THE MOST POPULAR TEXT BOOK OF

ENGINEERING MATHEMATICS - II

FOR : SEMESTER II

FIRST YEAR (F.E.) DEGREE COURSES IN ENGINEERING

COMMON FOR ALL DEGREE ENGINEERING COURSES
ACCORDING TO NEW REVISED CHOICE BASED CREDIT SYSTEM SYLLABUS
OF SAVITRIBAI PHULE PUNE UNIVERSITY, PUNE
(EFFECTIVE FROM ACADEMIC YEAR 2019-2020)

Dr. M. Y. GOKHALE
M. Sc. (Pure Maths.), M. Sc. (App. Maths.)
Ph. D. (I. I. T., Mumbai)
Professor and Head, Mathematics Deptt.
Maharashtra Institute of Technology,
PUNE.

Dr. N. S. MUJUMDAR
M. Sc., M. Phil., Ph. D. (Maths.)
Professor, Mathematics Deptt,
JSPM's Rajarshi Shahu
College of Engineering,
Tathawade, PUNE.

S. S. KULKARNI
M. Sc. (Maths.) (I. I. T. Mumbai)
Eminent & Experienced
Professor of Mathematics,
PUNE.

A. N. SINGH
M. A. (Mathematics Gold Medalist)
Formerly, Head Mathematics, Deptt.
D. Y. Patil College of Engineering
Pimpri, PUNE.

K. R. ATAL
M. Sc. (Mathematics)
Lecturer (Selection Grade) Deptt. of Applied Sciences,
Pune Institute of Computer Technology,
Dhankawdi, PUNE

N5161

ENGINEERING MATHEMATICS-II   **ISBN 978-93-89825-43-5**

First Edition   :   **January 2020**

©   :   **Authors**

Published By :
NIRALI PRAKASHAN

Abhyudaya Pragati, 1312 Shivaji Nagar

Off J.M. Road, PUNE 411005

Tel : (020) 25512336/37/39

Email : niralipune@pragationline.com

➤ DISTRIBUTION CENTRES

PUNE

Nirali Prakashan : 119, Budhwar Peth, Jogeshwari Mandir Lane, Pune 411002, Maharashtra
(For orders within Pune) Tel : (020) 2445 2044, Mobile : 9657703145
Email : niralilocal@pragationline.com

Nirali Prakashan : S. No. 28/27, Dhayari, Near Asian College Pune 411041
(For orders outside Pune) Tel : (020) 24690204 Fax : (020) 24690316; Mobile : 9657703143
Email : bookorder@pragationline.com

MUMBAI

Nirali Prakashan : 385, S.V.P. Road, Rasdhara Co-op. Hsg. Society Ltd.,
Girgaum, Mumbai 400004, Maharashtra; Mobile : 9320129587
Tel : (022) 2385 6339 / 2386 9976, Fax : (022) 2386 9976
Email : niralimumbai@pragationline.com

➤ DISTRIBUTION BRANCHES

JALGAON

Nirali Prakashan : 34, V. V. Golani Market, Navi Peth, Jalgaon 425001, Maharashtra,
Tel : (0257) 222 0395, Mob : 94234 91860; Email : niralijalgaon@pragationline.com

KOLHAPUR

Nirali Prakashan : New Mahadvar Road, Kedar Plaza, 1st Floor Opp. IDBI Bank, Kolhapur 416 012
Maharashtra. Mob : 9850046155; Email : niralikolhapur@pragationline.com

NAGPUR

Nirali Prakashan : Above Maratha Mandir, Shop No. 3, First Floor,
Rani Jhanshi Square, Sitabuldi, Nagpur 440012, Maharashtra
Tel : (0712) 254 7129; Email : niralinagpur@pragationline.com

DELHI

Nirali Prakashan : 4593/15, Basement, Agarwal Lane, Ansari Road, Daryaganj
Near Times of India Building, New Delhi 110002 Mob : 08505972553
Email : niralidelhi@pragationline.com

BENGALURU

Nirali Prakashan : Maitri Ground Floor, Jaya Apartments, No. 99, 6th Cross, 6th Main,
Malleswaram, Bengaluru 560003, Karnataka; Mob : 9449043034
Email: niralibangalore@pragationline.com

Other Branches : Hyderabad, Chennai

niralipune@pragationline.com | www.pragationline.com

Also find us on www.facebook.com/niralibooks

PREFACE TO THE FIRST EDITION

Our text books on **Engineering Mathematics-II** have occupied place of pride among engineering student's community for more than twenty five years now. All the teachers of this group of authors have been teaching mathematics in engineering colleges for the past several years. Difficulties of engineering students are well understood by the authors and that is reflected in the text material.

As per the policy of the University, Engineering Syllabi is revised every five years. Last revision was in the year 2015. New revision is coming little earlier, as Savitribai Phule Pune University has introduced new Choice Based Credit System from Academic Year 2019-2020.

As per the new Choice Based Credit System, the theory examination shall be conducted by university in two phases. Phase-I will be In-Semester Examination of 30 marks theory examination based on first and second units. Phase-II will be End-Semester Examination of 70 marks theory examination based on third, fourth, fifth and sixths units.

New text book on **Engineering Mathematics - II** is written, taking in to account all the new features that have been introduced. All the entrants to the engineering field will definitely find this book, complete in all respects. Students will find the subject matter presentation quite lucid. There are large number of illustrative examples and well graded exercises. Multiple Choice Questions based on the syllabus are appended at the end of the book and will be useful across universities/autonomous colleges and various competitive exams.

We take this opportunity to express our sincere thanks to Shri. Dineshbhai Furia of Nirali Prakashan, pioneer in all fields of education. Thanks are also due to Shri. Jignesh Furia, whose dynamic leadership is helpful to all the authors of Nirali Prakashan.

We specially appreciate the efforts of Shri. M. P. Munde, Mrs. Anagha Kaware and Mr. Santosh Bare and entire staff of Nirali Prakashan for making the publication of this book possible, well in time.

We have no doubt that like our earlier texts, student's community will respond favourably to this new venture.

The advice and suggestions of our esteemed readers to improve the text are most welcomed, and will be highly appreciated.

Pune   **Authors**

SYLLABUS

Unit - I : (09 Hrs.)

First Order Ordinary Differential Equations : Exact differential equations, Equations reducible to exact form. Linear differential equations, Equations reducible to linear form, Bernoulli's equation.

Unit - II : (09 Hrs.)

Applications of Differential Equations : Applications of differential equations to orthogonal trajectories, Newton's law of cooling, Kirchhoff's law of electrical circuits, Rectilinear motion, Simple harmonic motion, One dimensional conduction of heat.

Unit - III : (09 Hrs.)

Integral Calculus : Reduction formulae, Beta and Gamma functions, Differentiation under integral sign and error functions.

Unit - IV : (09 Hrs.)

Curve Tracing : Tracing of curves, Cartesian, Polar and Parametric curves. Rectification of curves.

Unit - V : (09 Hrs.)

Solid Geometry : Cartesian, Spherical, Polar and Cylindrical co-ordinate Systems. Sphere, Cone and Cylinder.

Unit - VI : (09 Hrs.)

Multiple Integrals and their Applications : Double and Triple integrations, Applications to find area, volume, mass, centre of gravity and moment of inertia.

•••

CONTENTS

•••

CHAPTER-1

FIRST ORDER ORDINARY DIFFERENTIAL EQUATIONS

1.1 INTRODUCTION

Study of differential equations is of prime importance because of its applications in almost all the engineering fields. Analysis of electrical networks, motion of fluid through porous media, heat transfer through spherical or cylindrical shells, decay of radioactive elements, projectile motion are some of the examples. Even medical sciences have not escaped the involvement of differential equations. Concentration of sugar in blood is analysed by a mathematical model involving differential equation. Studies of blood flow through the arteries, kidneys and the various organs of the body have attracted the attention of scientists working in the fields of Bio-medical engineering. Ordinary or Partial differential equations play an important role in different kinds of models representing various phenomena. Differential equations are also used in modeling of day-to-day life situations like concentration of traffic on roads in urban areas, arrival of customers in shopping mauls, landing of planes at the crowded airports and so on.

If an estimation of quantity depends upon only one factor, we come across a situation of one dependent variable and one independent variable and modeling of such phenomena will involve ordinary differential equation. In the case of problems involving more than one dependent variables depending on one independent variable, we come across system of ordinary differential equations. On the contrary if a variable for its variation depends upon two or more independent variables, models of such problems are represented by partial differential equations. Actual modeling takes into account many different factors as the situation demands. In what follows, methods to solve simple ordinary differential equations are discussed. In the next chapter, actual case studies are described.

1.2 DEFINITIONS

1. An equation involving the dependent variable, an independent variable and the differential coefficients of various orders is called a *Differential Equation (D.E.)*.

Any equation which contains a derivative is called a differential equation.

Following are some examples of differential equations :

(1) $\dfrac{dy}{dx} - y = 0$ (2) $\dfrac{dy}{dx} = \cos x$ (3) $\dfrac{d^2y}{dx^2} = 0$

(4) $(2x - y)\,dx + (y - x)\,dy = 0$ (5) $\sqrt{1 + \dfrac{dy}{dx}} = \dfrac{d^2y}{dx^2}$ (6) $\dfrac{dy}{dx} = \dfrac{x + y - 3}{2x - y + 1}$

(7) $\dfrac{dy}{dx} + \dfrac{2}{x}\,y = x^2$ (8) $x\dfrac{\partial z}{\partial x} + y\dfrac{\partial z}{\partial y} = nz$ (9) $\dfrac{\partial^2 y}{\partial t^2} = c^2\dfrac{\partial^2 y}{\partial x^2}$ (10) $\dfrac{\partial^2 u}{\partial x^2} + \dfrac{\partial^2 u}{\partial y^2} = 0$

2. **Ordinary Differential Equations :** Differential equations involving only one independent variable and one or more dependent variables and their derivatives w.r.t. independent variables are called *Ordinary Differential Equations (O.D.E.)*.

3. **Partial Differential Equations :** Differential equations involving two or more independent variables and one or more dependent variables and their partial derivatives w.r.t. independent variables are called *Partial Differential Equations (P.D.E.)*.

Thus, differential equations are divided into two parts.

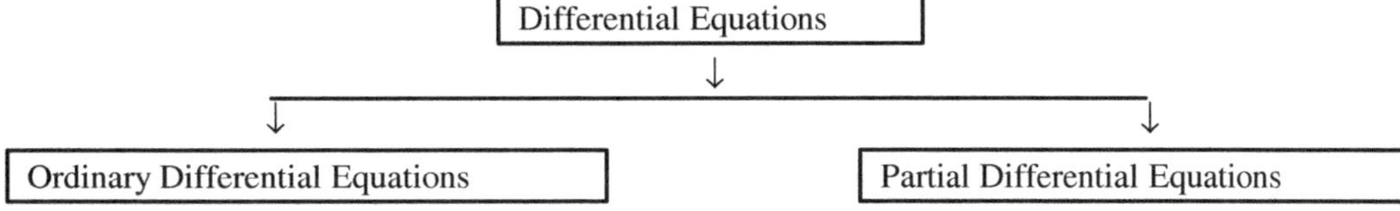

Equations (1) to (7) given in Article (1.2) are ordinary differential equations and equations (8) to (10) are partial differential equations. Differential equations are further classified according to their order and degree.

4. **Order of a Differential Equation**

Definition : The order of a differential equation is the order of the highest derivative that appears in the equation.

5. **Degree of a Differential Equation**

Definition : The degree of a differential equation is the degree of the highest order differential coefficient or derivative, when the differential coefficients are free from radicals and fractions.

Order and degree of equations (1), (2), (4), (6), (7), (8) of Article 1.2 is one.

Order of equations (3), (5), (9), (10) of Article 1.2 is two.

Degree of equations (3), (9), (10) of Article 1.2 is one.

To find degree of equation (5) i.e. $\sqrt{1 + \dfrac{dy}{dx}} = \dfrac{d^2y}{dx^2}$, squaring on both sides, we get

$$1 + \frac{dy}{dx} = \left(\frac{d^2y}{dx^2}\right)^2$$

$\therefore$ Degree of above equation is two.

This concept of order and degree of a differential equation is easily understood from the following illustrations.

1. $(2x - y + 3)\, dx + (y - 2x - 2)\, dy = 0$ $\therefore$ Order = 1, Degree = 1.

2. $\dfrac{dy}{dx} = \dfrac{x + 2y + 3}{x - y + 1}$ $\therefore$ Order = 1, Degree = 1.

3. $x^2\dfrac{d^2y}{dx^2} + x\,\dfrac{dy}{dx} = ny$ $\therefore$ Order = 2, Degree = 1.

4. $1 + \left(\dfrac{dy}{dx}\right)^2 + \dfrac{d^2y}{dx^2} = 0$ $\therefore$ Order = 2, Degree = 1.

5. $\left(\dfrac{dy}{dx}\right)^3 + 3y = \left(\dfrac{d^2y}{dx^2}\right)^2$ $\therefore$ Order = 2, Degree = 2.

6. $\left[1 + \left(\dfrac{dy}{dx}\right)^2\right]^{3/2} = \rho\,\dfrac{d^2y}{dx^2}$

Squaring on both sides, we get

$$\left[1 + \left(\frac{dy}{dx}\right)^2\right]^3 = \rho^2 \left(\frac{d^2y}{dx^2}\right)^2$$ $\therefore$ Order = 2, Degree = 2.

7. $\left(\dfrac{\partial z}{\partial x}\right)^2 + \dfrac{\partial^2 z}{\partial x^2} = \dfrac{\partial z}{\partial y}$ $\therefore$ Order = 2, Degree = 1

6. **Solution of a Differential Equation :** Consider the differential equation $\dfrac{dy}{dx} = \cos x$ and the relation $y = \sin x + C$. We observe that the derivative obtained from it satisfies the equation.

$\therefore$ $y = \sin x + C$ is a solution of the differential equation $\dfrac{dy}{dx} = \cos x,$ where C is an arbitrary constant.

Definition : A *solution* or *primitive* of a differential equation is any relation between the dependent and independent variables which is free from derivatives and which satisfies the differential equation.

7. **General Solution of a Differential Equation :** A relation between the dependent and independent variables, which is free from derivatives, which satisfies a given differential equation and which contains *arbitrary constants equal to the order of the differential equation* is called the *general solution* (G.S.) or *complete integral.*

   Hence, the *general solution of a differential equation of order n must involve n arbitrary constants.*

8. **Particular Solution of a Differential Equation :** The solution obtained by assigning particular values to the arbitrary constants in G.S. of a differential equation is called a *particular solution* or *particular integral.*

Note :

1. An arbitrary constant may be written in such a form as to make the answer simple. Thus, it may be written as C or log C, $\sin^{-1}$ C or $\tan^{-1}$ C or e^C etc.

2. Total number of arbitrary constants in the general solution is equal to the order of the equation.

1.3 REVISION : FORMATION OF ORDINARY DIFFERENTIAL EQUATIONS

In general, a general solution involving n arbitrary constants will give rise to a differential equation, of order n, free of arbitrary constants. This equation is obtained by eliminating the n constants between the (n + 1) equations consisting of the general solution and the n equations obtained by differentiating the general solution n times w.r.t. the independent variable.

Illustrations on Formation of Differential Equation

Ex. 1 : *Obtain the differential equation whose general solution is $y = a\, e^{-9t} \cos (3t + b)$ where a and b are arbitrary constants.*

$$\textit{(May 04)}$$

Sol. : Given G.S. is $y = a\, e^{-9t} \cos (3t + b)$, contains two arbitrary constants a and b, hence we will differentiate twice (w.r.t. 't').

$$\frac{dy}{dt} = a\, e^{-9t} [-9 \cos (3t + b) - 3 \sin (3t + b)]$$

$$= -9y - 3a\, e^{-9t} \sin (3t + b)$$

$$\frac{dy}{dt} + 9y = -3a\, e^{-9t} \sin (3t + b) \qquad \text{...(i)}$$

$$\frac{d^2y}{dt^2} + 9\frac{dy}{dt} = -3a\, e^{-9t} [-9 \sin (3t + b) + 3 \cos (3t + b)]$$

$$= -9 \left(\frac{dy}{dt} + 9y \right) - 9y \qquad \text{... by using (i)}$$

$$\frac{d^2y}{dt^2} + 18\frac{dy}{dt} + 90\, y = 0 \text{ is the required differential equation. Here order = 2, degree = 1.}$$

Ex. 2 : *Form the differential equation of which a general solution is $y = \log \cos (x - a) + b$, where a and b are arbitrary constants.*

$$\textit{(May 07, Dec. 05)}$$

Sol. : Given G.S. is $y = \log \cos (x - a) + b$, contains two arbitrary constants a and b, hence we will differentiate twice (w.r.t. x).

$$\frac{dy}{dx} = \frac{1}{\cos (x - a)} (-\sin (x - a)) = -\tan (x - a) \qquad \text{...(i)}$$

$$\frac{d^2y}{dx^2} = -\sec^2 (x - a) = -[1 + \tan^2 (x - a)]$$

$$\frac{d^2y}{dx^2} + 1 = -\left(\frac{dy}{dx} \right)^2 \qquad \text{... by using (i)}$$

$$\frac{d^2y}{dx^2} + \left(\frac{dy}{dx} \right)^2 + 1 = 0 \text{ is the required differential equation. Here order = 2, degree = 1.}$$

Ex. 3 : *Form the differential equation of which the primitive is $y = a \sin t + b \cos t$.*

Sol. : Here independent variable is t and primitive contains two arbitrary constants.

$\therefore$ Differentiating twice w.r.t. 't', we have

$$\frac{dy}{dt} = a \cos t - b \sin t$$

$$\frac{d^2y}{dt^2} = -a \sin t - b \cos t = -y$$

$\therefore$ $\dfrac{d^2y}{dt^2} + y = 0$ is the required differential equation. Here order = 2, degree = 1.

Ex. 4 : *Form the differential equation of all circles which touch the y-axis at the origin and centre lies on x-axis.*

Sol. : Consider the equation of circle as $(x - a)^2 + (y - 0)^2 = a^2$

or $x^2 + y^2 = 2ax$

or $2x + 2y \cdot y_1 = 2a$

or $x + y \cdot y_1 = a = \dfrac{x^2 + y^2}{2x}$

or $2x^2 + 2xyy_1 - x^2 - y^2 = 0$

i.e. $\dfrac{dy}{dx} = \dfrac{y^2 - x^2}{2xy}$ is the required differential equation.

Fig. 1.1

$$\text{Here order} = 1,\ \text{degree} = 1.$$

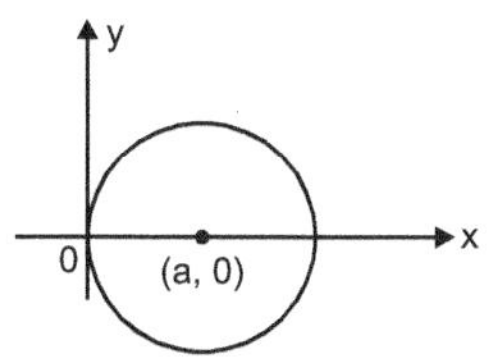

Ex. 5 : *Form the differential equation whose general solution is $y = a\, e^{-2x} + b\, e^{-3x}$.*   **(May 2006, Dec. 2004)**

Sol. : Since y contains two arbitrary constants, we have to differentiate twice

$$y = a\, e^{-2x} + b\, e^{-3x} \qquad \text{... (1)}$$

$$\frac{dy}{dx} = -2a\,e^{-2x} - 3b\,e^{-3x} \qquad \ldots (2)$$

$$\frac{d^2y}{dx^2} = 4a\,e^{-2x} + 9b\,e^{-3x} \qquad \ldots (3)$$

Eliminating a, b from (1), (2), (3), we get

$$\begin{vmatrix} y & -e^{-2x} & -e^{-3x} \\ \frac{dy}{dx} & 2e^{-2x} & 3e^{-3x} \\ \frac{d^2y}{dx^2} & -4e^{-2x} & -9e^{-3x} \end{vmatrix} = 0 \quad \text{or} \quad \begin{vmatrix} y & -1 & -1 \\ \frac{dy}{dx} & 2 & 3 \\ \frac{d^2y}{dx^2} & -4 & -9 \end{vmatrix} = 0$$

or
$$y(-18 + 12) - \frac{dy}{dx}(9 - 4) + \frac{d^2y}{dx^2}(-3 + 2) = 0$$

or
$$\frac{d^2y}{dx^2} + 5\frac{dy}{dx} + 6y = 0$$

Aliter : Multiply equation (2) by 5, equation (1) by 6 and add to equation (3).

$$\frac{d^2y}{dx^2} + 5\frac{dy}{dx} + 6y = 0$$

Note : When multipliers of equations (2) and (1) cannot be judged, first method is used.

EXERCISE 1.1

[A] State the order and degree of the following differential equations :

(1) $y\,dx - x\,dy = x$

(2) $(2x - y) + (x - y)\dfrac{dx}{dy} = 0$

(3) $\dfrac{dy}{dx} = \dfrac{2x + y + 1}{x - y + 2}$

(4) $(1 - x^2)\dfrac{d^2y}{dx^2} + x\dfrac{dy}{dx} = x^2 - y$

(5) $x^2\left(\dfrac{d^2y}{dx^2}\right)^3 + x\left(\dfrac{dy}{dx}\right)^4 = x^3$

(6) $\left(\dfrac{dy}{dx}\right)^2 = 16y$

(7) $\dfrac{y - x\,dy/dx}{dy/dx} = \left(\dfrac{dy}{dx}\right)^2$

(8) $y - x\dfrac{dy}{dx} = \sin^{-1}\left(\dfrac{dy}{dx}\right)$

(9) $\dfrac{dy}{dx} + \displaystyle\int y\,dx = x$

(10) $\dfrac{d^3y}{dx^3} = \left[k + \left(\dfrac{dy}{dx}\right)^2\right]^{3/2}$

(11) $\left[\dfrac{d^3y}{dx^3} + x\right]^{3/2} = \dfrac{dy}{dx}$

(12) $\sqrt[3]{\dfrac{d^2y}{dx^2}} = \sqrt{\dfrac{dy}{dx}}$

ANSWERS 1.1

(1) Order = 1, degree = 1

(2) Order = 1, degree = 1

(3) Order = 1, degree = 1

(4) Order = 2, degree = 1

(5) Order = 2, degree = 3

(6) Order = 1, degree = 2

(7) Multiply by $\dfrac{dy}{dx}$, order = 1, degree = 3

(8) Order = 1, degree cannot be determined

(9) First differentiate w.r.t. x, order = 2, degree = 1

(10) Order = 3, degree = 2

(11) Order = 3, degree = 3

(12) Order = 2, degree = 2

[B] Form the differential equations by eliminating the arbitrary constants from the following equations :

(1) $y = Ae^x + Be^{-x}$

(2) $y = A\cos nx + B\sin nx$

(3) $x = A\sin(kt + B)$

(4) $y = C^2 + \dfrac{C}{x}$ **(Dec. 2008)**

(5) $y = 4(x - A)^2$ **(May 2007)**

(6) $Ax^2 + By^2 = 1$ **(Dec. 2010, May 2004)**

(7) $y = A\cos(\log x) + B\sin(\log x)$ **(May 2006, Dec. 2004)**

(8) $xy = Ae^x + Be^{-x}$

(9) $(x - A)^2 + (y - B)^2 = r^2$ **(May 05)**

(10) $y = e^x (A \cos x + B \sin x)$ **(Dec. 2011, 2010, 2007, 2006)**

(11) $y = A \cos 4x + B \sin 4x + C$ **(May 2005)**

(12) $y = Ax + \dfrac{B}{x}$ **(Dec. 2009)**

(13) $xy = a\, e^x + b\, e^{-x} + x^2$ **(Dec. 2011; May 2010, 2009)**

(14) $y = e^x (ax^2 + bx + c)$ **(Dec. 2007)**

(15) $y = e^{at} (A \cos nt + B \sin nt)$

(16) $y = C_1 x^2 + C_2 x + C_3.$

(17) $y = a\, e^{2x} + b\, e^{3x}$

(18) $x = (C_1 + C_2 t)\, e^t$

(19) $\dfrac{x^2}{a^2} + \dfrac{y^2}{b^2} = 1$

(20) $x^2 + y^2 + 2gx + 2fy + c = 0$

(21) $y = Ax^2 + Bx^3$

(22) $(x - a)^2 = 4k (y - b)$, where a, b are arbitrary constants

(23) Form the differential equation of all circles having their centres on straight line $y = 5$ and touching the x-axis.

(24) Obtain differential equation of the family of confocal central conics $\dfrac{x^2}{a^2 + c} + \dfrac{y^2}{b^2 + c} = 1$ with a, b fixed, c arbitrary constants.

ANSWERS

(1) $\dfrac{d^2y}{dx^2} - y = 0$

(2) $\dfrac{d^2y}{dx^2} + n^2 y = 0$

(3) $\dfrac{d^2x}{dt^2} + k^2 x = 0$

(4) $y = x^4 \left(\dfrac{dy}{dx}\right)^2 - x\dfrac{dy}{dx}$

(5) $\left(\dfrac{dy}{dx}\right)^2 = 16 y$

(6) $xy\dfrac{d^2y}{dx^2} + x\left(\dfrac{dy}{dx}\right)^2 - y\dfrac{dy}{dx} = 0$

(7) $x^2\dfrac{d^2y}{dx^2} + x\dfrac{dy}{dx} + y = 0$

(8) $x\dfrac{d^2y}{dx^2} + 2\dfrac{dy}{dx} = xy$

(9) $\left[1 + \left(\dfrac{dy}{dx}\right)^2\right]^3 = r^2 \left(\dfrac{d^2y}{dx^2}\right)^2$

(10) $\dfrac{d^2y}{dx^2} - 2\dfrac{dy}{dx} + 2y = 0$

(11) $\dfrac{d^3y}{dx^3} + 16\dfrac{dy}{dx} = 0$

(12) $x^2\dfrac{d^2y}{dx^2} + x\dfrac{dy}{dx} - y = 0.$

(13) $x\dfrac{d^2y}{dx^2} + 2\dfrac{dy}{dx} - xy + x^2 - 2 = 0$

(14) $\dfrac{d^3y}{dx^3} - 3\dfrac{d^2y}{dx^2} + 3\dfrac{dy}{dx} - y = 0$

(15) $\dfrac{d^2y}{dx^2} - 2a\dfrac{dy}{dx} + (n^2 + a^2) y = 0$

(16) D.E. is $\dfrac{d^3y}{dx^3} = 0.$ Note that order = 3, degree = 1.

(17) $\dfrac{d^2y}{dx^2} - 5\dfrac{dy}{dx} + 6y = 0$

(18) $\dfrac{d^2x}{dt^2} - 2\dfrac{dx}{dt} + x = 0$

(19) $xy\dfrac{d^2y}{dx^2} + x\left(\dfrac{dy}{dx}\right)^2 - y\dfrac{dy}{dx} = 0$

(20) $y_3 (1 + y_1^2) - 3y_1 y_2^2 = 0$

(21) $x^2 y_2 - 4xy_1 + 6y = 0$

(22) $2k\dfrac{d^2x}{dy^2} + \left(\dfrac{dx}{dy}\right)^3 = 0$

(23) $(y - 5)^2 \left[1 + \left(\dfrac{dy}{dx}\right)^2\right] = 5^2$

(24) $(x + yy') (xy' - y) = (a^2 - b^2) y'$

1.4 ORDINARY DIFFERENTIAL EQUATIONS OF THE FIRST ORDER AND FIRST DEGREE

An ordinary differential equation of first order and first degree is of the form $M + N \dfrac{dy}{dx} = 0$ or $M\, dx + N\, dy = 0$, where M and N are functions of x and y or constants.

The general solution of such equation will contain only *one arbitrary constant.*

The method of solving a differential equation of the first order and first degree depends upon the type to which it belongs. These types are :

(1) Differential equations are of variables Separable form (V.S. form) or Reducible to Variable Separable form (Using substitutions, Homogeneous type and Non-homogeneous types).

(2) Exact differential equations or Reducible to Exact form (Using integrating factors).

(3) Linear differential equations or Reducible to Linear form (Bernoulli's Differential equation and equations of the form $f'(y) \dfrac{dy}{dx} + Pf(y) = Q$).

Students are already well acquainted with solutions of the differential equations of type (1) i.e. Variable Separable form or Reducible to Variable Separable forms. In what follows, we shall now briefly revise the solutions of differential equations of type (1) for sake of completeness and then discuss in detail type (2) and type (3).

1.4.1 Revision : Differential Equations are in Variable Separable Form or Reducible to Variable Separable Form

[A] Variables Separable Form (V.S. Form)

Many first-order differential equations can be reduced to the form $\dfrac{dy}{dx} = \dfrac{f(x)}{g(y)} \left(\text{or } \dfrac{dy}{dx} = \dfrac{g(y)}{f(x)} \right)$ by algebraic manipulations. We find it convenient to write, $g(y)\, dy = f(x)\, dx \left(\text{or } \dfrac{dy}{g(y)} = \dfrac{dx}{f(x)} \right)$.

Such an equation is in variables separable form (V.S. form) because the variables x and y are separated so that x appears only on one side of the equation and y appears only on the other side.

The solution is obtained by integrating on both sides giving,

$$\int g(y)\, dy = \int f(x)\, dx + C \qquad \text{or} \qquad \int \frac{dy}{g(y)} = \int \frac{dx}{f(x)} + C$$

It is of great importance to introduce the constant of integration immediately when the integration is carried out.

[B] Differential Equations Reducible to V.S. Form by using Substitution

1. **Linear Substitution :** If the given differential equation is of the form $\dfrac{dy}{dx} = f(ax + by + c)$ then we use the substitution $ax + by + c = u$. Given differential equation reduces to V.S. form in the variables u, x.

2. **Quotient Substitution :**

(i) Also differential equation of the form $\dfrac{dy}{dx} = f\left(\dfrac{y}{x}\right)$ reduces to V.S. form by using substitution $\boxed{\dfrac{y}{x} = u}$, $y = ux$, $\boxed{\dfrac{dy}{dx} = u + x\dfrac{du}{dx}}$

(ii) Also differential equation of the form $\dfrac{dx}{dy} = f\left(\dfrac{x}{y}\right)$ reduces to V.S. form by using substitution $\boxed{\dfrac{x}{y} = u}$, $x = uy$, $\boxed{\dfrac{dx}{dy} = u + y\dfrac{du}{dy}}$

Illustrations on V.S. Form

Ex. 1 : Solve $y - x\dfrac{dy}{dx} = a\left(y^2 + \dfrac{dy}{dx}\right)$.

Sol. : Given $\quad y - ay^2 = \dfrac{dy}{dx}(a + x)$

or $\quad \dfrac{dx}{a + x} = \dfrac{dy}{y(1 - ay)}$

G.S. $\quad \displaystyle\int \frac{dx}{a + x} = \int \left(\frac{1}{y} + \frac{a}{1 - ay}\right) dy + \log C \qquad$ (By using partial fraction)

$\log(a + x) = \log y - \log(1 - ay) + \log C$ or $\dfrac{(a + x)(1 - ay)}{y} = C$

$\boxed{(a + x)(1 - ay) = Cy}$ is the G.S.

Ex. 2 : Solve $\dfrac{dy}{dx} = e^{x-y} + 3x^2 e^{-y}$

Sol. : Multiply by e^y,

$$e^y \frac{dy}{dx} = e^x + 3x^2$$

$$e^y\, dy = (e^x + 3x^2)\, dx$$

$\therefore \qquad \boxed{e^y = e^x + x^3 + C}$ is the G.S.

Ex. 3 : *Solve* $\dfrac{dy}{dx} + x^2 = x^2 \cdot e^{3y}$

Sol. : Given $\qquad \dfrac{dy}{dx} = x^2 (e^{3y} - 1)$

G.S. $\qquad\qquad \displaystyle\int \dfrac{dy}{e^{3y} - 1} = \int x^2 \, dx + C$

or $\qquad\qquad \dfrac{1}{3} \displaystyle\int \dfrac{3 \, e^{-3y} \, dy}{1 - e^{-3y}} = \dfrac{x^3}{3} + C$

$\therefore \qquad\qquad \boxed{\log (1 - e^{-3y}) - x^3 = C_1}$ is the G.S.

Ex. 4 : *Solve* $\dfrac{dy}{dx} + \dfrac{1 + y^2}{1 + x^2} = 0$

Sol. : Given $\qquad \displaystyle\int \dfrac{dy}{1 + y^2} + \int \dfrac{dx}{1 + x^2} = 0$

G.S. $\qquad\qquad \tan^{-1} y + \tan^{-1} x = \tan^{-1} C$ (Note the selection of constant)

or $\qquad\qquad \tan^{-1} \left(\dfrac{x + y}{1 - xy} \right) = \tan^{-1} C$

$\therefore \qquad\qquad \boxed{\dfrac{x + y}{1 - xy} = C}$ is the G.S.

Ex. 5 : *Solve* $y \dfrac{dy}{dx} = \sqrt{1 + x^2 + y^2 + x^2 y^2}$ **(May 2004)**

Sol. : Given $\qquad y \dfrac{dy}{dx} = \sqrt{1 + x^2 + y^2 (1 + x^2)} = \sqrt{(1 + x^2)\,(1 + y^2)}$

G.S. $\qquad\qquad \displaystyle\int \dfrac{y \, dy}{\sqrt{1 + y^2}} = \int \sqrt{1 + x^2} \, dx + C$

$\therefore \qquad \boxed{\sqrt{1 + y^2} = \dfrac{x}{2}\sqrt{1 + x^2} + \dfrac{1}{2} \log (x + \sqrt{1 + x^2}) + C}$ is the G.S.

Illustrations on V.S. form by using substitutions :

Ex. 6 : *Solve* $\dfrac{dy}{dx} = 1 - x \tan (x - y).$ **(Dec. 2018)**

Sol. : Put $x - y = u \qquad\qquad \therefore \ 1 - \dfrac{dy}{dx} = \dfrac{du}{dx}$

$\therefore \qquad\qquad \dfrac{du}{dx} = x \tan u$

$\therefore \qquad\qquad \dfrac{du}{\tan u} = x \, dx$

G.S. $\qquad\qquad \displaystyle\int x \, dx - \int \cot u \, du = 0$

or $\qquad\qquad \dfrac{x^2}{2} - \log \sin u = C$

or $\qquad\qquad \boxed{\dfrac{x^2}{2} - \log \, \sin (x - y) = C}$ is the G.S.

Ex. 7 : *Solve* $x^4 \dfrac{dy}{dx} + x^3 y - \sec (xy) = 0.$ **(Dec. 2009, 2004, May 2017)**

Sol. : Put $xy = u \qquad\qquad \therefore \ x \dfrac{dy}{dx} + y = \dfrac{du}{dx}$

$\therefore \qquad\qquad x^3 \left(x \dfrac{dy}{dx} + y \right) - \sec (xy) = 0$

or $\qquad x^3 \dfrac{du}{dx} - \sec u = 0$

$\therefore \qquad x^3 \dfrac{du}{dx} = \sec u$

G.S. $\qquad \displaystyle\int \cos u \, du = \int \dfrac{dx}{x^3} \ \text{(V.S. form)}$

or $\qquad \sin u + \dfrac{1}{2x^2} = C$

$$\boxed{\sin (xy) + \dfrac{1}{2x^2} = C} \ \text{ is the G.S.}$$

Ex. 8 : *Solve* $\ y\left(x \cos \dfrac{y}{x} + y \sin \dfrac{y}{x}\right) - \left(y \sin \dfrac{y}{x} - x \cos \dfrac{y}{x}\right) x \dfrac{dy}{dx} = 0$

Sol. : We divide by x^2,

$$\dfrac{y}{x}\left(\cos \dfrac{y}{x} + \dfrac{y}{x} \sin \dfrac{y}{x}\right) - \left(\dfrac{y}{x}\sin \dfrac{y}{x} - \cos \dfrac{y}{x}\right)\dfrac{dy}{dx} = 0 \qquad \text{or} \qquad \dfrac{dy}{dx} = \dfrac{\dfrac{y}{x}\left(\cos \dfrac{y}{x} + \dfrac{y}{x}\sin \dfrac{y}{x}\right)}{\left(\dfrac{y}{x}\sin \dfrac{y}{x} - \cos \dfrac{y}{x}\right)}$$

Put $\dfrac{y}{x} = v$ or $y = xv \quad \therefore \dfrac{dy}{dx} = x \dfrac{dv}{dx} + v$

$\therefore \qquad x \dfrac{dv}{dx} + v = \dfrac{v(\cos v + v \sin v)}{(v \sin v - \cos v)} \qquad \text{or} \qquad x \dfrac{dv}{dx} = \dfrac{v(\cos v + v \sin v)}{v \sin v - \cos v} - v$

or $\qquad x \dfrac{dv}{dx} = \dfrac{2v \cos v}{v \sin v - \cos v} \qquad \text{or} \qquad \left(\dfrac{v \sin v - \cos v}{v \cos v}\right) dv = \dfrac{2}{x} dx \ \text{(V.S. form)}$

G.S. $\qquad 2 \displaystyle\int \dfrac{1}{x} dx + \int \dfrac{\cos v - v \sin v}{v \cos v} dv = \log C$

$\therefore \qquad 2 \log x + \log (v \cos v) = \log C$

or $\qquad \log (x^2 \, v \cos v) = \log C \qquad\qquad \text{or} \qquad x^2 \dfrac{y}{x} \cos \dfrac{y}{x} = C$

$$\boxed{xy \cos \dfrac{y}{x} = C} \ \text{ is the G.S.}$$

Ex. 9 : *Solve* $\ y \, e^{x/y} dx = (x \, e^{x/y} + y^2) dy$ **(May 2007)**

Sol. : Given $\qquad y \, e^{x/y} \dfrac{dx}{dy} - x \, e^{x/y} = y^2$ or $e^{x/y}\left(\dfrac{dx}{dy} - \dfrac{x}{y}\right) = y$

Put $\ x = yv \quad \therefore \dfrac{dx}{dy} = y \dfrac{dv}{dy} + v$

$\therefore \qquad e^v \left(y \dfrac{dv}{dy} + v - v\right) = y$ or $e^v \cdot y \cdot \dfrac{dv}{dy} = y$

G.S. $\qquad \displaystyle\int e^v \, dv = \int dy \Rightarrow e^v - y = C$

$$\boxed{e^{x/y} - y = C} \ \text{ is the G.S.}$$

Ex. 10 : *Solve* $\ x^2 \dfrac{dy}{dx} + xy + \sqrt{1 - x^2 y^2} = 0$

Sol. : Given $\ x\left(x \dfrac{dy}{dx} + y\right) + \sqrt{1 - x^2 y^2} = 0$

Put $\ xy = u \quad \therefore x \dfrac{dy}{dx} + y = \dfrac{du}{dx}$

$$\therefore \qquad x\frac{du}{dx} + \sqrt{1 - u^2} = 0$$

G.S. $\quad \displaystyle\int \frac{du}{\sqrt{1 - u^2}} + \int \frac{dx}{x} = C$ or $\sin^{-1} u + \log x = C$ $\boxed{\sin^{-1}(xy) + \log x = C}$ is the G.S.

EXERCISE 1.2

[A] Solve the Following Differential Equations :

1. $\dfrac{dy}{dx} = \dfrac{y}{x}$ **Ans. :** $y = Cx$

2. $x \cos x \cos y + \sin y \dfrac{dy}{dx} = 0$ **(May 2007)**

     Ans. : $x \sin x + \cos x + \log \sec y = C$

3. $(xy^2 - x)\, dx = (y + x^2 y)\, dy$ **Ans. :** $(1 + x^2) = C(1 - y^2)$

4. $(1 + x)\dfrac{dy}{dx} + 1 = 2e^{-}$ **Ans. :** $(1 + x)(2 - e^y) = C$

5. $3e^x \tan y\, dx + (1 + e^x) \sec^2 y\, dy = 0$

     Ans. : $(1 + e^x)^3 \tan y = C$

6. $\dfrac{dy}{dx} = \dfrac{x(2 \log x + 1)}{\sin y + y \cos y}$ **(Dec. 2011)**

     Ans. : $x^2 \log x = y \sin y + C$

7. $a\left(x\dfrac{dy}{dx} + 2y\right) = 2xy\dfrac{dy}{dx}$ **Ans. :** $\log x + \dfrac{1}{2}\log y - \dfrac{y}{a} = C$

8. $\dfrac{dy}{dx} = e^{x-y} + x^2 e^{-y}$ **(Dec. 2016) Ans. :** $e^y = e^x + \dfrac{x^3}{3} + C$

9. $xy^3 \dfrac{dy}{dx} = (1 - x^2)(1 + y^2)$ **(Nov. 15)**

     Ans. : $\log x - \dfrac{1}{2}(x^2 + y^2) + \dfrac{1}{2}\log(1 + y^2) = C$

10. $y - x\dfrac{dy}{dx} = 3\left(1 + x^2\dfrac{dy}{dx}\right)$ **Ans. :** $(y - 3)(3x + 1) = Cx$

11. $(1 - x^2)(1 + y)\, dx = xy(1 - y)\, dy$

     Ans. : $\log x - \dfrac{x^2 - y^2}{2} = 2y - 2\log(1 + y) + C$

12. $(4 + e^{2x})\dfrac{dy}{dx} = ye^{2x}$ **(May 2018) Ans. :** $y^2 = C(4 + e^{2x})$

13. $\dfrac{dy}{dx} = \dfrac{1 + y^2}{(1 + x^2)xy}$ **Ans. :** $(1 + x^2)(1 + y^2) = Cx^2$

14. $(3 + 2\sin x + \cos x)\, dy = (1 + 2\sin y + \cos y)\, dx$

     [Hint : Use half-angle formulae.]

     Ans. : $\dfrac{1}{2}\log(1 + 2\tan y/2) = \tan^{-1}(\tan x/2 + 1) + C$

15. $\dfrac{dy}{dx} + \dfrac{y^2 + 1}{x^2 + 1} = 0$ **Ans. :** $\dfrac{x + y}{1 - xy} = C$

16. $\dfrac{dy}{dx} + \dfrac{y^2 + y + 1}{x^2 + x + 1} = 0$

     [Hint : $\dfrac{dy}{(y + 1/2)^2 + (\sqrt{3}/2)^2} + \dfrac{dx}{(x + 1/2)^2 + (\sqrt{3}/2)^2}$]

     Ans. : $2\sqrt{3}(x + y + 1) = C[3 - (2y + 1)(2x + 1)]$

17. $\dfrac{dy}{dx} = \dfrac{x \sin x}{2e^y \sinh y}$

     [Hint : Use $\sinh y = \dfrac{e^y - e^{-y}}{2}$]

     Ans. $\dfrac{e^{2y}}{2} - y + x \cos x - \sin x = C$

18. $y^2 \cos \sqrt{x}\, dx - 2\sqrt{x}\, e^{1/y}\, dy = 0.$

     [Hint : $\displaystyle\int \dfrac{\cos \sqrt{x}}{2\sqrt{x}}\, dx - \int \dfrac{e^{1/y}}{y^2}\, dy = C.$

     Note that :

     $\displaystyle\int \cos f(x)\, f'(x)\, dx = \sin f(x)$ and $\displaystyle\int e^{f(y)}\, f'(y)\, dy = e^{f(y)}$]

     Ans. $\sin \sqrt{x} + e^{1/y} = C$

[B] Solve Each of the Following Differential Equations by Reducing it to Variables Separable form by Proper Substitution :

1. $\dfrac{dy}{dx} = \tan^2(x + y)$ **(Dec. 2005)**

     Ans. : $y - x + \sin(x + y)\cos(x + y) = C$

2. $(x - y)^2 \dfrac{dy}{dx} + a^2 = 0$ **Ans. :** $y - x = a \tan\left(\dfrac{c + y}{a}\right)$

3. $\cos(x + y)\, dy = dx$ **Ans. :** $\tan\left(\dfrac{x + y}{2}\right) = y + c$

4. $\dfrac{dy}{dx} = (x - y + 1)^2 + (x - y)$ **(Dec. 2004)**

     [Hint : Put $x - y + 1 = u$]   **Ans. :** $\dfrac{x - y + 3}{x - y} = ce^{3x}$

5. $\dfrac{dy}{dx} + x \tan(y - x) = 1$ **(Dec. 2008)**

     Ans. : $\cos(y - x) = ce^{x^2/2}$

6. $\dfrac{dy}{dx} = \cos x \cos y + \sin x \sin y$ **(May 2015)**

     Ans. : $x + \cot\left(\dfrac{x - y}{2}\right) = C$

7. $\dfrac{dy}{dx} = \left(\dfrac{x + y + 1}{x + y + 3}\right)^2$

     [Hint : $x + y = u$] **Ans. :** $(x + y)^2 + 4(x + y) + 5 = ce^{x-y}$

8. $(x - 2y)\, dy = (3x + y)(3x^2 - 5xy - 2y^2)\, dx$

     [Hint : $3x^2 - 5xy - 2y^2 = (3x + y)(x - 2y)$, Put $3x + y = u$]

     Ans. : $x = \dfrac{1}{\sqrt{3}}\tan^{-1}\dfrac{3x + y}{\sqrt{3}} + C$

9. $\dfrac{dy}{dx} = \sin(y - x)$ **Ans. :** $x + \tan(y - x) + \sec(y - x) = C$

10. $(1 + e^{x/y})\,dx + e^{x/y}\left(1 - \dfrac{x}{y}\right) dy = 0$ **Ans. :** $x + ye^{x/y} = C$

11. $x\,dx = y\,(x^2 + y^2 - 1)\,dy$. Put $x^2 + y^2 = u$
   Ans. : $\log(x^2 + y^2) = y^2 + C$

12. $e^{x+y}\left(x\dfrac{dy}{dx} + y\right) - e^{xy}\left(1 + \dfrac{dy}{dx}\right) = 0$ **(May 2007)**
   Ans. : $e^{-xy} = e^{-(x+y)} + C$

13. $\dfrac{x + y - a}{x + y + b}\,\dfrac{dy}{dx} = \dfrac{x + y + a}{x + y - b}$ **(May 2004)**

   [Hint : $\left(\dfrac{u^2 + (a + b)\,u + ab}{u^2 + ab}\right) du = 2dx]$

   Ans. : $(x + y) - \left(\dfrac{a + b}{2}\right) \log\,[(x + y)^2 + ab] = 2x + C$

14. $(4x + y)^2 \dfrac{dx}{dy} = 1$ **Ans. :** $4x + y = 2\tan(2x + C)$

15. $(x + y)^2\left(x\dfrac{dy}{dx} + y\right) = xy\left(1 + \dfrac{dy}{dx}\right)$ **(Dec. 2010)**

   [Hint : Put $x + y = u$, $xy = v]$ **Ans.** $\log xy + \dfrac{1}{x + y} = C$

16. $\left(\tan\dfrac{y}{x} - \dfrac{y}{x}\sec^2\dfrac{y}{x}\right) dx + \sec^2\dfrac{y}{x}\,dy = 0.$

   Ans. $x\tan\left(\dfrac{y}{x}\right) = C$

17. Solve $\dfrac{dy}{dx} = \sin(x + y) + \cos(x + y)$

   [Hint : Put $x + y = u$ $\therefore$ $1 + \dfrac{dy}{dx} = \dfrac{du}{dx}]$

   Ans. $\log\left[1 + \tan\left(\dfrac{x + y}{2}\right)\right] - x = C$

18. Solve $(\cos x \cos y - \sin x \sin y)\,dy = dx$

   [Hint : Put $x + y = u$ $\therefore$ $1 + \dfrac{dy}{dx} = \dfrac{du}{dx}]$

   Ans. : $y - \tan\left(\dfrac{x + y}{2}\right) = C$

19. Solve $xy\log\left(\dfrac{x}{y}\right) dx + \left(y^2 - x^2\log\left(\dfrac{x}{y}\right)\right) dy = 0$

   [Hint : Put $\dfrac{x}{y} = v,\ x = yv]$

   Ans. $\dfrac{x^2}{2y^2}\log\dfrac{x}{y} - \dfrac{x^2}{4y^2} + \log y = C$

20. Solve $x\dfrac{dy}{dx} - y = x\sqrt{x^2 + y^2}$

   [Hint : Put $\dfrac{y}{x} = u,\ y = ux$ $\therefore$ $\dfrac{dy}{dx} = u + x\dfrac{du}{dx}]$

   Ans. : $\sinh^{-1}\left(\dfrac{y}{x}\right) = x + C$

21. Solve $(ye^{xy} - \tan x)\,dx + (xe^{xy} - \sec y)\,dy = 0$

   Ans. : $e^{xy} - \log\sec x - \log\,[\sec y + \tan y] = C$

22. Solve $(x - 2\sin y + 3)\,dx + (2x - 4\sin y - 3)\cos y\,dy = 0$

   [Hint : Put $x - 2\sin y = u$ $\therefore$ $1 - 2\cos y \cdot \dfrac{dy}{dx} = \dfrac{du}{dx}]$

   Ans. : $x + 2\sin y + \dfrac{9}{4}\log(4x - 8\sin y + 3) = C$

23. If the tangent to a curve at any point (x, y) is known to be inclined at an angle $\tan^{-1}\left[(4x + y + 1)^2\right]$ to x-axis, find the equation of the curve.

   [Hint : Put $4x + y + 1 = u$ $\therefore$ $4 + \dfrac{dy}{dx} = \dfrac{du}{dx}]$

   Ans. : $\tan^{-1}\left(\dfrac{4x + y + 1}{2}\right) - 2x = C_1$

24. Solve $\dfrac{dy}{dx} = \sqrt{y - x}$

   [Hint : Put $y - x = t^2$
   $\therefore$ $\dfrac{dy}{dx} - 1 = 2t\dfrac{dt}{dx}$ or $\dfrac{dy}{dx} = 2t\dfrac{dt}{dx} + 1]$

   Ans. : $2\sqrt{y - x} + 2\log(\sqrt{y - x} - 1) - x = C$

25. Solve $\dfrac{dy}{dx} = \dfrac{2}{x + 2y - 3}$

   [Hint : Put $x + 2y - 3 = u]$

   Ans. : $2y - 3 - 4\log(x + 2y + 1) = C$

[C] Homogeneous Differential Equation :

Consider a differential equation in the form $M(x, y)\,dx + N(x, y)\,dy = 0$ or $\dfrac{dy}{dx} = \dfrac{M(x, y)}{N(x, y)}$.

The differential equation of the above form is said to be homogeneous if $M(x, y)$ and $N(x, y)$ are homogeneous functions in x and y of the *same degree*.

These equations are reducible to V.S. form by changing the dependent variable from y to u by the substitution $y = ux$

$\therefore$ $\dfrac{dy}{dx} = u + x\dfrac{du}{dx}.$

Illustrations on Homogeneous Differential Equation

Ex. 1 : *Solve* $x\dfrac{dy}{dx} + \dfrac{y^2}{x} = y$

Sol. : Given differential equation is homogeneous and can be written as

$$\dfrac{dy}{dx} + \dfrac{y^2}{x^2} = \dfrac{y}{x}$$

Put $\dfrac{y}{x} = u$ or $y = ux$ $\therefore$ $\dfrac{dy}{dx} = u + x\dfrac{du}{dx}$

$$x\dfrac{du}{dx} + u + u^2 = u \Rightarrow x\dfrac{du}{dx} + u^2 = 0$$

G.S. $\displaystyle\int \dfrac{du}{u^2} + \int \dfrac{dx}{x} = 0 \therefore -\dfrac{1}{u} + \log x = C$

$$\boxed{\log x - \dfrac{x}{y} = C}$$

Ex. 2 : Solve $x^3\, dx - y^3\, dy = 3xy\,(y\, dx - x\, dy)$

Sol. : Given $(x^3 - 3xy^2)\, dx = (y^3 - 3x^2y)\, dy$ which is homogeneous and can be written as

$$\dfrac{dy}{dx} = \dfrac{x^3 - 3xy^2}{y^3 - 3x^2\, y} \qquad\text{or}\qquad \dfrac{dy}{dx} = \dfrac{1 - 3\dfrac{y^2}{x^2}}{\dfrac{y^3}{x^3} - 3\dfrac{y}{x}}$$

Put $y = xv$ $\therefore$ $\dfrac{dy}{dx} = x\dfrac{dv}{dx} + v$

$$x\dfrac{dv}{dx} + v = \dfrac{1 - 3v^2}{v^3 - 3v} \qquad\text{or}\qquad x\dfrac{dv}{dx} = \dfrac{1 - 3v^2}{v^3 - 3v} - v = \dfrac{1 - v^4}{v^3 - 3v}$$

G.S. $\displaystyle\int \dfrac{v^3 - 3v}{1 - v^4}\, dv = \int \dfrac{dx}{x} + C$ or $-\dfrac{1}{4}\displaystyle\int \dfrac{-4v^3}{1 - v^4}\, dv - 3\int \dfrac{v\, dv}{1 - v^4} - \log x = C$

Put $v^2 = t - \dfrac{1}{4}\log(1 - v^4) - \dfrac{3}{2}\displaystyle\int \dfrac{dt}{1 - t^2} - \log x = C$ or $-\dfrac{1}{4}\log(1 - v^4) - \dfrac{3}{4}\log\left(\dfrac{1 + t}{1 - t}\right) - \log x = C$

$$\log(1 - v^4) + \log\left(\dfrac{1 + t}{1 - t}\right)^3 + \log x^4 = \log C_1$$

$$\dfrac{(1 - v^4)\,(1 + v^2)^3\, x^4}{(1 - v^2)^3} = C_1 \qquad\text{or}\qquad \dfrac{(1 + v^2)^4\, x^4}{(1 - v^2)^2} = C_1 \ \text{or} \ (1 + v^2)^2\, x^2 = C_2\,(1 - v^2)$$

$$\boxed{(x^2 + y^2)^2 = C_2\,(x^2 - y^2)} \text{ is the G.S.}$$

Ex. 3 : Solve $(y^4 - 2x^3\, y)\, dx + (x^4 - 2xy^3)\, dy = 0.$ $\hfill$ **(Dec. 09, 05, 04, 13; May 07)**

Sol. : Given differential equation can be written as

$$\left[\left(\dfrac{y}{x}\right)^4 - 2\left(\dfrac{y}{x}\right)\right] + \left[1 - 2\left(\dfrac{y}{x}\right)^3\right]\dfrac{dy}{dx} = 0$$

Put $y = ux$ $\therefore$ $\dfrac{dy}{dx} = u + x\dfrac{du}{dx}$

$$u^4 - 2u + (1 - 2u^3)\left(u + x\dfrac{du}{dx}\right) = 0 \Rightarrow (-u^4 - u) + (1 - 2u^3)\, x\dfrac{du}{dx} = 0$$

$$-\left(\dfrac{dx}{x}\right) + \left(\dfrac{1 - 2u^3}{u^4 + u}\right)du = 0 \Rightarrow \dfrac{dx}{x} = \left(\dfrac{1}{u} - \dfrac{3u^2}{1 + u^3}\right)du$$

G.S. $\displaystyle\int \dfrac{dx}{x} = \int \dfrac{1}{u}\, du - \int \dfrac{3u^2\, du}{1 + u^3} + \log C$

$$\log x = \log u - \log(1 + u^3) + \log C \Rightarrow \dfrac{x\,(1 + u^3)}{u} = C$$

$\therefore$ $\boxed{x^3 + y^3 = C\, xy}$ is the G.S.

Ex. 4 : Solve $x^3 \dfrac{dy}{dx} = y^3 + y^2 \sqrt{y^2 - x^2}$

Sol. : Given

$$\frac{dy}{dx} = \frac{y^3}{x^3} + \frac{y^2}{x^2} \sqrt{\frac{y^2}{x^2} - 1}$$

Put $y = xv$

$$\therefore \quad \frac{dy}{dx} = x \frac{dv}{dx} + v$$

$$x \frac{dv}{dx} + v = v^3 + v^2 \sqrt{v^2 - 1} \quad \text{or} \quad x \frac{dv}{dx} = v(v^2 - 1) + v^2 \sqrt{v^2 - 1}$$

$$x \frac{dv}{dx} = v \sqrt{v^2 - 1}\,(\sqrt{v^2 - 1} + v) = v \sqrt{v^2 - 1}\, \frac{(\sqrt{v^2 - 1} + v)\,(\sqrt{v^2 - 1} - v)}{(\sqrt{v^2 - 1} - v)} = \frac{v \sqrt{v^2 - 1}\,(v^2 - 1 - v^2)}{\sqrt{v^2 - 1} - v}$$

$$\therefore \quad \frac{v - \sqrt{v^2 - 1}}{v \sqrt{v^2 - 1}}\, dv = \frac{dx}{x}$$

G.S.

$$\int \frac{1}{\sqrt{v^2 - 1}}\, dv - \int \frac{1}{v}\, dv = \int \frac{dx}{x} + \log C$$

$$\log(v + \sqrt{v^2 - 1}) - \log v - \log x = \log C \quad \text{or} \quad \frac{(v + \sqrt{v^2 - 1})}{xv} = C$$

$$\boxed{y + \sqrt{y^2 - x^2} = Cxy} \quad \text{is the G.S.}$$

EXERCISE 1.3

Solve the following differential equations :

1. $(xy - x^2) \dfrac{dy}{dx} = y^2$ **(Dec. 2008, May 2011)**

      Ans. : $cy = e^{y/x}$

2. $yx^2\, dx = (x^3 - y^3)\, dy$ **Ans. :** $cy = e^{-x^3/3y^3}$

3. $\left(x + y \cot \dfrac{x}{y}\right) dy - y\, dx = 0$ **(Dec. 2016)**

      Ans. : $y \cdot \cos \dfrac{x}{y} = C$

4. $(x^4 + y^4)\, dx - 2x^3 y\, dy = 0$ **Ans. :** $\log x + \dfrac{x^2}{y^2 - x^2} = C$

5. $x\, dy - y\, dx = \sqrt{x^2 + y^2}\, dx$ **Ans. :** $y + \sqrt{x^2 + y^2} = cx^2$

6. $\left(x \tan \dfrac{y}{x} - y \sec^2 \dfrac{y}{x}\right) dx + x \sec^2 \dfrac{y}{x}\, dy = 0$

      (Dec. 2006, May 13) Ans. : $x \tan \dfrac{y}{x} = C$

7. $x(x - y)\, dy = y(x + y)\, dx$ **Ans. :** $x + y \log xy = cy$

8. $(x^2 + y^2)\, dx + 8xy\, dy = 0$ **Ans. :** $x(x^2 + 9y^2)^4 = C$

9. $2xy\, dy = (3y^2 + x^2)\, dx$ **Ans. :** $x^3 = c(x^2 + y^2)$

10. $2xy\, dx + (y^2 - x^2)\, dy = 0$ **Ans. :** $x^2 + y^2 = cy$

11. $(x^2 y - 2xy^2)\, dx = (x^3 - 3x^2 y)\, dy$

      (Dec. 2007, May 2010) Ans. : $cy^3 = x^2 \cdot e^{-x/y}$

12. $\dfrac{dy}{dx} = \dfrac{y^3 + 3x^2 y}{x^3 + 3xy^2}$ **Ans. :** $xy = c(x^2 - y^2)^2$

13. $\dfrac{dy}{dx} = \dfrac{x^2 - 3xy + 2y^2}{2xy - x^2}$ **(Dec. 2005)**   **Ans. :** $cx = e^{-y/x}$

14. $\dfrac{dy}{dx} = \dfrac{y}{x} + \tan\left(\dfrac{y}{x}\right)$ **(Nov. 2014)**   **Ans. :** $\sin \dfrac{y}{x} = cx$

15. $x \dfrac{dy}{dx} = y(\log y - \log x + 1)$ **Ans. :** $y = xe^{cx}$

[D] Non-Homogeneous Differential Equations Reducible to Homogeneous Form

A differential equation of the form $\boxed{\dfrac{dy}{dx} = \dfrac{a_1 x + b_1 y + c_1}{a_2 x + b_2 y + c_2}}$ is called non-homogeneous differential equation.   ...(1)

Case (i) : If $\dfrac{a_1}{a_2} = \dfrac{b_1}{b_2}$

In this case, the expressions $a_1 x + b_1 y$ and $a_2 x + b_2 y$ will always have a common factor of the form $lx + my$.

We put $lx + my = u$, then equation (1) reduces to V.S. form in the variables u, x.

Case (ii) : If $\dfrac{a_1}{a_2} \neq \dfrac{b_1}{b_2}$

In this case to reduce equation (1) to homogeneous form, we substitute $\boxed{x = X + h}$, $\boxed{y = Y + k}$ where h and k are constants to be determined.

Also $dx = dX$, $dy = dY$ $\therefore \dfrac{dy}{dx} = \dfrac{dY}{dX}$. Equation (1) becomes,

$$\frac{dY}{dX} = \frac{a_1 X + b_1 Y + (a_1 h + b_1 k + c_1)}{a_2 X + b_2 Y + (a_2 h + b_2 k + c_2)}$$

Choose h and k such that equation will become homogeneous in X and Y

i.e. $a_1 h + b_1 k + c_1 = 0$ and $a_2 h + b_2 k + c_2 = 0$

We get, $$\frac{dY}{dX} = \frac{a_1 X + b_1 Y}{a_2 X + b_2 Y}$$

which is a homogeneous equation in X and Y. Put $Y = VX$, $\dfrac{dY}{dX} = V + X\dfrac{dV}{dX}$. Finally equation reduces to V.S. form.

Illustrations on Non-homogeneous D.E.

Examples based on Case (i)

Ex. 1 : *Solve* $\dfrac{dy}{dx} = \dfrac{x + y + 1}{2x + 2y + 1}$ *(Dec. 2006)*

Sol. : $\dfrac{dy}{dx} = \dfrac{(x + y) + 1}{2(x + y) + 1}$

Put $x + y = u$ $\therefore 1 + \dfrac{dy}{dx} = \dfrac{du}{dx}$

$\therefore \qquad \dfrac{du}{dx} - 1 = \dfrac{u + 1}{2u + 1}$ or $\dfrac{du}{dx} = \dfrac{u + 1}{2u + 1} + 1 = \dfrac{3u + 2}{2u + 1}$

or $\dfrac{2u + 1}{3u + 2} du = dx$ or $\dfrac{2}{3}\left(\dfrac{3u + 2 - 1/2}{3u + 2}\right) du = dx$

G.S. $\dfrac{2}{3}\displaystyle\int du - \dfrac{1}{3}\int \dfrac{du}{3u + 2} = \int dx + C$

$\dfrac{2}{3} u - \dfrac{1}{9} \log (3u + 2) = x + C$ or $6u - 9x - \log (3u + 2) = C_1$

$6 (x + y) - 9x - \log (3x + 3y + 2) = C_1$

$\boxed{6y - 3x - \log (3x + 3y + 2) = C_1}$ is the G.S.

Ex. 2 : Solve $(x + 2y + 1)\, dx - (2x + 4y + 3)\, dy = 0$

Sol. : Given differential equation can be written as,

$$\frac{dy}{dx} = \frac{x + 2y + 1}{2x + 4y + 3} \quad \text{Here } \frac{a_1}{a_2} = \frac{1}{2} \,;\, \frac{b_1}{b_2} = \frac{2}{4} \therefore \frac{a_1}{a_2} = \frac{b_1}{b_2}$$

$$\frac{dy}{dx} = \frac{(x + 2y) + 1}{2(x + 2y) + 3}$$

Put $x + 2y = u$ $\therefore 1 + 2\dfrac{dy}{dx} = \dfrac{du}{dx} \Rightarrow \dfrac{dy}{dx} = \dfrac{1}{2}\left[\dfrac{du}{dx} - 1\right]$

$\dfrac{1}{2}\left(\dfrac{du}{dx} - 1\right) = \dfrac{u + 1}{2u + 3} \Rightarrow \dfrac{du}{dx} = \dfrac{2u + 2}{2u + 3} + 1 = \dfrac{4u + 5}{2u + 3}$

G.S. $\displaystyle\int \dfrac{2u + 3}{4u + 5} du = \int dx + C \Rightarrow \int \dfrac{4u + 5 + 1}{4u + 5} du = 2x + C_1$

$\displaystyle\int \left(1 + \dfrac{1}{4u + 5}\right) du - 2x = C_1 \Rightarrow u + \dfrac{1}{4} \log (4u + 5) - 2x = C_1$

$\boxed{2y - x + \dfrac{1}{4} \log (4x + 8y + 5) = C_1}$ is the G.S.

Examples based on Case (ii)

Ex. 1 : Solve $\dfrac{dy}{dx} + \dfrac{2x + 3y}{y + 2} = 0$

Sol. : Given differential equation can be written as,

$$\frac{dy}{dx} = \frac{-2x - 3y}{y + 2} \quad \text{Here} \quad \frac{a_1}{a_2} \neq \frac{b_1}{b_2}$$

Put $x = X + h$; $y = Y + k$

$\therefore \qquad dx = dX; \ dy = dY \ \text{and} \ \dfrac{dy}{dx} = \dfrac{dY}{dX}$

$\therefore \qquad \dfrac{dY}{dX} = \dfrac{-2X - 3Y - 2h - 3k}{Y + k + 2}$

We choose h, k such that $-2h - 3k = 0$, $k + 2 = 0$ giving $h = 3, k = -2$.

$$\therefore \qquad \frac{dY}{dX} = \frac{-2X - 3Y}{Y} = \frac{-2 - 3Y/X}{Y/X} \quad \text{Put} \ Y = XV, \ \frac{dY}{dX} = X\frac{dV}{dX} + V$$

$$X\frac{dV}{dX} + V = \frac{-2 - 3V}{V} \qquad \text{or} \qquad X\frac{dV}{dX} = \frac{-2 - 3V}{V} - V = \frac{-2 - 3V - V^2}{V}$$

or $\qquad \dfrac{V \, dV}{V^2 + 3V + 2} = -\dfrac{dX}{X}$

G.S. $\qquad \displaystyle\int \frac{V}{(V + 2)(V + 1)} \, dV + \int \frac{dX}{X} = \log C$

$$\int \frac{2}{V + 2} \, dV - \int \frac{1}{V + 1} \, dV + \log X = \log C \ \text{(by partial fraction)}$$

$$2 \log (V + 2) - \log (V + 1) + \log X = \log C$$

$$\log \frac{(V + 2)^2}{V + 1} X = \log C$$

$$\log \frac{(Y + 2X)^2}{(Y + X)} = \log C \qquad\qquad \because X = x - 3 \, ; \ Y = y + 2$$

$$\left[\frac{(y + 2 + 2x - 6)^2}{(x + y - 1)}\right] = C$$

$$\boxed{(2x + y - 4)^2 = C(x + y - 1)} \ \text{is the G.S.}$$

Ex. 2 : Solve $(2x - y + 1) \, dy - (x + 2y + 3) \, dx = 0$

Sol. : Given differential equation can be written as,

$$\frac{dy}{dx} = \frac{x + 2y + 3}{2x - y + 1}, \quad \frac{a_1}{a_2} = \frac{1}{2} \, ; \ \frac{b_1}{b_2} = \frac{2}{-1} \ \therefore \ \frac{a_1}{a_2} \neq \frac{b_1}{b_2}$$

We use the substitutions $\boxed{x = X + h}$, $\boxed{y = Y + k}$, $dx = dX$, $dy = dY$

$\therefore \qquad \dfrac{dy}{dx} = \dfrac{dY}{dX}$

Equation (1) becomes, $\dfrac{dY}{dX} = \dfrac{X + 2Y + (h + 2k + 3)}{2X - Y + (2h - k + 1)}$

We choose h, k such that this equation reduces to homogeneous form provided

$$h + 2k + 3 = 0, \ 2h - k + 1 = 0.$$

Solving for h, k we get, $h = -1, \ k = -1$

$$\frac{dY}{dX} = \frac{X + 2Y}{2X - Y} = \frac{1 + 2\dfrac{Y}{X}}{2 - \dfrac{Y}{X}} \qquad \text{Put} \ Y = XV \ \therefore \ \frac{dY}{dX} = X\frac{dV}{dX} + V$$

$$\therefore \qquad X\frac{dV}{dX} + V = \frac{1 + 2V}{2 - V} \qquad \text{or} \qquad X\frac{dV}{dX} = \frac{1 + 2V}{2 - V} - V$$

$$X\frac{dV}{dX} = \frac{1 + V^2}{2 - V} \qquad \text{or} \qquad \frac{2 - V}{V^2 + 1} \, dV = \frac{dX}{X}$$

G.S. $2 \int \dfrac{dV}{V^2 + 1} - \dfrac{1}{2} \int \dfrac{2V\, dV}{V^2 + 1} - \int \dfrac{dX}{X} = C$

$2 \tan^{-1} V - \dfrac{1}{2} \log (V^2 + 1) - \log X = C$

$2 \tan^{-1} \dfrac{Y}{X} - \dfrac{1}{2} \log \left(\dfrac{Y^2}{X^2} + 1 \right) X^2 = C$ But $Y = y + 1 \, ; \; X = x + 1$

$\boxed{2 \tan^{-1} \left(\dfrac{y + 1}{x + 1} \right) - \dfrac{1}{2} \log \left[(y + 1)^2 + (x + 1)^2 \right] = C}$ is the G.S.

EXERCISE 1.4

Problems on Case (i) where $\dfrac{a_1}{a_2} = \dfrac{b_1}{b_2}$

Solve the following differential equations :

1. $\dfrac{dy}{dx} = \dfrac{6x - 4y + 3}{3x - 2y + 1}$, Put $3x - 2y = u$ **(May 06)**

   Ans. : $2x - y - \log (3x - 2y + 3) = C$

2. $\dfrac{4x + 6y + 5}{3y + 2x + 4} \cdot \dfrac{dy}{dx} = 1,$

   [**Hint :** Put $2x + 3y = u$]   **(Dec. 2011)**

   Ans. : $7x - 14y + 3 \log (14x + 21y + 22) = C$

3. $\dfrac{dy}{dx} = \dfrac{2x + 3y - 1}{6x + 9y + 6}$ **(May 2014)**

   Ans. : $x - 3y = \log (2x + 3y + 1) + C$

4. $\dfrac{dy}{dx} = \dfrac{3x - 6y + 1}{6x - 12y + 5}$ **Ans. :** $2\,(5y - x) = 3\,(x - 2y)^2 + C$

5. $\dfrac{dy}{dx} = \dfrac{x + 2y - 3}{3x + 6y - 1}$ **(Dec. 2010, 2004; May 09)**

   Ans. : $5\,(x - 3y) = 8 \log (5x + 10y - 7) + C$

6. $\dfrac{dy}{dx} = \dfrac{x + 2y + 2}{2x + 4y - 1}$

   Ans. : $8y - 4x - 5 \log [4x + 8y + 3] = C$

7. $\dfrac{dy}{dx} = \dfrac{4x - 6y + 3}{6x - 9y - 1}$ **Ans. :** $\dfrac{3}{2}\,(2x - 3y)^2 - (2x - 3y) + 11x = C$

8. $\dfrac{dy}{dx} = \dfrac{x - y + 3}{2x - 2y + 5}$ **(Dec. 2009, 2005, 2017)**

   Ans. : $x - 2y + \log (x - y + 2) = C$

9. $\dfrac{dy}{dx} = \dfrac{x + y + 1}{2x + 2y + 3}$ **Ans. :** $6y - 3x + \log (3x + 3y + 4) = C$

10. $\dfrac{dy}{dx} = \dfrac{8x + 6y + 12}{4x + 3y + 2}$

    Ans. : $5\,(3y - 6x) - 12 \log (20x + 15y + 22) = C$

11. $\dfrac{dy}{dx} = \dfrac{3x - 4y - 2}{6x - 8y - 5}$ **(May 05, Dec. 07)**

    Ans. : $\log (6x - 8y - 7) = C - 2x + 4y$

Problems on Case (ii) where $\dfrac{a_1}{a_2} \neq \dfrac{b_1}{b_2}$

1. $\dfrac{dy}{dx} = \dfrac{x + 2y - 3}{2x + y - 3}$ **(Dec. 08, May 07)**

   Ans. : $(x - y)^3 = C\,(x + y - 2)$

2. $(4x + 3y + 1)\, dx + (3x + 2y + 1)\, dy = 0$

   Ans. : $2x^2 + 3xy + y^2 + x + y = C$

3. $(3y - 7x + 7)\, dx + (7y - 3x + 3)\, dy = 0$

   Ans. : $(x - y - 1)^2 (x + y - 1)^5 = C$

4. $\dfrac{dy}{dx} = \dfrac{2x + 2y + 1}{3x + y - 2}$

   Ans. : $(y - x + 3)^4 = C \left(y + 2x - \dfrac{3}{4} \right)$

5. $(x + 2y + 1)\, dx - (2x - 3)\, dy = 0$

   Ans. : $4y + 5 = (2x - 3) \log [C\,(2x - 3)]$

6. $\dfrac{dy}{dx} = \dfrac{y - x + 1}{y + x - 5}$

   Ans. : $\tan^{-1} \left(\dfrac{y - 2}{x - 3} \right) + \dfrac{1}{2} \log [(x - 3)^2 + (y - 2)^2] = C$

7. $\dfrac{dy}{dx} = \dfrac{x - y + 5}{x + y - 1}$ **Ans. :** $y^2 - x^2 + 2xy - 10x - 2y - 8 = C$

8. $\dfrac{dy}{dx} = \dfrac{2x + y + 3}{2y + x + 1}$ **Ans. :** $\left(x + y + \dfrac{4}{3} \right) (x - y + 2)^3 = C$

9. $\dfrac{dy}{dx} = \dfrac{y + 2}{x + y + 1}$ **Ans. :** $\log (y + 2) - \dfrac{x - 1}{y + 2} = C$

10. Solve $(y - 2x)\, dx + (2y - 3x + 1)\, dy = 0$ **Ans.** $\left[(y + 2)^2 - (x + 1)(y + 2) - (x + 1)^2 \right] \left(\dfrac{2y + 4 - x - 1 + \sqrt{5}\, x + \sqrt{5}}{2y + 4 - x - 1 - \sqrt{5}\, x - \sqrt{5}} \right)^{\frac{2}{\sqrt{5}}} = C$

1.4.2 Exact Differential Equations

Consider a differential equation of the form $M (x, y)\ dx + N (x, y)\ dy = 0$ ...(1)

If there exists a function $u (x, y)$ such that $M\,dx + N\,dy = du$ then the differential equation is called as an *exact differential equation*.

Condition of Exactness :

The necessary and sufficient condition that $M\,dx + N\,dy = 0$ be exact is,

$$\boxed{\frac{\partial M}{\partial y} = \frac{\partial N}{\partial x}}$$

When the condition of exactness is satisfied, the *general solution* can be obtained by the following rules.

Rule 1 : $\boxed{\int\limits_{y=constant} M\,dx + \int [\text{Terms of N not containing x}]\,dy = C}$

i.e. Integrate $M\,dx$ w.r.t. x treating y constant, integrate only those terms in $N\,dy$ which are free from x w.r.t. y and equate their sum to a constant.

Rule 2 : If N has no term which is free from x then $\boxed{\int\limits_{y=constant} M\,dx = C}$ is the general solution.

Rule 3 : Sometimes we may write the G.S. by using the following rule.

$$\boxed{\int\limits_{x=constant} N\,dy + \int [\text{Terms of M not containing y}]\,dx = C}$$

Remark : Sometimes an equation of the form $\dfrac{dy}{dx} = \dfrac{a_1 x + b_1 y + C_1}{a_2 x + b_2 y + C_2}$ becomes exact if $\boxed{b_1 = -a_2}$ because the equation can be written as,

$(a_1 x + b_1 y + C_1)\,dx - (a_2 x + b_2 y + C_2)\,dy = 0$ with $\dfrac{\partial M}{\partial y} = b_1,\ \dfrac{\partial N}{\partial x} = -a_2$ Accordingly, the solution is given by

$$\int (a_1 x + b_1 y + C_1)\,dx - \int (b_2 y + C_2)\,dy = C \ \text{(Ref. Solved Ex. No. 1 and 7)}$$

(treat y = constant)

Illustrations on Exact Differential Equation

Ex. 1 : *Solve* $(x + y - 2)\,dx + (x - y + 4)\,dy = 0$ **(Nov. 2014)**

Sol. : $(x + y - 2)\,dx + (x - y + 4)\,dy = 0$ ...(1)

This is of the type $M\,dx + N\,dy = 0$.

Here $M = x + y - 2\,;\ N = x - y + 4.\ \dfrac{\partial M}{\partial y} = 1 = \dfrac{\partial N}{\partial x}\ \therefore$ Equation (1) is exact differential equation. Its G.S. is given by

$$\text{G.S.} \quad \int\limits_{y=constant} (x + y - 2)\,dx + \int\limits_{No\ x} (-y + 4)\,dy = C$$

$$\frac{x^2}{2} + xy - 2x - \frac{y^2}{2} + 4y = C$$

or $\boxed{x^2 - y^2 + 2xy - 4x + 8y = C_1}$ is the required G.S.

Ex. 2 : *Solve* $\left(\dfrac{y^2}{(y - x)^2} - \dfrac{1}{x}\right) dx + \left(\dfrac{1}{y} - \dfrac{x^2}{(x - y)^2}\right) dy = 0$

Sol. : Here $M = \dfrac{y^2}{(y - x)^2} - \dfrac{1}{x}$ $\qquad \therefore \dfrac{\partial M}{\partial y} = \dfrac{2y}{(y - x)^2} - \dfrac{2y^2}{(y - x)^3} = \dfrac{-2xy}{(y - x)^3}$

and $N = \dfrac{1}{y} - \dfrac{x^2}{(x - y)^2}$ $\qquad \therefore \dfrac{\partial N}{\partial x} = \dfrac{-2x}{(x - y)^2} + \dfrac{2x^2}{(x - y)^3} = \dfrac{-2xy}{(y - x)^3}$

Therefore given equation is exact,

G.S. $\quad y^2 \int \dfrac{1}{(y-x)^2}\, dx - \int \dfrac{1}{x}\, dx + \int \dfrac{1}{y}\, dy = C$

$$y^2 \dfrac{1}{y-x} - \log x + \log y = C$$

or $\quad \boxed{\dfrac{y^2}{y-x} + \log \dfrac{y}{x} = C}$ is the G.S.

Ex. 3 : *Solve* $\dfrac{dy}{dx} = \dfrac{\tan y - 2xy - y}{x^2 - x\tan^2 y + \sec^2 y}$ **(May 2004)**

Sol. : Given differential equation can be written as

$$(\tan y - 2xy - y)\, dx + (x\tan^2 y - x^2 - \sec^2 y)\, dy = 0 \qquad \text{...(1)}$$

$$\dfrac{\partial M}{\partial y} = \sec^2 y - 2x - 1 = \tan^2 y - 2x ; \quad \dfrac{\partial N}{\partial x} = \tan^2 y - 2x$$

Equation (1) is exact.

G.S. $\quad \tan y \int dx - 2y \int x\, dx - y \int dx - \int \sec^2 y\, dy = C$

$$\boxed{x\tan y - x^2 y - xy - \tan y = C}$$ is the G.S.

Ex. 4 : *Solve* $\left(\dfrac{y}{(x-y)^2} - \dfrac{1}{2\sqrt{1-x^2}}\right) dx - \dfrac{x}{(x-y)^2}\, dy = 0$ **(May 2009)**

Sol. : Here $\quad M = \dfrac{y}{(x-y)^2} - \dfrac{1}{2\sqrt{1-x^2}}$ and $N = \dfrac{-x}{(x-y)^2}$

$$\dfrac{\partial M}{\partial y} = \dfrac{1}{(x-y)^2} + \dfrac{2y}{(x-y)^3} = \dfrac{x+y}{(x-y)^3}$$

and $\quad \dfrac{\partial N}{\partial x} = \dfrac{-1}{(x-y)^2} + \dfrac{2x}{(x-y)^3} = \dfrac{x+y}{(x-y)^3}$

$$\therefore \quad \dfrac{\partial M}{\partial y} = \dfrac{\partial N}{\partial x}$$

$\therefore$ Given differential equation is exact and the G.S. is given by,

G.S. $\quad y \int \dfrac{1}{(x-y)^2}\, dx - \dfrac{1}{2} \int \dfrac{dx}{\sqrt{1-x^2}} = C$

$$\dfrac{-y}{x-y} - \dfrac{1}{2}\sin^{-1} x = C$$

$$\boxed{\dfrac{2y}{x-y} + \sin^{-1} x = C_1}$$ is the G.S.

Ex. 5 : *Solve* $(y^2 \cdot e^{xy^2} + 4x^3)\, dx + (2xy\, e^{xy^2} - 3y^2)\, dy = 0$ **(Dec. 2010, 2013)**

Sol. : Here $\quad M = y^2 e^{xy^2} + 4x^3;$ $\quad \dfrac{\partial M}{\partial y} = 2y\, e^{xy^2} + y^2 e^{xy^2} (2xy)$

and $\quad N = 2xy\, e^{xy^2} - 3y^2;$ $\quad \dfrac{\partial N}{\partial x} = 2y\, e^{xy^2} + 2xy\, e^{xy^2} (y^2)$

$\therefore \quad \dfrac{\partial M}{\partial y} = \dfrac{\partial N}{\partial x}$ $\therefore$ Given differential equation is exact. G.S. is,

G.S. $\quad y^2 \int e^{xy^2}\, dx + 4 \int x^3\, dx - 3 \int y^2\, dy = C$

$$y^2 \dfrac{e^{xy^2}}{y^2} + x^4 - y^3 = C$$

$$\boxed{e^{xy^2} + x^4 - y^3 = C}$$ is the G.S.

Ex. 6 : *Solve* $\left(\dfrac{2x}{y^3}\right) dx + \left(\dfrac{y^2 - 3x^2}{y^4}\right) dy = 0$

Sol. : Here $M = \dfrac{2x}{y^3}$, $N = \dfrac{y^2 - 3x^2}{y^4}$ $\quad \therefore \dfrac{\partial M}{\partial y} = -\dfrac{6x}{y^4} = \dfrac{\partial N}{\partial x}$

$\therefore$ Given differential equation is exact. Its G.S. is

G.S. $\dfrac{2}{y^3} \displaystyle\int x\, dx + \int \dfrac{1}{y^2} dy = C$

$$\boxed{\dfrac{x^2}{y^3} - \dfrac{1}{y} = C}$$

Ex. 7 : *Solve* $\dfrac{dy}{dx} = \dfrac{2x - 3y + 1}{3x + 4y - 5}$.

Sol. : Given equation is of the form

$$\dfrac{dy}{dx} = \dfrac{a_1 x + b_1 y + C_1}{a_2 x + b_2 y + C_2} \text{ where } b_1 = -a_2 \text{ this is } -3 = -(3).$$

Therefore, we shall try exactness.

$\therefore \qquad (2x - 3y + 1)\, dx + (-3x - 4y + 5)\, dy = 0$

$\therefore \qquad \dfrac{\partial M}{\partial y} = -3, \quad \dfrac{\partial N}{\partial x} = -3.$

The given equation is exact and its G.S. is

G.S. $\qquad \displaystyle\int (2x - 3y + 1)\, dx + \int (-4y + 5)\, dy = C$

$\boxed{x^2 - 3xy + x - 2y^2 + 5y = C}$ is the G.S.

EXERCISE 1.5

Show that the following differential equations are exact and solve each equation :

1. $\dfrac{dy}{dx} = -\dfrac{4x^3 y^2 + y \cos xy}{2x^4 y + x \cos xy}$ **(May 2006, 2005)**

    Ans. : $x^4 y^2 + \sin xy = C$

2. $\dfrac{dy}{dx} = \dfrac{x - 2y + 5}{2x + y - 1}$ **Ans. :** $x^2 - 4xy + 10x - y^2 + 2y = C$

3. $\dfrac{1}{2x} \dfrac{dy}{dx} + \dfrac{x + y}{x^2 + y^2} = 0$ **Ans. :** $2x^3 + 3x^2 y + y^3 = C$

4. $x(x^2 + y^2 + a^2)\, dx + y(x^2 + y^2 - b^2)\, dy = 0$

    Ans. : $x^4 + 2x^2 y^2 + 2a^2 x^2 + y^4 - 2b^2 y^2 = C$

5. $\dfrac{dy}{dx} + \dfrac{y \cos x + \sin y + y}{\sin x + x \cos y + x} = 0$ **(Dec. 04)**

    Ans. : $y \sin x + x \sin y + xy = C$

6. $\left[y\left(1 + \dfrac{1}{x}\right) + \cos y\right] dx + [x + \log x - x \sin y]\, dy = 0$

    (Dec. 2006) Ans. : $y(x + \log x) + x \cos y = C$

7. $(x\sqrt{1 - x^2 y^2} - y)\, dy + (x + y\sqrt{1 - x^2 y^2})\, dx = 0$

    Ans. : $x^2 + xy\sqrt{1 - x^2 y^2} + \sin^{-1} xy = C$

8. $\dfrac{dy}{dx} = \dfrac{2x^2 + 3x - 2y + 3}{2y^2 - 3y + 2x - 3}$

    Ans. : $4x^3 + 9x^2 - 12xy + 18x - 4y^3 + 9y^2 + 18y = C$

9. $(x\sqrt{x^2 + y^2} - y)\, dx + (y\sqrt{x^2 + y^2} - x)\, dy = 0$

    Ans. : $(x^2 + y^2)^{3/2} - 3xy = C$

10. $\dfrac{dy}{dx} = \dfrac{1 + y^2 + 3x^2 y}{1 - 2xy - x^3}$ **(May 2017)**

    Ans. : $x(1 + y^2) + x^3 y - y = C$

11. $\dfrac{dy}{dx} = \dfrac{y + 1}{(y + 2)\, e^y - x}$ **(Dec. 2018)**

    Ans. : $(y + 1)(x - e^y) = C$

12. $\left(\dfrac{y^2}{1 + x^2} - 2y\right) dx + (2y \tan^{-1} x - 2x + \sinh y)\, dy = 0$

    Ans. : $y^2 \tan^{-1} x - 2xy + \cosh y = C$

13. $(1 + xy^2)\, dx + (1 + x^2 y)\, dy = 0$ **Ans. :** $x + \dfrac{x^2 y^2}{2} + y = C$

14. $(x^2 - 4xy - 2y^2)\, dx + (y^2 - 4xy - 2x^2)\, dy = 0$

    Ans. : $x^3 - 6x^2 y - 6xy^2 + y^3 = C$

15. $(1 + e^{x/y})\, dx + e^{x/y}\left(1 - \dfrac{x}{y}\right) dy = 0$ given $y(0) = 4$

    Ans. : $x + ye^{x/y} = 4$

16. $(2xy^4 + \sin y)\, dx + (4x^2 y^3 + x \cos y)\, dy = 0$

    Ans. : $x^2 y^4 + x \sin y = C$

17. $(1 + \log xy)\, dx + \left(1 + \dfrac{x}{y}\right) dy = 0$ **Ans. :** $y + x \log (xy) = C$

18. $(2xy + e^y)\, dx + (x^2 + xe^y)\, dy = 0$, given $y\,(1) = 0$

Ans. : $x^2 y + xe^y = 1$

19. $\left(\dfrac{1}{x^2} + \dfrac{3y^2}{x^4}\right) dx = \dfrac{2y}{x^3}\, dy$, given $y\,(1) = \sqrt{3}$

Ans. : $\left(\dfrac{1}{x} + \dfrac{y^2}{x^3}\right) = 4$

20. $y\, dx = (\sin y - x)\, dy$ **Ans. :** $xy + \cos y = C$

21. $(1 + x^2)\,(x\, dy + y\, dx) = -2y\, x^2\, dx$ **Ans. :** $xy\,(1 + x^2) = C$

22. $\cos y - x \sin y\, \dfrac{dy}{dx} = \sec^2 x$ **Ans. :** $\tan x - x \cos y = C$

23. $\dfrac{dy}{dx} = \dfrac{5 - 3x - 2y}{2x + 3y - 5}$

Ans. : $4xy + 3\,(x^2 + y^2) - 10\,(x + y) = C$

24. $\dfrac{dy}{dx} = \dfrac{4x - 2y + 1}{2x - 6y + 2}$ **Ans. :** $2xy + 2y - 3y^2 - x - 2x^2 = C$

25. $(2x^2y + 4x^3 - 12xy^2 + 3y^2 + xe^y + e^{2x})\, dy + (12x^2y + 2xy^2 + 4x^3 - 4y^3 + 2ye^{2x} + e^y)\, dx = 0$

Ans. : $4x^3y + x^2y^2 + x^4 - 4y^3x + ye^{2x} + xe^y + y^3 = C$

26. $(\sin x \cos y + e^{2x})\, dx + (\cos x \sin y + \tan y)\, dy = 0$

Ans. : $-\cos x \cos y + \dfrac{e^{2x}}{2} + \log \sec y = C$

27. $\left[\log\,(x^2 + y^2) + \dfrac{2x^2}{x^2 + y^2}\right] dx + \dfrac{2xy}{x^2 + y^2}\, dy = 0$

(May 08, 13; Dec. 07) Ans. : $x \log\,(x^2 + y^2) = C$

Hint : Use Rule 3 for G.S.

For G.S. take $\displaystyle\int N\, dy + \int M\, dx = C$

1.4.3 Equations Reducible to Exact form by using Integrating Factor

Integrating Factor : Definition : A function k (x, y) is said to be an Integrating Factor (I.F.) of the equation M dx + N dy = 0. If it is possible to obtain a function u (x, y) such that k (M dx + N dy) = du. In other words, an I.F. is a multiplying factor by which the equation can be made exact.

Rules for finding integrating factors of the equation M dx + N dy = 0 when it is not exact :

Rule 1 : If $x \cdot M + y \cdot N \neq 0$ and the given differential equation is homogeneous,

then, $\boxed{\text{I.F.} = \dfrac{1}{x \cdot M + y \cdot N}}$

Rule 2 : If $x \cdot M - y \cdot N \neq 0$ and the given D.E. has the form

$y \cdot f_1(xy)\, dx + x \cdot f_2\,(xy)\, dy = 0$ then, $\boxed{\text{I.F.} = \dfrac{1}{x \cdot M - y \cdot N}}$

Rule 3 : If $\dfrac{\dfrac{\partial M}{\partial y} - \dfrac{\partial N}{\partial x}}{N} = f\,(x)$ (say)   then, $\boxed{\text{I.F.} = e^{\int f\,(x)\, dx}}$

Rule 4 : If $\dfrac{\dfrac{\partial N}{\partial x} - \dfrac{\partial M}{\partial y}}{M} = \phi\,(y)$ (say)   then, $\boxed{\text{I.F.} = e^{\int \phi\,(y)\, dy}}$

Rule 5 : If the equation M dx + N dy = 0 can be written as,

$x^a\, y^b\,(my\, dx + nx\, dy) + x^r\, y^s\,(py\, dx + qx\, dy) = 0$

where a, b, m, n, r, s, p and q are all constants having any value, then the $\boxed{\text{I.F.} = x^h\, y^k}$, where h and k are such that after multiplying the integrating factor, the condition of exactness is satisfied.

Note : h, k can be determined from the following two equations :

$$nh - mk = (m - n) + (mb - na)$$
$$qh - pk = (p - q) + (ps - qr), \text{ provided } mq - np \neq 0$$

Illustrations on Equations Reducible to Exact Form

Ex. 1 : *Solve* $(xy - 2y^2)\, dx - (x^2 - 3xy)\, dy = 0$

Sol. : Here $M = xy - 2y^2$ and $N = -x^2 + 3xy$

$$\frac{\partial M}{\partial y} \neq \frac{\partial N}{\partial x}$$

∴ Given D.E. is not exact. But it is homogeneous.

Also, $x \cdot M + y \cdot N = x^2 y - 2xy^2 - x^2y + 3xy^2 = xy^2 \neq 0$

∴ (By using rule 1), I.F. $= \dfrac{1}{x \cdot M + N \cdot y} = \dfrac{1}{xy^2}$

Given equation becomes $\left(\dfrac{1}{y} - \dfrac{2}{x}\right) dx + \left(-\dfrac{x}{y^2} + \dfrac{3}{y}\right) dy = 0$ which is exact. Its G.S. is

$$\boxed{\dfrac{x}{y} - 2 \log x + 3 \log y \;=\; C}$$ is the G.S.

Ex. 2 : Solve $(x^2 y^2 + 2) \, y \, dx + (2 - 2x^2 y^2) \, x \, dy = 0$

Sol. : It can be seen that given equation is non-exact. It is in the form,

$f_1 (xy) \cdot y \, dx + f_2 (xy) \cdot x \, dy = 0$ (by using rule 2)

$$\therefore \qquad \text{I.F.} = \dfrac{1}{Mx - Ny} = \dfrac{1}{x^3 y^3 + 2xy - 2xy + 2x^3 y^3} = \dfrac{1}{3x^3 y^3}$$

Now, multiplying by $\dfrac{1}{3x^3 y^3}$ $\left(\text{or } \dfrac{1}{x^3 y^3}\right)$ to given equation, we get,

$$\dfrac{1}{3x^3 y^3} (x^2 y^3 + 2y) \, dx + \dfrac{1}{3x^3 y^3} (2x - 2x^3 y^2) \, dy = 0$$

$$\left(\dfrac{1}{3x} + \dfrac{2}{3x^3 y^2}\right) dx + \left(\dfrac{2}{3x^2 y^3} - \dfrac{2}{3y}\right) dy = 0 \qquad\qquad \text{...(1)}$$

Equation (1) is exact. Its G.S. is,

$$\int \dfrac{1}{3x} \, dx + \dfrac{2}{3y^2} \int \dfrac{1}{x^3} \, dx - \dfrac{2}{3} \int \dfrac{1}{y} \, dy = C$$

$$\log x - \dfrac{1}{y^2 x^2} - 2 \log y = C_1$$

$$\boxed{\log \left(\dfrac{x}{y^2}\right) - \dfrac{1}{x^2 y^2} \;=\; C_1}$$ is the required G.S.

Ex. 3 : Solve $y (1 + xy) \, dx + x (1 + xy + x^2 y^2) \, dy = 0$

Sol. : Given equation is not exact. To find I.F. (by rule 2), we have

$$\text{I.F.} = \dfrac{1}{Mx - Ny} = \dfrac{1}{xy + x^2 y^2 - xy - x^2 y^2 - x^3 y^3} = \dfrac{1}{-x^3 y^3}$$

Multiplying by $\dfrac{-1}{x^3 y^3}$ to given equation,

$$\left(-\dfrac{1}{x^3 y^2} - \dfrac{1}{x^2 y}\right) dx + \left(\dfrac{-1}{x^2 y^3} - \dfrac{1}{xy^2} - \dfrac{1}{y}\right) dy = 0 \qquad\qquad \text{...(1)}$$

For equation (1), $\dfrac{\partial M}{\partial y} = \dfrac{2}{x^3 y^3} + \dfrac{1}{x^2 y^2} = \dfrac{\partial N}{\partial x}$

Equation (1) is exact. Its G.S. is given by,

$$-\dfrac{1}{y^2} \int \dfrac{1}{x^3} \, dx - \dfrac{1}{y} \int \dfrac{1}{x^2} \, dx - \int \dfrac{1}{y} \, dy = C$$

$$\boxed{\dfrac{1}{2y^2 x^2} + \dfrac{1}{xy} - \log y \;=\; C}$$ is the G.S.

Ex. 4 : Solve $(x^2 + y^2 + x) \, dx + (xy) \, dy = 0$ **(Dec. 2013)**

Sol. : We have $(x^2 + y^2 + x) \, dx + (xy) \, dy = 0$ is of the type $M \, dx + N \, dy = 0$ giving us   ... (1)

$$M = x^2 + y^2 + x, \; N = xy \;\; \therefore \;\; \dfrac{\partial M}{\partial y} = 2y \,;\; \dfrac{\partial N}{\partial x} = y \,;\; \therefore \;\; \text{Equation (1) is not exact.}$$

To find the I.F. (by using rule 3),

Consider, $\quad \dfrac{\dfrac{\partial M}{\partial y} - \dfrac{\partial N}{\partial x}}{N} = \dfrac{2y - y}{xy} = \dfrac{1}{x} = f(x),$

$$\therefore \qquad \text{I.F.} = e^{\int f(x) \, dx} = e^{\int \frac{1}{x} \, dx} = e^{\log x} = x$$

Now multiplying by I.F. = x to equation (1), we get,

$$(x^3 + xy^2 + x^2)\ dx + (x^2 y)\ dy\ = 0\quad ...(2)$$

Thus, equation (2) is exact. Its G.S. is given by,

G.S. $\displaystyle\int (x^3 + xy^2 + x^2)\ dx = C \Rightarrow \dfrac{x^4}{4} + \dfrac{x^2 y^2}{2} + \dfrac{x^3}{3} = C$

or $\boxed{3x^4 + 6x^2 y^2 + 4x^3 = C_1}$ is the required G.S.

Ex. 5 : *Solve* $(x^4 e^x - 2mxy^2)\ dx\ +\ (2mx^2 y)\ dy\ =\ 0$ **(May 2011, 2013)**

Sol. : Given equation is

$$(x^4 e^x - 2mxy^2)\ dx + (2mx^2 y)\ dy\ = 0 \qquad ...(1)$$

is of the type $M\ dx + N\ dy = 0$

$$M = x^4 e^x - 2mxy^2 \qquad N = 2mx^2 y$$

$$\frac{\partial M}{\partial y} = -4mxy \qquad\qquad \frac{\partial N}{\partial x} = 4mxy$$

Equation (1) is non-exact. To find I.F. (by using Rule 3), we consider,

$$\frac{\dfrac{\partial M}{\partial y} - \dfrac{\partial N}{\partial x}}{N} = \frac{-8mxy}{2mx^2 y} = \frac{-4}{x}\ .$$

$$\text{I.F.} = e^{-4 \int \frac{1}{x}\ dx} = e^{-4 \log x} = \frac{1}{x^4}$$

Now multiplying to equation (1) by $\dfrac{1}{x^4}$,

$$\left(e^x - \frac{2my^2}{x^3}\right)\ dx + \left(\frac{2my}{x^2}\right)\ dy = 0 \qquad ...(2)$$

Equation (2) is exact. Its G.S. is,

$$\int \left(e^x - \frac{2my^2}{x^3}\right)\ dx = C$$

$$\boxed{e^x + \frac{my^2}{x^2} = C}\ \text{is the G.S.}$$

Ex. 6 : *Solve* $y\ (2x^2 y + e^x)\ dx\ =\ (e^x + y^3)\ dy$ **(May 2009, Dec. 2004)**

Sol. : Given equation is $(2x^2 y^2 + ye^x)\ dx + (-e^x - y^3)\ dy\ = 0$ $\qquad ...(1)$

Here $\dfrac{\partial M}{\partial y} = 4x^2 y + e^x, \quad \dfrac{\partial N}{\partial x} = -e^x \qquad \therefore$ (1) is not exact.

(By using Rule 4), consider,

$$\frac{\dfrac{\partial N}{\partial x} - \dfrac{\partial M}{\partial y}}{M} = \frac{-e^x - 4x^2 y - e^x}{y\ (2x^2 y + e^x)} = \frac{-2\ (2x^2 y + e^x)}{y\ (2x^2 y + e^x)} = -\frac{2}{y}\ .$$

$\therefore\qquad$ I.F. $= e^{-2 \int \frac{1}{y}\ dy} = e^{-2 \log y} = \dfrac{1}{y^2}$

$\therefore$ Equation (1) becomes

$$\left(2x^2 + \frac{e^x}{y}\right)\ dx + \left(-\frac{e^x}{y^2} - y\right)\ dy = 0 \qquad ...(2)$$

Equation (2) is exact and G.S. is

G.S. $\displaystyle\int 2x^2\ dx + \frac{1}{y} \int e^x\ dx - \int y\ dy = C$

$$\boxed{\frac{2x^3}{3} + \frac{e^x}{y} - \frac{y^2}{2} = C}\ \text{is the G.S.}$$

Ex. 7 : Solve $(x \sec^2 y - x^2 \cos y)\, dy = (\tan y - 3x^4)\, dx$ *(Dec. 2007)*

Sol. : Given equation is $(\tan y - 3x^4)\, dx + (x^2 \cos y - x \sec^2 y)\, dy = 0$ ...(1)

Equation (1) is not exact. To find I.F. consider (by using Rule 3)

$$\frac{\frac{\partial M}{\partial y} - \frac{\partial N}{\partial x}}{N} = \frac{\sec^2 y - 2x \cos y + \sec^2 y}{x\,(x \cos y - \sec^2 y)} = \frac{-2\,[x \cos y - \sec^2 y]}{x\,[x \cos y - \sec^2 y]} = -\frac{2}{x} = f(x)$$

$\therefore \qquad$ I.F. $= e^{\int -\frac{2}{x}\, dx} = e^{-2\log x}$

$\therefore \qquad$ I.F. $= \dfrac{1}{x^2}$

Multiplying to equation (1)

$$\left(\frac{\tan y}{x^2} - 3x^2\right) dx + \left(\cos y - \frac{\sec^2 y}{x}\right) dy = 0 \qquad \qquad \text{...(2)}$$

Equation (2) is exact. Its G.S. is,

$$\text{G.S.} \qquad \tan y \int \frac{1}{x^2}\, dx - 3\int x^2\, dx + \int \cos y\, dy = C$$

$$-\frac{\tan y}{x} - x^3 + \sin y = C$$

or $\qquad \boxed{\dfrac{\tan y}{x} + x^3 - \sin y = C_1}$ is the G.S.

Ex. 8 : Solve $\left(y + \dfrac{y^3}{3} + \dfrac{x^2}{2}\right) dx + \left(\dfrac{x + xy^2}{4}\right) dy = 0$

Sol. : Here $M = y + \dfrac{y^3}{3} + \dfrac{x^2}{2}$; $N = \dfrac{x + xy^2}{4}$

and $\qquad \dfrac{\partial M}{\partial y} = 1 + y^2$; $\dfrac{\partial N}{\partial x} = \dfrac{1 + y^2}{4}$

Given equation is not exact. To find I.F. (By using rule 3), consider,

$$\frac{\frac{\partial M}{\partial y} - \frac{\partial N}{\partial x}}{N} = \frac{(1 + y^2) - \left(\dfrac{1 + y^2}{4}\right)}{x\left(\dfrac{1 + y^2}{4}\right)} = \frac{3}{x} = f(x)$$

$\therefore \qquad$ I.F. $= e^{\int f(x)\, dx} = e^{\int \frac{3}{x}\, dx} = e^{3\log x} = x^3$

Multiplying by x^3 to given equation,

$$\left(yx^3 + \frac{x^3 y^3}{3} + \frac{x^5}{2}\right) dx + \left(\frac{x^4 + x^4 y^2}{4}\right) dy = 0 \qquad \qquad \text{...(1)}$$

Equation (1) is exact and its G.S. is,

$$\text{G.S.} \qquad y \int x^3\, dx + \frac{y^3}{3} \int x^3\, dx + \frac{1}{2} \int x^5\, dx = C$$

or $\qquad \boxed{3y\, x^4 + x^4 y^3 + x^6 = C_1}$ is the G.S.

Ex. 9 : Solve $(y^4 + 2y)\, dx + (xy^3 + 2y^4 - 4x)\, dy = 0$ *(Dec. 2009, 2007)*

Sol. : Here $M = y^4 + 2y$; $\qquad N = xy^3 + 2y^4 - 4x$

$\qquad \dfrac{\partial M}{\partial y} = 4y^3 + 2$; $\qquad \dfrac{\partial N}{\partial x} = y^3 - 4$

Given equation is not exact. To find I.F. (By using rule 4), consider,

$$\frac{\frac{\partial N}{\partial x} - \frac{\partial M}{\partial y}}{M} = \frac{y^3 - 4 - 4y^3 - 2}{y\,(y^3 + 2)} = \frac{-3\,(y^3 + 2)}{y\,(y^3 + 2)} = -\frac{3}{y}$$

$$\therefore \quad \text{I.F.} = e^{-3\int \frac{1}{y}dy} = e^{-3\log y} = \frac{1}{y^3}$$

Multiplying to given equation by $\dfrac{1}{y^3}$,

$$\left(y + \frac{2}{y^2}\right)\, dx \; + \; \left(x + 2y - \frac{4x}{y^3}\right)\, dy = 0 \qquad\qquad …(1)$$

$\therefore$ Equation (1) is exact and its G.S. is,

$$\boxed{\left(y + \frac{2}{y^2}\right)\int dx + 2\int y\, dy = C \Rightarrow \left(y + \frac{2}{y^2}\right)x + y^2 = C}\ \text{ is the G.S.}$$

Ex. 10 : *Solve* $(x^2 y + y^4)\, dx + (2x^3 + 4xy^3)\, dy = 0$ **(Dec. 2010, 2005; May 2010)**

Sol. : Here $M = x^2 y + y^4$; $N = 2x^3 + 4xy^3$ and $\dfrac{\partial M}{\partial y} \neq \dfrac{\partial N}{\partial x}$

Given equation is not exact. To find I.F., the above equation can be put in the form

$$(x^2 y\, dx + 2x^3\, dy) + (y^4\, dx + 4xy^3\, dy) = 0 \quad \text{i.e. } x^2\,(y\, dx + 2x\, dy) + y^3\,(y\, dx + 4x\, dy) = 0$$

which is of the form $x^a y^b\, (my\, dx + nx\, dy) + x^r y^s\, (py\, dx + qx\, dy) = 0$

where $a = 2$, $b = 0$, $m = 1$, $n = 2$, $r = 0$, $s = 3$, $p = 1$, $q = 4$; (use rule 5)

Hence, the integrating factor must be $x^h y^k$.

Multiplying the original equation by the I.F. $= x^h y^k$,

$$x^h y^k\,(x^2 y + y^4)\, dx + x^h y^k\,(2x^3 + 4xy^3)\, dy = 0$$

or $\qquad (x^{h+2} y^{k+1} + x^h y^{k+4})\, dx + (2x^{h+3} y^k + 4x^{h+1} y^{k+3})\, dy = 0$

$$\frac{\partial M}{\partial y} = (k + 1)\, x^{h+2} y^k + (k + 4)\, x^h y^{k+3} ;$$

$$\frac{\partial N}{\partial x} = 2\,(h + 3)\, x^{h+2} y^k + 4\,(h + 1)\, x^h y^{k+3}$$

But $\qquad \dfrac{\partial M}{\partial y} = \dfrac{\partial N}{\partial x}$ for exactness, which requires,

$$k + 1 = 2\,(h + 3) \text{ and } k + 4 = 4\,(h + 1) \text{ giving } k = 10, \ h = \frac{5}{2}$$

$\therefore \qquad$ I.F. $= x^h y^k = x^{5/2} y^{10}$.

Multiplying by the I.F. to the original differential equation, we get,

$$(x^2 y + y^4)\, x^{5/2} y^{10}\, dx + (2x^3 + 4xy^3)\, x^{5/2} y^{10}\, dy = 0$$

or $\qquad (x^{9/2} y^{11} + x^{5/2} y^{14})\, dx + (2x^{11/2} y^{10} + 4x^{7/2} y^{13})\, dy = 0 \qquad\qquad …(1)$

Hence, equation (1) becomes exact. Its G.S. is,

G.S. $\qquad y^{11}\displaystyle\int x^{9/2}\, dx + y^{14}\int x^{5/2}\, dx = C \Rightarrow \frac{2y^{11}}{11}\, x^{11/2} + \frac{2y^{14}}{7}\, x^{7/2} = C$

or $\qquad \boxed{7\, x^{11/2} y^{11} + 11\, x^{7/2} y^{14} = C_1}$ is the required G.S.

Ex. 11 : *Solve* $(y^3 - 2x^2 y)\, dx + (2xy^2 - x^3)\, dy = 0$ **(May 2008, 2004)**

Sol. : Here $M = y^3 - 2x^2 y$; $N = 2xy^2 - x^3$ and $\dfrac{\partial M}{\partial y} \neq \dfrac{\partial N}{\partial x}$

Given equation can be written as,

$$(y^3\, dx + 2xy^2\, dy) - (2x^2 y\, dx + x^3\, dy) = 0$$

$$\Rightarrow \qquad y^2\,(y\, dx + 2x\, dy) + x^2\,(-2y\, dx - x\, dy) = 0 \qquad\qquad \text{(use Rule 5)}$$

Here $a = 0$, $b = 2$, $m = 1$, $n = 2$, $r = 2$, $s = 0$, $p = -2$, $q = -1$.

$\therefore$ I.F. $= x^h y^k$

$\therefore$ $x^h y^k (y^3 - 2x^2 y)\ dx + x^h y^k (2xy^2 - x^3)\ dy = 0$

or $(x^h y^{k+3} - 2x^{h+2} y^{k+1})\ dx + (2x^{h+1} y^{k+2} - x^{h+3} y^k)\ dy = 0$

$$\frac{\partial M}{\partial y} = (k + 3)\ x^h y^{k+2} - 2\ (k + 1)\ x^{h+2} y^k\ ;$$

$$\frac{\partial N}{\partial x} = 2\ (h + 1)\ x^h y^{k+2} - (h + 3)\ x^{h+2} y^k$$

But $\dfrac{\partial M}{\partial y} = \dfrac{\partial N}{\partial x}$ for exactness, which requires,

$k + 3 = 2\ (h + 1)\ ;\ -2\ (k + 1)\ =\ -(h + 3)$ giving $h = 1$, $k = 1$.

$\therefore$ I.F. $= x^h y^k = xy$

Multiplying the given equation by xy, we get,

$$(xy^4 - 2x^3 y^2)\ dx + (2x^2 y^3 - x^4 y)\ dy = 0 \qquad \qquad \text{...(1)}$$

Hence, equation (1) is exact and its G.S. is

G.S. $y^4 \displaystyle\int x\ dx - 2y^2 \int x^3\ dx = C$

or $\boxed{y^4 x^2 - y^2 x^4 = C_1}$ is the required G.S.

Note : h, k can be determined by using short-cut method.

$\qquad nh - mk = m - n\ + (mb - na)$

$\Rightarrow \qquad 2h - k = -1 + (2 - 0)\ \Rightarrow\ 2h - k = 1.$ *(See imp. note in rule 5)*

Also $qh - pk\ + (ps - qr)\ \Rightarrow\ -h\ + 2k\ =\ -1 + (0 + 2) \Rightarrow -h + 2k = 1.$

Solving $h = 1$, $k = 1$. $\therefore$ I.F. $= xy$

EXERCISE 1.6

Solve the following differential equations :

Examples on Rule 1 :

1. $(x^2 y - 2xy^2)\ dx - (x^3 - 3x^2 y)\ dy = 0.$ $\left(\text{I.F.} = \dfrac{1}{x^2 y^2}\right)$

     (May 2005) Ans. : $\dfrac{x}{y} - 2 \log x + 3 \log y = C$

2. $(3xy^2 - y^3)\ dx + (xy^2 - 2x^2 y)\ dy = 0.$ $\left(\text{I.F.} = \dfrac{1}{x^2 y^2}\right)$

     Ans. : $\dfrac{Cy^2}{x^3} = e^{y/x}$

3. $(x^2 - 3xy + 2y^2)\ dx + x\ (3x - 2y)\ dy = 0.$ $\left(\text{I.F.} = \dfrac{1}{x^3}\right)$

     Ans. : $x^2 \log x + 3xy - y^2 = Cx^2$

4. $x\ (x - y)\ \dfrac{dy}{dx} = y\ (x + y).$ $\left(\text{I.F.} = \dfrac{1}{2xy^2}\right)$ **(May 2018)**

     Ans. : $xy^2 = C$

Examples on Rule 2 :

5. $(x^2 y^2\ + xy\ + 1)\ y\ dx + (x^2 y^2 - xy\ + 1)\ x\ dy = 0$

     $\left(\text{I.F.} = \dfrac{1}{2x^2 y^2}\right)$ **(May 2006)**

     Ans. : $xy + \log x - \dfrac{1}{xy} - \log y = C$

6. $(x^2 y^2\ + 5xy\ + 2)\ y\ dx + (x^2 y^2\ + 4xy\ + 2)\ x\ dy = 0$

     $\left(\text{I.F.} = \dfrac{1}{x^2 y^2}\right)$ **(Dec. 2006)**

     Ans. : $xy + 5 \log x + 4 \log y - \dfrac{2}{xy} = C$

7. $y\ (xy - 3)\ dx + x\ (3xy - 3)\ dy = 0.$ $\left(\text{I.F.} = \dfrac{-1}{2x^2 y^2}\right)$

     Ans. : $\log (xy^3)\ + \dfrac{3}{xy} = C$

8. $y\ (xy + 2x^2 y^2)\ dx + x\ (xy - x^2 y^2)\ dy = 0.$ $\left(\text{I.F.} = \dfrac{1}{3x^3 y^3}\right)$

     (May 2009) Ans. : $-\dfrac{1}{xy}\ + 2 \log x - \log y = C$

9. $(1 + xy)\ y\ dx + (1 - xy)\ x\ dy = 0.$ $\left(\text{I.F.} = \dfrac{1}{2x^2 y^2}\right)$

     (Dec. 2011, 2008, 2017) Ans. : $\log \dfrac{x}{y}\ - \dfrac{1}{xy} = C$

10. $(xy \sin xy + \cos xy)\ y\ dx + (xy \sin xy - \cos xy)\ x\ dy = 0.$

     $\left(\text{I.F.}\ = \dfrac{1}{2xy \cos xy}\right)$ **Ans. :** $y \cos xy = cx$

11. $(x^3y^3 + x^2y^2 + xy + 1)\, y\, dx + (x^3y^3 + x^2y^2 - xy - 1)\, x\, dy = 0$

$\left(\text{I.F.} = \dfrac{1}{2xy}\right)$ **Ans. :** $\dfrac{x^2y^2}{2} + \log\dfrac{x}{y} = C$

[First factorise M and N and cancel a common factor $(xy + 1)$]

Examples on Rule 3 :

12. $(2x \log x - xy)\, dy + 2y\, dx = 0.$ $\left(\text{I.F.} = \dfrac{1}{x}\right)$ **(Nov. 2015)**

Ans. : $2y \log x - \dfrac{y^2}{2} = C$

13. $(y - 2x^3)\, dx - x\,(1 - xy)\, dy = 0.$ $\left(\text{I.F.} = \dfrac{1}{x^2}\right)$ **(Dec. 2004)**

Ans. : $\dfrac{y}{x} + x^2 - \dfrac{y^2}{2} = C$

14. $\left[2x \sinh\dfrac{y}{x} + 3y \cosh\dfrac{y}{x}\right] dx - 3x \cosh\dfrac{y}{x}\, dy = 0.$

$\left(\text{I.F.} = x^{-8/3}\right)$ **Ans. :** $x^{-2/3} \sinh\dfrac{y}{x} = C$

15. $(x^4 e^x - 2mxy^2)\, dx + 2mx^2y\, dy = 0.$ $\left(\text{I.F.} = \dfrac{1}{x^4}\right)$

(May 2007)   **Ans. :** $e^x + \dfrac{my^2}{x^2} = C$

16. $(xy^2 - e^{1/x^3})\, dx - x^2y\, dy = 0.$ $\left(\text{I.F.} = \dfrac{1}{x^4}\right)$

Ans. : $2x^2 e^{1/x^3} - 3y^2 = cx^2$

17. $(x - y^2)\, dx + 2xy\, dy = 0.$ $\left(\text{I.F.} = \dfrac{1}{x^2}\right)$

Ans. : $\dfrac{y^2}{x} + \log x = C$

18. $(x^2 + y^2 + 1)\, dx - 2xy\, dy = 0.$ $\left(\text{I.F.} = \dfrac{1}{x^2}\right)$

(May 2014, 2015) Ans. : $x^2 - y^2 - 1 = cx$

19. $(20x^2 + 8xy + 4y^2 + 3y^3)y\, dx + 4(x^2 + xy + y^2 + y^3)x\, dy = 0$ $(\text{I.F.} = x^2)$ **(Dec. 2006)**

Ans. : $4x^5y + 2x^4y^2 + \dfrac{4}{3}x^3y^3 + y^4x^3 = C$

Examples on Rule 4 :

20. $(3x^2y^4 + 2xy)\, dx + (2x^3y^3 - x^2)\, dy = 0.$ $\left(\text{I.F.} = \dfrac{1}{y^2}\right)$

Ans. : $x^3 y^2 + \dfrac{x^2}{y} = C$

21. $y\,(x^2y + e^x)\, dx - e^x\, dy = 0.$ $\left(\text{I.F.} = \dfrac{1}{y^2}\right)$

Ans. : $\dfrac{x^3}{3} + \dfrac{e^x}{y} = C$

22. $y\,(2xy + e^x)\, dx = e^x\, dy.$ $\left(\text{I.F.} = \dfrac{1}{y^2}\right)$ **Ans. :** $x^2 + \dfrac{e^x}{y} = C$

23. $(2x + e^x \log y)\, y\, dx + e^x\, dy = 0.$ $\left(\text{I.F.} = \dfrac{1}{y}\right)$

(Dec. 2009, 2005; May 2008)   **Ans. :** $x^2 + e^x \log y = C$

24. $\dfrac{dy}{dx}(x + 2y^3) = y + 2x^3y^2.$ $\left(\text{I.F.} = \dfrac{1}{y^2}\right)$

Ans. : $\dfrac{x}{y} + \dfrac{x^4}{2} - y^2 = C$

25. $\left(\dfrac{y}{x}\sec y - \tan y\right) dx = (x - \sec y \log x)\, dy.$ $\left(\text{I.F.} = \dfrac{1}{\sec y}\right)$
(May 2005)   **Ans. :** $y \log x - x \sin y = C$

26. $y \log y\, dx + (x - \log y)\, dy = 0.$ $\left(\text{I.F.} = \dfrac{1}{y}\right)$

(Dec. 2010, May 2016)   **Ans. :** $2x \log y - (\log y)^2 = C$

Examples on Rule 5 :

27. $y\,(3y + 10x^2)\, dx - 2x\,(y + 3x^2)\, dy = 0.$ $(\text{I.F.} = x^2 y^{-4})$

Ans. : $x^3 y^{-2} + 2x^5 y^{-3} = C$

28. $(y^2 + 2yx^2)\, dx + (2x^3 - xy)\, dy = 0.$ $(\text{I.F.} = x^{-5/2} y^{-1/2})$

Ans. : $6\sqrt{xy} - x^{-3/2} y^{3/2} = C$

29. $(2x^2y^2 + y)\, dx - (x^3y - 3x)\, dy = 0.$ $(\text{I.F.} = x^{-11/7} y^{-19/7})$

Ans. : $4x^{10/7} y^{-5/7} - 5x^{-4/7} y^{-12/7} = C$

30. $(2x^2y - 3y^4)\, dx + (3x^3 + 2xy^3)\, dy = 0$ **(May 2007, 2006)**
[**Hint :** $(\text{I.F.} = x^{-49/13} \cdot y^{-23/13})$[

Ans. : $5x^{-36/13} y^{24/13} - 12x^{-10/13} y^{-15/13} = C$

31. $(2y + 6xy^2)\, dx + (3x + 8x^2y)\, dy = 0.$ $(\text{I.F.} = xy^2)$

(May 05) Ans. : $x^2 y^3 + 2x^2y^4 = C$

32. $(3x + 2y^2)\, y\, dx + 2x\,(2x + 3y^2)\, dy = 0.$ $(\text{I.F.} = xy^3)$

(May 09) Ans. : $x^3y^4 + x^2 y^6 = C$

33. $(3xy + 8y^5)\, dx + (2x^2 + 24xy^4)\, dy = 0.$ $(\text{I.F.} = xy)$

(Dec. 2006) Ans. : $x^3y^2 + 4x^2y^6 = C$

34. $(y^3 - 2x^2y)\, dx - (2xy^2 + x^3)\, dy = 0.$ $(\text{I.F.} = x^{-11/5} y^{-3/5})$

Ans. : $x^{-6/5} y^{12/5} + 3x^{4/5} y^{2/5} = C$

35. $(x^7y^2 + 3y)\, dx + (3x^8y - x)\, dy = 0.$ $(\text{I.F.} = x^{-7} y)$

Ans. : $2xy^3 - y^2x^{-6} = C$

1.4.4 Integrating Factors Found by Inspection

Integrating factors may sometimes be seen at a glance for some differential equations. For this purpose the following *integrable combinations* will be found useful in guessing the integrating factors.

1. $x\,dy + y\,dx = d\,(xy)$

2. $\dfrac{x\,dy + y\,dx}{xy} = d\,(\log\,(xy))$

3. $\dfrac{x\,dy - y\,dx}{x^2} = d\left(\dfrac{y}{x}\right)$

4. $\dfrac{x\,dy - y\,dx}{xy} = d\left[\log\left(\dfrac{y}{x}\right)\right]$

5. $\dfrac{x\,dy - y\,dx}{x^2 + y^2} = d\left[\tan^{-1}\left(\dfrac{y}{x}\right)\right]$

6. $\dfrac{x\,dy - y\,dx}{x^2 - y^2} = d\left(\dfrac{1}{2}\log\dfrac{x+y}{x-y}\right)$

7. $\dfrac{y\,dx - x\,dy}{y^2} = d\left(\dfrac{x}{y}\right)$

8. $\dfrac{y\,dx - x\,dy}{x^2 + y^2} = d\left(\tan^{-1}\left(\dfrac{x}{y}\right)\right)$

9. $\dfrac{y\,dx - x\,dy}{xy} = d\left(\log\left(\dfrac{x}{y}\right)\right)$

10. $\dfrac{x\,dx + y\,dy}{x^2 + y^2} = \dfrac{1}{2}\,d\,(\log\,(x^2 + y^2))$

11. $\dfrac{x\,dx + y\,dy}{\sqrt{x^2 + y^2}} = d\left(\sqrt{x^2 + y^2}\right)$

12. $x\,dx + y\,dy = \dfrac{1}{2}\,d\,(x^2 + y^2)$

13. $dx + dy = d\,(x + y)$

14. $\dfrac{dx + dy}{x + y} = d\,\log\,(x + y)$

15. $(x + y)^n\,(dx + dy) = d\left[\dfrac{(x + y)^{n+1}}{n + 1}\right]$ if $n \neq -1$

16. $\dfrac{x\,dy + y\,dx}{x^2\,y^2} = d\left(\dfrac{-1}{xy}\right)$

17. $\dfrac{y \cdot 2x\,dx - x^2\,dy}{y^2} = d\left(\dfrac{x^2}{y}\right)$

18. $\dfrac{x \cdot 2y\,dy - y^2\,dx}{x^2} = d\left(\dfrac{y^2}{x}\right)$

19. $\dfrac{2x^2\,y\,dy - 2y^2\,x\,dx}{x^4} = d\left(\dfrac{y^2}{x^2}\right)$

20. $\dfrac{2xy^2\,dx - 2yx^2\,dy}{y^4} = d\left(\dfrac{x^2}{y^2}\right)$

21. $\dfrac{y\,e^x\,dx - e^x\,dy}{y^2} = d\left(\dfrac{e^x}{y}\right)$

Illustrations on I.F. found by Inspection :

Ex. 1 : *Solve* $\quad (x + y)^2\left(x\dfrac{dy}{dx} + y\right) = xy\left(1 + \dfrac{dy}{dx}\right)$

Sol. : Given $\quad (x + y)^2\,\dfrac{(x\,dy + y\,dx)}{dx} = xy\,\dfrac{(dx + dy)}{dx}$

G.S. $\quad \displaystyle\int \dfrac{x\,dy + y\,dx}{xy} = \int \dfrac{dx + dy}{(x + y)^2}$

$\qquad \log xy = \dfrac{-1}{x + y} + C$

$$\boxed{\;\log\,(xy) + \dfrac{1}{x + y} = C\;}\;\text{ is the G.S.}$$

Ex. 2 : *Solve* $\;y\,dx - x\,dy + 3\,x^2\,y^2\,e^{x^3}\,dx = 0$

Sol. : Divide by y^2

$\therefore \qquad \dfrac{y\,dx - x\,dy}{y^2} + e^{x^3} \cdot 3x^2\,dx = 0 \;\text{ or }\; d\left(\dfrac{x}{y}\right) + d\,(e^{x^3}) = 0$

$$\boxed{\;\dfrac{x}{y} + e^{x^3} = C\;}\;\text{ is the G.S.}$$

Ex. 3 : *Solve* $x\,dy - y\,dx = (4x^2 + y^2)\,dy$

Sol. : Given : $\dfrac{x\,dy - y\,dx}{4x^2 + y^2} = dy \implies \dfrac{(x\,dy - y\,dx)/x^2}{4 + y^2/x^2} = dy$

G.S. $\displaystyle\int \dfrac{d\,(y/x)}{4 + (y/x)^2} = \int dy + C$

$\therefore$ $\boxed{\dfrac{1}{2}\tan^{-1}\left(\dfrac{y}{2x}\right) = y + C}$ is the required G.S.

Ex. 4 : *Solve* $e^{x+y}\left(x\dfrac{dy}{dx} + y\right) = e^{xy}\left(1 + \dfrac{dy}{dx}\right)$

Sol. : $e^{x+y}(x\,dy + y\,dx) = e^{xy}(dx + dy)$ or $\dfrac{x\,dy + y\,dx}{e^{xy}} = \dfrac{dx + dy}{e^{x+y}}$

G.S. $\displaystyle\int e^{-xy}\,d\,(xy) = \int e^{-(x+y)}\,d\,(x+y)$

$\boxed{e^{-xy} = e^{-(x+y)} + C}$ is the required G.S.

$$\boxed{\textbf{EXERCISE 1.7}}$$

Solve the following differential equations : [**Hint :** Use I.F. by inspection.]

1. $x\,dy - y\,dx = (x^2 + y^2)(dx + dy)$

    Ans. : $\tan^{-1}\dfrac{y}{x} = x + y + C$

2. $y(2xy + e^x)\,dx = e^x\,dy\ \left(\text{I.F.} = \dfrac{1}{y^2}\right)$ **Ans. :** $x^2 + \dfrac{e^x}{y} = C$

3. $(x^4 e^x - 2mx\,y^2)\,dx + 2mx^2\,y\,dy = 0\ \left(\text{I.F.} = \dfrac{1}{x^4}\right)$

    Ans. : $e^x + \dfrac{my^2}{x^2} = C$

4. $(y + x^3\,y^2)\,dx + x\,dy = 0$

    $\left(\textbf{Hint :}\ \dfrac{d\,(xy)}{x^2y^2} + d\left(\dfrac{x^2}{2}\right) = 0\right)$ **Ans. :** $-\dfrac{1}{xy} + \dfrac{x^2}{2} = C$

5. $x\dfrac{dy}{dx} = x^2\,y^2 - y$

    $\left(\textbf{Hint :}\ \dfrac{x\,dy + y\,dx}{x^2y^2} = dx\right)$ **Ans. :** $-\dfrac{1}{xy} = x + C$

6. $(xy^2 - e^{1/x^3})\,dx - x^2\,y\,dy = 0$

    $\left(\textbf{Hint :}\ 2x^2y\,dy - 2xy^2\,dx + 2e^{1/x^3}\,dx\ \text{and I.F.} = \dfrac{1}{x^4}\right)$

    Ans. : $\dfrac{y^2}{x^2} - \dfrac{2}{3}\,e^{1/x^3} = C$

7. $(y + y^2\cos x)\,dx - (x - y^3)\,dy = 0$

    $\left(\textbf{Hint :}\ \dfrac{y\,dx - x\,dy}{y^2} + \cos x\,dx + y\,dy = 0\right)$

    Ans. : $\dfrac{x}{y} + \sin x + \dfrac{y^2}{2} = C$

1.4.5 Linear Differential Equations of the First Order

Definition : A differential equation is said to be *linear* if the dependent variable and its derivatives appear only in the first degree. In other words, no term involves powers of derivatives or powers of dependent variables or product of derivatives and / or dependent variables.

A differential equation of the form $\boxed{\dfrac{dy}{dx} + Py = Q}$ where P, Q are functions of 'x' or constants, is called a linear differential equation of the first order.

Method of Solution : An I.F. of equation (1) is $\boxed{\text{I.F.} = e^{\int P\,dx}}$. Therefore, the G.S. is given by $\boxed{y \cdot e^{\int P\,dx} = \int Q \cdot e^{\int P\,dx}\,dx + C}$ ... (1)

Similarly, a differential equation $\boxed{\dfrac{dx}{dy} + Px = Q}$ where P and Q are functions of 'y' or constants, is also called a linear differential equation of the first order.

Method of Solution : We write I.F. of equation (2) as $\boxed{\text{I.F.} = e^{\int P\,dy}}$ and G.S. is given by $\boxed{x \cdot e^{\int P\,dy} = \int Q \cdot e^{\int P\,dy}\,dy + C}$ ... (2)

Important Note : The coefficient of $\dfrac{dy}{dx}\left(\text{or }\dfrac{dx}{dy}\right)$ in linear differential equation must be equal to ***one***.

Illustrations on Linear Differential Equation :

Ex. 1 : *Solve* $(1 + y^2) + (x - e^{-\tan^{-1}y}) \dfrac{dy}{dx} = 0$ *(Dec. 2010, 2017)*

Sol. : It can be written as $(1 + y^2) \dfrac{dx}{dy} + x = e^{-\tan^{-1}y}$

$$\frac{dx}{dy} + \left(\frac{1}{1 + y^2}\right) \cdot x = \frac{e^{-\tan^{-1}y}}{1 + y^2} \text{ which is linear in x.}$$

Here
$$\text{I.F.} = e^{\int \frac{1}{1 + y^2} dy} = e^{\tan^{-1}y}$$

G.S. is
$$x \cdot e^{\tan^{-1}y} = \int \frac{e^{-\tan^{-1}y}}{1 + y^2} \, e^{\tan^{-1}y} \, dy + C$$

$$= \int \frac{dy}{1 + y^2} + C$$

$$\boxed{x \, e^{\tan^{-1}y} = \tan^{-1}y + C}$$

Ex. 2 : *Solve* $\dfrac{dy}{dx} + \left(\dfrac{x}{(1-x^2)^{3/2}}\right) \cdot y = \dfrac{x(1 + \sqrt{1-x^2})}{(1-x^2)^2}$

Sol. : Given differential equation is of the type $\dfrac{dy}{dx} + Py = Q$, linear in y. Here $P = \dfrac{x}{(1-x^2)^{3/2}}$; $Q = \dfrac{x(1 + \sqrt{1-x^2})}{(1-x^2)^2}$

Consider
$$\int P \, dx = \int \frac{x}{(1-x^2)^{3/2}} \, dx = -\frac{1}{2} \int (1-x^2)^{-3/2} \, (-2x \, dx) = -\frac{1}{2} \left[\frac{(1-x^2)^{-1/2}}{-1/2}\right] = \frac{1}{\sqrt{1-x^2}}$$

Here,
$$\text{I.F.} = e^{\frac{1}{\sqrt{1-x^2}}}$$

G.S. is
$$y \, e^{\frac{1}{\sqrt{1-x^2}}} = \int \frac{x(1 + \sqrt{1-x^2})}{(1-x^2)^2} \, e^{\frac{1}{\sqrt{1-x^2}}} \, dx + C$$

Put $1 - x^2 = t^2$; $x \, dx = -t \cdot dt$

$$= \int \frac{(1 + t) \, e^{1/t} \, (-t \, dt)}{t^4} + C$$

Putting $\dfrac{1}{t} = u$, $-\dfrac{1}{t^2} dt = du$

$$= \int \left(1 + \frac{1}{u}\right) \cdot u \cdot e^u \, du + C = \int e^u \, (u + 1) \, du + C$$

$$= e^u \, u + C \qquad\qquad \left(\because \int e^u \, [f(u) + f'(u)] \, du = e^u \cdot f(u)\right)$$

$$\boxed{y \cdot e^{\frac{1}{\sqrt{1-x^2}}} = e^{\frac{1}{\sqrt{1-x^2}}} \frac{1}{\sqrt{1-x^2}} + C}$$

Ex. 3 : Solve $\left(\dfrac{e^{-2\sqrt{x}}}{\sqrt{x}} - \dfrac{y}{\sqrt{x}}\right) \dfrac{dx}{dy} = 1$ **(Dec. 2011)**

Sol. : Given equations is

$$\frac{dy}{dx} = \frac{e^{-2\sqrt{x}}}{\sqrt{x}} - \frac{y}{\sqrt{x}}$$

or $\dfrac{dy}{dx} + \left(\dfrac{1}{\sqrt{x}}\right) y = \dfrac{e^{-2\sqrt{x}}}{\sqrt{x}}$ Here $P = \dfrac{1}{\sqrt{x}}$; $Q = \dfrac{e^{-2\sqrt{x}}}{\sqrt{x}}$

$$\text{I.F.} = e^{\int \frac{1}{\sqrt{x}} dx} = e^{2\sqrt{x}}$$

$\therefore$ G.S. is $y \cdot e^{2\sqrt{x}} = \displaystyle\int \frac{e^{-2\sqrt{x}}}{\sqrt{x}} e^{2\sqrt{x}} \, dx + C = \int x^{-1/2} \, dx + C$

$$\boxed{y \cdot e^{2\sqrt{x}} = 2\sqrt{x} + C}$$

Ex. 4 : Solve $x^2(x^2-1)\dfrac{dy}{dx} + x(x^2+1)y = x^2 - 1$ **(Dec. 2006)**

Sol. : $\dfrac{dy}{dx} + \left(\dfrac{x^2+1}{x(x^2-1)}\right) y = \dfrac{1}{x^2}$ which is linear in y.

$$P = \frac{x^2+1}{x(x^2-1)}, \quad Q = \frac{1}{x^2}, \quad \int P \, dx = \int \frac{x^2+1}{x(x-1)(x+1)} \, dx$$

(Resolving into partial fractions)

$$\int P \, dx = \int \left(\frac{1}{x-1} + \frac{1}{x+1} - \frac{1}{x}\right) dx = \log(x-1) + \log(x+1) - \log x = \log\left(\frac{x^2-1}{x}\right)$$

$$\text{I.F.} = e^{\int P \, dx} = \frac{x^2-1}{x}$$

G.S. $y\left(\dfrac{x^2-1}{x}\right) = \displaystyle\int \frac{1}{x^2}\left(\frac{(x^2-1)}{x}\right) dx + C = \int \left(\frac{1}{x} - \frac{1}{x^3}\right) dx + C = \log x + \frac{1}{2x^2} + C$

$$\boxed{\frac{y(x^2-1)}{x} - \log x - \frac{1}{2x^2} = C} \text{ is the required G.S.}$$

Ex. 5 : Solve $(1 + \sin y)\dfrac{dx}{dy} = 2y \cos y - x(\sec y + \tan y)$. **(Dec. 2007)**

Sol. : Dividing by $1 + \sin y$,

$$\frac{dx}{dy} + \left(\frac{\sec y + \tan y}{1 + \sin y}\right) x = \frac{2y \cos y}{1 + \sin y}$$

$$\frac{dx}{dy} + \left(\frac{\frac{1 + \sin y}{\cos y}}{1 + \sin y}\right) x = \frac{2y \cos y}{1 + \sin y}$$

or $\dfrac{dx}{dy} + (\sec y) x = \dfrac{2y \cos y}{1 + \sin y}$ ...(1)

Equation (1) is linear in x with $P = \sec y$; $Q = \dfrac{2y \cos y}{1 + \sin y}$.

$$\text{I.F.} = e^{\int \sec y \, dy} = e^{\log(\sec y + \tan y)} = \sec y + \tan y$$

G.S. $x(\sec y + \tan y) = \displaystyle\int \frac{2y \cos y}{1 + \sin y} \cdot (\sec y + \tan y) \, dy + C = \int \left(\frac{2y \cos y}{1 + \sin y}\right) \cdot \frac{(1 + \sin y)}{\cos y} \, dy + C = \int 2y \, dy + C = y^2 + C$

$$\boxed{x(\sec y + \tan y) - y^2 = C} \text{ is the G.S.}$$

Ex. 6 : *Solve* $y^2 + \left(x - \dfrac{1}{y}\right)\dfrac{dy}{dx} = 0$

Sol. : Note that the equation contains y^2 and so it cannot be linear in y. We try to see whether it is linear in x. We write the given equation as

$$y^2 \frac{dx}{dy} + x = \frac{1}{y} \implies \frac{dx}{dy} + \frac{1}{y^2}\cdot x = \frac{1}{y^3} \text{ (Linear in x)}$$

$$\text{I.F.} = e^{\int \frac{1}{y^2} dy} = e^{-\frac{1}{y}}$$

Hence G.S. is, $x \cdot e^{-\frac{1}{y}} = \displaystyle\int \frac{1}{y^3} e^{-\frac{1}{y}} dy + C = \int \frac{1}{y} \cdot e^{-\frac{1}{y}} \left(\frac{1}{y^2} dy\right) + C$ (Note this step)

$$= \int (-t)\, e^t\, dt + C = -(t\, e^t - e^t) + C \qquad\qquad \left(\because\ t = -\frac{1}{y}\right)$$

$$= e^{-\frac{1}{y}}\left(1 + \frac{1}{y}\right) + C$$

$\therefore$ $\boxed{x = 1 + \dfrac{1}{y} + C\, e^{\frac{1}{y}}}$ is the required G.S.

EXERCISE 1.8

Solve the following linear differential equations :

1. $\sin x \dfrac{dy}{dx} + 2y = \tan^3 \dfrac{x}{2}$

        Ans. : $y \tan^2 \dfrac{x}{2} = \dfrac{\tan^5 x/2}{5} + C$

2. $x(1 - x^2)\dfrac{dy}{dx} + (2x^2 - 1)y = x^3$

        Ans. : $y = x + Cx\sqrt{1 - x^2}$

3. $\cosh x \dfrac{dy}{dx} + y \sinh x = 2 \cosh^2 x \sinh x$

        Ans. : $y \cosh x = \dfrac{2}{3}\cosh^3 x + C$

4. $(1 + x^2)\dfrac{dy}{dx} + xy = 1$ **(May 2007)**

        Ans. : $y\sqrt{1 + x^2} = \log(x + \sqrt{1 + x^2}) + C$

5. $x^2 \dfrac{dy}{dx} = 3x^2 - 2xy + 1$

        Ans. : $yx^2 = x^3 + x + C$ (Linear in y)

6. $(1 - x^2)\dfrac{dy}{dx} = 1 + xy$ **(May 2014)**

        Ans. : $y\sqrt{1 - x^2} = \sin^{-1} x + C$

7. $(x^2 + 1)\dfrac{dy}{dx} + 4xy = \dfrac{1}{(x^2 + 1)^2}$ **(May 2011, 2006)**

        Ans. : $y(x^2 + 1)^2 = \tan^{-1} x + C$

8. $\dfrac{dy}{dx} + y \cot x = \sin 2x$ **(Dec. 2018)**

        Ans. : $y = \dfrac{2}{3}\sin^3 x + C\ \mathrm{cosec}\, x$

9. $(x + 2y^3)\dfrac{dy}{dx} = y$

        Ans. : $x = y^3 + Cy$ (Linear in x)

10. $(2y + x^2)\, dx = x\, dy$

        Ans. : $y = x^2 \log(Cx)$ (Linear in y)

11. $(e^{-y} \sec^2 y - x)\, dy = dx$

        Ans. : $xe^y = C + \tan y$ (Linear in x)

12. $\dfrac{dy}{dx} + \dfrac{y}{1 - x} = x^2 - x$

        Ans. : $2y = (1 - x)(C_1 - x^2)$

13. $(1 + x^2)\, dy = (\tan^{-1} x - y)\, dx$ (Linear in y)

        Ans. : $y = \tan^{-1} x - 1 + ce^{-\tan^{-1} x}$

14. $\dfrac{dy}{dx} + (1 + 2x)y = e^{-x^2}$

        Ans. : $y \cdot e^{x^2 + x} = e^x + C$

15. $(1 + y^2)\, dx = (\tan^{-1} y - x)\, dy$ **(May 2008, 2005)**

Ans. : $x = \tan^{-1} y - 1 + c e^{-\tan^{-1} y}$

16. $(x^2 + 1)\dfrac{dy}{dx} = x^3 - 2xy + x$

Ans. : $y\,(x^2 + 1) = \dfrac{x^4}{4} + \dfrac{x^2}{2} + C$

17. $\sqrt{a^2 + x^2}\, \dfrac{dy}{dx} + y = \sqrt{a^2 + x^2} - x$

Ans. : $y\,(x + \sqrt{a^2 + x^2}) = a^2 \log(x + \sqrt{a^2 + x^2}) + C$

18. $\cos x \dfrac{dy}{dx} + y = \sin x$ **(May 2018)**

Ans. : $y = 1 + (C - x)(\sec x - \tan x)$

19. $y e^y = (y^3 + 2x e^y)\dfrac{dy}{dx}$

Ans. : $\dfrac{x}{y^2} + e^{-y} = C$ (Linear in x)

20. $x \cos x \dfrac{dy}{dx} + (\cos x - x \sin x)\, y = 1$ **(May 2016)**

Ans. : $xy \cos x - x = C$

21. $\dfrac{dy}{dx} = \dfrac{e^x - 3xy}{x^2}$

Ans. : $x^3 y = C + (x - 1) e^x$

22. $\dfrac{dy}{dx} + \dfrac{y}{(1 - x)\sqrt{x}} = 1 - \sqrt{x}$ **(May 2004)**

Ans. : $y\left(\dfrac{1 + \sqrt{x}}{1 - \sqrt{x}}\right) = x + \dfrac{2}{3} x^{3/2} + C$

1.4.6 Equations Reducible to the Linear Form

1. **Bernoulli's Differential Equation :** A differential equation of the form

$$\boxed{\dfrac{dy}{dx} + P\,y = Q \cdot y^n}$$ is called as *Bernoulli's differential equation.*

Method of Solution : We divide by y^n

$\therefore$ $y^{-n}\dfrac{dy}{dx} + P \cdot y^{1-n} = Q$ Put $y^{1-n} = u$ $\therefore$ $(1 - n)\, y^{-n}\dfrac{dy}{dx} = \dfrac{du}{dx}$

$\dfrac{1}{1 - n}\dfrac{du}{dx} + P \cdot u = Q$ or $\dfrac{du}{dx} + (1 - n)\, P \cdot u = (1 - n)\, Q$

which is linear and therefore, can be solved by using method discussed in type VI.

2. **Similarly, for equation** $\boxed{\dfrac{dx}{dy} + P\,x = Q \cdot x^n}$ is also called as *Bernoulli's differential equation.* Now dividing by x^n,

$x^{-n}\dfrac{dx}{dy} + P \cdot x^{1-n} = Q$ Put $x^{1-n} = u$ $\therefore$ $(1 - n)\, x^{-n}\dfrac{dx}{dy} = \dfrac{du}{dy}$

$\dfrac{du}{dy} + (1 - n)\, P \cdot u = (1 - n)\, Q$ which is linear and therefore, can be solved.

3. **Equations of the Form** $\boxed{f'(y)\dfrac{dy}{dx} + P\, f(y) = Q}$ are at once reducible to the linear form by the substitution of $f(y) = u$ and

$f'(y)\dfrac{dy}{dx} = \dfrac{du}{dx}$ transforming the equation to $\dfrac{du}{dx} + P \cdot u = Q$, which is linear.

4. **Similarly, for the equation** $\boxed{f'(x)\dfrac{dx}{dy} + P\, f(x) = Q}$, we substitute $f(x) = u$ and $f'(x)\dfrac{dx}{dy} = \dfrac{du}{dy}$

$\therefore$ Equation reduces to $\dfrac{du}{dy} + P \cdot u = Q$, which is linear.

Illustrations on Equations Reducible to Linear Form

Ex. 1 : *Solve* $\sin y \dfrac{dy}{dx} = \cos x\,(2 \cos y - \sin^2 x)$

Sol. : Here, we will use substitution $\cos y = u$.

$\sin y \dfrac{dy}{dx} = -\dfrac{du}{dx}$ and equation reduces to

$-\dfrac{du}{dx} = 2\, u \cos x - \sin^2 x \cos x$

or $\quad\quad \dfrac{du}{dx} + (2 \cos x)\, u = \sin^2 x \cos x$ which is linear in u.

Here $\quad P = 2 \cos x, \quad Q = \sin^2 x \, \cos x \quad$ I.F. $= e^{\int P\, dx} = e^{\int 2 \cos x\, dx} = e^{2 \sin x}$

G.S. is, $\quad u \cdot e^{2 \sin x} = \displaystyle\int \sin^2 x \cos x \cdot e^{2 \sin x}\, dx + C \quad (\because \ \sin x = t)$

$$= \int e^{2t} \cdot t^2 \cdot dt + C = t^2 \left(\dfrac{e^{2t}}{2}\right) - (2t)\left(\dfrac{e^{2t}}{4}\right) + (2)\left(\dfrac{e^{2t}}{8}\right) + C$$

$$u \cdot e^{2 \sin x} = \dfrac{e^{2t}}{4}\,(2t^2 - 2t + 1) + C$$

$$u = \dfrac{1}{4}\,(2 \sin^2 x - 2 \sin x + 1) + C \cdot e^{-2 \sin x}$$

or $\quad \boxed{4 \cos y = 2 \sin^2 x - 2 \sin x + 1 + C_1\, e^{-2 \sin x}}$ is the required G.S.

Ex. 2 : Solve $\dfrac{dy}{dx} - \dfrac{\tan y}{1 + x} = (1 + x)\, e^x \sec y$

Sol. : We divide by sec y,

$$\cos y \dfrac{dy}{dx} - \dfrac{\sin y}{1 + x} = (1 + x)\, e^x$$

This differential equation can be reduced to linear form by putting sin y = u

$\therefore \quad\quad \cos y \dfrac{dy}{dx} = \dfrac{du}{dx}$

$$\dfrac{du}{dx} - \dfrac{u}{1 + x} = (1 + x)\, e^x \ \left[\text{Linear in u, with } P = -\dfrac{1}{1 + x}, \ Q = e^x (1 + x)\right]$$

$$\text{I.F.} = e^{-\int \frac{1}{1 + x}\, dx} = e^{-\log (1 + x)} = \dfrac{1}{1 + x}$$

G.S. is, $\quad u\left(\dfrac{1}{1 + x}\right) = \displaystyle\int e^x (1 + x)\, \dfrac{1}{1 + x}\, dx + C \ \Rightarrow \ \dfrac{u}{1 + x} = e^x + C$

$\boxed{\sin y - e^x (1 + x) = C (1 + x)}$ is the required G.S.

Ex. 3 : Solve $\cos y - x \sin y \dfrac{dy}{dx} = \sec^2 x$ $\hfill$ *(May 2007, Dec. 2004, Nov. 2015)*

Sol. : Put $\cos y = u \Rightarrow -\sin y \dfrac{dy}{dx} = \dfrac{du}{dx}$

$$u + x \dfrac{du}{dx} = \sec^2 x \quad\quad\quad \therefore \quad \dfrac{du}{dx} + \dfrac{1}{x} \cdot u = \dfrac{\sec^2 x}{x} \quad\quad \text{(Linear in u)}$$

$$P = \dfrac{1}{x}, \quad Q = \dfrac{\sec^2 x}{x}$$

$$\text{I.F.} = e^{\int \frac{1}{x}\, dx} = e^{\log x} = x$$

$\therefore \quad\quad u \cdot x = \displaystyle\int \dfrac{\sec^2 x}{x}\, x\, dx + C = \int \sec^2 x\, dx + C$

$\boxed{x \cdot \cos y = \tan x + C}$ is the G.S.

Ex. 4 : *Solve* $\dfrac{dy}{dx} = \dfrac{y^3}{e^{2x} + y^2}$

Sol. : $\dfrac{e^{2x} + y^2}{y^3} = \dfrac{dx}{dy} \Rightarrow \dfrac{dx}{dy} - \dfrac{1}{y} = \dfrac{e^{2x}}{y^3}$ or $e^{-2x} \dfrac{dx}{dy} - \dfrac{e^{-2x}}{y} = \dfrac{1}{y^3}$

Put $\quad e^{-2x} = u \;\therefore\; e^{-2x} \dfrac{dx}{dy} = \dfrac{-1}{2} \dfrac{du}{dy}$

$\therefore \quad -\dfrac{1}{2} \dfrac{du}{dy} - \dfrac{u}{y} = \dfrac{1}{y^3} \quad$ or $\quad \dfrac{du}{dy} + \left(\dfrac{2}{y}\right) u = -\dfrac{2}{y^3}$ which is linear in u.

Here $\quad P = \dfrac{2}{y}, \quad Q = -\dfrac{2}{y^3}$

$\text{I.F.} = e^{\int \frac{2}{y}\, dy} = e^{2 \log y} = y^2$

G.S. $\quad u \cdot y^2 = \displaystyle\int -\dfrac{2}{y^3} \cdot y^2 \, dy + C = -2 \int \dfrac{1}{y} \, dy + C$

$u \cdot y^2 = -2 \log y + C$

$\boxed{e^{-2x} y^2 + \log y^2 = C}$ is the G.S.

Ex. 5 : *Solve* $xy - \dfrac{dy}{dx} = y^3 e^{-x^2}$ *(Dec. 2005)*

Sol. : Given differential equation can be written as $\dfrac{dy}{dx} - xy = -y^3 e^{-x^2}$.

This is Bernoulli's equation. Dividing by y^3,

$y^{-3} \dfrac{dy}{dx} - x \cdot y^{-2} = -e^{-x^2}$ Put $y^{-2} = u \quad \therefore -2y^{-3} \dfrac{dy}{dx} = \dfrac{du}{dx}$

$-\dfrac{1}{2} \dfrac{du}{dx} - x \cdot u = -e^{-x^2}$

or $\quad \dfrac{du}{dx} + (2x) u = 2 e^{-x^2}$

$P = 2x, \; Q = 2e^{-x^2}$

Here $\quad \text{I.F.} = e^{\int 2x\, dx} = e^{x^2}$

$\therefore \quad u \cdot e^{x^2} = \displaystyle\int 2 e^{-x^2} \cdot e^{x^2} \, dx + C = \int 2 \, dx + C = 2x + C$

$\boxed{\dfrac{e^{x^2}}{y^2} = 2x + C}$ is the required G.S.

Ex. 6 : *Solve* $xy + x^2 y^3 = \dfrac{dx}{dy}$

Sol. : Given $\dfrac{dx}{dy} - y \cdot x = x^2 y^3 \quad$ or $\quad x^{-2} \dfrac{dx}{dy} - y \cdot x^{-1} = y^3$

Put $\quad x^{-1} = u \quad \therefore \quad x^{-2} \dfrac{dx}{dy} = -\dfrac{du}{dy}$

$\therefore \quad -\dfrac{du}{dy} - yu = y^3 \quad$ or $\quad \dfrac{du}{dy} + y \cdot u = -y^3$ (Linear in u)

$\text{I.F.} = e^{\int y\, dy} = e^{y^2/2}$

G.S. $u \cdot e^{y^2/2} = \int -y^3\, e^{y^2/2} \cdot dy + C,$ Put $y^2 = 2t$ $\therefore$ $y\, dy = dt$

$$= -\int e^t\,(2t)\, dt + C = -2\,(t\, e^t - e^t) + C = 2 \cdot e^{y^2/2}\left(1 - \frac{y^2}{2}\right) C = e^{y^2/2}\,(2 - y^2) + C$$

$$\boxed{\dfrac{e^{y^2/2}}{x} = e^{y^2/2}\,(2 - y^2) + C}\ \text{ is the required G.S.}$$

Ex. 7 : *Solve* $(xy^2 + e^{-1/x^3})\, dx - x^2 y\, dy = 0$

Sol. : Given $xy^2 + e^{-1/x^3} - x^2 \cdot y\dfrac{dy}{dx} = 0$ or $y\dfrac{dy}{dx} - \dfrac{y^2}{x} = \dfrac{e^{-1/x^3}}{x^2}$

Put $y^2 = u$ $\therefore$ $y\dfrac{dy}{dx} = \dfrac{1}{2}\dfrac{du}{dx}$

$\therefore$ $\dfrac{1}{2}\dfrac{du}{dx} - \dfrac{u}{x} = \dfrac{e^{-\frac{1}{x^3}}}{x^2}$ or $\dfrac{du}{dx} - \dfrac{2}{x}\,u = \dfrac{2e^{-\frac{1}{x^3}}}{x^2}$ (Linear in u)

Here   I.F. $= e^{-2\int \frac{1}{x}\,dx} = e^{-2\log x} = \dfrac{1}{x^2}$

G.S. is $u \cdot \dfrac{1}{x^2} = \int \dfrac{2\,e^{-1/x^3}}{x^2} \cdot \dfrac{1}{x^2}\, dx + C,$ Put $-\dfrac{1}{x^3} = t \therefore \dfrac{1}{x^4}\, dx = \dfrac{1}{3}\, dt$

$$\dfrac{u}{x^2} = \int \dfrac{2}{3}\, e^t\, dt + C = \dfrac{2}{3}\, e^t + C$$

$$\boxed{\dfrac{3y^2}{x^2} - 2\,e^{-\frac{1}{x^3}} = C_1}\ \text{ is the required G.S.}$$

Ex. 8 : *Solve* $\cos x\,\dfrac{dy}{dx} + y \sin x = \sqrt{y \sec x}$ *(Dec. 2010)*

Sol. : Given $\dfrac{dy}{dx} + y \cdot \tan x = \sqrt{y}\, \sec^{3/2} x$ (Bernoulli's differential equation), divide by $\sqrt{y}$,

$$y^{-1/2}\,\dfrac{dy}{dx} + y^{1/2} \tan x = \sec^{3/2} x \qquad\qquad \text{Put } y^{1/2} = u \ \therefore\ y^{-1/2}\,\dfrac{dy}{dx} = 2 \cdot \dfrac{du}{dx}$$

$$2\dfrac{du}{dx} + u \cdot \tan x = \sec^{3/2} x \qquad \text{or} \qquad \dfrac{du}{dx} + \left(\dfrac{1}{2}\tan x\right) u = \dfrac{1}{2}\sec^{3/2} x$$

$$P = \dfrac{1}{2}\tan x,\ Q = \dfrac{1}{2}\sec^{3/2} x;\ \int P\, dx = \dfrac{1}{2}\int \tan x\, dx = \dfrac{1}{2}\log \sec x = \log\sqrt{\sec x}$$

Here   I.F. $= \sqrt{\sec x}$

G.S. is $u\sqrt{\sec x} = \int \dfrac{1}{2}\sec^{3/2} x\sqrt{\sec x}\, dx + C$

$$u\sqrt{\sec x} = \dfrac{1}{2}\int \sec^2 x\, dx + C = \dfrac{1}{2}\tan x + C$$

$$\boxed{\sqrt{y \sec x} = \dfrac{1}{2}\tan x + C}\ \text{ is the required G.S.}$$

Ex. 9 : *Solve* $\dfrac{dy}{dx} = 1 - x\,(y - x) - x^3\,(y - x)^2$

Sol. : Put $y - x = u$ $\therefore$ $\dfrac{dy}{dx} - 1 = \dfrac{du}{dx}$

$\therefore$ $\dfrac{du}{dx} = -x\,u - x^3\,u^2$ or $\dfrac{du}{dx} + x \cdot u = -x^3\,u^2$

This is Bernoulli's differential equation. Dividing by u^2,

$$u^{-2} \frac{du}{dx} + x \, u^{-1} = - x^3$$

Put $u^{-1} = t$ $\therefore$ $u^{-2} \frac{du}{dx} = - \frac{dt}{dx}$

$\therefore$ $\quad - \frac{dt}{dx} + xt = - x^3$ $\qquad$ or $\qquad$ $\frac{dt}{dx} - xt = x^3$ (Linear in t)

$\qquad P = -x, \quad Q = x^3;$

Here $\quad$ I.F. $= e^{\int -x \, dx} = e^{-\frac{x^2}{2}}$

G.S. $\quad t \cdot e^{-\frac{x^2}{2}} = \int x^3 \, e^{-\frac{x^2}{2}} \, dx + C$ $\qquad$ Put $x^2 = 2v$ $\therefore$ $x \, dx = dv$

$$= \int 2v \, e^{-v} \, dv + C = 2 \left[-v \, e^{-v} - e^{-v} \right] + C$$

$$t \, e^{-\frac{x^2}{2}} = - 2e^{-v} \, (v + 1) + C = - e^{-\frac{x^2}{2}} \, (x^2 + 2) + C$$

$$\frac{e^{-\frac{x^2}{2}}}{u} = - e^{-\frac{x^2}{2}} \, (x^2 + 2) + C$$

$$\boxed{\frac{1}{y - x} + x^2 + 2 = C \, e^{\frac{x^2}{2}}}$$ is the required G.S.

EXERCISE 1.9

Solve the Following Differential Equations Reducing it to Linear Form :

1. $xy (1 + xy^2) \frac{dy}{dx} = 1$

$\qquad$ **Ans. :** $\frac{1}{x} = - y^2 + 2 + c e^{-y^2/2}$

2. $\frac{dy}{dx} + \frac{1}{x} \tan y = \frac{1}{x^2} \tan y \sin y$

$\qquad$ **Ans. :** $\operatorname{cosec} y = \frac{1}{2x} + Cx$

3. $\frac{dr}{d\theta} = \frac{r \sin \theta - r^2}{\cos \theta}$

$\qquad$ **Ans. :** $\frac{1}{r} = \sin \theta + C \cos \theta$

4. $\frac{dy}{dx} + x \sin 2y = x^3 \cos^2 y$ $\qquad$ **(Nov. 2014, May 2017)**

$\qquad$ **Ans. :** $\tan y = \frac{x^2 - 1}{2} + C \, e^{-x^2}$

5. $e^y \left(1 + \frac{dy}{dx} \right) = e^x$ $\quad$ **Hint :** Put $e^y = u$

$\qquad$ **Ans. :** $e^y \cdot e^x = e^{2x}/2 + C$

6. $\frac{dy}{dx} = e^{x-y} (e^x - e^y)$ $\qquad$ **(May 2004, Dec. 2010)**

$\qquad$ **Ans. :** $e^y = e^x - 1 + C \, e^{-e^x}$

7. $(\sec x \tan x \tan y - e^x) \, dx + \sec^2 y \sec x \, dy = 0.$

$\qquad$ **Ans. :** $\tan y \sec x = e^x + C$

8. $\frac{dy}{dx} - \frac{f(y)}{f'(y)} \, \phi'(x) = \frac{\phi(x) \, \phi'(x)}{f'(y)}$

$\qquad$ **Ans. :** $f(y) + \phi(x) = 1 + C \, e^{\phi(x)}$

9. $\frac{dz}{dx} + \frac{z}{x} \log z = \frac{z}{x^2} (\log x)^2$

$\qquad$ **Ans. :** $x \log z = \frac{(\log x)^3 + C}{3}$, **Hint :** (Put $\log z = u$)

10. $x \frac{dy}{dx} + y = y^2 \log x$ $\qquad$ **(May 2015)**

$\qquad$ **Ans. :** $\frac{1}{y} = Cx + \log (ex)$

11. $3y^2 \frac{dy}{dx} + 2xy^3 = 4xe^{-x^2}$ (Put $y^3 = u$) $\qquad$ **(May 2006)**

$\qquad$ **Ans. :** $y^3 e^{x^2} = 2x^2 + C$

12. $xy^2 \frac{dy}{dx} - y^3 = x^2$

$\qquad$ **Ans. :** $y^3 = x^2 (C x - 3)$

13. $\frac{dy}{dx} + xy = y^2 e^{x^2/2} \log x$ $\qquad$ **(May 2009)**

$\qquad$ **Ans. :** $y^{-1} e^{-x^2/2} = - x \log x + x + C$

14. $y \, dy = (x - y^2) \, dx$

$\qquad$ **Ans. :** $y^2 = x - 1/2 + Ce^{-2x}$

15. $\dfrac{dy}{dx} - y \tan x = y^4 \sec x$ **(Dec. 2006)**

Ans. : $y^{-3} \sec^3 x = -3 \tan x - \tan^3 x + C$

16. $y + 2\dfrac{dy}{dx} = y^3 (x - 1)$

Ans. : $y^2 (x + ce^x) = 1$

17. $2x \dfrac{dy}{dx} + y - 2x (x + 1) y^3 = 0$

Ans. : $1/xy^2 = -2 (x + \log x) + C$

18. $(x^3 y^3 + xy) \dfrac{dy}{dx} = 1$ **(Dec. 2009, May 2005)**

Ans. : $x^{-2} e^{y^2} = C - e^{y^2} (y^2 - 1)$

19. $\dfrac{dx}{dy} - x \tan y = x^4 \sec y$

Ans. : $x^{-3} \sec^3 y = -3 \tan y + \tan^3 y + C$

20. $3x (1 - x^2) y^2 \dfrac{dy}{dx} + (2x^2 - 1) y^3 = ax^3$

Ans. : $y^3 = ax + cx \sqrt{1 - x^2}$

21. $xy - \dfrac{dy}{dx} = y^3 e^{-x^2}$

Ans. : $y^{-2} e^{x^2} = 2x + C$

22. $\tan x \cos y \, dy + \sin y \, dx + e^{\sin x} dx = 0$

Ans. : $\sin x \sin y = C - e^{\sin x}$

23. $x \dfrac{dy}{dx} + 3y = x^4 e^{1/x^2} y^3$ **(Dec. 2007)**

Ans. : $x^6 y^2 (C + e^{1/x^2}) = 1$

24. $\dfrac{dy}{dx} = \dfrac{x^2 + y^2 + 1}{2xy}$

Ans. : $y^2 = x^2 - 1 + Cx$

25. $\tan y \dfrac{dy}{dx} + \tan x = \cos y \cdot \cos^2 x$ **(May 2010)**

Ans. : $\sec x \sec y = C + \sin x$

26. $\dfrac{dy}{dx} + \dfrac{y}{x} \log y = \dfrac{y}{x^2} (\log y)^2$

Ans. : $\dfrac{1}{\log y} = \dfrac{1}{2x} + Cx$

27. $xy \dfrac{dy}{dx} + y^2 = gx$ if $y = 0$ when $x = 0$

Ans. : $y^2 = \dfrac{2}{3} gx$

28. $3\dfrac{dx}{dy} + \dfrac{x}{y + 1} = \dfrac{3 (y + 1)}{x^2}$

Ans. : $x^3 (y + 1) = C + (y + 1)^3$

29. $\dfrac{dy}{dx} = \dfrac{\sin^2 y}{x^2} - \dfrac{\sin y \cos y}{x}$

Ans. : $\cot y = \dfrac{1}{2x} + Cx$

30. $x^2 \dfrac{dy}{dx} = e^y - x$

Hint : Divide by $x^2 e^y$, Put $e^{-y} = u$

Ans. : $cx^2 + 2xe^{-y} = 1$

31. $\dfrac{dy}{dx} = 1 + 3x^2 e^{-y}$ **(May 2008)**

Ans. $e^y = -3 (x^2 + 2x + 2) + Ce^x$

32. $\tan y \dfrac{dy}{dx} + \tan x = \cos y \cos^3 x$ **(May 2009)**

Ans. : $\sec x \sec y = \dfrac{1}{2} \left(x + \dfrac{\sin 2x}{2} \right) + C$

1.4.7 Transformation to Polars

Sometimes it is not possible to solve a differential equation in variables x, y by usual methods, but when the equation is transformed to polars by using $x = r \cos \theta$, $y = r \sin \theta$, it may be convenient to solve the differential equation in the new variables r and θ.

In such cases, we note that, since $x = r \cos \theta$, $y = r \sin \theta$, $x^2 + y^2 = r^2$, $\theta = \tan^{-1} y/x$.

$\therefore$ $2x \, dx + 2y \, dy = 2r \, dr$ giving $\boxed{x \, dx + y \, dy = r \, dr}$

Also $x \, dy - y \, dx = r \cos \theta \, (r \cos \theta \, d\theta + \sin \theta \, dr) - r \sin \theta \, (\cos \theta \, dr - r \sin \theta \, d\theta)$

$\therefore$ $\boxed{x \, dy - y \, dx = r^2 \, d\theta}$

Ex. 1 : *Solve* $x^2(x \, dx + y \, dy) + y (x \, dy - y \, dx) = 0$

Sol. : Transforming to polar co-ordinates by using $x = r \cos \theta$, $y = r \sin \theta$ and therefore, $x^2 + y^2 = r^2$

$\therefore$ $\boxed{x \, dx + y \, dy = r \, dr}$

Also $\dfrac{y}{x} = \tan \theta$

$\therefore$ $\boxed{\dfrac{x \, dy - y \, dx}{x^2} = \sec^2 \theta \, d\theta}$

We have $\quad x\,dx + y\,dy + \dfrac{y(x\,dy - y\,dx)}{x^2} = 0$

$\therefore \qquad r\,dr + r\sin\theta \cdot \sec^2\theta\,d\theta = 0 \qquad$ or $\qquad dr + \tan\theta\,\sec\theta\,d\theta = 0$

$$\int dr + \int \sec\theta\,\tan\theta\,d\theta = C \qquad \text{or} \qquad r + \sec\theta = C$$

Hence, $\qquad \sqrt{x^2+y^2} + \dfrac{\sqrt{x^2+y^2}}{x} = C \qquad\qquad\qquad\qquad \left(\because\ \cos\theta = \dfrac{x}{r},\ \sec\theta = \dfrac{r}{x} \right)$

i.e. $\qquad \boxed{\sqrt{x^2+y^2}\,(x+1) = Cx}$ is the G.S.

Ex. 2 : *Solve* $\sqrt{x^2+y^2}\,(x\,dx + y\,dy) + (y\,dx - x\,dy)\,\sqrt{a^2 - x^2 - y^2} = 0$

Sol. : Put $x = r\cos\theta,\ y = r\sin\theta,\ x^2 + y^2 = r^2;\ x\,dx + y\,dy = r\,dr$

$\qquad x\,dy - y\,dx = r\cos\theta\,(r\cos\theta\,d\theta + \sin\theta\,dr) - r\sin\theta\,(\cos\theta\,dr - r\sin\theta\,d\theta) = r^2\,d\theta$

$\therefore$ Given differential equation reduces to $r(r\,dr) - \sqrt{a^2 - r^2}\,r^2\,d\theta = 0$

G.S. $\qquad\qquad \displaystyle\int \dfrac{dr}{\sqrt{a^2 - r^2}} - \int d\theta = C, \quad \sin^{-1}\dfrac{r}{a} - \theta = C$

$$\boxed{\sin^{-1}\left(\dfrac{\sqrt{x^2+y^2}}{a} \right) - \tan^{-1}\left(\dfrac{y}{x} \right) = C} \text{ is the G.S.}$$

EXERCISE 1.10

1. Solve $x\,dy - y\,dx = (x^2 + y^2)\,(x\,dx + y\,dy)$ **(May 2010, 2006)** $\qquad$ **Ans. :** $\tan^{-1}\dfrac{y}{x} = \dfrac{x^2+y^2}{2} + C$

2. $(x\,dy - y\,dx) + \dfrac{1}{x^2+y^2}\,(x\,dx + y\,dy) = 0$ **(May 2008)** $\qquad$ **Ans. :** $\tan^{-1}\dfrac{y}{x} = \dfrac{1}{2(x^2+y^2)} + C$

1.4.8 Equations which are both Homogeneous and Exact

Suppose that M and N are homogeneous functions of x and y of degree n (n ≠ –1). Further, suppose that Mdx + Ndy = 0 is exact. Then the general solution is Mx + Ny = C.

e.g. (1) $(x^3 + 3y^2\,x)\,dx + (y^3 + 3x^2\,y)\,dy = 0$

Here given equation is homogeneous and exact.

$\therefore$ The G.S. is $x\,(x^3 + 3y^2 x) + y\,(y^3 + 3x^2 y) = C$

or $\qquad x^4 + 6x^2\,y^2 + y^4 = C$

EXERCISE 1.10

Solve the following differential equations :

1. $3x^2\,y\,dx + (x^3 + y^3)\,dy = 0$ $\qquad\qquad\qquad$ 2. $(x^2 - 4xy - 2y^2)\,dx + (y^2 - 4xy - 2x^2)\,dy = 0$

$\qquad\qquad\qquad$ **Ans. :** $4x^3\,y + y^4 = C$ $\qquad\qquad\qquad$ **Ans. :** $x^3 - 6x^2\,y - 6y^2\,x + y^3 = C$

1.4.9 Miscellaneous Examples

In conclusion, the art of recognising the relevant form of the equation of first order and first degree equation and obtaining the solution can be achieved by proper observation and working a number of problems systematically and at random. To enable the student to master this art, the following examples are worked in detail.

Type I : V.S. Form and Reducible to V.S. Form (Revision) :

Ex. 1 : *Solve* $x\cos^2 y\,dx + (x^2 + 1)\,\tan y\,dy = 0$

Sol. : [Here the terms have factors x, $\cos^2 y$, $x^2 + 1$, tan y each of which is a function of either x only or y only. Therefore, we will use V.S. form.]

On dividing by $\cos^2 y$ and $(x^2 + 1)$, we get

$$\dfrac{x}{x^2+1}\,dx + \dfrac{\tan y}{\cos^2 y}\,dy = 0 \qquad\qquad \text{(V.S. form)}$$

G.S. $\quad \dfrac{1}{2} \displaystyle\int \dfrac{2x}{x^2 + 1}\, dx + \int \tan y \cdot \sec^2 y\, dy = C$

$\qquad \dfrac{1}{2} \log (x^2 + 1) + \tan^2 y = C$

or $\qquad \boxed{\log (x^2 + 1) + \tan^2 y = C_1}$

Ex. 2 : *Solve sin (x – y) dy = dx*

Sol. : [Here the factor sin (x – y) is not a function of x alone or y alone. Hence, we will use substitution method for reducing into V.S. form.]

Put $\quad x - y = u \quad \therefore \dfrac{dy}{dx} = 1 - \dfrac{du}{dx}$

$\therefore \qquad \dfrac{dy}{dx} = \dfrac{1}{\sin (x - y)} \quad \Rightarrow \quad 1 - \dfrac{du}{dx} = \dfrac{1}{\sin u}$

$\therefore \qquad \dfrac{\sin u}{\sin u - 1}\, du = dx \quad \Rightarrow \quad \displaystyle\int 1 \cdot du + \int \dfrac{du}{\sin u - 1} = x + C$

$\qquad u + \displaystyle\int \dfrac{\sin u + 1}{-\cos^2 u}\, du = x + C \ \text{ or } \ u - \int \sec u \tan u\, du - \int \sec^2 u\, du - y = C$

$\qquad \sec u + \tan u + y = C_1$

or $\qquad \boxed{\sec (x - y) + \tan (x - y) + y = C_1}$

Ex. 3 : *Solve x (x – y) dy + y² dx = 0*   ***(Dec. 2008)***

Sol. : [Here the variables are not separable, because though the factors x and y^2 are functions of single variable, the factor (x – y) is a mixed function of x and y. We also see that $\dfrac{\partial M}{\partial y} = 2y,\ \dfrac{\partial N}{\partial x} = 2x - y$ which are not equal. Hence not exact. Therefore, we will try homogeneous type.]

$\qquad (x^2 - xy)\dfrac{dy}{dx} + y^2 = 0 \qquad \text{Put} \quad y = vx \quad \therefore \quad \dfrac{dy}{dx} = x\dfrac{dv}{dx} + v$

But $\qquad (x^2 - x^2 v)\left(v + x\dfrac{dv}{dx}\right) + v^2 x^2 = 0$

or $\qquad (1 - v)\, v + (1 - v)\, x \cdot \dfrac{dv}{dx} + v^2 = 0$

$\qquad v + (1 - v)\, x\, \dfrac{dv}{dx} = 0$

G.S. $\qquad \displaystyle\int \dfrac{dx}{x} + \int \dfrac{1 - v}{v}\, dv = 0$

$\qquad \log x + \log v - v = C \quad \text{or} \quad \log\left(x \cdot \dfrac{y}{x}\right) - \dfrac{y}{x} = C$

$\therefore \qquad \boxed{\log y - \dfrac{y}{x} = C}$

Ex. 4 : *Solve* $\dfrac{dy}{dx} = \dfrac{x + y + 2}{-x + 2y + 1}$

Sol. : It is of the form $\dfrac{dy}{dx} = \dfrac{a_1 x + b_1 y + C_1}{a_2 x + b_2 y + C_2}$ where $b_1 = -a_2 \ \therefore$ we shall try exactness.

Given $\qquad (x + y + 2)\, dx + (x - 2y - 1)\, dy = 0$ ... (1)

$\qquad \dfrac{\partial M}{\partial y} = 1,\ \dfrac{\partial N}{\partial x} = 1,\ \therefore\ \text{Equation (1) is exact. Its G.S. is}$

G.S. $\qquad \displaystyle\int (x + y + 2)\, dx + \int (-2y - 1)\, dy = C$

$\qquad \boxed{\dfrac{x^2}{2} + xy + 2x - y^2 - y = C}$

Ex. 5 : *Solve $(x + y + 1)\ dx + (2x + 2y + 4)\ dy = 0$*

Sol. : We see that the coefficients of x and y in the two expressions are proportional. Hence we write

$$\frac{dy}{dx} = -\frac{(x + y + 1)}{2(x + y) + 4} \quad \text{Put } x + y = u \quad \therefore \quad \frac{dy}{dx} = \frac{du}{dx} - 1$$

$$\frac{du}{dx} - 1 = \frac{-(u + 1)}{2u + 4}$$

$$\frac{du}{dx} = \frac{2u + 4 - u - 1}{2u + 4} = \frac{u + 3}{2u + 4}$$

G.S. $\qquad \displaystyle\int \frac{u + 2}{u + 3}\ du = \frac{1}{2} \int dx + C \ \text{ or } \int \left(1 - \frac{1}{u + 3}\right) du = \frac{1}{2} \int dx + C$

$$u - \log(u + 3) = \frac{x}{2} + C$$

$\therefore \qquad \boxed{\dfrac{x}{2} + y - \log(x + y + 3) = C}$

Ex. 6 : *Solve $\dfrac{dy}{dx} = \dfrac{x + 2y - 3}{2x + y - 3}$.*   *(Dec. 2008, May 2007)*

Sol. : Given equation is of the form

$$\frac{dy}{dx} = \frac{a_1 x + b_1 y + C_1}{a_2 x + b_2 y + C_2}, \quad \text{where } \frac{a_1}{a_2} \neq \frac{b_1}{b_2}$$

We put $\qquad x = X + h,\ y = Y + k\ $ and choose h and k such that

$$a_1 h + b_1 k + C_1 = 0 \qquad \text{i.e.} \qquad h + 2k - 3 = 0$$
$$a_2 h + b_2 k + C_2 = 0 \qquad \text{i.e.} \qquad 2h + k - 3 = 0$$

Solving for h and k, $\ h = 1,\ k = 1$.

$\therefore\quad$ Given equation becomes

$$\frac{dY}{dX} = \frac{X + 2Y}{2X + Y} \qquad \text{Put } Y = VX \quad \therefore \quad \frac{dY}{dX} = X\frac{dV}{dX} + V$$

$$X\frac{dV}{dX} + V = \frac{X + 2VX}{2X + VX}$$

$$X\frac{dV}{dX} = \frac{1 + 2V}{2 + V} - V = \frac{1 - V^2}{2 + V}$$

$$\frac{2 + V}{1 - V^2}\ dV = \frac{dX}{X}$$

$$\left(\frac{1/2}{1 + V} + \frac{3/2}{1 - V}\right) dV = \frac{dX}{X}$$

$$\frac{1}{2}\log(1 + V) - \frac{3}{2}\log(1 - V) = \log X + \log C$$

$$\log(1 + V) - \log(1 - V)^3 - \log X^2 = \log C_1$$

$$(1 + V) = C_1 X^2 (1 - V)^3$$

$$(X + Y) = C_1 (X - Y)^3 \quad \text{But} \quad X = x - 1,\ Y = y - 1$$

$\therefore \qquad \boxed{(x + y - 2) = C_1 (x - y)^3}$

Ex. 7 : *Solve $(x^2 + y^2)\ dx - 2xy\ dy = 0$*

Sol. : Here variables are not separable. But it is homogeneous.

Dividing by x^2,

$$\left(1 + \frac{y^2}{x^2}\right) - \left(2\ \frac{y}{x}\right) \frac{dy}{dx} = 0$$

Put $y = xv \quad \therefore \quad \dfrac{dy}{dx} = x\dfrac{dv}{dx} + v$

$\therefore \qquad (1 + v^2) - (2v) \left(x\ \dfrac{dv}{dx} + v\right) = 0$

$$1 + v^2 - 2v^2 - 2v \cdot x \cdot \frac{dv}{dx} = 0$$

$$1 - v^2 - 2v \cdot x \frac{dv}{dx} = 0$$

$$\therefore \quad \int \frac{dx}{x} + \int \frac{-2v}{1-v^2} \, dv = 0 \quad \text{(V.S. form)}$$

$$\log x + \log (1 - v^2) = \log C$$

$$x (1 - v^2) = C$$

$$x \left(1 - \frac{y^2}{x^2}\right) = C$$

$$\boxed{(x^2 - y^2) = Cx}$$

Ex. 8 : *Solve $y^2 (x^2 + 2) \, dx + (x^3 + y^3) \, (y \, dx - x \, dy) = 0$* **(Dec. 2008)**

Sol. : Put $\dfrac{y}{x} = V \Rightarrow y = Vx \therefore \dfrac{dy}{dx} = V + x \dfrac{dV}{dx}$. Then given differential equation is,

$$y^2 (x^2 + 2) + (x^3 + y^3) \left(y - x \frac{dy}{dx}\right) = 0$$

$$\Rightarrow \quad V^2 x^2 (x^2 + 2) + (x^3 + V^3 x^3) \left[Vx - x \left(V + x \frac{dV}{dx}\right)\right] = 0$$

$$\Rightarrow \quad V^2 x^2 (x^2 + 2) + x^3 (1 + V^3) \left(-x^2 \frac{dV}{dx}\right) = 0 \Rightarrow \frac{x^2 + 2}{x^3} \, dx - \frac{1 + V^3}{V^2} \, dV = 0 \quad \text{[V.S. form]}$$

G.S. is
$$\int \frac{x^2 + 2}{x^3} \, dx - \int \frac{1 + V^3}{V^2} \, dV = C$$

$$\int \left(\frac{1}{x} + 2x^{-3}\right) dx - \int (V^{-2} + V) \, dV = C \Rightarrow \log x - \frac{1}{x^2} + \frac{1}{V} - \frac{V^2}{2} = C$$

$$\Rightarrow \quad \boxed{\log x - \frac{1}{x^2} + \frac{x}{y} - \frac{y^2}{2x^2} = C} \text{ is the required G.S.}$$

Type II : Exact and Reducible to Exact Form :

Ex. 9 : *Solve $(x^3 + xy^2 + y) \, dx + (y^3 + x^2y + x) \, dy = 0$*

Sol. : [Here the variables are not separable, because the terms cannot be factorised into functions of one variable alone, it is not also of the form $\dfrac{dy}{dx} = f(ax + by)$. Therefore, we will try exactness.]

$$M = x^3 + xy^2 + y \qquad\qquad N = y^3 + x^2y + x$$

$$\frac{\partial M}{\partial y} = 2xy + 1 \qquad\qquad \frac{\partial N}{\partial x} = 2xy + 1$$

Given equation is exact.

G.S.
$$\int (x^3 + xy^2 + y) \, dx + \int y^3 \, dy = C$$

$$\boxed{\frac{x^4}{4} + \frac{y^2 x^2}{2} + xy + \frac{y^4}{4} = C}$$

Ex. 10 : *Solve* $x^2y\, dx - (x^3 + y^3)\, dy = 0$

Sol. : Here the variables are not separable. The equation is not exact. But it is homogeneous. We shall try for finding integrating factor (Refer rule 1, Non-exact differential equation).

$$\text{I.F.} = \frac{1}{M\cdot x + N\cdot y} = \frac{1}{(x^2y)\, x + y\,(-x^3 - y^3)} = -\frac{1}{y^4}$$

$$(x^2y)\left(-\frac{1}{y^4}\right) dx - (x^3 + y^3)\left(-\frac{1}{y^4}\right) dy = 0$$

$$-\frac{x^2}{y^3}\, dx + \left(\frac{x^3}{y^4} + \frac{1}{y}\right) dy = 0 \qquad \qquad \dots (1)$$

Here equation (1) is exact.

G.S. $\qquad -\displaystyle\int \frac{x^2}{y^3}\, dx + \int \frac{1}{y}\, dy = C$

$$\boxed{-\frac{x^3}{3y^3} + \log y = C}$$

Ex. 11 : *Solve* $(xy \sin xy + \cos xy)\, y\, dx + (xy \sin xy - \cos xy)\, x\, dy = 0$

Sol. : The equation is not exact. But the equation is of the form

$$y\, f_1(xy)\, dx + x\, f_2(xy)\, dy = 0$$

Hence, $\qquad \text{I.F.} = \dfrac{1}{Mx - Ny} = \dfrac{1}{x^2 y^2 \sin xy + xy \cos xy - x^2 y^2 \sin xy + xy \cos xy}$

$$\text{I.F.} = \frac{1}{2xy \cos xy}$$

∴ Equation becomes

$$\left(\frac{y \sin xy}{2 \cos xy} + \frac{1}{2x}\right) dx + \left(\frac{x \sin xy}{2 \cos xy} - \frac{1}{2y}\right) dy = 0$$

or $\qquad \left(\dfrac{y}{2} \tan xy + \dfrac{1}{2x}\right) dx + \left(\dfrac{x}{2} \tan xy - \dfrac{1}{2y}\right) dy = 0$

which is exact and its G.S. is

$$\frac{y}{2} \int \tan xy\, dx + \frac{1}{2} \int \frac{1}{x}\, dx - \frac{1}{2} \int \frac{1}{y}\, dy = C$$

$$\frac{y}{2} \cdot \left(\frac{\log \sec xy}{y}\right) + \frac{1}{2} \log x - \frac{1}{2} \log y = C$$

$$\log \sec xy + \log x - \log y = \log C_1$$

$$\boxed{\frac{x}{y} \sec xy = C_1}$$

Ex. 12 : *Solve* $(3y^2 x^2 + y^2 x - 2y^3 + x^2)\, dx + (yx^2 - y^2)\, dy = 0$

Sol. : Here variables are not separable. It is not homogeneous.

$$\frac{\partial M}{\partial y} = 6yx^2 + 2xy - 6y^2 \ ; \quad \frac{\partial N}{\partial x} = 2xy$$

∴ It is not exact. We shall try for finding integrating factor.

Consider $\qquad \dfrac{\dfrac{\partial M}{\partial y} - \dfrac{\partial N}{\partial x}}{N} = \dfrac{6yx^2 + 2xy - 6y^2 - 2xy}{y\,(x^2 - y)} = \dfrac{6y\,(x^2 - y)}{y\,(x^2 - y)} = 6 = $ a function of x alone.

∴ $\qquad \text{I.F.} = e^{\int 6\cdot dx} = e^{6x}$ $\qquad\qquad\qquad$ (Note this step carefully)

∴ $\qquad e^{6x}\,(3y^2 x^2 + xy^2 - 2y^3 + x^2)\, dx + e^{6x}\,(yx^2 - y^2)\, dy = 0$

which is exact and its G.S. is

$$(3y^2 + 1) \int e^{6x} \cdot x^2 \, dx + y^2 \int e^{6x} x \, dx - 2y^3 \int e^{6x} \, dx = C$$

$$(1 + 3y^2) \left[x^2 \cdot \frac{e^{6x}}{6} - 2x \cdot \frac{e^{6x}}{36} + 2 \cdot \frac{e^{6x}}{(216)} \right] + y^2 \left[x \frac{e^{6x}}{6} - \frac{e^{6x}}{36} \right] - 2y^3 \left(\frac{e^{6x}}{6} \right) = C$$

$$\boxed{(1 + 3y^2) \left(\frac{x^2}{6} - \frac{x}{18} + \frac{1}{108} \right) + y^2 \left(\frac{x}{6} - \frac{1}{36} \right) - \frac{y^3}{3} = C.e^{-6x}}$$

Ex. 13 : *Solve* $(xy^3 + y) \, dx + 2 (x^2 y^2 + x + y^4) \, dy = 0$ *(Dec. 2008, 2011)*

Sol. : Here the variables are not separable.

$$\frac{\partial M}{\partial y} = 3xy^2 + 1, \quad \frac{\partial N}{\partial x} = 4xy^2 + 2.$$

Hence the equation is neither exact nor is it homogeneous. The equation is also not linear in x and y. We shall try for finding integrating factor.

Consider $\quad \dfrac{\dfrac{\partial N}{\partial x} - \dfrac{\partial M}{\partial y}}{M} = \dfrac{4xy^2 + 2 - 3xy^2 - 1}{y (xy^2 + 1)} = \dfrac{1}{y}$

$\therefore \qquad$ I.F. $= e^{\int 1/y \, dy} = e^{\log y} = y$

$\therefore \qquad y (xy^3 + y) \, dx + 2y (x^2 y^2 + x + y^4) \, dy = 0$

or $\qquad (xy^4 + y^2) \, dx + (2x^2 y^3 + 2xy + 2y^5) \, dy = 0$

which is exact and its G.S. is

$$y^4 \int x \, dx + y^2 \int dx + 2 \int y^5 \, dy = C$$

$$\boxed{\frac{y^4 x^2}{2} + y^2 x + \frac{y^6}{3} = C.}$$

Ex. 14 : *Solve* $(y^2 + 2yx^2) \, dx + (2x^3 - xy) \, dy = 0$ *(Dec. 2008)*

Sol. : Here variables are not separable. It is not homogeneous. It is also not exact.   It is also not linear type. We shall try for finding integrating factor. Regrouping the terms as

$$y^2 \, dx - xy \, dy + 2yx^2 \, dx + 2x^3 \, dy = 0$$
$$y (y \, dx - x \, dy) + x^2 (2y \, dx + 2x \, dy) = 0$$

This is of the form

$$x^a y^b (my \, dx + nx \, dy) + x^r y^s (py \, dx + q \cdot x \, dy) = 0$$

where $\qquad a = 0, \ b = 1, \ m = 1, \ n = -1, \ r = 2, \ s = 0, \ p = 2, q = 2.$

Let $\qquad$ I.F. $= x^h y^k,$ where h, k can be determined from

$$nh - mk = m - n + (mb - na)$$
$$qh - pk = p - q + (ps - qr)$$

i.e. $\qquad -h - k = 2 + (1 - 0) \qquad\qquad \therefore \ h + k = -3$

$\qquad\qquad 2h - 2k = 0 + (0 - 4) \qquad\qquad \therefore \ h - k = -2$

$\qquad\qquad h = \dfrac{-5}{2}, \quad k = \dfrac{-1}{2}$

$\therefore \qquad$ I.F. $= x^{-5/2} \, y^{-1/2}$

$\therefore \qquad x^{-5/2} y^{-1/2} (y^2 + 2yx^2) \, dx + x^{-5/2} y^{-1/2} (2x^3 - xy) \, dy = 0$

$\therefore \qquad (x^{-5/2} y^{3/2} + 2x^{-1/2} y^{1/2}) \, dx + (2x^{1/2} y^{-1/2} - x^{-3/2} y^{1/2}) \, dy = 0$

$$\frac{\partial M}{\partial y} = \frac{3}{2} x^{-5/2} y^{1/2} + x^{-1/2} y^{-1/2} = \frac{\partial N}{\partial x}$$

$\therefore \ $ It is exact and its G.S. is

$$y^{3/2} \int x^{-5/2} \, dx + 2y^{1/2} \int x^{-1/2} \, dx = C$$

$$y^{3/2} \frac{x^{-3/2}}{-3/2} + 2y^{1/2} \frac{x^{1/2}}{1/2} = C$$

$\therefore \qquad \boxed{6x^{1/2} y^{1/2} - x^{-3/2} y^{3/2} = C_1}$

Ex. 15 : *Solve* $\dfrac{dy}{dx} = \dfrac{2x - y}{x - y}$

Sol. : Method 1 : $(2x - y)\, dx + (y - x)\, dy = 0$. It is exact and its G.S. is

$$x^2 - xy + \frac{y^2}{2} \;=\; C \qquad\qquad \text{or} \qquad 2x^2 - 2xy + y^2 \;=\; C_1$$

Method 2 : It is homogeneous. Dividing by x,

$$\frac{dy}{dx} = \frac{2 - y/x}{1 - y/x} \qquad \text{Put } \frac{y}{x} = V,\; y = xV \quad \therefore \quad \frac{dy}{dx} = x\frac{dV}{dx} + V$$

$$x\frac{dV}{dx} + V = \frac{2 - V}{1 - V}$$

$$\therefore \qquad x\frac{dV}{dx} = \frac{2 - V}{1 - V} - V \quad \text{or} \quad x\frac{dV}{dx} = \frac{2 - 2V + V^2}{1 - V}$$

or $\qquad \dfrac{1 - V}{V^2 - 2V + 2}\, dV = \dfrac{dx}{x}$

G.S. $\quad \displaystyle\int \frac{dx}{x} + \frac{1}{2} \int \frac{2V - 2}{V^2 - 2V + 2}\, dV = \log C$

$$2 \log x + \log (V^2 - 2V + 2) = \log C_1 \Rightarrow x^2\, (V^2 - 2V + 2) = C_1$$

$$x^2 \left(\frac{y^2}{x^2} - 2\frac{y}{x} + 2\right) = C_1 \qquad \text{i.e.} \quad y^2 - 2xy + 2x^2 = C_1$$

Method 3 : $\qquad (x - y)\, dy = (2x - y)\, dx \qquad$ or $\qquad x\, dy + y\, dx = 2x\, dx + y$. By inspection,

$$d\,(xy) - 2x\, dx - y\, dy = 0 \qquad \therefore \qquad xy - x^2 - \frac{y^2}{2} = C \quad \text{or} \quad 2xy - 2x^2 - y^2 = C_1$$

Method 4 : $\qquad \dfrac{dy}{dx} = \dfrac{x + (x - y)}{x - y} \qquad$ Put $x - y = u \;\therefore\; 1 - \dfrac{dy}{dx} = \dfrac{du}{dx}$

$\therefore \qquad 1 - \dfrac{du}{dx} = \dfrac{x + u}{u} \qquad\qquad$ or $\qquad 1 - \dfrac{x + u}{u} = \dfrac{du}{dx}$

or $\qquad -\dfrac{x}{u} = \dfrac{du}{dx} \qquad\qquad$ or $\quad u\, du + x\, dx = 0$

G.S. $\qquad \dfrac{u^2}{2} + \dfrac{x^2}{2} = C \qquad\qquad\qquad u^2 + x^2 = C_1$

$$x^2 + (x - y)^2 = C_1 \qquad\qquad\qquad 2x^2 - 2xy + y^2 = C_1$$

Method 5 : *We know that if M dx + N dy = 0 is both homogeneous and exact, its solutions are given by,* $\;x\,M\,(x,\,y) + y\,N\,(x,\,y) = C.$
Here given equation can be written as $(2x - y)\, dx + (y - x)\, dy = 0,$
which is homogeneous and exact.

$\therefore\;$ Its G.S. is $\;x\,(2x - y) + y\,(y - x) = C$

or $\qquad \boxed{2x^2 - 2xy + y^2 = C}$

Type III : Linear and Reducible to Linear Form :

Ex. 16 : *Solve $x \cos x \dfrac{dy}{dx} + y\,(x \sin x + \cos x) = 1$* $\hfill$ *(May 2004; Dec. 2009, 2008)*

Sol. : Dividing by x cos x,

$$\frac{dy}{dx} + \left(\tan x + \frac{1}{x}\right)\cdot y = \frac{1}{x \cos x}$$

This is in the standard linear form.

$$\frac{dy}{dx} + P\cdot y = Q, \quad \text{where } P = \tan x + \frac{1}{x} \quad \text{and} \quad Q = \frac{1}{x \cos x}$$

$$\text{I.F.} = e^{\displaystyle\int \left(\tan x + \frac{1}{x}\right) dx} = e^{\log \sec x + \log x} = \frac{x}{\cos x}$$

G.S. is

$$y \cdot \frac{x}{\cos x} = \int \frac{1}{x \cos x} \cdot \frac{x}{\cos x} \, dx + C$$

$$\frac{xy}{\cos x} = \int \sec^2 x \, dx + C$$

$$\boxed{xy = \sin x + C \cdot \cos x.}$$

Ex. 17 : *Solve* $(x + \tan y) \, dy = \sin 2y \, dx$

Sol. : Here the variables are not separable, because $(x + \tan y)$ cannot be factorized into functions of single variable. The equation is not exact nor is it homogeneous. Further, it is not linear in y because the function multiplying dy is not a function of x alone. But the multiplying factor of dx is a function of y alone. Hence, we try for linear equation in x.

$$x + \tan y = \sin 2y \, \frac{dx}{dy}$$

$$\frac{dx}{dy} - \frac{x}{\sin 2y} = \frac{\tan y}{\sin 2y}$$

This is of the form

$$\frac{dx}{dy} + Px = Q, \text{ where } P = \frac{-1}{\sin 2y} \text{ and } Q = \frac{\tan y}{\sin 2y} \text{ are functions of y alone.}$$

$$\text{I.F.} = e^{\int P \, dy} = e^{\int -\frac{1}{\sin 2y} \, dy}$$

But

$$-\int \frac{1}{\sin 2y} \, dy = -\int \frac{1}{2 \sin y \cos y} \, dy = -\int \frac{dy}{2 \tan y \cos^2 y} = -\frac{1}{2} \int \frac{\sec^2 y \, dy}{\tan y} = -\frac{1}{2} \log \tan y$$

$$\therefore \quad \text{I.F.} = e^{-1/2 \log \tan y} = \frac{1}{\sqrt{\tan y}}$$

G.S. is

$$x \cdot \frac{1}{\sqrt{\tan y}} = \int \frac{\tan y}{\sin 2y} \cdot \frac{1}{\sqrt{\tan y}} \, dy + C = \int \frac{\sqrt{\tan y}}{2 \sin y \cos y} \, dy + C = \int \frac{\sqrt{\tan y} \, dy}{2 \tan y \cdot \cos^2 y} + C$$

$$= \frac{1}{2} \int \frac{\sec^2 y}{\sqrt{\tan y}} \, dy + C = \sqrt{\tan y} + C$$

$$\therefore \quad \boxed{x = \tan y + C \sqrt{\tan y}}$$

Ex. 18 : *Solve* $\dfrac{dy}{dx} = \dfrac{y}{2y \log y + y - x}$

Sol. : Here the variables are not separable. The equation is not homogeneous. Further, it is not linear in y, because the function multiplying dy is not a function of x alone. But the multiplying factor of dx is a function of y alone. Hence, we try for linear equation in x.

Given equation can be written as

$$\frac{dx}{dy} = \frac{2y \log y + y - x}{y} \qquad \text{(Note this step carefully)}$$

$$\frac{dx}{dy} = 2 \log y + 1 - \frac{x}{y}$$

$$\frac{dx}{dy} + \frac{1}{y} \cdot x = 2 \log y + 1$$

where $P = \dfrac{1}{y}$ and $Q = 1 + 2 \log y$

$$\text{I.F.} = e^{\int \frac{1}{y} \, dy} = e^{\log y} = y$$

G.S. is $\qquad x \cdot (y) = \int (1 + 2 \log y) \cdot y \cdot dy + C$

$$xy = \int y \, dy + 2 \int y \cdot \log y \, dy + C$$

$$= \frac{y^2}{2} + 2\left[\log y \cdot \frac{y^2}{2} - \int \frac{1}{y} \cdot \frac{y^2}{2} \, dy\right] + C = \frac{y^2}{2} + y^2 \log y - \frac{y^2}{2} + C$$

$$\boxed{xy - y^2 \log y = C}$$

Note : Given equation is also an exact equation and it can be solved by standard formula.

Ex. 19 : *Solve* $(y + e^y - e^{-x}) \, dx + x (1 + e^y) \, dy = 0$

Sol. : Put $\quad y + e^y = V \therefore (1 + e^y)\dfrac{dy}{dx} = \dfrac{dV}{dx}$. The given differential equation is,

$$(y + e^y - e^{-x}) + x (1 + e^y)\frac{dy}{dx} = 0 \Rightarrow V - e^{-x} + x\frac{dV}{dx} = 0$$

$$\therefore \qquad \frac{dV}{dx} + \frac{1}{x} \cdot V = \frac{1}{x} e^{-x} \text{ which is linear.}$$

$$\text{I.F.} = e^{\int \frac{1}{x} dx} = e^{\log x} = x$$

Its G.S. is $\qquad V x = \int \dfrac{1}{x} e^{-x} x \, dx + C \Rightarrow$

$$\boxed{(y + e^y) x = - e^{-x} + C}$$

Note : The given differential equation is also exact.

Ex. 20 : *Solve* $\dfrac{x^{n+1}}{y^{n+2}} \, dy - \dfrac{x^n}{y^{n+1}} \, dx + x^m \, dx = 0$

Sol. : Write the differential equation as, $\quad y^{-n-2}\dfrac{dy}{dx} - \dfrac{1}{x} y^{-n-1} = - x^{m-n-1}$

Put $\qquad y^{-n-1} = V$. Then $(-n-1) y^{-n-2} \dfrac{dy}{dx} = \dfrac{dV}{dx}$ or $y^{-n-2} \dfrac{dy}{dx} = \left(\dfrac{-1}{n+1}\right)\dfrac{dV}{dx}$

Then the equation becomes,

$$-\left(\frac{1}{n+1}\right)\frac{dV}{dx} - \frac{1}{x} \cdot V = - x^{m-n-1} \Rightarrow \frac{dV}{dx} + \frac{n+1}{x} V = (n+1) x^{m-n-1} \text{ which is linear.}$$

$$\text{I.F.} = e^{\int (n+1)/x \, dx} = e^{(n+1) \log x} = x^{n+1}$$

G.S. is $\qquad Vx^{n+1} = C + \int (n+1) x^{m-n-1} \cdot x^{n+1} \, dx = C + (n+1) \int x^m \, dx$

$$\Rightarrow \qquad y^{-n-1} \cdot x^{n+1} = C + (n+1)\frac{x^{m+1}}{m+1}$$

$$\boxed{\frac{x^{n+1}}{y^{n+1}} = C + \frac{n+1}{m+1} x^{m+1}}$$

Ex. 21 : *Solve* $4x^2 y \dfrac{dy}{dx} = 3x (3y^2 + 2) + (3y^2 + 2)^3$

Sol. : $\qquad$ Put $3y^2 + 2 = v$. Then, $6y \dfrac{dy}{dx} = \dfrac{dv}{dx} \Rightarrow y\dfrac{dy}{dx} = \dfrac{1}{6}\dfrac{dv}{dx}$

$\therefore$ The differential equation is

$$4x^2 \cdot \frac{1}{6}\frac{dv}{dx} = 3xv + v^3$$

$$\Rightarrow \qquad \frac{dv}{dx} - \frac{9}{2x} v = \frac{3}{2x^2} v^3 \text{ (Bernoulli's equation)}$$

Dividing by v^3

$$v^{-3}\frac{dv}{dx} - \frac{9}{2x}\, v^{-2} = \frac{3}{2x^2} \quad \text{Now put} \quad v^{-2} = t.$$

$\therefore \qquad -2v^{-3}\frac{dv}{dx} = \frac{dt}{dx} \Rightarrow v^{-3}\frac{dv}{dx} = -\frac{1}{2}\frac{dt}{dx}$

Then the equation becomes,

$$-\frac{1}{2}\frac{dt}{dx} - \frac{9}{2x}\, t = \frac{3}{2x^2} \quad \text{or} \quad \frac{dt}{dx} + \frac{9}{x}\, t = -\frac{3}{x^2} \ \text{(linear)}$$

$$\text{I.F.} = e^{\int \frac{9}{x}\, dx} = e^{9\log x} = x^9$$

G.S. is $\qquad t\cdot x^9 = C + \int -\frac{3}{x^2}\cdot x^9\, dx \Rightarrow x^9\cdot v^{-2} = C - 3\int x^7\, dx$

$$\boxed{\ \frac{x^9}{(3y^2+2)^2} = C - \frac{3}{8}\, x^8\ } \text{ is the required G.S.}$$

Ex. 22 : *Solve* $x\dfrac{dy}{dx} + 3y = x^4\, e^{1/x^2}\cdot y^3$ $\hfill$ **(Dec. 2008, 2007)**

Sol. : Here variables are not separable. Equation is not homogeneous. Also it is not exact. We shall try for linear type in y.

Given equation can be written as

$$\frac{dy}{dx} + \frac{3}{x}\cdot y = x^3\, e^{1/x^2}\cdot y^3$$

This is of Bernoulli's form with the factor y^3 on the R.H.S. $\therefore$ Dividing by y^3, we get

$$y^{-3}\frac{dy}{dx} + \frac{3}{x}\cdot y^{-2} = x^3\, e^{1/x^2}$$

Put $\ y^{-2} = u \ \therefore \ y^{-3}\dfrac{dy}{dx} = \dfrac{-1}{2}\cdot\dfrac{du}{dx}$

$\therefore \qquad -\dfrac{1}{2}\dfrac{du}{dx} + \dfrac{3}{x}\cdot u = x^3\, e^{1/x^2}$

or $\qquad \dfrac{du}{dx} - \dfrac{6}{x}\, u = -2x^3\, e^{1/x^2}$

which is linear in u with $P = \dfrac{-6}{x}$ and $Q = -2x^3\, e^{1/x^2}$

$$\text{I.F.} = e^{-6\int 1/x\, dx} = e^{-6\log x} = \frac{1}{x^6}$$

G.S. is $\qquad u\cdot\dfrac{1}{x^6} = \int -\dfrac{2x^3\, e^{1/x^2}}{x^6}\cdot dx + C$

$$\frac{u}{x^6} = \int e^{1/x^2}\left(-\frac{2}{x^3}\, dx\right) + C$$

$$\frac{u}{x^6} = e^{1/x^2} + C \hfill \left[\because \int e^f\, f'\, dx = e^f\right]$$

$$\boxed{\ \frac{1}{y^2 x^6} = e^{1/x^2} + C\ }$$

$$\boxed{\textbf{EXERCISE 1.13}}$$

Solve the following differential equations.

1. $\dfrac{dy}{dx} + \dfrac{2}{x}\,y = \dfrac{y^3}{x^3}$

   Ans. : $-\dfrac{1}{2y^2\,x^4} = \dfrac{-1}{6x^6} + C_1 \cdot \left(-\dfrac{1}{2y^2} = V\right)$

2. $y \sin 2x\, dx - (1 + y^2 + \cos^2 x)\, dy = 0$

   Ans. $-y \cos^2 x = y + \dfrac{y^3}{3} + C\ (-\cos^2 x = V)$

3. $\dfrac{dy}{dx} = \dfrac{-ax - hy - g}{hx + by + f}$

   Ans. : $ax^2 + 2hxy + by^2 + 2gx + 2fy = C_1$, (exact)

4. $(x^3 + 3xy^2)\, dx + (3x^2 y + y^3)\, dy = 0$

   Ans. : $x^4 + y^4 + 6x^2 y^2 = C$, (exact)

5. $x\,\dfrac{dy}{dx} = y + \cos\dfrac{1}{x}$

   Ans. : $\dfrac{y}{x} = -\sin\dfrac{1}{x} + C$ (linear in y)

6. $\dfrac{dy}{dx} - \dfrac{1}{x} = \dfrac{e^{2y}}{x^2}$

   Ans. : $-e^{-2y}\,x^2 = 2x + C$

7. $(1 - x^2 y^2)\, dx = y\, dx + x\, dy$

   Ans. : $\dfrac{1 - xy}{1 + xy} = C \cdot e^{2x}$ (Put $xy = t$)

8. $(x^2 y^2 + xy)\, y\, dx + (x^2 y^2 - 1)\, x\, dy = 0$ **Ans. :** $xy - \log y = C$

9. $x\left(\dfrac{dy}{dx} + y\right) = 1 - y$ **Ans. :** $(xy - 1)\, e^x = C$

10. $dr + (2r \cot\theta + \sin 2\theta)\, d\theta = 0$

    Ans. : $r \sin^2\theta = -\dfrac{\sin^4\theta}{2} + C$

11. $\dfrac{dy}{dx} = 2(y + e^{2x})$ **Ans.** $y = [C + 2x]\, e^{2x}$

12. $\dfrac{dy}{dx} = \dfrac{2y - x - 4}{y - 2x + 3}$ **Ans. :** $x^2 + y^2 - 4xy + 8x + 6y = C$

13. $(1 - x^2)\,\dfrac{dy}{dx} = 1 + xy$

    Ans. : $y\,\sqrt{1 - x^2} = C + \sin^{-1} x$

14. $2\,\dfrac{dy}{dx} + \cos^2(x - 2y) = 1$ **Ans. :** $\tan(x - 2y) - x = C$

15. $(3x + y - 7)\, dy = (2x + 2y - 6)\, dx$

    Ans. : $(y - x + 1)^4 = C\,(2x + y - 5)$

16. $(x^3 + xy^4)\, dx + 2y^3\, dy = 0$ **Ans. :** $e^{x^2}\,[x^2 - 1 + y^4] = C_1$

17. $3\,\dfrac{dy}{dx} + \dfrac{y}{2(x+1)} = \dfrac{x^3}{y^4}$

    Ans. : $y^{-3}\,\sqrt{x+1} = \dfrac{2}{9}(x+1)^{9/2} - \dfrac{6}{7}(x+1)^{7/2}$
    $+ \dfrac{6}{5}(x+1)^{5/2} - \dfrac{2}{3}(x+1)^{3/2} + C$

18. $(x^2 - xy + y^2)\, dx - xy\, dy = 0$ **Ans. :** $\log(y - x) + \dfrac{y}{x} = C$

19. $\dfrac{dy}{dx} = x^3 y^3 - xy$ **Ans. :** $\dfrac{1}{y^2} = x^2 + 1 + C \cdot e^{x^2}$

20. $\dfrac{1}{y}\,\dfrac{dy}{dx} + \dfrac{x}{1 - x^2} = xy^{-1/2}$

    Ans. : $3\sqrt{y} + 1 - x^2 = C_1\,(1 - x^2)^{1/4}$

21. $y\, dy - (1 - x^2 - y^2)\, x\, dx = 0$

    Ans. : $2y^2 = 2 - x^2 + C \cdot e^{-x^2}$

22. $(3y - 7x + 7)\, dx + (7y - 3x + 3)\, dy = 0$

    Ans. : $(x - y - 1)^2\,(x + y - 1)^5 = C$

23. $y - x\,\dfrac{dy}{dx} = x + y\,\dfrac{dy}{dx}$

    Ans. : $\tan^{-1}\dfrac{y}{x} + \dfrac{1}{2}\log\left(1 + \dfrac{y^2}{x^2}\right) + \log x = C$

24. $y\,(8x - 9y)\, dx + 2x\,(x - 3y)\, dy = 0$

    Ans. : $3x^3 y^2 - 2x^4 y = C$

25. $y + 2\,\dfrac{dy}{dx} = y^3\,(x - 1)$

    Ans. : $y^2\,(x + Ce^x) = 1$

26. $y\, dx + (2x + 1 - xy)\, dy = 0$

    Ans. : $xy^2 - y - 1 = C \cdot e^y$

27. $x\, dx + \sin^2\left(\dfrac{y}{x}\right)(y\, dx - x\, dy) = 0$

    Ans. : $\log x = \dfrac{y}{2x} - \dfrac{1}{4}\sin\left(\dfrac{2y}{x}\right) + C$

28. $e^{x-y}\, dx + y\, dy = ye^{2x}\, dy$

    Ans. : $\dfrac{1}{2}\log\left(\dfrac{1 + e^x}{1 - e^x}\right) + e^y\,(y - 1) = C$

29. $2(2x^2 + y^2)\, dx - xy\, dy = 0$ **Ans. :** $4x^2 + y^2 = Cx^4$

30. $\left(x - \sqrt{xy}\right) dy = y \cdot dx$

$$\textbf{Ans.:} \ 2\sqrt{\dfrac{x}{y}} + \log y + C = 0$$

31. $3x (xy - 2) dx + (x^3 + 2y) dy = 0$ **Ans.:** $x^3 y - 3x^2 + y^2 = C$

32. $(2xy - 3x^2) dx + (x^2 + y) dy = 0$

$$\textbf{Ans.:} \ x^2 y - x^3 + \dfrac{y^2}{2} = C$$

33. $(2x + y)^2 = xy \dfrac{dy}{dx}$ **Ans.:** $\log (x^4 + x^3 y) = \dfrac{y}{x} + C$

34. $\left(y - \sqrt{x^2 + y^2}\right) dx = x \, dy$

$$\textbf{Ans.:} \ \log x = - \sinh^{-1} \dfrac{y}{x} + C$$

35. $(x + y - 1) dx + (2x + 2y + 1) dy = 0$

$$\textbf{Ans.:} \ x + 2y - 3 \log (x + y + 2) + C = 0$$

36. $\dfrac{dy}{dx} = xy^3 - xy$ **Ans.:** $\dfrac{1}{y^2} = 1 + C e^{x^2}$

37. $\dfrac{dy}{dx} - \dfrac{1}{2}\left(1 + \dfrac{1}{x}\right) y + \dfrac{3y^3}{x} = 0$ **Ans.:** $\dfrac{x}{y^2} = 6 + C e^{-x}$

38. $(4xy + 3y^2 - x) dx + x (x + 2y) dy = 0$

$$\textbf{Ans.:} \ x^4 y + x^3 y^2 - \dfrac{x^4}{4} = C$$

39. $\tan y \dfrac{dy}{dx} + \tan x = \cos y \cos^2 x$ **Ans.:** $\sec y \ \sec x = \sin x + C$

40. $y (x + y + 1) dx + x (x + 3y + 2) dy = 0$ **(May 2007)**

$$\textbf{Ans.:} \ \dfrac{x^2 y^2}{2} + xy^3 + xy^2 = C$$

41. $xy \, dx + (x^2 - 3y) dy = 0$ **Ans.:** $\dfrac{x^2 y^2}{2} - y^3 = C$

42. $\dfrac{dy}{dx} = \dfrac{e^{-y}\left(2 \log x + \dfrac{1}{x} + \dfrac{1}{x^2}\right)}{e^{-x} [\cos e^{-y} + e^{-y} \sin e^{-y}]}$

$$\textbf{Ans.:} \ e^x \left(2 \log x - \dfrac{1}{x}\right) = e^y \cos e^{-y} + C$$

43. $(e^x + 1) y \, dy - (y + 1) e^x \, dx = 0$

$$\textbf{Ans.:} \ y - \log (y + 1) + \log (e^x + 1) - x = C$$

44. $e^{-y} \sec^2 y \, dy - dx = x \, dy$ **Ans.:** $\tan y = x e^y + C$

45. $\dfrac{\cos^2 y}{x} dy + \dfrac{\cos^2 x}{y} dx = 0$

$$\textbf{Ans.:} \ (x^2 + y^2) + x \sin 2x + y \sin 2y + \dfrac{\cos 2x}{2} + \dfrac{\cos 2y}{2} = C$$

46. $y \sqrt{1 + x^2} \, dx + x \sqrt{1 + y^2} \, dy = 0$

$$\textbf{Ans.:} \ \sqrt{1 + x^2} + \sqrt{1 + y^2} - \log \left[\dfrac{\left(\sqrt{1 + x^2} + 1\right)\left(\sqrt{1 + y^2} + 1\right)}{xy}\right] = C$$

47. $(y^2 + 2xy) dx + (2x^2 + 3xy) dy = 0$ **(Dec. 2007)**

$$\textbf{Ans.:} \ xy^3 + x^2 y^2 = C$$

48. $(y^2 + xy) dx - x^2 dy = 0$ **Ans.:** $\log x + \dfrac{x}{y} = C$

49. $\dfrac{(6x - 4y + 1)}{(3x - 2y + 1)} \cdot \dfrac{dy}{dx} = 1$

$$\textbf{Ans.:} \ x - 2y + \dfrac{1}{4} \log (12x - 8y + 1) = C$$

50. $(2x + 3y - 5) \dfrac{dy}{dx} + 3x + 2y - 5 = 0$

$$\textbf{Ans.:} \ 3 [(x - 1)^2 + (y - 1)^2] + 4 (x - 1) (y - 1) = C_1$$

51. $(x - y)^2 \left(x \dfrac{dy}{dx} - y\right) = x^2 \left(1 - \dfrac{dy}{dx}\right)$

$$\textbf{Ans.:} \ \dfrac{y}{x} + \dfrac{1}{x - y} = C_1$$

52. $\dfrac{dy}{dx} = \dfrac{\log x - 3x^3 y}{x^4}$ **Ans.:** $x^3 y = C + \dfrac{1}{2} (\log x)^2$

53. $x^2 \dfrac{dy}{dx} = e^y - x$ **Ans.:** $2x e^{-y} = 1 + C_1 \cdot x^2$

54. $x^2 dy = (5x^3 + 2 - 3 xy) \, dx$.   **Ans.:** $yx^3 = C + x^2 + x^5$

CHAPTER-2
APPLICATIONS OF FIRST ORDER DIFFERENTIAL EQUATIONS

2.1 INTRODUCTION

Differential equations are of great importance in engineering, because many physical laws and relations appear mathematically in the form of differential equations.

In the first chapter we discussed methods to solve ordinary differential equations of the first order and first degree. In this chapter, we shall show how these methods enable us to study many interesting problems, such as orthogonal trajectories, decay of radio- active materials, cooling or heating up of bodies, rectilinear motion, motion under gravity, analysis of electrical circuits, simple harmonic motion, vibrating string, heat conduction along a pipe or spherical shells and some problems related to chemical engineering.

It will be quite interesting to see, how methods discussed in first chapter are useful to solve the above mentioned problems of practical importance. Taking into account the importance of differential equations in mathematical modeling, we briefly discuss the technique of modeling.

2.2 TECHNIQUE OF MATHEMATICAL MODELLING

Mathematical modelling essentially consists of translating real world problems into mathematical problems, solving the mathematical problems and interpreting these solutions in the language of the real world. i.e.

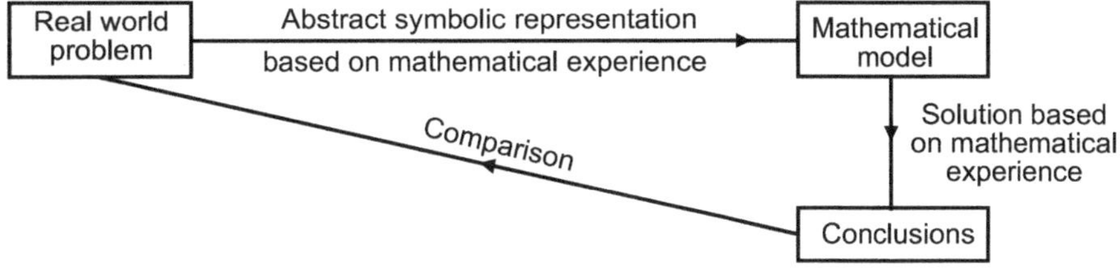

Fig. 2.1

Differential equations arise in many engineering and other applications as mathematical models of various physical and other systems.

For example, if we drop a stone then its acceleration $y'' = \dfrac{d^2y}{dt^2}$ is equal to the acceleration of gravity g (a constant). Hence the model of this problem of "free fall" is $y'' = g$ (neglecting air resistance). We have velocity $y' = \dfrac{dy}{dt} = gt + v_0$, where v_0 is the initial velocity with which the motion is started (e.g. $v_0 = 0$).

We get the distance traveled $y = \dfrac{g}{2}t^2 + v_0\ t + y_0$, where y_0 is the distance from 0 at the beginning (e.g. $y_0 = 0$).

Fig. 2.2

We shall consider physical problems which lead to a differential equation of first order and first degree and that of second order which reduces to first order.

2.3 ORTHOGONAL TRAJECTORIES

1.   **Trajectory :** A curve which cuts every member of a given family of curves according to some definite law is called a *trajectory* of the family.

2.   **Orthogonal Trajectory :** A curve which cuts every member of a given family of curves at *right angles* is called as *orthogonal trajectory* of the family.

3.   **Orthogonal Trajectories :** Two families of curves are said to be orthogonal if every member of the either family cuts each member of the other family at right angles.

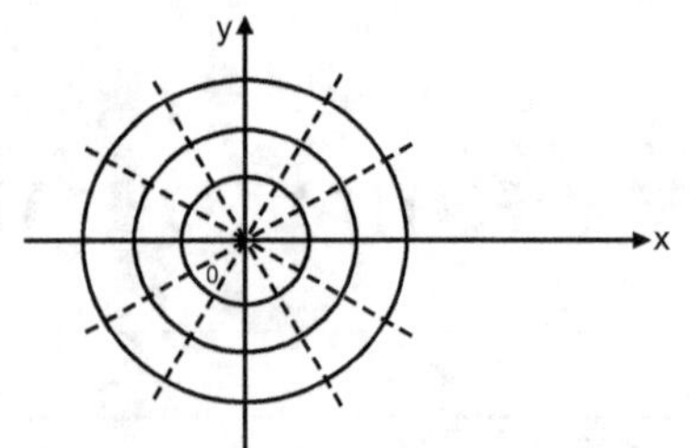

Fig. 2.3

Thus, if the given family consists of straight lines $y = mx$ (m = constant) representing family of straight lines all passing through the origin (shown by dotted lines in Fig. 2.3), then the family of circles $x^2 + y^2 = a^2$, (a is a parameter), with centres at $(0, 0)$ represents a family of orthogonal trajectories to the family $y = mx$ (Refer Fig. 2.3).

2.4 WORKING RULE TO FIND THE EQUATION OF ORTHOGONAL TRAJECTORIES

I. For Rectangular Cartesian Co-ordinates :

Step 1 : Given $f(x, y, a) = 0$, where a is a variable parameter.

Step 2 : Differentiate $f(x, y, a) = 0$ w.r.t. x and eliminate 'a'. We thus form a differential equation of the family of the form

$$\phi\left(x, y, \frac{dy}{dx}\right) = 0.$$

Step 3 : Replace $\dfrac{dy}{dx}$ by $-\dfrac{dx}{dy}$. Then the differential equation of the family of orthogonal trajectories will be : $\phi\left(x, y, -\dfrac{dx}{dy}\right) = 0$

Step 4 : The solution of step 3 is the family of orthogonal trajectories.

Ex. 1 : *Find the orthogonal trajectories of the family of straight lines $y = mx$.*

Sol. : Given $y = mx$ ... (1)

Differentiate (1) w.r.t. x, $\dfrac{dy}{dx} = m$

Eliminating m using $m = \dfrac{y}{x}$ from (1)

$\therefore$ $\dfrac{dy}{dx} = \dfrac{y}{x}$ ... (2)

which is the differential equation of the given family (1).

Replacing $\dfrac{dy}{dx}$ by $-\dfrac{dx}{dy}$ in (2), we get

$$-\frac{dx}{dy} = \frac{y}{x} \qquad \text{or} \qquad x\, dx + y\, dy = 0 \qquad\qquad \text{... (3)}$$

which is the differential equation of the orthogonal trajectories.

Integrating (3), $\displaystyle\int x\, dx + \int y\, dy = b$

$$\frac{x^2}{2} + \frac{y^2}{2} = b$$

or $\boxed{x^2 + y^2 = c^2}$

which is the equation of the required orthogonal trajectories of (1).

Ex. 2 : *Find the orthogonal trajectories of the curves given by $x^2 + 2y^2 = c^2$.*

Sol. : Given $x^2 + 2y^2 = c^2$ ... (1)

Differentiating (1) w.r.t. x, $x + 2y\dfrac{dy}{dx} = 0$ ... (2)

Replacing $\dfrac{dy}{dx}$ by $-\dfrac{dx}{dy}$ in (2), we obtain differential equation of orthogonal trajectories as

$$x + 2y\left(-\frac{dx}{dy}\right) = 0 \qquad \text{or} \qquad 2\frac{dx}{x} = \frac{dy}{y} \qquad\qquad \text{... (3)}$$

Integrating (3), $2\log x = \log y + \log k$ or $x^2 = ky$

Thus, $\boxed{x^2 = ky}$ is the required orthogonal trajectories of (1).

Ex. 3 : *Show that the family of confocal conics $\dfrac{x^2}{C} + \dfrac{y^2}{C - \lambda} = 1$, where C is arbitrary constant, is self orthogonal.*

Sol. : Given

$$\frac{x^2}{C} + \frac{y^2}{C - \lambda} = 1 \qquad \ldots (1)$$

Differentiating (1) w.r.t. x,

$$\frac{x}{C} + \frac{y\,(dy/dx)}{C - \lambda} = 0 \qquad \ldots (2)$$

Simplifying (2),

$$-\frac{x}{C} = \frac{y\,(dy/dx)}{C - \lambda} = \frac{x + y\,(dy/dx)}{-\lambda} \qquad \ldots (3)$$

From (3),

$$C = \frac{\lambda x}{x + y\,(dy/dx)} \quad \text{and} \quad C - \lambda = \frac{-\lambda y\,(dy/dx)}{x + y\,(dy/dx)} \qquad \ldots (4)$$

To obtain the differential equation for the family of curves (1), we eliminate C from (1) using C, C – λ from (4), thus

$$x^2\,\frac{[x + y\,(dy/dx)]}{\lambda x} + y^2\,\frac{[x + y\,(dy/dx)]}{-\lambda y\,(dy/dx)} = 1$$

or

$$[x + y\,(dy/dx)]\,[x\,(dy/dx) - y] - \lambda\,(dy/dx) = 0 \qquad \ldots (5)$$

Replacing dy/dx by (–dx/dy) in (5), we obtain differential equation of orthogonal family of curves as

$$[x - y\,(dx/dy)]\,[-x\,(dx/dy) - y] + \lambda\,(dx/dy) = 0$$

which on simplification gives the differential equation as

$$[x + y\,(dy/dx)]\,[x\,(dy/dx) - y] - \lambda\,(dy/dx) = 0 \qquad \ldots (6)$$

which is same as (5).

    Thus orthogonal family of curves of system given by (6) is the same as given by (1). Hence the confocal conics (1) is self orthogonal.

II. For Polar Co-ordinates :

Step 1 : Given $f\,(r, \theta, a) = 0$, where a is a variable parameter.

Step 2 : Form a differential equation of the family of the form $\phi\left(r, \theta, \dfrac{dr}{d\theta}\right) = 0$ by eliminating 'a'.

Step 3 : Replace $\dfrac{dr}{d\theta}$ by $\left(-r^2\,\dfrac{d\theta}{dr}\right)$ whereby the differential equation of the family of orthogonal trajectories become

$$\phi\left(r, \theta, -r^2\,\frac{d\theta}{dr}\right) = 0.$$

Step 4 : Solve step 3 which is the family of orthogonal trajectories.

Ex. 1 : *Find the orthogonal trajectories of the circles defined by $r = a\cos\theta$ which all pass through the origin and have their centres on the initial line, a being the variable diameter.*   ***(May 2005)***

Sol. : Given

$$r = a\cos\theta \qquad \ldots (1)$$

$$\log r = \log a + \log\cos\theta$$

$$\frac{1}{r}\,\frac{dr}{d\theta} = 0 + \frac{1}{\cos\theta}\,(-\sin\theta)$$

$$\frac{1}{r}\,\frac{dr}{d\theta} = -\tan\theta \qquad \ldots (2)$$

which is the differential equation of the given family (1).

Replacing $\dfrac{dr}{d\theta}$ by $-r^2\,\dfrac{d\theta}{dr}$ in (2), we get

$$\frac{1}{r}\left(-r^2\,\frac{d\theta}{dr}\right) = -\tan\theta$$

∴

$$r\,\frac{d\theta}{dr} = \tan\theta \quad \therefore \quad \frac{dr}{r} = \frac{d\theta}{\tan\theta}$$

$$\frac{dr}{r} = \cot\theta \cdot d\theta \qquad \ldots (3)$$

which is the differential equation of the family of orthogonal trajectories.

Integrating (3),

$$\int \frac{dr}{r} = \int \cot\theta\,d\theta + \log C$$

$$\log r = \log\sin\theta + \log C$$

$$\log r = \log (C\sin\theta)$$

$$\boxed{r = C \cdot \sin\theta}$$

which is the required equation of orthogonal trajectories of (1).

Ex. 2 : *Find orthogonal trajectories of the family of curve $r^2 = a \sin 2\theta$.*

Sol. : Given : $r^2 = a \sin 2\theta$ ... (1)

Differentiating (1), w.r.t. θ, $2r \dfrac{dr}{d\theta} = 2a \cos 2\theta$ ... (2)

Eliminating a, by putting $a = \dfrac{r^2}{\sin 2\theta}$ in (2), we get

$$\frac{dr}{d\theta} = r \cot 2\theta \qquad \text{... (3)}$$

Replacing $\dfrac{dr}{d\theta}$ by $-r^2 \dfrac{d\theta}{dr}$ in (3), we obtain

$$-r^2 \frac{d\theta}{dr} = r \cot 2\theta$$

or $\dfrac{dr}{r} = -\tan 2\theta \, d\theta$ ... (4)

which is the differential equation of the family of orthogonal trajectories.

Integrating (4), $\log r = \dfrac{1}{2} \log \cos 2\theta + \dfrac{1}{2} \log C$ (Note : $C_1 = \dfrac{1}{2} \log C$)

or $\boxed{r^2 = C \cos 2\theta.}$

which is the required equation of orthogonal trajectories of (1).

EXERCISE 2.1

Find the orthogonal trajectories of the family of

1. $xy = c$ **(May 2008)**

         Ans. : $x^2 - y^2 = c^2$

2. $2x^2 + y^2 = cx$

         Ans. : $x^2 = -y^2 \log (cy)$

3. $y^2 = 4ax$ **(Dec. 2010, May 07)**

         Ans. : $2x^2 + y^2 = c$

4. $\dfrac{x^2}{a^2} + \dfrac{y^2}{b^2 + \lambda} = 1$, λ is a parameter

         Ans. : $x^2 + y^2 = 2a^2 \, \lambda o\gamma \, x + c$

5. $e^x + e^{-y} = c$

         Ans. : $e^y - e^{-x} = c$

6. $x^2 + cy^2 = 1$ **(May 2009, Nov. 2014)**

         Ans. : $x^2 = c \, e^{x^2 + y^2}$

7. $r = a(1 - \cos \theta)$ **(Dec. 2008)**

         Ans. : $r = c(1 + \cos \theta)$

8. $r = \dfrac{2a}{1 + \cos \theta}$

         Ans. : $r = \dfrac{2c}{1 - \cos \theta}$

9. $r^2 = a^2 \cos 2\theta$

         Ans. : $r^2 = c^2 \sin 2\theta$

10. $r = a \cos^2 \theta$

         Ans. : $r^2 = c \sin \theta$

11. $y = ax^2$

         Ans. : $\dfrac{x^2}{(\sqrt{2}c)^2} + \dfrac{y^2}{c^2} = 1$

12. $r = a(1 + \cos \theta)$

         Ans. : $r = a(1 - \cos \theta)$

2.5 RATE OF DECAY OF RADIOACTIVE MATERIALS

This law states that disintegration at any instant, is proportional to the amount of material present.

If u is the amount of material at any time t, then $\boxed{\dfrac{du}{dt} = -ku}$ *where k is a constant.*

     Ex. 1 : *Uranium disintegrates at a rate proportional to the amount that is present at any instant. If M_1 and M_2 grams of uranium are present at times T_1 and T_2 respectively, find the half-life of uranium.*   **(May 2004)**

     Sol. : Let the mass of uranium at any time t be m grams.

Then the equation of disintegration of uranium is

$$\frac{dm}{dt} = -\mu m, \text{ where } \mu \text{ is a constant}$$

Integrating, we get
$$\int \frac{dm}{m} = -\mu \int dt + c$$

or
$$\log m = c - \mu t \qquad \ldots \text{(i)}$$

Initially, when $t = 0$, $m = M$ (say) so that $c = \log M$

$\therefore$ (i) becomes
$$\mu t = \log M - \log m \qquad \ldots \text{(ii)}$$

Also, when $t = T_1$, $m = M_1$ and when $t = T_2$, $m = M_2$

$\therefore$ From (ii), we get
$$\mu T_1 = \log M - \log M_1 \qquad \ldots \text{(iii)}$$
$$\mu T_2 = \log M - \log M_2 \qquad \ldots \text{(iv)}$$

Subtracting (iii) from (iv), we get,
$$\mu (T_2 - T_1) = \log M_1 - \log M_2 = \log (M_1/M_2)$$

Hence
$$\mu = \frac{\log (M_1/M_2)}{T_2 - T_1}$$

Let the mass reduce to half its initial value in time T.

i.e. when $t = T$, $m = \dfrac{1}{2} M$

$\therefore$ From (ii), we get
$$\mu T = \log M - \log (M/2) = \log 2$$

Thus,
$$\boxed{T = \frac{1}{\mu} \log 2 = \frac{(T_2 - T_1) \log 2}{\log (M_1/M_2)}}$$

EXERCISE 2.2

1. Radium decomposes at the rate proportional to the amount present. If 5% of the original amount disappears in 50 years, how much will remain after 100 years ?

   Ans. 90.25%

2. If 30% of a radioactive substance disappeared in 10 days, how long will it take for 90% of it to disappear ?

   (Dec. 2009, 2008) Ans. 64.5 days

3. Radium decomposes at the rate proportional to the quantity of radium present. Suppose that it is found that in 25 years approximately 1.1% of certain quantity of radium has decomposed. Determine approximately how long will it take for one half of the original amount of radium to decompose. **Ans.** $1564.66 \approx 1565$ years.

4. Radium decomposes at the rate proportional to the amount present. If a fraction M of the original amount disappears in 1 year, how much will remain at the end of 21 years ? **Ans.** $(1 - 1/M)^{21}$ times the original amount.

2.6 NEWTON'S LAW OF COOLING

According to this law, the *temperature of a body changes at a rate which is proportional to the difference in temperature between that of the surrounding medium and that of the body itself.*

If θ_0 is the temperature of the surroundings and θ that of the body at any time t, then

$$\boxed{\frac{d\theta}{dt} = -k (\theta - \theta_0)} \quad \text{where } k \text{ is a constant.}$$

Illustrations on Newton's Law of Cooling :

Ex. 1 : *A metal ball is heated to a temperature of 100°C and at time $t = 0$ it is placed in water which is maintained at 40°C. If the temperature of the ball is reduced to 60°C in 4 minutes, find the time at which the temperature of the ball is 50°C.* **(May 2011, 2006, 2018)**

Sol. : Let the temperature of the ball be T°C at time t min. Then the differential equation is given by

$$\frac{dT}{dt} = -k (T - 40) \quad \text{or} \quad \frac{dT}{T - 40} = -k\, dT \qquad \ldots \text{(1)}$$

Integration gives
$$-kt = \log (T - 40) + \log C \qquad \ldots \text{(2)}$$

At $t = 0$, $T = 100$. This gives $\log C = -\log 60$

and hence (2) becomes :
$$-kt = \log \frac{T - 40}{60} \qquad \ldots \text{(3)}$$

But $T = 60$ at $t = 4$. Substituting these values in (3), we obtain

$$-4k = \log \frac{1}{3} \quad \text{or} \quad k = \frac{1}{4} \log 3$$

Hence, equation (3) gives :

$$-\frac{t}{4} \log 3 = \log \frac{T-40}{60}$$

When T = 50, we obtain

$$\boxed{t = \frac{4 \log 6}{\log 3} = 6.5 \text{ minutes}}$$

Ex. 2 : *A body originally at 80°C cools down to 60°C in 20 minutes, the temperature of the air being 40°C. What will be the temperature of the body after 40 minutes from the original ?*   **(May 2008, 2007, 2006; Dec. 2011, Nov. 2014)**

Sol. : If θ is the temperature of the body at any time t, then

$$\frac{d\theta}{dt} = -k\,(\theta - 40), \quad \text{where k is a constant.}$$

Integrating,

$$\int \frac{d\theta}{\theta - 40} = -k \int dt + \log C, \quad \text{where C is a constant}$$

or

$$\log(\theta - 40) = -kt + \log C$$

i.e.

$$\theta - 40 = Ce^{-kt} \qquad \qquad \dots (1)$$

When t = 0, $\theta = 80°$ and when t = 20, $\theta = 60°$

$\therefore$

$$40 = C, \text{ and } 20 = Ce^{-20k}$$

$\therefore$

$$k = \frac{1}{20} \log 2 \qquad \qquad \dots (2)$$

Thus (1) becomes

$$\theta - 40 = 40\, e^{\left(-\frac{1}{20}\log 2\right)t}$$

When t = 40 min.,

$$\theta = 40 + 40\, e^{-2\log 2} = 40 + 40\, e^{\log(1/4)} = 40 + 40 \times \frac{1}{4}$$

$$\boxed{\theta = 50°C}$$

Ex. 3 : *According to Newton's law of cooling, the rate at which a substance cools in moving air is proportional to the difference between the temperature of the substance and that of the air. If the temperature of the air is 300°C and the substance cools from 370°C to 340°C in 15 minutes, find when the temperature will be 310°C.*

Sol. : Let T be the temperature of the substance at the time t minutes.

Then, by Newton's law of cooling the differential equation is,

$$\frac{dT}{dt} = -k\,(T - 300) \quad \text{or} \quad \frac{dT}{T - 300} = -k\,dt \qquad \qquad \dots (1)$$

Integrating between the limits t = 0, T = 370 and t = 15, T = 340,

$$\int_{370}^{340} \frac{dT}{T - 300} = -k \int_{0}^{15} dt \Rightarrow [\log(T - 300)]_{370}^{340} = -k[t]_{0}^{15}$$

or

$$\log 40 - \log 70 = -15k$$

$\therefore$

$$k = \frac{1}{15} \log \frac{7}{4} \qquad \qquad \dots (2)$$

Integrating between the limits t = 0, T = 370 and t = t, T = 310

$$\int_{370}^{310} \frac{dT}{T - 300} = -k \int_{0}^{t} dt \Rightarrow [\log(T - 300)]_{370}^{310} = -k\,[t]_{0}^{t}$$

$$\log 10 - \log 70 = -kt = -\left(\frac{1}{15} \log \frac{7}{4}\right)t$$

$\therefore$

$$\boxed{t = \frac{15\,(\log 7)}{\left(\log \dfrac{7}{4}\right)} = 52.2 \text{ minutes}}$$

Ex. 4 : *Water at temperature 100°C cools in 10 minutes to 60°C in a room temperature of 20°C. Find when the temperature will be 30°C.*

Sol. : Let T be the temperature of the water at any time t minutes. Then by Newton's law of cooling, the differential equation is

$$\frac{dT}{dt} = -k(T - 20)$$

or

$$\frac{ddT}{T - 20} = -k\,dt \qquad \qquad \dots (1)$$

Integrating (i) between the limits t = 0, T = 100°C and t = 10, T = 60°C

$$\int_{100}^{60} \frac{dT}{T - 20} = -k \int_{0}^{10} dt$$

∴

$$[\log(T - 20)]_{100}^{60} = -k\,[t]_{0}^{10}$$

or

$$\log 40 - \log 80 = -10k$$

∴

$$k = \frac{1}{10}\log 2 \qquad \qquad \dots (2)$$

Next, Integrating (1) between the limits t = 0, T = 100°C and t = t, T = 30°C.

$$\int_{100}^{30} \frac{dT}{T - 20} = -k \int_{0}^{t} dT$$

∴

$$[\log(T - 20)]_{100}^{30} = -k\,[t]_{0}^{t}$$

or

$$\log 10 - \log 80 = -kt = -\left(\frac{1}{10}\log 2\right)t \qquad \qquad \left(\because k = \frac{1}{10}\log 2\right)$$

∴

$$t = \frac{\log 8}{\left(\frac{1}{10}\log 2\right)} = 10\frac{\log 8}{\log 2}$$

or

$$\boxed{t = 30 \text{ minutes}}$$

Ex. 5 : *According to Newton's law of cooling, the rate at which a substance cools in moving air is proportional to the difference between the temperature of the substance and that of the air. If the temperature of the air is 30°C and the substance cools from 100°C to 70°C in 15 minutes, find when the temperature will be 40°C.*   **(Dec. 2008, 2006, 2005, 2016)**

Sol. : Let the unit of time be a minute and T be the temperature of the substance at any instant t. Then by Newton's law of cooling, we have

$$\frac{dT}{dt} = -k(T - 30)$$

or

$$\frac{dT}{T - 30} = -k\,dt$$

Integrating, $\log(T - 30) = -kt + C$ $\qquad \qquad \dots (1)$

Initially, when t = 0, T = 100

∴   from (1), $C = \log 70$

Substituting the value of C in (1), we have

$$\log(T - 30) = -kt + \log 70$$

or $kt = \log 70 - \log(T - 30)$ $\qquad \qquad \dots(2)$

Also, when t = 15, T = 70, gives

$$15k = \log 70 - \log 40 \qquad \qquad \dots (3)$$

Dividing (2) by (3), we have $\dfrac{t}{15} = \dfrac{\log 70 - \log(T - 30)}{\log 70 - \log 40}$ $\qquad \qquad \dots (4)$

Now, when T = 40, we have from (4),

$$\frac{t}{15} = \frac{\log 70 - \log 10}{\log 70 - \log 40} = \frac{\log_e 7}{\log_e 7/4} = \frac{\log_{10} 7}{\log_{10}(7/4)} = 3.48$$

∴ $\boxed{t = 15 \times 3.48 = 52.20}$

Hence the temperature will be 40°C after 52.2 minutes.

Ex. 6 : *If the temperature of the body drops from 100°C to 60°C in one minute when the temperature of the surrounding is 20°C, what will be the temperature of the body at the end of the second minute ?* **(May 2005; Dec. 2006, 2005, 2013, 2017)**

Sol. : Let T be the temperature of the body at time t minutes. The differential equation is given by

$$\frac{dT}{dt} = -k(T - 20) \quad \text{or} \quad \frac{dT}{T - 20} = -k\,dt \qquad \qquad \text{... (1)}$$

Integrating, $\log(T - 20) = -kt + C$

Initially at $t = 0$, $T = 100$, gives $C = \log 80$

$\therefore$ $kt = \log 80 - \log(T - 20)$ ... (2)

Also at $t = 1$, $T = 60$, gives $k = \log 80 - \log 40$

$\therefore$ $t = \dfrac{\log 80 - \log(T - 20)}{\log 80 - \log 40}$

For $t = 2$, $2 = \dfrac{\left(\log \dfrac{80}{T - 20}\right)}{\log 2}$ or $\log 2^2 = \log \dfrac{80}{T - 20}$

$\therefore$ $4 = \dfrac{80}{T - 20}$ or $T - 20 = 20$

$\therefore$ $\boxed{T = 40°C}$

Alternatively, Integrating (1) between the limits $t = 0$, $T = 100$ and $t = 1$, $T = 60$, we obtain

$$\int_{100}^{60} \frac{dT}{T - 20} = -k \int_{0}^{1} dt \Rightarrow [\log(T - 20)]_{100}^{60} = -k\,[t]_{0}^{1}$$

$\therefore$ $k = \log 80 - \log 40 = \log 2$

Next, integrate (1) between the limits $t = 0$, $T = 100$ and $t = 2$, $T = T$,

$$\int_{100}^{T} \frac{dT}{T - 20} = -k \int_{0}^{1} dt \Rightarrow [\log(T - 20)]_{100}^{T} = -k\,[t]_{0}^{2}$$

$\therefore$ $\log(T - 20) - \log 80 = -2k \Rightarrow \log\left(\dfrac{T - 20}{80}\right) = -2\log 2$ $(\because k = \log 2)$

$\therefore$ $\dfrac{T - 20}{80} = \dfrac{1}{4}$

or $\boxed{T = 40°C}$

EXERCISE 2.3

1. If a thermometer is taken outdoors where the temperature is 0°C, from a room in which the temperature is 21°C and the reading drops to 10°C in 1 minute, how long after its removal will the reading be 5°C ?

[**Hint :** $\dfrac{dT}{dt} = -k(T - 0)$; $t = 0$, $T = 21$ and $t = 1$, $T = 10$; $k = \log_e 2.1 = \ln 2.1$] **Ans.** 1.93424 min = 1 min 56 sec.

2. Water at temperature 100°C cools in 10 minutes to 88°C in a room of temperature 25°C. Find the temperature of water after 20 minutes.

      (Dec. 2008, May 2017)

[**Hint :** $\dfrac{d\theta}{dt} = -k(\theta - 25)$; $t = 0$, $\theta = 100$ and $t = 10$, $\theta = 88$; $k = \dfrac{1}{10} \log \dfrac{25}{21}$] **Ans.** when $t = 20$, $\theta = 77.9°C$

3. A body at temperature 100°C is placed in a room whose temperature is 20°C and cools to 60°C in 5 minutes. Find its temperature after a further interval of 3 minutes.

  (Nov./Dec. 19, Dec. 2009, 2006, 18; May 2009, 15)

[**Hint :** $t = 0$, $T = 100$; $t = 5$, $T = 60$; $k = \dfrac{1}{5}\log 2$; find T when $t = 8$.] **Ans.** 46.4°C

4. When a thermometer is placed in a hot liquid bath at temperature T, the temperature θ indicated by the thermometer rises at the rate of $T - \theta$. For a bath at 95°C, the temperature reads 15°C at a certain instant $(t = 0)$ and 35° at $t = 10$ second. What will be its temperature at $t = 20$ sec. ? **(Dec. 2007)**

[**Hint :** θ = Temperature of thermometer, $T = 95$ = temperature of hot liquid bath; $t = 0$, $\theta = 15$ and $t = 10$, $\theta = 35$; $k = \dfrac{1}{10}\log \dfrac{4}{3}$.] **Ans.** 50°C

5. A copper ball is heated to a temperature of 100°C. Then at time t = 0 it is placed in water which is maintained at a temperature of 30°C. At the end of 3 minutes, the temperature of the ball is reduced to 70°C. Find the time at which the temperature of the ball drops to 31°C. [**Hint :** t = 0, T = 100 and t = 3, T = 70.]

Ans. 22.78 ≈ 23 min

6. Two friends A and B order coffee and receive cups of equal temperature at the same time. A adds a small amount of cool cream immediately but does not drink his coffee until 10 minutes later, B waits for 10 minutes and adds the same amount of cool cream and begins to drink. Assuming the Newton's law of cooling, decide who drinks the hotter coffee ?

Ans. B drinks hotter coffee.

2.7 SIMPLE ELECTRIC CIRCUITS

We shall consider circuits made up of

(i) three passive elements – resistance, inductance, capacitance and

(ii) an active element – voltage source which may be a battery or a generator.

1. Table of Elements, Symbols and Units :

Sr. No.	Element	Symbol	Unit
1.	Time	t	second
2.	Quantity of electricity (electric charge)	q	coulomb
3.	Current (= time rate flow of charge)	$i = \dfrac{dq}{dt}$	ampere (A)
4.	Resistance, R	(R)	ohm (Ω)
5.	Inductance, L	(L)	henry (H)
6.	Capacitance, C	(C)	farad (F)
7.	Electromotive force or voltage (constant), E	Battery, E = constant	volt
8.	Variable voltage generator	Generator, E = Variable voltage	volt

2. Basic Relations :

(i) $i = \dfrac{dq}{dt}$ or $q = \int i\, dt$ [$\because$ current is the rate of flow of electricity]

(ii) Voltage drop across resistance R = Ri (Ohm's law)

(iii) Voltage drop across inductance L = $L\dfrac{di}{dt}$.

(iv) Voltage drop across capacitance C = $\dfrac{q}{C}$.

3. Kirchhoff's law : The formulation of differential equations for an electrical circuit depends on the following two Kirchhoff's laws which are of cardinal importance :

(I) The algebraic sum of the **voltage** drops around any closed circuit is equal to the resultant electromotive force in the circuit.

(II) The algebraic sum of the **currents** flowing into (or from) any node is zero.

4. Differential Equations :

(i) Circuit involving L and R along with a Voltage Source (battery) E, all in Series : Consider a circuit containing resistance R and inductance L in series with a voltage source (battery) E.

Let i be the current flowing in the circuit at any time t. Then by Kirchhoff's first law, we have

Sum of voltage drops across R and L = E

i.e. $\quad Ri + L\dfrac{di}{dt} = E$

or $\quad \dfrac{di}{dt} + \dfrac{R}{L}i = \dfrac{E}{L}$ which is a linear differential equation. ... (1)

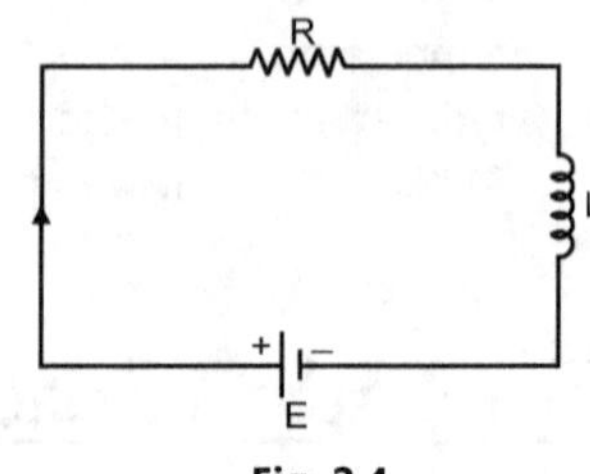

Fig. 2.4

$$\text{I.F.} = e^{\int \frac{R}{L}\,dt} = e^{Rt/L}$$

and therefore its solution is

$\therefore \qquad i \cdot e^{Rt/L} = \int \dfrac{E}{L}\, e^{Rt/L}\, dt + c = \dfrac{E}{L} \cdot \dfrac{L}{R} \cdot e^{Rt/L} + C$

Here, $\qquad i = \dfrac{E}{R} + Ce^{-Rt/L}$ ... (2)

If initially there is no current in the circuit, i.e. $i = 0$, when $t = 0$, we have $C = -\dfrac{E}{R}$. Thus, (2) becomes

$$\boxed{\; i = \dfrac{E}{R}\,(1 - e^{-Rt/L}) \;}\;.$$

As $t \to \infty$, $i = \dfrac{E}{R}$, which shows that i increases with t and attains the maximum value $\dfrac{E}{R}$.

(ii) Circuits involving R and C along with a Voltage Source (battery) E all in Series : Consider a circuit containing resistance R and capacitance C in series with a voltage source (battery) E.

Let i be the current flowing in the circuit at any time t. Then by Kirchhoff's first law, we have

Sum of voltage drops across R and C = E (e.m.f.)

i.e. $\qquad Ri + \dfrac{q}{C} = E.$

Since $i = \dfrac{dq}{dt}$, this equation in terms of q can be written as

$$R\dfrac{dq}{dt} + \dfrac{q}{C} = E \quad \text{or} \quad \dfrac{dq}{dt} + \dfrac{q}{RC} = \dfrac{E}{R}$$

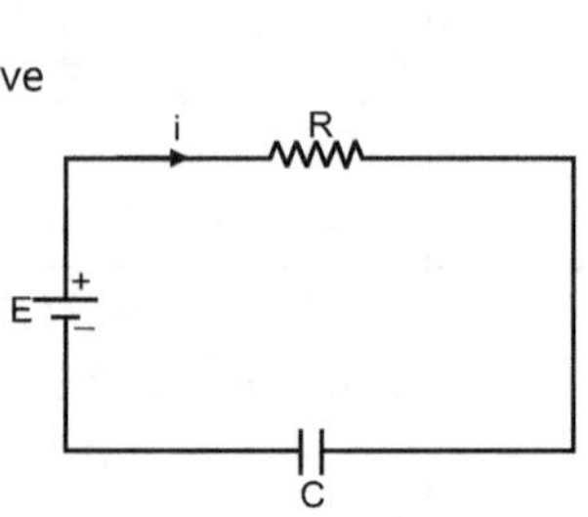

Fig. 2.5

which is a linear differential equation.

Here $\qquad \text{I.F.} = e^{\int \frac{1}{RC}\,dt} = e^{\frac{t}{RC}}$

and therefore its solution is

$$q\, e^{\frac{t}{RC}} = \int \dfrac{E}{R}\, e^{\frac{t}{RC}}\, dt + B = \dfrac{E}{R}\left(RC\, e^{\frac{t}{RC}}\right) + B$$

$$q = EC + B\, e^{-\frac{t}{RC}}$$

Assuming $\qquad q = q_0 \quad \text{when} \quad t = 0$

$\qquad\qquad q_0 = EC + B \quad \text{giving} \quad B = q_0 - EC$

$\therefore \qquad q = EC + (q_0 - EC)\, e^{-t/RC}$

or $\qquad q = EC\,(1 - e^{-t/RC}) + q_0\, e^{-t/RC}$

$\therefore \qquad i = \dfrac{dq}{dt} = EC\dfrac{1}{RC}\, e^{-t/RC} - \dfrac{q_0}{RC}\, e^{-t/RC}$

or $\qquad \boxed{\; i = \left(\dfrac{E}{R} - \dfrac{q_0}{RC}\right) e^{-t/RC} \;}$

(iii) Circuit involving L and C both in Series, after Removing Source Applied e.m.f. : Consider a circuit containing inductance L and capacitance C in series without applied e.m.f.

Let i be the current flowing in the circuit at any time t. Then by Kirchhoff's first law, we have sum of voltage drops across L and C = 0

i.e.
$$L \frac{di}{dt} + \frac{q}{C} = 0$$

$$\frac{di}{dt} = -\frac{q}{LC}$$

$\therefore$
$$\frac{di}{dq} \frac{dq}{dt} = -\frac{q}{LC}$$

$$i \frac{di}{dq} = -\frac{q}{LC}$$

$$\int i\, di = -\int \frac{q}{LC}\, dq + A \quad \Rightarrow \quad \frac{i^2}{2} = -\frac{q^2}{2LC} + A$$

$$i^2 = -\frac{q^2}{LC} + B$$

Assuming i = 0,   q = q_0 when t = 0 $\therefore$ B = $\dfrac{q_0^{\,2}}{LC}$

$$i^2 = \frac{\left(q_0^{\,2} - q^2\right)}{LC}$$

$$i = \pm \frac{\sqrt{\left(q_0^{\,2} - q^2\right)}}{\sqrt{LC}}$$

Since q decreases as t increases, i $= \dfrac{dq}{dt} = -\dfrac{1}{\sqrt{LC}} \sqrt{q_0^{\,2} - q^2}$

$$-\frac{dq}{\sqrt{q_0^{\,2} - q^2}} = \frac{dt}{\sqrt{LC}}$$

Integrating
$$\cos^{-1}\left(\frac{q}{q_0}\right) = \frac{t}{\sqrt{LC}} + C$$

Assuming q = q_0, when t = 0 $\therefore$ C = 0

$\therefore$
$$\frac{t}{\sqrt{LC}} = \cos^{-1}\left(\frac{q}{q_0}\right)$$

$$\boxed{q = q_0 \cos\left(\frac{t}{\sqrt{LC}}\right)}$$

USEFUL FORMULAE

1.	$\displaystyle\int e^{at} \sin bt\, dt = \frac{e^{at}}{a^2 + b^2} (a \sin bt - b \cos bt)$	
2.	$\displaystyle\int e^{at} \cos bt\, dt = \frac{e^{at}}{a^2 + b^2} (a \cos bt + b \sin bt)$	
3.	$\displaystyle\int e^{at} \sin bt\, dt = \frac{e^{at}}{\sqrt{a^2 + b^2}} \sin (bt - \phi)$, where $\phi = \tan^{-1}\left(\dfrac{b}{a}\right)$	
4.	$\displaystyle\int e^{at} \cos bt\, dt = \frac{e^{at}}{\sqrt{a^2 + b^2}} \cos (bt - \phi)$, where $\phi = \tan^{-1}\left(\dfrac{b}{a}\right)$	

Illustrations on Electrical Circuits :

Ex. 1 : *A resistance of 100 ohms, an inductance of 0.5 henry are connected in series with a battery of 20 volts. Find the current in a circuit as a function of t.*

(Nov./Dec. 2019, May 2005, 2014, 2017, 2018)

Sol. : By Kirchhoff's law, we have $L\dfrac{dI}{dt} + RI = E.$ (Read i = I)

$$\dfrac{dI}{dt} + \dfrac{R}{L}I = \dfrac{E}{L} \text{ which is linear. Here } P = \dfrac{R}{L}, Q = \dfrac{E}{L}.$$

$$\text{I.F.} = e^{\int \frac{R}{L}dt} = e^{\frac{Rt}{L}}$$

G.S. is

$$I e^{\frac{Rt}{L}} = \int \dfrac{E}{L} \cdot e^{\frac{Rt}{L}} \, dt + A = \dfrac{E}{R} e^{\frac{Rt}{L}} + A$$

$\therefore$ But at t = 0, I = 0 $0 = \dfrac{E}{R} + A$ giving A $= -\dfrac{E}{R}$

$\therefore$ G.S. becomes

$$I e^{\frac{Rt}{L}} = \dfrac{E}{R}\left(-1 + e^{\frac{Rt}{L}}\right) \text{ or } I = \dfrac{E}{R}\left(1 - e^{-\frac{Rt}{L}}\right)$$

Given : R = 100 ohms,   L = 0.5 henry,   E = 20 volts

$\therefore$

$$\boxed{I = \dfrac{20}{100}\left(1 - e^{-\frac{100t}{0.5}}\right) = \dfrac{1}{5}\left(1 - e^{-200\,t}\right)}$$

Ex. 2 : *A voltage $E\,e^{-at}$ is applied at t = 0 to a circuit containing inductance L and resistance R. Show that the current at any time t is* $\dfrac{E}{R - aL}\left(e^{-at} - e^{-\frac{Rt}{L}}\right)$.

(May 11, 09, 07, Dec. 16)

Sol. : Let i be the current in the circuit at any time t. Also a voltage Ee^{-at} is applied at t = 0 to a circuit containing inductance L and resistance R. Then by Kirchhoff's law, we have,

$$L\dfrac{di}{dt} + Ri = Ee^{-at} \quad \Rightarrow \quad \dfrac{di}{dt} + \dfrac{R}{L} i = \dfrac{E}{L} e^{-at} \qquad \text{... (1)}$$

which is a linear differential equation.

$\therefore$ $$\text{I.F.} = e^{\frac{Rt}{L}}$$

Solution of (1) is given by

$$i. e^{\frac{R}{L}t} = \dfrac{E}{L} \int e^{-at} \cdot e^{\frac{R}{L}t} \, dt + B = \dfrac{E}{L} \int e^{\left(\frac{R}{L} - a\right)t} \, dt + B = \dfrac{E}{L} \cdot \dfrac{e^{\left(\frac{R}{L} - a\right)t}}{(R/L - a)} + B$$

$$i \cdot e^{\frac{R}{L}t} = \dfrac{E}{R - aL} \cdot e^{\left(\frac{R}{L} - a\right)t} + B$$

Given i = 0, t = 0, $B = -\dfrac{E}{R - aL}$

$\therefore$

$$\boxed{i = \dfrac{E}{R - aL}\left(e^{-at} - e^{-\frac{R}{L}t}\right)}$$

Ex. 3 : *In a circuit containing inductance L, resistance R and voltage E, the current I is given by : $E = RI + L\dfrac{dI}{dt}$. Given L = 640 H, R = 250 Ω and E = 500 volts. I being zero when t = 0. Find the time that elapses, before it reaches 90% of its maximum value.*

(May 2008, 2006, 2004, 2015)

Sol. : Given that $E = RI + L\dfrac{dI}{dt} \Rightarrow \dfrac{dI}{dt} + \dfrac{R}{L} I = \dfrac{E}{L}$ which is a linear differential equation.

$$\text{I.F.} = e^{\frac{R}{L} \int dt} = e^{\frac{Rt}{L}}$$

G.S.
$$I e^{\frac{Rt}{L}} = \frac{E}{L} \int e^{\frac{Rt}{L}} \, dt + A = \frac{E}{L} \frac{e^{\frac{Rt}{L}}}{R/L} + A = \frac{E}{R} e^{\frac{Rt}{L}} + A$$

$$I = \frac{E}{R} + A e^{\frac{-Rt}{L}}$$

I being zero when $t = 0$ $\therefore$ $0 = \dfrac{E}{R} + A e^0$ $\therefore$ $A = -\dfrac{E}{R}$

$$I = \frac{E}{R} - \frac{E}{R} e^{-\frac{Rt}{L}}, \quad I = \frac{E}{R} \left(1 - e^{-\frac{Rt}{L}} \right) \qquad \ldots (1)$$

Maximum value of I be I_{max} which is obtained when $t \to \infty$

$$I_{max} = \frac{E}{R} (1 - e^{-\infty}) = \frac{E}{R} (1 - 0) = \frac{E}{R}$$

$$I_{max} = \frac{90}{100} \frac{E}{R} \quad (90\% \text{ of max.})$$

Putting in (1) if $t = t_1$ for 90% I_{max},

$$\frac{9}{10} \frac{E}{R} = \frac{E}{R} \left(1 - e^{-\frac{Rt_1}{L}} \right)$$

$$\frac{9}{10} = 1 - e^{-\frac{Rt_1}{L}}$$

$$e^{-\frac{Rt_1}{L}} = 1 - \frac{9}{10} = \frac{1}{10}$$

$$-\frac{Rt_1}{L} = -\log_e 10$$

$$t_1 = \frac{L}{R} \log_e 10$$

Given $L = 640$, $R = 250$ $\therefore$ $t_1 = \dfrac{640}{250} \log_e 10$

$\therefore$ $\boxed{t_1 = \dfrac{64}{25} \log_e 10 = 5.89 \text{ sec}}$ which is the required time.

Ex. 4 : *Show that the differential equation for the current 'i' in an electrical circuit containing an inductance L and a resistance R in series and acted on by an electromotive force E sin ωt satisfies the equation* $L \dfrac{di}{dt} + Ri = E \sin \omega t$. **(Dec. 2011, May 2019)**

Find the value of the current at an time t, if initially there is no current in the circuit.

Sol. : Given $\quad Ri + L\dfrac{di}{dt} = E \sin \omega t$ $\quad$ or $\quad \dfrac{di}{dt} + \dfrac{R}{L} i = \dfrac{E}{L} \sin \omega t$ which is a linear differential equation.

Here $\quad$ I.F. $= e^{\int \frac{R}{L} dt} = e^{\frac{Rt}{L}}$

The general solution is $\quad i \, (\text{I.F.}) = \int \dfrac{E}{L} \sin \omega t \cdot (\text{I.F.}) \, dt + c$

i.e. $\quad i \, e^{\frac{Rt}{L}} = \dfrac{E}{L} \int e^{\frac{Rt}{L}} \sin \omega t \, dt + c = \dfrac{E}{L} \dfrac{e^{\frac{Rt}{L}}}{\sqrt{[(R/L)^2 + \omega^2]}} \sin \left(\omega t - \tan^{-1} \dfrac{L\omega}{R} \right) + c$

or $\quad i = \dfrac{E}{\sqrt{(R^2 + \omega^2 L^2)}} \sin (\omega t - \phi) + c e^{-Rt/L}$ $\qquad \ldots (1)$

where, $\tan \phi = L\omega/R$

Initially when $t = 0$, $i = 0$ $\therefore$ $0 = \dfrac{E \sin (-\phi)}{\sqrt{(R^2 + \omega^2 L^2)}} + c$ i.e. $c = \dfrac{E \sin \phi}{\sqrt{(R^2 + \omega^2 L^2)}}$

Thus, (1) takes the form $i = \dfrac{E \sin (\omega t - \phi)}{\sqrt{(R^2 + \omega^2 L^2)}} + \dfrac{E \sin \phi}{\sqrt{(R^2 + \omega^2 L^2)}} \cdot e^{-Rt/L}$

$$\boxed{i = \dfrac{E}{\sqrt{(R^2 + \omega^2 L^2)}} \left[\sin (\omega t - \phi) + \sin \phi \cdot e^{-Rt/L}\right]}$$ which gives the current at any time t.

Ex. 5 : *The equation of an L-R circuit is given by $L\dfrac{dI}{dt} + RI = 10 \sin t$. If $I = 0$, at $t = 0$, express I as a function of t.* **(Dec. 2008, 2005)**

Sol. : Given $L\dfrac{dI}{dt} + RI = 10 \sin t$

$\therefore$ $\dfrac{dI}{dt} + \dfrac{R}{L}I = \dfrac{10}{L} \sin t$ is a linear differential equation

Here   I.F. $= e^{\frac{Rt}{L}}$

G.S. is $I \cdot e^{Rt/L} = \dfrac{10}{L} \displaystyle\int e^{Rt/L} \sin t \, dt + B = \dfrac{10}{L} \dfrac{e^{Rt/L}}{\sqrt{\dfrac{R^2}{L^2} + 1}} \sin (t - \phi) + B$ where $\tan \phi = \dfrac{L}{R}$

$$I = \dfrac{10}{\sqrt{R^2 + L^2}} \cdot \sin (t - \phi) + B\, e^{-\frac{Rt}{L}}$$

When $t = 0$, $I = 0$, $\therefore B = \dfrac{10}{\sqrt{R^2 + L^2}} \sin \phi$

$$\boxed{I = \dfrac{10}{\sqrt{R^2 + L^2}} \left[\sin (t - \phi) + \sin \phi \, e^{-\frac{Rt}{L}}\right]}$$

Ex. 6 : *A constant electromotive force E volts is applied to a circuit containing a constant resistance R ohms in series and a constant inductance L henries. If the initial current is zero, show that the current builds up to half its theoretical maximum in $(L \log 2)/R$ seconds.*

(Dec. 2011, 2009, 2008, 2006, 2017; May 2013)

Sol. : Let i be the current in the circuit at any time t. By Kirchhoff's law, we have

$$L\dfrac{di}{dt} + Ri = E \qquad \text{or} \qquad \dfrac{di}{dt} + \dfrac{R}{L} i = \dfrac{E}{L} \qquad \qquad \dots (1)$$

which is a linear differential equation.

Here   I.F. $= e^{\int \frac{R}{L} dt} = e^{\frac{Rt}{L}}$

The general solution of equation (1) is

$$i \,(\text{I.F.}) = \int \dfrac{E}{L} (\text{I.F.}) \, dt + c$$

or $i \cdot e^{\frac{Rt}{L}} = \displaystyle\int \dfrac{E}{L} \cdot e^{\frac{Rt}{L}} \, dt + c = \dfrac{E}{L} \cdot \dfrac{L}{R} e^{\frac{Rt}{L}} + c$

or $i = \dfrac{E}{R} + c\, e^{-\frac{Rt}{L}}$ $\qquad \qquad \dots (2)$

Initially, when $t = 0$, $i = 0$ so that $c = -\dfrac{E}{R}$

Thus, (2) becomes, $i = \dfrac{E}{R} \left(1 - e^{-\frac{Rt}{L}}\right)$ $\qquad \qquad \dots (3)$

This equation gives the current in the circuit at any time t. Clearly, i increases with t and attains the maximum value $\dfrac{E}{R}$.

Let the current in the circuit be half its theoretical maximum after a time T seconds.

Then,
$$\frac{1}{2} \cdot \frac{E}{R} = \frac{E}{R} \left(1 - e^{-\frac{RT}{L}} \right)$$

or
$$e^{-\frac{RT}{L}} = \frac{1}{2}$$

or
$$-\frac{RT}{L} = \log \frac{1}{2} = - \log 2$$

$\therefore$
$$\boxed{T = (L \log 2)/R}$$

Ex. 7 : *An electrical circuit contains an inductance of 5 henries and a resistance of 12 ohms in series with an e.m.f. 120 sin (20t) volts. Find the current at t = 0.01, if it is zero when t = 0.*   **(Dec. 2006)**

Sol. :
$$\frac{di}{dt} + \frac{R}{L} i = \frac{120 \sin (20 t)}{L} \quad \text{which is a linear differential equation.}$$

$$\text{I.F.} = e^{\frac{Rt}{L}} \quad \text{and its G.S. is}$$

$$i \cdot e^{\frac{Rt}{L}} = \frac{120}{L} \int e^{\frac{Rt}{L}} \sin (20 t) \, dt + A$$

$$= \frac{120}{L} \frac{e^{\frac{Rt}{L}}}{\sqrt{R^2/L^2 + 400}} \sin (20 t - \phi) + A \quad \text{where} \quad \tan \phi = \frac{20 L}{R}$$

$$i = \frac{120}{\sqrt{R^2 + 400 L^2}} \sin (20 t - \phi) + A e^{-\frac{Rt}{L}}$$

To find the current i at t $= 0.01$, given L = 5 H, R $= 12 \, \Omega$ and

$$\phi = \tan^{-1}\left[\frac{20 (5)}{12}\right] = 1.451367401$$

we get
$$\boxed{i = 0.023729634 \text{ ampere.}}$$

Ex. 8 : *The equation of electromotive force in terms of current i for an electrical circuit having resistance R and a condenser of capacity C, in series, is $E = Ri + \int \frac{i}{C} \, dt$. Find the current i at any time t, when $E = E_0 \sin \omega t$.*

Sol. : The given equation can be written as $Ri + \int \frac{i}{C} \, dt = E_0 \sin \omega t$.

Differentiating both sides w.r.t. t, we have

$$R \frac{di}{dt} + \frac{i}{C} = \omega E_0 \cos \omega t \quad \text{or} \quad \frac{di}{dt} + \frac{i}{RC} = \frac{\omega E_0}{R} \cos \omega t \qquad \text{... (1)}$$

which is linear differential equation.

$$\text{I.F.} = e^{\int \frac{1}{RC} dt} = e^{\frac{t}{RC}}$$

Here the general solution of equation (1) is

$$i. e^{\frac{t}{RC}} = \int \frac{\omega E_0}{R} \cos \omega t \cdot e^{\frac{t}{RC}} \, dt + k = \frac{\omega E_0}{R} \cdot \frac{e^{\frac{t}{RC}}}{\sqrt{\left(\frac{1}{RC}\right)^2 + \omega^2}} \cos \left(\omega t - \tan^{-1}\frac{\omega}{\frac{1}{RC}} \right) + k$$

$$= \frac{\omega C E_0}{\sqrt{1 + R^2 C^2 \omega^2}} e^{\frac{t}{RC}} \cos (\omega t - \phi) + k \quad \text{where} \quad \tan \phi = RC\omega$$

or
$$\boxed{i = \frac{\omega C E_0}{\sqrt{1 + R^2 C^2 \omega^2}} \cos (\omega t - \phi) + k e^{-\frac{t}{RC}}} \quad \text{which gives the current at any time t.}$$

Ex. 9 : *Find the current i in the circuit having resistance R and condenser of capacity C in series with emf E sin ωt.*

(May 2006; Dec. 2010, 2007)

Sol. : We have

$$Ri + \frac{q}{C} = E \sin \omega t \;\Rightarrow\; R \frac{dq}{dt} + \frac{q}{C} = E \sin \omega t.$$

or

$$\frac{dq}{dt} + \frac{q}{RC} = \frac{E}{R} \sin \omega t \quad \text{which is a linear differential equation}$$

Here

$$I.F. = e^{\frac{t}{RC}}$$

G.S. is

$$q \cdot e^{\frac{t}{RC}} = \int \frac{E}{R} \sin \omega t \; e^{\frac{t}{RC}} \, dt + A$$

$$= \frac{E}{R} \cdot \frac{e^{\frac{t}{RC}}}{\sqrt{1/R^2C^2 + \omega^2}} \sin (\omega t - \phi) + A, \quad \text{where } \tan \phi = RC\omega$$

$$= EC \frac{e^{\frac{t}{RC}}}{\sqrt{1 + R^2C^2\omega^2}} \sin (\omega t - \phi) + A$$

∴

$$q = \frac{EC}{\sqrt{1 + R^2C^2\omega^2}} \sin (\omega t - \phi) + A e^{-\frac{t}{RC}}$$

$$\boxed{\; i = \frac{dq}{dt} = \frac{EC\omega}{\sqrt{1 + R^2C^2\omega^2}} \cos (\omega t - \phi) - \frac{A}{RC} e^{-\frac{t}{RC}} \;}$$

Ex. 10 : *A circuit consists of resistance 'R' ohms and a condenser of 'C' farads connected to a constant e.m.f. E. If $\frac{q}{C}$ is the voltage of the condenser at time t after closing the circuit, show that the voltage at time t is $E\left(1 - e^{-\frac{t}{CR}}\right)$.* *(May 2007, 2005; Dec. 2009, 2013)*

Sol. : The differential equation for the circuit is $\;Ri + \frac{q}{C} = E \Rightarrow i + \frac{q}{RC} = \frac{E}{R}$

$$\frac{dq}{dt} + \left(\frac{1}{RC}\right) q = \frac{E}{R} \quad \text{which is a linear differential equation}$$

Here

$$I.F. = e^{\int \frac{1}{RC} dt} = e^{t/RC}$$

G.S. is

$$q \cdot e^{t/RC} = \int \frac{E}{R} \cdot e^{t/RC} \, dt + B \;\Rightarrow\; q \, e^{t/RC} = \frac{E e^{t/RC}}{R/RC} + B$$

or

$$\frac{q}{C} e^{t/RC} = E \, e^{t/RC} + B_1$$

At $t = 0$, $q = 0$, ∴

$$0 = E + B_1 \quad \therefore \quad B_1 = -E$$

∴

$$\frac{q}{C} e^{\frac{t}{RC}} = E \, e^{\frac{t}{RC}} - E$$

$$\boxed{\; \frac{q}{C} = E\left(1 - e^{-\frac{t}{RC}}\right) \;}$$

Ex. 11 : *The charge 'Q' on the plate of a condenser of capacity 'C' charged through a resistance 'R' by a steady voltage 'V' satisfies the differential equation $R \frac{dQ}{dt} + \frac{Q}{C} = V$. If Q = 0 at t = 0, show that Q = CV [1 − e^{−t/RC}]. Find the current flowing into the plate.*

(Dec. 2005; May 2008, 2004)

Sol. : The given equation is $\quad R \frac{dQ}{dt} + \frac{Q}{C} = V \quad$ i.e. $\quad \frac{dQ}{dt} + \frac{1}{RC} Q = \frac{V}{R} \quad$ which is a linear differential equation.

$$\text{I.F.} = e^{\int \frac{1}{RC} dt} = e^{t/RC}$$

G.S. is
$$Q\, e^{t/RC} = \int \frac{V}{R}\, e^{t/RC} \cdot dt + c_1 = \frac{V}{R} \cdot RC\, e^{t/RC} + c_1$$

$$Q = CV + c_1\, e^{-t/RC}$$

for $t = 0, Q = 0$ $\therefore$ $c_1 = -CV$

$\therefore$
$$0 = CV + c_1$$

$\therefore$
$$Q = CV - CV\, e^{-t/RC}$$

$$Q = CV\left[1 - e^{-\frac{t}{RC}}\right]$$

and
$$i = \frac{dQ}{dt} = CV\left[\frac{1}{RC}\, e^{-\frac{t}{RC}}\right]$$

$$\boxed{i = \frac{V}{R}\, e^{-\frac{t}{RC}}}$$

Ex. 12 : *An voltage $200\, e^{-5t}$ is applied to a circuit containing resistance $R = 20$ ohms and condenser of capacity $C = 0.01$ farads in series. Find the change and current at any time, assuming that $t = 0, q = 0$.*

Sol. : The differential equation for the R-C circuit is

$$Ri + \frac{q}{C} = E \quad \text{or} \quad R\frac{dq}{dt} + \frac{q}{C} = E \quad \text{(in terms of q)}$$

or
$$\frac{dq}{dt} + \frac{q}{RC} = \frac{E}{R} \qquad \ldots (1)$$

Given : $R = 20$, $C = 0.01$ and $E = 200\, e^{-5t}$.

$\therefore$
$$\frac{dq}{dt} + \frac{q}{20 \times 0.01} = \frac{200\, e^{-5t}}{20} \quad \text{or} \quad \frac{dq}{dt} + 5q = 10\, e^{-5t} \qquad \ldots (2)$$

which is a linear differential equation.

$$\text{I.F} = e^{\int 5dt} = e^{5t}$$

G.S. is
$$q\, e^{5t} = \int (10\, e^{-5t})\, e^{5t}\, dt + A = 10t + A \qquad \ldots (3)$$

Given : $t = 0, q = 0$ $\therefore$ $A = 0$

From G.S. (3),
$$q\, e^{5t} = 10t$$
or
$$q = 10t\, e^{-5t}$$

and
$$i = \frac{dq}{dt} = 10\, (e^{-5t} - 5t\, e^{-5t})$$

or
$$\boxed{i = 10\, (1 - 5t)\, e^{-5t}}$$

EXERCISE 2.5

1. In a circuit of resistance R, self inductance L, the current i is given by $L\frac{di}{dt} + Ri = E \cos pt$, where E, p are constants. Find the current at time t.

   Ans. $i = \dfrac{E}{L^2 p^2 + R^2}\, [pL \sin pt + R \cos pt] + ce^{-Rt/L}$

2. When a switch is closed in a circuit containing a battery E, a resistance R and an inductance L, the current i builds up at rate given by $L\frac{di}{dt} + Ri = E$. Find i as a function of t. How long will it be, before the current has reached one half its maximum value, if $E = 6$ volts, $R = 100$ ohms and $L = 0.1$ henry ?

   [Hint : $i = \dfrac{E}{R}\left(1 - e^{-\frac{R}{L}t}\right)$, $i_{max} = \dfrac{E}{R}$ and $t = \dfrac{L}{R} \log$**]**

   Ans. 0.0006931 sec.

3. Solve the equation, $L\frac{di}{dt} + Ri = 200 \cos (300\, t)$, where $R = 100$ ohms, $L = 0.05$ henry and find i, given that $i = 0$ when $t = 0$. What value does i approach after a long time ?

   Ans. $i = \dfrac{40}{409}\, (20 \cos 300\, t + 3 \sin 300\, t)$

   $$-\frac{800}{409}\, e^{-200t} \text{ and } \frac{40}{\sqrt{409}}$$

4. Find the current at any time $t > 0$ in a circuit having in series a constant electromotive force 40 volt, a resistor 10 ohms and an inductor 0.2 henry, given that initial current is zero. Also find the current when $E(t) = 150 \cos 200\, t$.   **Ans.** $i(t) = 4\,(1 - e^{-50\,t})$;

$$i(t) = \frac{3}{170}\,(50 \cos 200\, t + 200 \sin 200\, t) - \frac{15}{17}\,e^{-50t}$$

5. A capacitor $C = 0.01$ Farad in series with a resistor $R = 20$ ohms is charged from a battery $E = 10$ volts.

Assuming that initially the capacitor is completely uncharged, determine the charge $Q(t)$, voltage $v(t)$ on the capacitor and current $i(t)$ in the circuit.

(Nov. 2015, May 2016)

Ans. $Q(t) = 0.1\,(1 - e^{-5t})$, $v(t) = \dfrac{Q}{C} = 10\,(1 - e^{-5t})$,

$$i(t) = \frac{dQ}{dt} = 0.5\, e^{-5t}$$

2.8 RECTILINEAR MOTION

Rectilinear motion is a motion of a body along a straight line. Let a body of mass m start moving from a fixed point O along a straight line OX under the action of a force F. After any time t, let it be moving at P where OP = x, then

(i)   its velocity (v) $= \dfrac{dx}{dt}$

(ii)   its acceleration (a) $= \dfrac{dv}{dt}$ or $\dfrac{d}{dt}\left(\dfrac{dx}{dt}\right) = \dfrac{d^2x}{dt^2}$

Also, $a = \dfrac{dv}{dt} = \dfrac{dv}{dx} \cdot \dfrac{dx}{dt} = \dfrac{dv}{dx} \cdot v$

∴ $a = v\,\dfrac{dv}{dx}$

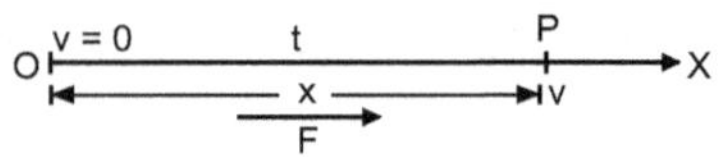

Fig. 2.6

∴   acceleration $= a = \dfrac{dv}{dt}$ or $v\,\dfrac{dv}{dx}$ or $\dfrac{d^2x}{dt^2}$

Newton's second law of motion states that $F = \dfrac{d}{dt}\,(mv)$.

If m is constant then $F = m\,\dfrac{dv}{dt} = ma$

∴ $F = m\,\dfrac{dv}{dt}$ or $mv\,\dfrac{dv}{dx}$ or $m\,\dfrac{d^2x}{dt^2}$, where F is the effective force.

D'Alembert's Principle : Algebraic sum of the forces acting on a body along a given direction is equal to the product of mass × acceleration in that direction.

i.e. | Net force $=$ Mass × Acceleration |

Note : The forces usually are : (i) vertically downward, (ii) tension in elastic string or spring, (iii) reactions or stresses at points in contact with other bodies, (iv) forces of attraction, (v) forces of resistance due to wind and friction etc.

Due precaution must be taken about the direction.

Force (Net) means *algebraic sum* of the forces acting along that direction i.e. direction of motion.

Illustrations on Rectilinear Motion :

Ex. 1 : *A body of mass m, falling from rest is subjected to the force of gravity and an air-resistance proportional to the square of the velocity 'kv^2'. If it falls through a distance x and possesses a velocity v at that instant, prove that* $\dfrac{2\,kx}{m} = \log\left(\dfrac{a^2}{a^2 - v^2}\right)$ *where mg $= ka^2$.*

(May 2006, 2005, 2004; Dec. 2011, 2016, 2017)

Sol. : The forces acting on the body are :

(i) the weight acting downwards $= mg$

(ii) the air resistance acting upwards $= -kv^2$

∴ Net force on the body $F = mg - kv^2$

Equation of motion is,

$$m\,v\,\frac{dv}{dx} = mg - kv^2$$

$$m\,v\,\frac{dv}{dx} = k\,(a^2 - v^2) \quad (\text{Given that } mg = ka^2)$$

Integrating,

$$\int \frac{v\,dv}{a^2 - v^2} = \int \frac{k}{m}\,dx + c$$

$$-\frac{1}{2} \log (a^2 - v^2) = \frac{kx}{m} + c$$

$\because\ x = 0,\ v = 0 \ \therefore\ c = -\frac{1}{2} \log a^2$

$\therefore$

$$\frac{1}{2} [\log a^2 - \log (a^2 - v^2)] = \frac{kx}{m}$$

$$\boxed{\frac{2\,kx}{m} = \log \left(\frac{a^2}{a^2 - v^2}\right)} \quad \text{which is the desired relation.}$$

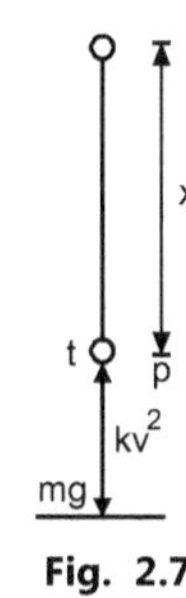

Fig. 2.7

Ex. 2 : *A body starts moving from rest is opposed by a force per unit mass of value 'cx' and resistance per unit mass of value 'bv²', where x and v are the displacement and velocity of the body at that instant. Show that the velocity of the body is given by*

$$v^2 = \frac{c}{2b^2} (1 - e^{-2bx}) - \frac{cx}{b}$$

(Dec. 2009, 2005, Nov. 2015, May 2017)

Sol. : By Newton's second law of motion, the equation of motion of the body is

$$v\frac{dv}{dx} = -cx - bv^2$$

or

$$v\frac{dv}{dx} + bv^2 = -cx \qquad\qquad \text{... (1)}$$

This is Bernoulli's linear differential equation, put $v^2 = z$ and $2v\dfrac{dv}{dx} = \dfrac{dz}{dx}$, equation (1) becomes

$$\frac{1}{2}\frac{dz}{dx} + bz = -cx$$

or

$$\frac{dz}{dx} + 2\,bz = -2cx \qquad\qquad \text{... (2)}$$

which is linear equation.

The general solution of (2) is

$$\text{I.F.} = e^{\int 2b\,dx} = e^{2bx}$$

$$z \cdot e^{2bx} = \int -2cx \cdot e^{2bx}\,dx + c_1$$

$$= -2c \int x\,e^{2bx}\,dx + c_1 \quad \text{[Integrating by parts]}$$

$$= -2c \left[x \cdot \frac{e^{2bx}}{2b} - \int 1 \cdot \frac{e^{2bx}}{2b}\,dx \right] + c_1$$

$$= -\frac{cx}{b} e^{2bx} + \frac{c}{2b^2} e^{2bx} + c_1$$

or

$$v^2 \cdot e^{2bx} = -\frac{cx}{b} e^{2bx} + \frac{c}{2b^2} e^{2bx} + c_1$$

or

$$v^2 = -\frac{cx}{b} + \frac{c}{2b^2} + c_1 e^{-2bx} \qquad\qquad \text{... (3)}$$

Initially, when $x = 0$, $v = 0$ $\therefore \dfrac{c}{2b^2} + c_1 = 0$ or $c_1 = -\dfrac{c}{2b^2}$

Substituting the value of c_1 in (3), we have

$$v^2 = -\frac{cx}{b} + \frac{c}{2b^2} - \frac{c}{2b^2} e^{-2bx}$$

or

$$\boxed{v^2 = \frac{c}{2b^2} (1 - e^{-2bx}) - \frac{cx}{b}}$$

Ex. 3 : Velocity of escape from the earth. *Determine the least velocity with which a particle must be projected vertically upwards so that it does not return to the earth. Assume that it is acted upon by the gravitational attraction of the earth only.*   **(May 2008)**

Sol. : Let r be the variable distance of the particle from the earth's centre. By Newton's law of gravitation, the acceleration a of the particle is proportional to $\frac{1}{r^2}$ $\left(\text{i.e. } a \propto \frac{1}{r^2} \right)$.

$$\therefore \qquad a = v\frac{dv}{dr} = -\frac{k}{r^2} \qquad \qquad \text{... (1)}$$

where v is the velocity of the particle when its distance from the earth's centre is r, the acceleration is negative because v is decreasing.

On the surface of the earth, r = R, the radius of the earth and a = – g, the acceleration of gravity at the surface.

i.e. $\qquad \qquad -g = -\frac{k}{R^2} \quad \text{or} \quad k = gR^2$

$\therefore$ From (1), $\qquad \qquad v\frac{dv}{dr} = -\frac{gR^2}{r^2}$

or $\qquad \qquad v\,dv = -\frac{gR^2}{r^2}\,dr$

Integrating, $\qquad \qquad \int v\,dv = -gR^2 \int \frac{dr}{r^2} + c$

or $\qquad \qquad \frac{v^2}{2} = \frac{gR^2}{r} + c$

or $\qquad \qquad v^2 = \frac{2gR^2}{r} + 2c \qquad \qquad \text{... (2)}$

Let v_0 be the initial velocity of projection from the surface of the earth. Then,

$$v = v_0 \text{ when } r = R$$

$\therefore$ From (2), $\qquad \qquad v_0^2 = 2gR + 2c \text{... (3)}$

Substituting (3) from (2) (to eliminate c),

$$v^2 - v_0^2 = \frac{2gR^2}{r} - 2gR$$

or $\qquad \qquad v^2 = \frac{2gR^2}{r} + (v_0^2 - 2gR)$

The particle will never return to earth if its velocity v during ascent remains positive. For, if v vanishes, the particle comes to rest and then descends, so that v becomes negative.

Now, as the particle rises upwards, $\frac{2gR^2}{r}$ goes on decreasing. The velocity will remain positive if and only if $v_0^2 - 2gR \geq 0$, i.e. If $v_0 \geq \sqrt{2gR}$

$\therefore$ The least velocity of projection is $\boxed{v_0 = \sqrt{2gR}}$

A particle projected with this velocity will never return to the earth, i.e. will escape from the earth. This velocity is called the velocity of escape from the earth.

Ex. 4 : *A body of mass m falls from rest under gravity in a fluid whose resistance to motion at any instant is mK times its velocity, where K is a constant. Find the terminal velocity of the body and also the time taken to acquire one-half of its limiting speed.* **(Dec. 07, 18)**

Sol. : Let v be the velocity of the body at any time t. Then the equation of motion of the body is given by,

$$m\frac{dv}{dt} = mg - m\,Kv$$

or $\qquad \qquad \frac{dv}{dt} = g - Kv \qquad \qquad \text{... (1)}$

Separating the variables in (1) and then integrating it, we obtain

$$t = \frac{-1}{K} \log (g - Kv) + C \qquad \qquad \text{... (2)}$$

Since the body is falling from rest, we have $v = 0$ when $t = 0$. This gives $C = \dfrac{1}{K} \log g$,

and therefore (2) becomes, $\qquad t = \dfrac{1}{K} \log \dfrac{g}{g - Kv} = \dfrac{-1}{K} \log \left(1 - \dfrac{K}{g} v\right)$ $\qquad$ ... (3)

We know that the terminal velocity is attained when the weight of the body exactly balances the resistance to motion. If V is the terminal velocity attained by the body, then

$$mg = m\,KV$$

or $\qquad\qquad V = g/K$ $\qquad\qquad$ ... (4)

Substituting (4) in (3), we obtain

$$t = -\dfrac{V}{g} \log \left(1 - \dfrac{v}{V}\right) \qquad\qquad \text{... (5)}$$

Let $t = t_1$ when $v = \dfrac{1}{2} V$. Then (5) gives

$$\boxed{\,t_1 = -\dfrac{V}{g} \log \left(\dfrac{1}{2}\right) = \dfrac{V}{g} \log 2\,}$$

Hence, the body attains one-half of its limiting speed at time $\dfrac{V}{g} \log 2$ seconds, where V is the terminal velocity of the body.

Remark : When resistance becomes equal to the weight, the acceleration becomes zero and particle continues to fall with a constant velocity, called the *limiting or terminal velocity*.

Ex. 5 : *A particle moves in a straight line under the action of an attraction varying inversely as the $\dfrac{3}{2}$ th power of the distance. Show that the velocity acquired by falling from an infinite distance to a distance 'a' from the centre is equal to the velocity which would be acquired in moving from rest at a distance 'a' to a distance $\dfrac{a}{4}$.*

Sol. : The equation of motion is $v\dfrac{dv}{dx} = -\lambda\, x^{-3/2}$, where λ is a constant and taking unit mass.

This gives $\qquad\qquad v\, dv = -\lambda x^{-3/2}\, dx$

If v_1 is the velocity acquired in moving from rest at infinity to a distance a, we have

$$\int_0^{v_1} v\, dv = -\lambda \int_\infty^a x^{-3/2}\, dx \quad \text{or} \quad \dfrac{v_1^2}{2} = \dfrac{2\lambda}{\sqrt{a}} \qquad\qquad \text{... (1)}$$

If v_2 is the velocity acquired in moving from rest at a distance 'a' to a distance a/4, we have,

$$\int_0^{v_2} v\, dv = -\lambda \int_a^{a/4} x^{-3/2}\, dx$$

or $\qquad\qquad \dfrac{v_2^2}{2} = \lambda \cdot 2\left[\dfrac{1}{\sqrt{x}}\right]_a^{a/4} = \dfrac{2\lambda}{\sqrt{a}} \qquad\qquad \text{...(2)}$

From (1) and (2), $\qquad \boxed{v_1 = v_2.}$

Ex. 6 : *A particle of mass m is projected vertically upward under gravity, the resistance of the air being mk times the velocity. Show that the greatest height attained by the particle is $\dfrac{V^2}{g} [\lambda - \log(1 + \lambda)]$, where V is the greatest velocity which the above mass will attain when it falls freely and λV is the initial velocity.*

Sol. : Let v be the velocity of the particle at time t. The forces acting on the particle are :

(i) its weight mg acting vertically downwards,

(ii) the resistance mkv of the air acting vertically downwards.

Accelerating force on the particle $= -mg - mkv$

$\therefore$　By Newton's second law, the equation of motion of the particle is,

$$mv\frac{dv}{dx} = -mg - mkv$$

or　　　$$v\frac{dv}{dx} = -g - kv \qquad \qquad ...(1)$$

When the particle falls freely (under gravity), equation (1) becomes (changing g to –g)

$$v\frac{dv}{dx} = g - kv \qquad \qquad ...(2)$$

When the particle attains the greatest velocity V, its acceleration is zero.

$\therefore$　From (2),　　　　$$0 = g - kV \quad \text{or} \quad k = \frac{g}{V}$$

Putting this value of k in (1), we have,

$$v\frac{dv}{dx} = -g - \frac{g}{V}v = -\frac{g}{V}(V + v)$$

or　　　$$\frac{v}{V+v}\, dv = -\frac{g}{V}\, dx$$

Integrating,　　　　$$\int \frac{v}{V+v}\, dv = -\frac{g}{V}\int dx + c$$

or　　　$$\int \left(1 - \frac{V}{V+v}\right) dv = -\frac{g}{V}x + c$$

or　　　$$v - V \log (V + v) = -\frac{g}{V}x + c \qquad \qquad ...(3)$$

Initially, when x = 0, v = λV　$\therefore$　From (3), we have,　$\lambda V - V \log (V + \lambda V) = c$

or　　　　　　　$$c = V[\lambda - \log V (1 + \lambda)]$$

Substituting the value of c in (3),

$$v - V \log (V + v) = -\frac{g}{V}x + V[\lambda - \log V (1 + \lambda)] \qquad \qquad ...(4)$$

Let h be the greatest height attained by the particle. Then x = h when v = 0

$\therefore$　From (4), we have,　　　$$-V \log V = -\frac{g}{V}h + V[\lambda - \log V (1 + \lambda)]$$

or　　　$$\frac{g}{V}h = V\lambda - V[\log V (1 + \lambda) - \log V] = V\lambda - V \log \frac{V(1 + \lambda)}{V}$$

or　　　$$\boxed{h = \frac{V^2}{g}[\lambda - \log (1 + \lambda)]}$$

Ex. 7 : *A paratrooper and his parachute weigh 50 kg. At the instant parachute opens, he is travelling vertically downward at the speed of 20 m/s. If the air resistance varies directly as the instantaneous velocity and it is 20 newtons when the velocity is 10 m/s, find the limiting velocity, the position and the velocity of the paratrooper at any time t.*

Sol. : Let v m/s be the velocity of the paratrooper t seconds after the parachute opens. The forces acting on the paratrooper are :

(i)　the weight 50 kg acting vertically downwards,

(ii)　the air resistance kv acting vertically upwards.

　　　Accelerating force　=　(50 – kv) N

By Newton's second law, the accelerating force is,

$$m\frac{dv}{dt} = \frac{W}{g}\frac{dv}{dt} = \frac{50}{g}\frac{dv}{dt}$$

$\therefore$　The equation of motion is　　$$\frac{50}{g}\frac{dv}{dt} = 50 - kv \qquad \qquad ...(1)$$

When　v = 10 m/s, the air resistance kv = 20 N

$\therefore$　　　　　　　$$k \cdot 10 = 20 \Rightarrow k = 2$$

Substituting the value of k in (1), we have,

$$\frac{50}{g} \frac{dv}{dt} = 50 - 2v \quad \text{or} \quad \frac{dv}{25 - v} = \frac{g}{25} \, dt$$

Integrating, $\qquad - \log (25 - v) = \dfrac{g\,t}{25} + c$ $\qquad\qquad$ ...(2)

When t = 0, v = 20 $\therefore$ c = $- \log 5$

Substituting the value of c in (2), we have,

$$- \log (25 - v) = \frac{g\,t}{25} - \log 5$$

or $\qquad \log \dfrac{25 - v}{5} = -\dfrac{g\,t}{25} \quad \text{or} \quad \dfrac{25 - v}{5} = e^{-g\,t/25}$

or $\qquad v = 5 (5 - e^{-gt/25})$ $\qquad\qquad$ ...(3)

which gives the velocity of the paratrooper at any time t.

The limiting velocity is the velocity when $t \to \infty$.

$\therefore$ From (3), the limiting velocity = 25 m/s.

Now, from (3), we have, $\qquad \dfrac{dx}{dt} = 5 (5 - e^{-gt/25})$

or $\qquad dx = 5 (5 - e^{-gt/25}) \, dt$

Integrating, $\qquad x = 5 \left(5t + \dfrac{25}{g} e^{-gt/25} \right) + C'$ $\qquad\qquad$ ...(4)

Initially, when t = 0, x = 0 $\therefore$ C' = $-\dfrac{125}{g}$

Substituting the value of C' in (4), the position of the paratrooper at any time t is given by,

$$\boxed{x = 25\,t - \frac{125}{g} (1 - e^{-gt/25})}$$

Ex. 8 : *Assuming that the resistance to movement of a ship through water in the form of $a^2 + b^2 v^2$, where v is the velocity and a and b are constants, write down the differential equation for retardation of the ship moving with engine stopped. Prove that the time in which the speed falls to one half its original value u is given by, $\dfrac{W}{abg} \tan^{-1} \dfrac{abu}{2a^2 + b^2 u^2}$, where W is the weight of the ship.*

(May 2011, Dec. 2007, May 2016)

Sol. : $\qquad m \dfrac{dv}{dt} = - (a^2 + b^2 v^2) \qquad\qquad \text{but} \qquad m = \dfrac{W}{g}$

$\therefore \qquad \dfrac{W}{g} \dfrac{dv}{dt} = - (a^2 + b^2 v^2) \qquad\qquad \text{or} \qquad \dfrac{W}{g} \left(\dfrac{dv}{a^2 + b^2 v^2} \right) = - dt$

or $\qquad \dfrac{W}{gb^2} \displaystyle\int_{u}^{u/2} \dfrac{dv}{\dfrac{a^2}{b^2} + v^2} = - \int_{0}^{t} dt$

$\therefore \qquad \dfrac{W}{gb^2} \left[\dfrac{b}{a} \tan^{-1} \dfrac{bv}{a} \right]_{u}^{u/2} = - t$

or $\qquad t = \dfrac{W}{abg} \left[\tan^{-1} \dfrac{bv}{a} \right]_{u/2}^{u} = \dfrac{W}{abg} \left[\tan^{-1} \dfrac{bu}{a} - \tan^{-1} \dfrac{bu}{2a} \right]$

$$= \dfrac{W}{abg} \tan^{-1} \dfrac{\left(\dfrac{bu}{a} - \dfrac{bu}{2a} \right)}{\left(1 + \dfrac{b^2 u^2}{2a^2} \right)} = \dfrac{W}{abg} \tan^{-1} \left[\dfrac{bu/2a}{(2a^2 + b^2 u^2)/2a^2} \right]$$

$\therefore \qquad \boxed{t = \dfrac{W}{abg} \tan^{-1} \left(\dfrac{abu}{2a^2 + b^2 u^2} \right)}$

Ex. 9 : The distance x descended by a parachuter satisfies the differential equation $v\dfrac{dv}{dx} = g\left(1 - \dfrac{v^2}{k^2}\right)$ where v is velocity, k, g constants. If v = 0 and x = 0 at time t = 0, show that $x = \dfrac{k^2}{g}\log\cosh\left(\dfrac{gt}{k}\right)$.   **(May 20011, 2006, 2005; Dec. 2013, 2009)**

Sol. : Since, $\qquad v\dfrac{dv}{dx} = \dfrac{dv}{dt}$ $\qquad\qquad \therefore\qquad \dfrac{dv}{dt} = g\left(\dfrac{k^2 - v^2}{k^2}\right)$

$$\int \frac{dv}{k^2 - v^2} = \int \frac{g}{k^2}\,dt + A \qquad\qquad \therefore \qquad \frac{1}{k}\tanh^{-1}\frac{v}{k} = \frac{g}{k^2}t + A$$

Using v = 0, at t = 0 we get, A = 0

$$\therefore \qquad \tanh^{-1}\left(\frac{v}{k}\right) = \frac{g}{k}t \qquad\qquad \text{or} \qquad v = k\tanh\left(\frac{g}{k}t\right)$$

i.e. $\qquad \dfrac{dx}{dt} = k\tanh\left(\dfrac{g}{k}t\right) \qquad\qquad \therefore \qquad \int dx = k\int \tanh\left(\dfrac{g}{k}t\right)dt$

i.e. $\qquad x = \dfrac{k}{(g/k)}\left[\log\cosh\dfrac{gt}{k}\right] + B \qquad$ or $\qquad x = \dfrac{k^2}{g}\log\cosh\left(\dfrac{gt}{k}\right) + B$

Also x = 0 when t = 0. $\qquad\qquad \therefore\ B = 0$

Hence, $\qquad\qquad \boxed{x = \dfrac{k^2}{g}\log\cosh\left(\dfrac{gt}{k}\right)}$

Ex. 10 : A particle is moving in a straight line with an acceleration $k\left[x + \dfrac{a^4}{x^3}\right]$ directed towards origin. If it starts from rest at a distance a from the origin, prove that it will arrive at origin at the end of time $\dfrac{\pi}{4\sqrt{k}}$.   **(May 2004, 2007, 2008, 2013; Dec. 2006)**

Sol. : Equation of motion is $\qquad v\dfrac{dv}{dx} = -k\left(x + \dfrac{a^4}{x^3}\right)$

$$\int v\,dv = -k\int \left(x + \frac{a^4}{x^3}\right)dx$$

or $\qquad\qquad \dfrac{v^2}{2} = -k\left[\dfrac{x^2}{2} - \dfrac{a^4}{2x^2}\right] + C$

When x = a,   v = 0,   ∴ C = 0

$$\therefore \qquad v^2 = k\left(\frac{a^4 - x^4}{x^2}\right)$$

Since acceleration is directed towards origin,

$$\therefore \qquad v = -\sqrt{k}\,\sqrt{\frac{a^4 - x^4}{x^2}} \quad \text{or} \quad \frac{dx}{dt} = \frac{-\sqrt{k}\sqrt{a^4 - x^4}}{x}$$

$$\int \frac{x\,dx}{\sqrt{a^4 - x^4}} = -\sqrt{k}\int dt,$$

Put $x^2 = u$ ∴ $x\,dx = \dfrac{du}{2}$

$$\frac{1}{2}\int \frac{du}{\sqrt{a^4 - u^2}} = -\sqrt{k}\cdot t + C$$

$$\therefore \qquad \sin^{-1}\left(\frac{u}{a^2}\right) = -2\sqrt{k}\cdot t + C_1$$

or $\qquad\qquad \sin^{-1}\left(\dfrac{x^2}{a^2}\right) = -2\sqrt{k}\cdot t + C_1$

When $t = 0$, $x = a$ $\therefore$ $C_1 = \dfrac{\pi}{2}$

$\therefore$ $\qquad\qquad\qquad\qquad \sin^{-1}\left(\dfrac{x^2}{a^2}\right) = \dfrac{\pi}{2} - 2\sqrt{k} \cdot t$

At $x = 0$, $\qquad\qquad\qquad\qquad 0 = \dfrac{\pi}{2} - 2\sqrt{k} \cdot t$

$\therefore$ $\qquad\qquad\qquad\qquad \boxed{t = \dfrac{\pi}{4\sqrt{k}}}$

Ex. 11 : *The distance 'x' descended by a person falling by means of a parachute satisfies the differential equation*

$\left(\dfrac{dx}{dt}\right)^2 = k^2\left[1 - e^{-\frac{2gx}{k^2}}\right]$ *where 'k' and 'g' are constants and $x = 0$ when $t = 0$. Show that $x = \dfrac{k^2}{g} \log \cosh\left(\dfrac{gt}{k}\right)$.* **(May 2007)**

Sol. : The equation is,

$$\left(\dfrac{dx}{dt}\right)^2 = k^2\left[1 - e^{-\frac{2gx}{k^2}}\right]$$

$\therefore$
$$\dfrac{dx}{dt} = k\sqrt{1 - e^{-\frac{2gx}{k^2}}}$$

$$\dfrac{dx}{\sqrt{1 - e^{-\frac{2gx}{k^2}}}} = k\,dt \quad \text{or} \quad \int \dfrac{e^{\frac{gx}{k^2}}}{\sqrt{e^{\frac{2gx}{k^2}} - 1}}\,dx = k \int dt$$

$$\dfrac{k^2}{g} \int \dfrac{du}{\sqrt{u^2 - 1}} = kt + C \qquad\qquad \left(\because e^{-\frac{2gx}{k^2}} = u\right)$$

$\therefore$
$$\dfrac{k^2}{g} \cosh^{-1} u = kt + C \qquad\qquad \left(\because \int \dfrac{du}{\sqrt{u^2-1}} = \cosh^{-1} u\right)$$

When $t = 0$, $x = 0$. But $e^{\frac{gx}{k^2}} = u$ $\therefore$ Put $x = 0$, $\therefore$ $e^0 = u \Rightarrow u = 1$. Hence, when $t = 0$, $u = 1$.

$$\dfrac{k^2}{g} \cosh^{-1} 1 = 0 + C$$

We note here that $\qquad \cosh(0) = 1, \quad \cosh^{-1}(1) = 0 \qquad\qquad \therefore \qquad C = 0$

$\Rightarrow$
$$\dfrac{k^2}{g} \cosh^{-1} u = kt \Rightarrow \cosh^{-1} u = \dfrac{gt}{k}$$

$\therefore$
$$u = \cosh\left(\dfrac{gt}{k}\right) \Rightarrow e^{\frac{gx}{k^2}} = \cosh\left(\dfrac{gt}{k}\right) \quad \text{or} \quad \dfrac{gx}{k^2} = \log \cosh\left(\dfrac{gt}{k}\right)$$

$\therefore$
$$\boxed{x = \dfrac{k^2}{g} \log \cosh\left(\dfrac{gt}{k}\right)}$$

Ex. 12 : *A body of mass 'm' falls from rest under the influence of gravity and a retarding force due to air resistance proportional to square of the velocity. Find the velocity and distance described as a function of time. Hence, show that the velocity of the body approaches the limiting value.* **(May 2009, Dec. 2004)**

Sol. : The forces acting on the body are :

(i)   the weight mg acting downwards.

(ii)   Retarding force due to air resistance is mkv^2 (mk is the constant of proportionality).

$\therefore$ Net force by virtue of which body falls $= mg - mkv^2 = m(g - kv^2)$

By D'Alembert's principle, $\qquad m\dfrac{dv}{dt} = m(g - kv^2) \qquad\qquad\qquad \text{or} \qquad \dfrac{dv}{dt} = k\left(\dfrac{g}{k} - v^2\right)$

Take $\qquad\qquad\qquad\qquad \dfrac{g}{k} = a^2 \Rightarrow$

We have
$$\frac{dv}{dt} = \frac{g}{a^2}(a^2 - v^2)$$

$\therefore$
$$\int \frac{dv}{a^2 - v^2} = \frac{g}{a^2} \int dt + C$$

$$\frac{1}{a} \tanh^{-1} \frac{v}{a} = \frac{gt}{a^2} + C$$

When $t = 0$, $v = 0$. $\therefore$ $C = 0$

$$\tanh^{-1} \frac{v}{a} = \frac{gt}{a} \qquad \boxed{v = a \tanh\left(\frac{gt}{a}\right)} \qquad \qquad \ldots(1)$$

But
$$v = \frac{ds}{dt}$$

$\therefore$
$$\frac{dS}{dt} = a \tanh\left(\frac{gt}{a}\right)$$

$$dS = a \int \tanh\left(\frac{gt}{a}\right) dt + A \quad \therefore \quad S = \frac{a^2}{g} \log \cosh\left(\frac{gt}{a}\right) + A$$

When $t = 0$, $S = 0$, $\therefore$ $A = 0$

$$S = \frac{a^2}{g} \log \cosh\left(\frac{gt}{a}\right) \qquad \qquad \ldots(2)$$

Equations (1) and (2) describe velocity and distance as a function of time.

To find the limiting value, we use result (1).

$$v = a \tanh\left(\frac{gt}{a}\right) \qquad \text{or} \qquad v = a\left(\frac{e^{gt/a} - e^{-gt/a}}{e^{gt/a} + e^{-gt/a}}\right)$$

$$v = a\left(\frac{1 - e^{-2gt/a}}{1 + e^{-2gt/a}}\right)$$

As $t \to \infty$, $v \to v_\infty$ $\therefore$
$$v_\infty = a\left(\frac{1 - 0}{1 + 0}\right) = a$$

But $a^2 = \dfrac{g}{k}$ $\qquad$ $\therefore$ $\qquad a = \sqrt{\dfrac{g}{k}}$

$\therefore$
$$\boxed{\text{Limiting value of } v = v_\infty = a = \sqrt{\frac{g}{k}}}$$

Ex. 13 : *A particle of mass m is projected upward with velocity V. Assuming the air resistance k times its velocity, write the equation of motion and show that it will reach maximum height in time* $\dfrac{m}{k} \log\left(1 + \dfrac{kV}{mg}\right)$ *and distance travelled at any time t is*   **(Dec. 2008)**

$$\left(\frac{mV}{k} + \frac{m^2 g}{k^2}\right)\left(1 - e^{-\frac{kt}{m}}\right) - \frac{gmt}{k}.$$

Sol. : Net force acting on the body $= -mg - kv$. By D'Alembert's principle,

$$m\frac{dv}{dt} = -mg - kv \qquad\qquad \text{or} \qquad\qquad \frac{dv}{dt} = -g - \frac{k}{m}v$$

$\therefore$
$$\int \frac{dv}{g + \frac{k}{m}v} = -\int dt + C \qquad\qquad \text{or} \qquad\qquad \frac{m}{k} \log\left(g + \frac{kv}{m}\right) = -t + C$$

When $t = 0$, $v = V$
$$\therefore \quad C = \frac{m}{k} \log\left(g + \frac{kV}{m}\right)$$

$\therefore$
$$t = \frac{m}{k} \log\left(\frac{g + \dfrac{kV}{m}}{g + \dfrac{kv}{m}}\right) \qquad\qquad \ldots(1)$$

It will attain maximum height when $v = 0$, $t = t_1$

$$\therefore \qquad t_1 = \frac{m}{k} \log \left(\frac{g + \frac{kV}{m}}{g} \right) \qquad \text{or} \qquad t_1 = \frac{m}{k} \log \left(1 + \frac{kV}{mg} \right)$$

which is the required time.

From (1),
$$\frac{k}{m} t = \log \left(\frac{g + \frac{kV}{m}}{g + \frac{kv}{m}} \right) \qquad \text{or} \qquad e^{\frac{kt}{m}} = \frac{g + \frac{kV}{m}}{g + \frac{kv}{m}}$$

$$\therefore \qquad g + \frac{k}{m} v = \left(g + \frac{kV}{m} \right) e^{-\frac{kt}{m}}$$

$$\therefore \qquad v = -\frac{gm}{k} + \left(V + \frac{mg}{k} \right) e^{-\frac{kt}{m}}$$

But
$$v = \frac{ds}{dt} = -\frac{gm}{k} + \left(V + \frac{mg}{k} \right) e^{-\frac{kt}{m}}$$

Integrating,
$$s = -\frac{gm}{k} \int dt + \left(V + \frac{mg}{k} \right) \int e^{-\frac{kt}{m}} dt + A$$

$$s = -\frac{gm}{k} t - \frac{m}{k} \left(V + \frac{mg}{k} \right) e^{-\frac{kt}{m}} + A$$

When $t = 0$, $s = 0$, $A = \frac{m}{k} \left(V + \frac{mg}{k} \right)$

$$\boxed{s = -\frac{gm}{k} t + \frac{m}{k} \left(V + \frac{mg}{k} \right) \left(1 - e^{-\frac{kt}{m}} \right)}$$

Ex. 14 : *A particle of unit mass is projected vertically upward with velocity u. Assuming that the air resistance is k times the instantaneous velocity of the particle, show that the particle will return to point of projection with velocity V given by,* **(Dec. 2006, 2005)**

$$V + u = \frac{g}{k} \log \left(\frac{g + ku}{g - kV} \right)$$

Sol. : The equation of motion is,

$$v \frac{dv}{dx} = -kv - g \qquad \text{or} \qquad \frac{v\, dv}{kv + g} = -dx$$

$$\therefore \qquad \int \frac{kv\, dv}{kv + g} = -k \int dx + C \qquad \text{or} \qquad \int \left(1 - \frac{g}{kv + g} \right) dv = -kx + C$$

$$\therefore \qquad v - \frac{g}{k} \log (kv + g) = -kx + C$$

When $x = 0$, $v = u$, $\therefore C = u - \frac{g}{k} \log (ku + g)$

$$v - \frac{g}{k} \log (kv + g) + \frac{g}{k} \log (ku + g) = -kx + u$$

$$v + \frac{g}{k} \log \left(\frac{ku + g}{kv + g} \right) = -kx + u$$

Let $x = x_1$ be maximum height attained. At this instant, $v = 0$.

$$\frac{g}{k} \log \left(\frac{ku + g}{g} \right) = -kx_1 + u \qquad \qquad \text{...(1)}$$

From this instant, body will start moving down. Algebraic sum of net forces $= -kv + g$

$\therefore$ The equation of motion is,

$$v \frac{dv}{dx} = -kv + g \qquad \text{or} \qquad \frac{v\, dv}{g - kv} = dx$$

or
$$\frac{-kv\, dv}{g - kv} = -k\, dx \qquad \text{or} \qquad \int \left(1 - \frac{g}{g - kv} \right) dv = -k \int dx$$

or $\qquad v + \dfrac{g}{k} \log (g - kv) = - kx + C_1$

When $x = 0$, $v = 0$ $\therefore C_1 = \dfrac{g}{k} \log g$

$\therefore \qquad v + \dfrac{g}{k} \log (g - kv) - \dfrac{g}{k} \log g = - kx$

or $\qquad v + \dfrac{g}{k} \log \left(\dfrac{g - kv}{g} \right) = - kx$

Body when reaches the point of projection, $x = x_1$, $\quad v = V$

$\therefore \qquad V + \dfrac{g}{k} \log \left(\dfrac{g - kV}{g} \right) = - kx_1 \qquad \qquad ...(2)$

By using equation (1),

$$V + \dfrac{g}{k} \log \left(\dfrac{g - kV}{g} \right) = \dfrac{g}{k} \log \left(\dfrac{ku + g}{g} \right) - u$$

$\therefore \qquad \boxed{\,V + u = \dfrac{g}{k} \log \left(\dfrac{g + ku}{g - kV} \right)\,}$

Ex. 15 : *A particle of unit mass moves in a horizontal straight line OA with an acceleration $\dfrac{k}{r^3}$ at a distance r and directed towards 0.*

If initially the particle was at rest at a distance a from 0, show that it will be at a distance $\dfrac{a}{2}$ from 0 at the end of time $\dfrac{a^2}{2} \sqrt{\dfrac{3}{k}}$.

(May 2009, Dec. 2004)

Sol. : Equation of motion is,

$$\dfrac{dv}{dr} = - \dfrac{k}{r^3}$$

$\therefore \qquad \displaystyle\int v \, dv = - \int \dfrac{k}{r^3} \, dr + C$

$\therefore \qquad \dfrac{v^2}{2} = \dfrac{k}{2r^2} + C$

When $r = a$, $v = 0$ $\therefore C = \dfrac{-k}{2a^2}$

$\therefore \qquad \dfrac{v^2}{2} = \dfrac{k}{2r^2} - \dfrac{k}{2a^2}$

$$v^2 = k \left(\dfrac{a^2 - r^2}{a^2 r^2} \right) \Rightarrow v = \dfrac{\sqrt{k}}{a} \dfrac{\sqrt{a^2 - r^2}}{r}$$

But $\qquad v = \dfrac{dr}{dt} = \dfrac{\sqrt{k}}{a} \dfrac{\sqrt{a^2 - r^2}}{r}$

$$\int \dfrac{r}{\sqrt{a^2 - r^2}} \, dr = \dfrac{\sqrt{k}}{a} \int dt + C_1$$

$\therefore \qquad - \sqrt{a^2 - r^2} = \dfrac{\sqrt{k}}{a} t + C_1$

Also $r = a$, $t = 0 \Rightarrow C_1 = 0$

$$- \sqrt{a^2 - r^2} = \dfrac{\sqrt{k}}{a} t \quad \text{or} \quad a^2 - r^2 = \dfrac{k}{a^2} t^2$$

When $t = \dfrac{a^2}{2} \sqrt{\dfrac{3}{k}}$ $a^2 - r^2 = \dfrac{k}{a^2} \dfrac{a^4}{4} \left(\dfrac{3}{k}\right) = \dfrac{3a^2}{4}$

$$r^2 = a^2 - \dfrac{3a^2}{4}$$

$\therefore$ $r^2 = \dfrac{a^2}{4} \Rightarrow \boxed{r = \dfrac{a}{2}}$

Ex. 16 : *A particle is projected vertically upwards with velocity V_1 and the resistance of the air produces a retardation Kv^2, where v is the velocity. Find the velocity V_2 with which the particle will return to the point of projection.*

Sol. : During the upward motion, let x be the distance of the particle at time t.

Taking the resistance to be mKv^2, the equation of motion may be written as :

$$m v \frac{dv}{dx} = - mg - mKv^2$$

or $v \dfrac{dv}{dx} = - g - Kv^2 \quad \text{... (1)}$

Separating the variables in (1) and then integrating it, we obtain

$$\int dx = - \int \frac{v\, dv}{g + Kv^2} \qquad \text{... (2)}$$

Suppose that the greatest height attained is H. At this time the velocity is zero. Initially, i.e. at $x = 0$, we have $v = V_1$. Hence equation (2) is written as :

$$\int_0^H dx = - \int_{V_1}^0 \frac{v\, dv}{g + Kv^2} = \int_0^{V_1} \frac{v\, dv}{g + Kv^2}$$

i.e. $H = \dfrac{1}{2K} \log \left(1 + \dfrac{K}{g} V_1^2\right) \qquad \text{... (3)}$

During the downward motion, let y be the distance fallen in time t seconds after starting from the highest point. The equation of motion is now,

$$m v \frac{dv}{dy} = mg - mKv^2$$

or $v \dfrac{dv}{dy} = g - Kv^2 \quad \text{... (4)}$

Separating the variables, we find

$$dy = \frac{v\, dv}{g - Kv^2} \quad \text{... (5)}$$

Now, y varies from $y = 0$ to $y = H$ and v varies from $v = 0$ to $v = V_2$. Hence, we obtain from equation (5),

$$\int_0^H dy = \int_0^{V_2} \frac{v\, dv}{g - Kv^2} = - \frac{1}{2K} \left[\log (g - Kv^2)\right]_0^{V_2}$$

i.e. $H = - \dfrac{1}{2K} \log \left(1 - \dfrac{K}{g} V_2^2\right) \qquad \text{... (6)}$

Equating (3) and (6), we obtain

$$\frac{1}{1 - \dfrac{K}{g} V_2^2} = 1 + \frac{K}{g} V_1^2$$

i.e. $V_2^2 = \dfrac{V_1^2}{1 + \dfrac{K}{g} V_1^2}$

We therefore have $\boxed{\dfrac{1}{V_2^2} = \dfrac{1 + \dfrac{K}{g} V_1^2}{V_1^2} = \dfrac{1}{V_1^2} + \dfrac{K}{g}}$

which gives the required velocity V_2.

Ex. 17 : *The resistance to the motion of a car of mass m varies as the square of its speed (i.e. kv^2) and the effective horse-power exerted at the road wheels is constant and equal to P. Show that the distance in which the car can accelerate from speed v_0 to v_1 is given by*

$$\frac{m}{3k} \log \frac{P - k\,v_0^3}{P - k\,v_1^3}$$

Sol. : If F is the pull exerted by the engine, then it is given that $F\,v = \text{power} = P$ i.e. $F = \dfrac{P}{v}$.

∴   The equation of motion of the car is

$$mv \frac{dv}{dx} = F - kv^2 = \frac{P}{v} - kv^2 \ \text{ or } \ \frac{dx}{m} = \frac{v^2\,dv}{P - kv^3}$$

Hence, on integrating, the required distance x is given by,

$$\frac{x}{m} = \int_{v_0}^{v_1} \frac{v^2\,dv}{P - kv^3} = -\frac{1}{3k}\,[\log(P - kv^3)]_{v_0}^{v_1} = -\frac{1}{3k}\,[\log(P - kv_1^3) - \log(P - kv_0^3)]$$

or $$x = \frac{m}{3k} \log \frac{P - kv_0^3}{P - kv_1^3}$$

Ex. 18 : *The acceleration of a moving particle being proportional to the cube of its velocity and negative, show that the distance passed over in time t is given by,*

$$S = \frac{\left(\sqrt{2k\,v_0^2\,t + 1} - 1\right)}{kv_0} .$$

the initial velocity being v_0, and the distance being measured from the position of the particle at time t = 0.

Sol. : The equation of motion is, $m\dfrac{dv}{dt} = -mk\,v^3$ where mk is the constant of proportionality.

$$\frac{dv}{dt} = -k\,v^3 \qquad\qquad \text{or} \qquad\qquad \int \frac{dv}{v^3} = -k \int dt + C$$

$$-\frac{1}{2v^2} = -\,kt + C$$

But $v = v_0$ when t = 0,   ∴ $C = \dfrac{-1}{2\,v_0^2}$ $-\dfrac{1}{2v^2} = -kt - \dfrac{1}{2v_0^2}$

$$\frac{1}{v^2} = 2kt + \frac{1}{v_0^2} = \frac{2kt\,v_0^2 + 1}{v_0^2} \qquad\qquad v = \frac{v_0}{\sqrt{2kt\,v_0^2 + 1}}$$

But $v = \dfrac{ds}{dt}$; $\displaystyle\int ds = v_0 \int \frac{dt}{\sqrt{2kt\,v_0^2 + 1}} + d$ $S = \dfrac{v_0}{2kv_0^2} \displaystyle\int \frac{2kv_0^2\,dt}{\sqrt{2kt\,v_0^2 + 1}} + d$ (Note this step)

$$S = \frac{1}{2kv_0}\left(2\sqrt{2kt\,v_0^2 + 1}\right) + d = \frac{\sqrt{2kt\,v_0^2 + 1}}{k\,v_0} + d$$

Since S = 0, when t = 0, ∴ $d = -\dfrac{1}{kv_0}$ ∴ $\boxed{\,S = \dfrac{1}{kv_0}\left(\sqrt{1 + 2kt\,v_0^2} - 1\right)\,}$

Ex. 19 : *A ship of mass 45,000 Mg starts from rest under the force of a constant propeller thrust of 9,00,000 N. (a) Find its velocity as a function of time t given that the resistance in newtons is 1,50,000 v with v = velocity measured in ms^{-1}, (b) Find the terminal velocity (i.e. v when $t \to \infty$) in kilometers per hour.*

Sol. : Since mass (kg) × acceleration (ms^{-2}) = net force (N)

then, $45 \times 10^6 \dfrac{dv}{dt} = 9{,}00{,}000 - 15 \times 10^4 v$

or $\qquad \dfrac{dv}{dt} + \dfrac{v}{300} = \dfrac{1}{50}$ ... (1)

$\therefore \qquad$ I.F. $= e^{\frac{t}{300}}$

G.S. is $\qquad v e^{\frac{t}{300}} = \dfrac{1}{50} \displaystyle\int e^{\frac{t}{300}} \, dt + C = 6 e^{\frac{t}{300}} + C$

(a) When $t = 0$, $v = 0$, $C = -6$ and $\boxed{v = 6\left(1 - e^{-\frac{t}{300}}\right)}$

(b) As $t \to \infty$, $v \to 6$, the terminal velocity is $\boxed{v = 6 \text{ ms}^{-1} = 21.6 \text{ km per hour}}$. This may also be obtained from (1) since, as v approaches a limiting value, $\dfrac{dv}{dt} \to 0$. Then $v = 6$ as before.

Ex. 20 : *A boat is being towed at the rate 20 km per hour. At the instant (t = 0) that the towing line is cast off, a man in the boat begins to row in the direction of motion exerting a force of 9 N. If the combined mass of the man and boat is 225 kg and the resistance (N) is equal to 26.25 v, where v is measured in ms^{-1}, find the speed of the boat after 1/2 minute.*

Sol. : Since $\qquad$ mass (kg) $\times$ acceleration (ms^{-2}) $=$ net force (N)

then, $\qquad 225 \dfrac{dv}{dt} = 90 - 26.25 v \qquad$ or $\qquad \dfrac{dv}{dt} + \dfrac{7}{60} v = \dfrac{2}{5}$

Integrating, $\qquad v e^{7t/60} = \dfrac{2}{5} \displaystyle\int e^{7t/60} \, dt + C = \dfrac{120}{35} e^{7t/60} + C$

When $t = 0$, $\qquad v = \dfrac{20000}{3600} = \dfrac{50}{9}$, $C = \dfrac{134}{63}$

and $\qquad \boxed{v = \dfrac{24}{7} + \dfrac{134}{63} e^{-7t/60}}$

When $t = 30$, $\qquad \boxed{v = \dfrac{24}{7} + \dfrac{134}{63} e^{-35} = 3.5 \text{ ms}^{-1}}$

EXERCISE 2.4

1. A vehicle starts from rest and its acceleration is given by $k\left(1 - \dfrac{t}{T}\right)$, where k is a constant and T is time taken to attain highest speed. Find the highest speed and distance travelled till the speed is attained.

   Ans. : $\dfrac{kT}{2}$, $\dfrac{kT^2}{3}$

2. A chain is coiled up near the edge of a smooth table and it just starts to fall over the edge. When a length x has fallen, its velocity v is given by, $xv \dfrac{dv}{dx} + v^2 = gx$. Show that if $v = 0$ at $x = 0$, then $v^2 = \dfrac{2}{3} gx$. **Hint :** Put $v^2 = t$

3. If $m \dfrac{dv}{dt} = X$ and $I \dfrac{dw}{dt} = a X$, where m, a, I are constants and X, v, w are functions of t, show that v_0, w_0 are the initial values of v, w respectively. $v = v_0 + \dfrac{I}{am} (w - w_0.)$

   Hint : Eliminate X.

4. A bullet is fired into a sand tank, its retardation is proportional to the square root of its velocity is $(k\sqrt{v})$. How long will it take to come to rest if it enters the sand tank with an initial velocity v_o. **(May 2014)**

   Hint : $\dfrac{dv}{dt} = -k\sqrt{v}$; $t = 0$, $v = v_o$. **Ans.** $t = \dfrac{2}{k} \sqrt{v_o}$

2.9 SIMPLE HARMONIC MOTION

If a particle moves on a straight line, so that the force acting on it is always directed towards a fixed point on the line and proportional to its distance from the point, the particle is said to move in Simple Harmonic Motion.

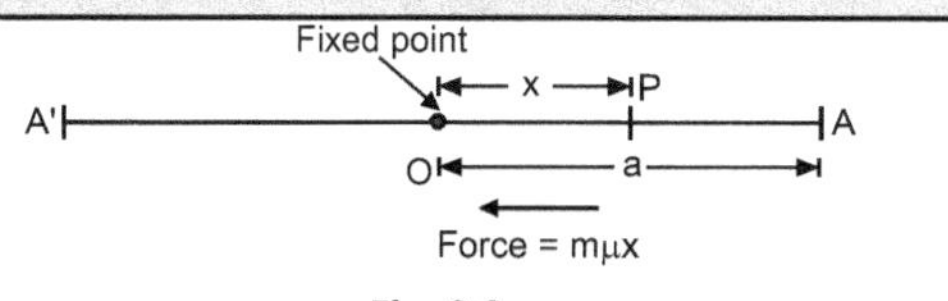

Fig. 2.8

(I) Let O be the fixed point and P be the position of particle at any time t. Let OP = x. Force acting on particle is mμx, where m is mass of the particle and μ a constant considered positive.

Equation of motion is

$$m \frac{d^2x}{dt^2} = -m\mu x \quad \text{or} \quad v \frac{dv}{dx} = -\mu x$$

$$v \, dv = -\mu x \, dx$$

Integrating, we get

$$v^2 = -\mu x^2 + A$$

Assuming that particle starts from point A (OA = a) and its initial velocity is zero.

When $x = a$, $v = 0$, $\therefore$ $A = \mu a^2$ $\therefore$ $v^2 = \mu (a^2 - x^2)$

or

$$v = \frac{dx}{dt} = -\sqrt{\mu} \, \sqrt{(a^2 - x^2)}$$

–ve sign is attached because x decreases as t increases.

$\therefore$

$$-\frac{dx}{\sqrt{a^2 - x^2}} = \sqrt{\mu} \, dt$$

Integrating,

$$\cos^{-1}\frac{x}{a} = \sqrt{\mu} \, t + B \qquad\qquad \left(\because \int -\frac{dx}{\sqrt{a^2 - x^2}} = \cos^{-1}\frac{x}{a} \right)$$

When $t = 0$, $x = a$ $\therefore$ $B = 0$

we get,

$$x = a \cos \left(\sqrt{\mu} \cdot t \right)$$

Particle will reach O in time t_1, given by

$$0 = a \cos \left(\sqrt{\mu} \cdot t_1 \right)$$

$\therefore$

$$\boxed{t_1 = \frac{\pi}{2\sqrt{\mu}}}$$

Its velocity at that time will be $\sqrt{\mu} \, a$.

As soon as it will cross O, the direction of force will change, however, particle will move with velocity $\sqrt{\mu} \, a$ and ultimately come to rest at A'. OA = OA' and time taken by the particle to travel from O and A' will be $\dfrac{\pi}{2\sqrt{\mu}}$.

Due to attraction, particle will start moving towards O. It is a to-fro motion i.e. oscillatory motion. Due to this, it is called Simple Harmonic Motion. Period of oscillation is $\dfrac{2\pi}{\sqrt{\mu}}$.

(II) Hooke's Law : Suppose that an elastic string of negligible weight hangs vertically with its upper end fixed and with a mass of M kg attached to its lower end. Let l metres be the natural length of the string and let T be its tension when it is stretched to a length x metres. Then the extension of the string beyond its natural length is given by the ratio $(x - l)/l$. The tension and the extension are related by Hooke's law :

The tension of an elastic string or spring is proportional to the extension of the string beyond its natural length. We thus have

$$T = \frac{\lambda}{l} (x - l)$$

where λ is a constant called the *modulus of elasticity* of the string. The value of λ depends on the material of which the string is composed. Clearly, $\lambda = T$ when $x = 2l$, i.e., λ is that force which would stretch the string to twice its natural length.

To determine motion, let v metres/sec. be the velocity of mass M when the extension is x metres. Then the tension is λx. The acceleration of the mass is $\dfrac{dv}{dt}$ or $v \dfrac{dv}{dx}$ so that the equation of motion may be written as,

$$Mv \frac{dv}{dx} = Mg - \lambda x$$

Separating the variables, we find :

$$v \, dv = \left(g - \frac{\lambda}{M} x \right) dx$$

Integrating both sides, we obtain :

$$\frac{v^2}{2} = gx - \frac{\lambda}{M} \frac{x^2}{2} + C \qquad\qquad\qquad \dots (1)$$

To calculate C, we assume that $v = V_0$ when $x = 0$. This gives $C = V_0^2/2$. Hence, equation (1) becomes :

$$\frac{v^2}{2} = gx - \frac{\lambda}{M}\frac{x^2}{2} + \frac{V_0^2}{2}$$

or

$$\boxed{v^2 = 2gx - \frac{\lambda}{M}x^2 + V_0^2}$$

which gives the velocity of the mass when the extension is x metres.

(III) Motion of a Particle of Mass m Suspended by an Elastic String (or Extensible Spring) : Let OA be an elastic string of natural length 'l' with fixed end 'O'. Let a mass 'm' be attached to 'A' so that the string elongates and the mass 'm' reaches a position of equilibrium at B. Consider the equilibrium of the mass 'm' at 'B'. It is acted on by force 'mg' downwards due to gravity and tension 'T' upwards due to extension. In equilibrium,

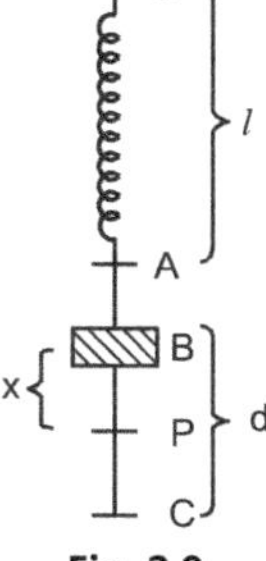

$$mg = T = \frac{\lambda AB}{l} \text{ (by Hooke's law)}$$

If $\qquad AB = c$ then $c = \dfrac{mgl}{\lambda}$ $\qquad\qquad$ ... (1)

Now, let the particle be pulled down upto 'C' and then let go. Let BC = d (say). The mass will go up due to tension in the 'spring'. Let the mass be at 'P' between B and C and BP = x (say).

Fig. 2.9

The force acting on the mass is mg – T downwards.

$\therefore$ Equation of motion is, $\qquad m \cdot \dfrac{d^2x}{dt^2} = mg - T = mg - \dfrac{\lambda \times \text{Extension}}{\text{Natural length}}$

$$= mg - \frac{\lambda\,(AB + x)}{l} = mg - \frac{\lambda}{l}\left(\frac{mgl}{\lambda} + x\right) = -\frac{\lambda x}{l}$$

$\therefore$ $\qquad\qquad \boxed{\dfrac{d^2x}{dt^2} = -\dfrac{\lambda}{lm}\,x}$ $\qquad\qquad$ ... (2)

This equation is same as that of S.H.M. except $\mu = \dfrac{\lambda}{lm}$.

Hence the particle of mass m will execute S.H.M. of period $\dfrac{2\pi}{\sqrt{\dfrac{\lambda}{lm}}}$. The expression $\dfrac{mgl}{\lambda}$ is called static extension of the spring and

denoted by c. Under this case, period of oscillation $2\pi\sqrt{\dfrac{lm}{\lambda}} = 2\pi\sqrt{\dfrac{c}{g}}$

$\therefore$ $\qquad\qquad$ Period $= 2\pi\sqrt{\dfrac{\text{Static extension}}{g}}$

Importance of c is obvious if period of oscillation is compared with that of simple pendulum of length l i.e. period of oscillation of mass is that of simple pendulum of length $\dfrac{lmg}{\lambda}$, i.e. static extension of spring.

Illustrations on Simple Harmonic Motion :

Ex. 1 : *A spring of negligible weight hangs vertically. A mass m is attached to the other end. If the mass is moving with velocity V_0 when the spring is unstretched, find the velocity v as a function of the stretch x (Take λ as Young's modulus of the spring).*

Sol. : If x is the increase in length of the spring when velocity of the mass m is v; then the equation of motion is

$$mv\frac{dv}{dx} = mg - \lambda x$$

$$\int mv\, dv = \int (mg - \lambda x)\, dx$$

$$m\frac{v^2}{2} = mgx - \lambda\frac{x^2}{2} + c$$

$x = 0,\ v = v_0\ \therefore\ \dfrac{mv_0^2}{2} = c$

$\therefore$ $\qquad\qquad \boxed{mv^2 = 2mgx - \lambda x^2 + mv_0^2}$

Ex. 2 : *Two identical loads are suspended from the end of a spring. Find the motion imparted to one load if the other breaks loose in case the increase in length of spring under action of one load at rest is a.*

Sol. :

When both loads are attached, we have,

$$2mg = \lambda \cdot \frac{2a}{l}$$

where λ is Young's modulus and l is natural length of spring.

$$\lambda = \frac{mgl}{a}$$

Equation of motion is

$$mv\frac{dv}{dx} = mg - \lambda\frac{(x + a)}{l}$$

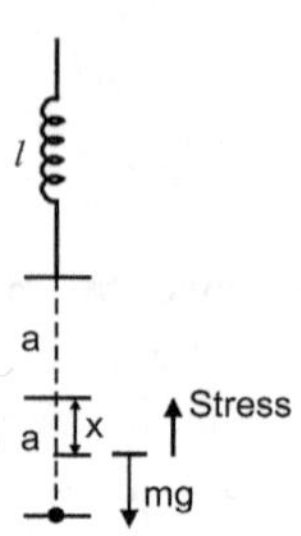

Fig. 2.10

$$\therefore \quad \frac{1}{2}m\frac{d}{dx}(v^2) = mg - \frac{mg}{a}\cdot(x + a) = -\frac{mgx}{a}$$

$$\therefore \quad \frac{d}{dx}(v^2) = -\frac{2g}{a}x$$

$$\therefore \quad v^2 = -\frac{g}{a}x^2 + A$$

When $x = a$, $v = 0$ $\therefore$ $A = ga$

$$\therefore \quad v^2 = \frac{g}{a}(a^2 - x^2)$$

$$v = \frac{dx}{dt} = \pm\sqrt{\frac{g}{a}}\sqrt{a^2 - x^2}$$

As t increases, x decreases.

Hence

$$\frac{dx}{dt} = -\sqrt{\frac{g}{a}}\sqrt{a^2 - x^2} \quad \text{or} \quad \frac{-dx}{\sqrt{a^2 - x^2}} = \sqrt{\frac{g}{a}}\,dt$$

$$\cos^{-1}\frac{x}{a} = \sqrt{\frac{g}{a}}\,t + B$$

When $t = 0$, $x = a$, $\therefore$ $B = 0$

$$\therefore \quad \boxed{x = a\cos\left(\sqrt{\frac{g}{a}}\cdot t\right)}$$

Ex. 3 : *A particle of mass m is attached to one end of a light elastic string of natural length a and modulus $\frac{mg}{k}$. The other end of the string is fixed to a point O and the particle is allowed to fall from rest at O. Obtain velocity of the particle and show that the highest magnitude is $\sqrt{ag\,(2 + k)}$.*   **(Dec. 2007)**

Sol. : Here OA is the natural length of the string. If P is a subsequent position of the particle such that AP = x (Refer Fig. 2.11) then the forces acting on the mass m are

(i)   the weight mg acting downwards (ii) the tension T in the string acting upwards.

By D-Alembert's principle, the equation of motion is $mv\frac{dv}{dx} = mg - T$, where $T = \frac{\lambda \cdot x}{l}$.

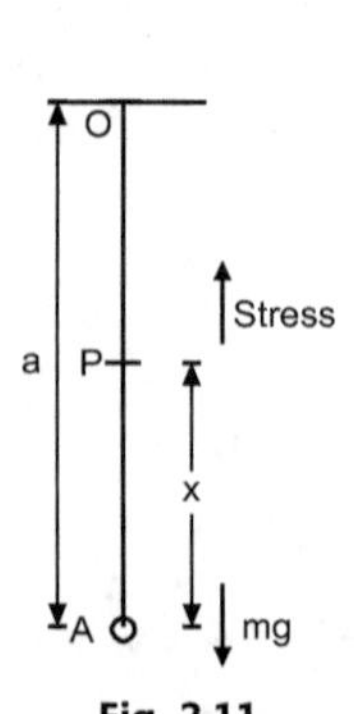

Fig. 2.11

Given λ = modulus of elasticity $= \frac{mg}{k}$ · x = extended length = x, l = natural length = a

$$\therefore \quad T = \frac{mg}{k}\cdot x\frac{1}{a} = \frac{mg}{ak}x$$

$\therefore$ Equation of motion is

$$mv\frac{dv}{dx} = mg - \frac{mg}{ak}x \Rightarrow v\frac{dv}{dx} = g - \frac{g}{ak}x = \frac{g}{ak}(ak - x)$$

$$\int v\,dv = \frac{g}{ak}\int(ak - x)\,dx + C \Rightarrow \frac{v^2}{2} = \frac{g}{ak}\frac{(ak - x)^2}{-2} + C \qquad \dots (1)$$

$\Big[$Since particle falls from rest with initial velocity u, then equation of motion is

$$mv \frac{dv}{dx} = mg$$

$$\int v\, dv = g \int dx \qquad \text{or} \qquad \frac{v^2}{2} = gx + C_1$$

$$v^2 = 2gx + C_2$$

When $x = 0$, $v = u$ $\quad \therefore\ C_2 = u^2$

$$v^2 = 2gx + u^2$$

$$x = 0 \Rightarrow \text{ it is a free fall through height a}$$

$\therefore$ $\qquad \boxed{v^2 = 2ag + u^2}$

When $x = 0$, $u = 0 \Rightarrow v^2 = 2ag$

$\therefore$ $\qquad \boxed{v = \sqrt{2ag}}$

Equation (1) becomes, $\quad \dfrac{2ag}{2} = \dfrac{g}{ak}\dfrac{(ak-0)^2}{-2} + C$

$\therefore$ $\qquad C = ag + \dfrac{agk}{2} = \dfrac{ag}{2}(2+k)$

From equation (1), $\quad \dfrac{v^2}{2} = \dfrac{g}{ak}\dfrac{(ak-x)^2}{-2} + \dfrac{ag}{2}(2+k)$

$$v^2 = -\frac{g}{ak}(ak-x)^2 + ag(2+k)$$

v is maximum when $ak - x = 0 \Rightarrow x = ak$

$$v_{max}^2 = ag(2+k)$$

$\therefore$ $\qquad \boxed{v_{max} = \sqrt{ag(2+k)}}$

Ex. 4 : *An elastic string without weight of natural length l and modulus of elasticity being weight of n-grams, is suspended by one end, and a mass m is attached to the other, show that the time of oscillations is $2\pi \sqrt{\dfrac{ml}{ng}}$.*

Sol. : Let the weight be at a distance x below the position of equilibrium then equation of motion will be,

$$m\frac{d^2x}{dt^2} = mg - \frac{\lambda(x + S_0)}{l} = mg - \frac{\lambda S_0}{l} - \frac{\lambda x}{l}$$

$$= mg - mg - \frac{\lambda x}{l} \qquad \left(\because mg = \frac{\lambda S_0}{l} \right)$$

$$\frac{d^2x}{dt^2} = \frac{-\lambda x}{lm}$$

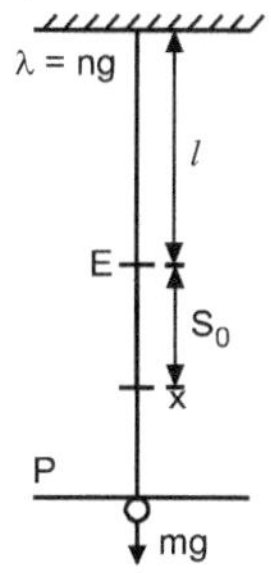

This equation is same as that of S.H.M. except

$$\mu = \frac{\lambda}{lm}$$

Hence the particle of mass m will execute S.H.M. of period $= \dfrac{2\pi}{\sqrt{\lambda/lm}}$.

$$\boxed{\text{Period } = 2\pi \cdot \sqrt{\frac{lm}{\lambda}} = 2\pi \sqrt{\frac{lm}{ng}}}$$

Fig. 2.12

Ex. 5 : *A point executing simple harmonic motion has velocities v_1 and v_2 and accelerations a_1 and a_2 in two positions respectively.*

Show that the distance between the two positions is $\left| \dfrac{v_1^2 - v_2^2}{a_1 + a_2} \right|$ **(May 2006, 2005, 2004)**

Sol. :

$$\frac{d^2 x_1}{dt^2} = -n^2 x_1 \qquad \therefore \qquad a_1 = -n^2 x_1 \qquad \dots (1)$$

$$\frac{d^2 x_2}{dt^2} = -n^2 x_2 \qquad \therefore \qquad a_2 = -n^2 x_2 \qquad \dots (2)$$

We have

$$v \frac{dv}{dx} = -n^2 x$$

$$v_1 \frac{dv}{dx_1} = -n^2 x_1 \qquad\qquad\qquad v_2 \frac{dv}{dx_2} = -n^2 x_2$$

$$\int v_1\, dv = -n^2 \int x_1\, dx_1 \qquad\qquad \int v_2\, dv = -n^2 \int x_2\, dx_2$$

$$\frac{v_1^2}{2} = -\frac{n^2 x_1^2}{2} + A \qquad\qquad \frac{v_2^2}{2} = -\frac{n^2 x_2^2}{2} + B$$

When $x_1 = a$, $v_1 = 0$ When $x_2 = a,\ v_2 = 0$

$$\therefore \qquad A = \frac{n^2 a^2}{2} \qquad\qquad\qquad B = \frac{n^2 a^2}{2}$$

$$v_1^2 = -n^2 x_1^2 + n^2 a^2 \qquad\qquad v_2^2 = -n^2 x_2^2 + n^2 a^2$$

$$v_1^2 = n^2 (a^2 - x_1^2) \qquad\qquad v_2^2 = n^2 (a^2 - x_2^2)$$

$$\therefore \qquad v_1 = n \sqrt{a^2 - x_1^2} \qquad\qquad v_2 = n \sqrt{a^2 - x_2^2}$$

a is the amplitude of S.H.M.

$$v_1^2 - v_2^2 = n^2 \left(x_2^2 - x_1^2 \right) = n^2 \left(\frac{a_2^2}{n^4} - \frac{a_1^2}{n^4} \right) \qquad\qquad \text{by using (1) and (2)}$$

$$v_1^2 - v_2^2 = \frac{\left(a_2^2 - a_1^2 \right)}{n^2}$$

$$n^2 = \frac{\left(a_2^2 - a_1^2 \right)}{\left(v_1^2 - v_2^2 \right)}$$

Also,

$$x_1 - x_2 = \frac{a_2 - a_1}{n^2} = \frac{(a_2 - a_1)\left(v_1^2 - v_2^2 \right)}{\left(a_2^2 - a_1^2 \right)}$$

$$\therefore \qquad \boxed{\ x_1 - x_2 = \frac{v_1^2 - v_2^2}{a_1 + a_2}\ }$$

Ex. 6 : *A mass of 2 kg is hung on a light spiral spring and produces a static deflection of 1/4 m. A mass of 2 kg is suddenly added to the original mass. Show that the maximum elongation produced is 0.75 m.*

Sol. : Let k be the spring constant. Then in the equilibrium position for the 2 kg mass, we have

$$2g = k\left(\frac{1}{4}\right) \quad \text{or} \quad k = 8g \qquad\qquad \dots (1)$$

Let x m be the extension of the spring when a mass of 2 kg is suddenly added to the original mass. Then $\left(x + \dfrac{1}{4}\right)$ m is the total elongation produced by the 4 kg mass and the equation of motion of the 4 kg mass is given by

$$4 \cdot v \cdot \frac{dv}{dx} = 4g - k\left(x + \frac{1}{4}\right) \qquad\qquad \dots (2)$$

where v m/sec is its velocity at any time t secs. Substituting (1) in (2), we obtain

$$4 \cdot v \cdot \frac{dv}{dx} = 4g - 8g \left(x + \frac{1}{4}\right) = 2g - 8gx$$

or
$$v \cdot \frac{dv}{dx} = \frac{g}{2} - 2gx$$

Separating the variables and integrating, we obtain $\frac{v^2}{2} = \frac{g}{2} x - 2g \cdot \frac{x^2}{2} + C.$ ... (3)

From the initial conditions, we have x = 0 and v = 0 at t = 0. This gives C = 0.

Hence, equation (3) becomes : $v^2 = gx - 2gx^2$ ... (4)

The elongation will be maximum when v = 0. Equation (4) then gives :

x = 0 or x = 1/2.

Hence $\boxed{\text{the maximum elongation produced} = 0.5 + 0.25 = 0.75 \text{ m.}}$

Ex. 7 : *An elastic spring of natural length l is fixed at a point A. To the lower end is attached a particle of mass m so that the spring stretches to a length 2l. If the particle is dropped from A, show that it descends a distance $l\left(2 + \sqrt{3}\right)$ before coming to rest.*

(Dec. 2006, 2005)

Sol. : The tension in the spring is given by $T = \frac{\lambda (2l - l)}{l} = \lambda.$

Since this balances the weight of the particle, we have $\lambda = mg$ ... (1)

At time t, let x be the extension of the spring beyond its natural length. Then the equation of motion of the particle is given by

$$m \cdot v \cdot \frac{dv}{dx} = mg - T = mg - \lambda \frac{x}{l} = mg - mg \cdot \frac{x}{l} = mg \left(1 - \frac{x}{l}\right)$$

or
$$v \, dv = g \left(1 - \frac{x}{l}\right) dx$$ ... (2)

Integrating both sides of (2), we find $\frac{v^2}{2} = gx - \frac{gx^2}{2l} + C$ ... (3)

Since the particle is dropped from A, we have $v = \sqrt{2gl}$ when x = 0. Substituting these values in (3), we find C = lg. Hence,

equation (iii) becomes : $\frac{v^2}{2} = gx - \frac{g}{2l} x^2 + gl$ ... (4)

When v = 0, x is given by $x^2 - 2lx - 2l^2 = 0$

from which we obtain $x = l + l\sqrt{3}$. Hence, the distance by which the particle descends before coming to rest is given by

$$\boxed{x + l = 2l + l\sqrt{3} = l\left(2 + \sqrt{3}\right)}.$$

Ex. 8 : *In the case of a stretched elastic spring which has one end fixed and a particle of mass m attached at the other end, the equation of motion is $m \frac{d^2x}{dt^2} = -\frac{mg}{e} (x - l)$, where l is the natural length of the string and e is the elongation due to weight mg. Find x and v under the condition that at t = 0, $x = x_0$ and v = 0.*

Sol. : Let O be the fixed end of the string and A the end to which particle is attached. If P is the position of the particle during the motion such that OP = x, then the equation of motion is $m \frac{d^2x}{dt^2} = -T$, where T is the tension in the string in that position.

If λ is the modulus of elasticity of the string, we have, as given

$$mg = \lambda \frac{e}{l} \quad \text{or} \quad \lambda = \frac{mg \, l}{e}$$

Now,
$$T = \lambda \frac{AP}{l} = \frac{\lambda}{l} (x - l) \quad \therefore \quad T = \frac{mg}{e} (x - l)$$

∴ The equation of motion is
$$m \frac{d^2x}{dt^2} = -\frac{mg}{e} (x - l)$$

or
$$v \frac{dv}{dx} = -\frac{g}{e} (x - l)$$

$x = x_0$ for v = 0 ∴
$$\int_0^v v \, dv = -\frac{g}{e} \int_{x_0}^x (x - l) \, dx$$

$$\frac{v^2}{2} = -\frac{g}{e}\left[\frac{x^2}{2} - lx\right]_{x_0}^{x}$$

$$\frac{v^2}{2} = -\frac{g}{e}\left[\frac{x^2}{2} - lx - \frac{x_0^2}{2} + lx_0\right]$$

$$v^2 = \frac{g}{e}\left[x_0^2 - 2lx_0 + l^2 - l^2 + 2lx - x^2\right] = \frac{g}{e}\left[(l - x_0)^2 - (l - x)^2\right]$$

$$v = \frac{dx}{dt} = \sqrt{\frac{g}{e}}\,[(l - x_0)^2 - (l - x)^2]^{1/2}$$

$$\int \frac{dx}{\sqrt{(l - x_0)^2 - (l - x)^2}} = \sqrt{\frac{g}{e}}\int dt$$

$$-\sin^{-1}\left(\frac{l - x}{l - x_0}\right) = \sqrt{\frac{g}{e}}\,t + C$$

At $\quad t = 0, \quad x = x_0 \quad \therefore \quad C = -\frac{\pi}{2}$

$\therefore$

$$\frac{\pi}{2} - \sin^{-1}\left(\frac{l - x}{l - x_0}\right) = \sqrt{\frac{g}{e}}\,t$$

$$\cos^{-1}\left(\frac{l - x}{l - x_0}\right) = \sqrt{\frac{g}{e}}\,t$$

$$\frac{l - x}{l - x_0} = \cos\sqrt{\frac{g}{e}}\,t$$

$$\boxed{x = l + (x_0 - l)\cos\left(\sqrt{\frac{g}{e}}\,t\right)}$$

and

$$\boxed{v = \frac{dx}{dt} = -(x_0 - l)\sqrt{\frac{g}{e}}\sin\left(\sqrt{\frac{g}{e}}\,t\right)}$$

EXERCISE 2.6

1. A particle is oscillating in a straight line about a centre of force O, towards which when at a distance x the force is $mn^2 x$ and a is the amplitude of the oscillation. When at a distance $\dfrac{a\sqrt{3}}{2}$ from O, the particle receives a blow in the direction of motion which generates a velocity na. If this velocity be away from O, show that the new amplitude is $a\sqrt{3}$.

   [**Hint :** $v\dfrac{dv}{dx} = -n^2 x$ when $x = \dfrac{a\sqrt{3}}{2}, v = \dfrac{3na}{2}$.

   For new amplitude, put $v = 0$. $\quad x = a\sqrt{3}$.]

2. A particle executes S.H.M. When it is 2 cm from mid path, its velocity is 10 cm/sec. and when it is 6 cm from centre of its path, its velocity is 2 cm/sec. Find its period and its greatest acceleration.

   (Dec. 2004; May 2008, 2007) Ans. $\dfrac{2\pi}{\sqrt{3}}$, $\sqrt{336}$ cm/sec^2

3. A particle of mass m is suspended from one end of spring whose other end is attached to a fixed point. If the extension in the length due to the mass of the particle is e, find the period of oscillation.

   (Dec. 2008) Ans. $2\pi\sqrt{e/g}$

4. A mass hangs from a fixed point by means of a tight elastic spring which obeys Hooke's law. The mass being given a small vertical displacement. If n is the number of oscillations per second in the ensuring S.H.M. and l is the length of the spring when the system is in equilibrium, show that the natural length of the spring is $l - \dfrac{g}{4\pi^2 n^2}$.

2.10 HEAT FLOW

The fundamental principles involved in the problems of heat conduction are :

(i)   Heat flows from a higher temperature to the lower temperature.

(ii)   The quantity of heat in a body is proportional to its mass and temperature.

(iii)   **Fourier's Law of Heat Conduction :** The rate of heat flow across an area is proportional to the area and to the rate of change of temperature with respect to its distance normal to the area.

If q (cal/sec.) be the quantity of heat that flows across a slab of area A (cm^2) and thickness δx in one second, where the difference of temperature at the faces is δT, then by (iii) above

$$q = \text{Thermal conductivity} \times \text{Area} \times \text{Temperature gradient}$$

$$\boxed{q = -kA \frac{dT}{dx}}$$

where k is a constant depending upon the material of the body and is called the thermal conductivity.

Negative sign is attached because T decreases as x increases.

Ex. 1 : *Obtain a formula for the steady-state heat loss per unit time from a unit length of pipe of radius r_0 carrying steam at temperature T_0 if the pipe is covered with insulation of thickness w, the outer surface of which remains at the constant temperature T_1. What is the temperature distribution through the insulation; i.e., what is the temperature in the insulation as a function of the radius ?*

Sol. : Since the problem tells us that steady-state conditions have been reached, it follows that the heat loss per unit time from a unit length of the pipe is a constant independent of time, say Q.

Let us now consider a typical cross-section of the pipe and insulation, as suggested in Fig. 2.14.

Let T denote the temperature in the insulation at the radius r, it follows that dT/dr is the temperature gradient (or temperature change per unit length) in the direction perpendicular to the cylindrical area of radius r. Hence, by Fourier's law, we have for the amount of heat Q flowing through this general area per unit time,

$$Q = \text{Thermal conductivity} \times \text{Area} \times \text{Temperature gradient}$$

$$Q = -k\,(1 \times 2\pi r)\,\frac{dT}{dr}$$

$$dT = \frac{-Q}{2\pi k}\,\frac{dr}{r} \Rightarrow T = \frac{-Q}{2\pi k}\,\log r + c$$

To determine the constant c, we use the fact that $T = T_0$ when $r = r_0$, from which

$$T_0 = \frac{-Q}{2\pi k}\,\log r_0 + c \quad \text{or} \quad c = T_0 + \frac{Q}{2\pi k}\,\log r_0$$

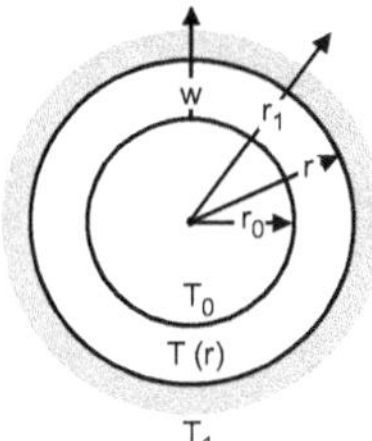

Fig. 2.13 : A typical cross-section of an insulated pipe

Substituting the value of c,

$$T = T_0 - \frac{Q}{2\pi k}\,(\log r - \log r_0) \qquad \ldots (1)$$

Furthermore, $T = T_1$ when $r = r_0 + w = r_1$.

Hence

$$T_1 = T_0 - \frac{Q}{2\pi k}\,(\log r_1 - \log r_0)$$

$\therefore$

$$Q = \frac{(T_0 - T_1)\,2\pi k}{\log r_1 - \log r_0} \qquad \ldots (2)$$

$\therefore$

$$\frac{Q}{2\pi k} = \frac{T_0 - T_1}{\log r_1 - \log r_0}$$

From (1), we get

$$T = T_0 - \left(\frac{T_0 - T_1}{\log r_1 - \log r_0}\right)(\log r - \log r_0)$$

$$T = T_0 - (T_0 - T_1)\,\frac{\log (r/r_0)}{\log (r_1/r_0)}$$

i.e.

$$\boxed{\frac{T_0 - T}{T_0 - T_1} = \frac{\log (r/r_0)}{\log (r_1/r_0)}}$$

Ex. 2 : *A pipe 20 cm in diameter contains steam at 150°C and is protected with a covering 5 cm thick for which k = 0.0025. If the temperature of the outer surface of the covering is 40°C, find the temperature half-way through the covering under steady-state conditions.*
(May 2011, 2009, 2005, 2018; Dec. 2009, May 2014)

Sol. : Let q cal/sec be the constant quantity of heat flowing out radially through a surface of the pipe having radius x cm and length 1 cm. Then the area of the lateral (belt) surface = $2\pi x$.

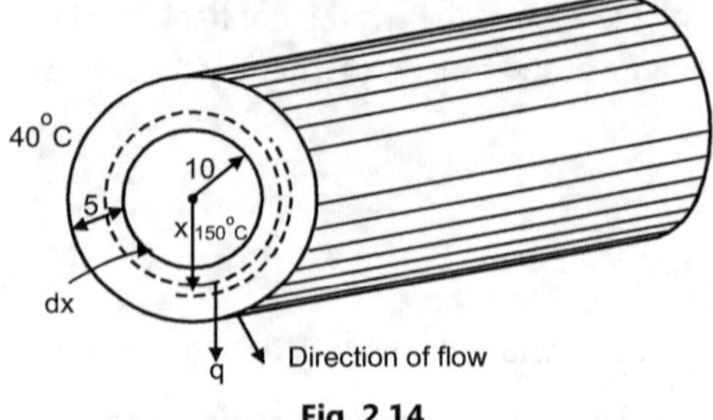

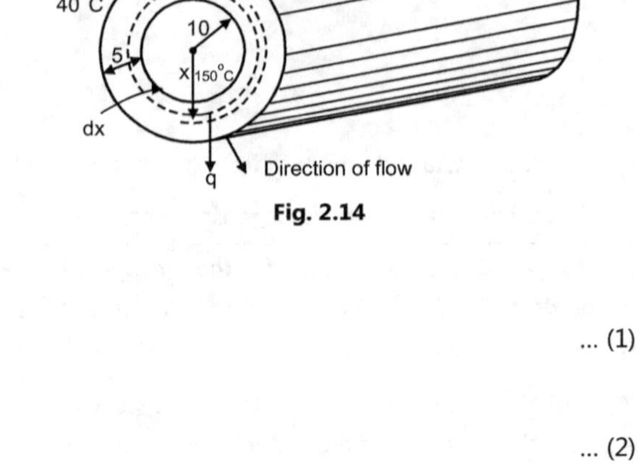

Fig. 2.14

Hence by Fourier's law, $\qquad q = -k \cdot 2\pi x \cdot \dfrac{dT}{dx}$

or $\qquad dT = -\dfrac{q}{2\pi k} \cdot \dfrac{dx}{x}$

Integrating, we have $\qquad T = -\dfrac{q}{2\pi k} \log_e x + c$

Since, $\qquad T = 150, \text{ when } x = 10$

$\therefore \qquad 150 = -\dfrac{q}{2\pi k} \log_e 10 + c \qquad \ldots (1)$

Again since $\qquad T = 40, \text{ when } x = 15$

$\qquad 40 = -\dfrac{q}{2\pi k} \log_e 15 + c \qquad \ldots (2)$

Subtracting (2) from (1), $\qquad 110 = \dfrac{q}{2\pi k} \log_e 1.5 \ldots (3)$

Let $\qquad T = t, \text{ when } x = 12.5$

$\therefore \qquad t = -\dfrac{q}{2\pi k} \log_e 12.5 + c \qquad \ldots (4)$

Subtracting (1) from (4), $\qquad t - 150 = -\dfrac{q}{2\pi k} \log_e 1.25 \qquad \ldots (5)$

Dividing (5) by (3),

$$\frac{t-150}{110} = -\frac{\log_e 1.25}{\log_e 1.5},$$

hence $\qquad \boxed{t = 89.5°C}$

Ex. 3 : *A long hollow pipe has an inner diameter of 10 cm and outer diameter of 20 cm. The inner surface is kept at 200°C and the outer surface at 50°C. The thermal conductivity is 0.12. How much heat is lost per minute from a portion of the pipe 20 metres long ? Find the temperature at a distance x = 7.5 cm from the centre of the pipe.* *(Nov.,/Dec. 2019, Dec. 2011, 06, 05; May 07)*

Sol. : Here the isothermal surfaces are cylinders, the axis of each one of them is the axis of the pipe. Consider one such cylinder of radius x cm and length 1 cm. The surface area of this cylinder is A = $2\pi x$ sq. cm. Let Q cal/sec be the quantity of heat flowing across this surface, then

Fig. 2.15

$$Q = -kA\frac{dT}{dx} = -k \cdot 2\pi x \frac{dT}{dx}$$

or $\qquad dT = -\dfrac{Q}{2\pi k} \cdot \dfrac{dx}{x}$

Integrating, we have $\qquad T = -\dfrac{Q}{2\pi k} \log_e x + c \qquad \ldots (1)$

Since $\qquad T = 200, \text{ when } x = 5$

$\therefore \qquad 200 = -\dfrac{Q}{2\pi k} \log_e 5 + c \qquad \ldots (2)$

Also $\qquad T = 50, \text{ when } x = 10$

$\therefore \qquad 50 = -\dfrac{Q}{2\pi k} \log_e 10 + c \qquad \ldots (3)$

Subtracting (3) from (2), we have

$$150 = \frac{Q}{2\pi k} (\log_e 10 - \log_e 5)$$

or $\qquad 150 = \dfrac{Q}{2\pi k} \log_e 2 \quad \ldots (4)$

$\therefore \qquad Q = \dfrac{2\pi k \times 150}{\log_e 2} = \dfrac{300\pi \times 0.12}{\log_e 2} = 163 \text{ cal/sec}$

Hence the heat lost per minute through 20 metre length of the pipe = $60 \times 2000\, Q = 120000 \times 163 = 1956000$ cal

Now, let $\qquad\qquad\qquad\qquad T = t,\quad$ when $\quad x = 7.5$

From (1), $\qquad\qquad\qquad\qquad t = -\dfrac{Q}{2\pi k}\, \log_e 7.5 + c$ $\qquad\qquad\qquad$ ... (5)

Subtracting (2) from (5), we have

$$t - 200 = -\frac{Q}{2\pi k}\,(\log_e 7.5 - \log_e 5)$$

or $\qquad\qquad\qquad\qquad t - 200 = -\dfrac{Q}{2\pi k}\,\log_e 1.5$ $\qquad\qquad\qquad$ ... (6)

Dividing (6) by (4), we have

$$\frac{t - 200}{150} = -\frac{\log_e 1.5}{\log_e 2}$$

or $\qquad\qquad\qquad\qquad t = 200 - 150 \times 0.58 = 113$

$\therefore\quad$ When $x = 7.5$ cm, $\qquad\boxed{T = 113°C}$

Ex. 4 : *A steam pipe 20 cm in diameter is protected with a covering 6 cm thick for which the coefficient of thermal conductivity is k = 0.0003 cal/cm deg. sec. in steady state. Find the heat lost per hour through a meter length of the pipe, if the surface of the pipe is at 200°C and the outer surface of the covering is at 30°C.* **(May 2006, 2005, 2015)**

Sol. : We have

$$q = -kA\,\frac{dt}{dx}$$

$\therefore$

$$q = -2\pi x \cdot k \cdot \frac{dt}{dx}$$

$$q \cdot \frac{dx}{x} = -2\pi k\, dt$$

$$\frac{q}{2\pi k} \int_{10}^{16} \frac{dx}{x} = -\int_{200}^{30} dt$$

$$\frac{q}{2\pi k} \log\left(\frac{16}{10}\right) = 200 - 30$$

$$q = \frac{170\,(2\pi k)}{\log (1.6)} \text{ cal/sec.}$$

Required heat loss $= \dfrac{340 \times (3.14) \times 0.0003}{\log 1.6} \times 100 \times 60 \times 60$

$$\boxed{\text{Heat loss } = 245443.3861 \text{ cal}}$$

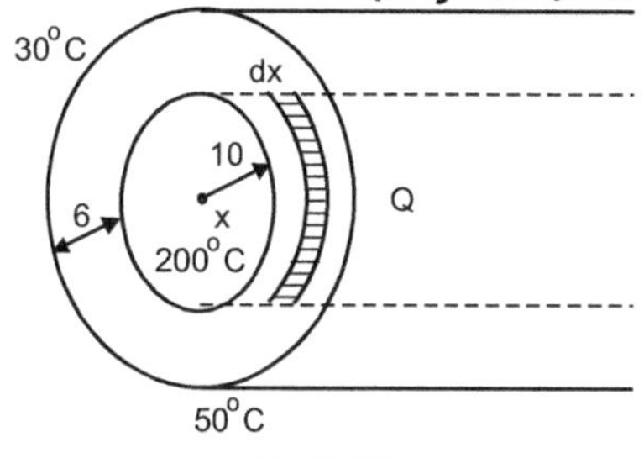

Fig. 2.16

Ex. 5 : *The inner and outer surfaces of a spherical shell are maintained at T_0 and T_1 temperatures respectively. If the inner and outer radii of the shell are r_0 and r_1 respectively and thermal conductivity of the shell is k, find the amount of heat lost from the shell per unit time. Find also the temperature distribution through the shell.* **(Dec. 2007, Nov. 2015)**

Sol. : Let x be the thickness of the spherical shell. Then heat flowing through the section is

$$q = -k\,(4\pi x^2)\,\frac{dT}{dx}$$

$$q\,\frac{dx}{x^2} = -4\pi k\, dT$$

$$q \int_{r_0}^{r_1} \frac{1}{x^2}\, dx = -4\pi k \int_{T_0}^{T_1} dT$$

$$q\left(\frac{1}{r_0} - \frac{1}{r_1}\right) = -4\pi k\,(T_1 - T_0)$$

$$q = \frac{4\pi k\,(T_0 - T_1)\, r_0\, r_1}{r_1 - r_0} \qquad\qquad\qquad\qquad \text{... (1)}$$

$$dT = -\frac{q}{4\pi k}\,\frac{dx}{x^2} \qquad T = \frac{q}{4\pi k}\,\frac{1}{x} + c$$

When $T = T_0,\ x = r_0,$ $\qquad c = T_0 - \dfrac{q}{4\pi k}\,\dfrac{1}{r_0}$

$$T = \frac{q}{4\pi k}\frac{1}{x} + T_0 - \frac{q}{4\pi k}\frac{1}{r_0}$$

by using (1)

$$= \frac{(T_0 - T_1)\,r_0\,r_1}{(r_1 - r_0)\,x} - \frac{(T_0 - T_1)\,r_0\,r_1}{(r_1 - r_0)\,r_0} + T_0$$

$$\Rightarrow \qquad \boxed{T = \frac{1}{r_1 - r_0}\left[\frac{(T_0 - T_1)\,r_0\,r_1}{x} + T_1 r_1 - T_0 r_0\right]}$$

Ex. 6 : One-dimensional steady-state heat conduction for a hollow cylinder with constant thermal conductivity k in the region $a \le r \le b$, the temperature T_r at a distance r, $(a \le r \le b)$ is given by $\dfrac{d}{dr}\left[r\dfrac{dT_r}{dr}\right] = 0$ with $T_r = T_1$, when $r = a$, and $T_r = T_2$ when $r = b$. Use this to determine steady-state temperature distribution T_r in the cylinder in terms of r.   **(May 2008)**

Sol. : Initial conditions are

$$T_r = T_1 \quad \text{for} \quad r = a$$
$$T_r = T_2 \quad \text{for} \quad r = b$$

We have

$$\frac{d}{dr}\left[r\frac{dT_r}{dr}\right] = 0$$

Integrating $\quad r\dfrac{dT_r}{dr} = c_1$, where, c_1 is a constant.

$\therefore$

$$dT_r = c_1 \frac{dr}{r}$$

Integrating both sides, $\qquad T_r = c_1 \log r + c_2$

We put initial conditions,

(1)

$$T_r = T_1 \quad \text{for} \quad r = a$$
$$T_1 = c_1 \log a + c_2 \qquad\qquad\qquad \text{... (1)}$$

(2)

$$T_r = T_2 \quad \text{for} \quad r = b$$
$$T_2 = c_1 \log b + c_2 \qquad\qquad\qquad \text{... (2)}$$

(1) – (2) gives

$$T_1 - T_2 = c_1 \log\frac{a}{b}$$

$\therefore$

$$c_1 = \frac{T_1 - T_2}{\log\dfrac{a}{b}}$$

and

$$c_2 = T_1 - \frac{T_1 - T_2}{\log\dfrac{a}{b}}\log a = T_1 - \frac{(T_1 - T_2)\log a}{\log a - \log b}$$

i.e.

$$c_2 = \frac{T_1(\log a - \log b) - (T_1 - T_2)\log a}{\log a - \log b} = -\frac{T_1 \log b + T_2 \log a}{\log a - \log b} = -\frac{T_2 \log a - T_1 \log b}{\log (a/b)}$$

$\therefore$

$$T_r = \left(\frac{T_1 - T_2}{\log (a/b)}\right)\log r + \frac{T_2 \log a - T_1 \log b}{\log (a/b)}$$

$\therefore$

$$\boxed{\left[\log\frac{a}{b}\right] T_r = (T_1 - T_2)\log r + T_2 \log a - T_1 \log b}$$

Ex. 7 : For steady heat flow through the wall of a spherical shell of inner and outer radii r_1 and r_2 respectively, the temperature T at a distance r from the centre of the sphere is given by $r\dfrac{d^2 T}{dr^2} + 2\dfrac{dT}{dr} = 0$. Integrate for T by substituting $\dfrac{dT}{dr} = y$ if u_1 and u_2 are temperatures at inner and outer surfaces. Find T in terms of r.   **(Dec. 2010)**

Sol. :

$$r\frac{d^2 T}{dr^2} + 2\frac{dT}{dr} = 0$$

Put

$$\frac{dT}{dr} = y$$

$\therefore$

$$r\frac{dy}{dr} + 2y = 0 \qquad\qquad \text{or} \qquad\qquad \frac{dy}{y} + 2\frac{dr}{r} = 0$$

Integrating $\log y + \log r^2 = \log c$ $\therefore$ $y = \dfrac{c}{r^2}$

Using; $\dfrac{dT}{dr} = \dfrac{c}{r^2}$ $\therefore$ $T = -\dfrac{c}{r} + d$

$T = u_1, \quad r = r_1, \quad T = u_2, \quad r = r_2,$

$$u_1 = -\frac{c}{r_1} + d, \quad u_2 = -\frac{c}{r_2} + d$$

$$u_1 - u_2 = c\left(\frac{1}{r_2} - \frac{1}{r_1}\right) = \left(\frac{r_1 - r_2}{r_1 r_2}\right) c$$

$$c = \frac{r_1 r_2}{r_1 - r_2}(u_1 - u_2)$$

$$d = u_1 + \frac{c}{r_1} = u_1 + \frac{1}{r_1}\frac{r_1 r_2}{r_1 - r_2}(u_1 - u_2) = \frac{u_2 r_2 - u_1 r_1}{r_2 - r_1}$$

$$\boxed{T = \frac{1}{r_2 - r_1}\left[(u_2 r_2 - u_1 r_1) - \frac{r_1 r_2 (u_2 - u_1)}{r}\right]}$$

EXERCISE 2.8

1.   A pipe 10 cm in diameter contains steam at 100°C. It is covered with asbestos, 5 cm thick, for which k = 0.0006 and the outside surface is at 30°C. Find the amount of heat lost per hour from a meter long pipe.   **(Dec. 2008, May 2013)**
     Ans. 14,0000 cal/hr.

2.11 MISCELLANEOUS EXAMPLES

Ex. 1 : *In a chemical reaction in which two substances A and B initially of amount a and b respectively are concerned, the velocity of transformation $\dfrac{dx}{dt}$ at any time t is known to be equal to the product $(a - x)(b - x)$ of the amounts of the two substances then remaining untransformed. Find t in terms of x if a = 0.7, b = 0.5 and x = 0.3 when t = 300 seconds.*

Sol. : We have $\dfrac{dx}{dt} = (a - x)(b - x) = \left(\dfrac{7}{10} - x\right)\left(\dfrac{1}{2} - x\right)$

$$\frac{dx}{\left(\dfrac{7}{10} - x\right)\left(\dfrac{1}{2} - x\right)} = dt \quad \text{or} \quad 5\left[\frac{dx}{\dfrac{1}{2} - x} - \frac{dx}{\dfrac{7}{10} - x}\right] = dt$$

Integrating, $5\left[\log\left(\dfrac{7}{10} - x\right) - \log\left(\dfrac{1}{2} - x\right)\right] = t + c$

But $x = \dfrac{3}{10}$ when   t = 300.

$\therefore$ $5\left(\log\dfrac{2}{5} - \log\dfrac{1}{5}\right) = 300 + c \Rightarrow c = 5\log 2 - 300$

$\therefore$ $t = 5\left[\log\left(\dfrac{7}{10} - x\right) - \log\left(\dfrac{1}{2} - x\right)\right] - 5\log 2 + 300$

$$\boxed{t = 5\left[\log\left(\frac{7}{10} - x\right) - \log\left(\frac{1}{2} - x\right) - \log 2\right] + 300}.$$

Ex. 2 : *When investigating the stress in the material of a thick cylinder subjected to internal pressure, the following relations are found to exist; $p + r\dfrac{dp}{dr} = q$ and $p + q = 2a$, where p and q are the radial stress and the circumferential stress respectively and r is the radius. Express p as a function of r.*

Sol. : We have to eliminate q between the given equations.

$\because$ $q = 2a - p$ $\therefore$ $p + r\dfrac{dp}{dr} = 2a - p$

i.e. $r\dfrac{dp}{dr} + 2p - 2a = 0$ or $\displaystyle\int \frac{dp}{p - a} + 2\int \frac{dr}{r} = 0$

$\therefore$ $\log(p - a) + 2\log r = \log c \Rightarrow (p - a) r^2 = c.$

$$\boxed{p = \frac{c}{r^2} + a}$$

Ex. 3 : *For a thick cylinder under internal pressure, if 'p' is the compressive stress and 'f' the tensile stress at a distance 'r' from the axis of the cylinder, the differential equation is $r \dfrac{dp}{dr} + p + f = 0$. Assuming $f + ap = b$, $p = 0$. When $r = r_2$ and $p = p_1$, when $r = r_1$, show that $\left(\dfrac{r_1}{r_2}\right)^{a-1} = \left(\dfrac{1-a}{b}\right) p_1 + 1$.*

Sol. : $r \dfrac{dp}{dr} + p + f = 0$ and $f = b - ap$.

$$r \dfrac{dp}{dr} + (p - ap + b) = 0$$

$$\frac{dp}{b - (a-1)\,p} + \frac{dr}{r} = 0 \quad \text{(by using V.S. form)}$$

But $p = 0$ when $r = r_2$; $p = p_1$ when $r = r_1$ and multiplying by $(a-1)$ throughout, we have,

$$(a-1) \int_{r_2}^{r_1} \frac{dr}{r} = \int_0^{p_1} \frac{-(a-1)\,dp}{b - (a-1)\,p}$$

$$\log \left(\frac{r_1}{r_2}\right)^{a-1} = \log \left(\frac{b - (a-1)\,p_1}{b}\right)$$

$$\left(\frac{r_1}{r_2}\right)^{a-1} = 1 - \left(\frac{a-1}{b}\right) p_1 = 1 + \left(\frac{1-a}{b}\right) p_1$$

Ex. 4 : *An equation in the theory of stability of an aeroplane is $\dfrac{dv}{dt} = g \cos \alpha - kv$, v being the velocity, g and k are constants. It is observed that at time t = 0, velocity is also zero. Solve completely.*

Sol. : We have $\qquad \dfrac{dv}{dt} = g \cos \alpha - kv$ at $t = 0$, $v = 0$

We have $\qquad \dfrac{dv}{dt} + kv = g \cos \alpha$ is a linear equation

$$\text{I.F.} = e^{kt}$$

G.S. is $\qquad v\, e^{kt} = g \cos \alpha \int e^{kt}\, dt + C$

$$v\, e^{kt} = \frac{g}{k} \cos \alpha\, e^{kt} + C$$

$\therefore \qquad v = \dfrac{g}{k} \cos \alpha + C e^{-kt}$ $\qquad\qquad$ ... (1)

At $t = 0$, $v = 0$, $0 = \dfrac{g}{k} \cos \alpha + C$ $\quad \therefore C = -\dfrac{g}{k} \cos \alpha$

$\therefore$ (1) becomes $\qquad v = \dfrac{g}{k} \cos \alpha - \dfrac{g}{k} \cos \alpha\, e^{-kt}$

$$\boxed{v = \frac{g}{k} \cos \alpha\, (1 - e^{-kt})}$$

Ex. 5 : *The amount x of a substance present in a certain chemical reaction at time t is given by $\dfrac{dx}{dt} + \dfrac{x}{10} = 2 - 1.5\, e^{-\frac{t}{10}}$. If at t = 0, x = 0.5, find x at t = 10.*

(Dec. 2009)

Sol. : $\qquad \dfrac{dx}{dt} + \dfrac{x}{10} = 2 - 1.5\, e^{-\frac{t}{10}}$ linear equation

$$\text{I.F.} = e^{t/10}$$

G.S. is $\qquad x\, e^{t/10} = \int e^{t/10}\, [2 - 1.5\, e^{-\frac{t}{10}}]\, dt + c$

$\therefore \qquad x\, e^{t/10} = 2 \times 10 \times e^{t/10} - \dfrac{3}{2} t + c$

$$x = 20 - \frac{3}{2} t\, e^{-\frac{t}{10}} + c\, e^{-t/10}, \text{ at } t = 0, \ x = 0.5$$

$$\frac{1}{2} = 20 + c \therefore c = -\frac{39}{2}$$

$$x = 20 - \frac{3}{2} t\, e^{-\frac{t}{10}} - \frac{39}{2}\, e^{-\frac{t}{10}}$$

Now at $t = 10$,
$$x = 20 - 15\, e^{-1} - \frac{39}{2}\, e^{-1}$$

$$\boxed{x = 20 - \frac{69}{2e}}$$

Ex. 6 : *A man invests Rs. 5000 at the rate of 6 percent per annum, interest being compounded continuously. When will the sum double itself ?* (Take $\log_e 2 = 0.693$)

Sol. : Let x be the amount after t years. Then

$$\frac{dx}{dt} = \frac{6}{100}\, x \qquad \text{or} \qquad \int \frac{dx}{x} = \frac{6}{100} \int dt$$

$$\log x = \frac{6\,t}{100} + c$$

But when $t = 0$, $x = 5000$ $\therefore c = \log 5000$

$\therefore$
$$\log x - \log 5000 = \frac{6\,t}{100}$$

$$\frac{6}{100}\, t = \log \left(\frac{x}{5000}\right)$$

To find t when $x = 10{,}000$
$$\frac{6}{100}\, t = \log \left(\frac{10{,}000}{5{,}000}\right) = \log 2$$

$$\boxed{t = \frac{100}{6}\ (0.693) = 11.55 \text{ years}}$$

Ex. 7 : *Motion of a boat across a stream* : *A boat is rowed with a velocity u directly across a stream of width a. If the velocity of the current is directly proportional to the product of the distances from the two banks, find the path of the boat and the distance downstream to the point where it lands.*

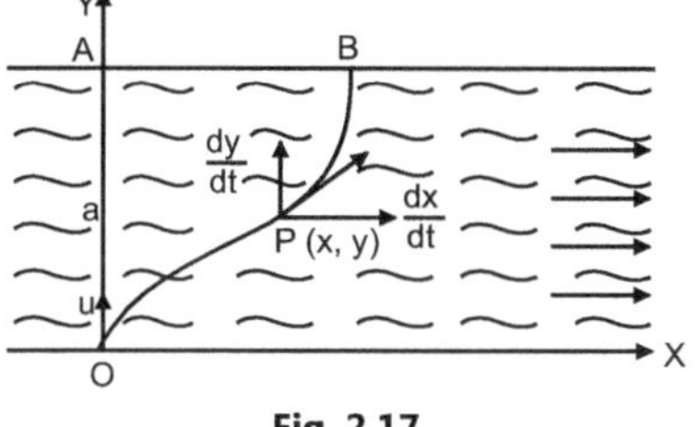

Fig. 2.17

Sol. : Taking the origin at the point from where the boat starts, let the axes be chosen as in Fig. 2.17. At any time t after its start from O, let the boat be at P (x, y), so that

$$\frac{dx}{dt} = \text{velocity of the current} = ky\,(a - y)$$

$$\frac{dy}{dt} = \text{velocity with which the boat is being rowed} = u$$

$$\frac{dy}{dx} = \frac{dy}{dt} \div \frac{dx}{dt} = \frac{u}{ky\,(a - y)} \qquad \qquad \text{... (1)}$$

This gives the direction of the resultant velocity of the boat which is also the direction of the tangent to the path of the boat.

Now (1) is of variables separable form and we can write it as

$$y\,(a - y)\, dy = \frac{u}{k}\, dx$$

Integrating, we get $\dfrac{ay^2}{2} - \dfrac{y^3}{3} = \dfrac{u}{k}\, x + c$

Since $y = 0$ when $x = 0$, $\therefore c = 0$

Hence the equation of the path of the boat is

$$\boxed{x = \frac{k}{6u}\, y^2\, (3a - 2y)}$$

Putting $y = a$, we get the distance AB, downstream where the boat lands $= ka^3/6u$.

Ex. 8 : *If the population of a country doubles in 50 years, in how many years will it treble under the assumption that the rate of increase is proportional to the number of inhabitants ?*

Sol. : Let y denote the population at time t years and y_0 the population at time $t = 0$. Then

$$\frac{dy}{dt} = ky \qquad \text{... (1)}$$

or $\qquad \dfrac{dy}{y} = k\,dt,$ where k is the proportionality factor.

Integrating, we have $\qquad \log y = kt + \log C$ or $y = C e^{kt}.$ ... (2)

At time $t = 0$, $\quad y = y_0$ and from (2), $C = y_0$. Thus, $y = y_0\, e^{kt}$ ... (3)

At $\quad t = 50$, $y = 2y_0$. From (3), $\quad 2y_0 = y_0\, e^{(50\,k)}$ $\quad$ or $\quad e^{(50\,k)} = 2.$

When $\quad y = 3y_0$, (3) gives $3 = e^{kt}.$

Then $\qquad\qquad 3^{50} = e^{(50\,k)t} = e^{(50\,k)\,t} = 2^t$

$\qquad\qquad 50 \log(3) = t \log(2)$

$\therefore \qquad \boxed{t = 79 \text{ years}}$

Ex. 9 : *In a certain culture of bacteria, the rate of increase is proportional to the number present. If it is found that the number doubles in 4 hours, how many may be expected at the end of 12 hours ?*

Sol. : Let x denote the number of bacteria at time t hours. Then

$$\frac{dx}{dt} = kx \quad \text{or} \quad \frac{dx}{x} = k\,dt. \qquad \text{...(1)}$$

Integrating (1), we have $\log x = kt + \log C \qquad$ or $\quad x = C\, e^{kt}$ ... (2)

Assuming that $x = x_0$ at time $t = 0$, $C = x_0$ and $x = x_0\, e^{kt}.$

At time $t = 4$, $\quad x = 2x_0$. Then $2x_0 = x_0\, e^{4k}$ and $e^{4k} = 2.$

When $t = 12$, $\boxed{x = x_0\, e^{12k} = x_0\,(e^{4k})^3 = x_0\,(2)^3 = 8x_0}$, that is, there are 8 times the original number.

Ex. 10 : *The temperature of a body decreases at a rate $k\theta$, where $\theta°$ is the amount of temperature of the body hotter than the surrounding air. The body is heated by a source which makes the body's temperature increase at a rate "at" where 't' is the time and 'a' is a constant. If this source is applied at $t = 0$, and the body is then at the temperature of the surrounding air, show that* $\theta = \dfrac{a}{k}\left(t - \dfrac{1}{k} + \dfrac{1}{k}e^{-kt}\right).$

Sol. : Let T be the temperature of the body at any time t, and T_1 be the constant temperature of the surrounding; then $\theta = T - T_1$

or $\dfrac{d\theta}{dt} = \dfrac{dT}{dt}.$ ... (1)

Now due to cooling alone, the temperature of the body decreases at the rate $\dfrac{dT}{dt}$

which equals $-k\theta$ $\therefore$ $\dfrac{dT}{dt} = -k\theta$ ... (2)

And due to the source of heat, the body's temperature increases at the rate $\dfrac{dT}{dt} = at$ ... (3)

$\therefore$ The rate at which the body's temperature changes due to both these effects is given by $(-k\theta + at)$.

$\therefore \qquad \dfrac{dT}{dt} = -k\theta + at$ $\quad$ or $\quad \dfrac{d\theta}{dt} = -k\theta + at$ $\left[\because \text{ by (1)}\right]$

$\qquad \dfrac{d\theta}{dt} + k\theta = at$ $\quad \therefore$ I.F. $= e^{kt}$

and G.S. is $\qquad \theta \cdot e^{kt} = a \int t\, e^{kt}\, dt + C = \dfrac{a}{k}\, e^{kt}\left(t - \dfrac{1}{k}\right) + C$

When $t = 0$, $\quad \theta = 0$ $\therefore$ $C = \dfrac{a}{k^2}$

$\therefore \qquad \boxed{\theta = \dfrac{a}{k}\left(t - \dfrac{1}{k} + \dfrac{1}{k}e^{-k/t}\right)}$

□□□

CHAPTER-3
REDUCTION FORMULAE, BETA AND GAMMA FUNCTIONS

3.1 INTRODUCTION

Students at this stage are well versed with elementary methods of integration and evaluation of real definite integrals. The aim of this chapter is to take them to higher level, by introducing them to some advance techniques. Reduction formula enables to deal with solution of some problems which are not immediately solvable. For example, $\int \sin^n x \, dx$ is connected with $\int \sin^{n-2} x \, dx$. This enables us to evaluate integrals of any power of sine or cosine and other trigonometric functions. Details will be taken up in next sections. Beta and Gamma integrals or typically called Beta and Gamma functions are the special kind of integrals which find their applications in theory of probability, integral transforms, fluid mechanics and so on. Certain kind of real definite integrals can be evaluated by using Beta and Gamma functions. Their use is prominent in evaluation of multiple integrals, which will be discussed in later chapters. In this chapter, we shall discuss Reduction formulae and some properties of Beta and Gamma functions.

3.2 REDUCTION FORMULAE

Many functions occur whose integrals are not immediately reducible to one or other standard forms and whose integrals are not directly obtainable. In some cases, however, such integrals may be linearly connected by some algebraic formulae with the integral of another expression, which itself may be either immediately integrable or easier to integrate than the original function.

For example, $\int (a^2 + x^2)^{5/2} \, dx$ may be connected with $\int (a^2 + x^2)^{3/2} \, dx$ and this latter may be expressed in terms of $\int (a^2 + x^2)^{1/2} \, dx$ which is a standard form.

Similarly, $\int \sin^n x \, dx$ may be ultimately connected with $\int \sin^2 x \, dx$ or $\int \sin x \, dx$ depending upon whether n is even or odd integer. Many such examples can be cited.

An algebraic relation connecting two integrals is termed as a *Reduction formula.*

3.3 REDUCTION FORMULAE FOR SINUSOIDAL FUNCTIONS

1. To find a reduction formula for $\int \sin^n x \, dx$, where n is a positive integer ≥ 2 and to evaluate completely $\int_0^{\pi/2} \sin^n x \, dx$

Let

$$I_n = \int \sin^n x \, dx = \int \sin^{n-1} x \cdot \sin x \, dx, \text{ on integrating by parts,}$$

$$I_n = \sin^{n-1} x \cdot (-\cos x) - \int (n-1) \sin^{n-2} x \cdot \cos x \, (-\cos x) \, dx$$

$$= -\sin^{n-1} x \cdot \cos x + (n-1) \int \sin^{n-2} x \cdot (1 - \sin^2 x) \, dx$$

$$= -\sin^{n-1} x \cos x + (n-1) \int \sin^{n-2} x \, dx - (n-1) \int \sin^n x \, dx$$

$$I_n = -\sin^{n-1} x \cos x + (n-1) I_{n-2} - (n-1) I_n$$

$$(1 + n - 1) I_n = -\sin^{n-1} x \cos x + (n-1) I_{n-2}$$

$$n I_n = -\sin^{n-1} x \cos x + (n-1) I_{n-2}$$

$$I_n = -\frac{1}{n} \sin^{n-1} x \cos x + \frac{n-1}{n} I_{n-2}$$

Thus the required reduction formula is

$$\int \sin^n x \, dx \;=\; -\frac{1}{n} \sin^{n-1} x \, \cos x + \frac{n-1}{n} \int \sin^{n-2} x \, dx \qquad \ldots(1)$$

From (1),

$$\int_0^{\pi/2} \sin^n x \, dx \;=\; \left[-\frac{1}{n} \sin^{n-1} x \, \cos x \right]_0^{\pi/2} + \frac{n-1}{n} \int_0^{\pi/2} \sin^{n-2} x \, dx$$

$$=\; \frac{n-1}{n} \int_0^{\pi/2} \sin^{n-2} x \, dx$$

Thus,
$$\boxed{I_n \;=\; \frac{n-1}{n} \, I_{n-2}} \qquad \ldots(2)$$

Changing n to n – 2 in equation (2) successively, we get,

$$I_{n-2} \;=\; \frac{n-3}{n-2} \, I_{n-4} \; ; \; I_{n-4} \;=\; \frac{n-5}{n-4} \, I_{n-6} \cdots$$

$$\therefore \qquad I_n \;=\; \frac{n-1}{n} \cdot \frac{n-3}{n-2} \cdot \frac{n-5}{n-4} \, I_{n-6} \text{ and so on.}$$

Next, consider two cases.

Case I : Let n be a positive even integer.

If n is an even integer, putting n = 4 in equation (2), we get,

$$I_4 \;=\; \frac{3}{4} I_2 ; \text{ Similarly } I_2 = \frac{1}{2} \cdot I_0$$

$$I_0 \;=\; \int_0^{\pi/2} \sin^0 x \cdot dx \;=\; \frac{\pi}{2}$$

$$I_n \;=\; \frac{n-1}{n} \cdot \frac{n-3}{n-2} \cdot \frac{n-5}{n-4} \cdots\cdots \frac{3}{4} \cdot \frac{1}{2} \cdot \frac{\pi}{2}$$

Case II : Let n be an odd positive integer.

Put n = 5 in equation (2),

$$I_5 \;=\; \frac{4}{5} I_3 ; \quad I_3 \;=\; \frac{2}{3} \cdot I_1$$

$$I_1 \;=\; \int_0^{\pi/2} \sin x \, dx = \left[-\cos x \right]_0^{\pi/2} \;=\; 1$$

$$\therefore \qquad I_n \;=\; \frac{n-1}{n} \cdot \frac{n-3}{n-2} \cdot \frac{n-5}{n-4} \cdots\cdots \frac{4}{5} \cdot \frac{2}{3} \cdot 1$$

Hence
$$\int_0^{\pi/2} \sin^n x \, dx \;=\; \frac{n-1}{n} \cdot \frac{n-3}{n-2} \cdot \frac{n-5}{n-4} \cdots\cdots \frac{3}{4} \cdot \frac{1}{2} \cdot \frac{\pi}{2} \; ; \text{ if n is even}$$

$$=\; \frac{n-1}{n} \cdot \frac{n-3}{n-2} \cdot \frac{n-5}{n-4} \cdots\cdots \frac{4}{5} \cdot \frac{2}{3} \cdot 1; \text{ if n is odd.} \qquad \ldots(3)$$

Note : Using the property
$$\int_0^a f(x) \, dx \;=\; \int_0^a f(a-x) \, dx.$$

$$\int_0^{\pi/2} \sin^n x \, dx \;=\; \int_0^{\pi/2} \sin^n \left(\frac{\pi}{2} - x \right) dx \;=\; \int_0^{\pi/2} \cos^n x \cdot dx$$

From equation (3),

$$\int_0^{\pi/2} \cos^n x \, dx = \frac{n-1}{n} \cdot \frac{n-3}{n-2} \cdot \frac{n-5}{n-4} \cdots \frac{3}{4} \cdot \frac{1}{2} \cdot \frac{\pi}{2} \text{ , if } n \text{ is even}$$

$$= \frac{n-1}{n} \cdot \frac{n-3}{n-2} \cdot \frac{n-5}{n-4} \cdots \frac{4}{5} \cdot \frac{2}{3} \cdot 1, \text{ if } n \text{ is odd} \qquad \ldots (4)$$

For example,

$$\int_0^{\pi/2} \sin^9 x \, dx = \frac{8}{9} \cdot \frac{6}{7} \cdot \frac{4}{5} \cdot \frac{2}{3} \cdot 1 = \frac{128}{315}$$

$$\int_0^{\pi/2} \cos^6 x \, dx = \frac{5}{6} \cdot \frac{3}{4} \cdot \frac{1}{2} \cdot \frac{\pi}{2} = \frac{5\pi}{32}$$

Example : Evaluate $\displaystyle\int_0^a \sqrt{a^2 - x^2} \, dx.$

Let $\qquad I = \displaystyle\int_0^a \sqrt{a^2 - x^2} \, dx, \quad$ Put $x = a \sin\theta, dx = a \cos\theta \, d\theta$

x	0	a
θ	0	π/2

$$= \int_0^{\pi/2} \sqrt{a^2 - a^2 \sin^2\theta} \cdot a \cos\theta \, d\theta = a^2 \int_0^{\pi/2} \cos^2\theta \, d\theta$$

$$= a^2 \cdot \frac{1}{2} \cdot \frac{\pi}{2} = \frac{\pi a^2}{4}$$

Additional Results :

I. $\qquad \displaystyle\int_0^{\pi} \sin^n x \, dx = 2 \int_0^{\pi/2} \sin^n x \, dx, \qquad$ for all n integral values of n.

II. $\qquad \displaystyle\int_0^{\pi} \cos^n x \, dx = 2 \int_0^{\pi/2} \cos^n x \, dx, \qquad$ if n is an even integer.

$\qquad\qquad\qquad\qquad = 0, \qquad\qquad\qquad$ if n is an odd integer.

III. $\qquad \displaystyle\int_0^{2\pi} \sin^n x \, dx = 4 \int_0^{\pi/2} \sin^n x \, dx, \qquad$ if n is an even integer.

$\qquad\qquad\qquad\qquad = 0, \qquad\qquad\qquad$ if n is an odd integer.

IV. $\qquad \displaystyle\int_0^{2\pi} \cos^n x \, dx = 4 \int_0^{\pi/2} \cos^n x \, dx, \qquad$ if n is an even integer.

$\qquad\qquad\qquad\qquad = 0, \qquad\qquad\qquad$ if n is an odd integer.

2. To find a reduction formula for $\int \sin^m x \cos^n x \, dx$**, where m and n are positive integers ≥ 2 and to completely evaluate** $\int_0^{\pi/2} \sin^m x \cos^n x \cdot dx$

Let

$$I_{m,n} = \int \sin^m x \cos^n x \, dx$$

$$= \int \sin^m x \cos^{n-1} x \cdot \cos x \, dx$$

$$= \int \cos^{n-1} x \cdot (\sin^m x \cdot \cos x) \, dx$$

Note that $\int \sin^m x \cos x \, dx = \dfrac{\sin^{m+1} x}{m+1}$ $\left\{ \because \int [f(x)]^m \cdot f'(x) \, dx = \dfrac{[f(x)]^{m+1}}{m+1} \right.$

Now applying integration by parts,

$$I_{m,n} = \cos^{n-1} x \cdot \frac{\sin^{m+1} x}{m+1} - \int (n-1) \cos^{n-2} x \, (-\sin x) \, \frac{\sin^{m+1} x}{m+1} \, dx.$$

$$I_{m,n} = \frac{\cos^{n-1} x \cdot \sin^{m+1} x}{m+1} + \frac{n-1}{m+1} \int \sin^{m+2} x \cdot \cos^{n-2} x \, dx$$

$$= \frac{\cos^{n-1} x \cdot \sin^{m+1} x}{m+1} + \frac{n-1}{m+1} \int \sin^m x \cdot \sin^2 x \, \cos^{n-2} x \, dx$$

$$= \frac{\cos^{n-1} x \sin^{m+1} x}{m+1} + \frac{n-1}{m+1} \int \sin^m x \cdot (1 - \cos^2 x) \, \cos^{n-2} x \, dx$$

$$= \frac{\cos^{n-1} x \sin^{m+1} x}{m+1} + \frac{n-1}{m+1} \int \sin^m x \cos^{n-2} x \, dx - \frac{n-1}{m+1} \int \sin^m x \cos^n x \, dx$$

$$I_{m,n} = \frac{\cos^{n-1} x \sin^{m+1} x}{m+1} + \frac{n-1}{m+1} I_{m,n-2} - \frac{n-1}{m+1} I_{m,n}$$

$$I_{m,n} + \frac{n-1}{m+1} I_{m,n} = \frac{\cos^{n-1} x \sin^{m+1} x}{m+1} + \frac{n-1}{m+1} I_{m,n-2}$$

$$I_{m,n} \left(\frac{m+1+n-1}{m+1} \right) = \frac{\cos^{n-1} x \sin^{m+1} x}{m+1} + \frac{n-1}{m+1} I_{m,n-2}$$

$$I_{m,n} = \frac{\cos^{n-1} x \sin^{m+1} x}{m+n} + \frac{n-1}{m+n} I_{m,n-2}$$

$$\int \sin^m x \cos^n x \, dx = \frac{\cos^{n-1} x \sin^{m+1} x}{m+n} + \frac{n-1}{m+n} \int \sin^m x \cdot \cos^{n-2} x \, dx \qquad \ldots(5)$$

which is the required reduction formula.

From equation (5),

$$\int_0^{\pi/2} \sin^m x \cos^n x \, dx = \left[\frac{\cos^{n-1} x \sin^{m+1} x}{m+n} \right]_0^{\pi/2} + \frac{n-1}{m+n} \int_0^{\pi/2} \sin^m x \cdot \cos^{n-2} x \, dx$$

$$= 0 + \frac{n-1}{m+n} I_{m,n-2}$$

$$\boxed{I_{m,n} = \frac{n-1}{m+n} I_{m,n-2}} \qquad \ldots(6)$$

Replacing n by n – 2 in equation (6) successively, we get,

$$I_{m,\,n-2} = \frac{n-3}{m+n-2}\, I_{m,\,n-4}$$

$$I_{m,\,n-4} = \frac{n-5}{m+n-4}\, I_{m,\,n-6} \cdots$$

$$\therefore \qquad I_{m,\,n} = \frac{n-1}{m+n} \cdot \frac{n-3}{m+n-2} \cdot \frac{n-5}{m+n-4}\, I_{m,\,n-6} \text{ and so on.}$$

We now have the following cases :

Case I : Let n be an even positive integer.

$$I_{m,\,4} = \frac{3}{m+4}\, I_{m,\,2} = \frac{3}{m+4} \cdot \frac{1}{m+2} \cdot I_{m,\,0}$$

$$I_{m,\,n} = \frac{n-1}{m+n} \cdot \frac{n-3}{m+n-2} \cdot \frac{n-5}{m+n-4} \cdots \frac{3}{m+4}\,\frac{1}{m+2} \cdot I_{m,\,0}$$

and

$$I_{m,\,0} = \int_0^{\pi/2} \sin^m x \cdot \cos^{o} x \cdot dx = \int_0^{\pi/2} \sin^m x \; dx$$

$$I_{m,\,0} = \frac{m-1}{m} \cdot \frac{m-3}{m-2}\,\frac{m-5}{m-4} \cdots \frac{3}{4}\,\frac{1}{2}\,\frac{\pi}{2} \qquad,\ \text{if } m = \text{even}$$

$$= \frac{m-1}{m} \cdot \frac{m-3}{m-2} \cdot \frac{m-5}{m-4} \cdots \frac{4}{5}\,\frac{2}{3} \cdot 1 \qquad,\ \text{if } m = \text{odd}$$

$\therefore$ If both m and n are even integers,

$$\therefore \qquad I_{m,\,n} = \frac{n-1}{m+n} \cdot \frac{n-3}{m+n-2} \cdots \frac{1}{m+2} \cdot \frac{m-1}{m} \cdot \frac{m-3}{m-2} \cdots \frac{3}{4}\,\frac{1}{2}\,\frac{\pi}{2}$$

which we write as,

$$\boxed{\; I_{m,\,n} = \frac{\{(n-1)\,(n-3)\,\ldots\ldots 3 \cdot 1\} \cdot \{(m-1)\,(m-3)\,\ldots\ldots 3\cdot 1\}}{\{(m+n)\,(m+n-2)\,\ldots\ldots 4.2\}} \cdot \frac{\pi}{2} \quad m\,,\,n \text{ both even} \;}$$

whereas, if m is odd and n is even, then

$$I_{m,\,n} = \frac{n-1}{m+n} \cdot \frac{n-3}{m+n-2} \cdots \frac{1}{m+2} \cdot \frac{m-1}{m} \cdot \frac{m-3}{m-2} \cdots \frac{4}{5}\,\frac{2}{3} \cdot 1$$

$$\boxed{\; I_{m,\,n} = \frac{\{(n-1)\,(n-3)\,\ldots\ldots 3 \cdot 1\} \cdot \{(m-1)\,(m-3)\,\ldots\ldots 4\cdot2\}}{\{(m+n)\,(m+n-2)\,\ldots\ldots 5.3.1\}} \quad m \text{ odd, } n \text{ even} \;}$$

Case II : Let n be an odd integer.

From equation (6),
$$I_{m,\,5} = \frac{4}{m+5}\, I_{m,\,3} = \frac{4}{m+5} \cdot \frac{2}{m+3} \cdot I_{m,\,1}$$

$$I_{m,\,n} = \frac{n-1}{m+n} \cdot \frac{n-3}{m+n-2} \cdot \frac{n-5}{m+n-4} \cdots \frac{4}{m+5} \cdot \frac{2}{m+3} \cdot I_{m,\,1}$$

and

$$I_{m,\,1} = \int_0^{\pi/2} \sin^m x \cdot \cos x \; dx = \left[\frac{\sin^{m+1} x}{m+1} \right]_0^{\pi/2} = \frac{1}{m+1}$$

$\therefore$ If n is odd and m may be even or odd, then

$$\int_0^{\pi/2} \sin^m x \cos^n x \; dx = \left[\frac{n-1}{m+n} \cdot \frac{n-3}{m+n-2} \cdots \frac{2}{m+3} \right] \frac{1}{m+1}$$

This is also written as,

$$\int_0^{\pi/2} \sin^m x \cos^n x \; dx = \left[\frac{\{(n-1)\,(n-3)\,\ldots\ldots 4.2\}\,\{(m-1)\,(m-3)\,\ldots\ldots 3.1\}}{(m+n)\,(m+n-2)\,\ldots\ldots (m+3)\,(m+1)\,(m-1)\,(m-3)\,\ldots 3.1} \right]$$

$$m = \text{even}\,;\ n = \text{odd}$$
$$\text{(Note the adjustment)}$$

$$\int_0^{\pi/2} \sin^m x \cos^n x \; dx = \left[\frac{\{(n-1)\,(n-3)\,\ldots\ldots 4.2\}\,\{(m-1)\,(m-3)\,\ldots\ldots 4.2\}}{(m+n)\,(m+n-2)\,\ldots\ldots (m+3)\,(m+1)\,(m-1)\,(m-3)\,\ldots 4.2} \right]$$

$$\text{if } m = \text{odd};\ n = \text{odd}$$

Note : From the above cases, it appears that the following *working rule* may be adopted for evaluation of integrals of the form

$$\int_0^{\pi/2} \sin^m x \cos^n x \, dx = \frac{\{(m-1)(m-3)\ldots 2 \text{ or } 1\}\{(n-1)(n-3)\ldots 2 \text{ or } 1\}}{(m+n)(m+n-2)(m+n-4)\ldots 2 \text{ or } 1} \times P$$

$$\text{where } P = \frac{\pi}{2}, \text{ if m and n are both even,}$$

$$= 1, \text{ for all other values of m and n.}$$

For example,

$$\int_0^{\pi/2} \sin^6 x \cos^4 x \, dx = \frac{(5 \cdot 3 \cdot 1)(3 \cdot 1)}{10 \cdot 8 \cdot 6 \cdot 4 \cdot 2} \times \frac{\pi}{2} = \frac{3\pi}{512}$$

$$\int_0^{\pi/2} \sin^5 x \cos^6 x \, dx = \frac{(4.2)(5.3.1)}{11.9.7.5.3.1} \times 1 = \frac{8}{693}$$

$$\int_0^{\pi/2} \sin^4 x \cdot \cos^5 x \, dx = \frac{(3.1)(4.2)}{9.7.5.3.1} \times 1 = \frac{8}{315}$$

$$\int_0^{\pi/2} \sin^3 x \cos^5 x \, dx = \frac{(2)(4.2)}{8.6.4.2} = \frac{1}{24}$$

Additional Results :

$$\text{I.} \quad \int_0^{\pi/2} \sin^P x \cos x \, dx = \frac{1}{P+1} = \int_0^{\pi/2} \cos^P x \cdot \sin x \cdot dx$$

$$\text{II.} \quad \int_0^{\pi} \sin^m x \cos^n x \, dx = 2 \int_0^{\pi/2} \sin^m x \cdot \cos^n x \, dx, \qquad \text{if } n = \text{even, } m = \text{even or odd}$$

$$= 0, \qquad \text{if } n = \text{odd, } m = \text{even or odd}$$

$$\text{III.} \quad \int_0^{2\pi} \sin^m x \cos^n x \, dx = 4 \int_0^{\pi/2} \sin^m x \cdot \cos^n x \, dx, \qquad \text{if } m, n = \text{even}$$

$$= 0 \qquad \text{otherwise}$$

3. Reduction formula for $\int \tan^n x \, dx$

Let

$$I_n = \int \tan^n x \, dx = \int \tan^{n-2} x \cdot \tan^2 x \, dx = \int \tan^{n-2} x \cdot (\sec^2 x - 1) \, dx$$

$$= \int \tan^{n-2} x \, \sec^2 x \, dx - \int \tan^{n-2} x \, dx$$

$$I_n = \frac{\tan^{n-1} x}{n-1} - I_{n-2}$$

$$\boxed{\int \tan^n x \, dx = \frac{\tan^{n-1} x}{n-1} - \int \tan^{n-2} x \, dx}$$

which is the required reduction formula.

4.   Reduction formula for $\int \sec^n \theta \, d\theta$

Let

$$I_n = \int \sec^n \theta \, d\theta$$

$$= \int \sec^{n-2} \theta \, \sec^2 \theta \, d\theta, \quad \text{integrate by parts}$$

$$= \sec^{n-2} \theta \cdot \tan \theta - \int (n-2) \sec^{n-3} \theta \cdot (\sec \theta \tan \theta) \, \tan \theta \, d\theta$$

$$= \sec^{n-2} \theta \tan \theta - (n-2) \int \sec^{n-2} \theta \, (\sec^2 \theta - 1) \, d\theta$$

$$= \sec^{n-2} \theta \tan \theta - (n-2) \int \sec^n \theta \, d\theta + (n-2) \int \sec^{n-2} \theta \, d\theta$$

$$= \sec^{n-2} \theta \tan \theta - (n-2) I_n + (n-2) I_{n-2}$$

$$(n-1) I_n = \sec^{n-2} \theta \, \tan \theta + (n-2) I_{n-2}$$

$$I_n = \frac{\sec^{n-2} \theta \tan \theta}{n-1} + \frac{n-2}{n-1} I_{n-2}$$

$$\boxed{\int \sec^n \theta \, d\theta = \frac{\sec^{n-2} \theta \tan \theta}{n-1} + \frac{n-2}{n-1} \int \sec^{n-2} \theta \, d\theta}$$

which is the required reduction formula.

3.4 ILLUSTRATIONS ON REDUCTION FORMULAE

Ex. 1 : *Evaluate* $\displaystyle\int_0^{2a} x\sqrt{2ax - x^2} \, dx.$ *(Dec. 2017)*

Sol. : Let

$$I = \int_0^{2a} x\sqrt{2ax - x^2} \, dx = \int_0^{2a} x \cdot \sqrt{x} \, \sqrt{2a - x} \, dx = \int_0^{2a} x^{3/2} (2a - x)^{1/2} \, dx.$$

Put $x = 2a \sin^2 \theta$, $dx = 4a \sin \theta \cos \theta \, d\theta$,

x	0	$2a$
θ	0	$\pi/2$

$$I = \int_0^{\pi/2} (2a)^{3/2} \sin^3 \theta \, (2a - 2a \sin^2 \theta)^{1/2} \, 4a \sin \theta \cos \theta \, d\theta$$

$$= 16 a^3 \int_0^{\pi/2} \sin^4 \theta \cos^2 \theta \, d\theta = 16a^3 \cdot \frac{(3.1) \, (1)}{6.4.2} \, \frac{\pi}{2} \quad \boxed{I = \frac{\pi a^3}{2}}$$

Ex. 2 : *Evaluate* $\displaystyle\int_0^{\pi/4} \cos^3 2\phi \, \sin^2 4\phi \, d\phi.$

Sol. : Let

$$I = \int_0^{\pi/4} \cos^3 2\phi \, \sin^2 4\phi \, d\phi$$

Put $2\phi = \theta$, $d\phi = \frac{1}{2} d\theta$,

ϕ	0	$\pi/4$
θ	0	$\pi/2$

$$\therefore \quad I = \int_0^{\pi/2} \cos^3\theta \, \sin^2 2\theta \left(\frac{1}{2} d\theta\right) = \frac{1}{2} \int_0^{\pi/2} \cos^3\theta \, (2\sin\theta\cos\theta)^2 \, d\theta$$

$$= 2 \int_0^{\pi/2} \sin^2\theta \, \cos^5\theta \, d\theta = 2 \cdot \frac{(1) \ (4.2)}{7.5.3.1} = \frac{16}{105}$$

$$\boxed{I = \frac{16}{105}}$$

Ex. 3 : *Evaluate* $\displaystyle\int_0^{\pi} x \sin^7 x \, \cos^4 x \, dx$ (Dec. 2010, May 2007)

Sol. : Let

$$I = \int_0^{\pi} x \sin^7 x \cos^4 x \, dx \qquad\qquad ...(1)$$

$$= \int_0^{\pi} (\pi - x) \, \sin^7 (\pi - x) \, \cos^4 (\pi - x) \, dx \qquad \left\{ \because \int_0^a f(x)\, dx = \int_0^a f(a - x) \, dx \right.$$

Also, $\sin(\pi - x) = \sin x, \ \cos(\pi - x) = -\cos x$

$$\therefore \quad I = \int_0^{\pi} (\pi - x) \, \sin^7 x \cos^4 x \, dx \qquad\qquad ...(2)$$

Adding (1) and (2),

$$2I = \int_0^{\pi} \pi \sin^7 x \cos^4 x \, dx = \pi \cdot 2 \int_0^{\pi/2} \sin^7 x \cos^4 x \, dx \qquad \text{(See additional result)}$$

$$I = \pi \cdot \frac{6.4.2.3.1}{11.9.7.5.3.1} = \frac{16\pi}{1155} \qquad \boxed{I = \frac{16\pi}{1155}}$$

Ex. 4 : *Evaluate* $\displaystyle\int_{-\pi/2}^{\pi/2} \cos^3\theta \, (1 + \sin\theta)^2 \, d\theta.$

Sol. : Let

$$I = \int_{-\pi/2}^{\pi/2} \cos^3\theta \, (1 + 2\sin\theta + \sin^2\theta) \, d\theta$$

$$= \int_{-\pi/2}^{\pi/2} \cos^3\theta \, d\theta + 2 \int_{-\pi/2}^{\pi/2} \cos^3\theta \sin\theta \, d\theta + \int_{-\pi/2}^{\pi/2} \cos^3\theta \sin^2\theta \, d\theta$$

$$= 2 \int_0^{\pi/2} \cos^3\theta \, d\theta + 0 + 2 \int_0^{\pi/2} \cos^3\theta \sin^2\theta \, d\theta.$$

$$I = 2\left(\frac{2}{3}\right) + 2 \cdot \frac{(2) \ (1)}{5.3.1} = \frac{8}{5} \qquad \boxed{I = \frac{8}{5}}$$

Note : Here we have used $\displaystyle\int_{-a}^{a} f(x) \, dx = 2 \int_0^a f(x) \, dx,$ if $f(x)$ is even

$$= 0, \qquad\qquad\qquad\qquad \text{if } f(x) \text{ is odd}$$

Ex. 5 : *Evaluate* $\displaystyle\int_0^{2\pi} \sin^2\theta\,(1+\cos\theta)^4\,d\theta.$

Sol. : Let

$$I = \int_0^{2\pi} \left(2\sin\frac{\theta}{2}\cos\frac{\theta}{2}\right)^2 \cdot \left(2\cos^2\frac{\theta}{2}\right)^4 d\theta$$

$$= 64 \int_0^{2\pi} \sin^2\frac{\theta}{2}\cos^{10}\frac{\theta}{2}\,d\theta \quad \text{Put } \theta = 2t,\ d\theta = 2\,dt$$

θ	0	2π
t	0	π

$$= 64 \int_0^{\pi} \sin^2 t\ \cos^{10} t \cdot 2dt$$

$$= 128 \cdot 2 \int_0^{\pi/2} \sin^2 t\ \cos^{10} t\,dt$$

$$= 256\ \frac{(1)\ (9.7.5.3.1)}{12.10.8.6.4.2}\ \frac{\pi}{2} = \frac{21\pi}{8}$$

$$\boxed{I = \frac{21\pi}{8}}$$

Ex. 6 : *Evaluate* $\displaystyle\int_0^{\infty} \frac{x^8 - x^5}{(1+x^3)^5}\,dx.$

Sol. : Let

$$I = \int_0^{\infty} \frac{x^6 - x^3}{(1+x^3)^5} \cdot x^2\,dx \quad \text{Put } x^3 = \tan^2\theta,\ 3x^2\,dx = 2\tan\theta\sec^2\theta\,d\theta$$

x	0	∞
θ	0	$\pi/2$

$$I = \int_0^{\pi/2} \frac{\tan^4\theta - \tan^2\theta}{(1+\tan^2\theta)^5}\,\frac{2}{3}\tan\theta\sec^2\theta\,d\theta$$

$$= \frac{2}{3} \int_0^{\pi/2} (\tan^5\theta - \tan^3\theta)\,\cos^8\theta\,d\theta$$

$$= \frac{2}{3}\left[\int_0^{\pi/2} \sin^5\theta\cos^3\theta\,d\theta - \int_0^{\pi/2} \sin^3\theta\cos^5\theta\,d\theta\right]$$

$$\boxed{I = 0}$$

Note : $\displaystyle\int_0^{\pi/2} \sin^5\theta\cos^3\theta\,d\theta = \int_0^{\pi/2} \sin^5(\pi/2-\theta)\cos^3(\pi/2-\theta)\,d\theta = \int_0^{\pi/2} \cos^5\theta\ \sin^3\theta\,d\theta)$

Ex. 7 : *Evaluate* $\displaystyle\int_0^1 x^5 \cdot \sin^{-1} x \, dx$

Sol. : Let $\displaystyle I = \int_0^1 \sin^{-1} x \cdot x^5 \, dx,$ integrating by parts

$$= \left[(\sin^{-1} x) \, \frac{x^6}{6} \right]_0^1 - \int_0^1 \frac{1}{\sqrt{1-x^2}} \cdot \frac{x^6}{6} \, dx$$

$$= \frac{1}{6} \cdot \frac{\pi}{2} - \frac{1}{6} \int_0^1 \frac{x^6}{\sqrt{1-x^2}} \, dx, \quad \text{Put } x = \sin\theta, \, dx = \cos\theta \, d\theta$$

x	0	1
θ	0	$\pi/2$

$$= \frac{\pi}{12} - \frac{1}{6} \int_0^{\pi/2} \frac{\sin^6\theta}{\cos\theta} \cos\theta \, d\theta$$

$$= \frac{\pi}{12} - \frac{1}{6} \left(\frac{5}{6} \cdot \frac{3}{4} \cdot \frac{1}{2} \cdot \frac{\pi}{2} \right) = \frac{11\pi}{192}$$

$$\boxed{I = \frac{11\pi}{192}}$$

Ex. 8 : *Evaluate* $\displaystyle\int_0^1 x^{4m+1} \sqrt{\frac{1-x^2}{1+x^2}} \, dx$

Sol. : Let $\displaystyle I = \int_0^1 x^{4m} \frac{(1-x^2)}{\sqrt{1-x^4}} \cdot x \, dx, \quad \text{Put } x^2 = \sin\theta, \, 2x \, dx = \cos\theta \, d\theta$

x	0	1
θ	0	$\pi/2$

$$= \int_0^{\pi/2} \sin^{2m}\theta \, \frac{(1-\sin\theta)}{\cos\theta} \, \frac{\cos\theta}{2} \, d\theta$$

$$= \frac{1}{2} \left[\int_0^{\pi/2} \sin^{2m}\theta \, d\theta - \int_0^{\pi/2} \sin^{2m+1}\theta \cdot d\theta \right]$$

$$\boxed{I = \frac{1}{2} \left[\left\{ \frac{2m-1}{2m} \cdot \frac{2m-3}{2m-2} \cdots \frac{3}{4} \cdot \frac{1}{2} \cdot \frac{\pi}{2} \right\} - \left\{ \frac{2m}{2m+1} \cdot \frac{2m-2}{2m-1} \cdots \frac{4}{5} \cdot \frac{2}{3} \cdot 1 \right\} \right]}$$

Ex. 9 : *Prove that* $\displaystyle\int_0^{\infty} \frac{dx}{(x^2+1)^n} = \frac{(2n-2)!}{2^{2n-2} \, [(n-1)!]^2} \cdot \frac{\pi}{2}$

Sol. : Let $\displaystyle I = \int_0^{\infty} \frac{dx}{(x^2+1)^n}, \quad \text{Put } x = \tan\theta, \, dx = \sec^2\theta \, d\theta$

x	0	∞
θ	0	$\pi/2$

$$I = \int_0^{\pi/2} \frac{\sec^2\theta \, d\theta}{(\tan^2\theta + 1)^n} = \int_0^{\pi/2} \cos^{2n-2}\theta \, d\theta$$

$$= \frac{2n-3}{2n-2} \cdot \frac{2n-5}{2n-4} \cdots \frac{3}{4} \cdot \frac{1}{2} \cdot \frac{\pi}{2}$$

Multiply numerator and denominator by $(2n - 2)\ (2n - 4)\ \dots\dots\ 4.2$.

$$I = \frac{(2n - 2)\ (2n - 3)\ (2n - 4)\ (2n - 5)\ \dots\dots\ 4.3.2.1}{[(2n - 2)\ (2n - 4)\ \dots\dots\ 4.2]^2} \cdot \frac{\pi}{2}$$

$$= \frac{(2n - 2)!}{[2\ (n - 1) \cdot 2\ (n - 2)\ \dots\dots\ 2(2) \cdot 2\ (1)]^2} \cdot \frac{\pi}{2}$$

$$= \frac{(2n - 2)\ !}{[2^{n-1}\ (n - 1)!]^2} \times \frac{\pi}{2}$$

$$\boxed{I = \frac{(2n - 2)\ !}{2^{2n-2}\ [(n - 1)!]^2} \cdot \frac{\pi}{2}}$$

Ex. 10 : Considering $\displaystyle\int_{0}^{1} x^{2n+1}\ (1 - x^2)^{-1/2}\ dx$, show that $\dfrac{1}{2n + 2} + \dfrac{1}{2} \cdot \dfrac{1}{2n + 4} + \dfrac{1}{2} \cdot \dfrac{3}{4} \cdot \dfrac{1}{2n + 6} + \ \dots\dots\ = \dfrac{2.4.6.\ \dots\dots\ 2n}{3.5.7.\ \dots\dots\ (2n + 1)}$

Sol. : We know from Binomial series, $(1 + z)^n = 1 + nz + \dfrac{n\ (n - 1)}{2!}\ z^2 + \ \dots\dots$

Put $z = -x^2,\ \ n = -\dfrac{1}{2},\ $ then $(1 - x^2)^{-1/2} = 1 + \dfrac{1}{2} \cdot x^2 + \dfrac{3}{8} x^4 + \ \dots\dots$

$$\int_{0}^{1} x^{2n+1}\ (1 - x^2)^{-1/2}\ dx = \int_{0}^{1} x^{2n+1} \left(1 + \frac{1}{2} x^2 + \frac{3}{8} x^4 + \ \dots\dots \right) dx$$

$$= \int_{0}^{1} \left(x^{2n+1} + \frac{1}{2} x^{2n+3} + \frac{3}{8} x^{2n+5} + \ \dots\dots \right) dx$$

$$= \left[\frac{x^{2n+2}}{2n + 2} + \frac{1}{2} \cdot \frac{x^{2n+4}}{2n + 4} + \frac{3}{8} \cdot \frac{x^{2n+6}}{2n + 6} + \ \dots\dots \right]_{0}^{1}$$

$$= \frac{1}{2n + 2} + \frac{1}{2}\ \frac{1}{2n + 4} + \frac{1}{2}\ \frac{3}{4}\ \frac{1}{2n + 6} + \ \dots\dots \qquad\qquad \dots(1)$$

Also,

$$\int_{0}^{1} x^{2n+1}\ (1 - x^2)^{-1/2}\ dx = \int_{0}^{1} \frac{x^{2n+1}}{\sqrt{1 - x^2}}\ dx, \ \ \text{Put } x = \sin\theta,\ dx = \cos\theta\ d\theta$$

x	0	1
θ	0	π/2

$$= \int_{0}^{\pi/2} \frac{\sin^{2n+1}\theta\ \cos\theta\ d\theta}{\cos\theta} = \int_{0}^{\pi/2} \sin^{2n+1}\theta\ d\theta$$

$$= \frac{2n}{2n + 1} \cdot \frac{2n - 2}{2n - 1}\ \dots\dots\ \frac{6}{7} \cdot \frac{4}{5} \cdot \frac{2}{3} \cdot 1 \qquad\qquad \dots(2)$$

Since (1) and (2) are the values of the same integral, it follows that,

$$\boxed{\frac{1}{2n + 2} + \frac{1}{2} \cdot \frac{1}{2n + 4} + \frac{1}{2} \cdot \frac{3}{4} \cdot \frac{1}{2n + 6} + \ \dots\dots = \frac{2n}{2n + 1} \cdot \frac{2n - 2}{2n - 1}\ \dots\dots\ \frac{4}{5} \cdot \frac{2}{3} \cdot 1}$$

Ex. 11 : If $I_n = \displaystyle\int_{0}^{\pi/4} \dfrac{\sin\ (2n - 1)\ x}{\sin x}\ dx$ then prove that, $n\ (I_{n+1} - I_n) = \sin\dfrac{n\pi}{2}$ and hence, find I_3.   **(Dec. 2011, 2010, 2007)**

Sol. : Given : $\qquad\qquad I_n = \displaystyle\int_{0}^{\pi/4} \dfrac{\sin\ (2n - 1)\ x}{\sin x}\ dx$

$$\therefore \qquad I_{n+1} = \int_0^{\pi/4} \frac{\sin(2n+1)x}{\sin x}\, dx$$

$$\therefore \qquad I_{n+1} - I_n = \int_0^{\pi/4} \frac{\sin(2n+1)x - \sin(2n-1)x}{\sin x}\, dx$$

$$= \int_0^{\pi/4} \frac{2\cos 2nx \sin x}{\sin x}\, dx \qquad\qquad \because\ \sin C - \sin D = 2\cos\frac{C+D}{2}\sin\frac{C-D}{2}$$

$$= 2\left[\frac{\sin 2nx}{2n}\right]_0^{\pi/4} = \frac{1}{n}\sin\left(n\frac{\pi}{2}\right)$$

$$\boxed{n\,(I_{n+1} - I_n) = \sin\frac{n\pi}{2}}$$

Put $n = 2$, $\qquad\qquad 2(I_3 - I_2) = 0;$

Put $n = 1$, $I_2 - I_1 = 1$

and $\qquad\qquad I_1 = \int_0^{\pi/4} \frac{\sin x}{\sin x}\, dx = \frac{\pi}{4}$

$$\therefore \qquad I_3 = I_2 = 1 + I_1 = 1 + \frac{\pi}{4}$$

$$\boxed{I_3 = 1 + \frac{\pi}{4}}$$

Ex. 12 : *If $I_{m,n} = \int \cos^m x \sin nx\, dx$, find the reduction formula connecting $I_{m,n}$ with $I_{m-1,\,n-1}$.* **(May 2008)**

Sol. : Given : $\qquad\qquad I_{m,\,n} = \int \cos^m x \sin nx\, dx$

Integrating by parts,

$$I_{m,\,n} = \cos^m x\left(-\frac{\cos nx}{n}\right) - \int m\cos^{m-1}x \cdot (-\sin x)\left(-\frac{\cos nx}{n}\right) dx$$

$$= -\frac{\cos^m x \cos nx}{n} - \frac{m}{n}\int \cos^{m-1} x \cdot \cos nx \cdot \sin x\, dx$$

Note : $\because\ \sin(n-1)x = \sin(nx - x) = \sin nx \cdot \cos x - \cos nx \cdot \sin x$

$\therefore \qquad\qquad \cos nx \cdot \sin x = \sin nx \cdot \cos x - \sin(n-1)x$

$$I_{m,\,n} = -\frac{\cos^m x \cos nx}{n} - \frac{m}{n}\int \cos^{m-1}x\,[\sin nx \cos x - \sin(n-1)x]\, dx$$

$$= -\frac{\cos^m x \cdot \cos nx}{n} - \frac{m}{n}\int \cos^m x \cdot \sin nx\, dx + \frac{m}{n}\int \cos^{m-1}x \sin(n-1)x\, dx$$

$$= -\frac{\cos^m x \cos nx}{n} - \frac{m}{n}\,I_{m,\,n} + \frac{m}{n}\,I_{m-1,\,n-1}$$

$$I_{m,n} + \frac{m}{n}\, I_{m,n} = -\frac{\cos^m x \cos nx}{n} + \frac{m}{n}\, I_{m-1,\,n-1}$$

$$I_{m,n}\left(\frac{n+m}{n}\right) = -\frac{\cos^m x \cos nx}{n} + \frac{m}{n}\, I_{m-1,\,n-1}$$

$$\boxed{I_{m,n} = -\frac{\cos^m x \cos nx}{m+n} + \frac{m}{m+n}\, I_{m-1,\,n-1}}$$

which is the required reduction formula.

Ex. 13 : If $I_n = \int [\log x]^n \, dx$ then prove that, $I_n + n\, I_{n-1} = x\,[\log x]^n$

Sol. : Given :

$$I_n = \int [\log x]^n\, 1 \cdot dx = [\log x]^n \cdot x - \int n \cdot [\log x]^{n-1} \cdot \frac{1}{x} \cdot x\, dx$$

$$I_n = x \cdot [\log x]^n - n\, I_{n-1}$$

$$\boxed{I_n + n\, I_{n-1} = x\,[\log x]^n}$$

Ex. 14 : If $U_n = \displaystyle\int_0^{\pi/4} \sin^{2n} x\, dx$, prove that $U_n = \left(1 - \dfrac{1}{2n}\right) U_{n-1} - \dfrac{1}{n \cdot 2^{n+1}}$ **(Dec. 2006, 2004; May 2007)**

Sol. : Given :

$$U_n = \int_0^{\pi/4} \sin^{2n} x\, dx = \int_0^{\pi/4} \sin^{2n-1} x \cdot \sin x\, dx$$

Integration by parts,

$$U_n = \left[\sin^{2n-1} x \cdot (-\cos x)\right]_0^{\pi/4} - \int_0^{\pi/4} (2n-1)\, \sin^{2n-2} x \cdot \cos x\, (-\cos x)\, dx$$

$$= -\left(\frac{1}{\sqrt{2}}\right)^{2n} + (2n-1) \int_0^{\pi/4} \sin^{2n-2} x\, (1 - \sin^2 x)\, dx$$

$$= -\frac{1}{2^n} + (2n-1) \int_0^{\pi/4} \sin^{2(n-1)} x\, dx - (2n-1) \int_0^{\pi/4} \sin^{2n} x\, dx$$

$$= -\frac{1}{2^n} + (2n-1)\, U_{n-1} - (2n-1)\, U_n$$

$$(1 + 2n - 1)\, U_n = -\frac{1}{2^n} + (2n-1)\, U_{n-1}$$

$$\boxed{U_n = -\frac{1}{n\,2^{n+1}} + \left(1 - \frac{1}{2n}\right) U_{n-1}}$$

Ex. 15 : If $I_n = \displaystyle\int_0^{\infty} e^{-x} \sin^n x\, dx$, obtain the relation between I_n and I_{n-2} and hence, find I_4.   **(Dec. 2011, May 2008, May 2014)**

Sol. : Given :

$$I_n = \int_0^{\infty} e^{-x} \sin^n x\, dx$$

Integration by parts

$$= \left[\sin^n x \cdot (-e^{-x})\right]_0^{\infty} - \int_0^{\infty} n \sin^{n-1} x \cdot \cos x\, (-e^{-x})\, dx = 0 + n \int_0^{\infty} e^{-x}\, (\sin^{n-1} x \cos x)\, dx$$

$$= \left[n \cdot \sin^{n-1} x \cos x \cdot (-e^{-x})\right]_0^\infty - n \int_0^\infty \left[\sin^{n-1} x \, (-\sin x) + (n-1) \sin^{n-2} x \cdot \cos^2 x\right] (-e^{-x}\, dx)$$

$$= 0 + n \int_0^\infty \left[-\sin^n x \, e^{-x} + (n-1) \sin^{n-2} x \, (1 - \sin^2 x) \cdot e^{-x}\right] dx$$

$$= -n \int_0^\infty e^{-x} \sin^n x \, dx + n(n-1) \int_0^\infty e^{-x} \sin^{n-2} x \, dx - n(n-1) \int_0^\infty e^{-x} \sin^n x \, dx$$

$$= -n^2 \int_0^\infty e^{-x} \sin^n x \, dx + n(n-1) I_{n-2}$$

$$I_n = -n^2 I_n + n(n-1) I_{n-2}$$

$$\boxed{I_n = \frac{n(n-1)}{n^2 + 1} \, I_{n-2}}$$

Put $n = 4$, $I_4 = \dfrac{4(3)}{17} \cdot I_2$,   Put $n = 2$, $I_2 = \dfrac{2(1)}{5} I_0$ and $I_0 = \displaystyle\int_0^\infty e^{-x} \, dx$

$\therefore$ $I_4 = \dfrac{12}{17} \cdot \dfrac{2}{5} I_0 = \dfrac{24}{85} \left[-e^{-x}\right]_0^\infty = \dfrac{24}{85} [0 + 1] = \dfrac{24}{85}$

$$\boxed{I_4 = \frac{24}{85}}$$

Ex. 16 : If $\;U_n = \displaystyle\int_0^{\pi/4} \tan^n \theta \, d\theta\;$ then show that, $\;n(U_{n+1} + U_{n-1}) = 1\;$ and hence, find $\displaystyle\int_0^{\pi/4} \tan^6 \theta \, d\theta\;$ and also evaluate

$$\int_0^a x^5 (2a^2 - x^2)^{-3} \, dx.$$

Sol. : Given : $U_n = \displaystyle\int_0^{\pi/4} \tan^n \theta \, d\theta$

$\therefore$ $U_{n+1} = \displaystyle\int_0^{\pi/4} \tan^{n+1} \theta \, d\theta$

$$U_{n+1} = \int_0^{\pi/4} \tan^{n-1} \theta \, \tan^2 \theta \, d\theta = \int_0^{\pi/4} \tan^{n-1} \theta \, (\sec^2 \theta - 1) \, d\theta$$

$$= \int_0^{\pi/4} \tan^{n-1} \theta \, \sec^2 \theta \, d\theta - \int_0^{\pi/4} \tan^{n-1} \theta \, d\theta$$

$$= \left[\frac{\tan^n \theta}{n}\right]_0^{\pi/4} - U_{n-1}$$

$$= \frac{1}{n} - U_{n-1}$$

$\therefore$ $\boxed{n(U_{n+1} + U_{n-1}) = 1}$

We have, $\quad U_{n+1} = \dfrac{1}{n} - U_{n-1}$ $\hspace{4cm}$ ... (1)

Put n = 5, 3, 1 : $U_6 = \left(\dfrac{1}{5} - U_4\right)$, $U_4 = \left(\dfrac{1}{3} - U_2\right)$ and $U_2 = (1 - U_0)$

$\therefore \qquad U_6 = \dfrac{1}{5} - U_4 = \dfrac{1}{5} - \left(\dfrac{1}{3} - U_2\right) = -\dfrac{2}{15} + U_2 = -\dfrac{2}{15} + (1 - U_0)$

$$= \frac{13}{15} - \int_0^{\pi/4} d\theta = \frac{13}{15} - \frac{\pi}{4}$$

$$\boxed{U_6 = \frac{13}{15} - \frac{\pi}{4}}$$

Next, consider $\qquad I = \displaystyle\int_0^a x^5 \left(2a^2 - x^2\right)^{-3} dx$ Put $x = \sqrt{2}\,a\,\sin\theta$, $dx = \sqrt{2}\cdot a\cos\theta\,d\theta$

x	0	a
θ	0	π/4

$$= \int_0^{\pi/4} (\sqrt{2})^5\, a^5 \sin^5\theta\ 2^{-3}\, a^{-6}\, (\cos^2\theta)^{-3}\ \sqrt{2}\cdot a\cos\theta\, d\theta$$

$$= \int_0^{\pi/4} \sin^5\theta \cos^{-5}\theta\, d\theta = \int_0^{\pi/4} \tan^5\theta\, d\theta = U_5$$

Put n = 4, $\qquad U_5 = \dfrac{1}{4} - U_3 = \dfrac{1}{4} - \left(\dfrac{1}{2} - U_1\right) = -\dfrac{1}{4} + \displaystyle\int_0^{\pi/4} \tan\theta\cdot d\theta = -\dfrac{1}{4} + \left[\log\sec\theta\right]_0^{\pi/4}$

$$= -\frac{1}{4} + \log\sqrt{2} = \frac{1}{2}\left[-\frac{1}{2} + \log 2\right]$$

$\therefore \qquad \boxed{\displaystyle\int_0^a x^5 \left(2a^2 - x^2\right)^{-3} dx = \frac{1}{2}\left[-\frac{1}{2} + \log 2\right]}$

Ex. 17 : *Establish the reduction formula connecting* $I_n = \displaystyle\int_0^{\pi/2} x\cos^n x\, dx$ *with* I_{n-2} $\hspace{1.5cm}$ ***(May 2006, 2005, 2004; Dec. 2009)***

Sol. : Given : $\qquad I_n = \displaystyle\int_0^{\pi/2} x\cos^n x\, dx = \int_0^{\pi/2} x\cos^{n-1} x \cdot \cos x\, dx$

$$I_n = \left[x\cos^{n-1} x \cdot (\sin x)\right]_0^{\pi/2} - \int_0^{\pi/2} \left[\cos^{(n-1)} x\,(1) + x\cdot(n-1)\cos^{n-2} x\,(-\sin x)\right] \sin x\, dx$$

$$= -\int_0^{\pi/2} \cos^{n-1} x \cdot \sin x\, dx + (n-1)\int_0^{\pi/2} x\cos^{n-2} x \cdot \sin^2 x\, dx$$

$$= \left[-\frac{\cos^n x}{n} \right]_0^{\pi/2} + (n-1) \int_0^{\pi/2} x \cos^{n-2} x \, (1 - \cos^2 x) \, dx$$

$$= -\frac{1}{n} + (n-1) \int_0^{\pi/2} x \cos^{n-2} x \, dx - (n-1) \int_0^{\pi/2} x \cdot \cos^n x \cdot dx$$

$$= -\frac{1}{n} + (n-1) \, I_{n-2} - (n-1) \, I_n$$

$$\boxed{I_n = -\frac{1}{n^2} + \frac{n-1}{n} \, I_{n-2}}$$

Ex. 18 : If $I_n = \int \dfrac{x^n}{(a^2 + x^2)^{3/2}} \, dx$, prove that $(n-2) \, I_n = \dfrac{x^{n-1}}{(a^2 + x^2)^{1/2}} - a^2 \, (n-1) \, I_{n-2}$

Sol. : Given :

$$I_n = \int x^n \, (a^2 + x^2)^{-3/2} \, dx$$

$$= \int x^{n-1} \, (a^2 + x^2)^{-3/2} \, x \, dx$$

$$I_n = x^{n-1} \frac{1}{2} \left(\frac{(a^2 + x^2)^{-1/2}}{-1/2} \right) - \int (n-1) \, x^{n-2} \, \frac{1}{2} \left(\frac{(a^2 + x^2)^{-1/2}}{-1/2} \right) dx$$

$$= - x^{n-1} \, (a^2 + x^2)^{-1/2} + (n-1) \int x^{n-2} \, (a^2 + x^2)^{-3/2} \, (a^2 + x^2) \, dx$$

$$= - x^{n-1} \, (a^2 + x^2)^{-1/2} + (n-1) \int x^{n-2} \cdot a^2 \, (a^2 + x^2)^{-3/2} \, dx + (n-1) \int x^{n-2} \cdot x^2 \, (a^2 + x^2)^{-3/2} \, dx$$

$$I_n = - x^{n-1} \, (a^2 + x^2)^{-1/2} + (n-1) \, a^2 \cdot I_{n-2} + (n-1) \, I_n$$

$$(2-n) \, I_n = -x^{n-1} \, (a^2 + x^2)^{-1/2} + (n-1) \, a^2 \, I_{n-2}$$

$$\boxed{(n-2) \, I_n = \frac{x^{n-1}}{(a^2 + x^2)^{1/2}} - (n-1) \, a^2 \, I_{n-2}}$$

Ex. 19 : If $I_n = \int x^n \, (a-x)^{1/2} \, dx$, prove that $(2n+3) \, I_n = 2an \, I_{n-1} - 2x^n \, (a-x)^{3/2}$. Hence evaluate $\displaystyle\int_0^a x^2 \, (a-x)^{1/2} \, dx$.

Sol. : Given :

$$I_n = \int x^n \, (a-x)^{1/2} \, dx$$

$$= \left[x^n \left(-\frac{2}{3} \, (a-x)^{3/2} \right) \right] - \int n x^{n-1} \left(-\frac{2}{3} \, (a-x)^{3/2} \right) dx$$

$$= -\frac{2}{3} \, x^n \, (a-x)^{3/2} + \frac{2n}{3} \int x^{n-1} \, (a-x)^{1/2} \, (a-x) \, dx$$

$$= -\frac{2}{3} \, x^n \, (a-x)^{3/2} + \frac{2n}{3} \int x^{n-1} \, (a-x)^{1/2} \cdot a \, dx - \frac{2n}{3} \int x^n \cdot (a-x)^{1/2} \, dx$$

$$= -\frac{2}{3} \, x^n \, (a-x)^{3/2} + \frac{2na}{3} \, I_{n-1} - \frac{2n}{3} \, I_n$$

$$\boxed{(2n+3) \, I_n = -2x^n \, (a-x)^{3/2} + 2an \, I_{n-1}}$$

Hence, consider, $(2n + 3) \int_0^a x^n (a - x)^{1/2}\, dx = [-2x^n(a - x)^{3/2}]_0^a + 2a\, n I_{n-1}$

or $(2n + 3) I_n = 2an\, I_{n-1}$

i.e. $I_n = \dfrac{2an}{2n + 3} I_{n-1}$

Put n = 2, $I_2 = \int_0^a x^2 (a - x)^{1/2}\, dx = \dfrac{2 \cdot a \cdot (2)}{7} I_1 = \dfrac{4a}{7} I_1$

x	0	a
θ	0	$\pi/2$

Here, $I_1 = \int_0^a x (a - x)^{1/2}\, dx$ Put $x = a \sin^2 \theta$, $dx = 2a \sin \theta \cos \theta\, d\theta$

$I_1 = \int_0^{\pi/2} a \sin^2 \theta (a - a \sin^2 \theta)^{1/2} \cdot 2a \sin \theta \cos \theta\, d\theta$

or $I_1 = 2 \cdot a^{5/2} \int_0^{\pi/2} \sin^3 \theta \cos^2 \theta\, d\theta$

$= 2a^{5/2} \dfrac{(2)\,(1)}{5 \cdot 3 \cdot 1} = \dfrac{4a^{5/2}}{15}$

$\therefore$ $I_2 = \int_0^a x^2 (a - x)^{1/2}\, dx = \dfrac{4a}{7} \cdot \left(\dfrac{4a^{5/2}}{15}\right)$

$$\boxed{I_2 = \dfrac{16a^{7/2}}{105}}$$

Ex. 20 : *Find reduction formula for* $\displaystyle\int_0^{\pi/3} \cos^n x\, dx$ *and using this evaluate* $\displaystyle\int_0^{\pi/3} \cos^6 x\, dx.$ *(May 2011, Dec. 2013)*

Sol. : Let $I_n = \int_0^{\pi/3} \cos^n x\, dx = \int_0^{\pi/3} \cos^{n-1} x \cdot \cos x\, dx$

$I_n = [\cos^{n-1} x \cdot \sin x]_0^{\pi/3} - \int_0^{\pi/3} (n - 1) \cos^{n-2} x\, (-\sin x)\, \sin x\, dx$

$= \left(\dfrac{1}{2}\right)^{n-1} \dfrac{\sqrt{3}}{2} + (n - 1) \int_0^{\pi/3} \cos^{n-2} x\, (1 - \cos^2 x)\, dx$

$= \dfrac{\sqrt{3}}{2^n} + (n - 1) \int_0^{\pi/3} \cos^{n-2} x\, dx - (n - 1) \int_0^{\pi/3} \cos^n x\, dx$

$I_n = \dfrac{\sqrt{3}}{2^n} + (n - 1) I_{n-2} - (n - 1) I_n$

$I_n = \dfrac{\sqrt{3}}{n \cdot 2^n} + \dfrac{n - 1}{n} I_{n-2}$

Now put n = 6,　　$I_6 = \dfrac{\sqrt{3}}{6 \cdot 2^6} + \dfrac{5}{6} I_4;$　when n = 4,　$I_4 = \dfrac{\sqrt{3}}{4 \cdot 2^4} + \dfrac{3}{4} \cdot I_2;$　when n = 2,　$I_2 = \dfrac{\sqrt{3}}{2 \cdot 2^2} + \dfrac{1}{2} I_0$

$\therefore$　　$I_6 = \dfrac{\sqrt{3}}{8} + \dfrac{1}{2} \displaystyle\int_0^{\pi/3} \cos^0 x \, dx = \dfrac{\sqrt{3}}{8} + \dfrac{\pi}{6} = \dfrac{5}{6} \left[\dfrac{\sqrt{3}}{64} + \dfrac{3}{4} \left(\dfrac{\sqrt{3}}{8} + \dfrac{\pi}{6} \right) \right]$

$\therefore$　　$\boxed{\,I_6 = \dfrac{3\sqrt{3}}{32} + \dfrac{5\pi}{48}\,}$

Ex. 21 : If $I_n = \displaystyle\int_0^{\pi/2} x^n (\sin x + \cos x) \, dx$　then show that $I_n = \left[\left(\dfrac{\pi}{2}\right)^n + n\left(\dfrac{\pi}{2}\right)^{n-1} \right] - n(n-1) I_{n-2}.$

Sol. : Given :　　$I_n = \displaystyle\int_0^{\pi/2} x^n \sin x \, dx + \int_0^{\pi/2} x^n \cos x \, dx$

$$= [x^n (-\cos x)]_0^{\pi/2} - \int_0^{\pi/2} nx^{n-1}(-\cos x) \, dx + [x^n \cdot \sin x]_0^{\pi/2} - \int_0^{\pi/2} nx^{n-1} \cdot \sin x \, dx$$

$$= 0 + n \int_0^{\pi/2} x^{n-1} \cos x \, dx + \left(\dfrac{\pi}{2}\right)^n \cdot (1) - 0 - n \int_0^{\pi/2} x^{n-1} \sin x \, dx$$

$$= \left(\dfrac{\pi}{2}\right)^n + [nx^{n-1} \sin x]_0^{\pi/2} - n \int_0^{\pi/2} (n-1)\, x^{n-2} \sin x \, dx - [n \cdot x^{n-1}(-\cos x)]_0^{\pi/2} + n \int_0^{\pi/2} (n-1)\, x^{n-2}(-\cos x) \, dx$$

$$= \left(\dfrac{\pi}{2}\right)^n + n\left(\dfrac{\pi}{2}\right)^{n-1} (1) - 0 - n(n-1) \int_0^{\pi/2} x^{n-2} (\sin x + \cos x) \, dx$$

$$\boxed{\,I_n = \left(\dfrac{\pi}{2}\right)^n + n\left(\dfrac{\pi}{2}\right)^{n-1} - n(n-1) I_{n-2}\,}$$

Ex. 22 : If $I_n = \displaystyle\int_0^{\pi/2} x^n \sin x \, dx$ then prove that $I_n = n\left(\dfrac{\pi}{2}\right)^{n-1} - n(n-1) I_{n-2}$

Sol. : Given :

$$I_n = \int_0^{\pi/2} x^n \sin x \, dx = [x^n (-\cos x)]_0^{\pi/2} - \int_0^{\pi/2} n\, x^{n-1}(-\cos x) \, dx$$

$$= 0 + n \int_0^{\pi/2} x^{n-1} \cos x \, dx = [n \cdot x^{n-1} \cdot \sin x]_0^{\pi/2} - n \int_0^{\pi/2} (n-1)\, x^{n-2} \sin x \, dx$$

$$= n \cdot \left(\dfrac{\pi}{2}\right)^{n-1} (1) - 0 - n(n-1) \int_0^{\pi/2} x^{n-2} \sin x \, dx$$

$$\boxed{\,I_n = n\left(\dfrac{\pi}{2}\right)^{n-1} - n(n-1) I_{n-2}\,}$$

EXERCISE 3.1

1. Evaluate $\displaystyle\int_0^{2a} x^{7/2} (2a - x)^{-1/2} \, dx$ **(May 2019)**

   Hint : Put $x = 2a \sin^2 \theta$ **Ans. :** $\dfrac{35 \pi a^4}{8}$

2. Evaluate $\displaystyle\int_0^{2a} x^3 (2ax - x^2)^{3/2} \, dx$

   Hint : Put $x = 2a \sin^2 \theta$ **Ans. :** $\dfrac{9\pi}{16} a^7$

3. Evaluate $\displaystyle\int_0^{\pi} x \sin^5 x \cos^8 x \, dx$. Refer Solved Ex. 3.   **Ans. :** $\dfrac{8\pi}{1287}$

4. Evaluate $\displaystyle\int_0^{\pi/2} \cos^3 2x \sin^4 4x \, dx$

5. Prove that $\displaystyle\int_0^1 \dfrac{x^8}{\sqrt{1 - x^2}} \, dx = \dfrac{35\pi}{256}$

   Hint : $\displaystyle\int_0^a f(x) \, dx = 0$ if $f(a - x) = -f(x)$ **Ans. :** 0

6. Prove that $\displaystyle\int_0^{\infty} \dfrac{t^4}{(1 + t^2)^3} \, dt = \dfrac{3\pi}{16}$

7. Prove that $\displaystyle\int_0^1 \dfrac{x^7}{\sqrt{1 - x^4}} \, dx = \dfrac{1}{3}$

8. Prove that $\displaystyle\int_0^3 \dfrac{x^{3/2}}{(3 - x)^{1/2}} \, dx = \dfrac{27\pi}{8}$

9. Prove that : $\displaystyle\int_0^1 \dfrac{x^{2n}}{\sqrt{1 - x^2}} \, dx = \dfrac{(2n) \, !}{2^{2n} \, (n!)^2} \dfrac{\pi}{2}$

10. Prove that $\displaystyle\int_0^{\infty} \dfrac{x^2}{(1 + x^2)^{7/2}} \, dx = \dfrac{2}{15}$

11. Prove that $\displaystyle\int_0^{\infty} \left(\dfrac{t}{1 + t^2}\right)^6 dt = \dfrac{3\pi}{512}$

12. Prove that $\displaystyle\int_0^{\infty} \dfrac{x^7 - x^8}{(1 + x)^{17}} \, dx = 0$

13. Prove that $\displaystyle\int_0^{\infty} \dfrac{x^7 (1 - x^{12})}{(1 + x)^{28}} \, dx = 0$

14. Prove that $\displaystyle\int_0^{2a} x \sqrt{2ax - x^2} \, dx = \dfrac{\pi a^3}{2}$

15. Prove that $\displaystyle\int_0^{2a} x^n \sqrt{2ax - x^2} \, dx = \dfrac{a^{n+2} \pi (2n + 1)!}{2^n \, n! \, (n + 2)!}$

16. Prove that $\displaystyle\int_{-1}^1 (1 + x)^m (1 - x)^n \, dx = 2^{m + n + 1} \dfrac{m! \, n!}{(m + n + 1)!}$ where m and n are positive integers.

    [**Hint :** Put $x = \cos 2\theta$.]

17. Considering $\displaystyle\int_0^1 (1 - x^2)^n \, dx$, show that : $1 - \dfrac{n}{1.3} + \dfrac{n(n - 1)}{1.2.5} - \dfrac{n(n - 1)(n - 2)}{1.2.3.7} + \, ... = \dfrac{2}{3} \cdot \dfrac{4}{5} \cdot \dfrac{6}{7} \, ... \dfrac{2n}{2n + 1}$

    [**Hint :** Refer solved example 10.]

18. Show that $\displaystyle\int_{-\pi/2}^{\pi/2} \sin^4 x \cos^2 x \, dx = \dfrac{\pi}{16}$

19. Show that $\displaystyle\int_{-\pi/2}^{\pi/2} \sin^5 x \, dx = 0$

20. Show that $\displaystyle\int_{-\pi}^{\pi} \sin^4 x \cos^2 x \, dx = \dfrac{\pi}{8}$

21. Show that $\displaystyle\int_0^{2\pi} \sin^4 x \cos^2 x \, dx = \dfrac{\pi}{8}$

22. Show that $\displaystyle\int_0^{\pi/4} \sin^7 2\theta \, d\theta = \dfrac{8}{35}$

23. Show that : $\displaystyle\int_0^{\pi} \sin^2 \theta (1 + \cos \theta)^4 \, d\theta = \dfrac{21\pi}{16}$

24. Show that $\displaystyle\int_0^\infty \frac{x^2}{(1+x^6)^{7/2}}\,dx = \frac{8}{45}$

     Hint : Put $x^3 = \tan\theta$

26. Prove that $\displaystyle\int_0^\pi x\sin^5 x\cos^4 x\,dx = \frac{8\pi}{315}$ **(May 2013)**

28. Prove that $\displaystyle\int_0^1 x^6\sqrt{1-x^2}\,dx = \frac{5\pi}{256}$

30. Prove that $\displaystyle\int_0^1 \left(1-x^{\frac{1}{n}}\right)^m dx = \frac{m!\,n!}{(m+n)!}$

     [**Hint :** Refer Q. 6, Exercise 4.3.]

31. If $f(m,n) = \displaystyle\int x^m(1-x)^n\,dx$, show that $f(m,n) = \dfrac{x^{m+1}(1-x)^n}{m+n+1} + \dfrac{n}{m+n+1}\,f(m,n-1)$.

     Hence show that $\displaystyle\int_0^1 x^m(1-x)^n\,dx = \frac{m!\,n!}{(m+n+1)!}$ **(May 2004)**

     [**Hint :** $(m+1)\,f(m,n) = (1-x)^n x^{m+1} - n\displaystyle\int (1-x)^{n-1} x^m\,[(1-x)-1]\,dx$]

32. Find reduction formulae for $\displaystyle\int_0^{\pi/4} \sin^n x\,dx$ and hence evaluate $\displaystyle\int_0^{\pi/4} \sin^6 x\,dx$.   **Ans. :** $\dfrac{5\pi}{64} - \dfrac{11}{48}$

33. If $I_n = \displaystyle\int_0^{\pi/4} \cos^{2n} x\,dx$, prove that $I_n = \dfrac{1}{n\,2^{n+1}} + \dfrac{2n-1}{2n}\,I_{n-1}$. Hence evaluate $\displaystyle\int_0^{\pi/4} \cos^6 x\,dx$.   **(Nov. 2014, Dec. 2016)**

     Ans. : $\dfrac{11}{48} + \dfrac{5\pi}{64}$. **Hint :** $I_n = \cos^{2n-1} x\,(\sin x) + (2n-1)\displaystyle\int \cos^{2n-2} x\,dx - (2n-1)\,I_n$

34. If $I_n = \displaystyle\int_0^{\pi/4} \sin^{2n} x\,dx$, prove that $I_n = \left(1-\dfrac{1}{2n}\right) I_{n-1} - \dfrac{1}{n2^{n+1}}$. Hence evaluate $\displaystyle\int_0^{\pi/4} \sin^6 x\cdot dx$

     Ans. : $\dfrac{5\pi}{64} - \dfrac{11}{48}$, **Hint :** For first part, refer solved example 14.

35. If $I_n = \displaystyle\int_{\pi/4}^{\pi/2} \cot^n\theta\,d\theta$, prove that $I_n = \dfrac{1}{n-1} - I_{n-2}$. Hence evaluate $\displaystyle\int_{\pi/4}^{\pi/2} \cot^6\theta\,d\theta$.   **(Dec. 2008, May 2005, 2015, 2017)**

     Ans. : $\dfrac{13}{15} - \dfrac{\pi}{4}$, **Hint :** $I_n = \displaystyle\int_{\pi/4}^{\pi/2} \cot^{n-2}\theta\,(\text{cosec}^2\theta - 1)\,d\theta,\ I_2 = 1 - \dfrac{\pi}{4}$.

25. Prove that : $\displaystyle\int_4^6 \sin^4 \pi x\,\cos^2 2\pi x\,dx = \frac{7}{16}$

     Hint : $\pi x = t,\ t - 4\pi = u$ or $\pi x - 4\pi = u$

27. Prove that $\displaystyle\int_0^\pi x\cos^6 x\,dx = \frac{5\pi^2}{32}$

29. Prove that $\displaystyle\int_0^{1/2} x^3\sqrt{1-4x^2}\,dx = \frac{1}{120}$

36. If $I_n = \int\limits_0^{\pi/4} \sec^n \theta\, d\theta$, prove that $I_n = \dfrac{(\sqrt{2})^{n-2}}{n-1} + \dfrac{n-2}{n-1}\, I_{n-2}$. Hence evaluate $\int\limits_0^{\pi/4} \sec^6 \theta\, d\theta$.   **(Dec. 09, 07, 04, 18; May 16)**

[**Hint :** $I_n = (\sqrt{2})^{n-2} - (n-2) \int\limits_0^{\pi/4} \sec^{n-2} \theta\, (\sec^2 \theta - 1)\, d\theta$]   **Ans. :** $\dfrac{28}{15}$

37. Establish a reduction formula for $\int \csc^n \theta\, d\theta$ and hence evaluate $\int\limits_{\pi/4}^{\pi/2} \csc^6 \theta\, d\theta$.   **Ans. :** $\dfrac{28}{15}$

38. If $I_n = \int\limits_0^{\infty} e^{-px} \sin^n x\, dx$ $(n \geq 2,\ p > 0)$, prove that $(p^2 + n^2)\, I_n = n\,(n-1)\, I_{n-2}$. Hence evaluate $\int\limits_0^{\infty} e^{-2x} \sin^4 x\, dx$

[**Hint :** Refer solved example 15.]   **Ans. :** $\dfrac{3}{40}$

39. If $I_n = \int\limits_0^{\pi/2} x^n \sin\,(2p+1)\, x \cdot dx$, prove that $(2p+1)^2\, I_n + n\,(n-1)\, I_{n-2} = (-1)^p \cdot n\left(\dfrac{\pi}{2}\right)^{n-1}$ where, n and p are positive integers.

Hence evaluate $\int\limits_0^{\pi/2} x^4 \sin 3x\, dx$.   **(Dec. 08, 05)**

[**Hint :** $\cos\,(2p+1)\,\dfrac{\pi}{2} = 0$; $\sin\,(2p+1)\,\dfrac{\pi}{2} = \sin\left(p\pi + \dfrac{\pi}{2}\right) = (-1)^p.$]   **Ans. :** $I_4 = \dfrac{8}{81} + \dfrac{4\pi}{27} - \dfrac{\pi^3}{18}$

40. If $I_n = \int\limits_0^{\pi/2} \theta \cdot \sin^n \theta\, d\theta$, prove that $I_n = \dfrac{n-1}{n}\, I_{n-2} + \dfrac{1}{n^2}$. Hence prove that $I_5 = \dfrac{149}{225}$.

Hint : Refer solved example 17.

41. If $I_n = \int\limits_0^{\pi/2} \cos^n x \cos nx\, dx$, prove that $I_n = \dfrac{1}{2}\, I_{n-1} = \dfrac{\pi}{2^{n+1}}$.   [**Hint :** $I_0 = \int\limits_0^{\pi/2} dx = \dfrac{\pi}{2}.$] **(May 2006)**

42. If $I_n = \int\limits_0^{\pi/2} x^n \cos x\, dx$, prove that $I_n = \left(\dfrac{\pi}{2}\right)^n - n\,(n-1)\, I_{n-2}$.   **(Nov. 2015)**

43. Find a reduction formula for $\int e^{mx} \cos^n x\, dx$.

44. Find a reduction formula for $I_{m,\,n} = \int x^m\, (\log x)^n\, dx$.   **Ans. :** $I_{m,\,n} = \dfrac{x^{m+1}}{m+1}\,(\log x)^n - \dfrac{n}{n+1}\, I_{m,\,n-1}$

45. Prove that $\int \cos^{2n} \phi\, d\phi = \dfrac{1}{2n} \tan \phi\, \cos^{2n} \phi + \left(1 - \dfrac{1}{2n}\right) \int \cos^{2n-2} \phi\, d\phi$.

46. Connect $I_{m,\,n} = \int \sin^m x \cos^n x\, dx$ with $I_{m-2,\,n}$.

47. If $I_n = \int x^n\, (a^2 + x^2)^{-3/2}\, dx$, show that $(n-2)\, I_n = x^{n-1}\, (a^2 + x^2)^{-1/2} - (n-1)\, a^2\, I_{n-2}$ and hence evaluate $\int\limits_0^1 \dfrac{x^5}{(3 + x^2)^{3/2}}\, dx$.

[**Hint :** For part (1) refer solved example (18); For part (2), put $a^2 = 3$ and find I_5.]   **Ans. :** $\dfrac{24}{\sqrt{3}} - \dfrac{83}{6}$.

48. If $I_n = \int x^n\, (a^2 - x^2)^{1/2}\, dx$, show that $(n+2)\, I_n = -x^{n-1}\, (a^2 - x^2)^{3/2} + a^2\,(n-1)\, I_{n-2}$.

3.5 GAMMA FUNCTIONS

Consider the definite integral $\int_0^\infty e^{-x} x^{n-1} dx$, it is denoted by the symbol $\overline{\lceil n}$ (we read it as Gamma 'n') and is called as *Gamma function of n.* Thus,

$$\overline{\lceil n} = \int_0^\infty e^{-x} x^{n-1} dx \quad (n > 0)$$ ...(1)

Gamma function is also called as Euler's integral of the second kind.

3.6 PROPERTIES OF GAMMA FUNCTIONS

1.
$$\overline{\lceil n} = 2 \int_0^\infty e^{-x^2} \cdot x^{2n-1} dx$$

Proof : We have,

$$\overline{\lceil n} = \int_0^\infty e^{-x} x^{n-1} dx. \qquad \text{Put } x = t^2, \ dx = 2t \, dt.$$

x	0	∞
t	0	∞

$$= \int_0^\infty e^{-t^2} t^{2n-2} 2t \, dt = 2 \int_0^\infty e^{-t^2} t^{2n-1} dt$$

$$\overline{\lceil n} = 2 \int_0^\infty e^{-x^2} x^{2n-1} dx$$ ... (2)

[It may be borne in mind that variable of integration is immaterial in a definite integral.]

Relations (1) and (2) are both considered as definitions of Gamma functions.

2.
$$\overline{\lceil 1} = 1$$

Proof : By def.
$$\overline{\lceil n} = \int_0^\infty e^{-x} x^{n-1} dx$$

Put $n = 1$
$$\overline{\lceil 1} = \int_0^\infty e^{-x} x^0 \, dx = [-e^{-x}]_0^\infty = (-e^{-\infty} + e^0) = 0 + 1 = 1$$

$$\overline{\lceil 1} = 1$$ ... (3)

3.   **Reduction Formula for Gamma Functions :**

$$\overline{\lceil (n+1)} = n \, \overline{\lceil n}$$

Proof : By definition

$$\overline{\lceil n} = \int_0^\infty e^{-x} x^{n-1} dx,$$

Replace n by $n + 1$.
$$\overline{\lceil (n+1)} = \int_0^\infty e^{-x} x^n \, dx.$$

Now, integrating by parts

$$\overline{|n+1} = \left\{x^n\,(-e^{-x})\right\}_0^\infty - \int_0^\infty n x^{n-1}\,(-e^{-x})\,dx.$$

Now, $\displaystyle\lim_{x\to\infty} \frac{x^n}{e^x} = 0.$ Also if $n > 0$, $\dfrac{x^n}{e^x} = 0$ for $x = 0$ $\therefore \left[\dfrac{x^n}{e^x}\right]_0^\infty = 0$

$\therefore$ $\overline{|n+1} = 0 + n\displaystyle\int_0^\infty e^{-x}\,x^{n-1}\,dx = n\,\overline{|n}$

$\therefore$ $\boxed{\overline{|n+1} = n\,\overline{|n}}$ ...(4)

If n is a positive integer,

$$\overline{|n+1} = n\,(n-1)\,\overline{|n-1}\qquad\qquad \because\ \overline{|n} = (n-1)\,\overline{|n-1}$$

$$= n\,(n-1)\,(n-2)\,\overline{|n-2}\ = n\,(n-1)\,(n-2)\,(n-3)\,(n-4)\dots 3.2.1\,\overline{|1}$$

$$= n\,(n-1)\,(n-2)\,(n-3)\dots\dots\dots 3.2.1\qquad\qquad \{\ \because\ \overline{|1} = 1$$

$$\overline{|n+1} = n! \quad\text{if n is a positive integer.}$$

Hence

$$\boxed{\begin{aligned}\overline{|n+1} &= n\,\overline{|n}\ ,\ \textbf{in general}\\ &= \textbf{n! if n is a positive integer.}\end{aligned}}$$

4. $\boxed{\overline{|0} = \infty}$ $\because\ \overline{|n} = \dfrac{\overline{|n+1}}{n}$ $\therefore\ \overline{|0} = \dfrac{\overline{|1}}{0} = \dfrac{1}{0} = \infty$

5. $\boxed{\overline{|\tfrac{1}{2}} = \sqrt{\pi}}$

6.   Since $\overline{|n+1} = n!$

$\therefore$ $\overline{|6} = 5!,\qquad \overline{|8} = 7!,\qquad \overline{|2} = 1! = 1$

$$\overline{|\tfrac{5}{2}} = \overline{|\tfrac{3}{2}+1} = \tfrac{3}{2}\,\overline{|\tfrac{3}{2}} = \tfrac{3}{2}\,\overline{|\tfrac{1}{2}+1} = \tfrac{3}{2}\cdot\tfrac{1}{2}\,\overline{|\tfrac{1}{2}} = \tfrac{3}{2}\cdot\tfrac{1}{2}\cdot\sqrt{\pi}$$

$$\overline{|\tfrac{11}{2}} = \tfrac{9}{2}\cdot\tfrac{7}{2}\cdot\tfrac{5}{2}\cdot\tfrac{3}{2}\cdot\tfrac{1}{2}\,\sqrt{\pi}$$

7. For negative fraction n, we use $\overline{|n} = \dfrac{\overline{|n+1}}{n}$

$$\overline{|-\tfrac{5}{3}} = \left(-\tfrac{3}{5}\right)\overline{|-\tfrac{2}{3}} = \left(-\tfrac{3}{5}\right)\left(-\tfrac{3}{2}\right)\overline{|\tfrac{1}{3}} = \tfrac{9}{10}\,\overline{|\tfrac{1}{3}}$$

3.7 TRANSFORMATION OF GAMMA FUNCTIONS

1. We know that

$$\overline{|n} = \int_0^\infty e^{-x}\,x^{n-1}\,dx \qquad\text{put } x = ky\ \therefore\ dx = k\,dy$$

x	0	∞
y	0	∞

$$= \int_0^\infty e^{-ky}\,k^{n-1}\cdot y^{n-1}\cdot k\cdot dy = k^n\int_0^\infty e^{-ky}\cdot y^{n-1}\,dy$$

$\therefore$ $\boxed{\displaystyle\int_0^\infty e^{-ky}\,y^{n-1}\,dy = \dfrac{\overline{|n}}{k^n}}$

2. We know that

$$\overline{n} = \int_0^\infty e^{-x} x^{n-1}\, dx \quad \text{Put } x^n = y \text{ or } x = y^{1/n} \quad \therefore nx^{n-1}\, dx = dy$$

x	0	∞
y	0	∞

$$= \int_0^\infty e^{-y^{1/n}} \cdot \frac{dy}{n}$$

$$\therefore \quad \boxed{\int_0^\infty e^{-y^{1/n}}\, dy = n\,\overline{n} = \overline{(n+1)}}$$

Put $n = \dfrac{1}{2}$,

$$\int_0^\infty e^{-y^2}\, dy = \frac{1}{2} \cdot \overline{\frac{1}{2}}, \quad \text{but} \quad \int_0^\infty e^{-y^2}\, dy = \frac{1}{2}\sqrt{\pi}$$

$$\therefore \qquad \frac{1}{2}\sqrt{\pi} = \frac{1}{2}\,\overline{\frac{1}{2}}$$

or

$$\overline{\frac{1}{2}} = \sqrt{\pi}$$

3. We know that

$$\overline{n} = \int_0^\infty e^{-x} x^{n-1}\, dx.$$

Put $e^{-x} = y$ $\therefore -e^{-x}\, dx = dy.$ $\quad e^{x} = \dfrac{1}{y}$ $\therefore x = \log\dfrac{1}{y}$

x	0	∞
y	1	0

$$= \int_1^0 \left(\log\frac{1}{y}\right)^{n-1} (-dy)$$

$$\boxed{\overline{n} = \int_0^1 \left(\log\frac{1}{y}\right)^{n-1} dy}$$

Additional Results :

$$\boxed{\overline{p}\ \overline{1-P} = \frac{\pi}{\sin p\pi} \quad \text{if } 0 < P < 1}$$

e.g.

$$\overline{\frac{1}{4}}\ \overline{\frac{3}{4}} = \overline{\frac{1}{4}}\ \overline{1 - \frac{1}{4}} \qquad \text{Let } P = \frac{1}{4} < 1$$

$$= \frac{\pi}{\sin\left(\frac{1}{4}\pi\right)} = \frac{\pi}{\dfrac{1}{\sqrt{2}}} = \sqrt{2}\cdot\pi$$

3.8 ILLUSTRATIONS ON GAMMA FUNCTION

Ex. 1 : *Evaluate* $\displaystyle\int_0^\infty \sqrt[4]{x}\, e^{-\sqrt{x}}\, dx.$ *(May 2008, Nov. 2014)*

Sol. : Let

$$I = \int_0^\infty x^{\frac{1}{4}} e^{-\sqrt{x}}\, dx \quad \text{Put } \sqrt{x} = t \text{ or } x = t^2;\ dx = 2t\, dt$$

x	0	∞
t	0	∞

$$= \int_0^\infty t^{1/2} \cdot e^{-t}\, 2t\, dt \;=\; 2\int_0^\infty e^{-t}\, t^{3/2}\, dt = 2\left|\frac{5}{2}\right.$$

$$= 2 \cdot \frac{3}{2} \cdot \frac{1}{2} \cdot \sqrt{\pi}$$

$$\boxed{I \;=\; \frac{3\sqrt{\pi}}{2}}$$

Ex. 2 : *Evaluate* $\displaystyle\int_0^\infty \frac{dx}{3^{4x^2}}$ *(May 2019)*

Sol. : Let

$$I \;=\; \int_0^\infty \frac{dx}{3^{4x^2}} \quad \text{Put } 3 = e^m,\; 4x^2 = t,\; x = \frac{\sqrt{t}}{2}\; dx = \frac{1}{4}\, t^{-\frac{1}{2}}\, dt$$

x	0	∞
t	0	∞

$$= \int_0^\infty \frac{1}{4}\, \frac{t^{-1/2}\, dt}{(e^m)^t} \;=\; \frac{1}{4}\int_0^\infty e^{-mt}\, t^{-1/2}\, dt$$

$$= \frac{1}{4}\, \frac{\left|\dfrac{1}{2}\right.}{m^{1/2}}$$

$$\left(\because \int_0^\infty e^{-ky}\, y^{n-1}\, dy = \frac{\left|\,n\right.}{k^n}\;\; k = m,\, n = \frac{1}{2} \right)$$

$$I \;=\; \frac{\sqrt{\pi}}{4\sqrt{\log 3}}$$

Alternatively, we can put $3^{4x^2} = e^t$ i.e. $x = \dfrac{t^{1/2}}{2\sqrt{\log 3}}$ $\quad\therefore\quad dx = \dfrac{t^{-1/2}}{4\sqrt{\log 3}}\, dt$

$$\therefore \qquad I \;=\; \int_0^\infty \frac{1}{t^4}\, \frac{t^{-1/2}}{4\sqrt{\log 3}}\, dt \;=\; \frac{\left|\dfrac{1}{2}\right.}{4\sqrt{\log 3}}$$

$$\boxed{I \;=\; \frac{\sqrt{\pi}}{4\sqrt{\log 3}}}$$

Ex. 3 : *Evaluate* $\displaystyle\int_0^\infty \frac{x^2}{3^{x^2}}\, dx.$

Sol. : Let

$$I \;=\; \int_0^\infty \frac{x^2}{3^{x^2}}\, dx, \quad \text{Put } 3^{x^2} = e^t \quad \text{i.e. } x = \frac{t^{1/2}}{\sqrt{\log 3}} \quad \therefore\; dx = \frac{t^{-1/2}}{2\sqrt{\log 3}}\, dt$$

x	0	∞
t	0	∞

$$\therefore \qquad I \;=\; \int_0^\infty \frac{t}{\log 3}\, \frac{1}{e^t}\, \frac{t^{-1/2}}{2\sqrt{\log 3}}\, dt$$

$$= \frac{1}{2\,(\log 3)^{3/2}} \int_0^\infty e^{-t}\, t^{1/2}\, dt = \frac{1}{2\,(\log 3)^{3/2}}\left|\frac{3}{2}\right.$$

or

$$\boxed{I \;=\; \frac{\sqrt{\pi}}{4\,(\log 3)^{3/2}}}$$

Ex. 4 : *Evaluate* $\displaystyle\int_0^\infty x^m e^{-ax^n}\,dx$ *(a > 0)*

Sol. : Let $\displaystyle I = \int_0^\infty x^m\, e^{-ax^n}\,dx$ Put $ax^n = t$, or $x = \dfrac{t^{1/n}}{a^{1/n}}$, $dx = \dfrac{1}{n}\dfrac{t^{(1/n)-1}}{a^{1/n}}\,dt$

x	0	∞
t	0	∞

$$= \int_0^\infty \frac{t^{m/n}}{a^{m/n}}\cdot e^{-t}\cdot\frac{t^{(1/n)-1}\,dt}{n\cdot a^{1/n}} = \frac{1}{n\cdot (a)^{\frac{m+1}{n}}}\int_0^\infty e^{-t}\cdot t^{\left(\frac{m+1}{n}-1\right)}\,dt$$

$$\boxed{\,I = \frac{1}{n}\cdot\frac{1}{(a)^{\frac{m+1}{n}}}\left\lceil\left(\frac{m+1}{n}\right)\right.}$$

Ex. 5 : *Evaluate* $\displaystyle\int_0^\infty \frac{x^a}{a^x}\,dx$ *(a > 1).*   **(May 2015)**

Sol. : Let $\displaystyle I = \int_0^\infty \frac{x^a}{a^x}\,dx$ Put $a^x = e^t$ or $x = \dfrac{t}{\log a}$, $dx = \dfrac{dt}{\log a}$

x	0	∞
t	0	∞

$$I = \int_0^\infty \frac{t^a}{(\log a)^a}\cdot\frac{1}{e^t}\cdot\frac{dt}{\log a} = \frac{1}{(\log a)^{a+1}}\int_0^\infty e^{-t}\cdot t^a\,dt$$

$$= \frac{1}{(\log a)^{a+1}}\left\lceil a+1\right.$$

$$\boxed{\,I = \frac{\left\lceil a+1\right.}{(\log a)^{a+1}}\,}$$

Ex. 6 : *Prove that* $\displaystyle\int_0^\infty e^{-h^2 x^2}\,dx = \frac{\sqrt{\pi}}{2h}$

Sol. : Let $\displaystyle I = \int_0^\infty e^{-h^2 x^2}\,dx$ Put $h^2 x^2 = t$ or $x = \dfrac{t^{1/2}}{h}$; $dx = \dfrac{1}{2h}\,t^{-1/2}\,dt$

x	0	∞
t	0	∞

$$= \int_0^\infty e^{-t}\cdot\frac{t^{-1/2}\,dt}{2h} = \frac{1}{2h}\int_0^\infty e^{-t}\,t^{-1/2}\,dt = \frac{1}{2h}\left\lceil\frac{1}{2}\right. \qquad \left(\because \left\lceil\frac{1}{2}\right. = \sqrt{\pi}\right)$$

$$\boxed{\,I = \frac{\sqrt{\pi}}{2h}\,}$$

Ex. 7 : *Evaluate* $\displaystyle\int_0^1 x^{a-1}\left(\log\frac{1}{x}\right)^{n-1}\,dx$ *(a > 0).*

Sol. : Let $\displaystyle I = \int_0^1 x^{a-1}\left(\log\frac{1}{x}\right)^{n-1}\,dx$, Put $\log\dfrac{1}{x} = t$ or $x = e^{-t}$, $dx = -e^{-t}\,dt$

x	0	1
t	∞	0

$$= \int_\infty^0 \left(e^{-t}\right)^{a-1} t^{n-1}\,(-e^{-t})\,dt$$

$$= \int_0^\infty e^{-at}\, e^t\, t^{n-1}\, e^{-t}\, dt = \int_0^\infty e^{-at}\, t^{n-1}\, dt \qquad \left(\because \int_0^\infty e^{-ky}\, y^{n-1}\, dy = \frac{\overline{|n}}{k^n} \right)$$

$$\boxed{I = \frac{1}{a^n}\, \overline{|n}} \ .$$

Ex. 8 : *Evaluate* $\displaystyle\int_0^\infty \sqrt{y}\ e^{-\sqrt{y}}\ dy$ $\hfill$ *(Nov./Dec. 2019, May 2017)*

Sol. : $\qquad I = \int_0^\infty \sqrt{y}\ e^{-\sqrt{y}}\, dy$ Put $\sqrt{y} = t$, $\dfrac{1}{2\sqrt{y}}\, dy = dt$ or $dy = 2\sqrt{y}\, dt = 2t\, dt$

y	0	∞
t	0	∞

$$I = \int_0^\infty t\, e^{-t}\, 2t\, dt = 2\int_0^\infty t^2\, e^{-t}\, dt = 2\int_0^\infty t^{3-1}\, e^{-t}\, dt = 2\,\overline{|3}$$

$$\boxed{I = 4}$$

Ex. 9 : *Show that* $\displaystyle\int_0^\infty \sqrt{x}\ e^{-x^3}\, dx = \frac{\sqrt{\pi}}{3}$

Sol. : $\qquad I = \int_0^\infty \sqrt{x}\ e^{-x^3}\, dx$ Put $x^3 = y$ $\therefore$ $3x^2\, dx = dy$ or $dx = \dfrac{dy}{3x^2} = \dfrac{dy}{3y^{2/3}}$

x	0	∞
y	0	∞

$$\therefore \qquad I = \int_0^\infty y^{1/6}\, e^{-y} \cdot \frac{dy}{3y^{2/3}}$$

$$= \frac{1}{3}\int_0^\infty y^{-1/2}\, e^{-y}\, dy = \frac{1}{3}\int_0^\infty y^{1/2-1}\, e^{-y}\, dy = \frac{1}{3}\,\overline{|1/2}$$

$$\boxed{I = \frac{\sqrt{\pi}}{3}}$$

Ex. 10 : *Evaluate* $\displaystyle\int_0^\infty x^n\, e^{-x^m}\, dx$ $\hfill$ *(Dec. 2004)*

Sol. : Put $x^m = t$ or $x = t^{1/m}$, $m x^{m-1}\, dx = dt$, $dx = \dfrac{dt}{m x^{m-1}}$

x	0	∞
t	0	∞

$$I = \int_0^\infty x^n\, e^{-x^m}\, dx = \int_0^\infty t^{n/m}\, e^{-t}\, \frac{dt}{m\, t^{(m-1)/m}}$$

$$I = \frac{1}{m}\int_0^\infty t^{\frac{n}{m}-1+\frac{1}{m}}\, e^{-t}\, dt = \frac{1}{m}\int_0^\infty e^{-t}\, t^{\frac{n+1}{m}-1}\, dt$$

$$\boxed{I = \frac{1}{m}\, \overline{\left|\frac{n+1}{m}\right.}}$$

Note : For $n = 4$, $m = 4$,

$$I = \int_0^\infty x^4\, e^{-x^4} = \frac{1}{4}\,\overline{\left|\frac{5}{4}\right.} = \frac{1}{16}\,\overline{\left|\frac{1}{4}\right.}$$

Ex. 11 : *Evaluate* $\displaystyle\int_0^\infty x^9\, e^{-2x^2}\, dx.$ **(Dec. 2009, 2018; May 2006)**

Sol. :
$$I = \int_0^\infty x^9\, e^{-2x^2}\, dx$$

Put $2x^2 = t$ or $x = \left(\dfrac{t}{2}\right)^{1/2}, \quad 4x\, dx = dt$

x	0	∞
t	0	∞

$$I = \int_0^\infty x^8\, e^{-2x^2}\cdot x\, dx = \int_0^\infty \left(\dfrac{t}{2}\right)^4 e^{-t}\,\dfrac{dt}{4}$$

$$= \dfrac{1}{64}\int_0^\infty t^4\, e^{-t}\, dt = \dfrac{1}{64}\int_0^\infty e^{-t}\, t^{5-1}\, dt$$

$$= \dfrac{1}{64}\,\overline{|5} = \dfrac{4!}{64} = \dfrac{24}{64}$$

$$\boxed{I = \dfrac{3}{8}}$$

Ex. 12 : *Show that* $\displaystyle\int_0^1 \left(\log\dfrac{1}{y}\right)^{n-1} dy = \overline{|n}$

Sol. : Let $\displaystyle I = \int_0^1 \left(\log\dfrac{1}{y}\right)^{n-1} dy$ Put $\log\dfrac{1}{y} = t$ or $\dfrac{1}{y} = e^t$ or $y = e^{-t}$ $dy = -e^{-t}\, dt$

y	0	1
t	∞	0

$$= \int_\infty^0 t^{n-1}\,(-e^{-t})\, dt = \int_0^\infty e^{-t}\, t^{n-1}\, dt$$

$$\boxed{I = \overline{|n}}$$

Ex. 13 : *Evaluate* $\displaystyle\int_0^1 (x\log x)^4\, dx.$ **(May 2005; Dec. 2011, 2007)**

Sol. : Let $\displaystyle I = \int_0^1 (x\log x)^4\, dx$ Put $\log x = -t$ or $x = e^{-t}$, $dx = -e^{-t}\, dt$

x	0	1
t	∞	0

$$= \int_\infty^0 (e^{-t})^4\,(-t)^4\,(-e^{-t}\, dt) = \int_0^\infty e^{-5t}\, t^4\, dt = \dfrac{\overline{|5}}{5^5}$$

$$\boxed{I = \dfrac{4!}{5^5}}$$

$$\left(\text{Put } k = 5, n = 5 \text{ in} \int_0^\infty e^{-ky}\, y^{n-1}\, dy = \dfrac{\overline{|n}}{k^n}\right)$$

Ex. 14 : *Show that* $\displaystyle\int_0^\infty x^{m-1}\cos ax\, dx = \dfrac{\overline{|m}}{a^m}\cos\dfrac{m\pi}{2}$ **(May 2004)**

Sol. :
$$e^{-i\,ax} = \cos ax - i\sin ax$$

$\therefore \qquad \cos ax = \text{Real part of } e^{-i\,ax}$

$$I = \int_0^\infty x^{m-1} \cos ax \, dx$$

$$= \text{Real part of} \int_0^\infty x^{m-1} \cdot e^{-i\,ax} \, dx \quad \text{Put } i\,ax = t \text{ or } x = \frac{t}{ia}, \; dx = \frac{dt}{ia}$$

x	0	∞
t	0	∞

(Note this step carefully)

$$= \text{Real part of} \int_0^\infty \frac{t^{m-1}}{(i\,a)^{m-1}} e^{-t} \cdot \frac{dt}{ia} = \text{Real part of} \frac{1}{i^m \, a^m} \int_0^\infty e^{-t} \, t^{m-1} \, dt$$

$$I = \text{Real part of} \frac{1}{i^m \, a^m} \overline{\lceil m} = \text{Real part of} \frac{\overline{\lceil m}}{a^m} \cdot \left(\frac{1}{i^m}\right)$$

But

$$i = \cos\frac{\pi}{2} + i\sin\frac{\pi}{2} \quad \therefore \quad i^m = \left(\cos\frac{\pi}{2} + i\sin\frac{\pi}{2}\right)^m$$

$$\therefore \quad i^m = \cos\frac{m\pi}{2} + i\sin\frac{m\pi}{2} \qquad \qquad \text{... By DeMoivre's theorem}$$

$$\therefore \quad \frac{1}{i^m} = \frac{1}{\cos\dfrac{m\pi}{2} + i\sin\dfrac{m\pi}{2}} = \cos\frac{m\pi}{2} - i\sin\frac{m\pi}{2}$$

$$\therefore \quad I = \text{Real part of} \frac{\overline{\lceil m}}{a^m}\left(\cos\frac{m\pi}{2} - i\sin\frac{m\pi}{2}\right)$$

$$= \text{Real part of} \left[\frac{\overline{\lceil m}}{a^m}\cos\frac{m\pi}{2} - i\frac{\overline{\lceil m}}{a^m}\sin\frac{m\pi}{2}\right]$$

$$\boxed{I = \frac{\overline{\lceil m}}{a^m}\cos\frac{m\pi}{2}}$$

Ex. 15 : If $\;I_n = \dfrac{\dfrac{\sqrt{\pi}}{2}\,\overline{\left\lceil\dfrac{n+1}{2}\right.}}{\overline{\left\lceil\dfrac{n}{2}+1\right.}}$, show that $I_{n+2} = \dfrac{n+1}{n+2}\,I_n$ and hence find I_5.

Sol. : Given :

$$I_n = \frac{\dfrac{\sqrt{\pi}}{2}\,\overline{\left\lceil\dfrac{n+1}{2}\right.}}{\overline{\left\lceil\dfrac{n}{2}+1\right.}}, \quad \text{Replace } n \text{ by } n+2 \text{ then,}$$

$$I_{n+2} = \frac{\dfrac{\sqrt{\pi}}{2}\,\overline{\left\lceil\dfrac{n+3}{2}\right.}}{\overline{\left\lceil\dfrac{n+2}{2}+1\right.}} = \frac{\dfrac{\sqrt{\pi}}{2}\,\overline{\left\lceil\dfrac{n+1}{2}+1\right.}}{\dfrac{n+2}{2}\,\overline{\left\lceil\dfrac{n+2}{2}\right.}}$$

$$= \frac{\sqrt{\pi}}{n+2}\,\frac{\dfrac{n+1}{2}\cdot\overline{\left\lceil\dfrac{n+1}{2}\right.}}{\overline{\left\lceil\dfrac{n}{2}+1\right.}} = \frac{n+1}{n+2}\,\frac{\dfrac{\sqrt{\pi}}{2}\cdot\overline{\left\lceil\dfrac{n+1}{2}\right.}}{\overline{\left\lceil\dfrac{n}{2}+1\right.}}$$

$$\boxed{I_{n+2} = \frac{n+1}{n+2}\cdot I_n}$$

Now put $n = 3$, then,

$$I_5 = \frac{4}{5} \cdot I_3 = \frac{4}{5} \cdot \frac{2}{3} \cdot I_1 = \frac{8}{15} \; \frac{\dfrac{\sqrt{\pi}}{2} \cdot \overline{|1}}{\overline{\left|\dfrac{1}{2} + 1\right.}} = \frac{4}{15} \; \frac{\sqrt{\pi}}{\dfrac{1}{2} \sqrt{\pi}} = \frac{8}{15}$$

$$\therefore \qquad \boxed{I_5 = \frac{8}{15}}$$

Ex. 16 : *Show that* $\dfrac{2^n \; \overline{\left|\left(n + \dfrac{1}{2}\right)\right.}}{\sqrt{\pi}} = 1 \cdot 3 \cdot 5 \; \ldots\ldots\ldots\ldots \; (2n - 1)$

Sol. :

$$\overline{\left|\left(n + \frac{1}{2}\right)\right.} = \left(n - \frac{1}{2}\right) \overline{\left|\left(n - \frac{1}{2}\right)\right.} \qquad \left(\because \; \overline{|(n+1)} = n \overline{|n} \right)$$

$$= \left(n - \frac{1}{2}\right) \left(n - \frac{3}{2}\right) \overline{\left|\left(n - \frac{3}{2}\right)\right.}$$

$$= \left(n - \frac{1}{2}\right) \left(n - \frac{3}{2}\right) \left(n - \frac{5}{2}\right) \overline{\left|\left(n - \frac{5}{2}\right)\right.} \; \ldots\ldots\ldots \; \frac{3}{2} \cdot \frac{1}{2} \; \overline{\left|\frac{1}{2}\right.}$$

$$= \frac{2n - 1}{2} \cdot \frac{2n - 3}{2} \cdot \frac{2n - 5}{2} \; \ldots \; \frac{3}{2} \cdot \frac{1}{2} \cdot \sqrt{\pi} \qquad \left\{ \because \; \overline{\left|\frac{1}{2}\right.} = \sqrt{\pi} \right\}$$

$$= \frac{(2n - 1) \; (2n - 3) \; \ldots\ldots \; 3.1}{2^n} \; \sqrt{\pi}$$

$$\therefore \qquad \boxed{\dfrac{2^n \; \overline{\left|\left(n + \dfrac{1}{2}\right)\right.}}{\sqrt{\pi}} = 1 \cdot 3 \cdot 5 \ldots\ldots\ldots\ldots \; (2n - 1)}$$

EXERCISE 3.2

Prove that,

1. $\displaystyle\int_0^\infty \sqrt{x} \; e^{-\sqrt[3]{x}} \; dx = \frac{315}{16} \sqrt{\pi}$ **[Hint :** $(x = t^3)$**]**

2. $\displaystyle\int_0^\infty x^7 \; e^{-2x^2} \; dx = \frac{3}{16}$ **[Hint :** $(2x^2 = t)$**] (May 2011, 2018)**

3. $\displaystyle\int_0^\infty x^2 \; e^{-h^2 x^2} \; dx = \frac{\sqrt{\pi}}{4h^3}$ **[Hint :** $h^2 x^2 = t$**] (Dec. 2010)**

4. $\displaystyle\int_0^\infty \sqrt{y} \; e^{-y^3} \; dy = \frac{\sqrt{\pi}}{3}$ **[Hint :** $y^3 = t$**]**

5. $\displaystyle\int_0^\infty x^{n-1} \; e^{-h^2 x^2} \; dx = \frac{\overline{|n/2}}{2h^n}$ **[Hint :** $h^2 x^2 = t$**]**

6. $\displaystyle\int_0^\infty e^{-x^4} \; dx = \frac{1}{4} \overline{\left|\frac{1}{4}\right.}$ **[Hint :** $x^4 = t$**]**

7. $\displaystyle\int_0^\infty \frac{x^4}{4^x} \; dx = \frac{24}{(\log 4)^5}$ **[Hint :** $(\because 4 = e^m)$**] (Dec. 2016)**

8. $\displaystyle\int_0^\infty \frac{x^5}{5^x} \; dx = \frac{120}{(\log 5)^6}$ **[Hint :** $(\because 5 = e^m)$**]**

9. $\displaystyle\int_0^\infty x^n \; e^{-\sqrt{ax}} \; dx = \frac{2 \, (2n + 1)!}{a^{n+1}}$ **[Hint :** $\sqrt{ax} = t$, n is integer**]**

10. $\displaystyle\int_0^\infty a^{-4x^2} \; dx = \frac{\sqrt{\pi}}{4 \sqrt{\log a}}$ **[Hint :** $(a = e^m, \; 4x^2 = t)$**]**

(May 2009)

11. $\displaystyle\int_0^\infty x^n\, e^{-x^m}\, dx = \frac{1}{m}\left|\overline{\left(\frac{n+1}{m}\right)}\right.$ **[Hint :** $x^m = t$]

12. $\displaystyle\int_0^\infty \sqrt[3]{x^2}\; e^{-\sqrt[3]{x}}\, dx = 72$ **[Hint :** $x = t^3$]

13. $\displaystyle\int_{-\infty}^\infty e^{-h^2 x^2}\, dx = \frac{\sqrt{\pi}}{h}$ **[Hint :** $I = 2\displaystyle\int_0^\infty e^{-h^2 x^2}\, dx,\ (h^2 x^2 = t)$]

14. $\displaystyle\int_0^1 (\log x)^n\, dx = (-1)^n\, \overline{\lceil n}$ **[Hint :** $\log x = -t$]

15. $\displaystyle\int_0^1 \frac{dx}{\sqrt{-\log x}} = \sqrt{\pi}$ **[Hint :** $(-\log x) = t$]

16. $\displaystyle\int_0^1 (x \log x)^3\, dx = -\frac{3}{128}$ **[Hint :** $\log x = -t$]

17. $\displaystyle\int_0^1 x^m (\log x)^n\, dx = (-1)^n\, \frac{\overline{\lceil(n+1)}}{(m+1)^{n+1}}$ **[Hint :** $(\log x = -t)$ **(Dec. 08, 05)**

18. $\displaystyle\int_0^1 \frac{dx}{\sqrt{x \log \frac{1}{x}}} = \sqrt{2\pi}$ **[Hint :** $\left(\log \frac{1}{x} = t\right)$]

19. $\displaystyle\int_0^1 \frac{x\, dx}{\sqrt{\log\left(\frac{1}{x}\right)}} = \sqrt{\frac{\pi}{2}}$ **(Dec. 2006)** **[Hint :** $\left(\log \frac{1}{x} = t\right)$]

20. $\displaystyle\int_0^\infty x^{n-1} e^{-ax} \cos bx\, dx = \frac{\overline{\lceil n}}{(a^2 + b^2)^{n/2}} \cos\left(n \tan^{-1}\frac{b}{a}\right)$

[Hint : $e^{ibx} = \cos bx + i \sin bx;\ (a - ib)\, x = t$

21. $\displaystyle\int_0^\infty x^{n-1} e^{-ax} \sin bx\, dx = \frac{\overline{\lceil n}}{(a^2 + b^2)^{n/2}} \sin\left(n \tan^{-1}\frac{b}{a}\right)$

[Hint : $e^{ibx} = \cos bx + i \sin bx$ and $(a - ib)\, x = t$]

22. $\displaystyle\int_0^\infty x\, e^{-ax} \sin bx\, dx = \frac{2ab}{(a^2 + b^2)^2}$

[Hint : $e^{ibx} = \cos bx + i \sin bx$ and $(a - ib)\, x = t^m$

$\therefore I = \text{Img} \dfrac{1}{(a - ib)^2} \displaystyle\int_0^\infty t\, e^{-t}\, dt = \dfrac{1}{(a - ib)^2}$]

23. $\displaystyle\int_0^\infty x^{n-1} \sin bx\ dx = \frac{\overline{\lceil n}}{b^n}\, \sin \frac{n\pi}{2}$

[Hint : $e^{-ibx} = \cos bx - i \sin bx\ ;\ ibx = t$]

3.9 BETA FUNCTION

Definition : Consider the definite integral $\displaystyle\int_0^1 x^{m-1} (1 - x)^{n-1}\, dx,\ m > 0,\ n > 0.$

It is denoted by the symbol B(m, n) (we read it as Beta (m, n)) and is called *Beta Function*.

$$\boxed{\ \mathbf{B\,(m,\ n)\ =\ \int_0^1 x^{m-1}\,(1-x)^{n-1}\,dx,\ \ m > 0\ \ n > 0}\ }$$ ...(1)

The Beta function is also called as *Euler's integral of the first kind*.

For example, (1) $B\left(3, \frac{3}{2}\right) = \displaystyle\int_0^1 x^2 (1 - x)^{1/2}\, dx$

(2) $\displaystyle\int_0^1 t^4 (1 - t)^{3/2}\, dt = B\left(5, \frac{5}{2}\right)$

3.10 PROPERTIES OF BETA FUNCTIONS

1. $\qquad B(m, n) = B(n, m)$

Proof : $\qquad B(m, n) = \displaystyle\int_0^1 x^{m-1}(1-x)^{n-1}\,dx = \int_0^1 (1-x)^{m-1}(1-(1-x))^{n-1}\,dx$

$\because \qquad \displaystyle\int_0^a f(x)\,dx = \int_0^a f(a-x)\,dx$

$\therefore \qquad B(m, n) = \displaystyle\int_0^1 (1-x)^{m-1}\cdot x^{n-1}\,dx = \int_0^1 x^{n-1}(1-x)^{m-1}\,dx = B(n, m)$

$\therefore \qquad \boxed{B(m, n) = B(n, m)}$

2. $\qquad \displaystyle\int_0^1 x^m (1-x)^n\,dx = B(m+1, n+1)$

3. $\qquad B(m, n) = 2\displaystyle\int_0^{\pi/2} \sin^{2m-1}\theta\,\cos^{2n-1}\theta\,d\theta$

Proof : $\qquad B(m, n) = \displaystyle\int_0^1 x^{m-1}(1-x)^{n-1}\,dx \qquad$ Put $x = \sin^2\theta, \quad dx = 2\sin\theta\cos\theta\,d\theta$

x	0	1
θ	0	π/2

$\qquad\qquad = \displaystyle\int_0^{\pi/2} \sin^{2m-2}\theta\,(1-\sin^2\theta)^{n-1}\,2\sin\theta\cos\theta\,d\theta$

$$\boxed{B(m, n) = 2\int_0^{\pi/2} \sin^{2m-1}\theta\cdot\cos^{2n-1}\theta\,d\theta}$$

We consider this as a definition of Beta function.

Further, let $2m - 1 = p,\ 2n - 1 = q \quad \therefore\ m = \dfrac{p+1}{2},\ n = \dfrac{q+1}{2}$ then

$$B\left(\frac{p+1}{2}, \frac{q+1}{2}\right) = 2\int_0^{\pi/2} \sin^p\theta\cos^q\theta\,d\theta$$

Standard Formula :

$$\boxed{\int_0^{\pi/2} \sin^p\theta\cos^q\theta\,d\theta = \frac{1}{2}B\left(\frac{p+1}{2}, \frac{q+1}{2}\right)}$$

4. Alternating definition :

$$B(m, n) = \int_0^\infty \frac{x^{m-1}}{(1+x)^{m+n}}\,dx$$

Proof : $\qquad B(m, n) = \displaystyle\int_0^1 x^{m-1}(1-x)^{n-1}\,dx$

Put $\boxed{x = \dfrac{t}{1+t}}$ i.e. $x(1+t) = t$ i.e. $x + xt = t$ $\therefore x = t - xt$ or $\boxed{t = \dfrac{x}{1-x}}$. Note this substitution

When $x = 0$, $t = \dfrac{0}{1-0} = 0$ and when $x = 1$, $t = \dfrac{1}{1-1} = \dfrac{1}{0} = \infty$ $\therefore$

x	0	1
t	0	∞

Also,
$$dx = \frac{(1+t)(1) - t(1)}{(1+t)^2}\, dt = \frac{1}{(1+t)^2}\, dt$$

$$B(m, n) = \int_0^\infty \frac{t^{m-1}}{(1+t)^{m-1}} \left(1 - \frac{t}{1+t}\right)^{n-1} \cdot \frac{dt}{(1+t)^2}$$

$$= \int_0^\infty \frac{t^{m-1}\, dt}{(t+1)^{m-1}(1+t)^{n-1}(1+t)^2}$$

$$= \int_0^\infty \frac{t^{m-1}\, dt}{(1+t)^{m+n}}$$

$$\boxed{\, B(m, n) = \int_0^\infty \frac{x^{m-1}}{(1+x)^{m+n}}\, dx \,}$$

We consider this result also as another definition of *Beta Function*.

5. Relation between Beta and Gamma Functions :

We have $\qquad \boxed{\, B(m, n) = \dfrac{\overline{|m}\ \overline{|n}}{\overline{|m+n}} \,} \qquad$ (For the proof, refer page 9.36)

6. $\boxed{\,\overline{\left|\dfrac{1}{2}\right.} = \sqrt{\pi}\,}$

We know, $\displaystyle\int_0^{\pi/2} \sin^p\theta \cos^q\theta\, d\theta = \frac{1}{2} B\left(\frac{p+1}{2}, \frac{q+1}{2}\right) = \frac{1}{2} \cdot \frac{\overline{\left|\dfrac{p+1}{2}\right.}\ \overline{\left|\dfrac{q+1}{2}\right.}}{\overline{\left|\dfrac{p+q+2}{2}\right.}}$

Put $p = q = 0$

$$\int_0^{\pi/2} d\theta = \frac{1}{2} \cdot \frac{\overline{\left|1/2\right.}\ \overline{\left|1/2\right.}}{\overline{|1}} \Rightarrow \frac{\pi}{2} = \left(\overline{\left|\dfrac{1}{2}\right.}\right)^2$$

$\therefore \qquad \boxed{\,\overline{\left|\dfrac{1}{2}\right.} = \sqrt{\pi}\,}$

3.11 DUPLICATION FORMULA OF GAMMA FUNCTIONS

$$\boxed{\Gamma m \; \Gamma\left(m + \frac{1}{2}\right) = \frac{\sqrt{\pi}}{2^{2m-1}} \; \Gamma 2m}$$

(Dec. 2005)

Proof : Consider

$$\frac{1}{2} \cdot \frac{\Gamma\left(\dfrac{p+1}{2}\right) \Gamma\left(\dfrac{q+1}{2}\right)}{\Gamma\left(\dfrac{p+q+2}{2}\right)} = \int_0^{\pi/2} \sin^p\theta \, \cos^q\theta \cdot d\theta$$

Put $p = 2m - 1$, $q = 2m - 1$ i.e. $\dfrac{p+1}{2} = m$, $\dfrac{q+1}{2} = m$

$$\frac{1}{2} \frac{\Gamma m \cdot \Gamma m}{\Gamma 2m} = \int_0^{\pi/2} \sin^{2m-1}\theta \, \cos^{2m-1}\theta \cdot d\theta$$

$$\frac{\Gamma m \, \Gamma m}{\Gamma 2m} = \frac{2}{2^{2m-1}} \int_0^{\pi/2} (2\sin\theta\cos\theta)^{2m-1} \, d\theta \qquad \text{(Note the adjustment)}$$

$$= \frac{2}{2^{2m-1}} \int_0^{\pi/2} (\sin 2\theta)^{2m-1} \, d\theta \quad \text{Put } 2\theta = t \,; \; d\theta = \frac{1}{2} dt$$

θ	0	$\pi/2$
t	0	π

$$= \frac{1}{2^{2m-1}} \int_0^{\pi} (\sin t)^{2m-1} \, dt = \frac{1}{2^{2m-1}} \, 2 \int_0^{\pi/2} \sin^{2m-1} t \cdot dt \quad [\because f(\pi - t) = f(t)]$$

$$= \frac{2}{2^{2m-1}} \int_0^{\pi/2} \sin^{2m-1} t \cdot \cos^0 t \cdot dt \qquad \text{(Note this step)}$$

$$\frac{\Gamma m \, \Gamma m}{\Gamma 2m} = \frac{2}{2^{2m-1}} \frac{1}{2} \frac{\Gamma\left(\dfrac{2m-1+1}{2}\right) \Gamma\left(\dfrac{0+1}{2}\right)}{\Gamma\left(\dfrac{2m-1+0+2}{2}\right)} = \frac{1}{2^{2m-1}} \frac{\Gamma m \; \sqrt{\pi}}{\Gamma\left(m + \dfrac{1}{2}\right)}$$

$$\therefore \quad \boxed{\Gamma m \; \Gamma\left(m + \frac{1}{2}\right) = \frac{\sqrt{\pi}}{2^{2m-1}} \; \Gamma 2m}$$

Additional Results :

Show that

$$\Gamma p \; \Gamma(1-p) = \frac{\pi}{\sin p\pi}, \; \text{given that} \int_0^\infty \frac{x^{p-1}}{1+x} \, dx = \frac{\pi}{\sin p\pi} \text{ for } 0 < p < 1$$

Proof : Consider

$$I = \int_0^\infty \frac{x^{p-1}}{1+x} \, dx \quad \text{Put } x = \tan^2\theta, \; dx = 2\tan\theta \sec^2\theta \, d\theta$$

x	0	∞
θ	0	$\pi/2$

$$= \int_0^{\pi/2} \frac{\tan^{2p-2}\theta \cdot 2\tan\theta \sec^2\theta \, d\theta}{1 + \tan^2\theta}$$

$$= 2 \int_0^{\pi/2} \tan^{2p-1}\theta \, d\theta = 2 \int_0^{\pi/2} \sin^{2p-1}\theta \cdot \cos^{1-2p}\theta \, d\theta$$

$$I = 2 \cdot \frac{1}{2} \, B\left(\frac{2p - 1 + 1}{2}, \frac{1 - 2p + 1}{2}\right)$$

$$= B(p, 1 - p) = \frac{\Gamma p \cdot \Gamma(1 - p)}{\Gamma(p + 1 - p)}$$

$$\boxed{\frac{\pi}{\sin p\pi} = \Gamma p \; \Gamma(1 - p)} \qquad \left(\text{given } I = \int_0^\infty \frac{x^{p-1}}{1 + x}\, dx = \frac{\pi}{\sin p\pi}\right)$$

(Note : This formula is to be used only when $0 < p < 1$.)

3.12 ILLUSTRATIONS ON BETA FUNCTION

Ex. 1 : *Prove that* $\Gamma(1/4)\; \Gamma(3/4) = \pi\sqrt{2}$

Sol. : By using duplication formulae, $\Gamma m \; \Gamma\left(m + \dfrac{1}{2}\right) = \dfrac{\sqrt{\pi}\; \Gamma(2m)}{2^{2m-1}}$

$$\Gamma(1/4)\; \Gamma(3/4) = \Gamma(1/4)\; \Gamma\left(\frac{1}{4} + \frac{1}{2}\right) = \frac{\sqrt{\pi}\; \Gamma\left(2\left(\frac{1}{4}\right)\right)}{2^{2\left(\frac{1}{4}\right) - 1}} = \frac{\sqrt{\pi}\; \Gamma\left(\frac{1}{2}\right)}{2^{-1/2}} \qquad \left(\text{Here } m = \frac{1}{4}\right)$$

$$= \sqrt{2}\; \sqrt{\pi}\; \sqrt{\pi} = \pi\sqrt{2}$$

$$\boxed{\Gamma\frac{1}{4}\; \Gamma\frac{3}{4} = n\sqrt{2}}$$

By using result $\qquad \Gamma p \; \Gamma(1 - p) = \dfrac{\pi}{\sin p\pi} \quad 0 < p < 1 \,;\; p = \dfrac{1}{4}$

Aliter : $\qquad \Gamma(1/4)\; \Gamma(3/4) = \Gamma\dfrac{1}{4}\; \Gamma\left(1 - \dfrac{1}{4}\right) = \dfrac{\pi}{\sin\dfrac{\pi}{4}} = \dfrac{\pi}{\dfrac{1}{\sqrt{2}}} = \pi\sqrt{2}$

Ex. 2 : Evaluate $\displaystyle\int_0^m x^m (m - x)^n \, dx$

Sol. : $\qquad I = \displaystyle\int_0^m x^m (m - x)^n \, dx \quad$ Put $\; x = my, \quad dx = m\, dy,$

x	0	m
y	0	1

$$I = \int_0^1 (my)^m (m - my)^n \, m\, dy = m^{m + n + 1} \int_0^1 y^m (1 - y)^n \, dy$$

$$= m^{m + n + 1}\, B(m + 1, n + 1)$$

$$\therefore \qquad \boxed{I = m^{m + n + 1} \, \frac{\Gamma(m + 1)\; \Gamma(n + 1)}{\Gamma(m + n + 2)}}$$

Ex. 3 : *Evaluate* $\displaystyle\int_a^b (x-a)^m (b-x)^n\, dx$ *(May 2014)*

Sol. :

$$I = \int_a^b (x-a)^m (b-x)^n\, dx \quad \text{Put } x-a = (b-a)\, t. \text{ Then } dx = (b-a)\, dt$$

x	a	b
t	0	1

$$= \int_0^1 (b-a)^m\, t^m\, [b-a-(b-a)\,t]^n\, (b-a)\, dt$$

$$= (b-a)^{m+1} \int_0^1 t^m\, [(b-a)(1-t)]^n\, dt$$

$$= (b-a)^{m+n+1} \int_0^1 t^m (1-t)^n\, dt$$

$$\boxed{I = (b-a)^{m+n+1}\, B\,(m+1,\, n+1)}$$

Ex. 4 : *Evaluate* $\displaystyle\int_3^7 (x-3)^{1/4} (7-x)^{1/4}\, dx$ *(Dec. 2009)*

Sol. :

$$I = \int_3^7 (x-3)^{1/4} (7-x)^{1/4}\, dx \quad \text{Put } x = 4t+3,\ dx = 4dt \text{ (Note this substitution)}$$

x	3	7
t	0	1

$$= \int_0^1 (4t)^{1/4} (7-4t-3)^{1/4}\, 4dt$$

$$= \int_0^1 4^{1/4}\, t^{1/4}\, [4(1-t)]^{1/4}\, 4\, dt = 4^{3/2} \int_0^1 t^{\frac{1}{4}} (1-t)^{1/4}\, dt = 8B\left(\frac{5}{4},\frac{5}{4}\right)$$

$$= 8\ \frac{\left\lfloor\frac{5}{4}\right. \left\lfloor\frac{5}{4}\right.}{\left\lfloor\frac{5}{2}\right.} = 8\ \frac{\frac{1}{4}\left\lfloor\frac{1}{4}\right. \frac{1}{4}\left\lfloor\frac{1}{4}\right.}{\frac{3}{2}\frac{1}{2}\sqrt{\pi}}$$

$$\boxed{I = \frac{2}{3\sqrt{\pi}}\left(\left\lfloor\frac{1}{4}\right.\right)^2}$$

Ex. 5 : *Show that* $\displaystyle\int_0^{\pi/2} \sqrt{\tan\theta}\ d\theta \int_0^{\pi/2} \sqrt{\cot\theta}\ d\theta = \dfrac{\pi^2}{2}$

Sol. :

$$\int_0^{\pi/2} \sqrt{\tan\theta}\ d\theta = \int_0^{\pi/2} \sin^{1/2}\theta\ \cos^{-1/2}\theta\ d\theta$$

$$= \frac{\dfrac{1}{2}\ \left|\!\!\dfrac{\frac{1}{2}+1}{2}\ \cdot\ \left|\!\!\dfrac{\frac{-1}{2}+1}{2}\right.\right.}{\left|\!\!\dfrac{\frac{1}{2}-\frac{1}{2}+2}{2}\right.} = \frac{\dfrac{1}{2}\cdot\left|\dfrac{3}{4}\cdot\left|\dfrac{1}{4}\right.\right.}{\left|1\right.}$$

$$= \frac{1}{2}\left|\frac{1}{4}\ \left|1-\frac{1}{4}\right.\right. = \frac{1}{2}\left(\frac{\pi}{\sin \pi/4}\right) \qquad \because\ \left|p\ \left|1-p\right.\right. = \frac{\pi}{\sin p\pi},\ 0 < p < 1$$

$$= \frac{\pi}{\sqrt{2}}$$

$$\int_0^{\pi/2} \sqrt{\cot\theta}\ d\theta = \int_0^{\pi/2} \sqrt{\cot\left(\frac{\pi}{2}-\theta\right)}\ d\theta \qquad \because\ \int_0^a f(x)\ dx = \int_0^a f(a-x)\ dx$$

$$= \int_0^{\pi/2} \sqrt{\tan\theta}\ d\theta \qquad \because\ \cot\left(\frac{\pi}{2}-\theta\right) = \tan\theta$$

$$= \frac{\pi}{\sqrt{2}} \text{ (by above result)}$$

$$\therefore \quad \boxed{\int_0^{\pi/2} \sqrt{\tan\theta}\ d\theta \int_0^{\pi/2} \sqrt{\cot\theta}\ d\theta = \frac{\pi}{\sqrt{2}}\cdot\frac{\pi}{\sqrt{2}} = \frac{\pi^2}{2}}$$

Ex. 6 : *Prove that* $\displaystyle\int_0^{\infty} \dfrac{dx}{1+x^4} = \dfrac{\pi}{2\sqrt{2}}$ **(Dec. 2017)**

Sol. : Let

$$I = \int_0^{\infty} \frac{dx}{1+x^4} \quad \text{Put } x^2 = \tan\theta \text{ or } x = \sqrt{\tan\theta},\ dx = \frac{1}{2}\tan^{-1/2}\theta\ \sec^2\theta\ d\theta$$

x	0	∞
θ	0	π/2

$$= \frac{1}{2}\int_0^{\pi/2} \tan^{-1/2}\theta\ d\theta = \frac{1}{2}\int_0^{\pi/2} \sin^{-1/2}\theta\ \cos^{1/2}\theta\ d\theta$$

$$= \frac{1}{2}\cdot\frac{1}{2}\ \frac{\left|\!\!\dfrac{\frac{-1}{2}+1}{2}\right.\ \left|\!\!\dfrac{\frac{1}{2}+1}{2}\right.}{\left|\!\!\dfrac{\frac{-1}{2}+\frac{1}{2}+2}{2}\right.} = \frac{1}{4}\ \frac{\left|\dfrac{1}{4}\ \left|\dfrac{3}{4}\right.\right.}{\left|1\right.} = \frac{1}{4}\left|\frac{1}{4}\ \left|1-\frac{1}{4}\right.\right. = \frac{1}{4}\ \frac{\pi}{\sin \pi/4}$$

$$\boxed{I = \frac{\pi}{2\sqrt{2}}}$$

Ex. 7 : *Show that* $\displaystyle\int_{0}^{\pi/2} \frac{\sin^{2m-1} x\, \cos^{2n-1} x}{(a \sin^2 x + b \cos^2 x)^{m+n}}\, dx = \frac{1}{2 a^m\, b^n}\, B\,(m, n)$

Sol. : Let

$$I = \int_{0}^{\pi/2} \frac{\sin^{2m-1} x\, \cos^{2n-1} x}{(a \sin^2 x + b \cos^2 x)^{m+n}}\, dx$$

$$= \int_{0}^{\pi/2} \frac{\sin^{2m-1} x\, \cos^{2n-1} x}{\cos^{2m+2n} x\, (a \tan^2 x + b)^{m+n}}\, dx$$

$$= \int_{0}^{\pi/2} \frac{\sin^{2m-1} x}{\cos^{2m-1} x}\, \frac{\cos^{2n-1} x}{\cos^{2n-1} x} \cdot \frac{1}{\cos^2 x} \cdot \frac{dx}{(a \tan^2 x + b)^{m+n}} \qquad \text{(Note this adjustment)}$$

$$= \int_{0}^{\pi/2} \frac{\tan^{2m-1} x \cdot \sec^2 x}{(a \tan^2 x + b)^{m+n}}\, dx$$

$$= \int_{0}^{\pi/2} \frac{\tan^{2m-2} x \cdot \tan x \sec^2 x}{(a \tan^2 x + b)^{m+n}}\, dx \quad \text{Put} \quad \boxed{a \tan^2 x = bt} \quad 2\, a \tan x \sec^2 x\, dx = b\, dt$$

x	0	$\pi/2$
t	0	∞

$$= \int_{0}^{\infty} \frac{\left(\dfrac{b\,t}{a}\right)^{m-1} \cdot \dfrac{b\,dt}{2a}}{b^{m+n}\,(1+t)^{m+n}} = \frac{1}{2 \cdot b^n\, a^m} \int_{0}^{\infty} \frac{t^{m-1}}{(1+t)^{m+n}}\, dt$$

$$\boxed{I = \frac{1}{2 b^n\, a^m}\, B\,(m, n)}$$

Ex. 8 : *Prove that* $\displaystyle\int_{0}^{\infty} \frac{dx}{(e^x + e^{-x})^n} = \frac{1}{4}\, B\left(\frac{n}{2}, \frac{n}{2}\right)$ *and hence evaluate* $\displaystyle\int_{0}^{\infty} \operatorname{sech}^8 x\, dx.$

Sol. : Let

$$I = \int_{0}^{\infty} \frac{dx}{(e^x + e^{-x})^n} = \frac{1}{2} \int_{-\infty}^{\infty} \frac{dx}{(e^x + e^{-x})^n} \qquad \text{(Note this step)}$$

Put $e^x = \tan \theta$, $e^x\, dx = \sec^2 \theta\, d\theta \Rightarrow dx = \dfrac{\sec^2 \theta}{\tan \theta}\, d\theta$

x	$-\infty$	∞
θ	0	$\pi/2$

$$\therefore \qquad I = \frac{1}{2} \int_{0}^{\pi/2} \frac{\sec^2 \theta\, / \tan \theta}{(\tan \theta + \cot \theta)^n}\, d\theta$$

$$= \frac{1}{2} \int_{0}^{\pi/2} \frac{\dfrac{1}{\sin \theta \cos \theta}}{\left(\dfrac{\sin \theta}{\cos \theta} + \dfrac{\cos \theta}{\sin \theta}\right)^n}\, d\theta = \frac{1}{2} \int_{0}^{\pi/2} \frac{\sin^n \theta \cos^n \theta}{\sin \theta \cos \theta}\, d\theta$$

$$= \frac{1}{2} \int_{0}^{\pi/2} \sin^{n-1} \theta \cos^{n-1} \theta\, d\theta = \frac{1}{2} \cdot \frac{1}{2}\, B\left(\frac{n-1+1}{2}, \frac{n-1+1}{2}\right)$$

$$\therefore \qquad \boxed{\int_{0}^{\infty} \frac{dx}{(e^x + e^{-x})^n} = \frac{1}{4}\, B\left(\frac{n}{2}, \frac{n}{2}\right)}$$

But $\cosh x = \dfrac{e^x + e^{-x}}{2}$ i.e. $e^x + e^{-x} = 2 \cosh x$

Hence
$$\int_0^\infty \frac{dx}{(2\cosh x)^n} = \frac{1}{4}\, B\left(\frac{n}{2}, \frac{n}{2}\right) \qquad \text{Put } n = 8.$$

$$\int_0^\infty \frac{dx}{(2\cosh x)^8} = \frac{1}{4}\, B(4,4) = \frac{1}{4}\, \frac{\overline{|4}\,\overline{|4}}{\overline{|8}} = \frac{1}{4}\, \frac{3!\,3!}{7!} = \frac{1}{560}$$

$$\therefore \qquad \boxed{\int_0^\infty \operatorname{sech}^8 x\, dx = \frac{2^8}{560} = \frac{16}{35}}$$

Ex. 9 : *Evaluate* $\displaystyle\int_0^{\pi/2} \frac{d\theta}{\sqrt{1 - \frac{1}{2}\sin^2\theta}}$ *(May 2007)*

Sol. :
$$I = \int_0^{\pi/2} \frac{d\theta}{\sqrt{1 - \frac{1}{2}\sin^2\theta}}$$

Put $\cos^2\theta = \sqrt{t}$ or $\cos\theta = t^{1/4}$, $\theta = \cos^{-1}(t^{1/4})$, $d\theta = \dfrac{-1}{\sqrt{1-t^{1/2}}} \cdot \dfrac{1}{4}\, t^{-3/4}\, dt$ Also $\sin^2\theta = 1 - \cos^2\theta = 1 - \sqrt{t}$.

When $\theta = 0$, $t = 1$; $\theta = \dfrac{\pi}{2}$, $t = 0$.

θ	0	$\pi/2$
t	1	0

$$I = \int_1^0 \frac{-1}{\sqrt{1-\sqrt{t}}} \cdot \frac{1}{4}\, \frac{t^{-3/4}\, dt}{\sqrt{1 - \frac{1}{2}(1-\sqrt{t})}} = \frac{1}{4}\int_0^1 \frac{t^{-3/4}\, dt}{\sqrt{1-\sqrt{t}}\,\sqrt{1 - \frac{1}{2} + \frac{1}{2}\sqrt{t}}}$$

$$= \frac{1}{4}\int_0^1 \frac{t^{-3/4}\, dt}{\sqrt{1-\sqrt{t}}\,\sqrt{\frac{1}{2}(1+\sqrt{t})}} = \frac{\sqrt{2}}{4}\int_0^1 \frac{t^{-3/4}\, dt}{\sqrt{1-t}}$$

$$= \frac{\sqrt{2}}{4}\int_0^1 t^{-3/4}(1-t)^{-1/2}\, dt$$

$$= \frac{\sqrt{2}}{4} \cdot B\left(\frac{1}{4}, \frac{1}{2}\right) = \frac{\sqrt{2}}{4}\, \frac{\overline{\left|\frac{1}{4}\right.}\,\overline{\left|\frac{1}{2}\right.}}{\overline{\left|\frac{3}{4}\right.}}$$

$$= \frac{\sqrt{2}\cdot\sqrt{\pi}}{4}\, \frac{(\overline{|1/4}\,)^2}{\overline{\left|\frac{1}{4}\right.}\,\overline{\left|1-\frac{1}{4}\right.}} = \frac{\sqrt{2\pi}\,(\overline{|1/4}\,)^2}{4\cdot\dfrac{\pi}{\sin \pi/4}} = \frac{2\sqrt{\pi}\,(\overline{|1/4}\,)^2}{4\pi}$$

$$\boxed{I = \frac{(\overline{|1/4}\,)^2}{2\sqrt{\pi}}}$$

Ex. 10 : *Using* $B(m, n) = \int_0^1 x^{m-1}(1-x)^{n-1}\,dx$, *show that* $B(m, n) = \int_0^\infty \dfrac{y^{m-1}}{(1+y)^{m+n}}\,dy$. *Also, evaluate* $\int_1^\infty \dfrac{dx}{x^{p+1}(x-1)^q}$.

Sol. : For the first part, refer property 4 of Beta function.

For the second part,

x	1	∞
t	1	0

Let
$$I = \int_1^\infty \frac{dx}{x^{p+1}(x-1)^q} \qquad \text{Put } x = \frac{1}{t}\ ,\ dx = -\frac{1}{t^2}\,dt$$

$$= \int_0^1 \frac{-1/t^2\,dt}{\dfrac{1}{t^{p+1}}\left(\dfrac{1}{t}-1\right)^q} = \int_0^1 \frac{1}{t^2}\cdot\frac{t^{p+1}\,t^q}{(1-t)^q}\,dt = \int_0^1 t^{p+q-1}(1-t)^{-q}\,dt$$

$$\boxed{I = B(p+q,\ 1-q)}$$

Ex. 11 : *Prove that* $\displaystyle\int_0^\infty \frac{x^{m-1}}{(a+bx)^{m+n}}\,dx = \frac{1}{a^n\,b^m}\,B(m, n)$ *(Dec. 2007)*

Sol. : Let
$$I = \int_0^\infty \frac{x^{m-1}}{(a+bx)^{m+n}}\,dx \qquad \text{Put } bx = at\ ;\ dx = \frac{a}{b}\,dt$$

x	0	∞
t	0	∞

$$= \int_0^\infty \frac{a^{m-1}\,t^{m-1}}{b^{m-1}}\cdot\frac{a}{b}\cdot dt\ \frac{1}{(a+at)^{m+n}}$$

$$= \int_0^\infty \frac{a^m\,t^{m-1}\,dt}{b^m\,a^{m+n}\,(1+t)^{m+n}} = \frac{1}{a^n b^m}\int_0^\infty \frac{t^{m-1}\,dt}{(1+t)^{m+n}}$$

$$\boxed{I = \frac{1}{a^n b^m}\,B(m, n)}$$

Ex. 12 : *Show that* $\displaystyle\int_0^1 \frac{y^{m-1}+y^{n-1}}{(1+y)^{m+n}}\,dy = B(m, n)$ *(May 2004)*

Sol. : We have,
$$B(m, n) = \int_0^\infty \frac{x^{m-1}}{(1+x)^{m+n}}\,dx$$

$$B(m, n) = \int_0^1 \frac{x^{m-1}}{(1+x)^{m+n}}\,dx + \int_1^\infty \frac{x^{m-1}}{(1+x)^{m+n}}\,dx = I_1 + I_2$$

Consider
$$I_2 = \int_1^\infty \frac{x^{m-1}}{(1+x)^{m+n}}\,dx \qquad \text{Put } x = \frac{1}{t} \text{ or } t = \frac{1}{x}\ \therefore\ dx = -\frac{dt}{t^2}$$

x	1	∞
t	1	0

$$= \int_1^0 \left(\frac{1}{t^{m-1}}\right)\frac{1}{\left(1+\dfrac{1}{t}\right)^{m+n}}\left(-\frac{dt}{t^2}\right)$$

$$= \int_0^1 \frac{t^{m+n}\,dt}{t^{m+1}\,(1+t)^{m+n}} = \int_0^1 \frac{t^{n-1}\,dt}{(1+t)^{m+n}}$$

$$I_2 = \int_0^1 \frac{x^{n-1}\,dx}{(1+x)^{m+n}}$$

$$\therefore \qquad B(m,n) = \int_0^1 \frac{x^{m-1}}{(1+x)^{m+n}}\,dx + \int_0^1 \frac{x^{n-1}}{(1+x)^{m+n}}\,dx$$

$$\boxed{\,B(m,n) = \int_0^1 \frac{x^{m-1} + x^{n-1}}{(1+x)^{m+n}}\,dx\,}$$

Ex. 12 : *Show that* $\displaystyle \int_0^1 \frac{x^{m-1}\,(1-x)^{n-1}}{(a+x)^{m+n}}\,dx = \frac{B(m,n)}{a^n\,(1+a)^m}$

Sol. : Let

$$I = \int_0^1 \frac{x^{m-1}\,(1-x)^{n-1}}{(a+x)^{m+n}}\,dx$$

Put

$$\frac{x}{a+x} = \frac{t}{a+1} \text{ (Note this substitution)}$$

$$\therefore \qquad x(a+1) = t(a+x) \quad \text{or} \quad x(a+1-t) = at$$

or

$$x = \frac{at}{a+1-t}$$

$$\therefore \qquad dx = \frac{(a+1-t)(a) - at(-1)}{(a+1-t)^2}\,dt = \frac{a(a+1)\,dt}{(a+1-t)^2}$$

Also,

$$1-x = 1 - \frac{at}{a+1-t} = \frac{a+1-t-at}{a+1-t}$$

$$= \frac{a+1-t(a+1)}{a+1-t} = \frac{(a+1)(1-t)}{a+1-t}$$

$$a+x = a + \frac{at}{a+1-t} = \frac{a(a+1) - at + at}{a+1-t} = \frac{a(a+1)}{(a+1-t)}$$

When $x = 0$, $0 = \dfrac{at}{a+1-t} \Rightarrow t = 0$; when $x = 1$, $\dfrac{1}{a+1} = \dfrac{t}{a+1} \Rightarrow t = 1$

x	0	1
t	0	1

$$\therefore \qquad I = \int_0^1 \frac{a^{m-1}\,t^{m-1}\,(a+1)^{n-1}\,(1-t)^{n-1}\,a(a+1)\,dt}{(a+1-t)^{m-1}\,(a+1-t)^{n-1}\,(a+1-t)^2} \cdot \frac{(a+1-t)^{m+n}}{a^{m+n}\,(a+1)^{m+n}}$$

$$= \frac{1}{a^n\,(1+a)^m} \int_0^1 t^{m-1}\,(1-t)^{n-1}\cdot dt$$

$$\boxed{\,I = \frac{B(m,n)}{a^n(a+1)^m}\,}$$

Ex. 14 : *Prove that* $\displaystyle\int_0^1 \frac{x^{p-1}(1-x)^{q-1}}{(a+bx)^{p+q}}\,dx = \frac{B(p,\,q)}{a^q\,(a+b)^p}$

Sol. : Let

$$I = \int_0^1 \frac{x^{p-1}(1-x)^{q-1}}{(a+bx)^{p+q}}\,dx$$

Put

$$\frac{x}{a+bx} = \frac{y}{b+a}$$

$$(b+a)\,x = ay + b\cdot x\cdot y,\quad (b+a-by)\,x = ay \quad\text{or}\quad x = \frac{ay}{b+a-by}$$

$$1-x = 1-\frac{ay}{b+a-by} = \frac{b+a-by-ay}{b+a-by} = \frac{(b+a)(1-y)}{b+a-by}$$

$$a+bx = a+\frac{aby}{a+b-by} = \frac{a(a+b)}{a+b-by}$$

Also,

$$dx = \frac{a(a+b)}{(b+a-by)^2}\,dy$$

When $x=0$, $0=\dfrac{ay}{b+a-by} \Rightarrow y=0$ and when $x=1$, $\dfrac{1}{a+b}=\dfrac{y}{b+a} \Rightarrow y=1$

x	0	1
y	0	1

$$I = \int_0^1 \frac{a^{p-1}\cdot y^{p-1}}{(b+a-by)^{p-1}}\cdot\frac{(b+a)^{q-1}(1-y)^{q-1}}{(b+a-by)^{q-1}}\cdot\frac{(a+b-by)^{p+q}}{a^{p+q}(a+b)^{p+q}}\,\frac{a(a+b)}{(b+a-by)^2}\,dy$$

$$= \frac{1}{a^q\,(a+b)^p}\int_0^1 y^{p-1}(1-y)^{q-1}\,dy$$

$$\boxed{\,I = \frac{1}{a^q\,(a+b)^p}\,B(p,\,q)\,}$$

Ex. 15 : *Prove that* $B(m,\,n) = B(m,\,n+1) + B(m+1,\,n)$

Sol. :

$$\text{R.H.S.} = B(m,\,n+1) + B(m+1,\,n)$$

$$= \frac{\overline{|m}\;\overline{|n+1}}{\overline{|m+n+1}} + \frac{\overline{|m+1}\;\overline{|n}}{\overline{|m+1+n}}$$

$$= \frac{\overline{|m}\;n\,\overline{|n}}{(m+n)\,\overline{|m+n}} + \frac{m\,\overline{|m}\;\overline{|n}}{(m+n)\,\overline{|m+n}}$$

$$= \frac{\overline{|m}\;\overline{|n}\,(n+m)}{(m+n)\,\overline{|m+n}} = \frac{\overline{|m}\;\overline{|n}}{\overline{|m+n}} = B(m,\,n) = \text{L.H.S.}$$

Ex. 16 : *Show that* $B(m,\,n)\;B(m+n,\,p) = \dfrac{\overline{|m}\;\overline{|n}\;\overline{|p}}{\overline{|m+n+p}}$

Sol. :

$$\text{L.H.S.} = B(m,\,n)\;B(m+n,\,p) = \frac{\overline{|m}\;\overline{|n}}{\overline{|m+n}}\cdot\frac{\overline{|m+n}\;\overline{|p}}{\overline{|m+n+p}} = \frac{\overline{|m}\;\overline{|n}\;\overline{|p}}{\overline{|m+n+p}} = \text{R.H.S.}$$

Ex. 17 : *Show that* $\quad B(m, m) = 2^{1-2m} B\left(m, \dfrac{1}{2}\right)$

Sol. : $\qquad B(m, m) = \dfrac{\overline{|m}\;\overline{|m}}{\overline{|2m}}$

We know by duplication formula, $\quad \overline{|m}\;\left|m + \dfrac{1}{2}\right. = \dfrac{\sqrt{\pi}\,\overline{|2m}}{2^{2m-1}}$

or $\qquad \dfrac{\overline{|m}}{\overline{|2m}} = \dfrac{\sqrt{\pi}}{2^{2m-1}\left|m + \dfrac{1}{2}\right.} = \dfrac{\left|\dfrac{1}{2}\right.}{2^{2m-1}\left|m + \dfrac{1}{2}\right.} = \dfrac{2^{1-2m}\left|\dfrac{1}{2}\right.}{\left|m + \dfrac{1}{2}\right.}$

$\therefore \qquad B(m, m) = \overline{|m}\left(\dfrac{\overline{|m}}{\overline{|2m}}\right) = \overline{|m}\;\dfrac{2^{1-2m}\left|\dfrac{1}{2}\right.}{\left|m + \dfrac{1}{2}\right.} = 2^{1-2m}\;\dfrac{\overline{|m}\;\left|\dfrac{1}{2}\right.}{\left|m + \dfrac{1}{2}\right.}$

$$\boxed{B(m, m) = 2^{1-2m} B\left(m, \dfrac{1}{2}\right)}$$

EXERCISE 3.3

1. Express in terms of gamma functions, $\displaystyle\int_0^1 x^m (1 - x^n)^p \, dx$

   (Dec. 2013, 2016) [**Hint :** $x^n = t$] **Ans. :** $\dfrac{1}{n}\dfrac{\left|\dfrac{m+1}{n}\right.\;\overline{|p+1}}{\left|\dfrac{m+1}{n} + p + 1\right.}$

2. Evaluate $\displaystyle\int_0^1 x^3 (1 - \sqrt{x})^5 \, dx$

   [**Hint :** $\sqrt{x} = t$] **(May 2011)** **Ans. :** $\dfrac{1}{5148}$

3. Evaluate $\displaystyle\int_0^n x^n (n - x)^p \, dx$

   [**Hint :** $x = nt$] **Ans. :** $n^{n+p+1} B(n+1, p+1)$

4. Prove that $B(m, n) = \displaystyle\int_0^\infty \dfrac{x^{n-1}}{(1+x)^{m+n}} \, dx$

   [**Hint :** Put $x = \dfrac{1}{1+t}$ in the definition of $B(m, n)$]

5. Evaluate $\displaystyle\int_0^1 \dfrac{x - 2x^2 + x^3}{(1+x)^5} \, dx$

   [**Hint :** $I = \displaystyle\int_0^1 \dfrac{x(1-x)^2}{(1+x)^5} \, dx$ Put $\dfrac{x}{1+x} = \dfrac{t}{2}$] **Ans. :** $\dfrac{1}{48}$

6. Prove that $\displaystyle\int_0^1 (1 - x^{1/n})^m \, dx = \dfrac{m!\,n!}{(m+n)!}$ **(May 2005)**

   [**Hint :** Put $x^{1/n} = t$, then $I = nB(n, m+1)$]

7. Express $\displaystyle\int_{-1}^1 (1+x)^m (1-x)^n \, dx$ in terms of gamma functions. **(May 2009)**

   [**Hint :** $x = \cos 2\theta$; $x = -1$; $\theta = \dfrac{\pi}{2}$; $x = 1$; $\theta = 0$]

   Ans. : $2^{m+n+1}\dfrac{\overline{|n+1}\;\overline{|m+1}}{\overline{|m+n+2}}$

8. Show that $\displaystyle\int_{-\pi/4}^{\pi/4} (\sin\theta + \cos\theta)^{1/3} \, d\theta = \dfrac{6\sqrt{\pi}}{2^{5/6}}\dfrac{\overline{|2/3}}{\overline{|1/6}}$

   [**Hint :** $(\sin\theta + \cos\theta)^{1/3} = \left[\sqrt{(\sin\theta + \cos\theta)^2}\right]^{\frac{1}{3}}$

   $= (1 + \sin 2\theta)^{\frac{1}{6}}$ Put $\theta = \dfrac{\pi}{4} - \phi$]

9. Prove that $\displaystyle\int_0^1 \dfrac{dx}{\sqrt{1 - x^m}} = \dfrac{\sqrt{\pi}}{m}\dfrac{\overline{|1/m}}{\left|\dfrac{1}{m} + \dfrac{1}{2}\right.}$ [**Hint :** $x^m = t$]

10. Show that (i) $B(m+1, n) = \dfrac{m}{m+n} B(m, n)$,

    (ii) $nB(m+1, n) = mB(m, n+1)$

11. Show that $\displaystyle\int_0^{\infty} \frac{x^{n-1}}{(a + bx)^{m+n}} \, dx = \frac{1}{a^m b^n} B(m, n)$

$$\left[\textbf{Hint :} \text{ Put } bx = \frac{a(1-t)}{t}\right]$$

12. Show that $\displaystyle B(n, n+1) = \frac{1}{2} \frac{(\overline{|n}\,)^2}{\overline{|2n}}$

13. Show that $\displaystyle\int_0^1 \frac{dx}{\sqrt[3]{1 - x^3}} = \frac{2\pi}{3\sqrt{3}}$ **(Dec. 2011)**

$$\left[\textbf{Hint :} \text{ Put } x^3 = t. \text{ Use } \overline{|p} \;\; \overline{|1-p} = \frac{\pi}{\sin p\pi} , 0 < p < 1\right]$$

14. Show that $\displaystyle\int_0^2 x (8 - x^3)^{1/3} \, dx = \frac{16\pi}{9\sqrt{3}}$

$$\left[\textbf{Hint :} \text{ Put } x^3 = 8t. \text{ Use } \overline{|p} \;\; \overline{|1-p} = \frac{\pi}{\sin p\pi}\right]$$

15. Show that $\displaystyle\int_0^{\infty} \frac{x^8 (1 - x^6) \, dx}{(1 + x)^{24}} = 0$ **(Dec. 2005, May 2013)**

$$\left[\textbf{Hint :} I = \int_0^{\infty} \frac{x^{9-1} \, dx}{(1 + x)^{9+15}} - \int_0^{\infty} \frac{x^{15-1} \, dx}{(1 + x)^{15+9}}\right.$$
$$\left. = B(9, 15) - B(15, 9) = 0\right]$$

16. Show that $\displaystyle\int_0^{\infty} \frac{x^6 - x^3}{(1 + x^3)^5} \, x^2 \, dx = 0$ Put $x^3 = t$

17. Show that $\displaystyle\int_0^1 \frac{x^2 + x^3}{(1 + x)^7} \, dx = \frac{1}{60}$

$$\left[\textbf{Hint :} I = \int_0^1 \frac{x^2 \, dx}{(1 + x)^7} + \int_0^1 \frac{x^3 \, dx}{(1 + x)^7} \left(\text{Put } x = \frac{1}{t} \text{ in } I_2\right)\right]$$

18. Prove that $\displaystyle\int_0^1 \frac{x^2 \, dx}{(1 - x^4)^{1/2}} \int_0^1 \frac{dx}{(1 + x^4)^{1/2}} = \frac{\pi}{4\sqrt{2}}$

[**Hint :** For I_1, put $x^2 = \sin\theta$; For I_2, put $x^2 = \tan\theta$, further $2\theta = t$]

19. Prove that $\displaystyle\int_0^{\pi/2} \frac{d\theta}{\sqrt{\sin\theta}} \int_0^{\pi/2} \sqrt{\sin\theta} \, d\theta = \pi$

$$\left[\textbf{Hint :} \text{Use } \int_0^{\pi/2} \sin^p\theta \cos^q\theta \, d\theta = \frac{1}{2} B\left(\frac{p+1}{2}, \frac{q+1}{2}\right)\right]$$

$$= \frac{\frac{1}{2} \left|\frac{p+1}{2}\right. \left|\frac{q+1}{2}\right.}{\left|\frac{p+q+2}{2}\right.}\,]$$

20. Prove that $\displaystyle B(m, m) = 2 \int_0^{1/2} (t - t^2)^{m-1} \, dt$

$\left[\textbf{Hint :} \; B(m, m) = \displaystyle\int_0^{1/2} t^{m-1} (1 - t)^{m-1} \, dt + \int_{1/2}^1 t^{m-1}\right.$

$(1 - t)^{m-1} \, dt$, put $t = 1 - x$ in $I_2]$

21. Prove that $\displaystyle\int_1^{\infty} \frac{x^{\frac{n}{2} - 1}}{(1 + x)^n} \, dx = \frac{1}{2} B\left(\frac{n}{2}, \frac{n}{2}\right)$

$\left[\textbf{Hint :} B\left(\dfrac{n}{2}, \dfrac{n}{2}\right) = \displaystyle\int_0^{\infty} \frac{x^{\frac{n}{2}-1}}{(1 + x)^n} = \int_0^1 \frac{x^{\frac{n}{2}-1}}{(1 + x)^n} + \right.$

$\displaystyle\int_1^{\infty} \frac{x^{\frac{n}{2}-1}}{(1 + x)^n} \left(\text{Put } x = \frac{1}{t} \text{ in } I_1\right)]$

22. Show that $\displaystyle\int_0^1 \frac{(1 - x^4)^{3/4}}{(1 + x^4)^2} \, dx = \frac{3\pi}{2^{15/4}}$

23. Show that $\displaystyle\frac{B(m, n+1)}{n} = \frac{B(m+1, n)}{m} = \frac{B(m, n)}{m + n}$

24. Show that $\displaystyle\int_0^1 x^{m-1} (1 - x^2)^{n-1} \, dx = \frac{1}{2} B\left(\frac{m}{2}, n\right)$

(Dec. 2010)

25. By putting $\dfrac{x}{1 - x} = \dfrac{at}{1 - t}$, where the constant a, is suitably selected, show that

$$\int_0^1 x^{-1/3} (1 - x)^{-2/3} (1 + 2x)^{-1} \, dx = \frac{1}{9^{1/3}} B\left(\frac{2}{3}, \frac{1}{3}\right)$$

26. $\displaystyle\int_0^1 \frac{x^{m-1} (1 - x)^{n-1}}{(1 + x)^{m+n}} \, dx = \frac{B(m, n)}{2^m}$ $\left[\textbf{Hint :} \text{ Put } x = \dfrac{t}{1 + t}\right]$

CHAPTER-4

DIFFERENTIATION UNDER THE INTEGRAL SIGN AND ERROR FUNCTIONS

4.1 INTRODUCTION

Differentiation under the integral sign or briefly DUIS is an effective technique used in evaluation of real definite integrals. When a definite integral $I = \int_a^b f(x, \alpha)\, dx$, which is to be integrated with respect to variable x and contains other parameter α, then we differentiate I, w.r.t. parameter α, by using DUIS. There are different rules when limits of integral are constants or functions of parameter α. When DUIS technique is used, the definite integral evaluation results into an ordinary differential equation, the solution of this equation results in the evaluation of definite integral. The technique is very useful in Laplace or Fourier transforms which students will be studying in subsequent years of their curriculum. Error function integral, which is another topic of this chapter is very close to Probability Integral and is used in probability distributions, particularly in Normal probability distribution. Complementary error functions are involved in finding inverse Laplace transforms of complicated functions. Advance software tools are used for this purpose. In subsequent sections, we will be discussing DUIS and Error function integrals in detail.

4.2 RULE - I : INTEGRAL WITH LIMITS (a, b) AS CONSTANTS

If
$$I(\alpha) = \int_a^b f(x, \alpha)\, dx, \text{ where a and b are constants, then}$$

$$\frac{dI}{d\alpha} = \int_a^b \frac{\partial}{\partial \alpha} f(x, \alpha)\, dx$$

Proof :
$$\frac{dI}{d\alpha} = \lim_{\delta\alpha \to 0} \frac{I(\alpha + \delta\alpha) - I(\alpha)}{\delta\alpha}$$

$$= \lim_{\delta\alpha \to 0} \frac{1}{\delta\alpha} \left[\int_a^b f(x, \alpha + \delta\alpha)\, dx - \int_a^b f(x, \alpha)\, dx \right]$$

$$= \lim_{\delta\alpha \to 0} \frac{1}{\delta\alpha} \int_a^b [f(x, \alpha + \delta\alpha) - f(x, \alpha)]\, dx$$

$$= \lim_{\delta\alpha \to 0} \int_a^b \left[\frac{f(x, \alpha + \delta\alpha) - f(x, \alpha)}{\delta\alpha} \right] dx$$

$$= \int_a^b \lim_{\delta\alpha \to 0} \left[\frac{f(x, \alpha + \delta\alpha) - f(x, \alpha)}{\delta\alpha} \right] dx$$

$$\frac{dI}{d\alpha} = \int_a^b \frac{\partial}{\partial \alpha} f(x, \alpha)\, dx \qquad \text{(by definition of partial derivative)}$$

Rule-I :

$$\text{If } I(\alpha) = \int_a^b f(x, \alpha)\, dx, \text{ then } \frac{dI}{d\alpha} = \int_a^b \frac{\partial}{\partial \alpha} f(x, \alpha)\, dx$$

It may be noted that if integral involves two parameters x and α, integration is to be carried out with respect to variable x treating α as a constant. Rule (I) gives method to differentiate integral with respect to parameter α.

Ex. : Verify the Rule-I of DUIS for $I(a) = \int_0^{\pi/2} \sin(ax)\, dx$.

Here integral sin (ax) involves two parameters x and a, and the integration is to be carried out with respect to x. According to rule (I), we carry out differentiation w.r.t. 'a' as,

$$\frac{dI}{da} = \int_0^{\pi/2} \frac{\partial}{\partial a} (\sin ax)\, dx = \int_0^{\pi/2} \cos(ax) \cdot x \cdot dx$$

$$= \left[x \cdot \left\{ \frac{\sin ax}{a} \right\} \right]_0^{\pi/2} - \int_0^{\pi/2} (1)\, \frac{\sin ax}{a}\, dx$$

$$= \frac{\pi}{2a} \sin \frac{\pi a}{2} - \frac{1}{a} \left(-\frac{\cos ax}{a} \right)_0^{\pi/2}$$

$$\boxed{ \frac{dI}{da} = \frac{\pi}{2a} \sin \frac{\pi a}{2} + \frac{1}{a^2} \cos \frac{\pi a}{2} - \frac{1}{a^2} } \qquad \ldots(1)$$

$$I = \int_0^{\pi/2} \sin ax\, dx = - \left. \frac{\cos ax}{a} \right|_0^{\pi/2}$$

$$I = -\frac{1}{a} \cos \frac{\pi a}{2} + \frac{1}{a}$$

$$\frac{dI}{da} = \frac{1}{a^2} \cos \frac{\pi a}{2} - \frac{1}{a} \left(-\sin \frac{\pi a}{2} \right) \frac{\pi}{2} - \frac{1}{a^2}$$

$$\boxed{ \frac{dI}{da} = \frac{1}{a^2} \cos \frac{\pi a}{2} + \frac{\pi}{2a} \sin \frac{\pi a}{2} - \frac{1}{a^2} } \qquad \ldots(2)$$

Results (1) and (2) are same which verify the rule I of differentiation under the sign of integration.

In above rule, limits of integration were constants i.e. they were independent of parameter 'a' with respect to which differentiation was carried out.

4.3 ILLUSTRATIONS ON RULE-I

Ex. 1 : *Verify the rule of differentiation under the integral sign for the integral* $\int_0^\infty e^{-at} \cos bt\, dt$, *where a is the parameter.*

Sol. : By direct integration

$$\phi(a) = \int_0^\infty e^{-at} \cos bt\, dt = \left[\frac{e^{-at}}{a^2 + b^2} (-a \cos bt + b \sin bt) \right]_0^\infty$$

$$= 0 - \frac{e^0}{a^2 + b^2} (-a + 0) = \frac{a}{a^2 + b^2}$$

Differentiating w.r.t. a,

$$\frac{d\phi}{da} = \frac{1}{a^2 + b^2} - \frac{2a^2}{(a^2 + b^2)^2} = \frac{b^2 - a^2}{(a^2 + b^2)^2} \qquad \ldots(1)$$

By using DUIS Rule - I,

$$\frac{d\phi}{da} = \int_0^\infty \frac{\partial}{\partial a} (e^{-at} \cos bt)\, dt = \int_0^\infty e^{-at} (-t) \cos bt\, dt = -\int_0^\infty t\, (e^{-at} \cos bt)\, dt$$

$$= -\left\{ \left[t\, \frac{e^{-at}}{a^2 + b^2} (-a \cos bt + b \sin bt) \right]_0^\infty - \int_0^\infty (1) \frac{e^{-at}}{a^2 + b^2} (-a \cos bt + b \sin bt)\, dt \right\}$$

$$= 0 + \frac{1}{a^2 + b^2} \left[\int_0^\infty -ae^{-at} \cos bt\, dt + \int_0^\infty b\, e^{-at} \sin bt\, dt \right]$$

$$\left\{ \because \int_0^\infty e^{-at} \cos bt\, dt = \frac{a}{a^2 + b^2}\, ; \int_0^\infty e^{-at} \sin bt\, dt = \frac{b}{a^2 + b^2} \right\}$$

$$= \frac{1}{a^2 + b^2} \left[-a \cdot \frac{a}{a^2 + b^2} + b \cdot \frac{b}{a^2 + b^2} \right] = \frac{b^2 - a^2}{(a^2 + b^2)^2} \qquad \ldots (2)$$

From (1) and (2), the theorem is verified.

Ex. 2 : *Evaluate* $\displaystyle\int_0^\infty \frac{e^{-x}}{x} (1 - e^{-ax})\, dx\ (a > -1).$

Sol. :

$$\phi(a) = \int_0^\infty \frac{e^{-x}}{x} (1 - e^{-ax})\, dx$$

Differentiating w.r.t. a,

$$\frac{d\phi}{da} = \int_0^\infty \frac{\partial}{\partial a} \left[\frac{e^{-x}}{x} (1 - e^{-ax}) \right] dx = \int_0^\infty \frac{e^{-x}}{x} (-e^{-ax})(-x)\, dx$$

$$= \int_0^\infty e^{-(a+1)x}\, dx = \left[\frac{e^{-(a+1)x}}{-(a+1)} \right]_0^\infty$$

$$= \frac{e^{-\infty}}{-(a+1)} - \frac{e^0}{-(a+1)} \qquad \{ \because a + 1 > 0 \text{ i.e. } a > -1$$

$$= 0 + \frac{1}{a+1}$$

$$\therefore \qquad d\phi = \frac{da}{a+1}$$

$$\phi(a) = \log(a+1) + C.$$

To determine C, put a = 0,

$$\therefore \qquad \phi(0) = 0 + C$$

But

$$\phi(0) = \int_0^\infty \frac{e^{-x}}{x} (1 - 1)\, dx = 0 \qquad \therefore\ C = 0$$

Hence,

$$\boxed{\phi(a) = \log(a+1)}$$

Ex. 3 : *Prove that* $\displaystyle\int_0^1 \frac{x^a - 1}{\log x}\, dx = \log(1 + a)\,;\, a \geq 0.$ **(Dec. 2004, 2018; May 2013)**

Sol. : Let

$$\phi(a) = \int_0^1 \frac{x^a - 1}{\log x}\, dx$$

Differentiating w.r.t. a,

$$\frac{d\phi}{da} = \int_0^1 \frac{\partial}{\partial a}\left(\frac{x^a - 1}{\log x}\right) dx = \int_0^1 \frac{1}{\log x} \cdot x^a \log x \cdot dx$$

$$= \int_0^1 x^a\, dx = \left[\frac{x^{a+1}}{a+1}\right]_0^1 \qquad \because a \geq 0$$

$$d\phi = \frac{1}{a+1}\, da$$

$$\therefore \qquad \phi(a) = \log(a+1) + C$$

To determine C, put $a = 0$,

$$\phi(0) = 0 + C$$

But

$$\phi(0) = \int_0^1 \frac{x^0 - 1}{\log x}\, dx = 0. \qquad \therefore C = 0$$

Hence,

$$\boxed{\phi(a) = \log(a+1)}$$

Ex. 4 : *Evaluate* $\displaystyle\int_0^\infty \frac{e^{-x}}{x}\left(a - \frac{1}{x} + \frac{1}{x} e^{-ax}\right) dx$

Sol. : Let

$$\phi(a) = \int_0^\infty \frac{e^{-x}}{x}\left(a - \frac{1}{x} + \frac{1}{x} e^{-ax}\right) dx$$

$$\phi'(a) = \int_0^\infty \frac{\partial}{\partial a}\left[\frac{e^{-x}}{x}\left(a - \frac{1}{x} + \frac{1}{x} e^{-ax}\right)\right] dx = \int_0^\infty \frac{e^{-x}}{x}\left[1 - 0 + \frac{1}{x} e^{-ax}(-x)\right] dx$$

$$\phi'(a) = \int_0^\infty \frac{e^{-x}}{x}\left[1 - e^{-ax}\right] dx$$

Again, differentiating, $\displaystyle\phi''(a) = \int_0^\infty \frac{e^{-x}}{x}\left[-e^{-ax}(-x)\right] dx = \int_0^\infty e^{-(1+a)x}\, dx$

$$= \left[\frac{e^{-(1+a)x}}{-(1+a)}\right]_0^\infty = \frac{1}{1+a}$$

$$\phi''(a) = \frac{1}{1+a},$$

Integration gives, $\phi'(a) = \log(1+a) + C_1$

Put $a = 0$ $\phi'(0) = \log 1 + C_1.$

But
$$\phi'(0) = \int_0^\infty \frac{e^{-x}}{x}\,[1-1]\,dx = 0 \quad \therefore\ C_1 = 0$$

$$\phi'(a) = \log(1+a)$$

Again, integrating,

$$\phi(a) = \int \log(1+a)\cdot 1\cdot da + C_2 = \log(1+a)\cdot(a) - \int \frac{1}{1+a}\cdot a\cdot da + C_2$$

$$= a\log(1+a) - \int \frac{a+1-1}{a+1}\,da\ + C_2 = a\log(1+a)\ - \int \left(1 - \frac{1}{a+1}\right)\,da + C_2$$

$$= a\log(1+a) - a + \log(a+1) + C_2$$

$$\therefore \qquad \phi(a) = (a+1)\,\log(a+1) - a + C_2$$

Put $a = 0$
$$\phi(0) = \log 1 - 0 + C_2$$

But
$$\phi(0) = \int_0^\infty \frac{e^{-x}}{x}\left(0 - \frac{1}{x} + \frac{1}{x}\,(1)\right)\,dx = 0 \quad \therefore\ C_2 = 0$$

Hence,
$$\boxed{\phi(a) = (a+1)\log(a+1) - a}$$

Ex. 5 : *Show that* $\displaystyle\int_0^\infty \frac{\tan^{-1}(ax)}{x(1+x^2)}\cdot dx = \frac{\pi}{2}\log(1+a).$

Sol. : Let
$$\phi(a) = \int_0^\infty \frac{\tan^{-1}(ax)}{x(1+x^2)}\,dx$$

$$\frac{d\phi}{da} = \int_0^\infty \frac{\partial}{\partial a}\frac{\tan^{-1}(ax)}{x(1+x^2)}\,dx = \int_0^\infty \frac{1\cdot(x)}{1+a^2x^2}\cdot\frac{1}{x(1+x^2)}\,dx$$

$$= \int_0^\infty \frac{dx}{(1+a^2x^2)(1+x^2)} = \int_0^\infty \left(\frac{\dfrac{1}{1-1/a^2}}{1+a^2x^2} + \frac{\dfrac{1}{1-a^2}}{1+x^2}\right)dx$$

$$= \frac{1}{1-a^2}\left[\int_0^\infty \frac{dx}{1+x^2} - \int_0^\infty \frac{a^2}{1+a^2x^2}\,dx\right] = \frac{1}{1-a^2}\left[\tan^{-1}x - a\tan^{-1}(ax)\right]_0^\infty$$

$$= \frac{1}{1-a^2}\left[\frac{\pi}{2} - a\cdot\frac{\pi}{2}\right] = \frac{\pi}{2}\frac{(1-a)}{(1-a)(1+a)} = \frac{\pi}{2}\cdot\frac{1}{1+a}$$

$$d\phi = \frac{\pi}{2}\cdot\frac{da}{1+a}$$

$$\therefore \qquad \phi(a) = \frac{\pi}{2}\log(1+a) + C$$

To determine C, we put $a = 0 \quad \therefore\ \phi(0) = C.$

But
$$\phi(0) = \int_0^\infty \frac{\tan^{-1}0}{x(1+x^2)}\,dx = 0 \quad \therefore\ C = 0$$

Hence
$$\boxed{\phi(a) = \frac{\pi}{2}\log(1+a)}$$

Ex. 6 : *Show that* $\displaystyle\int_0^\infty e^{-\left(x^2 + \frac{a^2}{x^2}\right)}\, dx = \frac{\sqrt{\pi}}{2}\, e^{-2a}$

Sol. : Let

$$\phi(a) = \int_0^\infty e^{-\left(x^2 + \frac{a^2}{x^2}\right)}\, dx \qquad \text{(Here 'a' is a parameter)}$$

$$\phi'(a) = \int_0^\infty \frac{\partial}{\partial a}\left[e^{-\left(x^2 + \frac{a^2}{x^2}\right)} \right] dx$$

$$= \int_0^\infty e^{-\left(x^2 + \frac{a^2}{x^2}\right)} \cdot \left(-\frac{2a}{x^2}\right) dx \qquad \text{Put } \frac{a}{x} = y \ ; \ -\frac{a}{x^2}\, dx = dy$$

x	0	∞
y	∞	0

$$= \int_\infty^0 e^{-\left(y^2 + \frac{a^2}{y^2}\right)}\, (2dy)$$

or

$$\phi'(a) = -2\int_0^\infty e^{-\left(x^2 + \frac{a^2}{x^2}\right)} dx = -2\,\phi(a)$$

or

$$\frac{\phi'(a)}{\phi(a)} = -2$$

$$\therefore \quad \log \phi(a) = -2a + C$$

or

$$\phi(a) = e^{-2a} \cdot C_1$$

Put $a = 0$;

$$\phi(0) = C_1$$

But

$$\phi(0) = \int_0^\infty e^{-x^2}\, dx = \frac{\sqrt{\pi}}{2} \quad \therefore \ C_1 = \frac{\sqrt{\pi}}{2}$$

$$\therefore \quad \phi(a) = e^{-2a}\frac{\sqrt{\pi}}{2}$$

i.e.

$$\boxed{\int_0^\infty e^{-\left(x^2 + \frac{a^2}{x^2}\right)} dx = \frac{\sqrt{\pi}}{2}\, e^{-2a}}$$

Ex. 7 : *Evaluate* $\displaystyle\int_0^\infty \frac{e^{-\beta x} \sin \alpha x}{x}\, dx$ *and hence deduce that* $\displaystyle\int_0^\infty \frac{\sin \alpha x}{x}\, dx = \begin{cases} -\dfrac{\pi}{2} & ; \text{ if } \alpha < 0 \\[2mm] 0 & ; \text{ if } \alpha = 0 \\[2mm] \dfrac{\pi}{2} & ; \text{ if } \alpha > 0 \end{cases}$

Sol. : Let

$$\phi(\alpha) = \int_0^\infty \frac{e^{-\beta x} \sin \alpha x}{x}\, dx$$

Differentiating w.r.t. α,

$$\frac{d\phi}{d\alpha} = \int_0^\infty \frac{e^{-\beta x} x \cdot \cos \alpha x}{x} \cdot dx = \int_0^\infty e^{-\beta x} \cos \alpha x \cdot dx$$

$$\therefore \qquad \frac{d\phi}{d\alpha} = \frac{\beta}{\beta^2 + \alpha^2} \qquad \left\{ \because \int_0^\infty e^{-at} \cos bt \, dt = \frac{a}{a^2 + b^2} \right.$$

$$d\phi = \frac{\beta}{\alpha^2 + \beta^2} \, d\alpha$$

Integrating,
$$\phi(\alpha) = \tan^{-1}\frac{\alpha}{\beta} + C \qquad \qquad \text{...(1)}$$

To determine C, put $\alpha = 0$ in (1)

$$\phi(0) = 0 + C$$

But
$$\phi(0) = \int_0^\infty \frac{e^{-\beta x} \sin 0}{x} \, dx = 0 \quad \therefore \ C = 0$$

and
$$\phi(\alpha) = \tan^{-1}\frac{\alpha}{\beta}$$

i.e.
$$\int_0^\infty \frac{e^{-\beta x} \sin \alpha x}{x} \, dx = \tan^{-1}\frac{\alpha}{\beta}$$

Put $\beta = 0$,
$$\int_0^\infty \frac{\sin \alpha x}{x} \, dx = \begin{cases} \tan^{-1}(-\infty) = -\dfrac{\pi}{2} & \text{if } \alpha < 0 \\[2mm] 0, & \text{if } \alpha = 0 \\[2mm] \tan^{-1}(\infty) = \dfrac{\pi}{2}, & \text{if } \alpha > 0 \end{cases}$$

Ex. 8 : *Show that* $\displaystyle\int_0^\pi \log(1 - a\cos x)\,dx = \pi \log\left(\frac{1 + \sqrt{1 - a^2}}{2}\right),\ |a| < 1.$

Sol. : Let
$$\phi(a) = \int_0^\pi \log(1 - a\cos x)\,dx$$

$$\phi'(a) = \int_0^\pi \frac{\partial}{\partial a}\{\log(1 - a\cos x)\}\,dx$$

$$= \int_0^\pi \frac{1\,(-\cos x)}{1 - a\cos x}\cdot dx = \frac{1}{a}\int_0^\pi \frac{-a\cos x}{1 - a\cos x}\cdot dx$$

$$= \frac{1}{a}\int_0^\pi \frac{1 - a\cos x - 1}{1 - a\cos x}\cdot dx = \frac{1}{a}\int_0^\pi \left(1 - \frac{1}{1 - a\cos x}\right)dx$$

$$= \frac{1}{a}\,[x]_0^\pi - \frac{1}{a}\int_0^\pi \frac{dx}{1 - a\cos x} \quad \text{Put } t = \tan x/2,\ dx = \frac{2\,dt}{1 + t^2},\ \cos x = \frac{1 - t^2}{1 + t^2}$$

x	0	π
t	0	∞

$$= \frac{1}{a}\,(\pi) - \frac{1}{a}\int_0^\infty \frac{\dfrac{2dt}{1 + t^2}}{1 - a\left(\dfrac{1 - t^2}{1 + t^2}\right)}$$

$$= \frac{\pi}{a} - \frac{2}{a} \int_0^\infty \frac{dt}{1 + t^2 - a + at^2}$$

$$= \frac{\pi}{a} - \frac{2}{a} \int_0^\infty \frac{dt}{1 - a + (1 + a)\, t^2} = \frac{\pi}{a} - \frac{2}{a\,(1 + a)} \int_0^\infty \frac{dt}{\dfrac{1 - a}{1 + a} + t^2}$$

$$= \frac{\pi}{a} - \frac{2}{a\,(1 + a)} \left[\frac{1}{\sqrt{\dfrac{1 - a}{1 + a}}} \tan^{-1}\left(\frac{t}{\sqrt{\dfrac{1 - a}{1 + a}}} \right) \right]_0^\infty = \frac{\pi}{a} - \frac{2}{a\,\sqrt{1 - a^2}} \left(\frac{\pi}{2} - 0 \right)$$

$$\phi'(a) = \frac{\pi}{a} - \frac{\pi}{a\,\sqrt{1 - a^2}}$$

Integrating w.r.t. 'a',

$$\phi(a) = \pi \log a - \pi \int \frac{1}{a\,\sqrt{1 - a^2}}\, da + C_1$$

$$= \pi \log a - \pi \log\left(\frac{1 - \sqrt{1 - a^2}}{a} \right) + C_1 \qquad \left\{ \because \int \frac{dx}{x\,\sqrt{1 - x^2}} = \log\left(\frac{1 - \sqrt{1 - x^2}}{x} \right) \right.$$

$$= \pi \log\left(\frac{a^2}{1 - \sqrt{1 - a^2}} \right) + C_1 = \pi \log\left[\frac{a^2\,(1 + \sqrt{1 - a^2})}{(1 - \sqrt{1 - a^2})\,(1 + \sqrt{1 - a^2})} \right] + C_1$$

$$= \pi \log\left[\frac{a^2\,(1 + \sqrt{1 - a^2})}{1 - 1 + a^2} \right] + C_1 = \pi \log(1 + \sqrt{1 - a^2}) + C_1$$

Put $a = 0$, $\qquad \phi(0) = \pi \log 2 + C_1.$

But $\phi(0) = 0 \qquad \therefore\ C_1 = -\pi \log 2$

$\therefore \qquad \phi(a) = \pi \log(1 + \sqrt{1 - a^2}) - \pi \log 2$

$\therefore$

$$\boxed{\phi(a) = \pi \log\left(\frac{1 + \sqrt{1 - a^2}}{2} \right)}$$

Ex. 9 : *Prove that* $\displaystyle\int_0^\infty e^{-bx^2} \cos(2ax)\, dx = \frac{1}{2}\sqrt{\frac{\pi}{b}}\, e^{-\frac{a^2}{b}},\ (b > 0)$

Sol. : Let

$$\phi(a) = \int_0^\infty e^{-bx^2} \cos(2ax)\, dx$$

$$\phi'(a) = \int_0^\infty \frac{\partial}{\partial a}\left[e^{-bx^2} \cos(2ax) \right] dx = \int_0^\infty e^{-bx^2} (-\sin 2ax)\,(2x\, dx)$$

$$= \frac{1}{b} \int_0^\infty \sin 2ax \left[e^{-bx^2} (-2bx\, dx) \right] \qquad\qquad \text{(Note this step)}$$

Integrating by parts using $\int e^{-bx^2} (-2bx)\, dx = e^{-bx^2}$,

$$\phi'(a) = \frac{1}{b}\left\{ [\sin 2ax\, (e^{-bx^2})]_0^\infty - \int_0^\infty \cos 2ax\,(2a)\, e^{-bx^2}\, dx \right\}$$

$$= \frac{1}{b}\left\{ 0 - 0 - 2a \int_0^\infty e^{-bx^2} \cos 2ax\, dx \right\} = -\frac{2a}{b}\, \phi(a)$$

$$\frac{\phi'(a)}{\phi(a)} = -\frac{2a}{b}$$

$\therefore \qquad \displaystyle\int \frac{\phi'(a)}{\phi(a)}\, da = -\frac{2}{b} \int a\, da$

$$\log \phi(a) = -\frac{2}{b} \cdot \frac{a^2}{2} + C = -\frac{a^2}{b} + C$$

$\therefore \qquad \phi(a) = e^{-\frac{a^2}{b}}\, C_1$

Put $a = 0$, $\phi(0) = C_1$

But $\displaystyle\phi(0) = \int_0^\infty e^{-bx^2} \cdot dx$ Put $bx^2 = t^2$ or $x = \dfrac{t}{\sqrt{b}}$, $dx = \dfrac{dt}{\sqrt{b}}$

x	0	∞
t	0	∞

$$\phi(0) = \int_0^\infty e^{-t^2} \frac{dt}{\sqrt{b}} = \frac{1}{\sqrt{b}} \int_0^\infty e^{-t^2}\, dt = \frac{1}{\sqrt{b}} \frac{\sqrt{\pi}}{2} = \frac{1}{2} \sqrt{\frac{\pi}{b}}$$

$\therefore \qquad C_1 = \dfrac{1}{2} \sqrt{\dfrac{\pi}{b}}$

Hence,

$$\boxed{\; \phi(a) = \frac{1}{2} \sqrt{\frac{\pi}{b}}\, e^{-\frac{a^2}{b}} \;}$$

Ex. 10 : *Show that* $\displaystyle\int_0^{\pi/2} \frac{\log(1 + a \sin^2 x)}{\sin^2 x}\, dx = \pi\left[\sqrt{a + 1} - 1\right]$ *(May 2007)*

Sol. : $\displaystyle\phi(a) = \int_0^{\pi/2} \frac{\log(1 + a \sin^2 x)}{\sin^2 x}\, dx$

$$\phi'(a) = \int_0^{\pi/2} \frac{1}{1 + a \sin^2 x} \sin^2 x \cdot \frac{1}{\sin^2 x}\, dx = \int_0^{\pi/2} \frac{1}{1 + a \sin^2 x}\, dx \qquad \text{(Divide by } \cos^2 x\text{)}$$

$$= \int_0^{\pi/2} \frac{\sec^2 x\, dx}{\sec^2 x + a \tan^2 x} = \int_0^{\pi/2} \frac{\sec^2 x\, dx}{1 + \tan^2 x + a \tan^2 x}$$

$$= \int_0^{\pi/2} \frac{\sec^2 x\, dx}{1 + (1 + a) \tan^2 x} = \frac{1}{a + 1} \int_0^{\pi/2} \frac{\sec^2 x\, dx}{\left(\dfrac{1}{\sqrt{a + 1}}\right)^2 + (\tan x)^2}$$

Put $\tan x = t$; $\sec^2 x\, dx = dt$

x	0	$\pi/2$
t	0	∞

$$\phi'(a) = \frac{1}{a+1} \int_0^{\infty} \frac{dt}{\left(\dfrac{1}{\sqrt{1+a}}\right)^2 + t^2} = \frac{1}{a+1} \left[\sqrt{a+1}\ \tan^{-1}\left(t\sqrt{a+1}\right)\right]_0^{\infty} \qquad \left(\because \int \frac{dx}{a^2 + x^2} = \frac{1}{a}\ \tan^{-1}\frac{x}{a}\right)$$

$$\phi'(a) = \frac{1}{\sqrt{a+1}} \left(\frac{\pi}{2} - 0\right) = \frac{\pi}{2} \cdot \frac{1}{\sqrt{a+1}}$$

Integrating w.r.t. 'a'

$$\phi(a) = \frac{\pi}{2} \int \frac{da}{\sqrt{a+1}} + C = \frac{\pi}{2} \left[2 \cdot (a+1)^{1/2}\right] + C$$

$$\phi(a) = \pi (a+1)^{1/2} + C$$

Put $a = 0$, $\phi(0) = \pi + C$

But $\phi(0) = 0 \quad \therefore \quad C = -\pi$

$$\phi(a) = \pi (a+1)^{1/2} - \pi$$

$$\boxed{\phi(a) = \pi \left(\sqrt{a+1} - 1\right)}$$

Ex. 11 : *Prove that* $\displaystyle \int_0^1 x^p (\log x)^n\, dx = \frac{(-1)^n\, n!}{(p+1)^{n+1}}$ *, where n is a positive integer and $p > -1$.*   *(May 2008)*

Sol. : On putting $n = 0$, $\displaystyle \phi(p) = \int_0^1 x^p (\log x)^0\, dx$

Simplification gives, $\displaystyle \phi(p) = \int_0^1 x^p\, dx = \left[\frac{x^{p+1}}{p+1}\right]_0^1 = \frac{1}{p+1}$ $(\because p > -1)$

Differentiating, $\displaystyle \frac{d\phi}{dp} = \int_0^1 \frac{\partial}{\partial p}(x^p)\, dx = -\frac{1}{(p+1)^2}$

or $\displaystyle \frac{d\phi}{dp} = \int_0^1 x^p \log x\, dx = -\frac{1}{(p+1)^2}$

Again differentiating, $\displaystyle \frac{d^2\phi}{dp^2} = \int_0^1 x^p (\log x)^2\, dx = \frac{(-1)^2\, 2!}{(p+1)^3} \cdot$

Similarly, on differentiating n times w.r.t. p,

$$\boxed{\frac{d^n \phi}{dp^n} = \int_0^1 x^p (\log x)^n\, dx = \frac{(-1)^n\, n!}{(p+1)^{n+1}}}$$

Ex. 12 : *The Bessel function of integral order n is defined as* $J_n(x) = \dfrac{1}{\pi} \displaystyle\int_0^{\pi} \cos(nt - x \sin t)\, dt.$ *Show that*

$$\frac{d}{dx} J_n(x) = \frac{1}{2}\left[J_{n-1}(x) - J_{n+1}(x)\right].$$

Sol. : We have

$$J_n(x) = \frac{1}{\pi} \int_0^{\pi} \cos(nt - x \sin t)\, dt \qquad \qquad \ldots(1)$$

$$\therefore \qquad J_{n-1}(x) = \frac{1}{\pi} \int_0^{\pi} \cos((n-1)t - x \sin t)\, dt \qquad \qquad \ldots(2)$$

$$J_{n+1}(x) = \frac{1}{\pi} \int_0^{\pi} \cos((n+1)t - x \sin t)\, dt \qquad \qquad \ldots(3)$$

Treating x as a parameter, differentiating (1) w.r.t. x,

$$\frac{d}{dx}(J_n(x)) = \frac{1}{\pi} \int_0^{\pi} \frac{\partial}{\partial x} \cos(nt - x \sin t)\, dt$$

$$= \frac{1}{\pi} \int_0^{\pi} (-\sin(nt - x \sin t))\,(-\sin t) \cdot dt$$

$$= \frac{1}{2\pi} \int_0^{\pi} 2 \sin(nt - x \sin t) \sin t \cdot dt$$

$$= \frac{1}{2\pi} \int_0^{\pi} [\cos(nt - x \sin t - t) - \cos(nt - x \sin t + t)]\, dt.$$

$$= \frac{1}{2\pi} \int_0^{\pi} \cos((n-1)t - x \sin t)\, dt - \frac{1}{2\pi} \int_0^{\pi} \cos((n+1)t - x \sin t)\, dt$$

From (2) and (3), $\boxed{\dfrac{d}{dx}(J_n(x)) = \dfrac{1}{2} \cdot \left[J_{n-1}(x) - J_{n+1}(x)\right]}$

Ex. 13 : *Prove that* $\displaystyle\int_0^{\infty} \frac{\cos \lambda x}{x}\,(e^{-ax} - e^{-bx})\, dx = \frac{1}{2} \log\left(\frac{b^2 + \lambda^2}{a^2 + \lambda^2}\right) ;\ a > 0,\ b > 0.$

Sol. : Let

$$\phi(a) = \int_0^{\infty} \frac{\cos \lambda x}{x}\,(e^{-ax} - e^{-bx}) \cdot dx$$

$$\frac{d\phi}{da} = \int_0^{\infty} \frac{\partial}{\partial a}\left[\frac{\cos \lambda x}{x}\,(e^{-ax} - e^{-bx})\right] dx$$

$$= \int_0^{\infty} \frac{\cos \lambda x}{x}\,[e^{-ax}\,(-x) - 0]\, dx = -\int_0^{\infty} e^{-ax} \cos \lambda x\, dx$$

$$= -\left[\frac{e^{-ax}}{a^2 + \lambda^2}\,(-a \cos \lambda x + \lambda \sin \lambda x)\right]_0^{\infty}$$

by using $\displaystyle\int e^{ax}\cos bx\,dx = \frac{e^{ax}}{a^2+b^2}(a\cos bx + b\sin bx)$

$$\frac{d\phi}{da} = -\left[0 - \frac{e^0}{a^2+\lambda^2}(-a+0)\right] = -\frac{a}{a^2+\lambda^2}$$

$$d\phi = -\frac{a}{a^2+\lambda^2}\,da$$

$\therefore\qquad$ $\displaystyle\phi(a) = -\frac{1}{2}\int \frac{2a}{a^2+\lambda^2}\,da$

$$\phi(a) = -\frac{1}{2}\left[\log(a^2+\lambda^2)\right] + C \qquad\qquad\qquad\ldots(1)$$

To find C, put $a = b$

$\therefore\qquad$ $\displaystyle\phi(b) = -\frac{1}{2}\log(b^2+\lambda^2) + C$

But $\qquad\displaystyle\phi(b) = \int_0^\infty \frac{\cos\lambda x}{x}(e^{-bx} - e^{-bx})\,dx = 0 \quad \therefore\ C = \frac{1}{2}\log(b^2+\lambda^2)$

$\therefore$ From (1) $\qquad\displaystyle\phi(a) = -\frac{1}{2}\log(a^2+\lambda^2) + \frac{1}{2}\log(b^2+\lambda^2)$

or $\qquad\boxed{\displaystyle\phi(a) = \frac{1}{2}\log\left(\frac{b^2+\lambda^2}{a^2+\lambda^2}\right)}$

Ex. 14 : *Evaluate* $\displaystyle\int_0^\pi \frac{dx}{a+b\cos x}$, $a > 0$, $|b| < a$. *Hence deduce that*

$$\int_0^\pi \frac{dx}{(a+b\cos x)^2} = \frac{\pi a}{(a^2-b^2)^{3/2}} \quad and \quad \int_0^\pi \frac{\cos x\,dx}{(a+b\cos x)^2} = \frac{-\pi b}{(a^2-b^2)^{3/2}}.$$

Sol. : Let $\qquad\displaystyle I = \int_0^\pi \frac{dx}{a+b\cos x}$ Put $t = \tan\dfrac{x}{2}$, $dx = \dfrac{2\,dt}{1+t^2}$, $\cos x = \dfrac{1-t^2}{1+t^2}$

x	0	π
t	0	∞

$$I = \int_0^\infty \frac{\dfrac{2\,dt}{1+t^2}}{a + \dfrac{b(1+t^2)}{(1+t^2)}}$$

$$= 2\int_0^\infty \frac{dt}{a + at^2 + b - bt^2} = 2\int_0^\infty \frac{dt}{(a+b) + (a-b)t^2}$$

$$= \frac{2}{a-b}\int_0^\infty \frac{dt}{\dfrac{(a+b)}{(a-b)} + t^2} = \frac{2}{a-b}\left[\frac{1}{\sqrt{\dfrac{a+b}{a-b}}}\tan^{-1}\left(\sqrt{\dfrac{a-b}{a+b}}\,t\right)\right]_0^\infty$$

$$= \frac{2}{\sqrt{a^2-b^2}}(\tan^{-1}\infty - \tan^{-1}0) = \frac{2}{\sqrt{a^2-b^2}}\frac{\pi}{2}$$

$\therefore\qquad\boxed{\displaystyle\int_0^\pi \frac{dx}{a+b\cos x} = \frac{\pi}{\sqrt{a^2-b^2}}}\qquad\qquad\ldots(1)$

Differentiating (1) both sides w.r.t. a,

$$\int_0^\pi \frac{\partial}{\partial a}\left(\frac{1}{a + b \cos x}\right) dx = -\frac{\pi}{2} (a^2 - b^2)^{-3/2} \, 2a$$

$$\int_0^\pi -\frac{1}{(a + b \cos x)^2} \, dx = -\pi a \, (a^2 - b^2)^{-3/2}$$

$$\boxed{\int_0^\pi \frac{dx}{(a + b \cos x)^2} = \frac{\pi a}{(a^2 - b^2)^{3/2}}}$$ ... (2)

Again, differentiating (1) w.r.t. b,

$$\int_0^\pi \frac{\partial}{\partial b}\left(\frac{1}{a + b \cos x}\right) dx = -\frac{\pi}{2} (a^2 - b^2)^{-3/2} (-2b)$$

$$\int_0^\pi \frac{-1 \cdot \cos x}{(a + b \cos x)^2} \, dx = \frac{\pi b}{(a^2 - b^2)^{3/2}}$$

i.e. $$\boxed{\int_0^\pi \frac{\cos x \cdot dx}{(a + b \cos x)^2} = \frac{-\pi b}{(a^2 - b^2)^{3/2}}}$$ ... (3)

Hence, (1), (2) and (3) are required results.

Ex. 15 : *Evaluate* $\displaystyle\int_0^{\pi/2} \log (a^2 \cos^2 \theta + b^2 \sin^2 \theta) \, d\theta.$

Sol. :

$$\phi(a) = \int_0^{\pi/2} \log (a^2 \cos^2 \theta + b^2 \sin^2 \theta) \, d\theta$$

$$\frac{d\phi}{da} = \int_0^{\pi/2} \frac{2a \cos^2 \theta \, d\theta}{a^2 \cos^2 \theta + b^2 \sin^2 \theta} = 2a \int_0^{\pi/2} \frac{\cos^2 \theta \, d\theta}{(a^2 - b^2) \cos^2 \theta + b^2}$$

$$= \frac{2a}{a^2 - b^2} \int_0^{\pi/2} \frac{(a^2 - b^2) \cos^2 \theta + b^2 - b^2}{(a^2 - b^2) \cos^2 \theta + b^2} \, d\theta = \frac{2a}{a^2 - b^2} \int_0^{\pi/2} \left[1 - \frac{b^2}{(a^2 - b^2) \cos^2 \theta + b^2}\right] d\theta$$

$$= \frac{2a}{a^2 - b^2} \left[\frac{\pi}{2} - b^2 \int_0^{\pi/2} \frac{\sec^2 \theta \, d\theta}{a^2 - b^2 + b^2 \sec^2 \theta}\right] = \frac{2a}{a^2 - b^2} \left[\frac{\pi}{2} - b^2 \int_0^{\pi/2} \frac{\sec^2 \theta \, d\theta}{a^2 + b^2 (\sec^2 \theta - 1)}\right]$$

$$= \frac{2a}{a^2 - b^2} \left[\frac{\pi}{2} - b^2 \int_0^{\pi/2} \frac{\sec^2 \theta \, d\theta}{a^2 + b^2 \tan^2 \theta}\right] = \frac{2a}{a^2 - b^2} \left[\frac{\pi}{2} - \int_0^{\pi/2} \frac{\sec^2 \theta \, d\theta}{\dfrac{a^2}{b^2} + \tan^2 \theta}\right]$$

Put $\tan \theta = t$, $\sec^2 \theta \, d\theta = dt$

θ	0	$\pi/2$
t	0	∞

$$= \frac{2a}{a^2 - b^2} \left[\frac{\pi}{2} - \int_0^\infty \frac{dt}{\left(\frac{a}{b}\right)^2 + t^2} \right] = \frac{2a}{a^2 - b^2} \left[\frac{\pi}{2} - \left(\frac{b}{a} \tan^{-1} \frac{bt}{a}\right)_0^\infty \right]$$

$$= \frac{2a}{a^2 - b^2} \left[\frac{\pi}{2} - \frac{b}{a} \cdot \frac{\pi}{2} \right] = (\pi) \frac{a}{a^2 - b^2} \cdot \left(\frac{a - b}{a}\right)$$

$$\int d\phi = \int \frac{\pi}{a + b} \, da$$

∴ $\phi(a) = \pi \log(a + b) + C$

To find C, put a = b,

∴ $\phi(b) = \pi \log 2b + C$

But $\phi(b) = \int_0^{\pi/2} \log(b^2 \cos^2 \theta + b^2 \sin^2 \theta) \, d\theta = \log b^2 \left(\frac{\pi}{2}\right) = \pi \log b$

∴ $\pi \log b = \pi \log 2b + C$ ∴ $C = \pi \log \frac{1}{2}$

Hence, $\phi(a) = \pi \log(a + b) + \pi \log \frac{1}{2}$

or $\boxed{\phi(a) = \pi \log \left(\frac{a + b}{2}\right)}$

EXERCISE 4.1

1. Prove that $\displaystyle\int_0^1 \frac{x^a - x^b}{\log x} \, dx = \log \left(\frac{a + 1}{b + 1}\right)$; a > 0, b > 0. **(May 2009)**

[**Hint :** $\dfrac{dI}{da} = \displaystyle\int_0^1 x^a \, dx = \dfrac{1}{a + 1}$, Integrate w.r.t. a and put a = b.]

2. Prove that $\displaystyle\int_0^\infty e^{-x^2} \cos 2ax \cdot dx = \frac{\sqrt{\pi}}{2} e^{-a^2}$. Assume $\displaystyle\int_0^\infty e^{-x^2} \, dx = \frac{\sqrt{\pi}}{2}$

[**Hint :** $\phi'(a) = \displaystyle\int_0^\infty (\sin 2ax) [e^{-x^2}(-2x)] \, dx = -2a \, \phi(a)$ ∴ $\phi(a) = c \, e^{-a^2}$, put a = 0, C = $\frac{\sqrt{\pi}}{2}$]

3. Assuming that $\displaystyle\int_0^\infty \frac{\sin x}{x} \, dx = \frac{\pi}{2}$, evaluate $\displaystyle\int_0^\infty \frac{1 - \cos ax}{x^2} \, dx$. **(May 2011) Ans. :** $\dfrac{\pi a}{2}$

[**Hint :** $\phi'(a) = \displaystyle\int_0^\infty \frac{\sin ax}{x} \, dx = \frac{\pi}{2}$ ∴ $\phi(a) = \frac{\pi a}{2} + c$, put a = 0, c = 0]

4. Prove that $\displaystyle\int_0^\infty \frac{e^{-ax} - e^{-bx}}{x} \, dx = \log \frac{b}{a}$; a > 0, b > 0. **(Nov./Dec. 2019, Dec. 2007, 2004, 2017)**

[**Hint :** $\phi'(a) = -\dfrac{1}{a}$ ∴ $\phi(a) = -\log a + C$, put a = b, C = log b]

5. Prove that $\displaystyle\int_0^\infty \frac{e^{-\alpha x} \sin x}{x} \, dx = \cot^{-1} \alpha$. Deduce that $\displaystyle\int_0^\infty \frac{\sin x}{x} \, dx = \frac{\pi}{2}$ **(Dec. 2013)**

[**Hint :** $\phi'(\alpha) = -\dfrac{1}{\alpha^2 + 1}$ ∴ $\phi(\alpha) = -\tan^{-1} \alpha + c$, put $\alpha = \infty$, C = $\frac{\pi}{2}$]

6. Prove that $a^2 f(y) - \dfrac{d^2 f}{dy^2} = \dfrac{\pi}{2}$, where $f(y) = \displaystyle\int_0^\infty \dfrac{\sin xy}{x(a^2 + x^2)}\, dx$. Assume $\displaystyle\int_0^\infty \dfrac{\sin x}{x}\, dx = \dfrac{\pi}{2}$.   **(Dec. 2005)**

[**Hint :** $f'(y) = \displaystyle\int_0^\infty \dfrac{\cos xy}{(a^2 + x^2)}\, dx$, $f''(y) = \displaystyle\int_0^\infty \dfrac{-x \sin xy}{(a^2 + x^2)} \cdot \dfrac{x}{x}\, dx = -\dfrac{\pi}{2} + a^2 f(y)$]

7. Prove that $\displaystyle\int_0^\infty \dfrac{1}{x^2} \log(1 + ax^2)\, dx = \pi\sqrt{a}\ (a > 0)$. Deduce $\displaystyle\int_0^\infty \dfrac{1}{x^2} \log(1 + x^2)\, dx = \pi$ **(Dec. 2008, 2006; May 2010)**

[**Hint :** $\phi'(a) = \dfrac{1}{a}\displaystyle\int_0^\infty \dfrac{dx}{x^2 + (1/\sqrt{a})^2} = \dfrac{\pi}{2\sqrt{a}}\ \therefore\ \phi(a) = \pi\sqrt{a} + C$, put $a = 0$, $C = 0$.]

8. Prove that $\displaystyle\int_0^\infty e^{-x^2} \cos \alpha x\, dx = \dfrac{\sqrt{\pi}}{2}\, e^{-\frac{\alpha^2}{4}}$.

[**Hint :** $\phi'(\alpha) = \displaystyle\int_0^\infty (\sin \alpha x)\, [e^{-x^2}(-x)]\, dx = -\dfrac{\alpha}{2}\phi(\alpha)\ \therefore\ \phi(\alpha) = C\, e^{-\frac{\alpha^2}{4}}$, put $\alpha = 0$, $C = \dfrac{\sqrt{\pi}}{2}$]

9. Prove that $\displaystyle\int_0^\infty \dfrac{e^{-x} - e^{-ax}}{x \sec x}\, dx = \dfrac{1}{2} \log\left(\dfrac{a^2 + 1}{2}\right)$; $a > 0$. **(May 2006, 2015, Dec. 2010, May 2019)**

[**Hint :** $\phi'(a) = \displaystyle\int_0^\infty e^{-ax} \cos x\, dx = \dfrac{a}{a^2 + 1}\ \therefore\ \phi(a) = \dfrac{1}{2} \log(a^2 + 1) + C$ then put $a = 1$ to find $C = -\dfrac{1}{2} \log 2.$]

10. Prove that $\displaystyle\int_0^{\pi/2} \dfrac{\log(1 + \cos \alpha \cos x)}{\cos x}\, dx = \dfrac{\pi^2}{8} - \dfrac{\alpha^2}{2}$

[**Hint :** $\phi'(\alpha) = -\alpha$, $\phi(\alpha) = -\alpha^2/2 + C$, put $\alpha = \pi/2$, $C = \pi^2/8$]

4.4 RULE - II : INTEGRAL WITH LIMITS AS FUNCTIONS OF THE PARAMETER : LEIBNITZ'S RULE

If $I(\alpha) = \displaystyle\int_{a(\alpha)}^{b(\alpha)} f(x, \alpha)\, dx$, where a and b are functions of the parameter α, then,

$$\dfrac{dI}{d\alpha} = \int_{a(\alpha)}^{b(\alpha)} \dfrac{\partial}{\partial \alpha} \{f(x, \alpha)\}\, dx + f(b, \alpha)\dfrac{db}{d\alpha} - f(a, \alpha)\dfrac{da}{d\alpha}$$

Proof : Since the parameter α enters into the function $I(\alpha)$ due to the integral $f(x, \alpha)$ and due to the limits a, b which are functions of α, we express this by denoting $I(\alpha)$ as $I(\alpha) = \phi(\alpha, b, a)$. From tree diagram 1,

$$\dfrac{dI}{d\alpha} = \dfrac{\partial I}{\partial \alpha}\ (1) + \dfrac{\partial I}{\partial b}\dfrac{db}{d\alpha} + \dfrac{\partial I}{\partial a}\dfrac{da}{d\alpha} \qquad \text{...(I)}$$

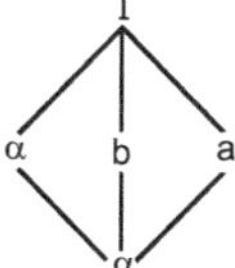

Tree diagram 1

Now, $I(\alpha) = \displaystyle\int_{a(\alpha)}^{b(\alpha)} f(x, \alpha)\, dx.$

By using Rule I, $\dfrac{\partial I}{\partial \alpha}$ is obtained by treating a, b as constants, we have,

$$\frac{\partial I}{\partial \alpha} = \int_{a(\alpha)}^{b(\alpha)} \frac{\partial}{\partial \alpha} f(x, \alpha)\, dx \qquad \ldots\text{(II)}$$

Let the indefinite integral be represented as

$$\int f(x, \alpha)\, dx = \Psi(x, \alpha)$$

i.e.
$$\frac{\partial}{\partial x}[\Psi(x, \alpha)] = f(x, \alpha) \qquad \ldots\text{(III)}$$

Hence
$$\phi(\alpha, b, a) = I(\alpha) = \int_{a(\alpha)}^{b(\alpha)} f(x, \alpha)\, dx = [\Psi(x, \alpha)]_{a(\alpha)}^{b(\alpha)} = \Psi(b, \alpha) - \Psi(a, \alpha) \qquad \ldots\text{(IV)}$$

Hence from (IV), we get,
$$\frac{\partial I}{\partial b} = \frac{\partial \phi}{\partial b} = \frac{\partial}{\partial b} \Psi(b, \alpha) = f(b, \alpha) \ \text{(from III)} \qquad \ldots\text{(V)}$$

$$\frac{\partial I}{\partial a} = \frac{\partial \phi}{\partial a} = -\frac{\partial}{\partial a} \Psi(a, \alpha) = -f(a, \alpha) \ \text{(from III)} \qquad \ldots\text{(VI)}$$

Hence substituting from equations (II), (V), (VI) in (I), we get,

Rule-II :
$$\boxed{\frac{dI}{d\alpha} = \frac{d}{d\alpha} \int_{a(\alpha)}^{b(\alpha)} f(x, \alpha)\, dx = \int_{a(\alpha)}^{b(\alpha)} \frac{\partial}{\partial \alpha} f(x, \alpha)\, dx + f(b, \alpha)\frac{db}{d\alpha} - f(a, \alpha)\frac{da}{d\alpha}}$$

4.5 ILLUSTRATIONS ON RULE-II

Ex. 1 : *Verify the Leibnit's rule of differentiation under integral sign for the integral* $\int_{a}^{a^2} \log(ax)\, dx$.

Sol. : Let
$$\phi(a) = \int_{a}^{a^2} \log(ax)\, dx$$

Using DUIS Rule II :
$$\frac{d\phi}{da} = \int_{a}^{a^2} \frac{\partial}{\partial a} \log(ax)\, dx + \left\{\frac{d}{da}(a^2)\right\} \log(a \cdot a^2) - \left\{\frac{d}{da}(a)\right\} \cdot \log a^2$$

$$= \int_{a}^{a^2} \frac{1}{ax} \cdot x \cdot dx + 2a \cdot \log a^3 - 2\log a = \left[\frac{1}{a}x\right]_{a}^{a^2} + 6a\log a - 2\log a$$

$$= \frac{1}{a}(a^2 - a) + 6a\log a - 2\log a$$

$$\boxed{\frac{d\phi}{da} = a - 1 + 6a\log a - 2\log a} \qquad \ldots\text{(1)}$$

By Actual Integration
$$\phi(a) = \int_{a}^{a^2} \log(ax) \cdot 1 \cdot dx = [\log(ax) \cdot x]_{a}^{a^2} - \int_{a}^{a^2} \frac{1}{ax} \cdot a \cdot x \cdot dx$$

$$= a^2 \log a^3 - a\log a^2 - [x]_{a}^{a^2} = 3a^2 \log a - 2a\log a - (a^2 - a)$$

$$\frac{d\phi}{da} = 6a \log a + 3a^2 \cdot \frac{1}{a} - 2 \log a - 2a \cdot \frac{1}{a} - (2a - 1)$$

$$\boxed{\frac{d\phi}{da} = 6a \log a - 2 \log a + a - 1} \qquad \text{...(2)}$$

From (1) and (2), the rule is verified.

Ex. 2 : *Verify the rule of differentiation under integral sign for the integral* $\displaystyle\int_0^{a^2} \tan^{-1}\frac{x}{a}\, dx$

Sol. : Let $\qquad \phi\,(a) = \displaystyle\int_0^{a^2} \tan^{-1}\frac{x}{a}\, dx$

By DUIS Rule II, $\qquad \phi'\,(a) = \displaystyle\int_0^{a^2} \frac{\partial}{\partial a}\left(\tan^{-1}\frac{x}{a}\right) dx + \left\{\frac{d}{da}(a^2)\right\} \tan^{-1}\left(\frac{a^2}{a}\right) - \left\{\frac{d}{dx}(0)\right\}(\tan^{-1} 0)$

$$= \int_0^{a^2} \frac{1}{1 + \frac{x^2}{a^2}}\left(-\frac{x}{a^2}\right) dx + 2a \tan^{-1} a = -\int_0^{a^2} \frac{x}{a^2 + x^2}\, dx + 2a \tan^{-1} a$$

$$= -\frac{1}{2}\int_0^{a^2} \frac{2x\, dx}{a^2 + x^2} + 2a \tan^{-1} a = -\frac{1}{2}\left[\log(a^2 + x^2)\right]_0^{a^2} + 2a \tan^{-1} a \qquad \text{(Note this step)}$$

$$= -\frac{1}{2}\left[\log(a^2 + a^4) - \log a^2\right] + 2a \tan^{-1} a = -\frac{1}{2}\log\frac{a^2(1 + a^2)}{a^2} + 2a \tan^{-1} a$$

$\therefore \qquad \boxed{\phi'\,(a) = -\frac{1}{2}\log(a^2 + 1) + 2a \tan^{-1} a} \qquad \text{...(1)}$

Next, by actual integration by parts,

$$\phi\,(a) = \int_0^{a^2} \tan^{-1}\left(\frac{x}{a}\right) \cdot 1 \cdot dx = \left[\tan^{-1}\left(\frac{x}{a}\right)(x)\right]_0^{a^2} - \int_0^{a^2} \frac{1}{1 + x^2/a^2} \cdot \frac{1}{a} \cdot x \cdot dx$$

$$= a^2 \tan^{-1} a - 0 - a\int_0^{a^2} \frac{x\, dx}{a^2 + x^2} = a^2 \tan^{-1} a - \frac{a}{2}\left[\log(a^2 + x^2)\right]_0^{a^2}$$

$$= a^2 \tan^{-1} a - \frac{a}{2}\log\frac{a^2(1 + a^2)}{a^2}$$

$$\phi\,(a) = a^2 \tan^{-1} a - \frac{a}{2}\log(a^2 + 1)$$

Differentiating $\qquad \phi'\,(a) = 2a \tan^{-1} a + a^2\frac{1}{1 + a^2} - \frac{1}{2}\log(a^2 + 1) - \frac{a}{2}\left(\frac{2a}{a^2 + 1}\right)$

$$\boxed{\phi'\,(a) = 2a \tan^{-1} a - \frac{1}{2}\log(a^2 + 1)} \qquad \text{...(2)}$$

From (1) and (2), the rule of differentiation under integral sign for the integral is verified.

Ex. 3 : *Show that* $\phi(a) = \displaystyle\int_{\pi/6a}^{\pi/2a} \dfrac{\sin ax}{x}\, dx$ *is independent of a.*   *(Dec. 2010, Nov. 2014)*

Sol. : To show that $\phi(a) = \displaystyle\int_{\pi/6a}^{\pi/2a} \dfrac{\sin ax}{x}\, dx$ is independent of a, we find $\phi'(a)$ using DUIS Rule II,

$$\frac{d\phi}{da} = \int_{\pi/6a}^{\pi/2a} \frac{\partial}{\partial a}\left(\frac{\sin ax}{x}\right) dx + \left\{\frac{d}{da}\left(\frac{\pi}{2a}\right)\right\} \frac{\sin\left(a\,\frac{\pi}{2a}\right)}{\left(\frac{\pi}{2a}\right)} - \left\{\frac{d}{da}\left(\frac{\pi}{6a}\right)\right\} \frac{\sin\left(a\,\frac{\pi}{6a}\right)}{\left(\frac{\pi}{6a}\right)}$$

$$= \int_{\pi/6a}^{\pi/2a} \frac{\cos ax \cdot x \cdot dx}{x} + \left(-\frac{\pi}{2a^2}\right)\frac{1}{(\pi/2a)} - \left(-\frac{\pi}{6a^2}\right)\frac{(1/2)}{(\pi/6a)}$$

$$= \left[\frac{\sin ax}{a}\right]_{\pi/6a}^{\pi/2a} - \frac{1}{a} + \frac{1}{2a} = \frac{1}{a}\left[\sin\frac{\pi}{2} - \sin\frac{\pi}{6}\right] - \frac{1}{a} + \frac{1}{2a}$$

$$= \frac{1}{a} - \frac{1}{2a} + \frac{1}{2a} - \frac{1}{a} = 0$$

Thus, $\qquad\boxed{\dfrac{d\phi}{da} = 0 \text{ implies that } \phi(a) \text{ is independent of a.}}$

Ex. 4 : If $\quad y = \displaystyle\int_0^x f(t)\,\sin a\,(x-t)\, dt$, show that $\dfrac{d^2 y}{dx^2} + a^2 y = a\, f(x)$ *(May 2008)*

Sol. : Given $\qquad y = \displaystyle\int_0^x f(t)\,\sin a\,(x-t)\, dt$

Differentiating by Rule II w.r.t. x,

$$\frac{dy}{dx} = \int_0^x \frac{\partial}{\partial x}[f(t)\,\sin a\,(x-t)]\, dt + \left\{\frac{d}{dx}(x)\right\} f(x)\,\sin 0 - \left\{\frac{d}{dx}(0)\right\} f(0)\cdot\sin 0$$

$$= \int_0^x a\, f(t)\,\cos a\,(x-t)\, dt + 0 - 0$$

Again, differentiating w.r.t. x,

$$\frac{d^2 y}{dx^2} = \int_0^x \frac{\partial}{\partial x}[a\, f(t)\,\cos a\,(x-t)]\, dt + \left[\frac{d}{dx}(x)\right] a\, f(x)\,\cos 0 - \frac{d}{dx}(0)\cdot a\, f(0)\cdot\cos 0$$

$$= \int_0^x a\, f(t)\,(-\sin a\,(x-t))\cdot a \cdot dt + a\cdot f(x) - 0$$

$$= -a^2 \int_0^x f(t)\,\sin a\,(x-t)\, dt + a\, f(x) = -a^2\, y + a\, f(x)$$

$$\therefore \qquad \boxed{\dfrac{d^2 y}{dx^2} + a^2 y = a\, f(x)}$$

Ex. 5 : *Evaluate* $\displaystyle\int_0^a \frac{\log(1+ax)}{1+x^2}\,dx$ *and show that* $\displaystyle\int_0^1 \frac{\log(1+x)}{1+x^2}\,dx = \frac{\pi}{8}\log 2$

Sol. :

$$\phi(a) = \int_0^a \frac{\log(1+ax)}{1+x^2}\,dx$$

$$\phi'(a) = \int_0^a \frac{\partial}{\partial a}\left[\frac{\log(1+ax)}{1+x^2}\right]dx + \left\{\frac{d}{da}(a)\right\}\left(\frac{\log(1+a^2)}{1+a^2}\right) - \left\{\frac{d}{da}(0)\right\}\left(\frac{\log(1+0)}{1+0}\right)$$

$$= \int_0^a \frac{(x)}{(1+ax)(1+x^2)}\,dx + \frac{\log(1+a^2)}{1+a^2} - 0 \qquad\qquad \dots (1)$$

Using partial fraction for

$$\frac{x}{(1+ax)(1+x^2)} = \frac{A}{1+ax} + \frac{Bx+C}{1+x^2}$$

$$\Rightarrow \qquad x = A + Ax^2 + Bx + C + aBx^2 + aCx$$

$$= (A + aB)x^2 + (B + aC) + x + (A + C)$$

Equating coefficient of x^2, x and constant terms

$$A + aB = 0, \quad B + aC = 1, \quad A + C = 0$$

Solving, $\qquad B = \dfrac{1}{a^2+1} \;;\; A = -\dfrac{a}{a^2+1} \;;\; C = \dfrac{a}{a^2+1}$

$$\therefore \qquad \int_0^a \frac{x}{(1+ax)(1+x^2)}\,dx = -\frac{a}{a^2+1}\int_0^a \frac{1}{1+ax}\,dx + \frac{1}{a^2+1}\int_0^a \frac{x+a}{1+x^2}\,dx.$$

$$= \left\{-\frac{1}{a^2+1}\log(1+ax) + \frac{1}{a^2+1}\left[\frac{1}{2}\log(1+x^2) + a\tan^{-1}x\right]\right\}_0^a$$

$$= \frac{1}{a^2+1}\left\{\frac{1}{2}\log(1+x^2) - \log(1+ax) + a\tan^{-1}x\right\}_0^a$$

$$= \frac{1}{a^2+1}\left\{\frac{1}{2}\log(1+a^2) - \log(1+a^2) + a\tan^{-1}a\right\}$$

$\therefore$ From (1), $\qquad \phi'(a) = \dfrac{1}{a^2+1}\left\{\dfrac{1}{2}\log(1+a^2) - \log(1+a^2) + a\tan^{-1}a\right\} + \dfrac{\log(1+a^2)}{1+a^2}$

$$= \frac{1}{a^2+1}\left\{\frac{1}{2}\log(1+a^2) + a\tan^{-1}a\right\}$$

$\therefore \qquad \phi'(a) = \dfrac{1}{2}\dfrac{\log(1+a^2)}{1+a^2} + \dfrac{a}{1+a^2}\cdot\tan^{-1}a$

Integrating both sides w.r.t. 'a'

$$\phi(a) = \frac{1}{2}\int \frac{\log(1+a^2)}{1+a^2}\,da + \int \frac{a}{1+a^2}\cdot\tan^{-1}a\cdot da + C. \qquad\qquad \dots (2)$$

Consider, $\quad \displaystyle\int \frac{a}{1+a^2}\tan^{-1}a\cdot da = \tan^{-1}a\left(\frac{1}{2}\log(1+a^2)\right) - \int \frac{1}{1+a^2}\cdot\frac{1}{2}\log(1+a^2)\cdot da$

Substituting in (2) $\quad \phi(a) = \dfrac{1}{2}\displaystyle\int \dfrac{\log(1+a^2)}{1+a^2}\,da + \dfrac{1}{2}\tan^{-1}a\,\log(1+a^2) - \dfrac{1}{2}\displaystyle\int \dfrac{\log(1+a^2)}{1+a^2}\,da + C$

$$\phi(a) = \frac{1}{2} \tan^{-1} a \, \log(1 + a^2) + C$$

Put $a = 0$,

$$\phi(0) = 0 + C$$

But $\phi(0) = 0 \qquad \therefore \; C = 0$

$\therefore \qquad \phi(a) = \frac{1}{2} \tan^{-1} a \, \log(1 + a^2)$

$$\boxed{\phi(a) = \int_0^a \frac{\log(1 + ax)}{1 + x^2} \, dx = \frac{1}{2} \tan^{-1} a \, \log(1 + a^2)}$$

Put $a = 1$,

$$\boxed{\int_0^1 \frac{\log(1 + x)}{1 + x^2} \, dx = \frac{1}{2} \cdot \frac{\pi}{4} \cdot \log(2) = \frac{\pi}{8} \log 2}$$

EXERCISE 4.2

1. Verify the rule of differentiation under integral sign for the integral $\displaystyle\int_a^{a^2} \frac{1}{x + a} \, dx$.

   [**Hint :** Given $\phi(a) = \log \dfrac{a + 1}{2}$ $\therefore$ $\phi'(a) = \dfrac{1}{a + 1}$. Also, Rule II : $\phi'(a) = \displaystyle\int_a^{a^2} - \dfrac{1}{(x + a)^2} \, dx + (2a)\dfrac{1}{a^2 + a} - (1)\dfrac{1}{2a} = \dfrac{1}{a + 1}$]

2. If $f(x) = \displaystyle\int_0^x (x - t)^2 \, G(t) \, dt$ then show that $\dfrac{d^3 f}{dx^3} - 2\,G(x) = 0$ *(May 2005, 2016, 2018)*

   [**Hint :** Here x is a parameter, $f'(x) = \displaystyle\int_0^x 2(x - t)\,G(t)\,dt$, $f''(x) = 2\displaystyle\int_0^x G(t)\,dt$ and $f'''(x) = 2\left[\displaystyle\int_0^x \frac{\partial}{\partial x}\,G(t)\,dt + \left\{\frac{dx}{dx}\right\}G(x) - \left\{\frac{da}{dx}\right\}G(a)\right]$]

3. If $F(t) = \displaystyle\int_t^{t^2} e^{tx^2} \, dx$, then show that $\dfrac{dF}{dt} = \dfrac{1}{2t}\left[5t^2 e^{t^5} - 3\,t\,e^{t^3} - F(t)\right]$

4. Show that $\dfrac{d}{da} \displaystyle\int_{\sqrt{a}}^{1/a} \cos ax^2 \, dx = -\displaystyle\int_{\sqrt{a}}^{1/a} x^2 \cdot \sin ax^2 \, dx - \dfrac{1}{a^2} \cos \dfrac{1}{a} - \dfrac{1}{2\sqrt{a}} \cos a^2$

5. If $\phi(a) = \displaystyle\int_a^{a^2} \frac{\sin ax}{x} \, dx$, find $\dfrac{d\phi}{da}$

6. Prove that $\displaystyle\int_0^x \frac{dx}{(x^2 + a^2)^2} = \frac{1}{2a^2} \tan^{-1} \frac{x}{a} + \frac{x}{2a^2 (x^2 + a^2)}$

   [**Hint :** starting from $I(x) = \displaystyle\int_0^x \frac{dx}{x^2 + a^2} = \frac{1}{a} \tan^{-1} \frac{x}{a}$]

4.6 ERROR FUNCTION   (Dec. 2009)

1. **Definition :** Error function x is defined as $\dfrac{2}{\sqrt{\pi}} \displaystyle\int_0^x e^{-u^2}\, du$ and is denoted by erf (x).

We write
$$\boxed{\; \text{erf (x)} \;=\; \frac{2}{\sqrt{\pi}} \int_0^x e^{-u^2}\, du \;}$$ ...(1)

This function or integral is also called *Error function integral* or *probability integral* and is encountered in many branches of Mathematics, Physics or Engineering.

2. **Complementary Error Function :** Complementary error function x is defined as $\dfrac{2}{\sqrt{\pi}} \displaystyle\int_x^{\infty} e^{-u^2}\, du$ and is denoted by erfc (x).

We write
$$\boxed{\; \text{erfc (x)} \;=\; \frac{2}{\sqrt{\pi}} \int_x^{\infty} e^{-u^2}\, du \;}$$ ...(2)

3. **Alternate definition of Error Function :** In integral of (1), if we put $u^2 = t$, $2u\, du = dt$ or $du = \dfrac{dt}{2\sqrt{t}}$;

u	0	x
t	0	x^2

$$\text{erf (x)} \;=\; \frac{2}{\sqrt{\pi}} \int_0^{x^2} e^{-t}\,\frac{dt}{2\sqrt{t}} \;=\; \frac{1}{\sqrt{\pi}} \int_0^{x^2} e^{-t}\, t^{-1/2}\, dt$$

$\therefore$
$$\boxed{\; \text{erf (x)} \;=\; \frac{1}{\sqrt{\pi}} \int_0^{x^2} e^{-t}\, t^{-1/2}\, dt \;}$$ ...(3)

This is also considered as definition of Error function x and either (1) or (3) can be used for erf (x) according to the need of the problem.

4.7 PROPERTIES OF ERROR FUNCTIONS

(1)
$$\text{erf } (\infty) \;=\; \frac{2}{\sqrt{\pi}} \int_0^{\infty} e^{-u^2}\, du \qquad\qquad (\text{Put } u^2 = y)$$

$$=\; \frac{2}{\sqrt{\pi}} \int_0^{\infty} e^{-y}\,\frac{1}{2}\, y^{-1/2}\, dy \;=\; \frac{1}{\sqrt{\pi}} \int_0^{\infty} e^{-y} \cdot y^{-1/2}\, dy$$

$$=\; \frac{1}{\sqrt{\pi}} \left|\underline{\frac{1}{2}}\right| \;=\; \frac{1}{\sqrt{\pi}}\, \sqrt{\pi} \;=\; 1$$

$\therefore$ $\boxed{\; \text{erf } (\infty) \;=\; 1 \;}$ ...(4)

(2)
$$\text{erf } (0) \;=\; \frac{2}{\sqrt{\pi}} \int_0^{0} e^{-u^2}\, du \;=\; 0$$

$\therefore$ $\boxed{\; \text{erf } (0) \;=\; 0 \;}$ ...(5)

(3)
$$\text{erf}(x) + \text{erfc}(x) = \frac{2}{\sqrt{\pi}}\left[\int_0^x e^{-u^2}\,du + \int_x^\infty e^{-u^2}\,du\right]$$

$$= \frac{2}{\sqrt{\pi}}\left[\int_0^\infty e^{-u^2}\,du\right] = \text{erf}(\infty) = 1$$

$$\boxed{\text{erf}(x) + \text{erfc}(x) = 1} \qquad \dots(6)$$

(4) Error function is an odd function :

Proof :
$$\text{erf}(x) = \frac{2}{\sqrt{\pi}}\int_0^x e^{-u^2}\,du$$

Replace x by $-x$

$$\therefore \quad \text{erf}(-x) = \frac{2}{\sqrt{\pi}}\int_0^{-x} e^{-u^2}\,du. \quad \text{Put } u = -y;\ du = -dy.$$

u	0	$-x$
y	0	x

$$\text{erf}(-x) = \frac{2}{\sqrt{\pi}}\int_0^x e^{-y^2}(-dy) = -\frac{2}{\sqrt{\pi}}\int_0^x e^{-y^2}\,dy$$

$$\therefore \quad \boxed{\text{erf}(-x) = -\text{erf}(x)} \qquad \dots(7)$$

Thus erf (x) is an odd function.

(5) Expression for erf (x) in series :

$$\text{erf}(x) = \frac{2}{\sqrt{\pi}}\int_0^x e^{-u^2}\,du$$

Since
$$e^{-t} = 1 - t + \frac{t^2}{2!} - \frac{t^3}{3!} + \dots$$

$$\therefore \quad \text{erf}(x) = \frac{2}{\sqrt{\pi}}\int_0^x \left[1 - u^2 + \frac{u^4}{2!} - \frac{u^6}{3!} + \dots\right] du \qquad \text{(By putting } t = -u^2 \text{ in } e^{-t})$$

$$= \frac{2}{\sqrt{\pi}}\left[u - \frac{u^3}{3} + \frac{u^5}{10} - \frac{u^7}{42} + \dots\right]_0^x$$

$$\boxed{\text{erf}(x) = \frac{2}{\sqrt{\pi}}\left[x - \frac{x^3}{3} + \frac{x^5}{10} - \frac{x^7}{42} + \dots\right]} \qquad \dots(8)$$

This series is uniformly convergent and hence erf (x) is a continuous function of x. Values of erf (x) can be tabulated using above series.

Remark : Using above results, we note that : erf$(0) = 0$, erf$(\infty) = 1$, erf$(-\infty) = -$ erf$(\infty) = -1$ and erf(x) is an odd function, graph of erf(x) is given in Fig. 4.1.

(6) Alternate definition of Complementary error function : By result (4), erf $(\infty) = 1$.

$$\text{erf}(\infty) = \frac{1}{\sqrt{\pi}}\int_0^\infty e^{-t}\,t^{-1/2}\,dt = 1$$

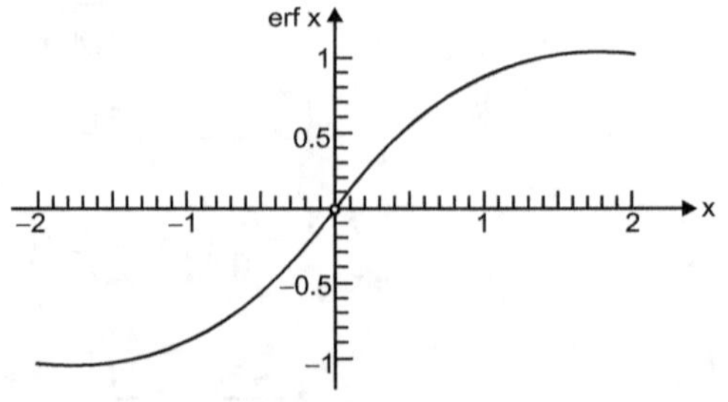

Fig. 4.1 : Error Function

This can be rewritten as,

$$\frac{1}{\sqrt{\pi}} \left\{ \int_{0}^{x^2} e^{-t}\, t^{-1/2}\, dt + \int_{x^2}^{\infty} e^{-t}\, t^{-1/2}\, dt \right\} = 1$$

$$\frac{1}{\sqrt{\pi}} \int_{0}^{x^2} e^{-t}\, t^{-1/2}\, dt + \frac{1}{\sqrt{\pi}} \int_{x^2}^{\infty} e^{-t}\, t^{-1/2}\, dt = 1 \qquad\qquad …(9)$$

Here first integral on L.H.S. of (9) is erf (x) [refer result (3)] and second integral $= \dfrac{1}{\sqrt{\pi}} \displaystyle\int_{x^2}^{\infty} e^{-t}\, t^{-1/2}\, dt$ is defined as *complementary error function x* or briefly written as erfc (x). This function also occurs in many branches of engineering and sciences. Hence we have,

$$\boxed{\; \text{erfc (x)} \;=\; \frac{1}{\sqrt{\pi}} \int_{x^2}^{\infty} e^{-t}\, t^{-1/2}\, dt \;} \qquad\qquad … (10)$$

Thus, from result (9), we note that, erf (x) + erfc (x) $=$ 1

4.8 DIFFERENTIATION OF ERROR FUNCTION

$$\text{erf (x)} \;=\; \frac{2}{\sqrt{\pi}} \int_{0}^{x} e^{-u^2}\, du$$

$$\therefore \qquad \text{erf (ax)} \;=\; \frac{2}{\sqrt{\pi}} \int_{0}^{ax} e^{-u^2}\, du$$

Using second rule of differentiation under the integral sign (Refer article 4.4), noting that integration is w.r.t. u and differentiation is to be carried out w.r.t. x.

$$\frac{d}{dx} \text{ erf (ax)} \;=\; \frac{2}{\sqrt{\pi}} \left[\int_{0}^{ax} \frac{\partial}{\partial x} e^{-u^2}\, du + \left\{\frac{d}{dx}\,(ax)\right\} e^{-a^2 x^2} - \left\{\frac{d}{dx}\,(0)\right\} e^{-0} \right] \qquad\textbf{(May 2008)}$$

$$= \frac{2}{\sqrt{\pi}} \left[0 + a \cdot e^{-a^2 x^2} - 0 \right] = \frac{2a\, e^{-a^2 x^2}}{\sqrt{\pi}}$$

$$\therefore \qquad \boxed{\; \frac{d}{dx} \text{ erf (ax)} \;=\; \frac{2\, a\, e^{-a^2 x^2}}{\sqrt{\pi}} \;} \qquad\qquad … (11)$$

4.9 INTEGRATION OF ERROR FUNCTION

$$\int_{0}^{t} \text{erf (ax)}\, dx \;=\; \int_{0}^{t} 1 \,.\, \text{erf (ax)}\, dx \qquad\qquad\textbf{(May 2008)}$$

Integrating by parts treating unity as second function and erf (ax) as first function

$$= \big[\text{erf (ax)} \cdot x \big]_{0}^{t} - \int_{0}^{t} \frac{d}{dx} \text{ erf (ax)} \cdot x \cdot dx$$

$$= t\, \text{erf (at)} - 0 - \int_{0}^{t} \frac{2\, ae^{-a^2 x^2}}{\sqrt{\pi}} \cdot x \cdot dx \qquad\qquad \left(\because\; \frac{d}{dx} \text{ erf (ax)} = \frac{2a\, e^{-a^2 x^2}}{\sqrt{\pi}} \right)$$

$$= t\ \mathrm{erf}\,(at) + \frac{1}{\sqrt{\pi}} \cdot \frac{1}{a} \int_0^t e^{-a^2 x^2}\,(-2a^2\,x\,dx) \qquad \text{(Note this step)}$$

$$= t \cdot \mathrm{erf}\,(at) + \frac{1}{a\sqrt{\pi}} \left[e^{-a^2 x^2} \right]_0^t$$

$$= t \cdot \mathrm{erf}\,(at) + \frac{1}{a\sqrt{\pi}} \left(e^{-a^2 t^2} - 1 \right)$$

$$\boxed{\int_0^t \mathrm{erf}\,(ax)\,dx \;=\; t\,\mathrm{erf}\,(at) + \frac{1}{a\sqrt{\pi}}\,e^{-a^2 t^2} - \frac{1}{a\sqrt{\pi}}} \qquad \ldots (12)$$

4.10 ILLUSTRATIONS ON ERROR FUNCTION

Ex. 1 : *Prove that* $\dfrac{d}{dx}\,[\mathrm{erf}\,(x)] = \dfrac{2}{\sqrt{\pi}}\,e^{-x^2}$ *and use it to show that* $\dfrac{d}{dx}\,[\mathrm{erf}\,(ax^n)] = \dfrac{2\,an}{\sqrt{\pi}}\,x^{n-1}\,e^{-a^2\,x^{2n}}$ **(Dec. 08, 07, 06, 04; May 2011)**

Sol. : We know that
$$\mathrm{erf}\,(x) = \frac{2}{\sqrt{\pi}} \int_0^x e^{-u^2}\,du$$

Using second rule of differentiation under the integral sign (Refer article 4.4), noting that integration is w.r.t. u and differentiation is to be carried out w.r.t. x,

$$\frac{d}{dx}\,[\mathrm{erf}\,(x)] = \frac{2}{\sqrt{\pi}} \left\{ \int_0^x \frac{\partial}{\partial x}\,(e^{-u^2})\,du + \frac{d}{dx}\,(x) \cdot e^{-x^2} - \frac{d}{dx}\,(0) \cdot e^{-0} \right\}$$

$$= \frac{2}{\sqrt{\pi}} \left\{ 0 + e^{-x^2} - 0 \right\}$$

$$\boxed{\frac{d}{dx}\,[\mathrm{erf}\,(x)] = \frac{2}{\sqrt{\pi}}\,e^{-x^2}}$$

Also,
$$\mathrm{erf}\,(ax^n) = \frac{2}{\sqrt{\pi}} \int_0^{ax^n} (e^{-u^2})\,du$$

$\therefore$
$$\frac{d}{dx}\,[\mathrm{erf}\,(ax^n)] = \frac{2}{\sqrt{\pi}} \left\{ \int_0^{ax^n} \frac{\partial}{\partial x}\,(e^{-u^2})\,du + \frac{d}{dx}\,(ax^n) \cdot e^{-a^2 x^{2n}} - \frac{d}{dx}\,(0) \cdot e^{-0} \right\}$$

$$= \frac{2}{\sqrt{\pi}} \left\{ 0 + nax^{n-1}\,e^{-a^2 x^{2n}} - 0 \right\}$$

$$\boxed{\frac{d}{dx}\,[\mathrm{erf}\,(ax^n)] = \frac{2an}{\sqrt{\pi}}\,x^{n-1}\,e^{-a^2 x^{2n}}}$$

Ex. 2 : *Show that* $\dfrac{d}{dx}\,[\mathrm{erfc}\,(ax)] = -\dfrac{2a}{\sqrt{\pi}}\,e^{-a^2 x^2}$ *and evaluate* $\displaystyle\int_0^t \mathrm{erfc}\,(au)\,du$

Sol. : We know that $\mathrm{erfc}\,(ax) = \dfrac{2}{\sqrt{\pi}} \displaystyle\int_{ax}^{\infty} e^{-u^2}\,du$

$$\frac{d}{dx}[\text{erfc}(ax)] = \frac{2}{\sqrt{\pi}}\left\{\int_{ax}^{\infty}\frac{\partial}{\partial x}(e^{-u^2})\,du + \frac{d}{dx}(\infty)\cdot e^{-\infty} - \frac{d}{dx}(ax)\cdot e^{-a^2x^2}\right\}$$

$$= \frac{2}{\sqrt{\pi}}\left\{0 + 0 - ae^{-a^2x^2}\right\}$$

$$\boxed{\frac{d}{dx}[\text{erfc}(ax)] = -\frac{2a}{\sqrt{\pi}}\,e^{-a^2x^2}} \qquad \qquad \dots (1)$$

Next, consider, $\displaystyle\int_0^t \text{erfc}(au)\,du = \int_0^t \text{erfc}(au)\cdot 1\cdot du$ (Note this step)

$$= [\text{erfc}(au)\cdot u]_0^t - \int_0^t \frac{d}{du}[\text{erfc}(au)\cdot u]\,du$$

$$= t\cdot\text{erfc}(at) - 0 - \int_0^t -\frac{2a}{\sqrt{\pi}}\,e^{-a^2u^2}\cdot u\cdot du \qquad\qquad [\text{(using above result (1)]}$$

$$= t\,\text{erfc}(at) - \frac{1}{a\sqrt{\pi}}\int_0^t e^{-a^2u^2}\cdot(-2a^2u)\cdot du \qquad\qquad (\text{Note this step})$$

$$= t\cdot\text{erfc}(at) - \frac{1}{a\sqrt{\pi}}\left[e^{-a^2u^2}\right]_0^t$$

$$\therefore \qquad \boxed{\int_0^t \text{erfc}(au)\,du = t\,\text{erfc}(at) - \frac{1}{a\sqrt{\pi}}\,e^{-a^2t^2} + \frac{1}{a\sqrt{\pi}}}$$

Ex. 3 : *Show that* $\displaystyle\int_0^t \text{erf}(ax)\,dx + \int_0^t \text{erfc}(ax)\,dx = t$

Sol. : $\displaystyle\int_0^t \text{erf}(ax)\,dx + \int_0^t \text{erfc}(ax)\,dx = \int_0^t [\text{erf}(ax) + \text{erfc}(ax)]\,dx.$

$$= \int_0^t (1)\cdot dx = [x]_0^t \qquad\qquad \{\because \text{erf}(ax) + \text{erfc}(ax) = 1\}$$

$$\boxed{\int_0^t \text{erf}(ax)\,dx + \int_0^t \text{erfc}(ax)\,dx = t}$$

Ex. 4 : *Prove that* $\text{erfc}(-x) + \text{erfc}(x) = 2$ **(Dec. 2008, 2007, 2004)**

Sol. : We have $\text{erf}(x) + \text{erfc}(x) = 1$

Replace x by –x $\text{erf}(-x) + \text{erfc}(-x) = 1$

$$-\text{erf}(x) + \text{erfc}(-x) = 1 \qquad\qquad \{\because \text{erf}(-x) = -\text{erf}(x)\}$$

$$\text{erfc}(-x) = 1 + \text{erf}(x)$$

$$\text{erfc}(-x) + \text{erfc}(x) = 1 + \text{erf}(x) + \text{erfc}(x) = 1 + 1$$

$$\boxed{\text{erfc}(-x) + \text{erfc}(x) = 2}$$

Ex. 5 : *Prove that* $\dfrac{1}{x}\dfrac{d}{da}\,erfc\,(ax) = -\dfrac{1}{a}\dfrac{d}{dx}\,erf\,(ax)$ **(May 2009, 2008)**

Sol. : Consider

$$erfc\,(ax) \;=\; \frac{2}{\sqrt{\pi}}\int_{ax}^{\infty} e^{-u^2}\,du$$

$$\frac{d}{da}\,erfc\,(ax) \;=\; \frac{2}{\sqrt{\pi}}\left\{\int_{ax}^{\infty}\frac{\partial}{\partial a}(e^{-u^2})\cdot du + \frac{d}{da}(\infty)\cdot e^{-\infty} - \frac{d}{da}(ax)\cdot e^{-a^2 x^2}\right\}$$

$$=\; \frac{2}{\sqrt{\pi}}\left\{0 + 0 - x\,e^{-a^2 x^2}\right\} \;=\; -\,\frac{2x\,e^{-a^2 x^2}}{\sqrt{\pi}}$$

$$\boxed{\;\frac{1}{x}\frac{d}{da}\,erfc\,(ax) \;=\; -\frac{2}{\sqrt{\pi}}\,e^{-a^2 x^2}\;}\qquad\qquad\text{...(1)}$$

Next, consider

$$\frac{d}{dx}\,erf\,(ax) \;=\; \frac{2}{\sqrt{\pi}}\left\{\int_{0}^{ax}\frac{\partial}{\partial x}(e^{-u^2})\,du + \frac{d}{dx}(ax)\cdot e^{-a^2 x^2} - \frac{d}{dx}(0)\cdot e^0\right\}$$

$$=\; \frac{2}{\sqrt{\pi}}\left\{0 + a\cdot e^{-a^2 x^2} - 0\right\} \;=\; \frac{2a\,e^{-a^2 x^2}}{\sqrt{\pi}}$$

$$\boxed{\;-\frac{1}{a}\frac{d}{dx}\,erf\,(ax) \;=\; -\frac{2}{\sqrt{\pi}}\,e^{-a^2 x^2}\;}\qquad\qquad\text{...(2)}$$

From (1) and (2), the required result follows.

Ex. 6 : *Show that* $\displaystyle\int_{a}^{b} e^{-x^2}\,dx = \frac{\sqrt{\pi}}{2}\,[erf\,(b) - erf\,(a)]$ **(May 2007, Dec. 2004, May 2014)**

Sol. : By definition,

$$erf\,(x) \;=\; \frac{2}{\sqrt{\pi}}\int_{0}^{x} e^{-u^2}\,du$$

If $x = \infty$, then

$$erf\,(\infty) \;=\; \frac{2}{\sqrt{\pi}}\int_{0}^{\infty} e^{-u^2}\,du$$

$\therefore$

$$1 \;=\; \frac{2}{\sqrt{\pi}}\int_{0}^{\infty} e^{-x^2}\,dx \quad\text{(Note this step)}\qquad\qquad \{\because erf\,(\infty) = 1\}$$

Assuming that $b > a$, we can write,

$$1 \;=\; \frac{2}{\sqrt{\pi}}\left\{\int_{0}^{a} e^{-x^2}\,dx + \int_{a}^{b} e^{-x^2}\,dx + \int_{b}^{\infty} e^{-x^2}\,dx\right\}$$

$$1 \;=\; \frac{2}{\sqrt{\pi}}\int_{0}^{a} e^{-x^2}\,dx + \frac{2}{\sqrt{\pi}}\int_{a}^{b} e^{-x^2}\,dx + \frac{2}{\sqrt{\pi}}\int_{b}^{\infty} e^{-x^2}\,dx$$

$$1 \;=\; erf\,(a) + \frac{2}{\sqrt{\pi}}\int_{a}^{b} e^{-x^2}\,dx + erfc\,(b).$$

$$1 - \text{erfc}(b) = \text{erf}(a) + \frac{2}{\sqrt{\pi}} \int_{a}^{b} e^{-x^2} \, dx.$$

$$\text{erf}(b) - \text{erf}(a) = \frac{2}{\sqrt{\pi}} \int_{a}^{b} e^{-x^2} \, dx. \qquad \{\because \text{erf}(b) + \text{erfc}(b) = 1\}$$

$$\therefore \qquad \boxed{\int_{a}^{b} e^{-x^2} \, dx = \frac{\sqrt{\pi}}{2} \left[\text{erf}(b) - \text{erf}(a)\right]}$$

Ex. 7 : *Show that* $\displaystyle\int_{0}^{\infty} e^{-x^2 - 2bx} \, dx = \frac{\sqrt{\pi}}{2} \cdot e^{b^2} [1 - \text{erf}(b)]$ **(May 05, 04; Dec. 09, Nov. 15)**

Sol. :

$$I = \int_{0}^{\infty} e^{-x^2 - 2bx} \, dx = \int_{0}^{\infty} e^{-x^2 - 2bx - b^2 + b^2} \, dx = e^{b^2} \int_{0}^{\infty} e^{-(x+b)^2} \, dx$$

Put $x + b = u$, $dx = du$ and

x	0	∞
u	b	∞

$$I = e^{b^2} \int_{b}^{\infty} e^{-u^2} \, du = e^{b^2} \frac{\sqrt{\pi}}{2} \cdot \frac{2}{\sqrt{\pi}} \int_{b}^{\infty} e^{-u^2} \, du$$

$$= \frac{\sqrt{\pi}}{2} e^{b^2} \cdot \text{erfc}(b)$$

$$\boxed{I = \frac{\sqrt{\pi}}{2} e^{b^2} [1 - \text{erf}(b)]}$$

Ex. 8 : *If* $\alpha(x) = \sqrt{\dfrac{2}{\pi}} \displaystyle\int_{0}^{x} e^{-\frac{t^2}{2}} \, dt$ *show that* $\text{erf}(x) = \alpha\left[x\sqrt{2}\right]$ **(Dec. 2010, 2006)**

Sol. :

$$\alpha(x) = \sqrt{\frac{2}{\pi}} \int_{0}^{x} e^{-\frac{t^2}{2}} \, dt$$

$$\therefore \qquad \alpha\left[x\sqrt{2}\right] = \sqrt{\frac{2}{\pi}} \int_{0}^{x\sqrt{2}} e^{-\frac{t^2}{2}} \, dt$$

Put $t^2 = 2u^2$, $2t \, dt = 4u \, du$, $dt = \dfrac{2u \, du}{t} = \dfrac{2u \, du}{\sqrt{2} \cdot u} = \sqrt{2} \, du$

t	0	$x\sqrt{2}$
u	0	x

$$= \sqrt{\frac{2}{\pi}} \int_{0}^{x} e^{-u^2} \sqrt{2} \cdot du$$

$$= \frac{2}{\sqrt{\pi}} \int_{0}^{x} e^{-u^2} \, du$$

$$\therefore \qquad \boxed{\alpha\left[x\sqrt{2}\right] = \text{erf}(x)}$$

EXERCISE 4.3

1. Define erf (x), erfc (x), erf $(\sqrt{t})$, erfc $(\sqrt{t})$.

2. Show that erf (x) is an odd function.   **(Dec. 2011)**

3. Find erf (0), erf (∞),   **Ans. :** erf $(0) = 0$, erf $(\infty) = 1$

4. Plot the graphs of erf (x) and erfc (x).

5. Find $\dfrac{d}{dx}$ erfc (ax^n).

   [**Hint :** Refer solved example 1.]

6. Express erf $(\sqrt{t})$ as an infinite series.   **Ans.** erf $(\sqrt{t}) = \dfrac{2}{\sqrt{\pi}}\left[\sqrt{t} - \dfrac{t^{3/2}}{3} + \dfrac{t^{5/2}}{2!\,.5} \cdots\cdots\right]$

7. Show that $\displaystyle\int_0^\infty e^{-(x+a)^2}\,dx = \dfrac{\sqrt{\pi}}{2}\,[1 - \text{erf}(a)]$

8. Show that $\dfrac{d}{dt}$ erf $(\sqrt{t}) = \dfrac{e^{-t}}{\sqrt{\pi t}}$ and hence evaluate $\displaystyle\int_0^\infty e^{-t}\,\text{erf}(\sqrt{t})\,dt$.   **(May 2010, 2006)**

   [**Hint :** $\dfrac{d}{dt}\text{erf}(\sqrt{t}) = \dfrac{2}{\sqrt{\pi}}\left[\displaystyle\int_0^{\sqrt{t}} \dfrac{\partial}{\partial t} e^{-u^2}\,du + \left\{\dfrac{d}{dt}(\sqrt{t})\right\}e^{-t} - \left\{\dfrac{d}{dt}(0)\right\}e^{-0}\right] = \dfrac{e^{-t}}{\sqrt{\pi t}}$

   and $\displaystyle\int_0^\infty e^{-t}\,(\text{erf}\sqrt{t})\,dt = [\text{erf}(\sqrt{t})\,(-e^{-t})]_0^\infty - \displaystyle\int_0^\infty \left\{\dfrac{d}{dt}\text{erf}(\sqrt{t})\right\}(-e^{-t})\,dt = \dfrac{1}{\sqrt{\pi}}\displaystyle\int_0^\infty e^{-2t}\,t^{-1/2}\,dt = \dfrac{1}{\sqrt{2}}$]

9. Find $\dfrac{d}{dx}$ erfc $(\sqrt{x})$

   [**Hint :** $\dfrac{2}{\sqrt{\pi}}\dfrac{d}{dx}\left[\displaystyle\int_{\sqrt{x}}^\infty e^{-u^2}\,du\right] = \dfrac{2}{\sqrt{\pi}}\left[\displaystyle\int_{\sqrt{x}}^\infty 0 + 0 - \left\{\dfrac{d}{dx}(\sqrt{x})\right\}e^{-x}\right] = -\dfrac{1}{\sqrt{\pi x}}\,e^{-x}$]

10. Show that $\displaystyle\int_0^\infty e^{-st}\,\text{erf}(\sqrt{t})\,dt = \dfrac{1}{s\sqrt{s+1}}$ **(May 2005)**

    [**Hint :** $\left[\text{erf}(\sqrt{t})\left(\dfrac{e^{-st}}{-s}\right)\right]_0^\infty - \displaystyle\int_0^\infty \left\{\dfrac{d}{dt}\text{erf}(\sqrt{t})\right\}\dfrac{e^{-st}}{-s}\,dt = \dfrac{1}{s}\displaystyle\int_0^\infty e^{-(s+1)t}\,t^{-1/2}\,dt = \dfrac{1}{s\sqrt{s+1}}$]

□□□

CHAPTER-5

CURVE TRACING AND RECTIFICATION OF CURVES

5.1 INTRODUCTION

A mathematical function can be better described or understood by its plot or graphical representation. Till now students are well acquainted with curves like straight line, parabola, hyperbola and ellipse. Standard mathematical equations of these curves and their properties are well known at this stage. Aim of this chapter is to introduce the students, with general principles of curve plotting, so that many more curves with their mathematical equations are well understood. Mathematical studies of various kinds of regions with different boundaries can then be taken-up. Curve tracing principles taken-up in this chapter enable us to know the approximate shape of the curve, without plotting large number of points on the curve. Method to trace a curve depends upon the representation of its equation in cartesian, polar or parametric form. Study of this chapter is based upon these three categories of curves along with certain common principles.

After discussing the methods to trace the curve, measurements of the lengths of the arcs of the curves are taken-up. After covering the formulae to measure the arcs of the curves, large number of problems are solved to demonstrate the use of the formulae and methods to trace the curve. This chapter forms the basis of the chapter describing many more applications of integration.

5.2 BASIC DEFINITIONS

1. **Convex Upwards :** If the portion of the curve on both sides of 'A' lies below the tangent at A, then the curve is *convex upwards* (or *concave downwards*).

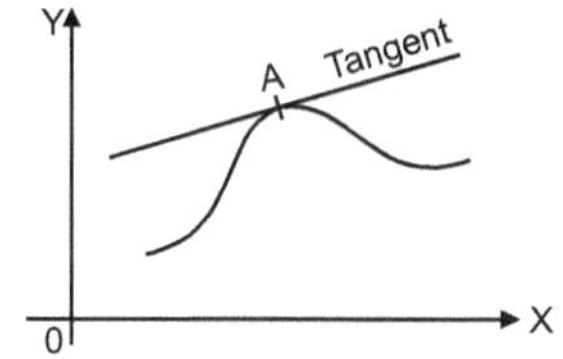

Fig. 5.1

2. **Convex Downwards :** If the portion of the curve on both sides of 'A' lies above the tangent at A, then the curve is *convex downwards* (or *concave upwards*).

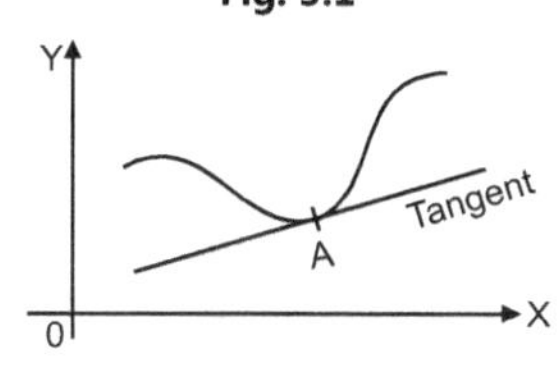

Fig. 5.2

3. **Singular Points :** An unusual point on a curve is called a *singular point* such as, a point of inflexion, a double point, a multiple point, cusp, node or a conjugate point.

4. **Point of Inflexion :** The point that separates the convex part of a continuous curve from the concave part is called the *point of inflexion* of the curve.

   It is obvious that at the point of inflexion the tangent line, if it exists, **cuts** the curve, because on one side the curve lies **under** the tangent and on the other side, **above** it.

   In other words, a point where the curve unusually crosses its tangent is called a *point of inflexion*.

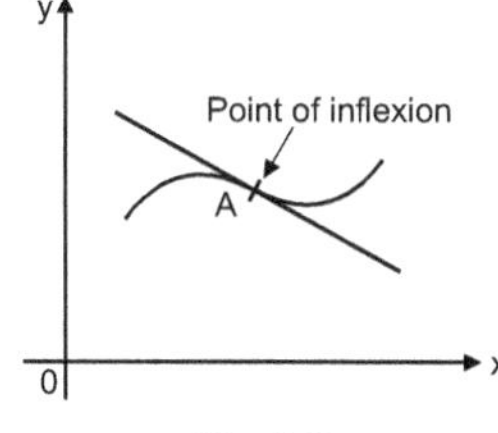

Fig. 5.3

5. **Multiple Point :** A point through which more than one branches of a curve pass is called a *multiple point* of the curve.

6. **A Double Point :** A point on a curve is called a *double point,* if two branches of the curve pass through it.

   A *triple point*, if three branches pass through it. If r branches pass through a point, the point is called a *multiple point* of rth order.

7. **Node :** A double point is called *node* if the branches of curve passing through it are real and the tangents at the common point of intersection are real and distinct (not coincident).

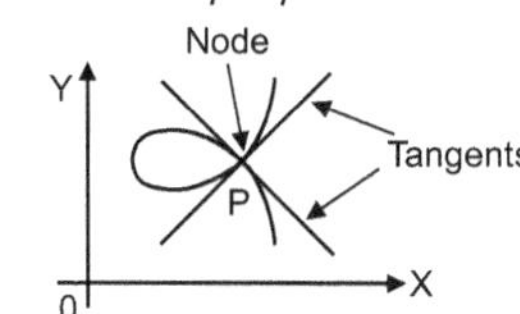

Fig. 5.4

8. **Cusp :** A double point is called *cusp* if the tangents at it to the two branches of the curve are coincident.

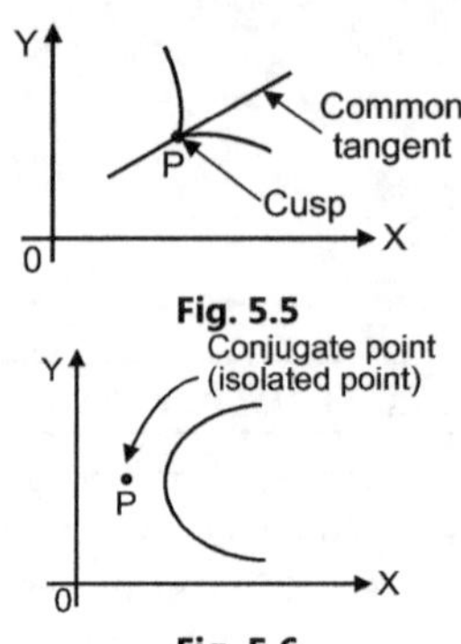

Fig. 5.5

9. **A Conjugate Point :** P is called a *conjugate point* on the curve if there are no real points on the curve in the vicinity of the point P. It is also sometimes called an *isolated point.*

Fig. 5.6

For illustrating methods of tracing, we divide the curves into five types as :

Type 1 : Curves given by Cartesian equations (explicit relations).

Type 2 : Curves given by parametric equations.

Type 3 : Curves given by polar equations.

Type 4 : Curves given by polar equations of the type $r = a \sin n\theta$ or $r = a \cos n\theta$.

Type 5 : Curves given by Cartesian equations (implicit relations).

TYPE 1 : CURVES GIVEN BY CARTESIAN EQUATIONS (EXPLICIT RELATIONS)

5.3 TRACING OF CARTESIAN CURVES

The following rules will help in tracing a Cartesian curve together with the definitions stated in article 5.2.

Rule 1 : Symmetry :

(a) **Symmetry About X-Axis :** If the equation of the curve is such that the powers of y are even everywhere, then the curve is symmetrical about X-axis.

For example, $y^2 = 4\ ax$.

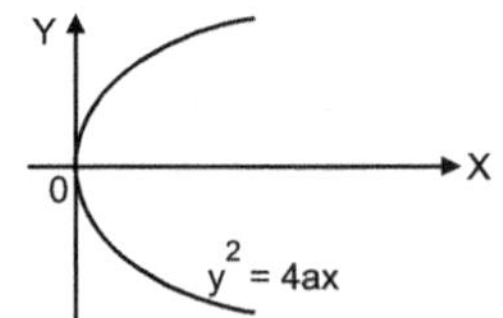

Fig. 5.7

(b) **Symmetry About Y-Axis :** If the equation of the curve is such that the powers of x are even everywhere, then the curve is symmetrical about Y-axis.

For example, $x^2 = 4\ ay$

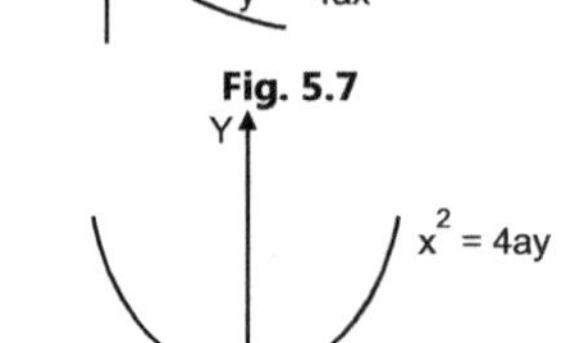

Fig. 5.8

(c) **Symmetry About Both X and Y Axes :** If the equation of the curve is such that the powers of x and y both are even everywhere, then the curve is symmetrical about both the axes.

For example, $\dfrac{x^2}{a^2} + \dfrac{y^2}{b^2} = 1.$

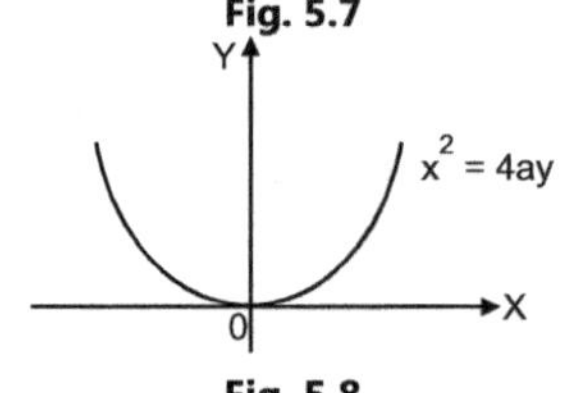

Fig. 5.9

(d) **Symmetry in Opposite Quadrants :** A curve is symmetrical in opposite quadrants if its equation remains unchanged if x and y are changed to $-x$ and $-y$ simultaneously.

For example, $y = x^3$

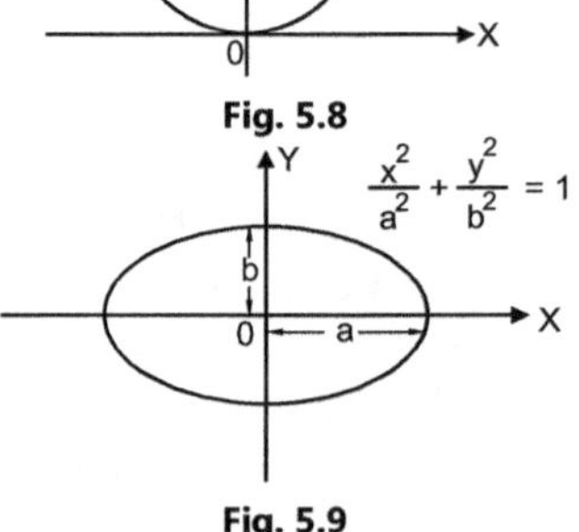

Fig. 5.10

(e) **Symmetry About the Line y = x :** If the equation of the curve remains unaltered when x is changed to y and y to x, the curve is said to be symmetrical about the line $y = x$.

For example, $xy = c^2$

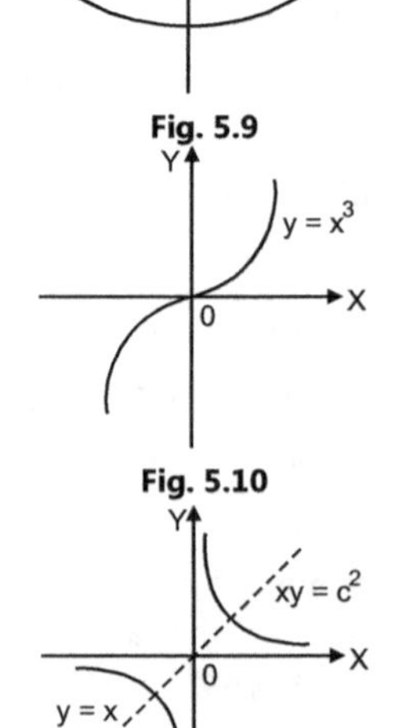

Fig. 5.11

(f) Symmetry About the Line y = – x : If the equation of the curve remains unaltered when x is changed to – y and y to – x, the curve is said to be symmetrical about the line y = – x.

Rule 2 : Points of Intersection :

(a) Origin : Find out whether the curve passes through the origin. It will pass through the origin if the equation is satisfied by (0, 0). In other words, if the equation of the curve does not contain any absolute constant, then it passes through the origin.

(b) Tangents at the Origin : To investigate the nature of a multiple point, it is necessary to find the tangent or tangents at that point.

Newton's Method : If a curve is given by a rational integral algebraic equation and passes through the origin; **the equation of the tangent or tangents at the origin,** can be obtained by equating to zero, the lowest degree terms taken together in the equation of the curve.

(c) Intersections with the Co-ordinate Axes : If possible, try to express the given equation in the explicit form say $y = f(x)$ or $x = f(y)$.

To find the intersection with X-axis, put $y = 0$.

To find the intersection with Y-axis, put $x = 0$.

Find the tangents at these points, if necessary and the position of the curve relative to these lines.

(d) If a curve is symmetrical about the line $y = x$ or $y = – x$, find the points of intersection of the curve with these lines and also the tangents at that point because *the tangent leads the curve.*

Rule 3 : Asymptotes :

Asymptotes are the tangents to the curve at infinity. Find the asymptotes and the position of the curve with respect to them.

Asymptotes Parallel to Co-ordinate Axes :

(a) To find the equation of the *asymptote parallel to X-axis*, equate to zero the coefficient of highest degree terms in x.

(b) To find the equation of the *asymptote parallel to Y-axis*, equate to zero the coefficient of highest degree terms in y.

Oblique Asymptotes : Asymptotes which are not parallel to co-ordinate axes are called as *oblique asymptotes.*

(c) **Method 1 :** Let $y = mx + c$ be the asymptote. The points of intersection with the curve $f(x, y) = 0$ are given by $f(x, mx + c) = 0$. Equate to zero the coefficients of two successive highest powers of x, giving equations to determine m and c.

(d) **Method 2 :**

    (i) Let $y = mx + c$ be the equation of the asymptote.

    (ii) Find $\phi_n (m)$ by putting $x = 1$ and $y = m$ in the highest degree (n) terms of the equation.

    (iii) Similarly, find $\phi_{n-1} (m)$.

    (iv) Solve $\phi_n (m) = 0$ to determine m.

    (v) Find 'c' by the formula $c = -\dfrac{\phi_{n-1} (m)}{\phi'_n (m)}$.

    (vi) If the roots of m are equal, then find c by $\dfrac{c^2}{2} \phi'' (m) + c \phi'' (m) + \phi_{n-2} (m) = 0$.

    (For details see solved example 1, Type 5)

(e) Find more finite points where asymptote meets any other branch of the curve anywhere.

Rule 4 : Special points on the curve :

(a) Find out such points on the curve whose presence can be readily detected.

(b) Find $\dfrac{dy}{dx}$ and the points where the tangent is parallel to either axes.

(c) If $\left(\dfrac{dy}{dx}\right)_{P\,(x_1,\,y_1)} = 0$ then the tangent at P (x_1, y_1) is parallel to X-axis.

(d) If $\left(\dfrac{dy}{dx}\right)_{P\,(x_1,\,y_1)} = \infty$ then the tangent at P (x_1, y_1) is parallel to Y-axis.

(e) Also find point of inflexion, if any, double or multiple points, node, cusp, conjugate or (isolated) point (see basic definitions, refer article 5.2).

Rule 5 : Region of absence of the curve :

(a) If possible, express the equation in the explicit form say $y = f(x)$ and examine how y varies as x varies continuously.

(b) For $y = f(x)$, if y becomes imaginary for some value of $x > a$ (say), then no part of the curve exists beyond $x = a$.

(c) For $x = f(y)$, if x becomes imaginary for some value of $y > b$ (say), then no part of the curve exists beyond $y = b$.

Some Useful Remarks :

(a) When we have to solve for $y = f(x)$, put $x = 0$ and see what is y. Also, observe how y varies as x increases from 0 to $+\infty$, paying special attention to the values of y for which $y = 0$ or $y \to +\infty$.

Also, observe how y behaves as x becomes negative and $x \to -\infty$; being more careful for y becoming zero or tending to $-\infty$.

(b) If $y \to \infty$ as $x \to a$ then $x = a$ must be an asymptote parallel to Y-axis.

If $x \to \infty$ as $y \to b$ then $y = b$ must be an asymptote parallel to X-axis.

(c) If we observe that $y \to \infty$ as $x \to \infty$ and there is approximately a linear relation between x and y for larger values of x, we may expect an oblique asymptote.

(d) If the curve is symmetrical about X-axis or in the opposite quadrants then only positive value of y may be considered. We may draw the curve for negative values of y by symmetry.

If the curve is symmetrical about Y-axis, we need not consider negative values of x.

At this time, we may observe the existence of a loop, which is the part of the curve.

Note : If there are two points of the curve on any axis and curve ceases to exist on either side of the two points, then there is always *a loop between the two points.*

(e) If necessary, we change the equation to polar form by using polar transformations $x = r\cos\theta$, $y = r\sin\theta$ to get the equation $r = f(\theta)$.

(f) Sometimes we change the equation to parametric form.

(g) Use only as many steps are necessary for tracing the curve.

(h) If possible prepare a table for certain values of x, y, $\dfrac{dy}{dx}$ which helps us to draw a rough shape of the curve.

ILLUSTRATIONS ON TYPE 1 : Cartesian Curves (Explicit Relations)

Ex. 1 : *Trace the curve $y^2 (2a - x) = x^3$.*   **(Nov./Dec. 2019, Dec. 2010, 2007, 2004)**

Sol. : This curve is known as *"The Cissoid of Diocle".*

The given equation of the curve can be written as $y^2 = \dfrac{x^3}{2a - x}$.   ... (1)

1. Equation of the curve contains only even powers of y, therefore, it is symmetrical about X-axis.

2. Equation does not contain any absolute constant therefore, it passes through the origin.

3. Tangents at the origin are obtained by equating to zero the lowest degree terms in the equation.

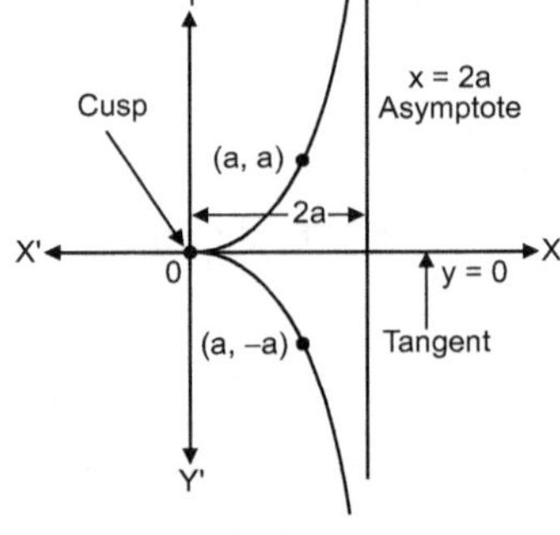

Fig. 5.12

From (1), we have

$y^2 (2a - x) = x^3$ i.e. $2ay^2 - xy^2 - x^3 = 0$

$\therefore$ $2ay^2 = 0 \Rightarrow y^2 = 0$, $y = 0$ is a double point.

$\therefore$ X-axis is a tangent at origin.

4. Since the two tangents at origin are coincident, therefore the origin is a *cusp.*

5. For intersection with X-axis, we put $y = 0$ $\therefore$ $\dfrac{x^3}{2a - x} = 0$ $\Rightarrow x = 0$ and for intersection with Y-axis, we put $x = 0$

$\therefore$ $y^2 = 0 \Rightarrow y = 0.$

Thus the curve meets the co-ordinate axis only at (0, 0).

6. The asymptote parallel to Y-axis can be obtained by equating to zero the coefficient of highest powers of y.

$y^2 (2a - x) - x^3 = 0$ i.e. $2a - x = 0 \Rightarrow x = 2a$ is the asymptote parallel to Y-axis.

7. From the equation of the curve, we observe that for $x < 0$ and $x > 2a$, y^2 becomes negative, hence y becomes imaginary, therefore the curve does not exist for $x < 0$ and $x > 2a$.

A rough sketch of the curve is as shown above.

Ex. 2 : *Trace the curve $x (x^2 + y^2) = a (x^2 - y^2)$ where $a > 0$.*   **(Dec. 2007, 2018, May 2005)**

Sol. : This curve is known as *"strophoid"*. The given equation can be written as

$$y^2 = \dfrac{x^2 (a - x)}{(a + x)}$$ ... (1)

1. Equation of the curve contains only even powers of y, therefore, it is symmetrical about X-axis.

2.   Equation of the curve does not contain any absolute constant, therefore, it passes through the origin.

3.   Tangents at origin are obtained by equating to zero the lowest degree terms in the equation. i.e. $ay^2 = ax^2 \Rightarrow y^2 = x^2$.

Hence $y = x$, $y = -x$ are tangents at the origin.

4.   Since tangents to the curve at origin are $y = x$, $y = -x$ which are real and different, origin is a node.

5.   For intersection with X-axis, we put $y = 0$,

$\therefore$ $x = 0$ or $x = a$.

For intersection with Y-axis, we put $x = 0$,

$\therefore$ $y = 0$. Thus the curve meets the co-ordinate axes at $(0, 0)$, $(a, 0)$.

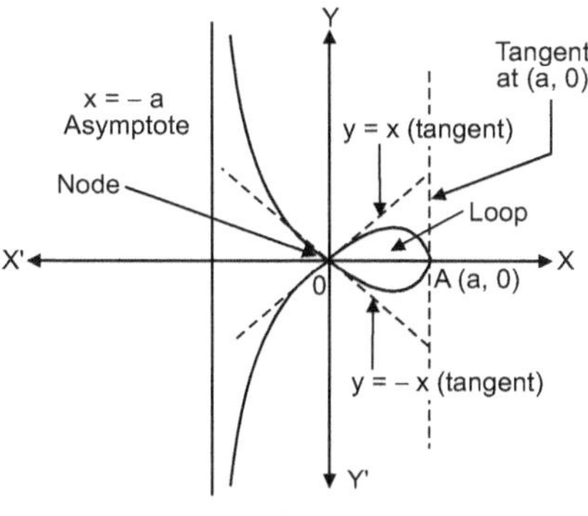

Fig. 5.13

6.   The asymptotes parallel to Y-axis can be obtained by equating to zero the coefficient of highest powers of y i.e. $a + x = 0 \Rightarrow x = -a$ is the asymptote parallel to Y-axis.

7.   For $x < -a$ and $x > a$, y becomes imaginary, therefore, the curve does not exist for $x < -a$ and $x > a$.

8.   Since the curve passes through the origin and no branch of the curve exists to the right of $x = a$, therefore, there exists a loop between $(0, 0)$ and $(a, 0)$.

A rough sketch of the curve is as shown above.

Ex. 3 : *Trace the curve $xy^2 = a^2 (a - x)$.*   **(May 2011, 2010, 2005, 2018; Dec. 2005, 2013)**

Sol. : This curve is known as *"Witch of Agnesi"*.

1.   **Symmetry** about X-axis (powers of y even).

2.   **Origin :** Curve does not pass through origin but cuts X-axis at $(a, 0)$. If we transfer the origin to the point $(a, 0)$, the new equation becomes $y^2 (x + a) + a^2 x = 0$.

Hence tangent at the new origin is $x = 0$, that is new Y-axis.

Hence at $(a, 0)$, tangent is parallel to old Y-axis.

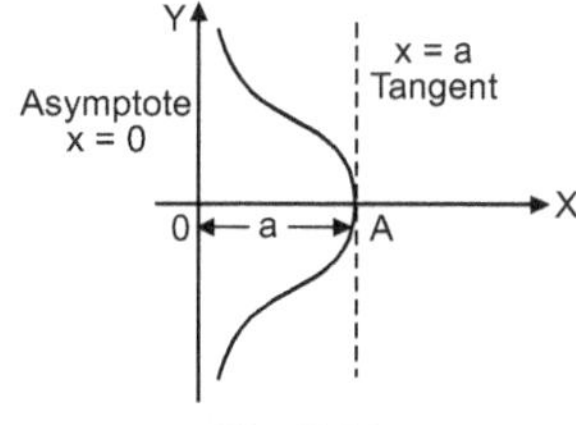

Fig. 5.14

3.   **Asymptote** $y^2 = \dfrac{a^2 (a - x)}{x}$.

As $x \to 0$, $y \to \infty$, hence the only asymptote is the line $x = 0$,   i.e.   Y-axis.

4.   **Region of Absence :** If we solve for y, $y = a\sqrt{\dfrac{a - x}{x}}$.

(i)   If $x < 0$, y becomes imaginary. Hence no part of curve lies to the left of Y-axis.

(ii)   As x increases from $x = 0$ to $x = a$, remaining positive y decreases from $y = \infty$ to $y = 0$ for the upper part.

Also if $x > a$, y becomes again imaginary, so curve is absent beyond $x = a$ to the right.

Ex. 4 : *Trace the curve $x = (y - 1) (y - 2) (y - 3)$.*

Sol. :

1.   **Symmetry :** No symmetry at all.

2.   **Origin :** It does not pass through $(0, 0)$.

3.   It is difficult to solve for y. But it is already solved for x. Here we shall regard y as independent variable and start tracing.

4.   **Special Points :** By observation we can easily find that the points $(0, 1)$, $(0, 2)$ and $(0, 3)$ do lie on the curve. All the points lie on Y-axis.

5.   Also when $y = 0$, $x = -6$. Hence it passes through $(-6, 0)$.

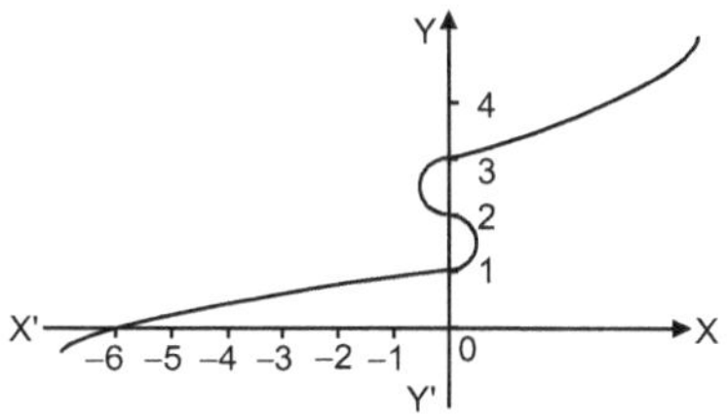

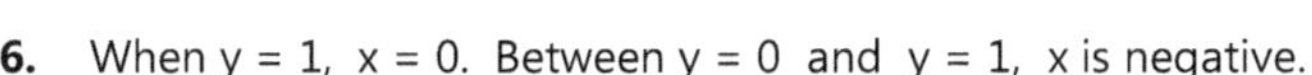

Fig. 5.15

6.   When $y = 1$, $x = 0$. Between $y = 0$ and $y = 1$, x is negative.

7.   When y lies between 1 and 2, x is positive and x is again zero at $y = 2$.

8.   Between $y = 2$ and $y = 3$, x is negative again. Again at $y = 3$, $x = 0$.

When $y > 3$, x is positive and remains positive continuously. As $y \to \infty$, $x \to \infty$. For very large value of y, x is almost equal to y^3. Hence no linear asymptote for this branch.

9.   x is negative when y is negative. As $y \to -\infty$, $x \to -\infty$. No linear asymptote for this branch too.

10.   When $y = \dfrac{3}{2}$, $x = \dfrac{3}{8}$ and when $y = \dfrac{5}{2}$, $x = \dfrac{-3}{8}$. Hence the curve is as traced above in the Fig. 5.15.

Ex. 5 : *Trace the curve $x^2 y^2 = a^2 (y^2 - x^2)$.*   *(Dec. 2009, 2004, May 2014)*

Sol. : It is symmetrical about both axes and in the opposite quadrants also. It passes through origin and $y = \pm x$ are tangents at $(0, 0)$. It intersects the co-ordinate axes only at $(0, 0)$.

Given equation can be written as $a^2 y^2 - x^2 y^2 = a^2 x^2$ or $y^2 = \dfrac{a^2 x^2}{a^2 - x^2}$.

$x = \pm a$ are the asymptotes to the curve. For $x > a$ and $x < -a$, the curve does not exist. A rough sketch of the curve is as shown below.

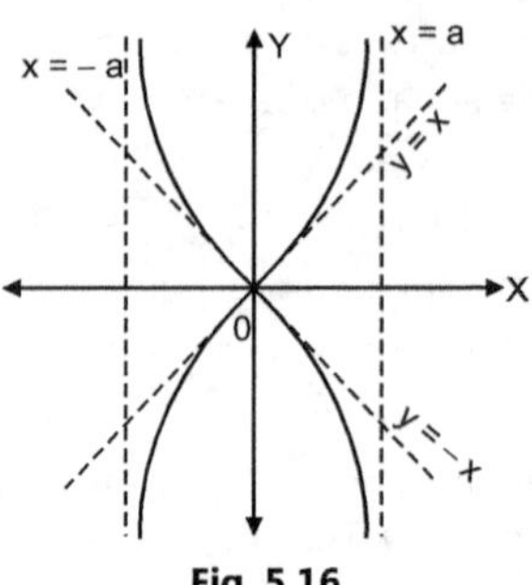

Fig. 5.16

Ex. 6 : *Trace the curve $y^2 (a^2 + x^2) = a^2 x^2$.*   *(May 2006)*

Sol. : It is symmetrical about both the axes and in the opposite quadrants also. It passes through origin and $y = \pm x$ are tangents at origin. It intersects the co-ordinate axes only at $(0, 0)$.

To find the equation of asymptote parallel to X-axis, we have $a^2 y^2 = a^2 x^2 - x^2 y^2$

or $x^2 (a^2 - y^2) = a^2 y^2$. $\therefore x^2 = \dfrac{a^2 y^2}{a^2 - y^2}$. Therefore $y = \pm a$ are the asymptotes to the curve.

For $y > a$ and $y < -a$, $x^2 = \dfrac{a^2 y^2}{a^2 - y^2}$ becomes negative, hence the curve does not exist for $y > a$, $y < -a$.

A rough sketch of the curve is as shown in the Fig. 5.17.

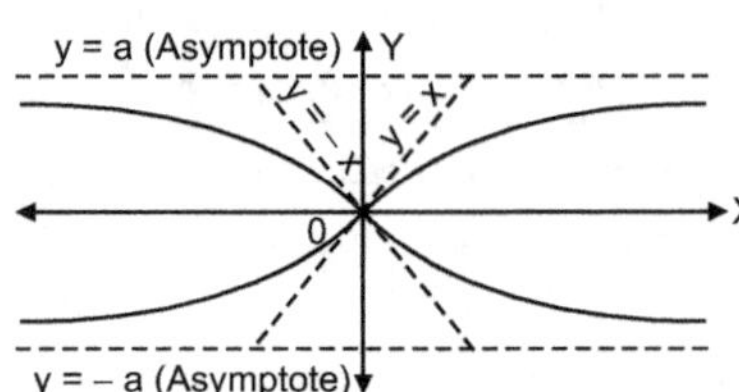

Fig. 5.17

Ex. 7 : *Trace the curve $y (1 + x^2) = x$.*   *(May 2009)*

Sol. : It is symmetrical in the opposite quadrants. It passes through origin and $y = x$ is the tangent at origin. It intersects the co-ordinate axes only at origin $y = 0$ i.e. X-axis is an asymptote to the given curve. Because $y (1 + x^2) - x = 0$. Equate coefficient of highest power of x to zero i.e. $y = 0$ is asymptote. The curve does not exist in the second and fourth quadrants.

We have $y = \dfrac{x}{1 + x^2}$

x	0	1	2	3	4	5
y	0	0.5	0.4	0.3	0.24	0.9

From the table it is clear that as x increases, y decreases.

A rough sketch of the curve is as shown in Fig. 5.18.

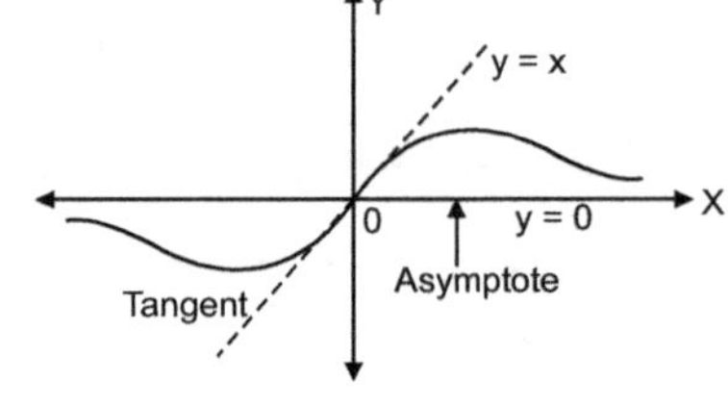

Fig. 5.18

Ex. 8 : *Trace the curve $a^2 y^2 = x^2 (2a - x)(x - a)$.*

Sol. : It is symmetrical about X-axis and passes through origin. Here origin is a conjugate point.

Intersection with X-axis are $(0, 0)$, $(2a, 0)$, $(a, 0)$.

For $x < 0, x > 2a$ curve does not exist.

For $0 < x < a$ curve does not exist.

For $a < x < 2a$ curve exists.

For $x = \dfrac{3a}{2}$, $y = \pm \dfrac{3a}{4}$

No asymptote to the curve. A rough sketch of the curve is as shown in the Fig. 5.19.

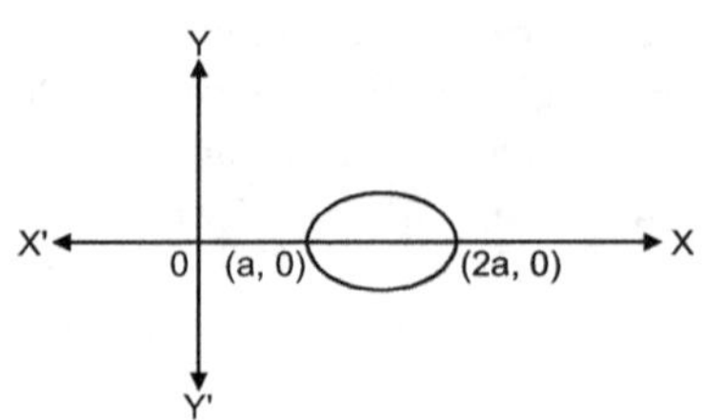

Fig. 5.19

Ex. 9 : *Trace the curve $y (x^2 + 4a^2) = 8a^3$.*

Sol. : It is symmetrical about Y-axis. It does not pass through origin. It intersects Y-axis at $(0, 2a)$.

We have $x^2 + 4a^2 = \dfrac{8a^3}{y}$ $\therefore \quad 2x \dfrac{dx}{dy} = -\dfrac{8a^3}{y^2}$

$\dfrac{dx}{dy} = -\dfrac{4a^3}{xy^2}$ $\therefore \quad \left(\dfrac{dx}{dy}\right)_{(0, 2a)} = -\infty$

Therefore tangent at (0, 2a) is parallel to X-axis.
$y = 0$ i.e. X-axis is asymptote to the given curve.
For $y < 0$, $y > 2a$, the curve does not exist.
For $y = a$, $x = \pm 2a$, $(2a, a)$ $(-2a, a)$ are the points on the curve.
A rough sketch of the curve is as shown in the Fig. 5.20.

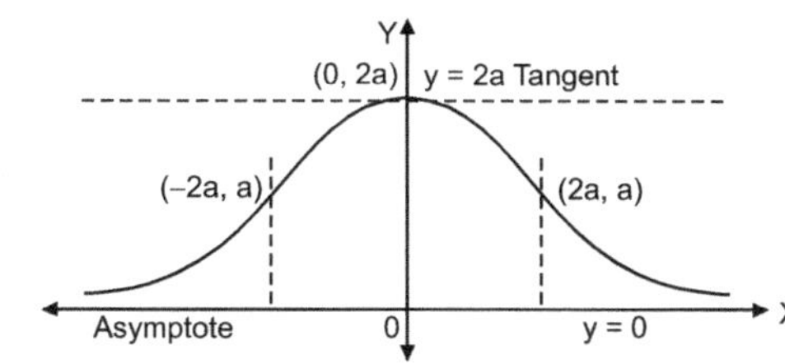

Fig. 5.20

Ex. 10 : *Trace the curve $y^2 (a^2 - x^2) = a^3 x$.*   *(May 2019, May 2008, 2004)*

Sol. : We have $y^2 = \dfrac{a^3 x}{a^2 - x^2}$

It is symmetrical about X-axis, passes through origin $x = 0$ i.e. Y-axis is tangent at origin. $x = \pm a$ are the asymptotes to the curve. Also $y = 0$ i.e. X-axis is an asymptote to the given curve.

For $x > a$ the curve does not exist.
For $x < -a$ the curve exists.
For $0 < x < a$ the curve exists.
For $-a < x < 0$ the curve does not exist.
A rough sketch of the curve is as shown below.

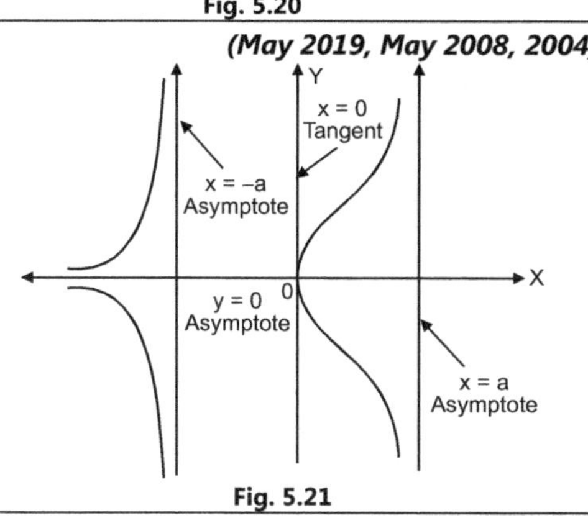

Fig. 5.21

Ex. 11 : *Trace the curve $x^{1/2} + y^{1/2} = a^{1/2}$.*   *(May 2007)*
Sol. : It is symmetrical about the line $y = x$. It does not pass through the origin.

It intersects X-axis at $(a, 0)$ and Y-axis at $(0, a)$. It also intersects the line $y = x$ at $\left(\dfrac{a}{4}, \dfrac{a}{4}\right)$.

The curve entirely lies in the first quadrant because x and y cannot be negative.

Also, $\dfrac{1}{2} x^{-1/2} + \dfrac{1}{2} y^{-1/2} \dfrac{dy}{dx} = 0$ $\therefore$ $\dfrac{dy}{dx} = -\sqrt{\dfrac{y}{x}}$

$\therefore$ At $(a, 0)$, $\dfrac{dy}{dx} = 0$ $\Rightarrow$ tangent at $(a, 0)$ is X-axis itself.

At $(0, a)$, $\dfrac{dy}{dx} = -\infty \Rightarrow$ tangent at $(0, a)$ is Y-axis itself.

A rough sketch of the curve is as shown in the following figure.

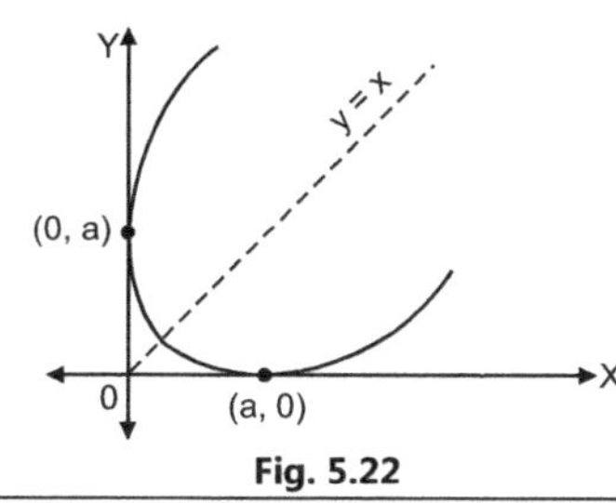

Fig. 5.22

Ex. 12 : *Trace the following curves :*
(i) $x^2 (x^2 - 4a^2) = y^2 (x^2 - a^2)$ *(May 2008)*
(ii) $a^2 x^2 = y^3 (2a - y)$ *(May 2009) (Dec. 2011)*

Sol. : (i) $y^2 = \dfrac{x^2 (x^2 - 4a^2)}{x^2 - a^2}$

A rough sketch of the curve is as shown in the following figure.
It is symmetrical about both x and y axes and also in the opposite quadrants. It passes through origin and $y = \pm 2x$ are tangents at origin. Intersection with the co-ordinate axes is at $(0, 0)$, $(2a, 0)$, $(-2a, 0)$. Also $x = \pm a$ are the asymptotes to the given curve.
For $a < x < 2a$, $-2a < x < -a$, the curve does not exist.
For $x > 2a$, $x < -2a$, the curve exists.

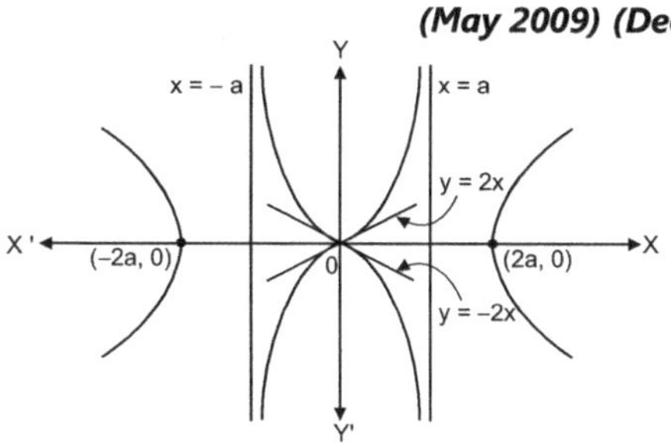

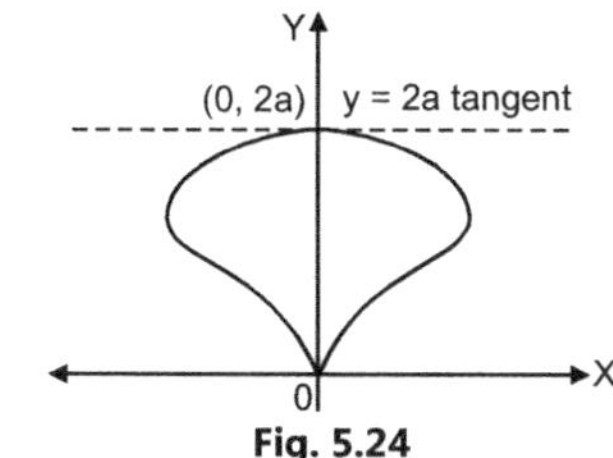

Fig. 5.23

(ii) $a^2 x^2 = y^3 (2a - y)$
It is symmetrical about Y-axis. It passes through the origin and $x = 0$ is the tangent at $(0, 0)$. It meets Y-axis at $(0, 0)$, $(0, 2a)$ and X-axis at $(0, 0)$.

$$2a^2 \cdot x \cdot \dfrac{dx}{dy} = 6ay^2 - 4y^3 \qquad \therefore \quad \dfrac{dx}{dy} = \dfrac{y^2 (3a - 2y)}{a^2 x}$$

$$\left(\dfrac{dx}{dy}\right)_{(0, 2a)} = -\infty \Rightarrow \text{tangent at } (0, 2a) \text{ is parallel to X-axis.}$$

There exists a cusp at $(0, 0)$. When $y = a$, $x = \pm a$. For $y < 0$, $y > 2a$, x^2 becomes negative, therefore, curve does not exist for $y > 2a$ and $y < 0$.
A rough sketch of the curve is as shown in the following figure.

**Fig. 5.24

EXERCISE 5.1

Trace the following curves :

1. $y = x^3$
2. $y = x (x^2 - 1)$ **(Dec. 2009)**
3. $y^2 (x - a) = x^2 (2a - x)$ **(Dec. 06, 08)**
4. $27ay^2 = 4 (x - 2a)^3$
5. $ay^2 = x^2 (a - x)$ **(May 06, 13, Nov. 15)**
6. $3ay^2 = x (x - a)^2$
7. $ay^2 = x (a^2 + x^2)$
8. $xy^2 = a (x^2 - a^2)$ **(Dec. 2010)**
9. $y^2 (a + x) = (x - a)^3$
10. $a^2 y^2 = x^2 (a^2 - x^2)$ **(May 2010)**
11. $a^2 y^2 = x^2 (x + 2a) (x - a)$
12. $a^2 y^2 = x^2 (a - x) (x - b)$ where $b < a$

13. $y^2 = (x - 1) (x - 2) (x - 3)$ **(Dec. 2008, 2006)**
14. $y^2 (x^2 - 1) = x$ **(May 2011, 2017)**
15. $y (x^2 - 1) = x^2 + 1$
16. $(x^2 - a^2) (y^2 - b^2) = a^2 b^2$
17. $x^2 y^2 = x^2 + 1$
18. $ay^2 = x (a^2 - x^2)$ **(Dec. 2011)**
19. $ay^2 = (x - a) (x - 5a)^2$
20. $y^2 = x^5 (2a - x)$ **(May 2015)**
21. $y^2 (x^2 + y^2) + a^2 (x^2 - y^2) = 0$
22. $y^2 (4 - x) = x (x - 2)^2$ **(May 2006, 2010)**
23. $y^2 (a - x) = x^3$. **(Dec. 2011, 2017)**

ANSWERS

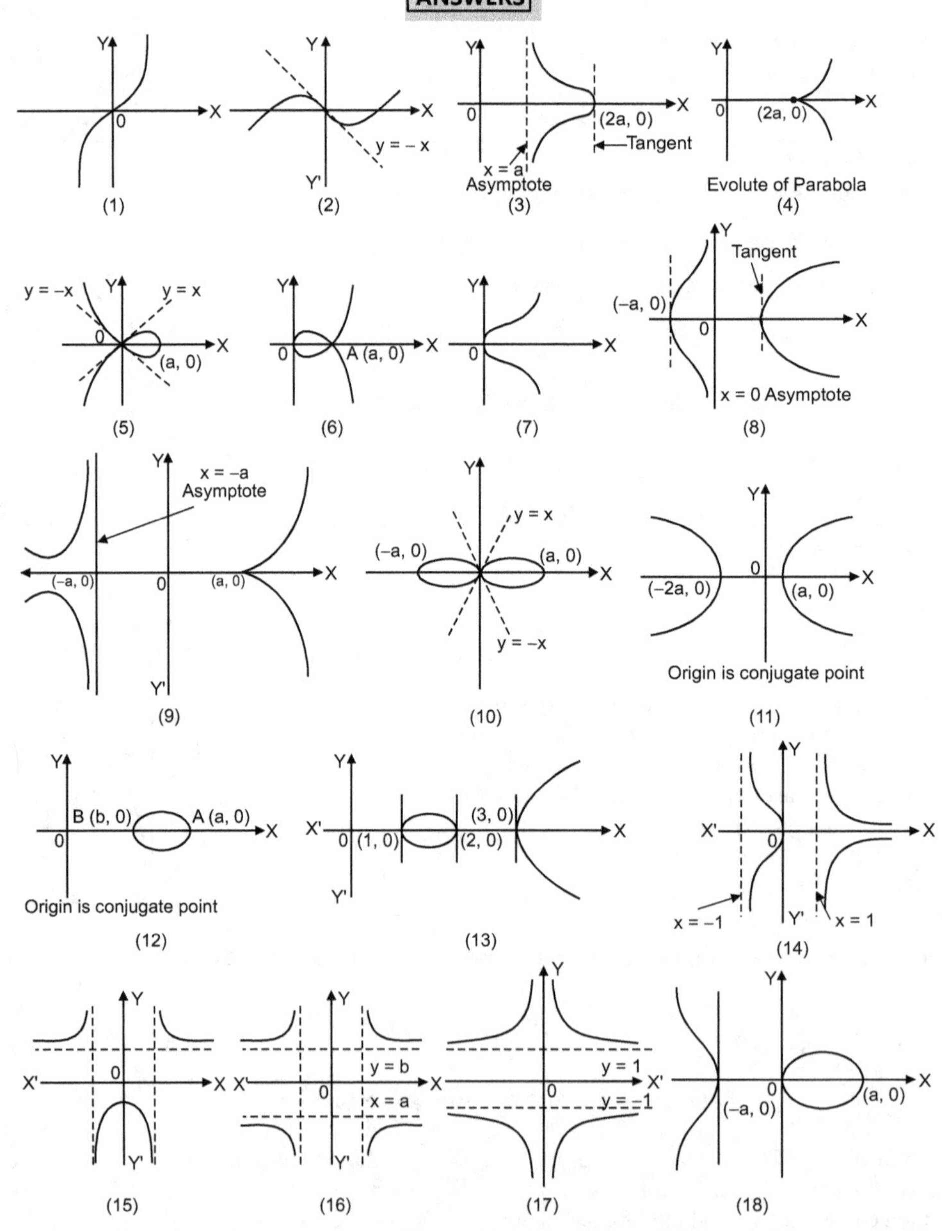

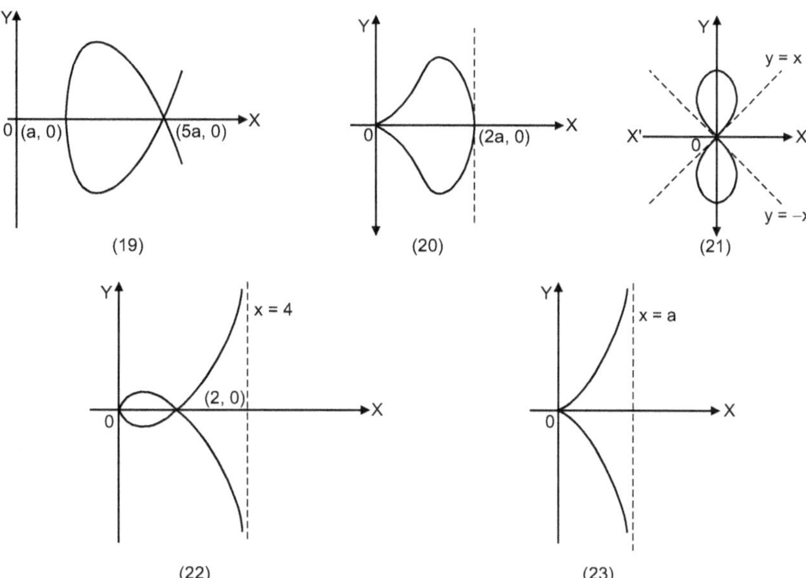

(19)   (20)   (21)

(22)   (23)

Fig. 5.25

TYPE 2 : CURVES GIVEN BY PARAMETRIC EQUATIONS

5.4 TRACING OF PARAMETRIC CURVES

Rule : (If possible chart to Cartesian and then trace).

1. Limitations of the Curve : Let the parametric equation be given by $x = f(t)$, $y = g(t)$. If possible find the greatest and least values of x and y for a proper value of t and therefore the boundary lines parallel to x and y axes between which the curve lies.

2. Symmetry :

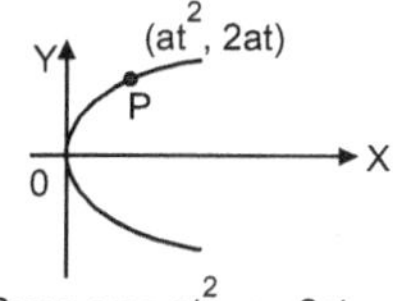

Curve : $x = at^2$, $y = 2at$

Fig. 5.26

(a) If $f(t)$ be even function of t and $g(t)$ an odd, the curve is symmetrical to X-axis. As the parabola $x = at^2$, $y = 2at$ is symmetrical about X-axis.

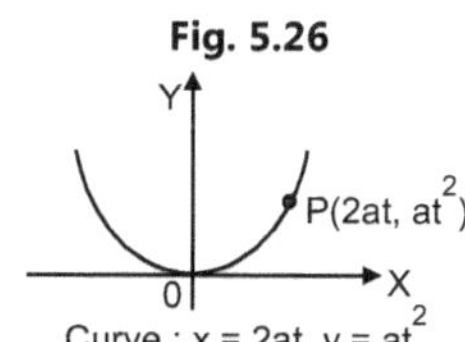

Curve : $x = 2at$, $y = at^2$

Fig. 5.27

(b) If $f(t)$ be odd and $g(t)$ an even function, symmetry about Y-axis. For example, the parabola $x = 2at$, $y = at^2$ has symmetry about Y-axis.

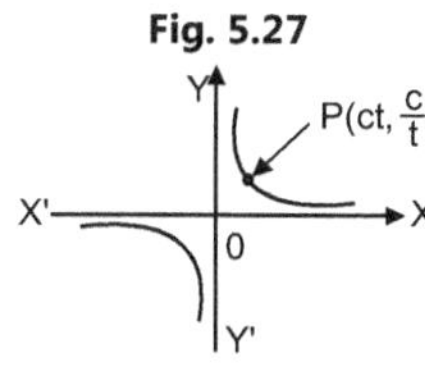

Fig. 5.28

(c) If both $f(t)$ and $g(t)$ are odd, the curve is symmetrical in opposite quadrants.

For example, $x = ct$, $y = \dfrac{c}{t}$, the rectangular hyperbola.

(d) Also we note that for values of t and $-t$, x remains unchanged but y has equal and opposite values, therefore, the curve is symmetrical about X-axis. For example, $x = at^2$, $y = 2at$.

(e) Also we note that for values of t and $\pi - t$, y remains unchanged but x has equal and opposite values, hence the curve is symmetrical about Y-axis.

For example, $x = a\cos^3 t$, $y = a\sin^3 t$.

3. Origin : If on putting $x = 0$, we obtain $y = 0$ for some value of t, then the curve passes through origin.

Also find the points of intersection of the curve and the axes.

Find asymptotes if any.

4. Special Points : Try to find few points on the curve by observation and also those points where $\dfrac{dy}{dx} = 0$ or ∞. Here $x = f(t)$,

$y = g(t)$, hence use the formula $\dfrac{dy}{dx} = \dfrac{dy/dt}{dx/dt}$.

5. **Region of Absence of Curve :**

    (a) Find those regions where curve does not exist.

    (b) Make a table of values of t, x, y, $\dfrac{dx}{dt}$ and $\dfrac{dy}{dt}$.

    (c) If both x and y are periodic functions of t, with a common period, we need to study the position of the curve for one period only.

6. **Cycloid :**

    When a circle rolls in a plane along given straight line, the locus traced out by a fixed point on the circumference of rolling circle is called as *cycloid.*

    There are four types of equations of the curve depending upon choice of axes and position of line along which the circle rolls.

    The sketching of the cycloid from its equation depends on the values of x, y and $\dfrac{dy}{dx}$ at t = 0.

ILLUSTRATION ON TYPE 2 : Parametric Curves

Ex. 1 : *Trace the cycloid, x = a (t + sin t), y = a (1 – cos t).*   ***(May 2011, Dec. 2017)***

Sol. : The curve is known as *cycloid.*

1. **Limitations :** The curve lies between the lines y = 0 and y = 2a, because the greatest value of y is 2a and least is 0.

2. **Symmetry :** x = a (t + sin t), being odd function of t and y = a (1 – cos t) an even function, the curve is symmetrical to Y-axis.

3. **Origin :** When t = 0, x = 0 and y = 0, hence curve passes through the origin. The curve cuts X-axis (putting y = 0) when y = a (1 – cos t) = 0, or t = 0 and then x = 0 i.e. at (0, 0) only. Similarly, it cuts Y-axis also at (0, 0).

4. **No Asymptotes.**

5. **Special points :** We have after differentiation

$$\frac{dy}{dx} \;=\; \frac{dy/dt}{dx/dt} \;=\; \frac{a \sin t}{a\,(1 + \cos t)} \;=\; \tan \frac{t}{2}$$

$$\Rightarrow \qquad \frac{dy}{dx} \;=\; 0 \ \text{ when } \ \frac{t}{2} = 0 \ \text{ or } \ \text{when } t = 0$$

$$\text{and} \qquad \frac{dy}{dx} \;=\; \infty \ \text{ at } \tan \frac{t}{2} = \infty \ \text{ or } \ \text{at } t = \pi.$$

    Hence at (0, 0), tangent is parallel to X-axis (rather X-axis itself)

    and at (aπ, 2a) tangent is parallel to Y-axis.

6. **Region of absence (a) :** Also y = 2a $\sin^2 \dfrac{t}{2}$ or $\sin \dfrac{t}{2} = \sqrt{y/2a}$, hence when y is negative, t is imaginary, i.e. no part of the curve lies below X-axis (i.e. in 3rd and 4th quadrants).

7. The table of values of t, x, y, $\dfrac{dy}{dx}$ is as follows.

Table 5.1

t	0	π/2	π	2π	– π/2	–π
x	0	a (π/2 + 1)	a π	2a π	– a (π/2 + 1)	–a π
y	0	a	2a	0	a	2a
$\dfrac{dy}{dx}$	0	1	∞	0	–1	– ∞

Hence some of the special points that lie on the curve are (0, 0), $\left[a \left(\dfrac{\pi}{2} + 1 \right), a \right]$, (aπ, 2a), (2aπ, 0), $\left[- a \left(\dfrac{\pi}{2} + 1 \right), a \right]$ and (– aπ, 2a).

From the table it is clear that as x increases from 0 to aπ, y also increases from 0 to 2a. But as x further increases to 2aπ, y decreases again to zero.

When t is positive and varies from 0 to 2π, we can sketch the portion OAC of the curve in the first quadrant.

As x increases from –aπ to 0, y is found to decrease from 2a to 0. Also the portion DBO can be traced by symmetry.

The curve consists of congruent arches extending to infinity in both directions of X-axis.

The curve is as traced below.

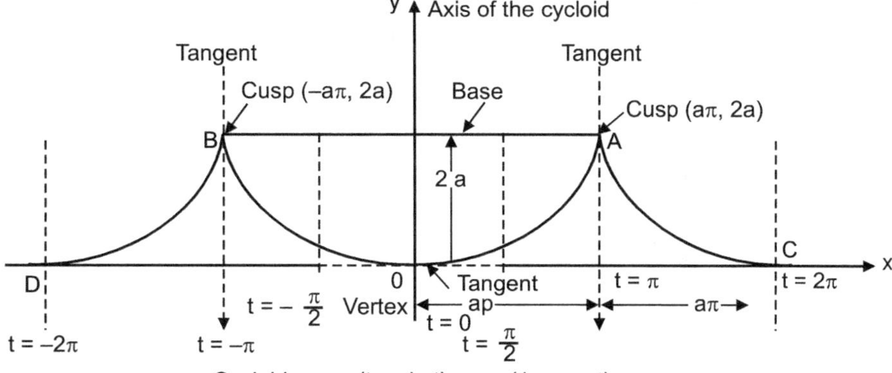

Fig. 5.29

The points A, B are called the *cusps* of the cycloid.

The line OY (i.e. Y-axis) about which the curve is symmetrical is called the *axis of the cycloid*.

The line AB joining the cusps is called the *base* of the cycloid.

The point O is called the *vertex* of the cycloid.

Ex. 2 : *Trace the cycloid $x = a\,(t - \sin t)$, $y = a\,(1 - \cos t)$.*   *(May 2009, 2004; Dec. 2005, May 2014, Nov. 2014, Dec. 2016)*

Sol. : The curve is known as cycloid.

The curve lies between the lines $y = 0$ and $y = 2a$ because the greatest value of y is 2a and least is 0. When t = 0, x = 0 and y = 0, hence the curve passes through the origin. It does not contain any asymptote.

We have $y = a\,(1 - \cos t) = a \cdot 2 \sin^2 \left(\dfrac{t}{2}\right)$

$\therefore \quad \sin \dfrac{t}{2} = \sqrt{\dfrac{y}{2a}}$. Hence when y is negative, t is imaginary i.e. no part of the curve lies below X-axis (i.e. in 3rd and 4th

quadrants).

Also, $\qquad \dfrac{dy}{dx} = \dfrac{a \sin t}{a\,(1 - \cos t)} = \dfrac{2 \sin \dfrac{t}{2} \cos \dfrac{t}{2}}{2 \sin^2 \dfrac{t}{2}} = \cot \left(\dfrac{t}{2}\right)$

$\qquad \dfrac{dy}{dx} = \infty \quad$ when $\quad t = 0$

$\qquad \dfrac{dy}{dx} = 0 \quad$ when $\quad t = \pi$

When t = 0, x = 0, y = 0 and when t = π, x = aπ, y = 2a. Therefore at (0, 0), tangent is parallel to Y-axis and at (aπ, 2a), tangent is parallel to X-axis (rather X-axis itself).

Table 5.2

t	0	π/2	π	2π
x	0	a (π/2 − 1)	aπ	2aπ
y	0	a	2a	0
$\dfrac{dy}{dx}$	∞	1	0	−∞

From the table it is clear that the curve is symmetrical about the line x = aπ.

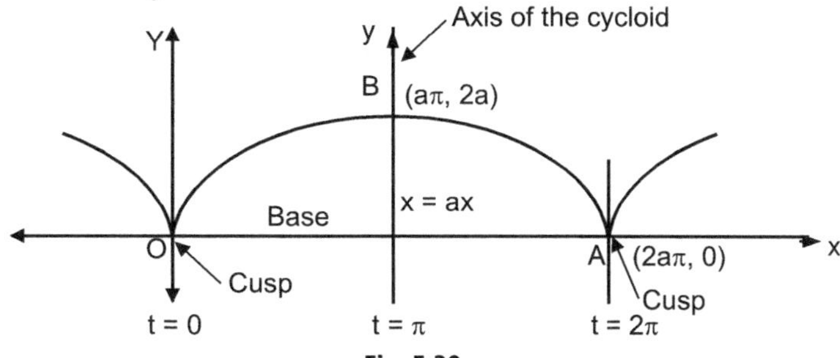

Fig. 5.30

As t increases from 0 to π, x increases from 0 to $a\pi$ and y also increases from 0 to 2a. But as x further increases to $2a\pi$, y decreases again to zero. The points O and A are called the *cusps* of the cycloid. The line $x = a\pi$ about which the curve is symmetrical is called the *axis* of the cycloid. The line OA joining the cusps is called the *base* of the cycloid. The curve is as traced in the figure.

Ex. 3 : *Trace the curve* $x^{2/3} + y^{2/3} = a^{2/3}$.   **(May 05; Dec. 2011, 2010, 2009, 2008, 2006)**

Sol. : This famous curve is called as *Astroid* (or *star-shaped curve*).

The parametric equations of the Astroid are $x = a \cos^3 t$, $y = a \sin^3 t$.

(Students are advised to remember these parametric equations of the Astroid.)

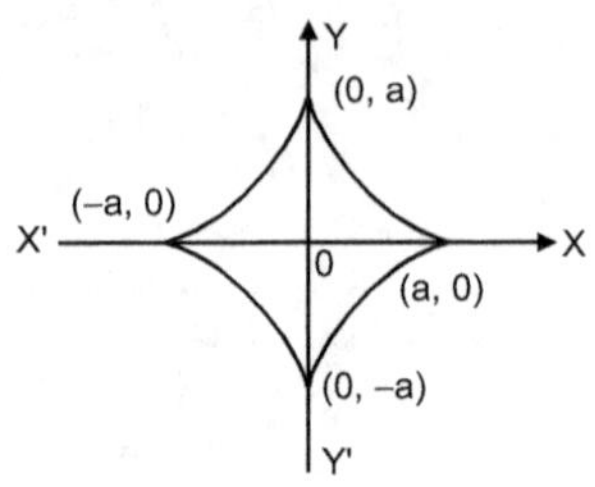

Fig. 5.32

1. $\therefore$ $a \cos^3 t$ is even and $a \sin^3 t$ is odd.

   $\therefore$ Symmetry about X-axis. Also we note that for t and –t, x has same value but y has equal and opposite values, hence curve is symmetrical about X-axis.

2. $\therefore$ for t and $\pi - t$, y has same value but x has opposite values. Therefore, symmetry about Y-axis.

3.   The table of values of x, y, t is as follows :

t	0	$\pi/2$	π	$3\pi/2$	2π
x	a	0	–a	0	a
y	0	a	0	–a	0

As t ranges from 0 to $\dfrac{\pi}{2}$, cos t decreases and sin t increases i.e. x decreases, y increases.

4.   For any value of t $|\cos t| < 1$ and $|\sin t| < 1$, the values of $|x| < a$ and $|y| < a$ i.e. values of x, y numerically cannot exceed a.

5.   The curve cuts X-axis at $(\pm a, 0)$ and Y-axis at $(0, \pm a)$. A rough sketch of the curve is as shown in the figure.

Ex. 4 : *Trace the curve* $x = a\left[\cos t + \dfrac{1}{2} \log\left(\tan^2 \dfrac{t}{2}\right)\right]$, $y = a \sin t$.   **(May 2009)**

Sol. : This curve is known as "The Tractrix".

1.   **Limitations of the curve :** Since $y = a \sin t$ and $-1 < \sin t < 1$, the greatest value of y is a and the least value of y is $- a$. x can change from $- \infty$ to $+ \infty$.

2.   **Symmetry :** $x = a\left[\cos t + \dfrac{1}{2} \log \tan^2 \dfrac{t}{2}\right]$, being even function and $y = a \sin t$, odd hence symmetry about X-axis.

Also we note that for values of t and –t, x remains unchanged but y has equal and opposite values. Thus the curve is symmetrical about X-axis.

For $\pi - t$, $\sin (\pi - t) = \sin t$ i.e. for t and $\pi - t$, value of y remains same.

Again for $\pi - t$, $x = a\left[\cos (\pi - t) + \dfrac{1}{2} \log \tan^2 \left(\dfrac{\pi}{2} - \dfrac{t}{2}\right)\right]$

$$= a\left[-\cos t - \dfrac{1}{2} \log \tan^2 \dfrac{t}{2}\right] = -a\left[\cos t + \dfrac{1}{2} \log \tan^2 \dfrac{t}{2}\right]$$

Thus for t and $\pi - t$, values of x are equal and opposite. Hence, the curve is symmetrical about Y-axis.

3.   **Origin :** Origin does not lie on the curve. But from the table we see that the curve concretely passes through the points $(0, a)$ and $(0, -a)$.

t	0	$\pi/2$	π	$3\pi/2$
x	$-\infty$	0	∞	0
y	0	a	0	–a

$$y = a \sin t, \ \frac{dy}{dt} = a \cos t, \ \frac{dx}{dt} = \frac{a \cos^2 t}{\sin t}$$

$$\therefore \quad \frac{dy}{dx} = \frac{dy/dt}{dx/dt} = \frac{a \cos t}{a \cos^2 t/\sin t} = \tan t$$

$$\frac{dy}{dx} = \tan \psi \Rightarrow \tan \psi = \tan t \Rightarrow \psi = t$$

For $t = \dfrac{\pi}{2}, \dfrac{3\pi}{2}, \dfrac{dy}{dx} = \pm \infty$ i.e. y-axis is tangent to the curve.

Fig. 5.31

The curve is as shown above.

EXERCISE 5.2

Trace the following curves :

1. $x = a(t + \sin t)$, $y = a(1 + \cos t)$

(May 2006, 2008, 2010; Dec. 2007, May 2016)

2. $x = a(t - \sin t)$, $y = a(1 + \cos t)$ **(May 2007)**

3. $\left(\dfrac{x}{a}\right)^{2/3} + \left(\dfrac{y}{b}\right)^{2/3} = 1$ **(Dec. 2007, May 2008)**

4. $x = t^2$, $y = t - \dfrac{t^3}{3}$ **(May 2007, Dec. 04, 05, 07)**

5. $x = at$, $y = \dfrac{a}{t}$

6. $ay^2 = x^3$

ANSWERS 5.2

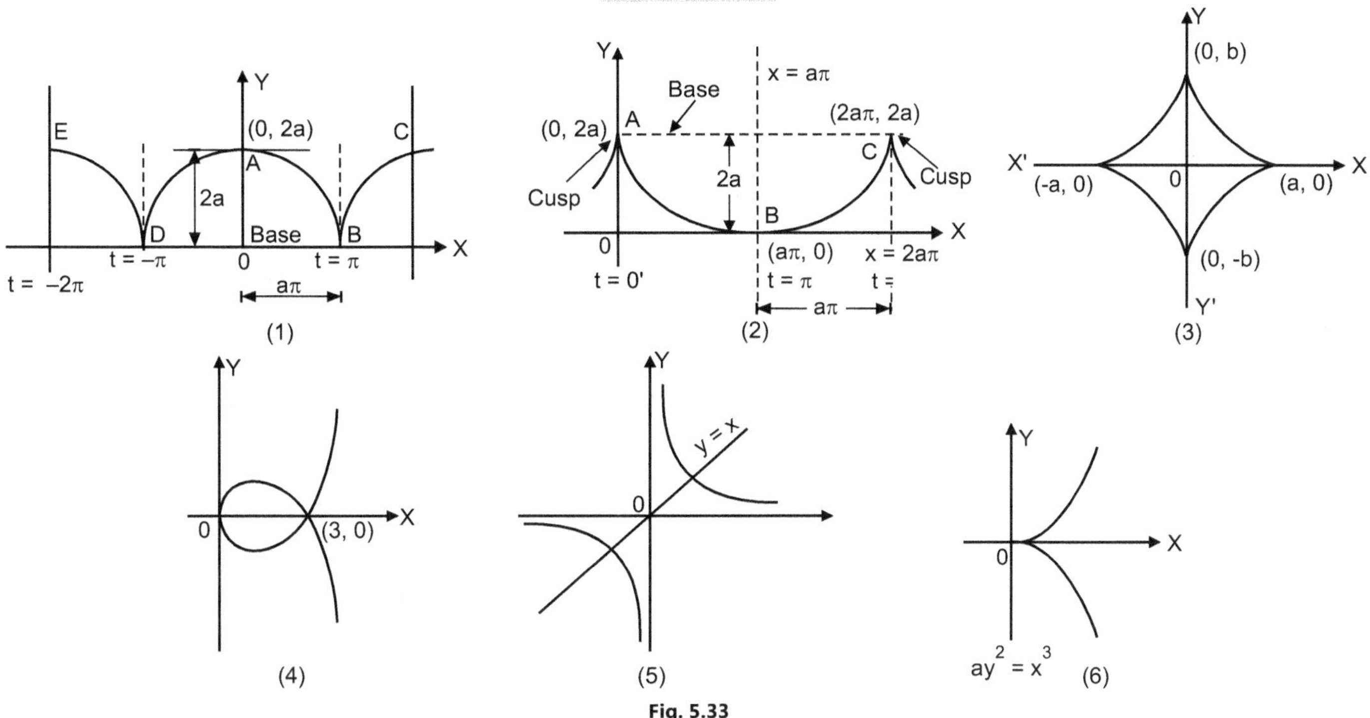

Fig. 5.33

TYPE 3 : CURVES GIVEN BY POLAR CO-ORDINATES

5.5 TRACING OF POLAR CURVE

Introduction : In polar system, the fixed point 'O' is called pole or origin. From fixed point O (pole), draw a straight line in any direction, say OX (X-axis), then this fixed straight line OX is called as initial line.

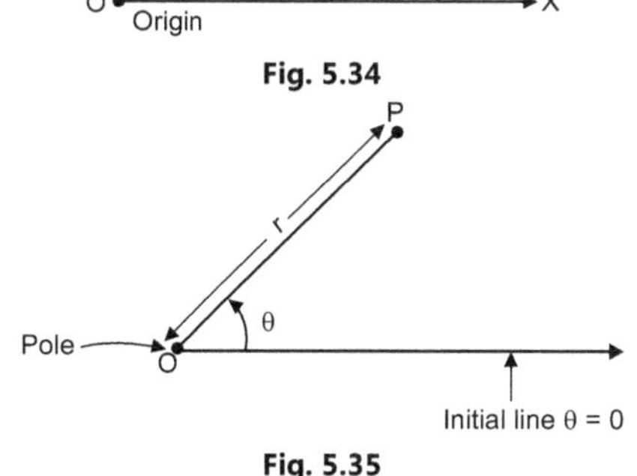

Fig. 5.34

Let P be any given point.

We note that the distance OP = r is called radius vector.

OP makes an angle θ with the initial line and is called vectorial angle measured positive in anticlockwise sense $\therefore$ $\angle$XOP = θ.

OP = radius vector, any point is P(r, θ)

Co-ordinates of P in polar co-ordinates are (r, θ) and r = f(θ) is referred as polar equation of the curve.

Fig. 5.35

The initial line OX represents $\theta = 0$. We draw $\theta = \dfrac{\pi}{2}$ as a line perpendicular to initial line and passing through the pole.

The following rules given below should enable the students to sketch a curve given by polar equation in simple cases.

Rule 1 : Symmetry :

In polar co-ordinates, the curve is often given by the equation r = f(θ).

(a) If the equation to the curve remains unchanged by changing θ to $-\theta$, it will be symmetrical to the initial line [viz. $r = a(1 + \cos \theta)$].

(b) If the equation of the curve remains unchanged by changing r to – r, the curve is symmetrical to the pole. In such a case, only even power of r will occur in the equation [$r^2 = a^2 \cos 2\theta$].

(c) If the equation remains unchanged by changing θ to $-\theta$ and r to – r, at the same time, the curve is symmetrical to the line through the pole, perpendicular to the initial line (i.e. Y-axis) [$r^2 = a^2 \cos 2\theta$].

The same symmetry also exists if the equation remains unchanged when θ is changed to $\pi - \theta$, as for example, the curve $r = (1 + \sin \theta)$.

Rule 2 : Pole :

(a) The pole will lie on the curve if for some value of θ, r becomes zero.

(b) Then find the equation of tangent or tangents at the pole. If we put r = 0, the value of θ gives the tangent at the pole.

Care should be taken of the points where the curve cuts the initial line and the line $\theta = \dfrac{\pi}{2}$.

Rule 3 : The table showing values of r for different values of θ is very useful in plotting a polar curve. Also find the values of θ at which r = 0 or r = ∞.

Rule 4 : Angle between the radius vector and tangent [ϕ] :

Use the formula $\tan \phi = r \dfrac{d\theta}{dr}$ and find ϕ and also the points where $\phi = 0$ or ∞.

i.e. find the points where the tangent coincides with the radius vector or is perpendicular to it.

Rule 5 : Asymptotes : Find asymptotes if any.

Rule 6 : Region of absence of the curve :

(a) Solve the equation for r and consider how r varies as θ increases from 0 to $+ \infty$ and also when θ decreases from 0 to $- \infty$. If necessary form a table of values of r and θ.

(b) If for some values of θ, say α and β, r^2 is negative, i.e. r imaginary, this means that no branch of the curve exists between lines $\theta = \alpha$ and $\theta = \beta$.

(c) If the maximum numerical value of r is a, the entire curve will lie within a circle of radius a (i.e. r = a). If least numerical value of r is b, the curve will lie outside the circle r = b.

(d) In most of the polar equations, only periodic functions $\sin \theta$ and $\cos \theta$ occur and hence values of θ from 0 to 2π should only be considered. The remaining values of θ, give no new branch of the curve.

Note : In some problems, changing to Cartesian proves more convenient.

ILLUSTRATIONS ON TYPE 3 : Polar Curves

Ex. 1 : *Trace the curve $r^2 = a^2 \cos 2\theta$.* *(May 2009, 2004; Dec. 2011, 2010, 2005)*

Sol. : This curve is known as *"Lemniscate of Bernoulli"*.

1. The curve is symmetrical to the initial line.
2. The curve is symmetrical to the pole.
3. The curve is symmetrical to the line perpendicular to the initial line passing through the pole.
4. When $\theta = \pi/4$ or $3\pi/4$, r becomes zero, hence the curve passes through the pole.
5. Tangents at pole are obtained by putting r = 0, we have $\cos 2\theta = 0$.

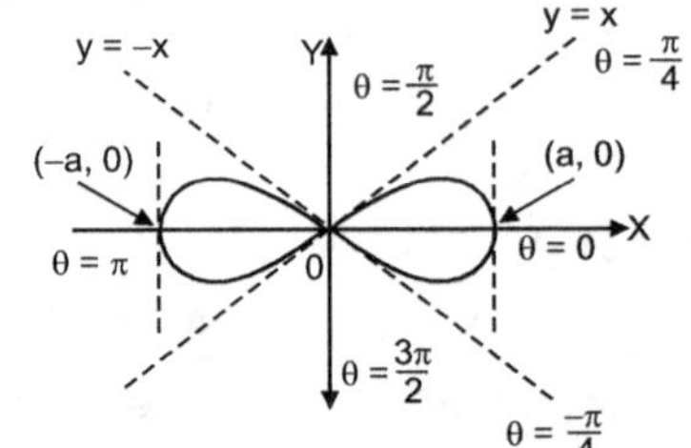

Fig. 5.36

$$\therefore \quad 2\theta = \frac{\pi}{2}, \frac{3\pi}{2}, \frac{5\pi}{2}, \frac{7\pi}{2}, \ \dots \quad \text{i.e.} \quad \theta = \frac{\pi}{4}, \frac{3\pi}{4}, \frac{5\pi}{4}, \frac{7\pi}{4}, \dots$$

6. Variation of r corresponding to θ is tabulated as follows :

θ	0	$\pi/4$	$3\pi/4$	π	$5\pi/4$	$7\pi/4$	2π
r	a	0	0	$-a$	0	0	a

Thus we see that maximum value of r is a (since maximum of $\cos 2\theta$ is +1).

When $\theta = 0$, $r = \pm a$ i.e. curve passes through the points (a, 0) and (–a, 0). Minimum value of r is 0.

7. As θ increases from $\dfrac{\pi}{4}$ to $\dfrac{3\pi}{4}$, r^2 remains negative, hence r becomes imaginary for $\dfrac{\pi}{4} < \theta < \dfrac{3\pi}{4}$ and the curve does not exist in this region. Similarly curve does not exist in $\dfrac{5\pi}{4} < \theta < \dfrac{7\pi}{4}$.

8. Angle ϕ : We use the formula $\tan \phi = r \dfrac{d\theta}{dr}$ given $r^2 = a^2 \cos 2\theta$, differentiating with respect to θ, $\therefore \ 2r \dfrac{dr}{d\theta} = -2a^2 \sin 2\theta$.

i.e.
$$\frac{dr}{d\theta} = -\frac{a^2 \sin 2\theta}{r} \quad \therefore \quad r \frac{d\theta}{dr} = \frac{-r^2}{a^2 \sin 2\theta} = -\frac{a^2 \cos 2\theta}{a^2 \sin 2\theta} = -\cot 2\theta$$

$$\tan \phi = r \frac{d\theta}{dr} = \tan\left(\frac{\pi}{2} + 2\theta\right) \quad \text{i.e.} \quad \phi = \frac{\pi}{2} + 2\theta$$

Hence at $\theta = 0$, $\phi = \dfrac{\pi}{2}$ i.e. tangent is perpendicular to the initial line at the points (a, 0) and (–a, 0).

9. The complete curve lies within the circle
   r = a since $r^2 \le a^2$.
   A rough sketch of the curve is as shown in Fig. 5.36.

Ex. 2 : *Trace the curve r = a + b cos θ for a > b, a < b and a = b.*   **(May 2011, 2017, Nov. 2015)**

Sol. : Case I : r = a + b cos θ, where a > b. The curve is known as *Pascal's Limacon.*

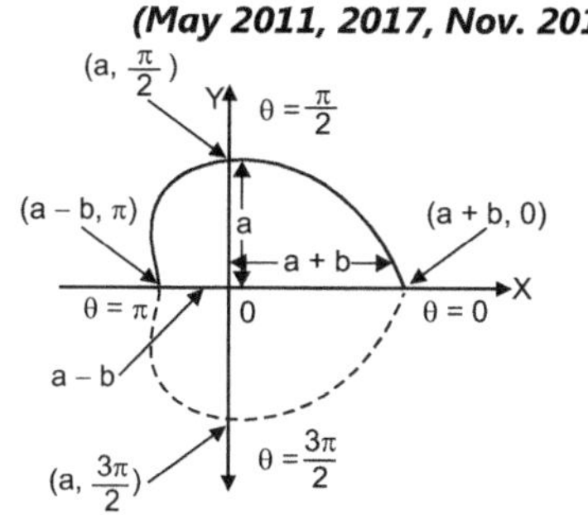

Fig. 5.37

1. The curve is symmetrical about the initial line.
2. It does not pass through the pole.
3. Maximum value of r is a + b at $\theta = 0$ and minimum value of r is a – b at $\theta = \pi$.
4. Table of values of r corresponding to θ :

θ	0	π/4	π/2	3π/4	π
r	a + b	a + b/$\sqrt{2}$	a	a – b/$\sqrt{2}$	a – b

5. Since $-1 \le \cos\theta \le 1$ and a > b, for this case r is never negative.
6. We trace the curve for values of θ between 0 and π. Remaining portion of the curve (i.e. for the values of θ between π and 2π) can be traced by symmetry about initial line.

Note :
(i)   Since a and b are arbitrary, approximate shape is as shown in Fig. 5.37.
(ii) Dotted portion represents the curve between θ = π and θ = 2π by symmetry about initial line.

Case II : r = a + b cos θ when a < b. This curve is known as *Pascal's Limacon.*   **(May 2008, 2006)**

1. The curve is symmetrical about the initial line.
2. Since a < b, the curve passes through the pole.
   [e.g. if we set a = 1, b = 2 i.e. r = 1 + 2 cos θ then r = 0 $\Rightarrow$ 2 cos θ = – 1.

   $\cos\theta = -\dfrac{1}{2}$ giving $\theta = \dfrac{2\pi}{3}$ $\therefore$ r = 1 + 2 cos θ passes through the pole.]

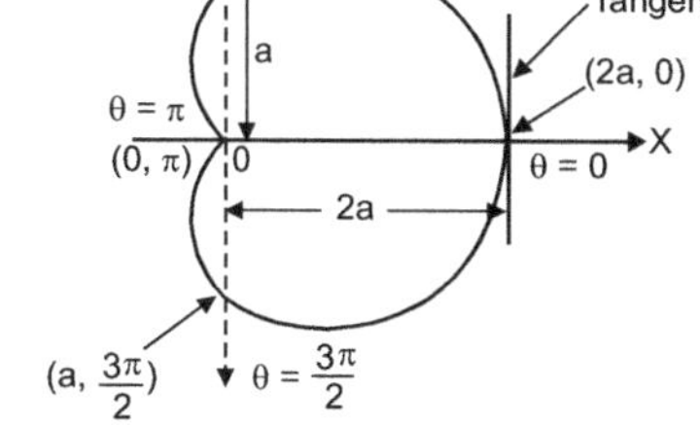

Fig. 5.38

3. Also note that for some value of θ, r can become negative also because for a < b, a – b < 0 hence when θ = π, r = a – b is negative. Hence we get an inner loop.
4. Table of values of r corresponding to θ. Maximum value of r is a + b, when θ = 0 and minimum value of r is a – b at θ = π.

θ	0	π/4	π/2	3π/4	π
r	a + b	a + b/$\sqrt{2}$	a	a – b/$\sqrt{2}$	a – b

The shape of the curve r = a + b cos θ, (a < b) is as shown in Fig. 5.38.
Note that the negative r = a – b for θ = π is denoted in the opposite direction.

Case III : r = a + b cos θ where a = b. For a = b, the curve becomes the cardioide

   r = a (1 + cos θ). *(Heart-shaped curve).*

1. The curve is symmetrical about the initial line.
2. The curve passes through the pole because for θ = π, r = 0.
3. Tangent at the pole is obtained by putting r = 0 in the equation.

   $\therefore$ 0 = a (1 + cos θ)

   i.e. cos θ = – 1 i.e. θ = π

   Hence tangent at the pole is the initial line itself.

4. If θ = 0, r = 2a, hence the curve cuts the initial line also at (2a, 0).
5. Table of values of r corresponding to θ is as given below :

θ	0	π/2	π	3π/2	2π
r	2a	a	0	a	2a

From table we get the information that, as θ increases from θ = 0 to θ = π, r decreases from 2a to 0 and also as θ increases from θ = π to θ = 2π, r increases from 0 to 2a. Maximum value of r is 2a and minimum value of r is 0.

6. Angle between radius vector r and tangent to the curve [ϕ] :

We use the formula $\tan \phi = r \dfrac{d\theta}{dr}$.

Now, given $r = a (1 + \cos \theta)$ $\therefore \dfrac{dr}{d\theta} = -a \sin \theta$

Hence $\tan \phi = r \cdot \dfrac{d\theta}{dr} = \dfrac{a (1 + \cos \theta)}{-a \sin \theta} = \dfrac{2 \cos^2 \dfrac{\theta}{2}}{-2 \sin \dfrac{\theta}{2} \cos \dfrac{\theta}{2}}$

or $\tan \phi = -\cot \dfrac{\theta}{2} = \tan \left(\dfrac{\pi}{2} + \dfrac{\theta}{2} \right)$

So that $\phi = \dfrac{\pi}{2} + \dfrac{\theta}{2}$

$\therefore$ at $\theta = 0$, $\phi = \dfrac{\pi}{2}$

Hence the tangent is perpendicular to the initial line at (2a, 0).

Note : The shape of the curve from $\theta = \pi$ to $\theta = 2\pi$ may also be traced by symmetry.

Spirals : There is an important class of curves called *spirals*.

A spiral is a curve usually given by an equation of the type $r = a\theta^m$.

In such curves, as θ increases indefinitely, r either goes on increasing indefinitely or decreases continuously, i.e. the curve goes on winding and winding round the pole. We shall trace some important spirals only.

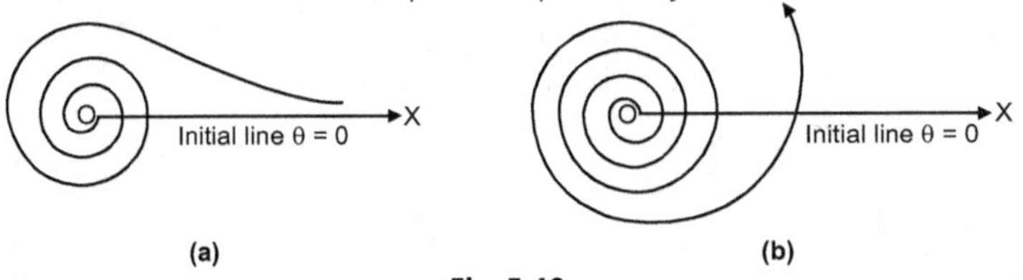

Fig. 5.40

Ex. 3 : Trace the curve r = ae^{mθ}, where a and m are positive (Equiangular spiral).

Sol. : The curve is known as Equiangular spiral.

1. **Symmetry :** No symmetry of any type.

2. **Pole :** Curve does not pass through the pole, because $r \neq 0$ for any value of θ.

3. **The angle** ϕ : $\dfrac{dr}{d\theta} = ame^{m\theta} = mr$.

$\Rightarrow$ $\tan \phi = r \dfrac{d\theta}{dr} = \dfrac{1}{mr} = \dfrac{1}{m}$ = constant.

Hence ϕ is a constant angle; i.e. the tangent always makes a constant angle with the radius vector.

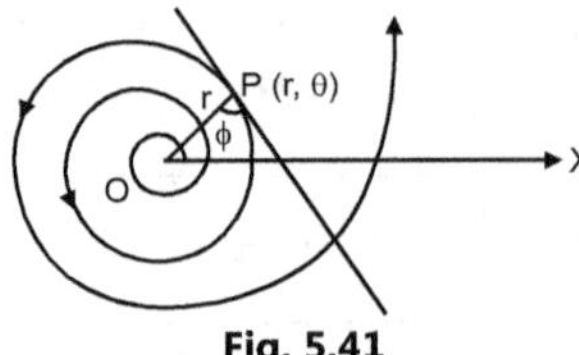

Fig. 5.41

4. **Region :**

    (a) When $\theta = 0$, $r = a$, i.e. the point (a, 0) lies on the curve.

    (b) As θ increases from 0 to ∞, r also increases from a to ∞, but remains positive.

    (c) In fact, r remains always positive.

    (d) If θ decreases from 0 to $-\infty$, r while remaining positive tends to 0. Hence the curve is as shown above.

Note : If we put $m = \cot \alpha$ in the curve $r = a e^{m\theta}$, it becomes the equiangular spiral $r = ae^{m \cot \alpha}$ and then $\phi = \alpha$. Hence the spiral where tangent m takes a fixed angle α with the radius vector always. This is also known as *logarithmic spiral.*

Ex. 4 : Trace the curve r = aθ.

Sol. : This curve is known as the *"Spiral of Archimidies"*.

$$r = a\theta \qquad \qquad \text{... (1)}$$

1. **Symmetry :** If θ and r have negative signs simultaneously, equation (1) does not change, hence symmetry about the line $\theta = \dfrac{\pi}{2}$.

2. **Pole :** If $\theta = 0$, r = 0, hence pole lies on the curve.

3. As θ increases, r also increases and as $\theta \to \infty$, $r \to \infty$. But when $\theta \to -\infty$, $r \to -\infty$.

    Hence the curve starting from the pole goes round and round the pole for an infinite number of times.

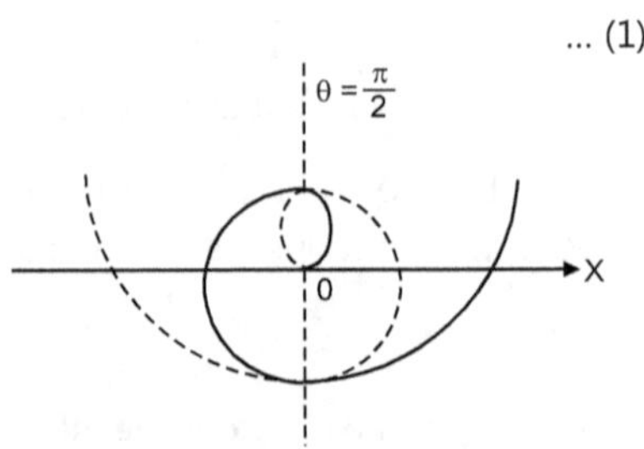

The dotted lines show the tracing for negative θ. The curve is as traced above.

Fig. 5.42

EXERCISE 5.3

Trace the following curves :

1. $r = a (1 + \sin \theta)$ **(Dec. 2008, 2006, 2018)**
2. $r = a (1 - \sin \theta)$ **(May 2009)**
3. $r (1 + \sin \theta) = 2a$ **(May 2011, 2007, 2005, 2004)**
4. $r (1 - \sin \theta) = 2a$
5. $r = a \operatorname{cosec} \theta \pm b$
6. $r^2 \theta = a^2$
7. $r\theta = a, \quad a > 0$

On type : $r = a + b \cos \theta$

8. $r = a \left(\dfrac{\sqrt{3}}{2} + \cos \dfrac{\theta}{2} \right)$

9. $r = a \left(\sqrt{3} + 2 \cos \theta \right)$

10. $r = a (1 + 2 \cos \theta)$ **(May 2010, 2004)**

11. $r = \dfrac{a}{2} (1 + \cos \theta)$ **(Dec. 2005)**

12. $r = 2a \cos \theta$

On Type : $r^n = a^n \cos n\theta$: For $n = \pm 1, \pm 2, \pm 1/2$

13. $r \cos \theta = a$
14. $r^2 = a^2 \cos 2\theta$
15. $r^2 \cos 2\theta = a^2$

16. $r = \dfrac{1}{2} a (1 + \cos \theta)$

17. $r (1 + \cos \theta) = 2a$

18. $r = \dfrac{a}{2} (1 - \cos \theta)$

19. $r = 2a \sin \theta$

On Type : $r^n = a^n \sin n\theta$: For $n = \pm 1, \pm 2, \pm \dfrac{1}{2}$

20. $r \sin \theta = a$
21. $r^2 = a^2 \sin 2\theta$
22. $r^2 \sin 2\theta = a^2$
23. $r (1 - \cos \theta) = 2a$

ANSWERS

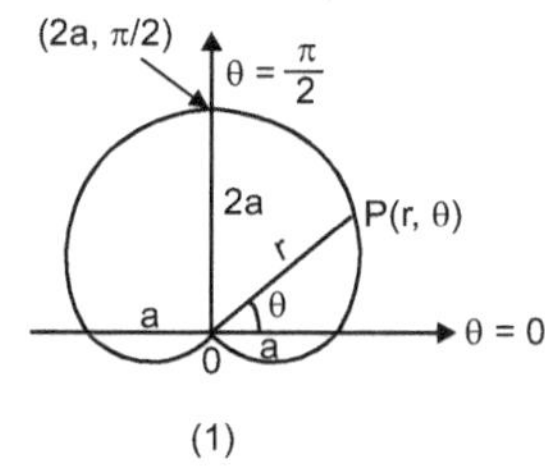

(1)

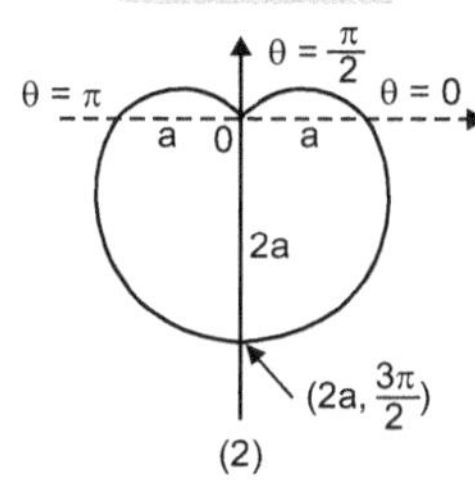

(2)

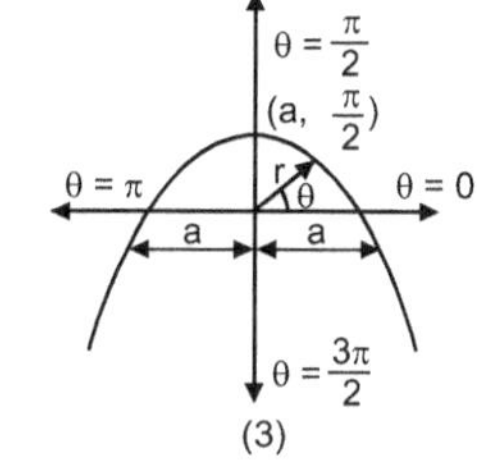

(3)

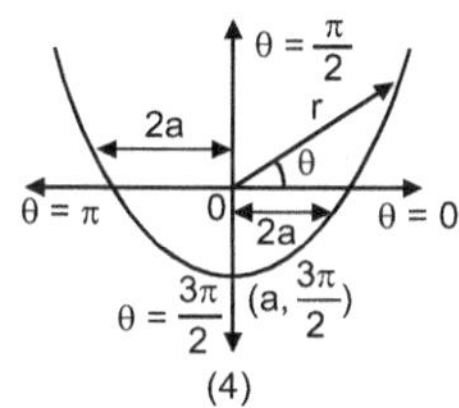

(4)

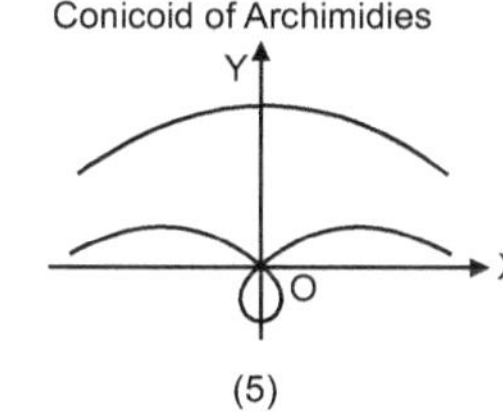

(5)

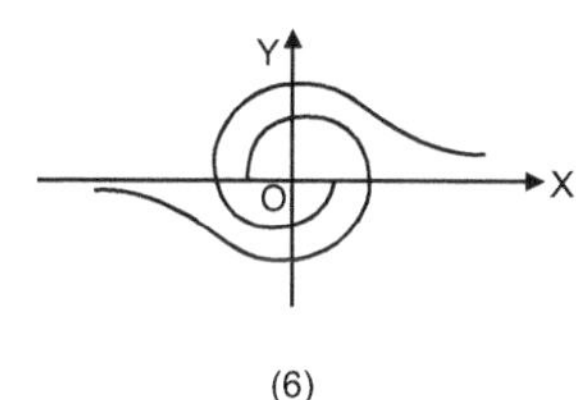

(6)

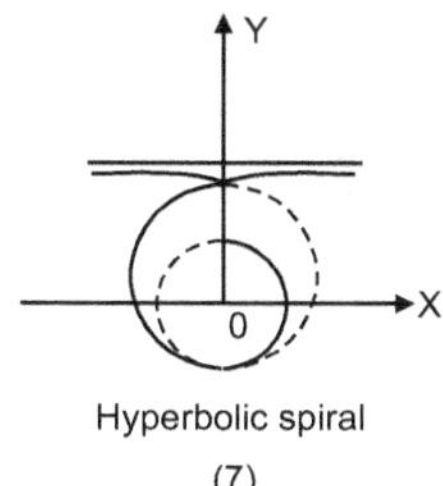

(7)

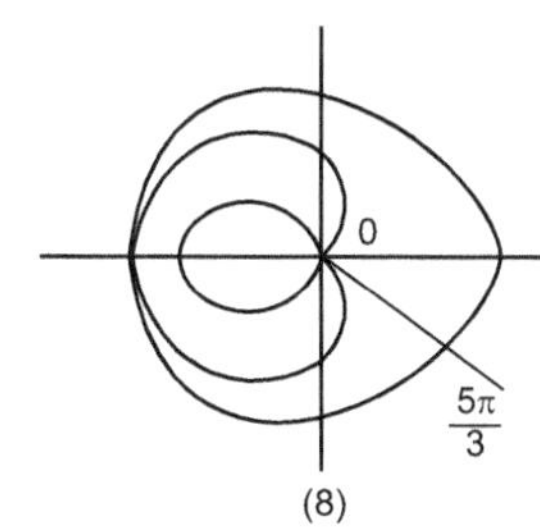

(8)

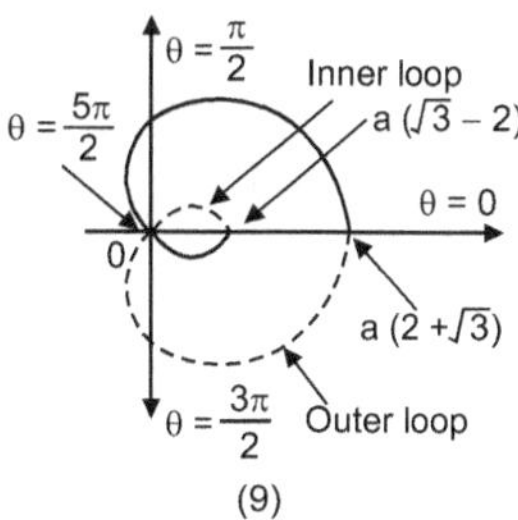

(9)

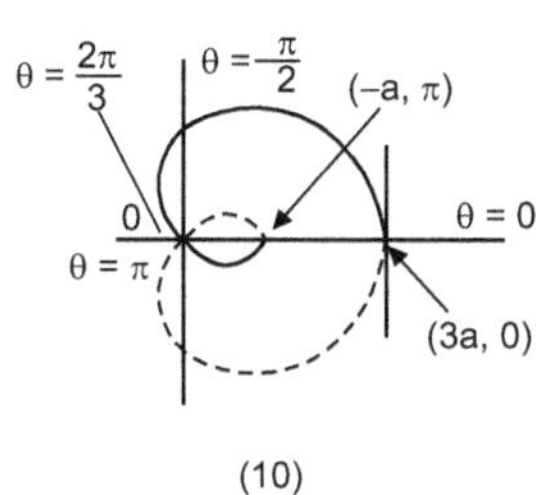

(10)

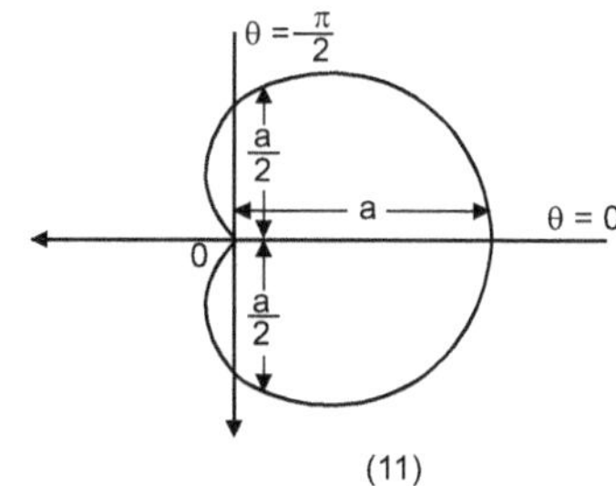

(11)

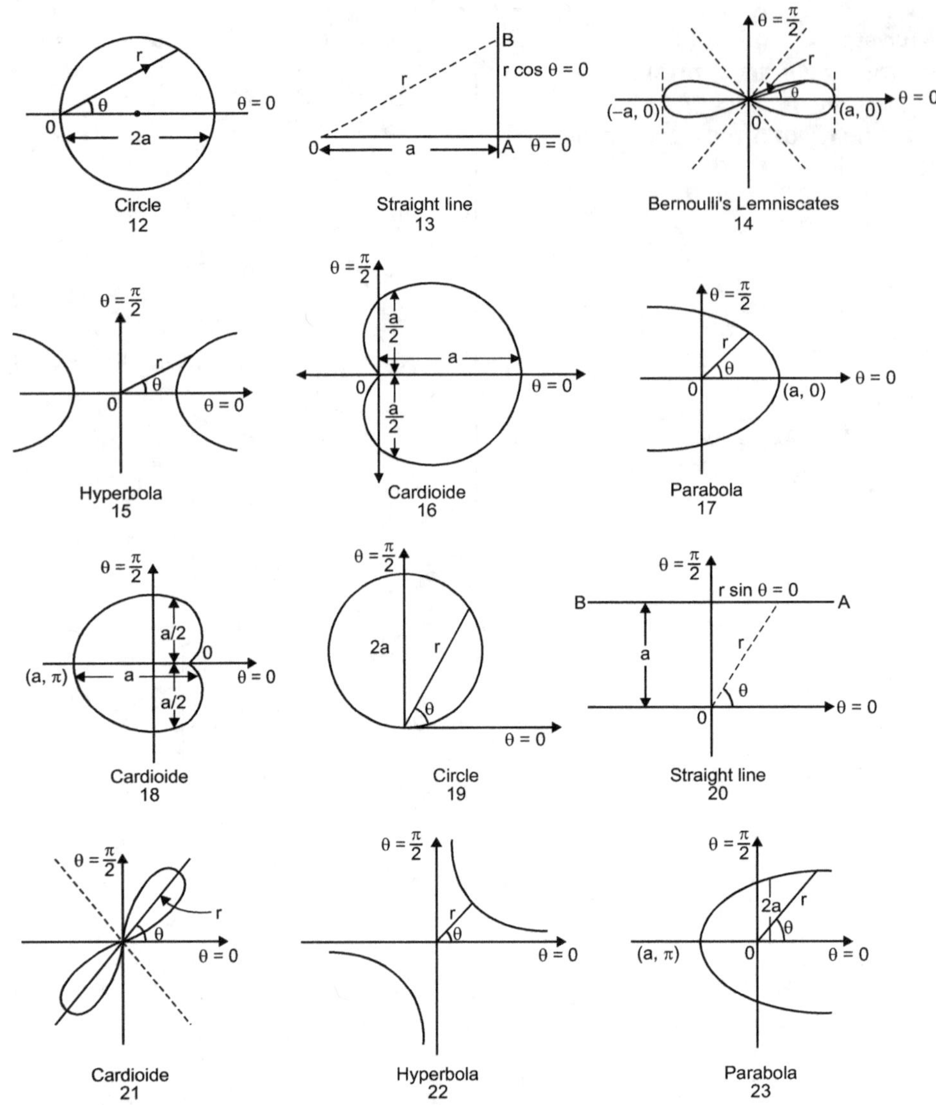

Fig. 5.43

TYPE 4 : Curves given by Polar equations of the type r = a sin nθ or r = a cos nθ.

5.6 TRACING OF ROSE CURVES

The following rules will help in tracing a polar curve given by r = a sin nθ or r = a cos nθ.

Rule 1 : Symmetry :

(a) If the substitution of $-\theta$ for θ in the equation leaves the equation unaltered, the curve is symmetrical about the initial line. For example, r = a cos 2θ.

(b) If the equation remains unchanged by changing θ to $-\theta$ and r to $-r$, at the same time, then the curve is symmetrical about the line θ = π/2 through the pole perpendicular to the initial line. e.g. r = a sin 3θ.

Rule 2 : Find in particular values of θ which give r = 0.

Rule 3 : If pole lies on the curve then find the equation of tangent or tangents at the pole. Put r = 0, the value of θ gives the tangent at the pole.

Rule 4 : For r = a sin nθ or r = a cos nθ, the maximum numerical value of r is a, hence the entire curve will lie within a circle of radius a (i.e. r = a).

Rule 5 : Since sin θ and cos θ are periodic functions, values of θ from 0 to 2π should only be considered. Values of θ > 2π give no new branch of the curve.

Rule 6 : The curve $r = a \sin n\theta$ or $r = a \cos n\theta$ consists of

    (a)   n equal loops if n is odd.

    (b)   2n equal loops if n is even.

Rule 7 : For drawing the loops, divide each quadrant into 'n' equal parts.

    For $r = a \sin n\theta$,

    (a)   First loop is drawn along $\theta = \dfrac{\pi/2}{n}$.

    (b)   If n is even, draw loops in two sectors consecutively from $\theta = 0$ to $\theta = 2\pi$.

    (c)   If n is odd, draw loops in two sectors alternatively keeping two sectors between the loops vacant.

Rule 8 : For drawing the loops, divide each quadrant into 'n' equal parts.

    For $r = a \cos n\theta$,

    (a)   First loop is drawn along $\theta = 0$.

    (b)   If n is even, draw loops in two sectors consecutively from $\theta = 0$ to $\theta = 2\pi$.

    (c)   If n is odd, draw loops in two sectors alternately keeping two sectors between the loops vacant.

Rule 9 : Angle between the radius vector and the tangent [ϕ] :

Use the formula $\tan \phi = r \dfrac{d\theta}{dr}$ and find ϕ and also the points where $\phi = 0$ or ∞ i.e. find the points where the tangent coincides with the radius vector or is perpendicular to it.

Rule 10 : Prepare the table of values of r and θ and observe how r varies as θ increases from 0 to 2π.

Note :

1. $\qquad \sin n\theta = 0 \quad$ for $n\theta = 0, \pi, 2\pi, 3\pi, 4\pi \dots$

    $\therefore \qquad \theta = 0, \dfrac{\pi}{n}, \dfrac{2\pi}{n}, \dfrac{3\pi}{n}, \dfrac{4\pi}{n}, \dfrac{5\pi}{n}, \dots$

2. $\qquad \cos n\theta = 0 \quad$ for $n\theta = \dfrac{-\pi}{2}, \dfrac{\pi}{2}, \dfrac{3\pi}{2}, \dfrac{5\pi}{2}, \dfrac{7\pi}{2}, \dfrac{9\pi}{2}, \dfrac{11\pi}{2} \dots$

    $\therefore \qquad \theta = \dfrac{-\pi}{2n}, \dfrac{\pi}{2n}, \dfrac{3\pi}{2n}, \dfrac{5\pi}{2n}, \dfrac{7\pi}{2n}, \dfrac{9\pi}{2n}, \dfrac{11\pi}{2n}$

3. $r = -a$ means radial distance to be taken in the opposite direction.

4.   Since the shapes of the curves given by $r = a \sin n\theta$ or $r = a \cos n\theta$ are like rose, these curves are known as *Rose curves.*

***Ex. 1 :** Trace the curve $r = a \sin 3\theta$.*   **(Dec. 2009, 2006; May 2013 Nov. 2014)**

Sol. : This curve is known as *Three leaved rose.*

1.   The curve $r = a \sin 3\theta$ consists of three equal loops (since the curve $r = a \sin n\theta$ consists of n equal loops if n is odd.)

2.   We see that the equation remains unchanged by changing θ to $-\theta$ and r to $-r$, at the same time. Hence the curve is symmetrical to line $\theta = \dfrac{\pi}{2}$ through the pole perpendicular to the initial line (i.e. Y-axis in cartesian co-ordinates).

3.   The pole lies on the curve (since for $\theta = 0, r = 0$).

4.   To obtain equation of tangents at pole we put $r = 0$,

    we get $\sin 3\theta = 0$. i.e. $3\theta = 0, \pi, 2\pi, 3\pi, 4\pi, 5\pi, \dots$ Hence $\theta = 0, \dfrac{\pi}{3}, \dfrac{2\pi}{3}, \pi, \dfrac{4\pi}{3}, \dfrac{5\pi}{3}, \dots$

    Draw these lines and place equal loops in alternate division, the first loop between $\theta = 0$, $\theta = \dfrac{\pi}{3}$, second between $\theta = \dfrac{2\pi}{3}$ and $\theta = \pi$ and third between $\theta = \dfrac{4\pi}{3}$ and $\dfrac{5\pi}{3}$.

5.   Hence we divide each quadrant into 3 parts and as such we have following table of values.

θ	0	$\dfrac{\pi}{6}$	$\dfrac{2\pi}{6}$	$\dfrac{\pi}{2}$	$\dfrac{4\pi}{6}$	$\dfrac{5\pi}{6}$	π	$\dfrac{7\pi}{6}$	$\dfrac{8\pi}{6}$	$\dfrac{3\pi}{2}$	$\dfrac{10\pi}{6}$	$\dfrac{11\pi}{6}$	2π
r	0	a	0	$-a$	0	a	0	$-a$	0	a	0	$-a$	0

From the table we see that r is never greater than a. Hence the curve lies within the circle of radius a.

6. Angle between radius vector and tangent to the curve [ϕ]. We use the formula :

$$\tan \phi = r\frac{d\theta}{dr} , \frac{dr}{d\theta} = 3a \cos 3\theta$$

$$\therefore \quad \tan \phi = r\frac{d\theta}{dr} = \frac{a \sin 3\theta}{3a \cos 3\theta} = \frac{1}{3} \tan 3\theta$$

$$\therefore \quad \tan \phi = r\frac{d\theta}{dr} = \frac{1}{3} \tan 3\theta$$

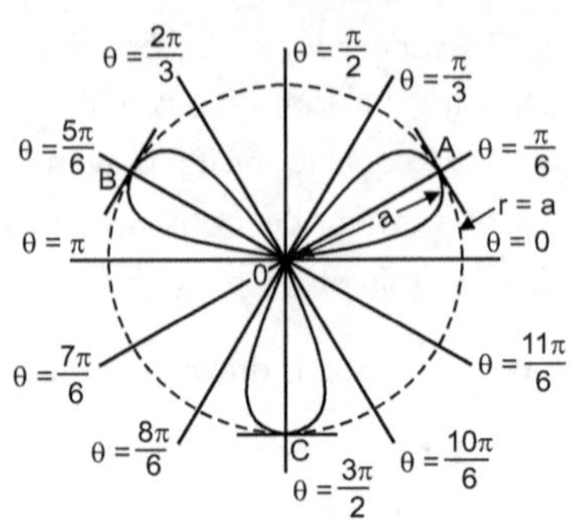

Fig. 5.44

For $\theta = 0, \dfrac{\pi}{3}, \dfrac{2\pi}{3}, \pi, \dfrac{4\pi}{3}, \dfrac{5\pi}{3}, \dots$ r = 0 (from table) and at these points

as $\phi = 0$ [from (1)], the tangent and radius vectors are coincident.

For $\theta = \dfrac{\pi}{6}, \dfrac{\pi}{2}, \dfrac{5\pi}{6}, \dfrac{7\pi}{6}, \dfrac{3\pi}{2}, \dfrac{11\pi}{6}$, the values of r (from table) are $\pm$ a (i.e. points A, B, C) and ϕ is $\dfrac{\pi}{2}$ or $\dfrac{3\pi}{2}$. Hence at points A, B, C, tangent is perpendicular to the radius vector.

7. Here we consider the value of θ from 0 to 2π since sin θ is periodic function and $\theta > 2\pi$ gives no new branch of the curve.

8. When θ increases from 0 to $\dfrac{\pi}{6}$, r remains positive and increases from 0 to a. Further when θ increases from $\dfrac{\pi}{6}$ to $\dfrac{\pi}{3}$, r remains positive but decreases from a to 0. Thus we get first loop of the curve between $\theta = 0$ to $\theta = \dfrac{\pi}{3}$ i.e. the loop OAO.

When θ increases from $\dfrac{2\pi}{6}$ to $\dfrac{\pi}{2}$ $\left(\text{i.e.} \dfrac{\pi}{3} \text{ to } \dfrac{\pi}{2}\right)$, r remains negative and numerically increases from 0 to a (i.e. portion of the curve below initial line).   As θ increases from $\dfrac{\pi}{2}$ to $\dfrac{4\pi}{6}$ $\left(\text{i.e.} \dfrac{\pi}{2} \text{ to } \dfrac{2\pi}{3}\right)$, r is negative and numerically decreases from a to 0.

Now, we get the second loop of the curve OCO. When θ increases from $\theta = \dfrac{2\pi}{3}$ to $\theta = \dfrac{5\pi}{6}$, r remains positive and increases from 0 to a. Further, when θ increases from $\dfrac{5\pi}{6}$ to π, r remains positive but decreases from a to 0. Finally, we get the third loop of the curve OBO. A rough sketch of the curve is as shown above.

Ex. 2 : *Trace the curve r = a cos 2θ.*   **(May 2019, May 2010, 2015, 2018; Dec. 2013)**

Sol. : This curve is known as *Four Leaved Rose*.

1. The curve r = a cos 2θ consists of four equal loops.

2. The curve has only symmetry about initial line.

3. The pole will lie on the curve (or curve passes through the pole).

4. Equations of tangents (or tangent) at the pole are obtained by putting r = 0, we have

$$\cos 2\theta = 0 \text{ i.e. } 2\theta = -\frac{\pi}{2}, \frac{\pi}{2}, \frac{3\pi}{2}, \frac{5\pi}{2}, \frac{7\pi}{2}. \text{ Hence } \theta = -\frac{\pi}{4}, \frac{\pi}{4}, \frac{3\pi}{4}, \frac{5\pi}{4}, \frac{7\pi}{4}, \dots$$

Draw these lines and place equal loops in each division. The first loop between $\theta = -\dfrac{\pi}{4}$ to $\theta = \dfrac{\pi}{4}$, second loop between $\theta = \dfrac{\pi}{4}$ to $\theta = \dfrac{3\pi}{4}$, third loop between $\theta = \dfrac{3\pi}{4}$ to $\theta = \dfrac{5\pi}{4}$ and fourth loop between $\theta = \dfrac{5\pi}{4}$ to $\theta = \dfrac{7\pi}{4}$.

5. Hence we divide each quadrant into two (since n = 2) parts and as such we have following table of values.

θ	0	$\dfrac{\pi}{4}$	$\dfrac{\pi}{2}$	$\dfrac{3\pi}{4}$	π	$\dfrac{5\pi}{4}$	$\dfrac{3\pi}{2}$	$\dfrac{7\pi}{4}$	2π
r	a	0	–a	0	a	0	–a	0	a

Thus we see that r is never greater than a (i.e. maximum value of r is a) and the curve lies wholly within a circle of radius a.

6. Angle between radius vector r and tangent to the curve [φ] :

We use the formula $\tan \phi = r \dfrac{d\theta}{dr}$, where value of φ will indicate the direction of tangent.

Given $r = a \cos 2\theta$

Differentiating with respect to θ, we have,

$$\frac{dr}{d\theta} = -2a \sin 2\theta$$

$$r \frac{d\theta}{dr} = \frac{a \cos 2\theta}{-2a \sin 2\theta} = -\frac{1}{2} \cot 2\theta$$

$$\tan \phi = r \frac{d\theta}{dr} = \frac{1}{2} \tan \left(\frac{\pi}{2} + 2\theta \right)$$

$$\tan \phi = 0 \text{ when } r = 0 \text{ i.e. when}$$

$$\theta = -\frac{\pi}{4} \left(\text{or} \frac{7\pi}{4} \right), \frac{\pi}{4}, \frac{3\pi}{4}, \frac{5\pi}{4}$$

... (1)

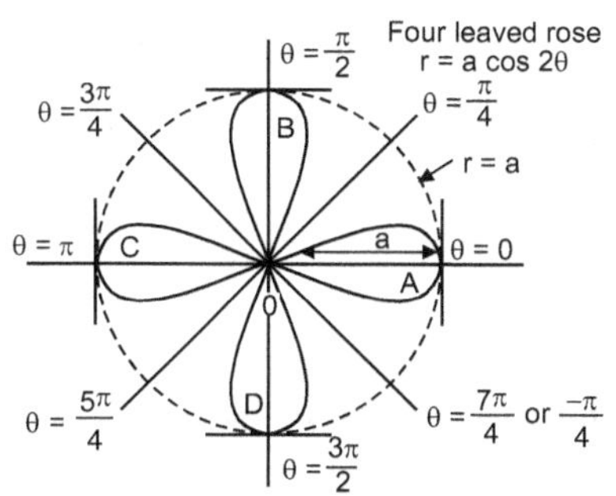

Fig. 5.45

Thus at points $\theta = -\frac{\pi}{4}, \frac{\pi}{4}, \frac{3\pi}{4}, \frac{5\pi}{4}$ $(r = 0)$, the tangents are coincident with radius vectors. Also tan φ is infinite when $r = \pm a$ i.e.

when $\theta = 0, \frac{\pi}{2}, \pi, \frac{3\pi}{2}$. Thus at points $\theta = 0, \frac{\pi}{2}, \pi, \frac{3\pi}{2}$ $(r = \pm a)$, the tangents are perpendicular to radius vectors. Hence at each of the points A, B, C, D, tangent is perpendicular to the radius vector.

7. A rough sketch of the curve is as shown above.

EXERCISE 5.4

Trace the following curves :

1. $r = a \cos 5\theta$

2. $r = a \sin 4\theta$ **(May 2008)**

3. $r = a \cos 4\theta$

4. $r = a \cos 3\theta$

5. $r = a \sin 5\theta$

(Nov./Dec. 2019, Dec. 2010, 2008, 2004; May 2007, 2006, 2016)

6. $r = a \sin 2\theta$ **(May 2005, Dec. 2007)**

ANSWERS

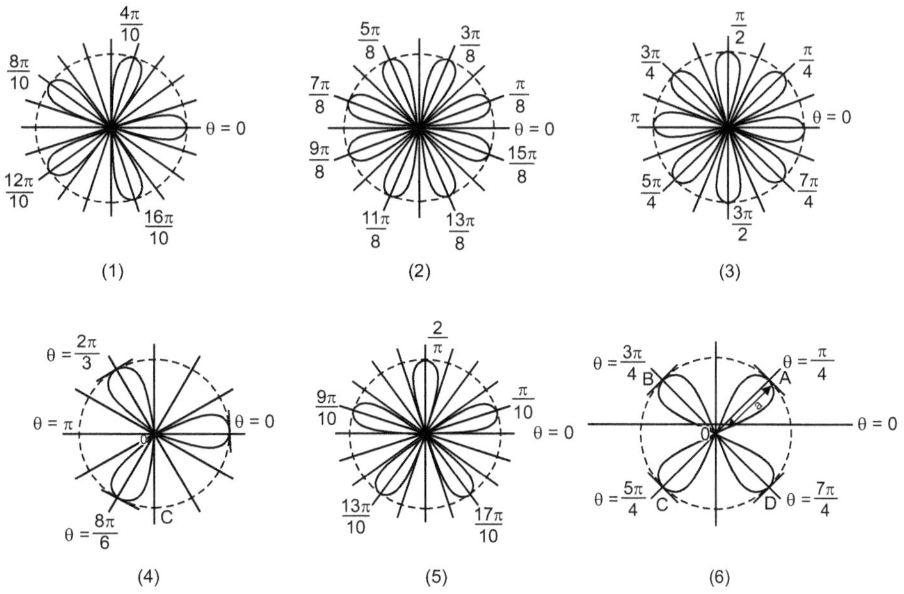

Fig. 5.46

5.7 TRACING OF CARTESIAN CURVES (WITH IMPLICIT RELATIONS)

For the curves expressed in implicit form, first we will apply all the rules already discussed under cartesian equation of the curve. Now, we transform the equation to polar form by using $x = r \cos \theta$, $y = r \sin \theta$ getting the polar equation of the curve in explicit form $r = f(\theta)$.

We note that if power of r is even then the curve is symmetrical about the pole. Also find the oblique asymptotes by using method discussed earlier. The transformation to polar helps us to determine that region where the part of the curve does not exist.

Hence to trace the curves given by implicit equations, it is convenient to make use of polar as well as cartesian forms of their equations.

Ex. 1 : *Trace the curve $x^3 + y^3 = 3axy$.*   **(May 2007, 2005; Dec. 2010, 2009, 2008, 2006)**

Sol. : This curve is known as *"Foulium of Descartes"*.

1. The curve has only symmetry about line $y = x$.

2. The curve passes through the origin.

3. Tangents at the origin are $x = 0$, $y = 0$ (i.e. X-axis and Y-axis). Hence we may expect a node at the origin.

4. The curve cuts the axes only at $(0, 0)$. It also cuts the line $y = x$ at $(0, 0)$ and $\left(\dfrac{3a}{2}, \dfrac{3a}{2}\right)$.

5. For the equation $x^3 + y^3 = 3axy$ if x and y both become negative then R.H.S. becomes positive but L.H.S. remains negative, which is absurd. Hence x and y both cannot be negative at the same time. This means curve has no branch in the third quadrant.

6. Oblique asymptote : $x^3 + y^3 - 3axy = 0$.

    Here third degree term $= x^3 + y^3$, second degree term $= - 3axy$.

    We put $x = 1$, $y = m$

    Let $\phi_3(m) = 1 + m^3$ and $\phi_2(m) = - 3am$

    Solve $\phi_3(m) = 0 \Rightarrow 1 + m^3 = 0$

    $(m + 1)(m^2 - m + 1) = 0$ giving only real value $m = -1$ because the remaining two roots are imaginary.

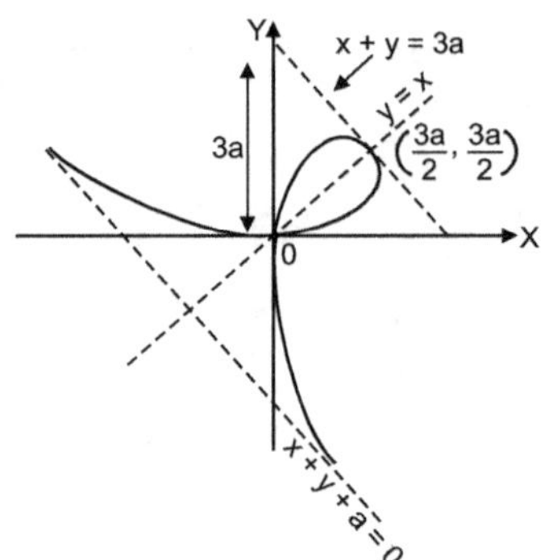

Fig. 5.47

To find c, we use the formula $c = -\dfrac{\phi_2(m)}{\phi_3'(m)}$

Here $\phi_3(m) = 1 + m^3$ ∴ $\phi_3'(m) = \dfrac{d}{dm}(1 + m^3) = 3m^2$ ∴ $c = -\left(\dfrac{-3am}{3m^2}\right) = \dfrac{a}{m}$

Now put $m = -1$ ∴ $c = -a$

By putting $m = -1$, $c = -a$ in $y = mx + c$, we have $y = -x - a$ i.e. $x + y + a = 0$ is an oblique asymptote to the curve.

7. Since tangents to the curve at origin are $x = 0$, $y = 0$ which are real and different, origin is a node.

8. Also $\dfrac{dy}{dx} = -\left(\dfrac{x^2 - ay}{y^2 - ax}\right)$; ∴ $\left[\dfrac{dy}{dx}\right]_{\left(\frac{3a}{2}, \frac{3a}{2}\right)} = - 1$ which shows that the tangent at $\left(\dfrac{3a}{2}, \dfrac{3a}{2}\right)$ is perpendicular to $y = x$. Now the

    equation of this tangent is $y - y_1 = m(x - x_1)$ where $m = - 1$ and (x_1, y_1) is $\left(\dfrac{3a}{2}, \dfrac{3a}{2}\right)$.

    ∴ $y - \dfrac{3a}{2} = -1\left(x - \dfrac{3a}{2}\right)$ i.e. $x + y = 3a$ is the tangent at $\left(\dfrac{3a}{2}, \dfrac{3a}{2}\right)$ which is perpendicular to $y = x$. A rough sketch of the curve is as shown above.

    We also note that the parametric equations of $x^3 + y^3 = 3axy$ are

    $x = \dfrac{3at}{1 + t^3}$, $y = \dfrac{3at^2}{1 + t^3}$ and polar equation is $r = \dfrac{3a \cos \theta \sin \theta}{\cos^3 \theta + \sin^3 \theta}$.

Ex. 2 : *Trace the curve* $x^5 + y^5 - 5a^2 x^2 y = 0$

Sol. :

1. It is symmetrical in the opposite quadrants.

2. It passes through the origin.

3. Tangents at origin are $x = 0$, $y = 0$.

4. Since tangents at the origin are $x = 0$, $y = 0$ which are real and different, origin is a node.

5. For oblique asymptote, fifth degree term = $x^5 + y^5$, fourth degree term = 0.

   We put $x = 1$, $y = m$.

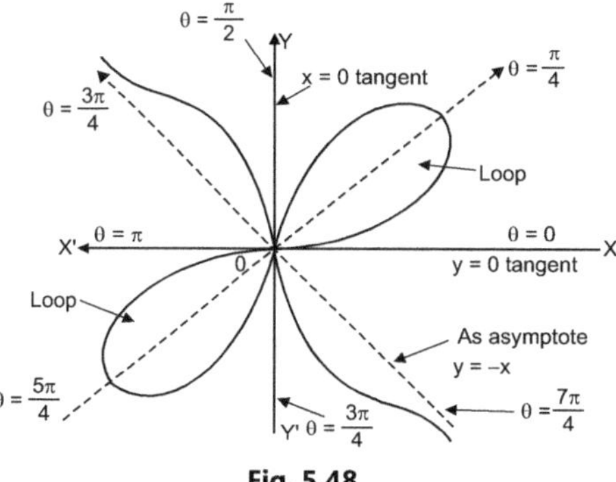

Fig. 5.48

$\phi_5(m) = 1 + m^5$, $\phi_4(m) = 0$. Solve $\phi_5(m) = 0$ i.e. $1 + m^5 = 0$ giving only $m = -1$ as real value and we note that $c = 0$.

$y = mx + c$ i.e. $y = -x$ is oblique asymptote to the curve.

6. Now, convert the equation to polar form

$$r^5(\cos^5\theta + \sin^5\theta) = 5a^2 r^3 \cos^2\theta \sin\theta \; r^2 = \frac{5a^2 \cos^2\theta \sin\theta}{\cos^5\theta + \sin^5\theta}.$$

when $\theta = 0$, $r = 0$, $\theta = \dfrac{\pi}{4}$, $r = \sqrt{5} \cdot a$.

$\theta = \pi/2$, $r = 0$. Thus the curve has one loop in first quadrant symmetrical about $y = x$. Because of symmetry in the opposite quadrants, we get one more loop in the 3rd quadrant.

When $\theta = \dfrac{3\pi}{4}$, $r \to \infty$ and for $\pi/2 < \theta < \dfrac{3\pi}{4}$, r^2 increases for 0 to ∞, curve exists.

Further for $\dfrac{3\pi}{4} < \theta < \pi$, r^2 is negative, the curve does not exist for $\dfrac{3\pi}{4} < \theta < \pi$.

Values of θ between π to 2π need not be considered because of symmetry in opposite quadrants.

7. Since power of r is even, curve is also symmetrical about pole. Hence the curve is as shown above.

EXERCISE 5.5

Trace the following curves :

1. $x^4 + y^4 = 4axy^2$
2. $x^4 + y^4 = a^2(x^2 - y^2)$
3. $x^6 + y^6 = a^2 x^2 y^2$
4. $x^5 + y^5 = 5ax^2 y^2$
5. $x^4 + y^4 = 2a^2 xy$ **(Dec. 2004)**
6. $y^4 - x^4 + xy = 0$

ANSWERS

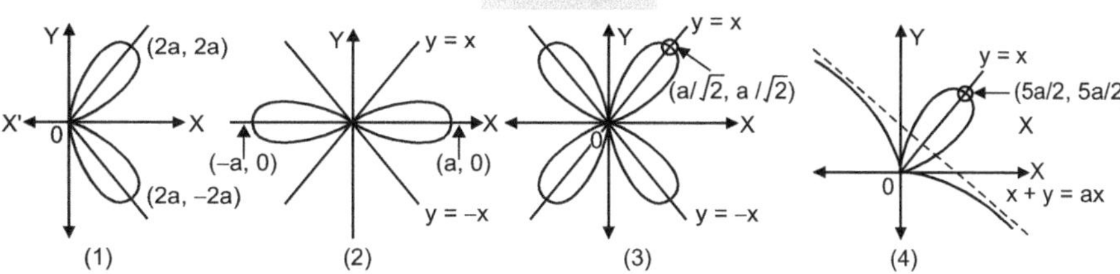

(1)   (2)   (3)   (4)

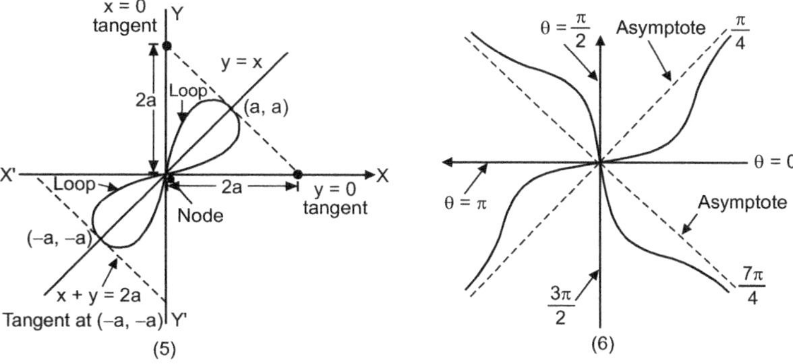

(5)   (6)

Fig. 5.49

5.8 SOME WELL-KNOWN STANDARD CURVES

Ex. 1 : *Trace the following curve $y = c\,\cosh\dfrac{x}{c}$.*

Sol. : This curve is known as "*The common catenary*".

A curve in which a perfectly flexible and uniform string hangs under gravity is called *catenary*. The equation of the curve is $y = c\,\cosh\dfrac{x}{c}$, where c is known as *parameter* of the curve. It is symmetrical about Y-axis.

It does not pass through the origin. When x = 0 we have y = c ($\because \cosh 0 = 1$).

It cuts Y-axis at A (0, c) and this point A (0, c) is called as *vertex* of the common catenary.

Here X-axis is called as the *directrix*. It does not cut X-axis at all.

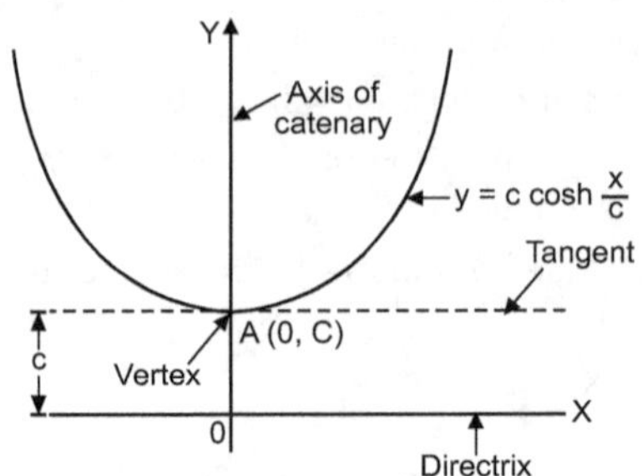

Fig. 5.50 (a)

$$\frac{dy}{dx} = c\cdot\sinh\frac{x}{c}\cdot\frac{1}{c}\left[\because \frac{d}{dx}\cosh\left(\frac{x}{c}\right) = \frac{1}{c}\sinh\frac{x}{c}\right]$$

$$\frac{dy}{dx} = \sinh\frac{x}{c} > 0 \ \text{ for } x > 0$$

$\therefore$　y increases as x increases from 0 to ∞

when x = 0, $\dfrac{dy}{dx} = \sinh 0 = 0$.

Therefore, the tangent at A (0, c) is parallel to X-axis. Here AY is called as the *axis of the common catenary*. A rough sketch of the curve is as shown in Fig. 5.50 (b)

SOME WELL-KNOWN CURVES

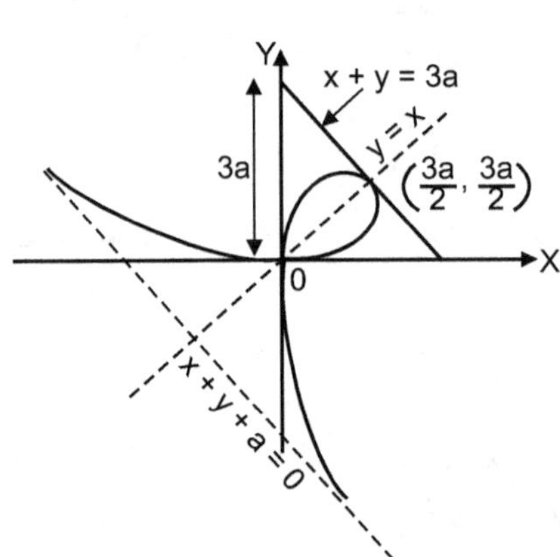

Foulium of Descart as $x^3 + y^3 = 3$ any

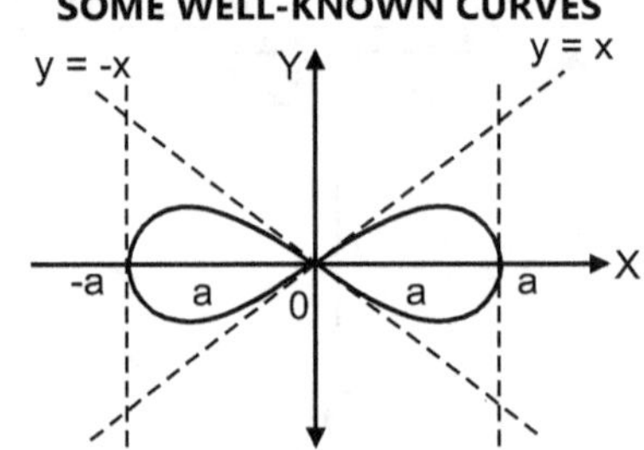

Bemoulli's Lemniscate $r^2 = a^2\cos 2\theta$

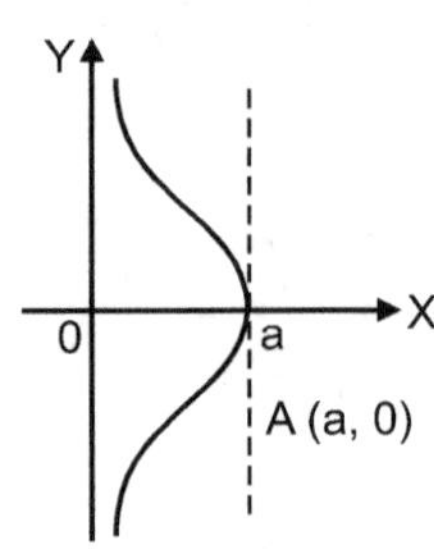

Witch of agnest $y^2 = 4a^2(a - x)/x$

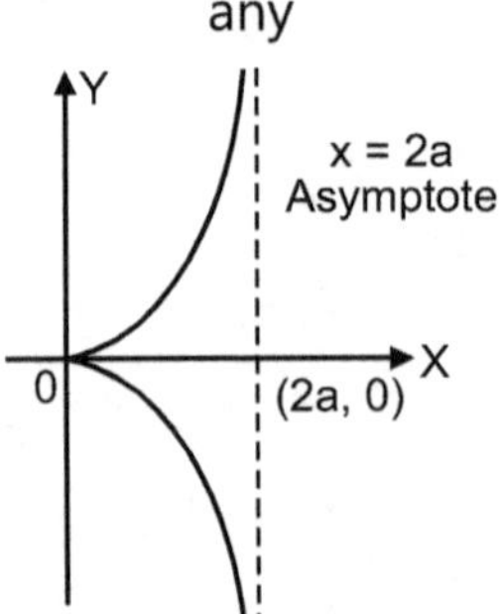

The Cissoid of Diocle $y^2(2a - x) = x^3$

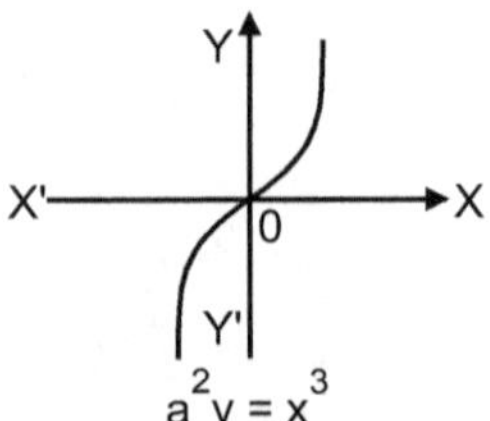

Semi-cubic parabola

Cubic parabola

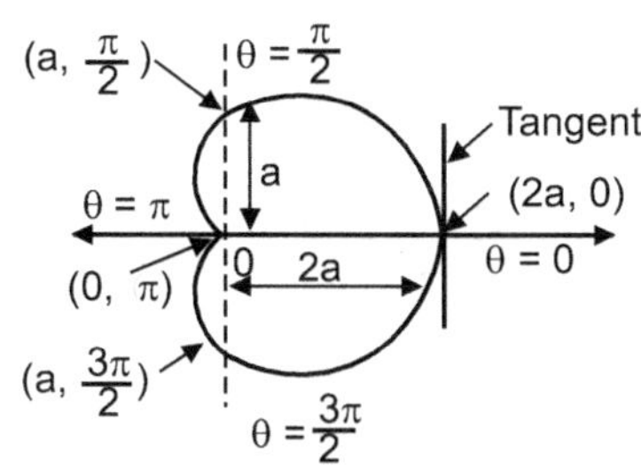

$r = a (1 + \cos \theta)$

cardiode

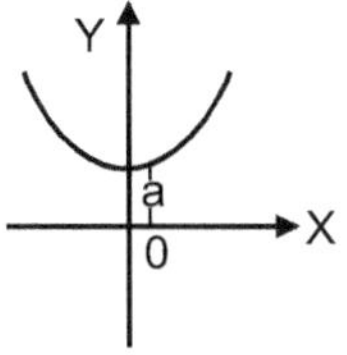

$$y = a \cos h \frac{x}{a}$$

Catenary

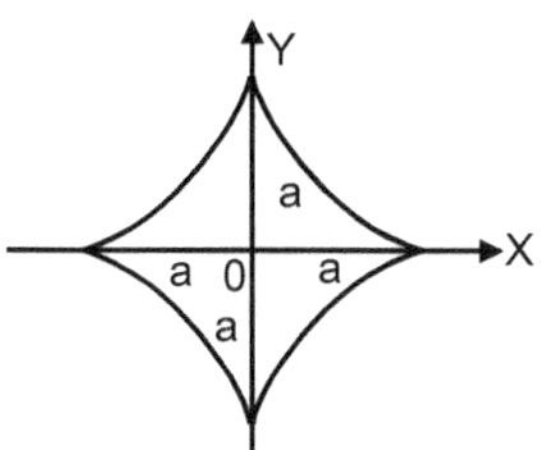

$x^{2/3} + y^{2/3} = a^{2/3}$ or

$x = a \cos^3\theta, \; y = a \sin^3\theta$

Asstroid

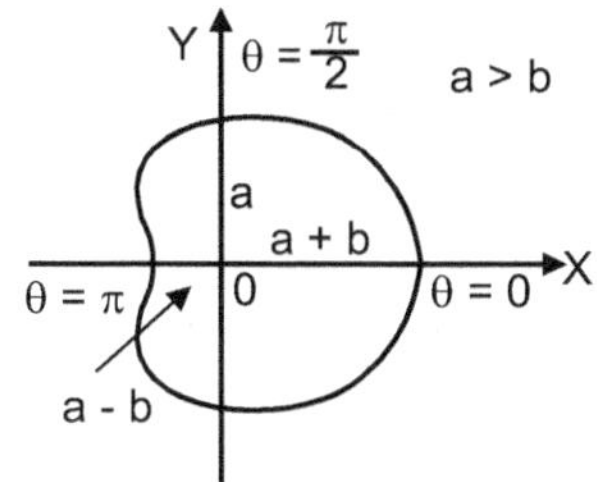

Pascal's Limacon

$r = a + b \cos \theta$

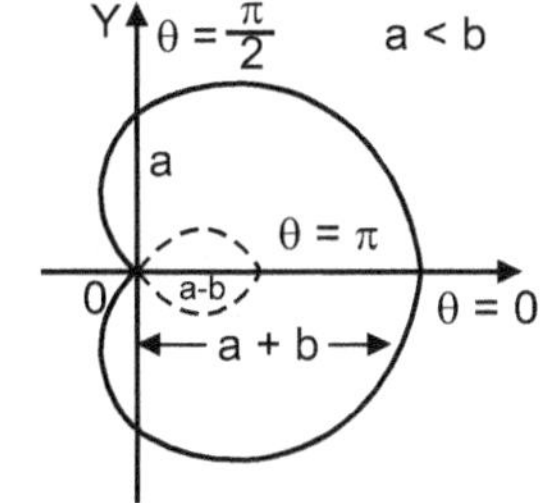

Pascal's Limacon

$r = a + b \cos \theta$

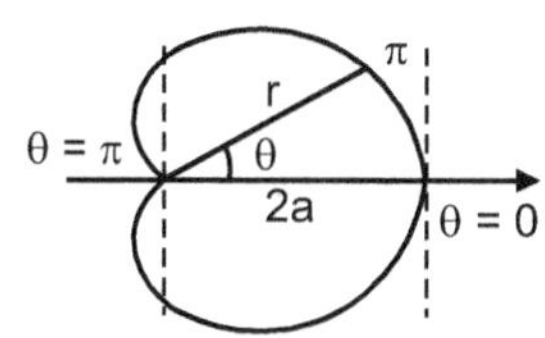

Cardioide

$r = (a + b \cos \theta), \, a = b$

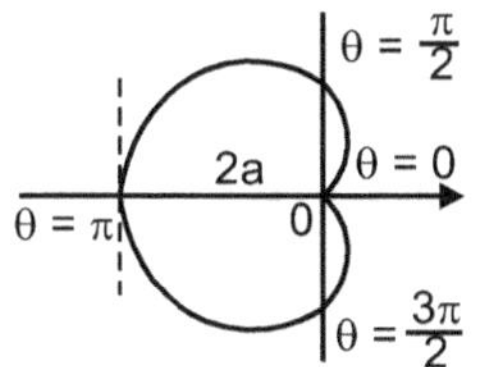

Cardioide

$r = a (1 - \cos \theta)$

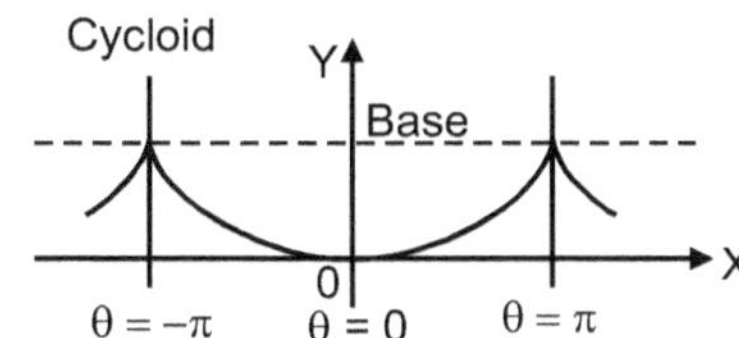

$x = a (\theta + \sin \theta), \; y = a (1 - \cos \theta)$

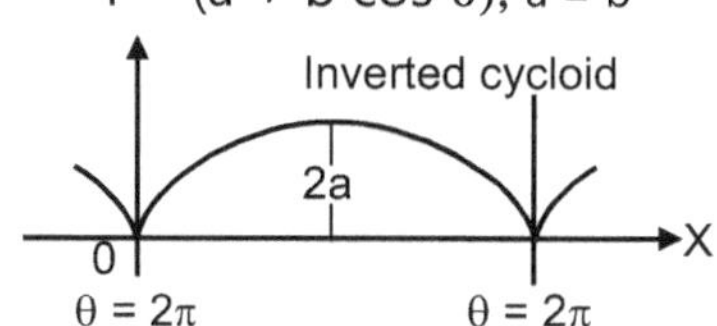

$X = a (\theta - \sin \theta), \; y = a(1 - \cos \theta)$

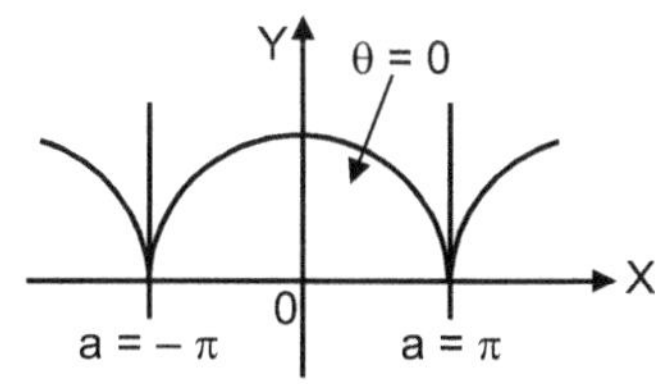

$x = a (\theta + \sin \theta),$

$y = a (1 + \cos \theta)$

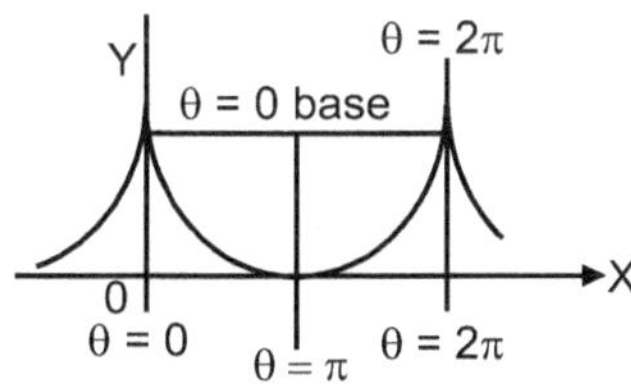

$x = a (\theta - \sin \theta),$

$y = a \; (1 - \cos \theta)$

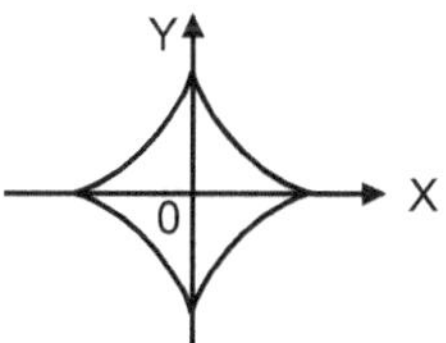

$(ax)^{2/3} + (by)^{2/3} = (a^2 - b^2)^{2/3}$

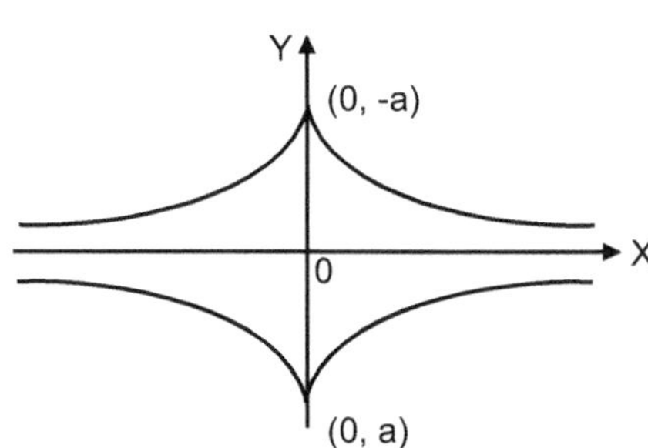

$x = a \cos \theta + \dfrac{a}{2} \log \tan^2 \dfrac{\theta}{2}, \; y = a \sin \theta$

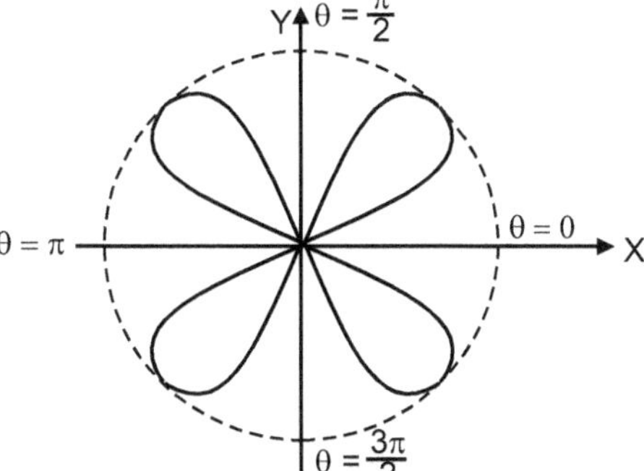

$r = a \sin 2\theta$

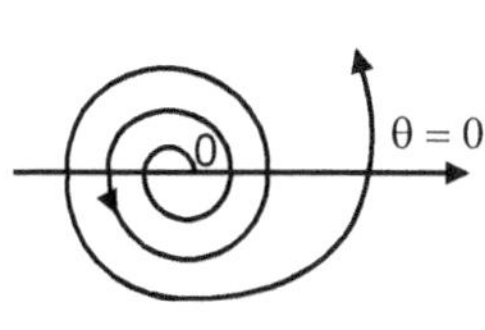

$r = ae^{m\theta}$

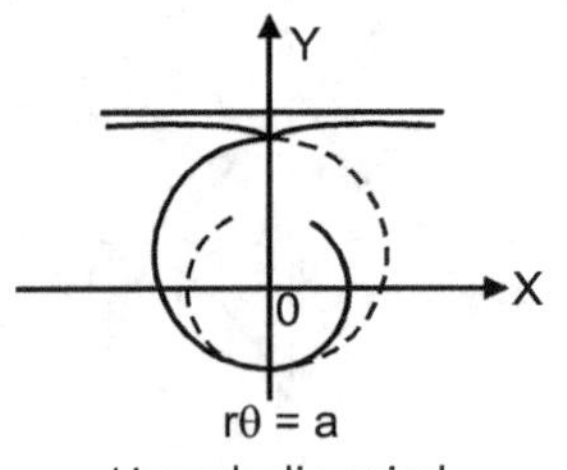

$r\theta = a$

Hyperbolic spiral

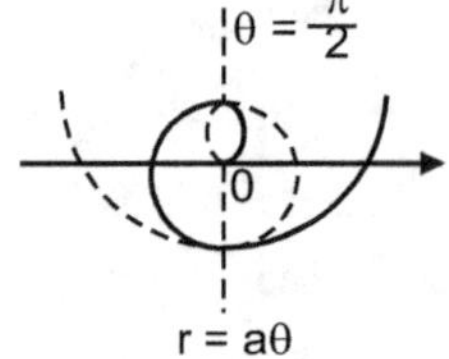

$r = a\theta$

Spiral of Archimidies

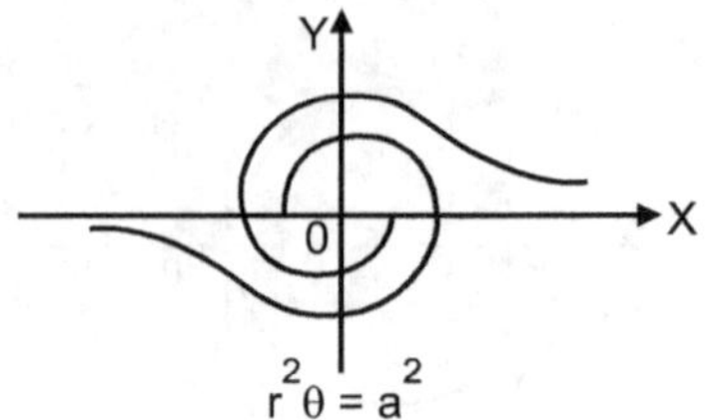

$r^2\theta = a^2$

Fig. 5.50 (b)

5.9 RECTIFICATION OF CURVES

Introduction :

In this section, we shall consider the applications of integration to measure the length of arc of plane curves. This procedure is called *Rectification*.

The curves that we come across, have their equations either in cartesian, parametric or polar form. We shall first develop formulae which are used when equations of curves are given in either cartesian or parametric form.

5.10 RECTIFICATION OF PLANE CURVE FOR CARTESIAN EQUATION

1. Cartesian Equation : y = f(x) : Consider the curve C as shown in figure below. Consider two points P (x, y), Q (x + δx, y + δy), where arc PQ = δs. PM, QN are perpendiculars on X-axis and PR is perpendicular on QN.

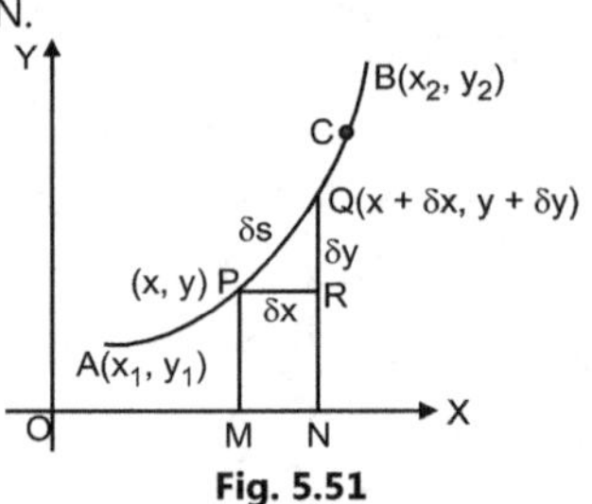

Fig. 5.51

Then   OM = x,   ON = x + δx,   MP = y,   NQ = y + δy

∴   PR = ON – OM = δx

        RQ = NQ – NR = NQ – MP = δy

From right angled triangle PQR,

$$PQ^2 = PR^2 + RQ^2$$

or $(\delta s)^2 = (\delta x)^2 + (\delta y)^2$ ...(1)

(When Q is indefinitely closer to P, arc PQ = δs = Chord PQ)

Dividing (1) by $(\delta x)^2$, we get,

$$\left(\frac{\delta s}{\delta x}\right)^2 = 1 + \left(\frac{\delta y}{\delta x}\right)^2$$

Taking the limit as Q → P or δx → 0

$$\left(\frac{ds}{dx}\right)^2 = 1 + \left(\frac{dy}{dx}\right)^2$$

or

$$\left(\frac{ds}{dx}\right) = \sqrt{1 + \left(\frac{dy}{dx}\right)^2}$$

i.e.

$$ds = \sqrt{1 + \left(\frac{dy}{dx}\right)^2}\, dx$$

Integrating

$$s = \int \sqrt{1 + \left(\frac{dy}{dx}\right)^2}\, dx$$

In particular, if we want to determine arc length AB, where $A(x_1, y_1)$, $B(x_2, y_2)$ are any two points on the curve, we substitute the limits of integration for x and we have,

$$s = \int_{x_1}^{x_2} \sqrt{1 + \left(\frac{dy}{dx}\right)^2}\, dx \qquad ...(a)$$

2. Cartesian Equation : x = f (y) : Again dividing (1) by $(\delta y)^2$ and taking limiting case as before,

$$\left(\frac{ds}{dy}\right)^2 = 1 + \left(\frac{dx}{dy}\right)^2 \quad \text{or} \quad \frac{ds}{dy} = \sqrt{1 + \left(\frac{dx}{dy}\right)^2}$$

i.e.

$$ds = \sqrt{1 + \left(\frac{dx}{dy}\right)^2}\, dy$$

Integrating between the limits y_1 to y_2,

$$s = \int_{y_1}^{y_2} \sqrt{1 + \left(\frac{dx}{dy}\right)^2}\, dy$$

...(b)

Formulae (a) and (b) both give the same arc length AB.

They are to be used when equation of the curve is given in cartesian form $f(x, y) = 0$. Choice between (a) and (b) is to be made so that integration becomes simple.

5.11 RECTIFICATION OF PLANE CURVE FOR PARAMETRIC EQUATIONS OF THE FORM X = φ(t), Y = Ψ(t)

When equation of the curve is given in parametric form $x = \phi(t)$, $y = \Psi(t)$, where t is the parameter. (As t varies we get different points on the curve).

Dividing (1) by $(\delta t)^2$ and taking limit as $Q \to P$, we get,

$$\left(\frac{ds}{dt}\right)^2 = \left(\frac{dx}{dt}\right)^2 + \left(\frac{dy}{dt}\right)^2 \quad \text{or} \quad \frac{ds}{dt} = \sqrt{\left(\frac{dx}{dt}\right)^2 + \left(\frac{dy}{dt}\right)^2}$$

i.e.
$$ds = \sqrt{\left(\frac{dx}{dt}\right)^2 + \left(\frac{dy}{dt}\right)^2}\, dt$$

Integrating,
$$s = \int \sqrt{\left(\frac{dx}{dt}\right)^2 + \left(\frac{dy}{dt}\right)^2}\, dt$$

If a point A corresponds to the parameter t_1 and point B corresponds to the parameter t_2, then integrating between the limits t_1 and t_2,

$$s = \int_{t_1}^{t_2} \sqrt{\left(\frac{dx}{dt}\right)^2 + \left(\frac{dy}{dt}\right)^2}\, dt$$

...(c)

This formula gives length of arc AB when equation of the curve is expressed in parametric form.

5.12 RECTIFICATION OF PLANE CURVE WHEN ITS EQUATION IS IN POLAR FORM r = f(θ)

1. Polar Equation : r = f(θ) : We shall now obtain two formulae which give length of arc of a curve when its equation is given in polar form $r = f(\theta)$.

P (r, θ), Q $(r + \delta r, \theta + \delta\theta)$ are two closed points on the curve C as shown in Fig. 5.53.

A (r_1, θ_1), B (r_2, θ_2) are any two points on the curve C and arc PQ = δs.

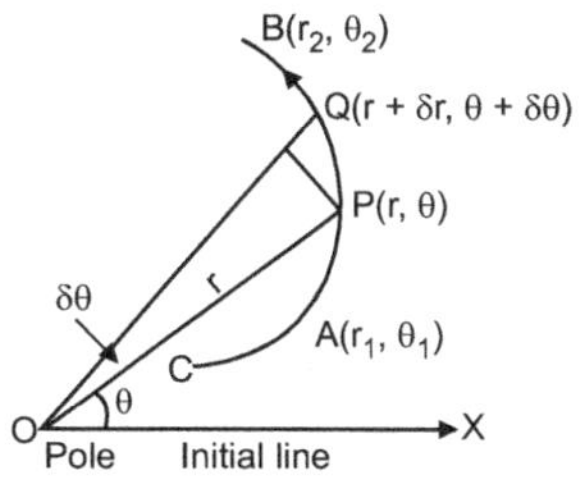

$$OP = r, \quad OQ = r + \delta r. \quad PR \text{ is } \perp \text{ on } OQ.$$
$$OR = OP \cos\delta\theta = r \cos\delta\theta \to r \text{ as } \delta\theta \to 0$$

and
$$PR = OP \sin\delta\theta = r \sin\delta\theta \to r\,\delta\theta$$
$$RQ = OQ - OR = r + \delta r - r = \delta r$$

When P and Q are indefinitely closer to each other, arc PQ = δs = chord PQ.

From $\triangle$ PQR, $\quad\quad PQ^2 = RQ^2 + PR^2 \quad\quad$ or $\quad (\delta s)^2 = (\delta r)^2 + (r\,\delta\theta)^2 \quad\quad$ ... (2)

Dividing by $(\delta\theta)^2$ and taking limit as $Q \to P$ or $\delta\theta \to 0$, we get,

$$\left(\frac{ds}{d\theta}\right)^2 = \left(\frac{dr}{d\theta}\right)^2 + r^2$$

or
$$\frac{ds}{d\theta} = \sqrt{r^2 + \left(\frac{dr}{d\theta}\right)^2} \quad\quad \text{or} \quad\quad ds = \sqrt{r^2 + \left(\frac{dr}{d\theta}\right)^2}\, d\theta$$

Integration gives,
$$s = \int \sqrt{r^2 + \left(\frac{dr}{d\theta}\right)^2}\, d\theta.$$

Fig. 5.52

If arc length AB is to be determined then we substitute the limits of θ as θ_1 and θ_2 and we get

$$s = \int_{\theta_1}^{\theta_2} \sqrt{r^2 + \left(\frac{dr}{d\theta}\right)^2} \, d\theta \qquad \ldots (d)$$

2. Polar Equation : $\theta = f(r)$: Similarly dividing (2) by $(\delta r)^2$ and taking the limit $Q \to P$ or $\delta r \to 0$

$$\left(\frac{ds}{dr}\right)^2 = 1 + r^2 \left(\frac{d\theta}{dr}\right)^2 \quad \text{or} \quad \frac{ds}{dr} = \sqrt{1 + r^2 \left(\frac{d\theta}{dr}\right)^2}$$

Integration gives
$$s = \int \sqrt{1 + r^2 \left(\frac{d\theta}{dr}\right)^2} \, dr$$

To determine arc length AB, substituting the limits of r from r_1 to r_2,

$$s = \int_{r_1}^{r_2} \sqrt{1 + r^2 \left(\frac{d\theta}{dr}\right)^2} \, dr \qquad \ldots (e)$$

Formulae (d) and (e) are applicable to polar curves. Formula (d) is to be used, if integration w.r.t. θ is simple and formula (e) is to be used, if integration w.r.t. r is simple.

1. Cartesian Curves :

	Equation of Curve	Formula in Differential Calculus	Formula in Integral Calculus
1.	$y = f(x)$	$ds = \sqrt{1 + \left(\frac{dy}{dx}\right)^2} \cdot dx$	$s = \int_{x_1}^{x_2} \sqrt{1 + \left(\frac{dy}{dx}\right)^2} \cdot dx$
2.	$x = g(y)$	$ds = \sqrt{1 + \left(\frac{dx}{dy}\right)^2} \cdot dy$	$s = \int_{y_1}^{y_2} \sqrt{1 + \left(\frac{dx}{dy}\right)^2} \cdot dy$
3.	$x = f_1(t), \; y = f_2(t)$	$ds = \sqrt{\left(\frac{dx}{dt}\right)^2 + \left(\frac{dy}{dt}\right)^2} \cdot dt$	$s = \int_{t_1}^{t_2} \sqrt{\left(\frac{dx}{dt}\right)^2 + \left(\frac{dy}{dt}\right)^2} \cdot dt$

2. Polar Curves :

	Equation of Curve	Formula in Differential Calculus	Formula in Integral Calculus
1.	$r = f(\theta)$	$ds = \sqrt{r^2 + \left(\frac{dr}{d\theta}\right)^2} \cdot d\theta$	$s = \int_{\theta_1}^{\theta_2} \sqrt{r^2 + \left(\frac{dr}{d\theta}\right)^2} \cdot d\theta$
2.	$\theta = f(r)$	$ds = \sqrt{1 + r^2 \left(\frac{d\theta}{dr}\right)^2} \cdot dr$	$s = \int_{r_1}^{r_2} \sqrt{1 + r^2 \left(\frac{d\theta}{dr}\right)^2} \cdot dr$
3.	$r = f_1(\alpha); \; \theta = f_2(\alpha)$ α is parameter	$ds = \sqrt{\left(\frac{dr}{d\alpha}\right)^2 + r^2 \left(\frac{d\theta}{d\alpha}\right)^2} \cdot d\alpha$	$s = \int_{\alpha_1}^{\alpha_2} \sqrt{\left(\frac{dr}{d\alpha}\right)^2 + r^2 \left(\frac{d\theta}{d\alpha}\right)^2} \cdot d\alpha$

ILLUSTRATIONS :

Ex. 1 : *Find the circumference of circle of radius 'a'.*

Sol. : Equation of circle of radius a with centre as origin as shown in the figure can be taken as,
$x^2 + y^2 = a^2$

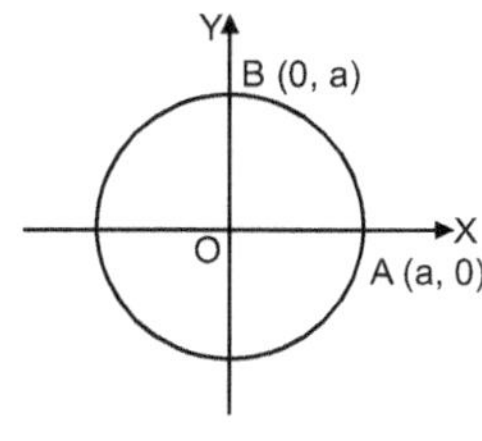

Fig. 5.53

$$\therefore \qquad y = \sqrt{a^2 - x^2}$$

$$\frac{dy}{dx} = \frac{1\,(-2x)}{2\sqrt{a^2 - x^2}} = \frac{-x}{\sqrt{a^2 - x^2}}$$

$$1 + \left(\frac{dy}{dx}\right)^2 = 1 + \frac{x^2}{a^2 - x^2} = \frac{a^2 - x^2 + x^2}{a^2 - x^2} = \frac{a^2}{a^2 - x^2}$$

Using formula (a), $\qquad S_{AB} = $ length of arc AB $= \int \sqrt{1 + \left(\frac{dy}{dx}\right)^2}\ dx$

Along arc AB, x varies from a to 0, but as x decreases, s increases and hence to get positive value, limits of x are taken from x = 0 to x = a.

$$\therefore \qquad S_{AB} = \int_0^a \sqrt{\frac{a^2}{a^2 - x^2}}\cdot dx = \int_0^a \frac{a}{\sqrt{a^2 - x^2}}\ dx = a\left[\sin^{-1}\frac{x}{a}\right]_0^a$$

$$= a\left[\sin^{-1}\frac{a}{a} - \sin^{-1}0\right] = a\left[\frac{\pi}{2} - 0\right] = \frac{\pi a}{2}$$

Circumference $= 4 \times S_{AB}$

$$\boxed{\text{Circumference} = 4 \times \frac{\pi a}{2} = 2\pi a}$$

Ex. 2 : *Find the length of arc of parabola $y^2 = 4ax$ from vertex to one extremity of latus rectum.*

Sol. : Latus rectum of a parabola ($y^2 = 4ax$) is a line segment perpendicular to the axis (y = 0) of the parabola, through the focus [S(a, 0)] and whose end points are [A(a, 2a) and B(a, – 2a)] lie on the parabola. Length of latus rectum is 4a.

Required length is of arc OA. Here we use formula (b).

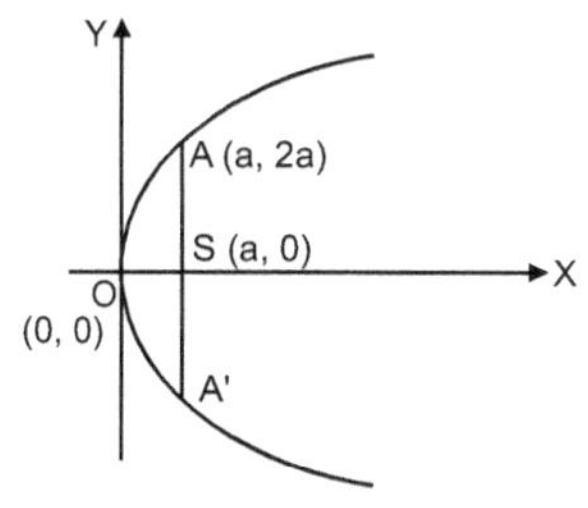

Fig. 5.54

$$S_{OA} = \int_0^{2a} \sqrt{1 + \left(\frac{dx}{dy}\right)^2}\ dy$$

$$x = \frac{y^2}{4a} \quad \therefore \quad \frac{dx}{dy} = \frac{2y}{4a} = \frac{y}{2a}$$

$$1 + \left(\frac{dx}{dy}\right)^2 = 1 + \frac{y^2}{4a^2} = \frac{4a^2 + y^2}{4a^2}$$

$$\therefore \qquad S_{OA} = \int_0^{2a} \sqrt{1 + \left(\frac{dx}{dy}\right)^2}\cdot dy = \frac{1}{2a}\int_0^{2a} \sqrt{4a^2 + y^2}\ dy$$

which is in a standard form. In this example if we had used formula (a), integral would not have been directly in standard form.

$$S_{OA} = \frac{1}{2a}\left[\frac{y\sqrt{4a^2 + y^2}}{2} + \frac{4a^2}{2}\log\left\{y + \sqrt{4a^2 + y^2}\right\}\right]_0^{2a}$$

$$= \frac{1}{2a}\left[\frac{2a\sqrt{4a^2 + 4a^2}}{2} + 2a^2 \log\left\{2a + \sqrt{4a^2 + 4a^2}\right\}\right] - \left[2a^2 \log\left\{\sqrt{4a^2}\right\}\right]$$

$$= \frac{1}{2a}\left[2\sqrt{2}\,a^2 + 2a^2 \log\frac{2a + 2\sqrt{2}\,a}{2a}\right]$$

$$\boxed{S_{OA} = a\left[\sqrt{2} + \log\left(1 + \sqrt{2}\right)\right]}$$

Ex. 3 : *Show that the length of the arc of the curve $ay^2 = x^3$ from origin to the point whose abscissa is b is*

$\dfrac{1}{27\sqrt{a}}\,(9b + 4a)^{3/2} - \dfrac{8a}{27}$. **(May 2009)**

Sol. : Curve is symmetrical about X-axis as shown in the figure.

P is the point on the curve whose abscissa is b.

Differentiating the equation of the curve,

$$2ay\frac{dy}{dx} = 3x^2 \quad \therefore \quad \frac{dy}{dx} = \frac{3x^2}{2ay}$$

$$\left(\frac{dy}{dx}\right)^2 = \frac{9x^4}{4a^2 y^2} = \frac{9x^4}{4a \cdot x^3} = \frac{9x}{4a}$$

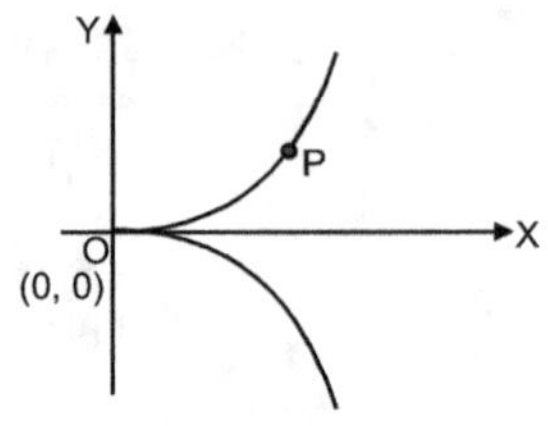

Fig. 5.55

Using the formula (a) with limits of x from 0 to b, we get,

$$S = \int_0^b \sqrt{1 + \left(\frac{dy}{dx}\right)^2}\, dx = \int_0^b \sqrt{1 + \frac{9x}{4a}}\, dx$$

$$= \frac{1}{2\sqrt{a}}\int_0^b \sqrt{4a + 9x}\, dx = \frac{1}{2\sqrt{a}}\left[\frac{(4a + 9x)^{3/2}}{\frac{3}{2}\cdot 9}\right]_0^b = \frac{1}{27\sqrt{a}}\,[(4a + 9b)^{3/2} - (4a)^{3/2}] = \frac{(4a + 9b)^{3/2}}{27\sqrt{a}} - \frac{8a^{3/2}}{27\sqrt{a}}$$

$$\boxed{S = \frac{(4a + 9b)^{3/2}}{27\sqrt{a}} - \frac{8a}{27}}$$

Ex. 4 : *Show that the whole length of the loop of the curve $9y^2 = (x + 7)(x + 4)^2$ is $4\sqrt{3}$.* **(May 2007)**

Sol. : From the equation of the curve it is clear that loop is around X-axis between $x = -7$ and $x = -4$. The curve is symmetrical about X-axis. Shape of the curve is as shown in Fig. 5.58.

We use formula (a), with appropriate limits to get whole length of the loop as,

$$S = 2\int_{-7}^{-4} \sqrt{1 + \left(\frac{dy}{dx}\right)^2}\, dx$$

(We multiply the integral by 2 to get the whole length of the loop.)

Differentiating equation of the curve w.r.t. 'x',

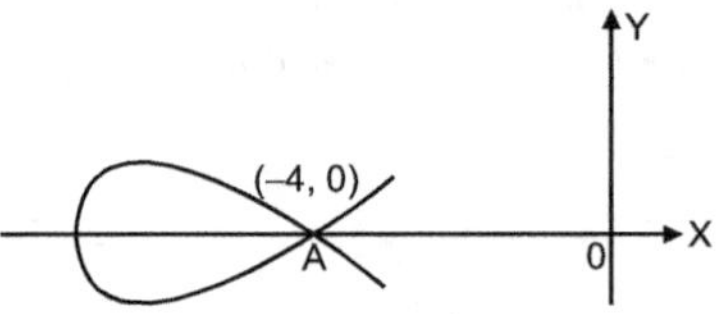

Fig. 5.56

$$18y\frac{dy}{dx} = (x + 4)^2 + 2(x + 4)(x + 7)$$

$$= (x + 4)(x + 4 + 2x + 14)$$

$$\frac{dy}{dx} = \frac{3(x + 4)(x + 6)}{18y} = \frac{(x + 4)(x + 6)}{6y}$$

$$\left(\frac{dy}{dx}\right)^2 = \frac{(x + 4)^2 (x + 6)^2}{36y^2} = \frac{(x + 4)^2 (x + 6)^2}{4(x + 7)(x + 4)^2}$$

$$1 + \left(\frac{dy}{dx}\right)^2 = 1 + \frac{(x + 6)^2}{4(x + 7)} = \frac{4x + 28 + x^2 + 12x + 36}{4(x + 7)} = \frac{x^2 + 16x + 64}{4(x + 7)} = \frac{(x + 8)^2}{4(x + 7)}$$

$$\therefore \quad S = \frac{2}{2}\int_{-7}^{-4} \frac{x + 8}{\sqrt{x + 7}}\, dx = \int_{-7}^{-4} \frac{x + 7 + 1}{\sqrt{x + 7}}\, dx = \int_{-7}^{-4} \left\{\sqrt{x + 7} + (x + 7)^{-1/2}\right\} dx$$

$$= \left[\frac{(x + 7)^{3/2}}{3/2} + 2(x + 7)^{1/2}\right]_{-7}^{-4} = \frac{2}{3}\,3\sqrt{3} + 2\sqrt{3} = 4\sqrt{3}$$

$$\boxed{\text{Length of the loop} = 4\sqrt{3}}$$

***Ex. 5 :** Find the length of the arc of the catenary $y = c \cosh \frac{x}{c}$ measured from its vertex to any point (x, y) and show that $S^2 = y^2 - c^2$*

Sol. : Here, $y = c \cosh \frac{x}{c}$ $\therefore$ $\frac{dy}{dx} = c \cdot \sinh \frac{x}{c} \cdot \frac{1}{c} = \sinh \frac{x}{c}$

$$1 + \left(\frac{dy}{dx}\right)^2 = 1 + \sinh^2 \frac{x}{c} = \cosh^2 \frac{x}{c}$$

$$\therefore \quad S = \int_0^x \sqrt{1 + \left(\frac{dy}{dx}\right)^2}\, dx = \int_0^x \cosh \frac{x}{c} \cdot dx$$

$$= \left[c \sinh \frac{x}{c}\right]_0^x = c \cdot \sinh \frac{x}{c}$$

$$S^2 = c^2 \sinh^2 \frac{x}{c} = c^2 \left[\cosh^2 \frac{x}{c} - 1\right] = c^2 \left[\frac{y^2}{c^2} - 1\right] = y^2 - c^2$$

$$\therefore \quad \boxed{S^2 = y^2 - c^2}$$

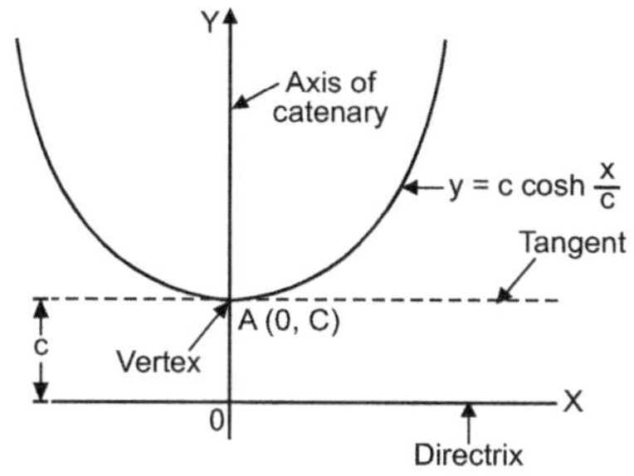

Fig. 5.57

***Ex. 6 :** Show that for the curve $8a^2 y^2 = x^2 (a^2 - x^2)$, $S = \frac{a}{2\sqrt{2}} [2\theta + \sin\theta \cos\theta]$ and the perimeter of one of the loops is $\frac{\pi a}{\sqrt{2}}$.*

(Dec. 2009)

Sol. : Here the curve is symmetrical about X-axis and has two loops around X-axis between x = 0 and x = a.

We first integrate between x = 0 and x = x using formula (a) i.e. we first find arc length of the curve from origin to any point (x, y) on the curve.

$$S = \int_0^x \sqrt{1 + \left(\frac{dy}{dx}\right)^2}\, dx$$

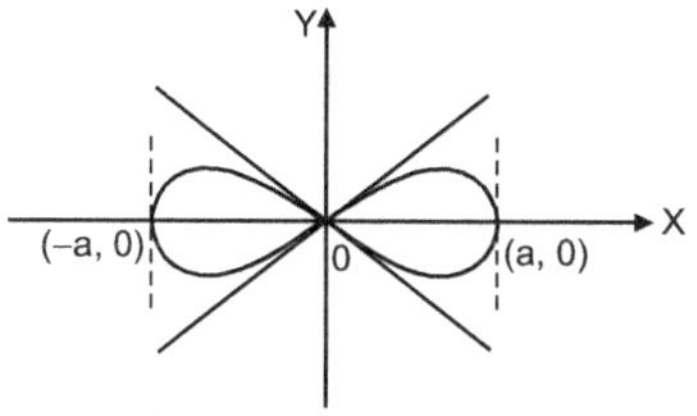

Fig. 5.58

Differentiating w.r.t. x, the equation of the curve, we get,

$$8a^2 \cdot 2y \frac{dy}{dx} = 2x (a^2 - x^2) - 2x^3$$

or $$\frac{dy}{dx} = \frac{a^2 x - 2x^3}{8a^2 y}$$

or $$\left(\frac{dy}{dx}\right)^2 = \frac{x^2 (a^2 - 2x^2)^2}{64 a^4 y^2} = \frac{x^2 (a^2 - 2x^2)^2}{8a^2 \cdot x^2 (a^2 - x^2)}$$

and $$1 + \left(\frac{dy}{dx}\right)^2 = 1 + \frac{(a^2 - 2x^2)^2}{8a^2 (a^2 - x^2)} = \frac{8a^4 - 8a^2 x^2 + a^4 - 4a^2 x^2 + 4x^4}{8a^2 (a^2 - x^2)}$$

$$= \frac{9a^4 - 12a^2 x^2 + 4x^4}{8a^2 (a^2 - x^2)} = \frac{(3a^2 - 2x^2)^2}{8a^2 (a^2 - x^2)}$$

$$\therefore \quad S = \int_0^x \sqrt{1 + \left(\frac{dy}{dx}\right)^2}\, dx = \int_0^x \frac{3a^2 - 2x^2}{2\sqrt{2}\, a \sqrt{a^2 - x^2}}\, dx$$

$$= \frac{1}{2\sqrt{2}\, a} \int_0^x \frac{3a^2 - 2x^2}{\sqrt{a^2 - x^2}}\, dx$$

Putting $x = a \sin\theta$, $dx = a \cos\theta \, d\theta$. When $x = 0$, $\theta = 0$, $x = x$, $\theta = \theta$

$$S = \frac{1}{2\sqrt{2}\,a} \int_0^\theta \frac{3a^2 - 2a^2 \sin^2\theta}{a\cos\theta} \cdot a\cos\theta \, d\theta$$

$$= \frac{1}{2\sqrt{2}\,a} \int_0^\theta [3a^2 - a^2(1 - \cos 2\theta)] \, d\theta$$

$$= \frac{1}{2\sqrt{2}\,a} \int_0^\theta (2a^2 + a^2 \cos 2\theta) \, d\theta = \frac{a}{2\sqrt{2}} \left[2\theta + \frac{\sin 2\theta}{2} \right]_0^\theta$$

$$\boxed{S = \frac{a}{2\sqrt{2}} [2\theta + \sin\theta \cdot \cos\theta]} \quad \text{which is the first required result.}$$

To get the perimeter of the loop, we put $\theta = \dfrac{\pi}{2}$ [When $x = a$, $\theta = \dfrac{\pi}{2}$ which gives length of upper half of the loop]

$$S = \frac{a}{2\sqrt{2}} \left[2 \cdot \frac{\pi}{2} \right] = \frac{\pi a}{2\sqrt{2}}$$

$\therefore$ $$\boxed{\text{Length of one loop} = 2 \cdot \frac{\pi a}{2\sqrt{2}} = \frac{\pi a}{\sqrt{2}}}$$

Ex. 7 : *Find the length of the loop of the curve* $x = t^2$; $y = t\left(1 - \dfrac{t^2}{3}\right)$. **(Dec. 2007)**

Sol. : To trace the curve, it is advisable in this problem to convert the equation into Cartesian form.

$$y^2 = t^2\left(1 - \frac{t^2}{3}\right)^2 \quad \text{or} \quad y^2 = x\left(1 - \frac{x}{3}\right)^2$$

Loop is around X-axis between $x = 0$ and $x = 3$ as shown in the figure. Corresponding to $x = 0$ and $x = 3$, we have $t = 0$ and $t = \sqrt{3}$.
To obtain length of loop, we use formula (c).

$$S = \int_0^{\sqrt{3}} \sqrt{\left(\frac{dx}{dt}\right)^2 + \left(\frac{dy}{dt}\right)^2} \, dt$$

which will give length of upper half of the loop

We have $$\frac{dx}{dt} = 2t, \quad \frac{dy}{dt} = 1 - t^2$$

and $$\left(\frac{dx}{dt}\right)^2 + \left(\frac{dy}{dt}\right)^2 = 4t^2 + 1 - 2t^2 + t^4 = (1 + t^2)^2$$

$\therefore$ $$S = \int_0^{\sqrt{3}} \sqrt{(1 + t^2)^2} \, dt = \int_0^{\sqrt{3}} (1 + t^2) \, dt$$

$$\text{Total length} = 2\int_0^{\sqrt{3}} [1 + t^2] \, dt = 2\left[t + \frac{t^3}{3} \right]_0^{\sqrt{3}} = 2\left[\sqrt{3} + \frac{3\sqrt{3}}{3} \right] = 4\sqrt{3}.$$

$$\boxed{\text{Length of the loop} = 4\sqrt{3}}$$

Ex. 8 : *Show that in the Astroid* $x^{2/3} + y^{2/3} = a^{2/3}$, $S^3 \alpha x^2$. *S being measured from the cusp which lies on Y-axis.*

(May 2010, 2005, 2016; Dec. 2010, 2018)

Sol. : $A(0, a)$ is a cusp on Y-axis. $P(x, y)$ is any point on the curve where arc $AP = S$ (See Fig. 5.57)
To obtain the arc length, we use parametric equations

$$x = a\cos^3\theta, \quad \frac{dx}{d\theta} = 3a\cos^2\theta\,(-\sin\theta)$$

$$y = a\sin^3\theta, \quad \frac{dy}{d\theta} = 3a\sin^2\theta\,\cos\theta$$

and

$$\left(\frac{dx}{d\theta}\right)^2 + \left(\frac{dy}{d\theta}\right)^2 = 9a^2\cos^4\theta\sin^2\theta + 9a^2\sin^4\theta\cos^2\theta$$

$$= 9a^2\sin^2\theta\cos^2\theta\,(\cos^2\theta + \sin^2\theta)$$

$$= 9a^2\sin^2\theta\cos^2\theta$$

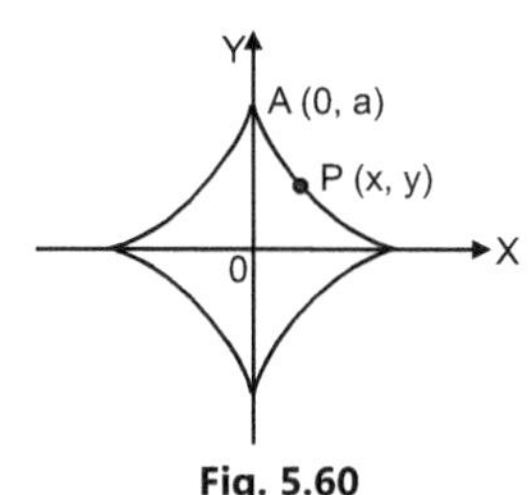

Fig. 5.60

Corresponding to $y = a$, $\sin^3\theta = 1$ $\therefore$ $\theta = \dfrac{\pi}{2}$

We use formula (c) with appropriate limits.

$$S = \int_{\pi/2}^{\theta} \sqrt{\left(\frac{dx}{d\theta}\right)^2 + \left(\frac{dy}{d\theta}\right)^2}\; d\theta = \int_{\pi/2}^{\theta} 3a\sin\theta\cos\theta\; d\theta$$

$$= \frac{3a}{2}\int_{\pi/2}^{\theta}\sin 2\theta\; d\theta = \frac{3a}{2}\left[-\frac{\cos 2\theta}{2}\right]_{\pi/2}^{\theta} = -\frac{3a}{4}[\cos 2\theta - \cos\pi] = -\frac{3a}{4}[1 + \cos 2\theta]$$

$$S = -\frac{3a}{4}[2\cos^2\theta] = -\frac{3a}{2}\cos^2\theta$$

$$S^3 = -\frac{27a^3}{8}\cos^6\theta = -\frac{27\,a^3}{8}(\cos^3\theta)^2$$

$$S^3 = -\frac{27\,a^3}{8}\left(\frac{x^2}{a^2}\right) = -\frac{27\,a}{8}x^2$$

$\therefore$ $\boxed{S^3 \propto x^2.}$

Ex. 9 : *Find the length of the arc of the curve* $x = e^{\theta}\left(\sin\dfrac{\theta}{2} + 2\cos\dfrac{\theta}{2}\right)$, $y = e^{\theta}\left(\cos\dfrac{\theta}{2} - 2\sin\dfrac{\theta}{2}\right)$ *from* $\theta = 0$ *to* $\theta = \pi$

Sol. : Here equation of the curve is given in parametric form, so we use formula (c) with t replaced by θ, limits from $\theta = 0$ to $\theta = \pi$.

$$S = \int_{0}^{\pi} \sqrt{\left(\frac{dx}{d\theta}\right)^2 + \left(\frac{dy}{d\theta}\right)^2}\; d\theta$$

Now,

$$\frac{dx}{d\theta} = e^{\theta}\left(\sin\frac{\theta}{2} + 2\cos\frac{\theta}{2}\right) + e^{\theta}\left(\frac{1}{2}\cos\frac{\theta}{2} - \sin\frac{\theta}{2}\right) = \frac{5}{2}e^{\theta}\cos\frac{\theta}{2}$$

$$\frac{dy}{d\theta} = e^{\theta}\left(\cos\frac{\theta}{2} - 2\sin\frac{\theta}{2}\right) + e^{\theta}\left(-\frac{1}{2}\sin\frac{\theta}{2} - \cos\frac{\theta}{2}\right) = -\frac{5}{2}e^{\theta}\sin\frac{\theta}{2}$$

$\therefore$

$$S = \int_{0}^{\pi} \sqrt{\frac{25}{4}e^{2\theta}\left(\cos^2\frac{\theta}{2} + \sin^2\frac{\theta}{2}\right)}\; d\theta = \frac{5}{2}\int_{0}^{\pi} e^{\theta}\; d\theta = \frac{5}{2}\left[e^{\theta}\right]_{0}^{\pi}$$

$$\boxed{S = \frac{5}{2}\,(e^{\pi} - 1)}$$

Ex. 10 : *A curve is described by the equations,*

$$x\sin\theta + y\cos\theta = f'(\theta)$$

$$x\cos\theta - y\sin\theta = f''(\theta)$$

Show that the length of arc expression is given by

$$S = f'(\theta) + f''(\theta) + c, \text{ where } c \text{ is a constant.}$$

Sol. : Consider $x\sin\theta + y\cos\theta = f'(\theta)$ ...(1)

$$x\cos\theta - y\sin\theta = f''(\theta) \qquad \qquad ...(2)$$

Multiplying equation (1) by $\sin\theta$, (2) by $\cos\theta$ and adding we get,

$$x = f'(\theta)\sin\theta + f''(\theta)\cos\theta \qquad \qquad ...(3)$$

Similarly multiplying (1) by $\cos\theta$, (2) by $\sin\theta$, and subtracting

$$y = f'(\theta) \cos\theta - f''(\theta) \sin\theta \qquad \text{...(4)}$$

(3) and (4) are parametric equations of the curve. To get arc length expression, we use formula (c).

$$S = \int \sqrt{\left(\frac{dx}{d\theta}\right)^2 + \left(\frac{dy}{d\theta}\right)^2} \, d\theta$$

From (3) and (4), by differentiating,

$$\frac{dx}{d\theta} = f''(\theta)\sin\theta + f'(\theta)\cos\theta + f'''(\theta)\cos\theta - f''(\theta)\sin\theta = \cos\theta\,\{f'(\theta) + f'''(\theta)\},$$

$$\frac{dy}{d\theta} = f''(\theta)\cos\theta - f'(\theta)\sin\theta - f'''(\theta)\sin\theta - f''(\theta)\cos\theta = -\sin\theta\,\{f'(\theta) + f'''(\theta)\}$$

Substituting in formula for S,

$$S = \int \sqrt{\{f'(\theta) + f'''(\theta)\}^2 (\cos^2\theta + \sin^2\theta)} \, d\theta = \int \{f'(\theta) + f'''(\theta)\} \, d\theta$$

Integrating we get, $\boxed{S = f(\theta) + f''(\theta) + c}$

which is the arc length expression for the given curve.

Ex. 11 : *Find the arc length of the cycloid $x = a\,(\theta + \sin\theta)$; $y = a\,(1 - \cos\theta)$ from one cusp to another cusp. If S is the length of the arc from origin to point P (x, y), show that $S^2 = 8ay$.* **(Dec. 2004, May 2017)**

Sol. : Part 1 : Here

$$x = a\,(\theta + \sin\theta); \quad \frac{dx}{d\theta} = a\,(1 + \cos\theta)$$

$$y = a\,(1 - \cos\theta); \quad \frac{dy}{d\theta} = a \sin\theta$$

$$\left(\frac{dx}{d\theta}\right)^2 + \left(\frac{dy}{d\theta}\right)^2 = a^2 (1 + \cos\theta)^2 + a^2 \sin^2\theta$$

$$= a^2 [1 + 2\cos\theta + \cos^2\theta + \sin^2\theta]$$

$$= 2a^2 (1 + \cos\theta) = 2a^2 \left(2 \cos^2\frac{\theta}{2}\right) = 4a^2 \cos^2\frac{\theta}{2}$$

Length of arc AB $= 2 \times$ length of OB.

$$S = 2\int_0^\pi 2a \cos\frac{\theta}{2} \cdot d\theta$$

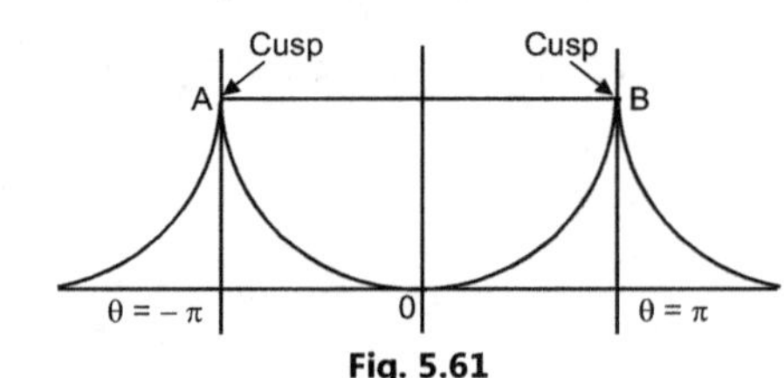

Fig. 5.61

$$= 4a\left[2 \sin\frac{\theta}{2}\right]_0^\pi = 8a$$

$$\boxed{S = 8a}$$

which is the required length from one cusp to another cusp.

Part 2 :

$$S = \int_0^\theta \sqrt{\left(\frac{dx}{d\theta}\right)^2 + \left(\frac{dy}{d\theta}\right)^2} \, d\theta = \int_0^\theta 2a \cos\frac{\theta}{2} \cdot d\theta = 2a\left[2 \sin\frac{\theta}{2}\right]_0^\theta$$

$$S = 4a \sin\frac{\theta}{2}$$

$$S^2 = 16a^2 \sin^2\frac{\theta}{2} = 16a^2 \left(\frac{1 - \cos\theta}{2}\right) = 8a\,(a\,(1 - \cos\theta)) = 8ay$$

$$\boxed{S^2 = 8ay}$$

Ex. 12 : *Evaluate $\int xy \, ds$ along the arc of the ellipse $\dfrac{x^2}{a^2} + \dfrac{y^2}{b^2} = 1$ in the first quadrant* **(Dec. 2005).**

Sol. : Parametric equations of the ellipse are $x = a \cos\theta$, $y = b \sin\theta$.

Along the arc in the positive quadrant, θ varies from 0 to $\dfrac{\pi}{2}$ as shown in Fig. 5.61.

$$\int xy \, ds = \int xy \frac{ds}{d\theta} \, d\theta = \int xy \sqrt{\left(\frac{dx}{d\theta}\right)^2 + \left(\frac{dy}{d\theta}\right)^2} \, d\theta$$

Substituting $x = a \cos \theta$, $y = b \sin \theta$, $\dfrac{dx}{d\theta} = -a \sin \theta$, $\dfrac{dy}{d\theta} = b \cos \theta$ and integrating between the limits $\theta = 0$ to $\theta = \dfrac{\pi}{2}$,

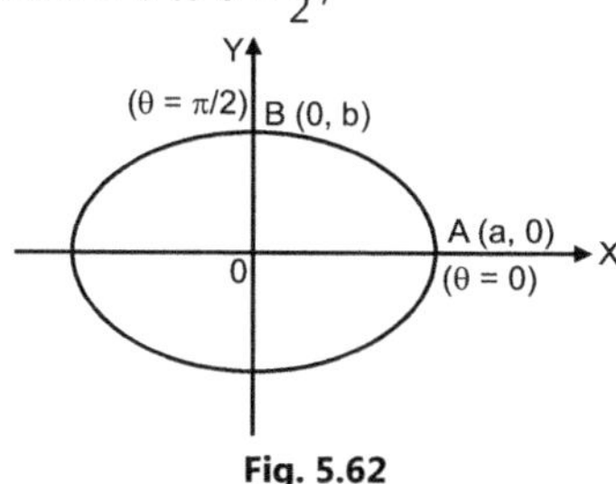

Fig. 5.62

$$I = \int_0^{\pi/2} xy \, ds = \int_0^{\pi/2} a \cos \theta \cdot b \sin \theta \sqrt{a^2 \sin^2 \theta + b^2 \cos^2 \theta} \; d\theta$$

$$= \frac{ab}{2} \int_0^{\pi/2} \sqrt{a^2 \sin^2 \theta + b^2 (1 - \sin^2 \theta)} \; 2 \sin \theta \cos \theta \, d\theta$$

$$= \frac{ab}{2} \int_0^{\pi/2} \sqrt{b^2 + (a^2 - b^2) \sin^2 \theta} \; d(\sin^2 \theta)$$

Integrating w.r.t. $\sin^2 \theta$

$$I = \frac{ab}{2} \left[\frac{(b^2 + (a^2 - b^2) \sin^2 \theta)^{3/2}}{\frac{3}{2} \cdot (a^2 - b^2)} \right]_0^{\pi/2} = \frac{ab}{3 (a^2 - b^2)} [(a^2)^{3/2} - (b^2)^{3/2}]$$

$$= \frac{ab}{3 (a^2 - b^2)} [a^3 - b^3] = \frac{ab}{3 (a^2 - b^2)} (a - b) (a^2 + ab + b^2)$$

$$\boxed{I = \frac{1}{3} \frac{ab (a^2 + ab + b^2)}{(a + b)}}$$

Ex. 13 : *For the curve* $x = (a + b) \cos \theta - b \cos \left(\dfrac{a + b}{b} \theta \right)$, $y = (a + b) \sin \theta - b \sin \left(\dfrac{a + b}{b} \theta \right)$

Show that $S = \dfrac{4b}{a} (a + b) \cos \left(\dfrac{a\theta}{2b} \right)$ *measured from* $\theta = \dfrac{\pi b}{a}$ *to* θ.

Sol. : Here,

$$x = (a + b) \cos \theta - b \cos \left(\frac{a + b}{b} \theta \right)$$

$$\frac{dx}{d\theta} = -(a + b) \sin \theta + (a + b) \sin \left(\frac{a + b}{b} \theta \right)$$

$$y = (a + b) \sin \theta - b \sin \left(\frac{a + b}{b} \theta \right)$$

$$\frac{dy}{d\theta} = (a + b) \cos \theta - (a + b) \cos \left(\frac{a + b}{b} \theta \right)$$

and

$$\left(\frac{dx}{d\theta} \right)^2 + \left(\frac{dy}{d\theta} \right)^2 = (a + b)^2 \left[\sin^2 \theta - 2 \sin \theta \sin \left(\frac{a + b}{b} \right) \theta + \sin^2 \left(\frac{a + b}{b} \right) \theta + \cos^2 \theta \right.$$

$$\left. - 2 \cos \theta \cos \left(\frac{a + b}{b} \right) \theta + \cos^2 \left(\frac{a + b}{b} \right) \theta \right]$$

$$= (a + b)^2 \left[2 - 2 \cdot \cos \left(\theta - \left(\frac{a + b}{b} \right) \theta \right) \right] = 2 (a + b)^2 \left[1 - \cos \left(- \frac{a\theta}{b} \right) \right] = 2 (a + b)^2 \left(1 - \cos \frac{a\theta}{b} \right)$$

$$= 2 (a + b)^2 \left(2 \sin^2 \frac{a\theta}{2b} \right) = 4 (a + b)^2 \sin^2 \frac{a\theta}{2b}$$

$\therefore$

$$S = \int_{\frac{\pi b}{a}}^{\theta} \sqrt{\left(\frac{dx}{d\theta} \right)^2 + \left(\frac{dy}{d\theta} \right)^2} \; d\theta = \int_{\frac{\pi b}{a}}^{\theta} 2 (a + b) \sin \left(\frac{a\theta}{2b} \right) d\theta = 2 (a + b) \left[-2 \frac{b}{a} \cos \left(\frac{a\theta}{2b} \right) \right]_{\frac{\pi b}{a}}^{\theta}$$

$$= \frac{4b}{a} (a + b) \left[\cos \frac{\pi}{2} - \cos \frac{a\theta}{2b} \right] = -\frac{4b}{a} (a + b) \cos \left(\frac{a\theta}{2b} \right)$$

$$\boxed{S = \frac{4b}{a} (a + b) \cos \left(\frac{a\theta}{2b} \right)}$$

Ex. 14 : *Find the perimeter of cardioide $r = a(1 + \cos\theta)$ and show that a line $\theta = \dfrac{\pi}{3}$ divides upper half of the cardioide.*

(May 2009, 2019, 2018, Dec. 2013)

Sol. : Curve is symmetrical about initial line OX. For upper half of the arc, θ varies from $\theta = 0$ to $\theta = \pi$.

Since the curve is given in Polar form, we use formula (d) with appropriate limits to find arc length,

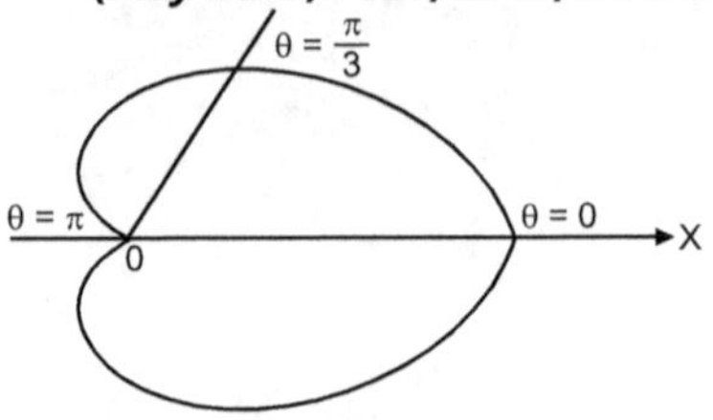

$$S = \int_0^\pi \sqrt{r^2 + \left(\frac{dr}{d\theta}\right)^2}\, d\theta \qquad ...(1)$$

Fig. 5.63

Here
$$r = a(1 + \cos\theta), \quad \frac{dr}{d\theta} = -a\sin\theta$$

and
$$r^2 + \left(\frac{dr}{d\theta}\right)^2 = a^2(1 + 2\cos\theta + \cos^2\theta + \sin^2\theta)$$
$$= 2a^2(1 + \cos\theta)$$
$$\sqrt{r^2 + \left(\frac{dr}{d\theta}\right)^2} = \sqrt{2}\, a\sqrt{1 + \cos\theta}$$

Substituting in (1),

$$S = \int_0^\pi \sqrt{2}\, a\sqrt{1 + \cos\theta}\, d\theta$$

$$= \sqrt{2}\, a \int_0^\pi \sqrt{2\cos^2\frac{\theta}{2}}\, d\theta = 2a \int_0^\pi \cos\frac{\theta}{2}\, d\theta = 4a\left[\sin\frac{\theta}{2}\right]_0^\pi = 4a \qquad ...(i)$$

This gives length of upper half. Because of symmetry of the curve, perimeter of the cardioide $= 2(4a) = 8a$.

To prove second part, we integrate (1) between the limits $\theta = 0$ to $\theta = \dfrac{\pi}{3}$.

$$S = \sqrt{2}\, a \int_0^{\pi/3} \sqrt{2\cos^2\frac{\theta}{2}}\, d\theta$$

$$= 2a \int_0^{\pi/3} \cos\frac{\theta}{2}\, d\theta = 4a\left[\sin\frac{\theta}{2}\right]_0^{\pi/3} = 4a\sin\frac{\pi}{6} = 4a\left(\frac{1}{2}\right) = 2a \qquad ...(ii)$$

From (i) and (ii), $\boxed{\text{A line } \theta = \dfrac{\pi}{3} \text{ divides upper half of cardioiode}}$

Ex. 15 : *Find the length of the upper arc of one loop of Lemniscate $r^2 = a^2\cos 2\theta$.*

Sol. : For upper arc of the curve, θ varies from $\theta = 0$ to $\theta = \dfrac{\pi}{4}$.   **(Nov./Dec. 2019, Dec. 2006, May 2011, May 2014)**

$$S = \int_0^{\pi/4} \sqrt{r^2 + \left(\frac{dr}{d\theta}\right)^2}\, d\theta \qquad\qquad r = a\sqrt{\cos 2\theta}$$

$$\frac{dr}{d\theta} = a\cdot\frac{1(-2\sin 2\theta)}{2\sqrt{\cos 2\theta}} \qquad r^2 + \left(\frac{dr}{d\theta}\right)^2 = a^2\cos 2\theta + \frac{a^2\sin^2 2\theta}{\cos 2\theta} = \frac{a^2}{\cos 2\theta}$$

$$S = \int_0^{\pi/4} \frac{a}{\sqrt{\cos 2\theta}}\, d\theta$$

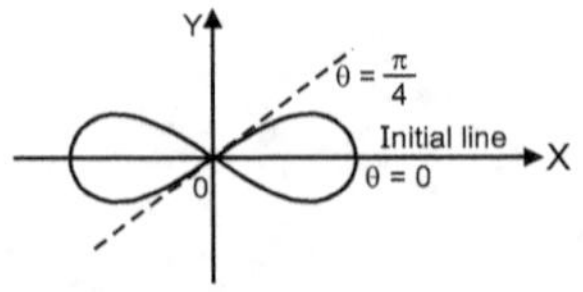

Fig. 5.64

Put $2\theta = t$ $\therefore$ $d\theta = \dfrac{1}{2}dt$ When $\theta = 0$, $t = 0$; $\theta = \dfrac{\pi}{4}$, $t = \dfrac{\pi}{2}$

$\therefore$

$$S = \frac{a}{2} \int_0^{\pi/2} \frac{dt}{\sqrt{\cos t}} = \frac{a}{2} \int_0^{\pi/2} \sin^0 t \, \cos^{-1/2} t \, dt$$

$$= \frac{a}{2} \left[\frac{\left\lfloor \frac{0+1}{2} \right. \left\lfloor \frac{-1/2+1}{2} \right.}{2 \left\lfloor \frac{0-1/2+2}{2} \right.} \right] = \frac{a}{4} \frac{\sqrt{\pi} \, \lfloor 1/4}{\lfloor 3/4}$$

$$= \frac{a}{4} \frac{\sqrt{\pi} \, (\lfloor 1/4 \,)^2}{\lfloor 1/4 \, \lfloor 3/4} = \frac{a\sqrt{\pi}}{4} \frac{(\lfloor 1/4 \,)^2}{\pi\sqrt{2}}$$

$$= \frac{a}{4\sqrt{2}} \frac{(\lfloor 1/4 \,)^2}{\sqrt{\pi}} \qquad \left[\because \lfloor 1/4 \; \lfloor 3/4 \; = \; \lfloor 1/4 \; \left\lfloor 1 - \frac{1}{4} \right. \; = \; \frac{\pi}{\sin \pi/4} = \; \pi\sqrt{2} \right]$$

$$\boxed{\text{Length of the upper arc of one loop} = \frac{a}{4\sqrt{2}} \frac{(\lfloor 1/4 \,)^2}{\sqrt{\pi}}}$$

Ex. 16 : *Find the length of arc of the curve* $r = a\, e^{m\theta}$ *intercepted between radii vectors* r_1 *and* r_2.   **(Dec. 2005)**

Sol. : In this problem, it is convenient to use formula (e).

$$S = \int_{r_1}^{r_2} \sqrt{1 + r^2 \left(\frac{d\theta}{dr}\right)^2} \cdot dr$$

Differentiating equation of given curve, $r = a\, e^{m\theta}$

$$\frac{dr}{d\theta} = a\, m e^{m\theta} = mr$$

$$r^2 \left(\frac{d\theta}{dr}\right)^2 = r^2 \cdot \frac{1}{m^2 r^2} = \frac{1}{m^2}$$

$\therefore$

$$S = \int_{r_1}^{r_2} \sqrt{1 + \frac{1}{m^2}} \, dr = \frac{\sqrt{1 + m^2}}{m} \, [r]_{r_1}^{r_2}$$

$$\boxed{\text{Length of arc of curve} = \frac{\sqrt{1 + m^2}}{m} \, (r_2 - r_1)}$$

Ex. 17 : *Find the length of the cardioide* $r = a\,(1 + \cos\theta)$ *which lies outside the circle* $r + a\cos\theta = 0$.

Sol. : The point of intersection of the curves is given by $r = -a\cos\theta$,

$r = a\,(1 + \cos\theta) - a\cos\theta = a\,(1 + \cos\theta)$, $\cos\theta = -\dfrac{1}{2}$, $\theta = \dfrac{2\pi}{3}$.

The arc length outside the circle is twice the arc length BA.

$$\text{Required length} = L = \int ds = \int \frac{ds}{d\theta} \, d\theta.$$

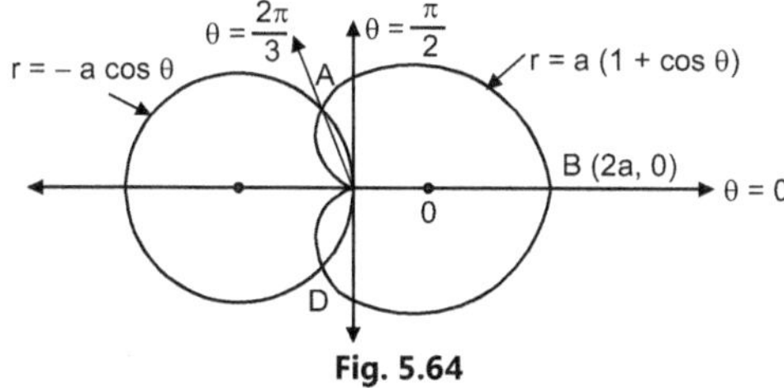

Here

$$\left(\frac{ds}{d\theta}\right)^2 = r^2 + \left(\frac{dr}{d\theta}\right)^2$$

$$= a^2 \, [1 + 2\cos\theta + \cos^2\theta + \sin^2\theta] = a^2 \,(1 + \cos\theta) = 2a^2 \cdot 2\cos^2\frac{\theta}{2} = 4a^2 \cos^2\frac{\theta}{2}$$

$\therefore$

$$L = 2 \int_0^{2\pi/3} 2a \cos\frac{\theta}{2} \, d\theta = 4a \left[2\sin\frac{\theta}{2} \right]_0^{2\pi/3} = 8a \left(\sin\frac{\pi}{3} \right) = 4\sqrt{3}\, a$$

$$\boxed{\text{Required length} = 8a \left(\sin\frac{\pi}{3} \right) = 4\sqrt{3}\, a}$$

EXERCISE 5.6

1. Show that in the catenary $y = a \cosh \frac{x}{a}$, the length of arc of the curve from the vertex to any point (x, y) is given by $S = a \sinh \frac{x}{a}$.

[**Hint :** Here $\sqrt{1 + \left(\frac{dy}{dx}\right)^2} = \cosh \frac{x}{a}$ and limits are x = 0 to x]

2. Find the whole length of the loop of the curve $3y^2 = x (x - 1)^2$.   (May 2006)

[**Hint :** Here $\sqrt{1 + \left(\frac{dy}{dx}\right)^2} = \frac{1 + 3x}{2\sqrt{3}\sqrt{x}}$ and limits are x = 0 to a]   **Ans. :** $\frac{4}{\sqrt{3}}$

3. Find the length of the arc of the parabola $y^2 = 4x$, cut off by the line $3y = 8x$.   **Ans. :** $\log 2 + \frac{15}{16}$

[**Hint :** Cutting points (0, 0) an $\left(\frac{9a}{16}, \frac{3a}{2}\right)$, Also $\sqrt{1 + \left(\frac{dx}{dy}\right)^2} = \frac{1}{2a} \sqrt{4a^2 + y^2}$ and limits are y = 0 to $\frac{3a}{2}$]

4. Show that the length of the arc of the curve $4ax = y^2 - 2a^2 \log \frac{y}{a} - a^2$ from (0, a) to any point (x, y) is given by $S = \frac{y^2}{2a} - \frac{a}{2} - x$.

[**Hint :** Here $\sqrt{1 + \left(\frac{dx}{dy}\right)^2} = \frac{1}{2a}\left(y + \frac{a^2}{y}\right)$ and limits are y = a to y.]

5. Find the length of the arc of the cycloid $x = a (\theta - \sin \theta)$, $y = a (1 - \cos \theta)$ between the two consecutive cusps.   (Dec. 09, May 15)
Ans. : 8a.

6. Find the length of the arc of the curve $x = e^\theta \cos \theta$, $y = e^\theta \sin \theta$ from $\theta = 0$ to $\theta = \frac{\pi}{2}$.   (May 2011, Dec. 2017)

[**Hint :** Here $\sqrt{\left(\frac{dx}{d\theta}\right)^2 + \left(\frac{dy}{d\theta}\right)^2} = \sqrt{2}\, e^\theta$ and limits are $\theta = 0$ to $\frac{\pi}{2}$]

Ans. : $\sqrt{2}\, (e^{\pi/2} - 1)$

7. Find the length of the arc of the curve $\left(\frac{x}{a}\right)^{2/3} + \left(\frac{y}{b}\right)^{2/3} = 1$ in the positive quadrant.   (Dec. 2011)

[**Hint :** Here $\sqrt{\left(\frac{dx}{dt}\right)^2 + \left(\frac{dy}{dt}\right)^2} = 3 \cos t \sin t \sqrt{a^2 \cos^2 t + b^2 \sin^2 t}$; t = 0 to $\frac{\pi}{2}$ Put $a^2 \cos^2 t + b^2 \sin^2 t = t$]   **Ans. :** $\frac{(a^2 + ab + b^2)}{a + b}$

8. Show that the length of the arc of the tractrix $x = a\left(\cos t + \log \tan \frac{t}{2}\right)$, $y = a \sin t$ from $t = \frac{\pi}{2}$ to any point t is "a log sin t".

9. Find the complete arc length of the curve $x^{2/3} + y^{2/3} = a^{2/3}$.   (Dec. 2018) **Ans. :** 6a

[**Hint :** Here $\sqrt{\left(\frac{dx}{dt}\right)^2 + \left(\frac{dy}{dt}\right)^2} = \frac{3a}{2} \sin 2t$ and limits are t = 0 to $\frac{\pi}{2}$]

10. Find the length of the arc of the curve $x = a (\cos t + t \sin t)$, $y = a (\sin t - t \cos t)$ from t = 0 to t = 2π.   **Ans. :** $2\pi^2 a$

11. Evaluate $\int y^2 ds$ along the arc of the curve, $x = a (\cos \theta + \theta \sin \theta)$, $y = a (\sin \theta - \theta \cos \theta)$.   **Ans. :** $\frac{256 a^3}{15}$

12. Find the length of the arc of the cardioide $r = a (1 - \cos \theta)$ which lies outside the circle, $r = a \cos \theta$.   (May 2010, Nov. 2014)
Ans. : $4a \sqrt{3}$

13. Show that the length of the arc of that part of the cardioide $r = a (1 + \cos \theta)$ which lies on the side of the line $4r = 3a \sec \theta$ remote from the pole, is equal to 4a.

14. Show that the whole length of the arc of the limacon $r = a \cos \theta + b$ is equal to that of an ellipse whose semiaxes are equal in length to the maximum and minimum radii vectors of the limacon.

15. Find the length of any arc of the curve $x^{2/3} - y^{2/3} = a^{2/3}$.   **Ans. :** $\left[\frac{1}{2} (x_2^{2/3} + y_2^{2/3})^{3/2} - (x_1^{2/3} + y_1^{2/3})^{3/2}\right]$

CHAPTER-6
CO-ORDINATE SYSTEM, PLANE, STRAIGHT LINE AND SOLIDS OF REVOLUTION

6.1 INTRODUCTION

Students are very well acquainted with the Co-ordinate Geometry of two dimensions, wherein the position of a point in a plane is described by means of an ordered pair of numbers (x, y) referred to two intersecting lines at right angles, called co-ordinate axes. The physical activity which is confined to a plane such as flow of heat along a bar or a plane, flow of fluid along a plane, etc. can be studied with the help of two-dimensional geometry, straight lines or curves confined to a plane require for their study the description of points lying on them which can be made with an ordered pair (x, y) called cartesian co-ordinates of a point or (r, θ) called polar co-ordinates of a point. In the study of solid objects, different points on it do not lie in the same plane and hence the reference system used in two-dimensional geometry cannot describe the position of various points on the solid. Reference system used in the geometry of two dimensions is thus required to be extended. In the study of solid objects we do not rule out hollowness. The outer boundaries of the solids are called surfaces. Study of the surfaces of solids is important from the physical point of view and constitutes the main subject matter of the geometry of three dimensions.

6.2 CO-ORDINATE SYSTEM

Consider three mutually perpendicular straight lines x'ox, y'oy, z'oz intersecting at the point 'o'. These straight lines constitute our reference system and the point 'o' is called the origin of reference. All the distances measured in the directions of ox, oy and oz are taken as positive and those along ox', oy' and oz' are taken as negative. ox, oy and oz are the three co-ordinate axes and are called x, y and z axes respectively. Three mutually perpendicular lines ox, oy and oz constitute three mutually perpendicular planes xoy, yoz and zox called co-ordinate planes.

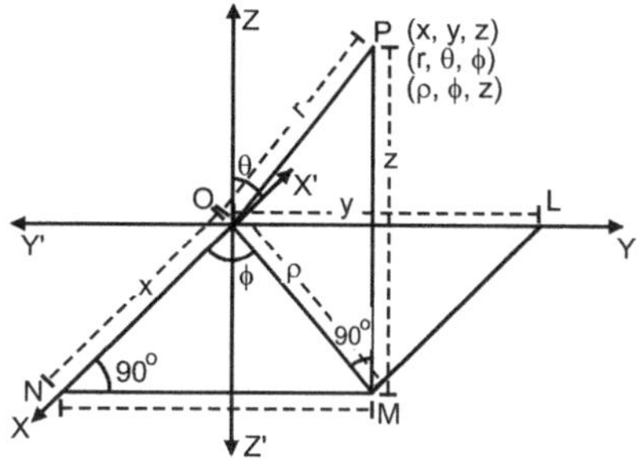

Fig. 6.1

With reference to Fig. 6.1, we now describe three co-ordinate systems.

1. Cartesian Co-ordinate System : Let P be any point in space. PM is perpendicular on xoy plane. From M draw parallel lines which cut ox and oy in N and L respectively. Let the distance ON = x, NM = OL = y and MP = z. Point P is associated with these numbers (x, y, z), which describe the position of the point P in space with respect to the reference system ox, oy, oz. (x, y, z) are called the *cartesian coordinates of the point P.*

2. Spherical Polar System : Let the distance OP be denoted by r, angle made by OP with positive direction of z-axis be denoted by θ and ∠ MOX = φ. The numbers (r, θ, φ) which can be associated with the point P are called *spherical polar co-ordinates* of *the point P.*

Remark : Here length OP = r can also be associated with radius vector of P, ∠ MOX = φ is called azimuth of P, being the angle between azimuthal plane POM and plane ZOX. ∠ ZOP = θ is vectorial angle or polar angle.

3. Cylindrical Co-ordinate System : Let the distance OM be denoted by ρ, φ and z are as defined in earlier cases. The numbers (ρ, φ, z) associated with the point P are called *cylindrical co-ordinates of the point P.*

6.3 RELATIONS BETWEEN THREE CO-ORDINATE SYSTEMS

With reference to Fig. 6.1,

$$PM = z = OP \cos θ = r \cos θ \qquad \therefore \ z = r \cos θ$$
$$OM = r \sin θ$$
$$ON = OM \cos φ \qquad \therefore \ x = r \sin θ \cos φ$$
$$NM = OM \sin φ \qquad \therefore \ y = r \sin θ \sin φ$$

Thus, the relations between cartesian and spherical polar system of co-ordinates are

$$\boxed{\begin{aligned} x &= r \sin\theta \cos\phi \\ y &= r \sin\theta \sin\phi \\ z &= r \cos\theta \end{aligned}} \qquad \text{... (a)}$$

Again with respect to Fig. 6.1, OM = ρ

$$ON = x = OM \cos\phi \qquad\qquad \therefore \ x = \rho \cos\phi$$
$$NM = y = OM \sin\phi \qquad\qquad \therefore \ y = \rho \sin\phi$$

and
$$z = z$$

which is defined in identical manner in cartesian and cylindrical system. Thus, the relations between cartesian and cylindrical system of co-ordinates are

$$\boxed{\begin{aligned} x &= \rho \cos\phi \\ y &= \rho \sin\phi \\ z &= z \end{aligned}} \qquad \text{... (b)}$$

Note :

(i)   In Cartesian Co-ordinate System :

   $-\infty < x < \infty, \ -\infty < y < \infty, \ -\infty < z < \infty$

(ii) In Spherical Polar System :

   $0 < r < \infty, \ 0 \le \theta \le \pi, \ 0 \le \phi \le 2\pi$

(iii) In Cylindrical Co-ordinate System :

   $0 < \rho < \infty, \ 0 \le \phi \le 2\pi, \ -\infty < z < \infty$

6.4 ILLUSTRATIONS ON TRANSFORMATION OF CO-ORDINATES

Ex. 1 : *Find the spherical polar and cylindrical co-ordinates of a point (1, 1, 1).*

Sol. : Here $\qquad\qquad x = 1, \quad y = 1, \quad z = 1$

From relations (a),

$$r^2 = x^2 + y^2 + z^2$$

$\therefore \qquad\qquad r = \sqrt{1+1+1} = \sqrt{3} \qquad\qquad (\because \ r > 0)$

Again $\qquad\qquad \cos\theta = \dfrac{z}{r} = \dfrac{1}{\sqrt{3}} \qquad\qquad \therefore \ \theta = \cos^{-1}\left(\dfrac{1}{\sqrt{3}}\right) = 54.74°$

and $\qquad\qquad \tan\phi = \dfrac{y}{x} = \dfrac{1}{1} \qquad\qquad \therefore \ \theta = \tan^{-1}(1) = \dfrac{\pi}{4} = 45°$

Thus the spherical polar co-ordinates of the given point are

$$(r, \theta, \phi) = \left(\sqrt{3}, \ 54.74°, \ 45°\right)$$

From relations (b),

$$\rho^2 = x^2 + y^2$$

$\therefore \qquad\qquad \rho = \sqrt{1+1} = \sqrt{2} \qquad\qquad (\because \ \rho > 0)$

Again $\qquad\qquad \tan\phi = \dfrac{y}{x} = 1 \qquad\qquad \therefore \ \phi = \tan^{-1}(1) = \dfrac{\pi}{4} = 45°$

and $\qquad\qquad z = 1$

Thus the cylindrical co-ordinates of the given point are

$$(\rho, \phi, z) = \left(\sqrt{2}, \ 45°, \ 1\right)$$

Ex. 2 : *Find the spherical polar and cylindrical co-ordinates of the point, whose cartesian co-ordinates are (3, 4, –5).*

Sol. : Here $\qquad\qquad x = 3, \ y = 4, \ z = -5$

From relations (a), $\qquad r = \sqrt{x^2 + y^2 + z^2} = \sqrt{9 + 16 + 25} = 5\sqrt{2}$

Also $\qquad\qquad \cos\theta = \dfrac{z}{r} = \dfrac{-5}{5\sqrt{2}} = -\dfrac{1}{\sqrt{2}} \qquad\qquad \therefore \ \theta = \dfrac{3\pi}{4} = 135° \text{ (Obtuse angle)}$

and $\qquad\qquad \tan\phi = \dfrac{y}{x} = \dfrac{4}{3} \qquad\qquad \therefore \ \phi = \tan^{-1}\dfrac{4}{3} = 53°\,8'$

Thus the spherical polar co-ordinates of the given point are

$$(r, \theta, \phi) \;=\; \left(5\sqrt{2},\, 135°,\, 53°\,8'\right)$$

From relations (b), $\qquad\qquad \rho \;=\; \sqrt{x^2 + y^2} \;=\; \sqrt{9 + 16} \;=\; 5$

Also $\qquad\qquad\qquad \tan\phi \;=\; \dfrac{y}{x} \;=\; \dfrac{4}{3} \qquad\qquad \therefore\;\; \phi \;=\; \tan^{-1}\dfrac{4}{3} \;=\; 53°\,8'$

and $\qquad\qquad\qquad z \;=\; -5$

Thus the cylindrical co-ordinates of the given point are

$$(\rho, \phi, z) \;=\; (5,\, 53°\,8',\, -5).$$

Ex. 3 : *Find the spherical polar co-ordinates of the points (i) (–3, 4, –5) (ii) (–1, 2, –3).*

Sol. : (i) Here $\qquad\qquad x \;=\; -3,\; y \;=\; 4,\; z = -5$

From relations (a), $\qquad r \;=\; \sqrt{x^2 + y^2 + z^2} \;=\; \sqrt{9 + 16 + 25} \;=\; 5\sqrt{2}$

and $\qquad\qquad\qquad \cos\theta \;=\; \dfrac{z}{r} \;=\; \dfrac{-5}{5\sqrt{2}} \;=\; \dfrac{-1}{\sqrt{2}} \qquad \therefore\;\; \theta = \dfrac{3\pi}{4} \;=\; 135° \qquad\qquad$ (Obtuse angle)

For angle ϕ, we note from $x = r\sin\theta\cos\phi$ and $y = r\sin\theta\sin\phi$,

$$-3 \;=\; 5\sqrt{2}\;\sin\dfrac{3\pi}{4}\;\cos\phi \;\text{ and }\; 4 = 5\sqrt{2}\;\sin\dfrac{3\pi}{4}\;\sin\phi.$$

$\therefore \qquad\qquad\qquad \cos\phi \;=\; -\dfrac{3}{5} \quad \text{ and } \quad \sin\phi \;=\; \dfrac{4}{5}$

This shows that ϕ is the angle in the second quadrant $\qquad\qquad (\because\; \cos\phi$ is $-$ve and $\sin\phi$ is $+$ve).

$\therefore \qquad\qquad\qquad \phi \;=\; 126.87°$

Thus the spherical polar co-ordinates of the given point are

$$(r, \theta, \phi) \;=\; \left(5\sqrt{2},\, 135°,\, 126.87°\right)$$

(ii) Here $\qquad x \;=\; -1,\; y \;=\; 2,\; z \;=\; -3$

From relations (a), $\qquad r \;=\; \sqrt{x^2 + y^2 + z^2} \;=\; \sqrt{1 + 4 + 9} \;=\; \sqrt{14}$

and $\qquad\qquad\qquad \cos\theta \;=\; \dfrac{z}{r} \;=\; \dfrac{-3}{\sqrt{14}} \;\therefore\; \theta = \cos^{-1}\left(-\dfrac{3}{\sqrt{14}}\right) = 143.30°$ (Obtuse angle)

For angle ϕ, we note from $x = r\sin\theta\cos\phi$ and $y = r\sin\theta\sin\phi$,

$$-1 \;=\; \sqrt{14}\left(\sqrt{\dfrac{5}{14}}\right)\cos\phi \;\text{ and } 2 \;=\; \sqrt{14}\left(\sqrt{\dfrac{5}{14}}\right)\sin\phi \qquad\qquad \left(\because\; \sin\theta = \sqrt{1 - \cos^2\theta}\right)$$

$\therefore \qquad\qquad\qquad \cos\phi \;=\; -\dfrac{1}{\sqrt{5}} \quad \text{ and } \quad \sin\phi \;=\; \dfrac{2}{\sqrt{5}}$

This shows that ϕ is the angle in second quadrant.

$\therefore \qquad\qquad\qquad \phi \;=\; 116.56°$

Thus the spherical polar co-ordinates of the given point are

$$(r, \theta, \phi) \;=\; \left(\sqrt{14},\, 143.30°,\, 116.56°\right)$$

Ex. 4 : *Find the cartesian co-ordinates of the points :* $\left(2, \dfrac{\pi}{3}, \dfrac{\pi}{4}\right)$, *(ii)* $\left(3, \dfrac{2\pi}{3}, \dfrac{\pi}{6}\right)$.

Sol. : (i) Here $\left(2, \dfrac{\pi}{3}, \dfrac{\pi}{4}\right)$ are actually the polar co-ordinates (r, θ, ϕ), hence $r = 2$, $\theta = \dfrac{\pi}{3}$ and $\phi = \dfrac{\pi}{4}$.

From relations (a), $\qquad x \;=\; r\sin\theta\cos\phi = 2\sin\left(\dfrac{\pi}{3}\right)\cos\left(\dfrac{\pi}{4}\right) = 2\left(\dfrac{\sqrt{3}}{2}\right)\left(\dfrac{1}{\sqrt{2}}\right) = \sqrt{\dfrac{3}{2}}$

$$y \;=\; r\sin\theta\sin\phi = 2\sin\left(\dfrac{\pi}{3}\right)\sin\left(\dfrac{\pi}{4}\right) = 2\left(\dfrac{\sqrt{3}}{2}\right)\left(\dfrac{1}{\sqrt{2}}\right) = \sqrt{\dfrac{3}{2}}$$

$$z \;=\; r\cos\theta = 2\cos\left(\dfrac{\pi}{3}\right) = 2\left(\dfrac{1}{2}\right) = 1$$

Thus the cartesian co-ordinates of the given point are $(x, y, z) = \left(\sqrt{\dfrac{3}{2}}, \sqrt{\dfrac{3}{2}}, 1\right)$.

(ii) Here $\left(3, \dfrac{2\pi}{3}, \dfrac{\pi}{6}\right)$ are actually the polar co-ordinates (r, θ, ϕ), hence $r = 3$, $\theta = \dfrac{2\pi}{3}$ and $\phi = \dfrac{\pi}{6}$.

From relations (a),

$$x = r\sin\theta\cos\phi = 3\sin\left(\frac{2\pi}{3}\right)\cos\left(\frac{\pi}{6}\right) = 3\left(\frac{\sqrt{3}}{2}\right)\left(\frac{\sqrt{3}}{2}\right) = \frac{9}{4}$$

$$y = r\sin\theta\sin\phi = 3\sin\left(\frac{2\pi}{3}\right)\sin\left(\frac{\pi}{6}\right) = 3\left(\frac{\sqrt{3}}{2}\right)\left(\frac{1}{2}\right) = \frac{3\sqrt{3}}{4}$$

$$z = r\cos\theta = 3\cos\left(\frac{2\pi}{3}\right) = 3\left(-\frac{1}{2}\right) = -\frac{3}{2}$$

Thus the cartesian co-ordinates of the given points are $(x, y, z) = \left(\dfrac{9}{4}, \dfrac{3\sqrt{3}}{4}, -\dfrac{3}{2}\right)$.

Ex. 5 : *Find (i) polar, (ii) cylindrical equation of the right circular cone, whose cartesian equation is given by* $x^2 + y^2 = z^2\tan^2\alpha$

Sol. : (i) Using spherical polar transformations $x = r\sin\theta\cos\phi$, $y = r\sin\theta\sin\phi$ and $z = r\cos\theta$, polar equation of the cone is given by

$$(r\sin\theta\cos\phi)^2 + (r\sin\theta\sin\phi)^2 = (r\cos\theta)^2\tan^2\alpha$$

$\therefore \qquad r^2\sin^2\theta(\cos^2\phi + \sin^2\phi) = r^2\cos^2\theta\tan^2\alpha$

$\therefore \qquad \tan^2\theta = \tan^2\alpha$

$\therefore \qquad \theta = \alpha$

(ii) Using cylindrical transformations $x = \rho\cos\phi$, $y = \rho\sin\phi$ and $z = z$, cylindrical equation of the cone is given by

$$(\rho\cos\phi)^2 + (\rho\sin\phi)^2 = z^2\tan^2\alpha$$

$\therefore \qquad \rho^2(\cos^2\phi + \sin^2\phi) = z^2\tan^2\alpha$

$\therefore \qquad \rho = z\tan\alpha$

Exercise 6.1

1. Find the spherical polar co-ordinates of the points (i) (3, 4, 5) (ii) (−1, 1, −1).

   Ans. : (i) $\left(5\sqrt{2}, 45°, 53.13°\right)$ (ii) $\left(\sqrt{3}, 125.26°, 135°\right)$.

2. Find the spherical polar co-ordinates of the points (i) (− 2, 1, − 2), (ii) (− 1, − 2, − 3)

   Ans. : (i) $(3, 131.81°, 154.43°)$ or $\left(3, \dfrac{\pi}{2} + \tan^{-1}\dfrac{2}{\sqrt{5}}, \dfrac{\pi}{2} + \tan^{-1}2\right)$.

   (ii) $\left(\sqrt{14}, 143.30°, 243.43°\right)$, third quadrant.

3. Find the cylindrical co-ordinates of the point, whose cartesian coordinates are (−3, 4, 12).   **Ans. :** (5, 126.87°, 12).

4. Find the cartesian co-ordinates of the points $\left(2, \dfrac{5\pi}{6}, \dfrac{3\pi}{4}\right)$.

   (May 2003) Ans. : $\left(-\dfrac{1}{\sqrt{2}}, \dfrac{1}{\sqrt{2}}, -\sqrt{3}\right)$.

5. Find (i) polar, (ii) cylindrical equation of surfaces whose cartesian equations are

   (a) $x^2 + y^2 = a^2$ (b) $x^2 + y^2 + z^2 = 16$

   Ans. : (a) $r\sin\theta = a$; $\rho = a$ (b) $r = 4$; $\rho^2 + z^2 = 16$.

We shall now review the preliminary results related to Co-ordinate Systems, Direction Cosines, Planes and Straight Lines which will be useful throughout our work.

6.5 SHIFT OF ORIGIN

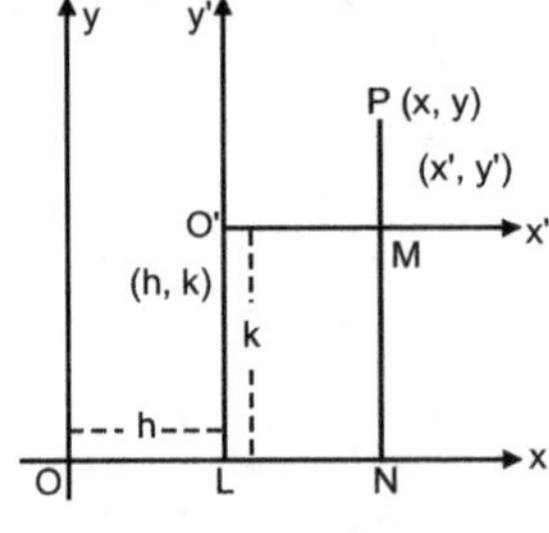

Fig. 6.2

Consider the reference system ox, oy and o'x', o'y' in the same plane. Let the point P has co-ordinates (x, y) with reference to ox, oy as co-ordinate axes and (x', y') with respect to o'x' and o'y' as axes.

Let the co-ordinates of O' be (h, k) with respect to ox, oy. From Fig. 6.2, it is clear   that :

$$ON = x = OL + LN = OL + O'M = h + x'$$

and

$$NP = y = NM + MP = LO' + MP = k + y'$$

∴

$$\boxed{\begin{aligned} x' &= x - h \\ y' &= y - k \end{aligned}}$$ ... (1)

Thus when the origin is shifted to O'(h, k) with axes transferred parallel to themselves, co-ordinates of the point P with respect to new system are given by relations (1).

Similarly, if we consider the reference systems ox, oy, oz and o'x', o'y', o'z' in three dimensional geometry and if P, any point in space has co-ordinates (x, y, z) and (x', y', z') with respect to these systems, where O' has co-ordinates (h, k, l) with respect to ox, oy, oz, then it can be established that

$$\boxed{\begin{aligned} x' &= x - h \\ y' &= y - k \\ z' &= z - l \end{aligned}}$$ ... (2)

6.6 DISTANCE FORMULA

(a) Distance of a Point P(x, y, z) from the Origin :

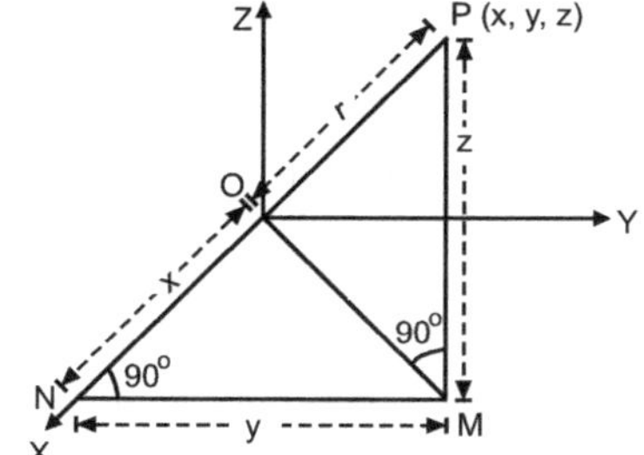

Fig. 6.3

Let P(x, y, z) be any point in space. M is the foot of the perpendicular from P on XOY plane. MN is perpendicular on OX and ON = x, NM = y, MP = z. Also, let OP = r.

Then

$$OP^2 = OM^2 + MP^2$$

$$= (ON^2 + NM^2) + MP^2 \qquad (\because OM^2 = ON^2 + NM^2)$$

∴

$$OP^2 = r^2 = x^2 + y^2 + z^2$$

or

$$\boxed{OP = r = \sqrt{x^2 + y^2 + z^2}}$$ ... (3)

which gives distance between origin and any point P (x, y, z).

(b) Distance between Two Points : Consider any two points $P_1(x_1, y_1, z_1)$ and $Q_1(x_2, y_2, z_2)$. If the origin is shifted to $P_1(x_1, y_1, z_1)$ with axes transferred parallel to themselves, then by results (2) of an article 6.5, co-ordinates of Q_1 with respect to P_1 as origin are $(x_2 - x_1, y_2 - y_1, z_2 - z_1)$. Hence by result (3),

$$\boxed{P_1Q_1 = \sqrt{(x_2 - x_1)^2 + (y_2 - y_1)^2 + (z_2 - z_1)^2}}$$

which gives distance between any two points in space.

Ex. 1 *Find the distance between the points P_1 (1, 2, – 3) and Q_1 (2, – 1, 4).*

Sol. : $P_1Q_1 = \sqrt{(2 - 1)^2 + (-1 - 2)^2 + (4 + 3)^2} = \sqrt{1 + 9 + 49} = \sqrt{59}$

6.7 DIVISION OF THE JOIN OF TWO GIVEN POINTS

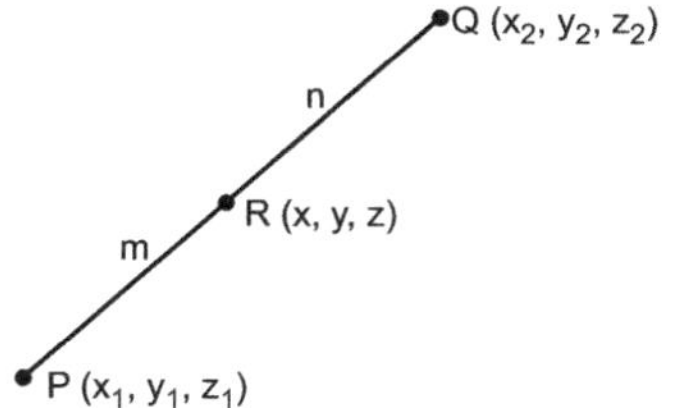

If PQ joins the points (x_1, y_1, z_1) and (x_2, y_2, z_2) (Fig. 6.4) and the point R internally divides PQ in the ratio m : n, then co-ordinates of the point R are given by :

Fig. 6.4

$$\boxed{x = \frac{mx_2 + nx_1}{m + n}, \quad y = \frac{my_2 + ny_1}{m + n}, \quad z = \frac{mz_2 + nz_1}{m + n}}$$ ... (5)

Note : (i) If $\dfrac{m}{n} = \lambda$ or if R divides PQ in the ratio $\lambda : 1$, then the general co-ordinates of the point R are given by

$$x = \frac{\lambda\, x_2 + x_1}{\lambda + 1}, \quad y = \frac{\lambda\, y_2 + y_1}{\lambda + 1}, \quad z = \frac{\lambda\, z_2 + z_1}{\lambda + 1} \qquad \text{... (6)}$$

(ii) If $\dfrac{m}{n} = \lambda$ is +ve, then the point R divides the line PQ internally and if it is – ve, then externally.

Ex. 1 : *Find the co-ordinates of the point which divides the join of two points (2, – 4, 3), (– 4, 5, 6) (i) internally in the ratio 1 : 2, (ii) externally in the ratio 1 : 4.*

Sol. :   (i) Here $\lambda = \dfrac{1}{2}$ (internal division) $\therefore$ $x = 0,\ y = -1,\ z = 4.$

(ii) Here $\lambda = -\dfrac{1}{4}$ (external division) $\therefore$ $x = 4,\ y = -7,\ z = 2.$

6.8 DIRECTION COSINES

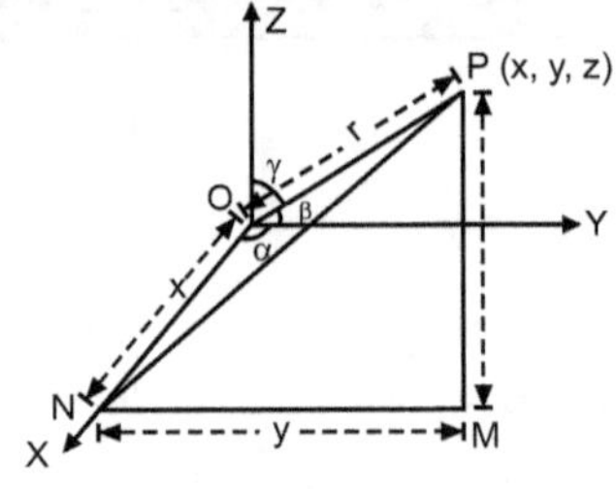

Fig. 6.5

Let the line OP in space be inclined at angles α, β, γ with ox, oy and oz respectively (Refer Fig. 6.5). The quantities cos α, cos β, cos γ which provide the information about certain direction in space are called direction cosines (d.c.'s) of the line OP and are denoted by l, m, n respectively. Thus, l = cos α, m = cos β and n = cos γ.

Cor. : Direction Cosines of Co-ordinate Axes : Since the line x-axis, makes with the co-ordinate axes the angle 0°, 90°, 90° respectively, therefore its direction cosines are cos 0°, cos 90°, cos 90° i.e. 1, 0, 0. Similarly, the d.c.'s of y-axis are 0, 1, 0 and that of z-axis are 0, 0, 1.

6.8.1 A Useful Result

Let P (x, y, z) be any point at distance r from the origin O and l, m, n be the d.c.'s of the line OP (Refer Fig. 6.5). Through P draw PN perpendicular on x-axis, then ON = x, then from right angled triangle PNO,

$$\frac{ON}{OP} = \cos \alpha \quad \text{or} \quad x = lr$$

Similarly by dropping perpendicular from the point P on y-axis and z-axis, we get   y = mr and z = nr.

Thus, if a line joining origin to a point P (x, y, z) has d.c.'s. l, m, n, then co-ordinates of P are

$$x = lr,\ y = mr,\ z = nr \qquad \text{... (7)}$$

6.8.2 Relation Between Direction Cosines

From result (7), on squaring and adding, we obtain

$$x^2 + y^2 + z^2 = r^2\,(l^2 + m^2 + n^2)$$

But $x^2 + y^2 + z^2 = OP^2 = r^2$ (by result (3) of article 6.6)

$\therefore$ $\text{or} \begin{array}{l} l^2 + m^2 + n^2 = 1 \\ \cos^2 \alpha + \cos^2 \beta + \cos^2 \gamma = 1 \end{array}$ ... (8)

6.8.3 Direction Ratios

Any three numbers a, b, c which are proportional to the direction cosines l, m, n respectively of a given line are called direction numbers or direction ratios (or briefly d.r.'s) of the given line.

Now, from definition of direction ratios, when a, b, c are proportional to l, m, n, we have

$$\frac{l}{a} = \frac{m}{b} = \frac{n}{c} = \pm \frac{\sqrt{l^2 + m^2 + n^2}}{\sqrt{a^2 + b^2 + c^2}} = \pm \frac{1}{\sqrt{a^2 + b^2 + c^2}}$$

Hence, if a, b, c are the d.r.'s of a given line, then the actual d.c.'s are

$$l = \frac{a}{\sqrt{a^2 + b^2 + c^2}}, \quad m = \frac{b}{\sqrt{a^2 + b^2 + c^2}}, \quad n = \frac{c}{\sqrt{a^2 + b^2 + c^2}} \qquad \text{... (9)}$$

Cor. : We have, from result (7), $\dfrac{x}{l} = \dfrac{y}{m} = \dfrac{z}{n} = r$ which means x, y, z are proportional to l, m, n or x, y, z are d.r.'s of the line OP.

Now, P has co-ordinates (x, y, z) and O has (0, 0, 0). The **d.r.'s of OP** can be written as **(x – 0, y – 0, z – 0) i.e. (x, y, z).** Similarly, if the line P_1P_2 joins the points P_1 (x_1, y_1, z_1) and P_2 (x_2, y_2, z_2), then **d.r.'s of P_1P_2** can be taken as **$(x_2 - x_1, y_2 - y_1, z_2 - z_1)$.** Thus, if two points on a line are known, we can first find its d.r.'s and then d.c.'s.

ILLUSTRATIONS :

Ex. 1 : *A line makes angles 45° and 60° with positive axes of x and y respectively. What angle does it make with the positive axis of z ?*

Sol. : Let the line makes an angle γ with the positive z-axis. Thus, line makes angles 45°, 60° and γ with the axes; its d.c.'s are $\cos 45°$, $\cos 60°$ and $\cos γ$. i.e. $\dfrac{1}{\sqrt{2}}, \dfrac{1}{2}, \cos γ$. But we know that $l^2 + m^2 + n^2 = 1$.

$$\therefore \quad \frac{1}{2} + \frac{1}{4} + \cos^2 γ = 1 \quad \text{or} \quad \cos γ = \pm \frac{1}{2}$$

$$\therefore \quad \boxed{γ = 60° \text{ or } 120°.}$$

Ex. 2 : *What are the d.c.'s of the lines equally inclined to the axes ?*

Sol. : If a line makes angles α, β, γ with the axes, then we have given that α = β = γ

$$\therefore \qquad \cos α = \cos β = \cos γ \text{ or } l = m = n$$

Again we know that $l^2 + m^2 + n^2 = 1$

$$\therefore \qquad l^2 + l^2 + l^2 = 1 \quad \text{or} \quad 3 l^2 = 1 \quad \therefore l = \pm \frac{1}{\sqrt{3}}$$

$$\therefore \quad \text{The d.c.'s of the lines are} \quad \boxed{\left(\pm \frac{1}{\sqrt{3}}, \pm \frac{1}{\sqrt{3}}, \pm \frac{1}{\sqrt{3}} \right)} .$$

Ex. 3 : *1, – 2, 2 are proportional to the direction cosines of a line. What are their actual values ?*

Sol. : From result (9), the actual direction cosines are given by

$$\frac{1}{\sqrt{1 + 4 + 4}}, \quad \frac{-2}{\sqrt{1 + 4 + 4}}, \quad \frac{2}{\sqrt{1 + 4 + 4}} \quad \text{or} \quad \left(\frac{1}{3}, -\frac{2}{3}, \frac{2}{3} \right) .$$

Ex. 4 : *P and Q are (4, 3, – 5) and (–2, 1, – 8). Find the d.c.'s of OP, OQ and PQ.*

Sol. : The d.c.'s of OP are proportional to 4, 3, – 5

$$\therefore \quad \text{actual d.c.'s of OP are} \quad \frac{4}{5\sqrt{2}}, \quad \frac{3}{5\sqrt{2}}, \quad -\frac{5}{5\sqrt{2}}$$

The d.c.'s of OQ are proportional to – 2, 1, – 8

$$\therefore \quad \text{actual d.c.'s of OQ are} \quad -\frac{2}{\sqrt{69}}, \quad -\frac{1}{\sqrt{69}}, \quad -\frac{8}{\sqrt{69}}$$

and d.c.'s of PQ are proportional to (– 2 – 4), (1 – 3), (– 8 + 5) or – 6, – 2, – 3

$$\therefore \quad \boxed{\text{actual d.c.'s of PQ are} \quad -\frac{6}{7}, -\frac{2}{7}, -\frac{3}{7}} .$$

6.9 ANGLE BETWEEN TWO LINES

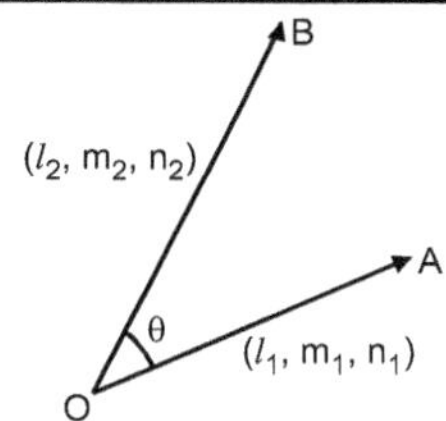

Fig. 6.6

If (l_1, m_1, n_1) and (l_2, m_2, n_2) are the direction cosines of the two lines inclined at an angle θ (Refer Fig. 6.6),

then
$$\boxed{\cos\theta = l_1 l_2 + m_1 m_2 + n_1 n_2}$$
   ... (10)

If direction ratios of two lines are given as a_1, b_1, c_1 and a_2, b_2, c_2,

then
$$\boxed{\cos\theta = \frac{a_1 a_2 + b_1 b_2 + c_1 c_2}{\sqrt{a_1^2 + b_1^2 + c_1^2}\ \sqrt{a_2^2 + b_2^2 + c_2^2}}}$$
   ... (11)

Lagrange's Identity :

$$\left(l_1^2 + m_1^2 + n_1^2\right)\left(l_2^2 + m_2^2 + n_2^2\right) - (l_1 l_2 + m_1 m_2 + n_1 n_2)^2$$
$$= (l_1 m_2 - l_2 m_1)^2 + (m_1 n_2 - m_2 n_1)^2 + (n_1 l_2 - n_2 l_1)^2$$

Cor. 1 : Expression for sin θ :

$$\sin^2\theta = 1 - \cos^2\theta = 1 - (l_1 l_2 + m_1 m_2 + n_1 n_2)^2$$
$$= (l_1 m_2 - l_2 m_1)^2 + (m_1 n_2 - m_2 n_1)^2 + (n_1 l_2 - n_2 l_1)^2$$

Hence
$$\boxed{\sin\theta = \sqrt{(l_1 m_2 - l_2 m_1)^2 + (m_1 n_2 - m_2 n_1)^2 + (n_1 l_2 - n_2 l_1)^2}}$$
   ... (12)

Cor. 2 : Conditions for Perpendicularity and Parallelism :

(a) When given lines are perpendicular,

$\theta = 90°$ i.e. $\cos\theta = 0$, hence we have from (10) and (11),

$$\text{or}\quad \boxed{\begin{array}{c} l_1 l_2 + m_1 m_2 + n_1 n_2 = 0 \\ a_1 a_2 + b_1 b_2 + c_1 c_2 = 0 \end{array}}$$
   ... (13)

(b) When given lines are parallel,

$\theta = 0$ i.e. $\sin\theta = 0$, hence we have from (12),

$$l_1 m_2 - l_2 m_1 = m_1 n_2 - m_2 n_1 = n_1 l_2 - n_2 l_1 = 0$$

Similarly, $\quad a_1 b_2 - a_2 b_1 = b_1 c_2 - b_2 c_1 = c_1 a_2 - c_2 a_1 = 0$

Hence
$$\text{or}\quad \boxed{\begin{array}{c} \dfrac{l_1}{l_2} = \dfrac{m_1}{m_2} = \dfrac{n_1}{n_2} = 1 \\[2mm] \dfrac{a_1}{a_2} = \dfrac{b_1}{b_2} = \dfrac{c_1}{c_2} \end{array}}$$
   ... (14)

ILLUSTRATIONS :

Ex. 1 : *Prove that the points (1, 2, 3), (4, 0, 4) and (– 2, 4, 2) are collinear.*

Sol. : The given points are A (1, 2, 3), B (4, 0, 4) and C (– 2, 4, 2).

The d.r.'s of AB, by Cor. 7.8.3, $x_2 - x_1$, $y_2 - y_1$, $z_2 - z_1$ are 3, –2, 1 and d.r.'s of AC are –3, 2, –1 or 3, –2, 1. Hence the two lines are parallel and since they both pass through the point A, therefore, the points A, B, C are collinear.

Ex. 2 : *Show that the join of the points (1, 2, 3), (4, 5, 7) is parallel to the join of the points (– 4, 3, – 6), (2, 9, 2).*

Sol. : The given points are A (1, 2, 3), B (4, 5, 7) and C (– 4, 3, – 6), D (2, 9, 2).

The d.r.'s of AB, by Cor. 7.8.3 by $x_2 - x_1$, $y_2 - y_1$, $z_2 - z_1$ are 3, 3, 4 and d.r.'s of CD are 6, 6, 8 which are proportional to 3, 3, 4. Hence the lines are parallel.

Ex. 3 : *If two lines have d.c.'s proportional to (1, 2, 3) and (–2, 1, 3) respectively, find the direction cosines of the line perpendicular to both of them.*

Sol. : Let the d.c.'s of a line perpendicular to them be l, m, n.

$\therefore$ we have $\quad\quad\quad\quad l + 2m + 3n = 0$ ... (i)

and $\quad\quad\quad\quad\quad -2l + m + 3n = 0$ ... (ii)

From (i) and (ii), we get $\quad \dfrac{l}{3} = \dfrac{m}{-9} = \dfrac{n}{5}$

$\therefore$ The d.c.'s of the line are proportional to 3, – 9, 5.

6.10 PROJECTION

(a) Projection of a Point :

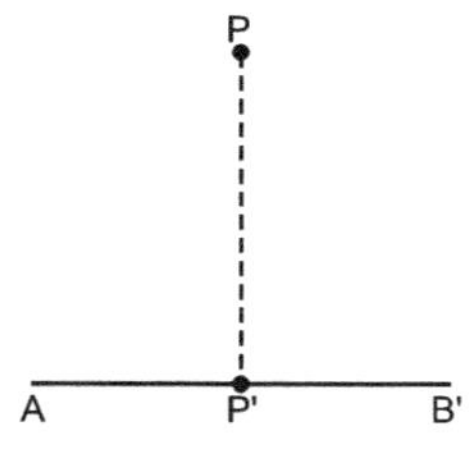

Fig. 6.7

The projection of a point P on a given line AB is the point P' which is the foot of the perpendicular from P on AB (Refer Fig. 6.7)

(b) Projection of a Segment of a Line :

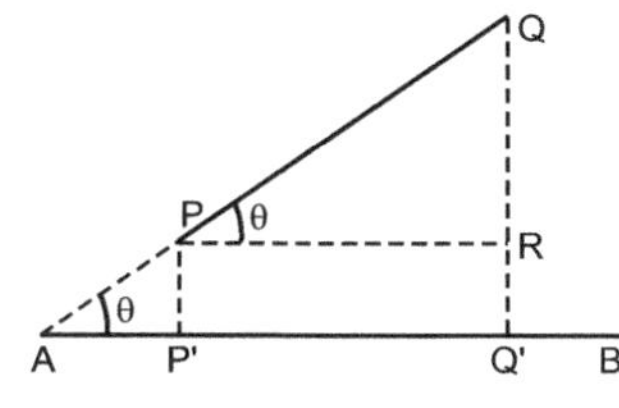

Fig. 6.8

The projection of a segment PQ of a line on another line AB is the segment P'Q' of AB where P' and Q' are the projections of the points P and Q on the line AB (Refer Fig. 6.8).

(c) Length of the Projection :

$$\boxed{P'Q' = PQ \cos \theta} \qquad \ldots (15)$$

where, θ is the angle which the line PQ makes with PR i.e. with AB.

Note : If PQ is perpendicular to AB, then projection of PQ on AB = 0, as $\theta = 90°$.

6.10.1 Projection of Join of Two Points on a Line whose Direction Cosines are l, m, n

Let $P(x_1, y_1, z_1)$ and $Q(x_2, y_2, z_2)$ be the given points and AB be the line whose d.c.'s are l, m, n. Draw PP' and QQ' perpendiculars on AB. Then P'Q' is the projection of PQ on AB so that,

$$P'Q' = PQ \cos \theta \qquad \ldots (i)$$

Now d.r.'s of PQ are $\quad x_2 - x_1,\ y_2 - y_1,\ z_2 - z_1$

$\therefore$ d.c.'s of PQ are $\quad \dfrac{x_2 - x_1}{PQ},\ \dfrac{y_2 - y_1}{PQ},\ \dfrac{z_2 - z_1}{PQ}\ \left[\because PQ = \sqrt{(x_2 - x_1)^2 + (y_2 - y_1)^2 + (z_2 - z_1)^2}\right]$

Also d.c.'s of AB are l, m, n.

$$\therefore \qquad \cos \theta = l\left(\frac{x_2 - x_1}{PQ}\right) + m\left(\frac{y_2 - y_1}{PQ}\right) + n\left(\frac{z_2 - z_1}{PQ}\right)$$

$$\therefore \qquad PQ \cos \theta = l(x_2 - x_1) + m(y_2 - y_1) + n(z_2 - z_1) \qquad \ldots (ii)$$

From (i) and (ii), we have

$$\boxed{P'Q' = l(x_2 - x_1) + m(y_2 - y_1) + n(z_2 - z_1)} \qquad \ldots (16)$$

ILLUSTRATIONS :

Ex. 1 : *If P, Q, A, B are (–2, 1, 3), (1, 2, 5), (2, 1, – 4), and (4, 4, 2), find the projection of PQ on AB.*

Sol. : Since the projection is on AB, we must know d.c.'s of AB.

The d.r.'s of AB are 2, 3, 6 $\qquad \therefore \qquad$ d.c.'s of AB are $\dfrac{2}{7}, \dfrac{3}{7}, \dfrac{6}{7}$.

From result (16), we get projection of PQ on AB

$$= \frac{2}{7}(1 + 2) + \frac{3}{7}(2 - 1) + \frac{6}{7}(5 - 3)$$

$$= \frac{1}{7}(6 + 3 + 12) = 3$$

Ex. 2 : *P, Q, R, S are the points (–2, 3, 4), (–4, 4, 6), (4, 3, 5) and (0, 1, 2). Prove by projections that PQ is at right angles to RS.*

Sol. : The d.r.'s of RS are – 4, – 2, – 3 or 4, 2, 3. $\therefore$ d.c.'s of RS are $\dfrac{4}{\sqrt{29}}, \dfrac{2}{\sqrt{29}}, \dfrac{3}{\sqrt{29}}$

$\therefore$ Projection of PQ on RS $= \dfrac{4}{\sqrt{29}} (-4 + 2) + \dfrac{2}{\sqrt{29}} (4 - 3) + \dfrac{3}{\sqrt{29}} (6 - 4)$

$$= \dfrac{1}{\sqrt{29}} (-8 + 2 + 6) = 0.$$

Hence PQ is perpendicular to RS.

After considering the preliminary results related to Co-ordinate systems, Direction cosines etc. we shall deal with the surface of common occurrence 'Plane'.

6.11 THE PLANE

Definition : A plane is a surface such that if any two points are taken on it, then the straight line joining them lies wholly in the surface.

Equation of Plane (Standard Forms) :

(a) General Form : The general equation of first degree in x, y, z represents a plane.

Thus, general form of the equation of plane is

$$\boxed{ax + by + cz + d = 0} \qquad \text{... (17)}$$

where a, b, c are d.r.'s of normal to the plane.

Cor. 1 : Equation of the plane passing through the point (x_1, y_1, z_1) : Let the equation of the plane be $ax + by + cz + d = 0$. Since it passes through (x_1, y_1, z_1),

then $ax_1 + by_1 + cz_1 + d = 0$

Subtracting, we get

$$\boxed{a\,(x - x_1) + b\,(y - y_1) + c\,(z - z_1) = 0} \qquad \text{... (18)}$$

Cor. 2 : General equation of a plane passing through origin is

$$\boxed{ax + by + cz = 0} \qquad \text{... (19)}$$

Since it passes through origin, $d = 0$.

Note : Number of arbitrary constants in the general equation of a plane

$$ax + by + cz + d = 0 \quad \text{or} \quad \dfrac{a}{d}\,x + \dfrac{b}{d}\,y + \dfrac{c}{d}\,z + 1 = 0$$

The above form shows that there are only three independent arbitrary constants in the equation of the plane. Thus a plane must satisfy three conditions.

(b) Intercept Form : The equation of the plane in terms of the intercepts a, b and c, which it makes on the axes is

$$\boxed{\dfrac{x}{a} + \dfrac{y}{b} + \dfrac{z}{c} = 1} \qquad \text{... (20)}$$

(c) Normal (Perpendicular) Form : If l, m, n are d.c.'s of normal to the plane and p perpendicular from origin to the plane, then normal form of the plane is

$$\boxed{lx + my + nz = p} \qquad \text{... (21)}$$

Note 1 : Any equation $lx + my + nz = p$ is in normal form if

(i)   (coeff. of x)^2 + (coeff. of y)^2 + (coeff. of z)^2 = 1

(ii)   Constant term on the R.H.S. is +ve.

Note 2 : If we divide general equation of the plane $ax + by + cz + d = 0$ throughout by $\sqrt{a^2 + b^2 + c^2}$, we get

$$\dfrac{a}{\sqrt{a^2 + b^2 + c^2}}\,x + \dfrac{b}{\sqrt{a^2 + b^2 + c^2}}\,y + \dfrac{c}{\sqrt{a^2 + b^2 + c^2}}\,z + \dfrac{d}{\sqrt{a^2 + b^2 + c^2}} = 0$$

Now comparing with $lx + my + nz = p$, we see that d.c.'s of normal to the plane are

$$\frac{a}{\sqrt{a^2 + b^2 + c^2}}, \frac{b}{\sqrt{a^2 + b^2 + c^2}}, \frac{c}{\sqrt{a^2 + b^2 + c^2}} \text{ and } p = +\frac{d}{\sqrt{a^2 + b^2 + c^2}} \qquad \text{(p is always taken as positive)}$$

Thus $p = +\dfrac{d}{\sqrt{a^2 + b^2 + c^2}}$ gives length of perpendicular from origin and coefficients of x, y, z i.e. a, b, c are d.r.'s of normal to the plane in the general equation of plane.

(d) Three Points Form : The equation of the plane passing through three points (x_1, y_1, z_1), (x_2, y_2, z_2) and (x_3, y_3, z_3) is

$$\begin{vmatrix} x & y & z & 1 \\ x_1 & y_1 & z_1 & 1 \\ x_2 & y_2 & z_2 & 1 \\ x_3 & y_3 & z_3 & 1 \end{vmatrix} = 0 \qquad \text{... (22)}$$

(e) Equation of the Plane Passing through the Common Section of Two Planes :

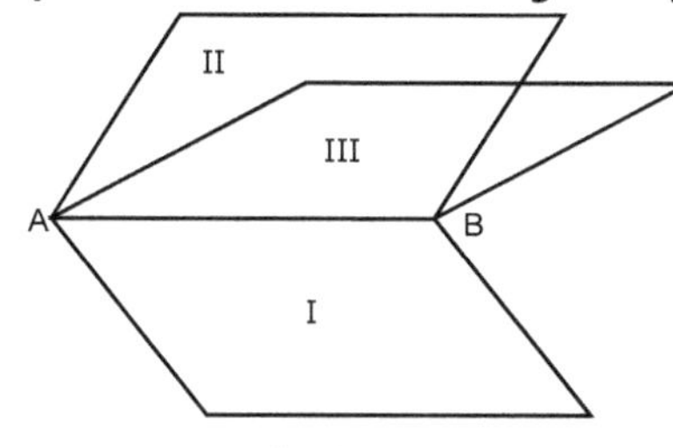

Fig. 6.9

Consider two planes :

$$a_1x + b_1y + c_1z + d_1 = 0 \qquad \text{... (I)}$$
$$a_2x + b_2y + c_2z + d_2 = 0 \qquad \text{... (II)}$$

If they are non-parallel, they will intersect in a line (say AB). Fig. 6.9 represents the common section of two planes.

Multiplying (II) by λ and adding to (I) we get,

$$a_1x + b_1 y + c_1 z + d_1 + \lambda (a_2x + b_2 y + c_2 z + d_2) = 0 \qquad \text{... (III)}$$

All the points which satisfy (I) and (II) both lie on the line of section AB, and will satisfy equation (III). But (III) is general equation of first degree in x, y, z and represents a plane which contains the points on the line of section AB. For different values of λ, we get different planes. Hence equation (III) represents a family of planes passing through the line of section of planes (I) and (II).

Hence $\boxed{a_1x + b_1 y + c_1 z + d_1 + \lambda (a_2x + b_2 y + c_2 z + d_2) = 0}$

6.12 ANGLE BETWEEN TWO PLANES

Consider two planes

$$a_1x + b_1 y + c_1 z + d_1 = 0 \qquad \text{... (i)}$$
$$a_2x + b_2 y + c_2 z + d_2 = 0 \qquad \text{... (ii)}$$

Angle between two planes is equal to the angle between their normals.

The d.r.'s of normals to the two planes are (a_1, b_1, c_1) and (a_2, b_2, c_2).

If θ is the angle between these two normals, then angle between two planes is given by

$$\boxed{\cos \theta = \frac{a_1 a_2 + b_1 b_2 + c_1 c_2}{\sqrt{a_1^2 + b_1^2 + c_1^2}\sqrt{a_2^2 + b_2^2 + c_2^2}}} \qquad \text{... (24)}$$

Cor. : If the planes are perpendicular, $\theta = 90°$ and hence by result (13),

$$\boxed{a_1 a_2 + b_1 b_2 + c_1 c_2 = 0} \qquad \text{... (25)}$$

Similarly, if the planes are parallel, $\theta = 0°$ i.e. $\sin \theta = 0$, hence by result (14),

$$\boxed{\frac{a_1}{a_2} = \frac{b_1}{b_2} = \frac{c_1}{c_2}} \qquad \text{... (26)}$$

From above result we can easily conclude that the equations of two planes differ only in the constant term.

6.13 LENGTH OF THE PERPENDICULAR

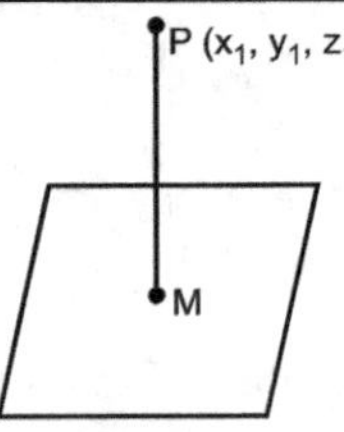

Fig. 6.10

The length of the perpendicular from the point $P(x_1, y_1, z_1)$ on the plane $ax + by + cz + d = 0$ is

$$p = \frac{ax_1 + by_1 + cz_1 + d}{\sqrt{a^2 + b^2 + c^2}} \qquad \ldots (27)$$

6.14 GENERAL EQUATION OF A PLANE IN SOME PARTICULAR CASES

(a) Equation to the Co-ordinate Planes : Every point in the xoy plane will have its z co-ordinate zero and hence its equation is $z = 0$. This can also be obtained by considering normal to the xoy plane which is z-axis whose D.C.S. are $(0, 0, 1)$ and it passes through origin. Hence equation of xoy plane is

$$0 \cdot x + 0 \cdot y + 1 \cdot z = 0 \quad \text{or} \quad z = 0$$

Similarly, equations of yoz and zox planes are $x = 0$ and $y = 0$ respectively.

(b) Planes Parallel to the Co-ordinate Planes : If we draw a plane parallel to xoy plane at a distance 'a' from it, then every point in the plane will have its z co-ordinate as 'a', hence equation of the plane can be taken as $z = a$. Similarly plane parallel to xoz plane at a distance b from it will have its equation as $y = b$ and plane parallel to yoz plane at a distance c from it will have its equation as $x = c$.

(c) Planes Parallel to Co-ordinate Axes : Consider the general plane

$$ax + by + cz + d = 0 \qquad \ldots (i)$$

If it is parallel to x-axis, normal to the plane (i) will be perpendicular to the x-axis. Normal to the plane (i) has D.R.S. (a, b, c) and D.R.S. of x-axis are $(1, 0, 0)$.

$\therefore$ By the condition of perpendicularity 6.12, (25)

$$a \cdot 1 + b \cdot 0 + c \cdot 0 = 0 \quad \text{or} \quad a = 0$$

Putting in (i), we get

$$by + cz + d = 0$$

Thus if x term is absent in the general equation of a plane, then it is parallel to x-axis. Similarly, planes parallel to y and z axes can be respectively taken as

$$ax + bz + d = 0 \quad \text{and} \quad ax + by + d = 0$$

(d) Planes Perpendicular to Co-ordinate Planes : Any plane perpendicular to yoz plane is parallel to x-axis and its equation by case (c) can be taken as $by + cz + d = 0$.

Similarly, we can take planes perpendicular to zox and xoy planes as

$$ax + cz + d = 0 \quad \text{and} \quad ax + by + d = 0 \text{ respectively.}$$

ILLUSTRATIONS :

Ex. 1 : *Find the intercepts made by the plane $2x + y - 2z = 3$ on the coordinate axes. What are the d.c.'s of normal to the plane ?*

Sol. : The given equation can be written as

$$\frac{x}{\frac{3}{2}} + \frac{y}{3} + \frac{z}{\frac{-3}{2}} = 1 \quad \therefore \text{ Intercepts are } \frac{3}{2}, 3, -\frac{3}{2}$$

Also d.r.'s of the normal to the plane are given by coefficients of x, y, z in the equation of the plane which are 2, 1, – 2 and hence d.c.'s are $\frac{2}{3}, \frac{1}{3}, -\frac{2}{3}$.

Ex. 2 : *Find the angle between the planes $2x - y + z = 6$ and $x + y + 2z = 7$.*

Sol. : If θ be the angle between the planes, then by result (24),

$$\cos \theta = \frac{2 \cdot 1 + (-1) \cdot 1 + 1 \cdot 2}{\sqrt{(4 + 1 + 1)} \sqrt{1 + 1 + 4}} = \frac{1}{2}$$

$\therefore$ $$\theta = \frac{\pi}{3}$$

Ex. 3 : *Find the equation of the plane through the point (4, 0, 1) and parallel to the plane $4x + 3y - 12z + 6 = 0$.*

Sol. : We know from result (26), equation of the plane parallel to

$4x + 3y - 12z + 6 = 0$ is $4x + 3y - 12z + k = 0$.

Since it passes through (4, 0, 1), we get $16 + 0 - 12 + k = 0$ or $k = -4$

Hence the required equation of the plane is $4x + 3y - 12z - 4 = 0$

Ex. 4 : *Find the angle between the plane $x + 2y - 3z + 4 = 0$ and the line whose direction cosines are $\dfrac{2}{\sqrt{14}}, \dfrac{3}{\sqrt{14}}, \dfrac{1}{\sqrt{14}}$.*

Sol. :

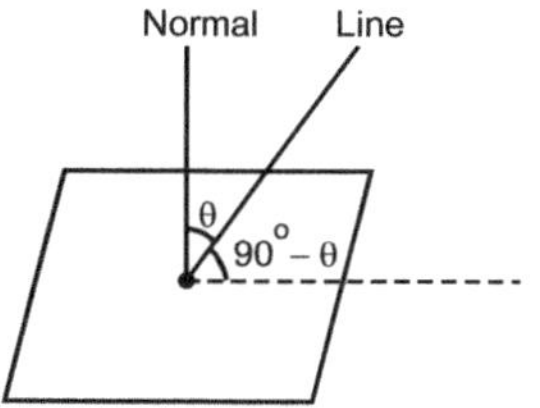

Fig. 6.11

The given plane is $x + 2y - 3z + 4 = 0$ … (i)

The d.r.'s of the normal to (i) are 1, 2, – 3

∴ d.c.'s of the normal to (i) are

$$\frac{1}{\sqrt{14}}, \frac{2}{\sqrt{14}}, -\frac{3}{\sqrt{14}}$$

Also d.c.'s of the line are given as

$$\frac{2}{\sqrt{14}}, \frac{3}{\sqrt{14}}, \frac{1}{\sqrt{14}}.$$

Let θ be the angle between the normal and the given line.

Then

$$\cos \theta = \frac{1}{\sqrt{14}} \cdot \frac{2}{\sqrt{14}} + \frac{2}{\sqrt{14}} \cdot \frac{3}{\sqrt{14}} - \frac{3}{\sqrt{14}} \cdot \frac{1}{\sqrt{14}} = \frac{5}{14}$$

∴

$$\theta = \cos^{-1}\left(\frac{5}{14}\right)$$

Note : $90° - \theta$ is the angle between the plane and the line.

Ex. 5 : *Find the equation of the plane which passes through the section of two planes*

$$2x - y + z - 3 = 0, \qquad x + y - z + 2 = 0$$

and (i) is parallel to the line with direction ratios (1, 1, 1), (ii) passes through the point (1, 0, – 1).

Sol. : Plane through the given section may be taken as (by result 23)

$$2x - y + z - 3 + \lambda (x + y - z + 2) = 0 \qquad … (1)$$

or $$x (2 + \lambda) + y (-1 + \lambda) + z (1 - \lambda) - 3 + 2\lambda = 0 \qquad … (2)$$

(i)   If plane (2) is parallel to the given line, then normal to the plane (2) is perpendicular to the given line.

∴   From the condition of perpendicularity,

$$(2 + \lambda) \cdot 1 + (\lambda - 1) \cdot 1 + (1 - \lambda) \cdot 1 = 0$$

∴

$$\lambda + 2 = 0 \quad \text{or} \quad \lambda = -2$$

Putting $\lambda = -2$ in equation (2), we get the required plane as

$$-3y + 3z - 7 = 0$$

or $$3y - 3z + 7 = 0$$

(ii)   If plane (2) passes through the point (1, 0, – 1), then

$$1 (2 + \lambda) + 0 (-1 + \lambda) - 1 (1 - \lambda) - 3 + 2\lambda = 0$$

∴

$$\lambda + 2 - 1 + \lambda - 3 + 2\lambda = 0$$

or $$4\lambda - 2 = 0 \qquad\qquad ∴ \quad \lambda = \frac{1}{2}$$

Putting $\lambda = \dfrac{1}{2}$ in (2) or in (1), we get

$$2x - y + z - 3 + \frac{1}{2}(x + y - z + 2) = 0$$

$$4x - 2y + 2z - 6 + x + y - z + 2 = 0$$

or
$$5x - y + z - 4 = 0$$

is the required plane.

We shall now present certain results related to the topic of straight line. These results will often be used when we deal with different surfaces in the geometry of three dimensions.

6.15 STRAIGHT LINE

(a) General Form : Any two equations of first degree in x, y, z taken together represent a line.

We know that every equation of first degree in x, y, z represents a plane [see article 6.11 (17)].

Consider two equations of first degree

$$\begin{array}{c} a_1 x + b_1 y + c_1 z + d_1 = 0 \\ a_2 x + b_2 y + c_2 z + d_2 = 0 \end{array}$$
... (28)

Now, if the co-ordinate of any point satisfies equations (28) individually then it lies on both the planes and therefore on their line of intersection. Hence these two equations taken together represent the curve of intersection of the two given planes which is a straight line. (since a plane cuts another plane in a straight line).

Thus equations (28) taken together represent a line.

Cor. : The x-axis is the line of intersection of the XY and ZX planes whose equations are z = 0, y = 0. Thus the equations of x-axis are y = 0, z = 0. Similarly, the equations of y-axis are z = 0, x = 0 and that of z-axis are x = 0, y = 0.

(b) Symmetrical Form :

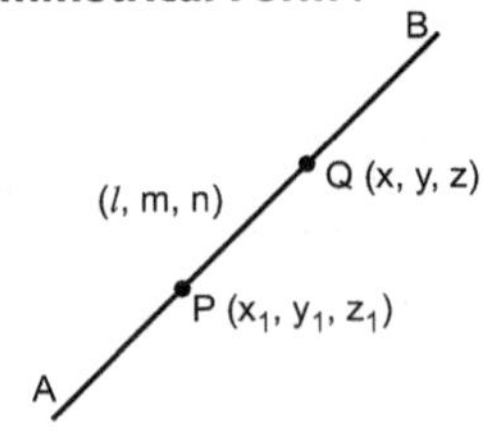

Fig. 6.12

Consider the straight line which passes through the fixed point $P(x_1, y_1, z_1)$ and has Direction Cosines (l, m, n). Let Q (x, y, z) be any point on the line (Refer Fig. 6.12). Now, direction ratios of the line PQ or of AB are $x - x_1, y - y_1, z - z_1$ which are proportional to (l, m, n).

$\therefore$
$$\frac{x - x_1}{l} = \frac{y - y_1}{m} = \frac{z - z_1}{n} = r$$
... (29)

where, r represents the actual distance of the point (x, y, z) from the given point (x_1, y_1, z_1).

Equation (29) is true for any point on the line AB, hence is the equation of straight line AB.

Similarly, if the line passes through a fixed point (x_1, y_1, z_1) and has d.r.'s (a, b, c), its equation can be taken as

$$\frac{x - x_1}{a} = \frac{y - y_1}{b} = \frac{z - z_1}{c} = r$$
... (30)

where, r is a constant of proportionality.

Cor. : From the relations (29) and (30), we find that the coordinates of any point on the line are

$$(x_1 + lr,\ y_1 + mr,\ z_1 + nr) \quad \text{or} \quad (x_1 + ar,\ y_1 + br,\ z_1 + cr)$$
... (31)

6.16 LINE THROUGH TWO POINTS

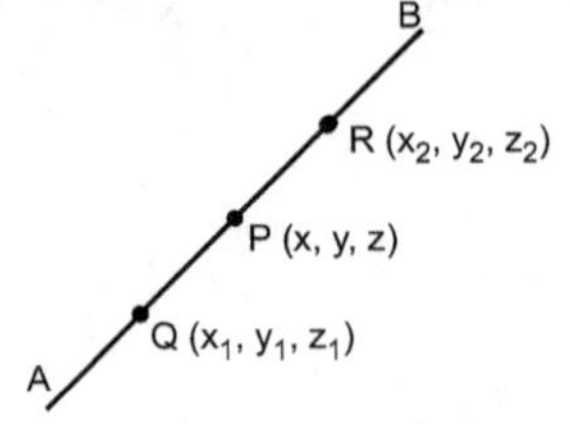

Fig. 6.13

Consider the straight line AB which passes through two fixed points Q (x_1, y_1, z_1) and R (x_2, y_2, z_2) (Refer Fig. 6.13). The d.r.'s of the line AB are $x_2 - x_1, y_2 - y_1, z_2 - z_1$.

Let P(x, y, z) be any point on the line QP. The d.r.'s of QP or those of AB are also $x - x_1, y - y_1, z - z_1$ which must be proportional to $x_2 - x_1, y_2 - y_1, z_2 - z_1$.

$$\therefore \qquad \boxed{\frac{x - x_1}{x_2 - x_1} = \frac{y - y_1}{y_2 - y_1} = \frac{z - z_1}{z_2 - z_1}} \qquad \text{... (32)}$$

which is the equation of the straight line AB passing through two fixed points. The equation can also be written as

$$\boxed{\frac{x - x_2}{x_2 - x_1} = \frac{y - y_2}{y_2 - y_1} = \frac{z - z_2}{z_2 - z_1}} \qquad \text{... (33)}$$

6.17 COPLANARITY OF LINES

Consider two lines

$$\frac{x - x_1}{l_1} = \frac{y - y_1}{m_1} = \frac{z - z_1}{n_1}$$

$$\frac{x - x_2}{l_2} = \frac{y - y_2}{m_2} = \frac{z - z_2}{n_2}$$

Equation of plane through the line (i) may be taken as

$$a\,(x - x_1) + b\,(y - y_1) + c\,(z - z_1) = 0 \qquad \text{... (iii)}$$

with the condition $\qquad\qquad al_1 + bm_1 + cn_1 = 0 \qquad \text{... (iv)}$

If the plane (iii) also contains the line (ii), then (x_2, y_2, z_2) will satisfy equation (iii).

$$\therefore \qquad a\,(x_2 - x_1) + b\,(y_2 - y_1) + c\,(z_2 - z_1) = 0 \qquad \text{... (v)}$$

Normal to the plane (iii) is perpendicular to the line (ii).

$$al_2 + bm_2 + cn_2 = 0 \qquad \text{... (vi)}$$

Eliminating a, b, c from (v), (iv) and (vi), we get

$$\begin{vmatrix} x_2 - x_1 & y_2 - y_1 & z_2 - z_1 \\ l_1 & m_1 & n_1 \\ l_2 & m_2 & n_2 \end{vmatrix} = 0 \qquad \text{... (34)}$$

which is the condition of coplanarity of the lines (i) and (ii).

Equation of plane containing the lines (i) and (ii) will be obtained by eliminating a, b, c from (iii), (iv) and (vi), we get

$$\begin{vmatrix} x - x_1 & y - y_1 & z - z_1 \\ l_1 & m_1 & n_1 \\ l_2 & m_2 & n_2 \end{vmatrix} = 0 \qquad \text{... (35)}$$

ILLUSTRATIONS :

Ex. 1 : *Find the equations of the line through (1, 2, 3) and making angles 45°, 60°, 120° with positive direction of axes.*

Sol. : The d.c.'s of the line are cos 45°, cos 60°, cos 120° i.e. $\dfrac{1}{\sqrt{2}}, \dfrac{1}{2}, \dfrac{-1}{2}$ and the line passes through (1, 2, 3).

$\therefore$ Equations of the line are

$$\frac{x - 1}{1/\sqrt{2}} = \frac{y - 2}{1/2} = \frac{z - 3}{-1/2} \quad \text{or} \quad \frac{x - 1}{\sqrt{2}} = \frac{y - 2}{1} = \frac{z - 3}{-1}$$

Ex. 2 : *Find the equations of the line joining P (2, 1, 1) and Q (1, 3, – 2), and find the point where it meets the plane*
x + 3y + 2z – 6 = 0.

Sol. : Equations to the line PQ, using result (33) are

$$\frac{x - 2}{2 - 1} = \frac{y - 1}{1 - 3} = \frac{z - 1}{1 + 2} \text{ i.e. } \frac{x - 2}{1} = \frac{y - 1}{-2} = \frac{z - 1}{3} = r, \text{ say}$$

$\therefore$ Any point on the line is given by r + 2, – 2r + 1, 3r + 1.

If this lies on the given plane, we must have

$$(r + 2) + 3\,(-2r + 1) + 2\,(3r + 1) - 6 = 0 \quad \therefore \quad r = -1$$

and thus the required point is (1, 3, – 2).

Ex. 3 : *Find the equation of the line through (1, –1, 2) and parallel to the planes x – 2y + z – 2 = 0 and 2x – 3y – 4z + 5 = 0.*

Sol. : Let (l, m, n) be d.r.'s of the line. Since it is parallel to the given planes, it is perpendicular to their normals. Using condition of perpendicularity,

$$l - 2m + n = 0 \quad \text{and} \quad 2l - 3m - 4n = 0$$

$$\therefore \qquad \frac{l}{8 + 3} = \frac{m}{2 + 4} = \frac{n}{-3 + 4} \quad \text{or} \quad \frac{l}{11} = \frac{m}{6} = \frac{n}{1}$$

Equation of the line, therefore can be written as,

$$\frac{x - 1}{11} = \frac{y + 1}{6} = \frac{z - 2}{1}$$

Ex. 4 : *Find the equation of the line through (0, 2, 1) and parallel to the line 3x + 2y – z – 4 = 0 = x – 2y – 2z – 5.*

Sol. : Let (l, m, n) be d.r.'s of the line. It is parallel to the line of section of given two planes and therefore perpendicular to the normals to the given two planes.

$$\therefore \qquad 3l + 2m - n = 0 \qquad \text{and} \qquad l - 2m - 2n = 0$$

$$\therefore \qquad \frac{l}{-4 - 2} = \frac{m}{-1 + 6} = \frac{n}{-6 - 2} \quad \text{or} \quad \frac{l}{6} = \frac{m}{-5} = \frac{n}{8}$$

Equation of the required line is

$$\frac{x}{6} = \frac{y - 2}{-5} = \frac{x - 1}{8}$$

Ex. 5 : *Show that the lines*

$$\frac{x - 1}{-1} = \frac{y - 8}{7} = \frac{z - 2}{2} \quad and \quad \frac{x + 1}{1} = \frac{y - 2}{-1} = \frac{z + 4}{1}$$

are coplanar and find the equation of plane containing them.

Sol. : The equation of the plane containing the first line and parallel to second by result (35) is

$$\begin{vmatrix} x - 1 & y - 8 & z - 2 \\ -1 & 7 & 2 \\ 1 & -1 & 1 \end{vmatrix} = 0$$

or $\qquad (x - 1)\,(9) - (y - 8)\,(-3) + (z - 2)\,(-6) = 0$

or $\qquad 3x + y - 2z - 7 = 0$

Clearly this plane is satisfied by the point (–1, 2, –4) which lies on the second line. Thus, the given lines are coplanar.

EXERCISE 6.2

1. Find the projection of the join of points (1, 2, 1) and (–1, 0, 2) on the line joining the points (0, 1, 3), (5, 0, 2). **Ans. :** $\left(-\dfrac{3}{\sqrt{3}}\right)$

2. Show that the triangle whose vertices are the points (1, 4, 2), (– 2, 1, 2), (2, – 3, 4) is a right angled triangle. Find also the other angles. **Ans. :** $\left(\cos^{-1}\left(\dfrac{1}{\sqrt{3}}\right);\ \cos^{-1}\left(\sqrt{\dfrac{2}{3}}\right)\right)$

3. A line makes angles $\alpha, \beta, \gamma, \delta$ with the four diagonals of a cube. Show that $\cos^2 \alpha + \cos^2 \beta + \cos^2 \gamma + \cos^2 \delta = \dfrac{4}{3}$

4. Find the ratio in which the line joining the points (2, 4, 5), (3, 5, – 4) is divided by the XY-plane. **Ans. :** ratio 5 : 4

5. Prove that the two lines whose direction cosines are connected by the two relations $al + bm + cn = 0$ and $ul^2 + vm^2 + wn^2 = 0$ are perpendicular if

$a^2 (v + w) + b^2 (w + u) + c^2 (u + v) = 0$ and parallel if $\dfrac{a^2}{u} + \dfrac{b^2}{v} + \dfrac{c^2}{w} = 0$

6. Find the equation of the plane through the points (0, 1, 1), (2, 1, 3), (1, – 2, – 2). **Ans. :** $3x + 4y - 3z - 1 = 0$

7. Find the equation of the plane passing through the line of intersection of the planes $x - 2y + 4z = 5$, $2x + y - 4z - 3 = 0$ and perpendicular to the plane $3x - 4y + 5z - 5 = 0$. **Ans. :** $80x - 5y - 52z - 183 = 0$

8. Find the condition that the lines with d.c.'s (l_1, m_1, n_1), (l_2, m_2, n_2) and (l_3, m_3, n_3) are coplanar.

Ans. : $\begin{vmatrix} l_1 & m_1 & n_1 \\ l_2 & m_2 & n_2 \\ l_3 & m_3 & n_3 \end{vmatrix} = 0$

6.18 INTERPRETATION OF SIMPLE LOCI

The Equation to a Surface : Any equation involving one or more of the current co-ordinates of a variable point represents a *surface* or system of surfaces which is the locus of the variable point.

We shall now discuss various simple loci represented by equations in three dimensions :

(A) Equation Containing One Variable :

The locus of all points whose x-coordinates are equal to a constant α, is a plane parallel to the plane yoz and the equation $x = \alpha$ represents that plane. If the equation $f(x) = 0$ has roots $\alpha_1, \alpha_2, \alpha_3, \dots \alpha_n$, it is equivalent to the equations $x = \alpha_1, x = \alpha_2, \dots x = \alpha_n$ and therefore, represents a system of planes (real or imaginary) parallel to the plane yoz. Similarly, $f(y) = 0$, $f(z) = 0$ represent system of planes parallel to zox, xoy respectively.

Note :

(1) Equations $x = \alpha$, $y = \beta$ together represent a line of intersection of these planes which is parallel to z-axis and passes through $(\alpha, \beta, 0)$.

(2) In spherical polar co-ordinates, $f(r) = 0$ represents a system of spheres with common centre at the origin, $f(\theta) = 0$, a system of coaxial right circular cone whose axis is the z-axis, $f(\phi) = 0$, a system of planes passing through z-axis.

(B) Equation Involving Two Variables :

Consider the equation $f(x, y) = 0$. This equation is satisfied by the co-ordinates of all points of the curve in the plane xoy whose two-dimensional equation is $f(x, y) = 0$.

Let $P(x_0, y_0, 0)$ be any point on the curve C (i.e. $f(x, y) = 0$). Through P, draw a line parallel to z-axis, and let Q (x_0, y_0, z) be any point on it. Then the co-ordinates of Q (x_0, y_0, z) satisfy the equation $f(x, y) = 0$ (since P is on the curve $f(x_0, y_0) = 0$) for any value of z. Therefore the co-ordinates of every point on PQ satisfy the equation and every point on PQ lies on the locus of the equation. But P is any point of the curve, therefore, the locus of the equation $f(x, y) = 0$ is the cylinder generated by straight lines drawn parallel to z-axis through points of the curve. Similarly, $f(y, z) = 0$, $f(z, x) = 0$ represent cylinders generated by lines parallel to x-axis and y-axis respectively.

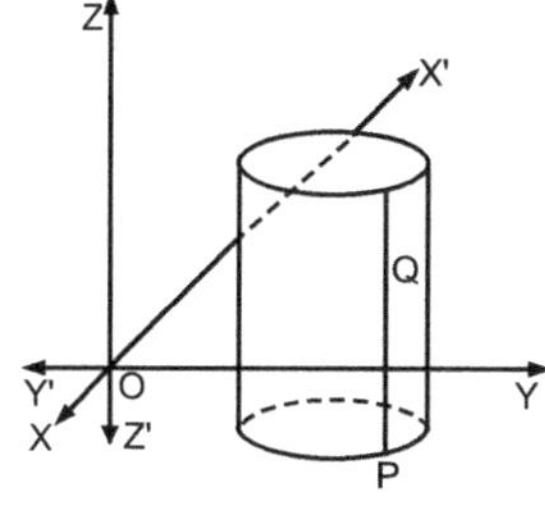

Fig. 6.14

Note : (1) Section of a cylinder by the plane xoy is the curve C.

(2) We note that in three dimensions, two equations are necessary to determine the curve in the xoy plane. The curve is on the cylinder whose equation is $f(x, y) = 0$ and on the plane whose equation is $z = 0$, and hence "the equations to the curve" are $f(x, y) = 0$, $z = 0$.

Ex. : (i) $x^2 + y^2 = a^2$, is a circular cylinder with generators parallel to z-axis whose section with xoy plane is a circle with a as its radius.

(ii) $y^2 = 4ax$, $\dfrac{x^2}{a^2} + \dfrac{y^2}{b^2} = 1$ are parabolic and elliptic cylinders.

(C) Equations Involving Three Variables :

Consider now the equation $f(x, y, z) = 0$. The equation $z = k$ represents a plane parallel to xoy. The equation $f(x, y, k) = 0$ represents, as   proved above, a cylinder generated by lines parallel to z-axis. Thus the equation $f(x, y, k) = 0$ is satisfied by all points which simultaneously satisfy the equation $f(x, y, z) = 0$ and $z = k$. (i.e. at all points common to the plane and the locus of the equation $f(x, y, z) = 0$).

Hence $f(x, y, k) = 0$ represents the cylinder generated by lines parallel to z-axis which pass through the common points (Refer Fig. 6.15) of $f(x, y, z) = 0$ and $z = k$.

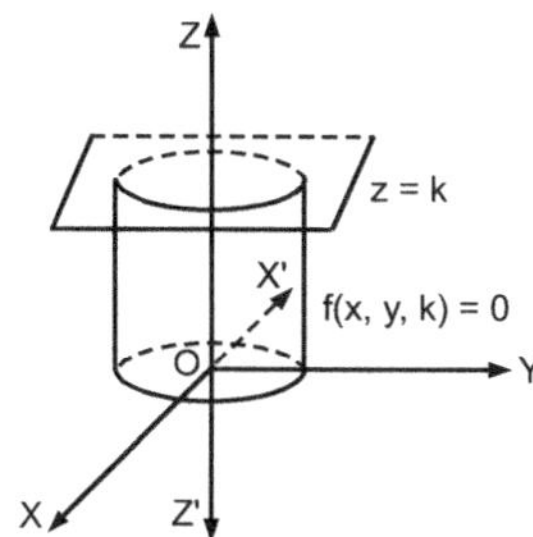

Fig. 6.15

The two equations $f(x, y, k) = 0$, $z = k$ represent the curve of section of cylinder by the plane. $z = k$, which is the curve of section of the locus by the plane $z = k$. If k varies from $-\infty$ to $+\infty$, the curve $f(x, y, k) = 0$, $z = k$, varies continuously and generates a surface.

The co-ordinates of every point on this surface satisfy the equation $f(x, y, z) = 0$, for they satisfy, for some value of k, $f(x, y, k) = 0$, $z = k$; and any point (x_1, y_1, z_1) whose co-ordinates satisfy $f(x, y, z) = 0$ lies on the surface for the co-ordinates satisfy $f(x, y, z_1) = 0$, $z = z_1$, and therefore, the point is on one of the curves which generate surface. Hence $f(x, y, z) = 0$ represents a surface and the surface is the locus of a variable point whose co-ordinates satisfy the equation. In what follows we shall next discuss some standard surfaces.

Quadratic Surfaces (or a Conicoid)

The surface represented by general equation of the second degree in x, y, z is called a **Quadratic Surface** or a **Conicoid**.

Thus the general equation of quadratic surface is of the form

$$ax^2 + by^2 + cz^2 + 2fyz + 2gx + 2hxy + 2ux + 2vy + 2wz + d = 0$$

which can be reduced to one of the following standard forms.

(a) The Ellipsoid :

The equation to the surface is given by,

$$\frac{x^2}{a^2} + \frac{y^2}{b^2} + \frac{z^2}{c^2} = 1 \qquad \qquad \text{... (36)}$$

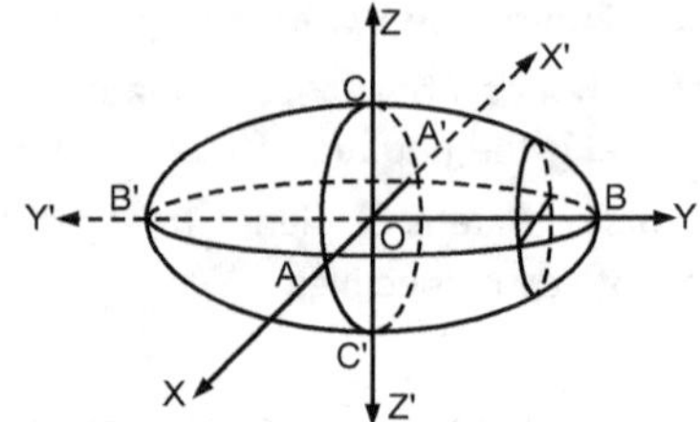

Fig. 6.16

The sections of the surface by the co-ordinate planes x = 0, y = 0, z = 0 are ellipses given by

$$\frac{y^2}{b^2} + \frac{z^2}{c^2} = 1, \quad \frac{x^2}{a^2} + \frac{z^2}{c^2} = 1, \quad \text{and} \quad \frac{x^2}{a^2} + \frac{y^2}{b^2} = 1$$

as shown in Fig. 6.16.

These are known as principal sections.

The surface meets the x-axis (i.e. y = 0, z = 0) at A(a, 0, 0), A'(–a, 0, 0); the y-axis (i.e. x = 0, z = 0) at B(0, b, 0), B'(0, –b, 0); the z-axis at C(0, 0, c), C'(0, 0, – c). The lengths OA (= a), OB (= b) and OC (= c) are known as principal semi-axes.

Next consider the section by the plane z = k which has equations

$$\frac{x^2}{a^2} + \frac{y^2}{b^2} = 1 - \frac{k^2}{c^2} \; ; \; z = k$$

The section is therefore a real ellipse if $k^2 < c^2$, is imaginary if $k^2 > c^2$ and reduces to a point if $k^2 = c^2$. The surface is therefore generated by a **Variable Ellipse** whose plane is parallel to xoy and whose centre is on z-axis called **Ellipsoid**. As k varies, centre of the ellipse moves along z-axis and the ellipse increases from point z = – c in the plane xoy which is given by $\frac{x^2}{a^2} + \frac{y^2}{b^2} = 1$, and then decreases to a point in the plane z = c. Similar interpretations hold for sections by planes parallel to yoz and zox.

Remarks :

1. Since by changing the signs of x, y, z, the surface equation is unaltered (or only even powers of x, y, z occur in surface equation) the surface is symmetrical about co-ordinate planes and the co-ordinate axes.

2. Section by any plane is ellipse.

3. If a = b, we have $\frac{x^2 + y^2}{b^2} + \frac{z^2}{c^2} = 1$. This is the equation of an ellipsoid of revolution generated by revolving the ellipse $\frac{y^2}{b^2} + \frac{z^2}{c^2} = 1$ about z-axis. If c > b, the axis of rotation (z-axis) is the major axis and the surface is termed as **Polate Spheroid** and if c < b, the axis of rotation is minor axis and the surface is known as **Oblate Spheroid.**

(b) The Hyperboloids :

The equations to hyperboloid of one sheet and two sheets are given by

$$\frac{x^2}{a^2} + \frac{y^2}{b^2} - \frac{z^2}{c^2} = 1 \qquad \qquad \text{... (37)}$$

$$\frac{x^2}{a^2} + \frac{y^2}{b^2} - \frac{z^2}{c^2} = -1 \qquad \qquad \text{... (38)}$$

respectively.

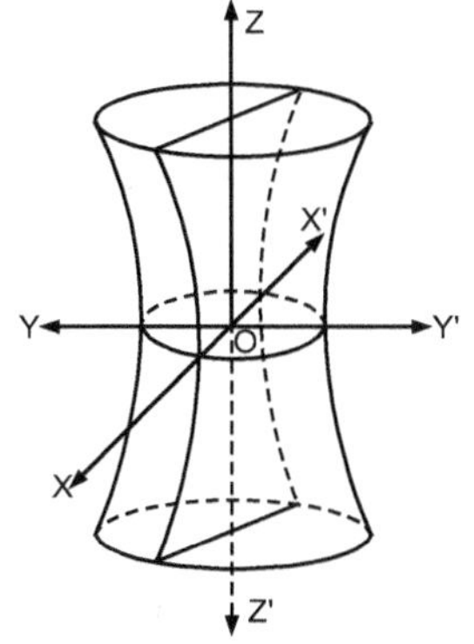

Fig. 6.17 : Hyperboloid of One Sheet

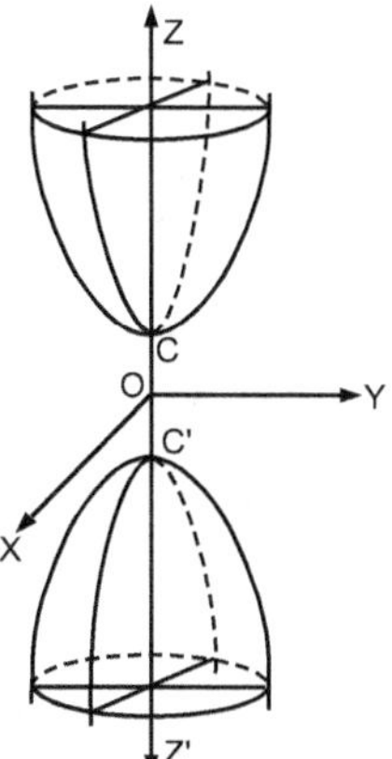

Fig. 6.18 : Hyperboloid of Two Sheets

The sections of the surface (37) by the co-ordinate plane (x = 0) is the hyperbola $\dfrac{y^2}{b^2} - \dfrac{z^2}{c^2} = 1$; co-ordinate plane (y = 0) is the hyperbola $\dfrac{x^2}{a^2} - \dfrac{z^2}{c^2} = 1$ and section by co-ordinate plane (z = 0) is the ellipse $\dfrac{x^2}{a^2} + \dfrac{y^2}{b^2} = 1$.

The surface meets the x-axis at A (a, 0, 0), A' (– a, 0, 0); the y-axis at B (0, b, 0), B' (0, – b, 0); and z-axis in imaginary points.

Taking section by any plane z = k, we have

$$\frac{x^2}{a^2} + \frac{y^2}{b^2} = 1 + \frac{k^2}{c^2} ; \quad z = k$$

which represents a real ellipse. Thus surface represented by equation (37) is generated by a variable ellipse (growing larger and larger as k varies from – ∞ to ∞) whose centre moves on z-axis, passing in turn through every point on it. Thus surface given by equation (37) is called **Hyperboloid of One Sheet** and is represented in Fig. 6.17, which is like that of Juggler's dabru.

The sections of the surface (38) by the co-ordinate plane yoz is the hyperbola $\dfrac{z^2}{c^2} - \dfrac{y^2}{b^2} = 1$; by the coordinate plane zox is the hyperbola $\dfrac{z^2}{c^2} - \dfrac{x^2}{a^2} = 1$ and section by plane xoy is the imaginary ellipse $\dfrac{x^2}{a^2} + \dfrac{y^2}{b^2} = -1$.

The surface meets the z-axis at C(0, 0, c), C'(0, 0, – c) and the x and y-axes in imaginary points.

The surface given by equation (38) is also generated by a variable ellipse whose centre moves on z-axis. The ellipse is given by

$$\frac{x^2}{a^2} + \frac{y^2}{b^2} = \frac{k^2}{c^2} - 1, \quad z = k.$$

which is real ellipse, if $|k| > c$ (i.e. k varies from – ∞ to – c and c to + ∞ and extends to infinity on both sides) and is imaginary ellipse if $|k| < c$ (i.e. c < k < c) and hence no part of the surface lies between the planes z = ± c. Thus surface given by equation (38) is called **Hyperboloid of Two Sheets** and is represented by two portions as shown in Fig. 6.18.

(c) Paraboloid :

The equations to the elliptic paraboloid and hyperbolic paraboloid are given by

$$\frac{x^2}{a^2} + \frac{y^2}{b^2} = \frac{2z}{c} \tag{39}$$

$$\frac{x^2}{a^2} - \frac{y^2}{b^2} = \frac{2z}{c} \tag{40}$$

respectively.

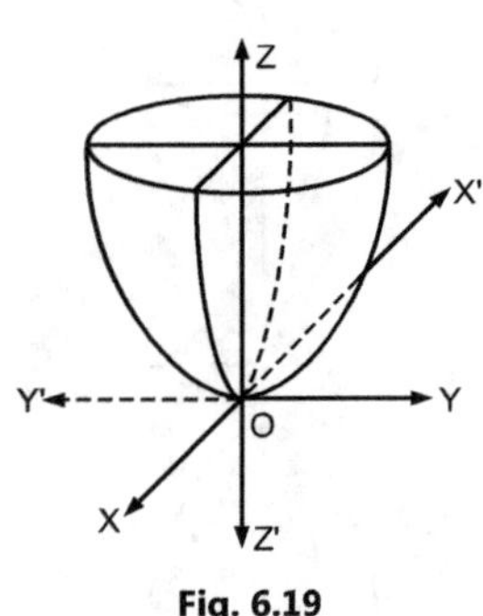

Fig. 6.19

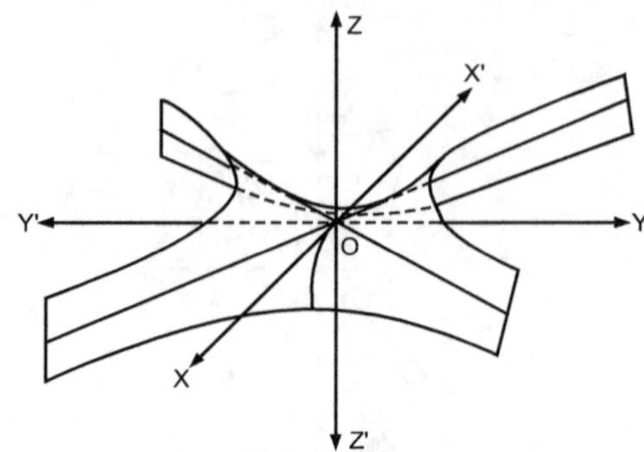

Fig. 6.20

The equation (39) represents the surface generated by the variable ellipse $\dfrac{x^2}{a^2} + \dfrac{y^2}{b^2} = \dfrac{2k}{c}$, z = k. This ellipse is imaginary unless k and c have the same sign, hence the centre of the ellipse lies on oz, if c > 0 and on oz' if c < 0. The sections of the surface by planes parallel to the coordinate planes yoz and zox are parabolas $y^2 = \dfrac{2b^2 z}{c}$, x = 0 and $x^2 = \dfrac{2a^2 z}{c}$, y = 0. Fig. 6.19 shows the form and position of the surface for a positive value of c. The surface is called the **Elliptic Paraboloid.**

The equation (40) represents the surface generated by the variable hyperbola $\dfrac{x^2}{a^2} - \dfrac{y^2}{b^2} = \dfrac{2k}{c}$, z = k, which is real for all real values of k and its centre passes in turn through every point on z'z. When k = 0, the hyperbola degenerates into the two lines $\dfrac{x^2}{a^2} - \dfrac{y^2}{b^2} = 0$, z = 0. The projections of sections of the surface by planes z = k, z = – k on xoy plane are conjugate hyperbolas with their asymptotes given by $\dfrac{x^2}{a^2} - \dfrac{y^2}{b^2} = 0$, z = 0. The sections by planes parallel to yoz and zox are parabolas. The surface (40) is called the **Hyperbolic Paraboloid** and Fig. 6.20 shows the form and position of the surface for a negative value of c.

7.1 DEFINITION

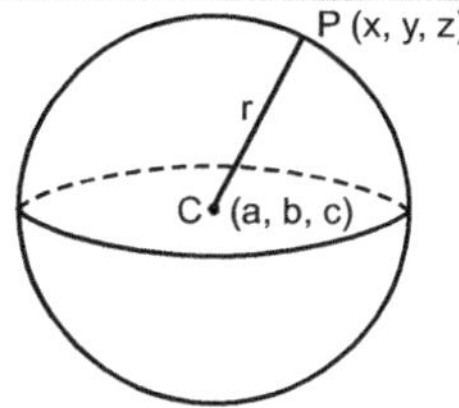

Fig. 7.1

The locus of a point which moves in space such that it remains at a constant distance from a fixed point defines a spherical surface or a sphere. The fixed point is called the **centre** of the sphere and the constant distance is called the **radius**.

7.2 EQUATIONS OF SPHERE IN DIFFERENT FORMS

(A) Centre and Radius Form : Let P(x, y, z) be any point on the locus (sphere), C(a, b, c) be the centre of the sphere and r be the radius.

Then by distance formula (Art. 6.6 (4)), we have

$$(x - a)^2 + (y - b)^2 + (z - c)^2 = r^2$$ ... (1)

This being true for every point on the sphere (Fig. 7.1) represents the equation of the sphere in centre and radius form.

Particular Case : If the centre is at the origin O, then the equation (1) takes the form

$$x^2 + y^2 + z^2 = r^2$$ ... (2)

which is called the **Standard Form** of the equation of the sphere.

(B) General Form : The equation of the sphere in centre and radius form is

$$(x - a)^2 + (y - b)^2 + (z - c)^2 = r^2$$

Rewriting above equation as

$$x^2 + y^2 + z^2 - 2ax - 2by - 2cz + (a^2 + b^2 + c^2 - r^2) = 0$$

Put $a = -u$, $b = -v$, $c = -w$ and $a^2 + b^2 + c^2 - r^2 = d$.

Then the equation of the sphere takes the form

$$x^2 + y^2 + z^2 + 2ux + 2vy + 2wz + d = 0$$ ... (3)

which is known as the general form of the equation of the sphere.

Its centre is $(-u, -v, -w)$ and radius is given by;

$$r^2 = a^2 + b^2 + c^2 - d$$
$$= u^2 + v^2 + w^2 - d$$
$$\therefore \quad r = \sqrt{u^2 + v^2 + w^2 - d}$$

Note :

1. If $u^2 + v^2 + w^2 > d$, equation (3) represents a sphere with centre $(-u, -v, -w)$ and real radius.

2. If $u^2 + v^2 + w^2 = d$, equation (3) represents a sphere with centre $(-u, -v, -w)$ and radius zero i.e. the sphere coincides with the centre. Such a sphere is called a *point sphere*.

3. If $u^2 + v^2 + w^2 < d$, equation (3) represents a sphere with centre $(-u, -v, -w)$ and imaginary radius.

4. Equation (3) is a second degree equation in x, y, z with coefficients of x^2, y^2, z^2 as equal and terms xy, yz, and zx are absent. It contains four arbitrary constants u, v, w and d, determination of which requires four conditions.

(C) Intercept Form : To Find the Equation of the Sphere Which Cuts-off Intercepts a, b, c From ox, oy, oz Axes :

Consider the sphere,

$$x^2 + y^2 + z^2 + 2ux + 2vy + 2wz + d = 0 \qquad \text{... (i)}$$

To determine the sphere we require four conditions. Let us take the fourth condition as sphere passing through the origin i.e. (0, 0, 0).

$\because$ (0, 0, 0) satisfies equation (i), putting x = 0, y = 0, z = 0 in (i), we get d = 0.

$\therefore$ Equation (i) takes the form

$$x^2 + y^2 + z^2 + 2ux + 2vy + 2wz = 0 \qquad \text{... (ii)}$$

which is the general equation of the sphere passing through the origin.

$\because$ It cuts-off intercept 'a' from x-axis, (a, 0, 0) should satisfy equation (ii).

Putting x = a, y = 0, z = 0 in (ii), we get $a^2 + 2ua = 0$ $\therefore$ $u = -\dfrac{a}{2}$

Similarly, since the sphere cuts-off intercepts b from y-axis and c from z-axis, we get $v = -\dfrac{b}{2}$, $w = -\dfrac{c}{2}$. Putting for u, v, w in (ii),

we get

$$\boxed{x^2 + y^2 + z^2 - ax - by - cz = 0} \qquad \text{... (4)}$$

which is the required equation of the sphere in intercept form.

(D) Diameter Form : To Find the Equation of the Sphere Described on the Join of Two Given Points as Diameter :

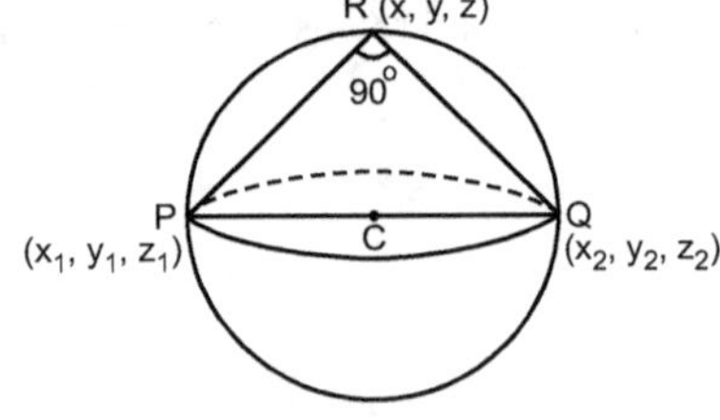

Fig. 7.2

Consider the sphere described on the line joining two points $P(x_1, y_1, z_1)$, $Q(x_2, y_2, z_2)$ as diameter (Fig. 7.2). Let R (x, y, z) be any point on the sphere, then d.r.'s of PR are $(x - x_1, y - y_1, z - z_1)$ and of QR are $(x - x_2, y - y_2, z - z_2)$.

$\because$ $\angle$ PRQ = Angle in a semicircle = 90°

$\therefore$ PR $\perp$ QR

Using condition of perpendicularity, we get

$$\boxed{(x - x_1)(x - x_2) + (y - y_1)(y - y_2) + (z - z_1)(z - z_2) = 0} \qquad \text{... (5)}$$

which is the required equation of the sphere in diameter form.

Note : (1) Expansion of above equation (5) leads to general form of equation (3) of the sphere.

(2) Equation of the sphere can also be obtained by finding the coordinates of centre 'C' as $\left(\dfrac{x_1 + x_2}{2}, \dfrac{y_1 + y_2}{2}, \dfrac{z_1 + z_2}{2}\right)$ and finding the radius i.e. the distance CP or CQ. Once the centre and radius are obtained, form (1) of 7.2 can be used to obtain the equation of the sphere.

(E) Four-Point Form : To Find Equation of the Sphere Passing through Four Given Points : Let the co-ordinates of the four given points be (x_1, y_1, z_1), (x_2, y_2, z_2), (x_3, y_3, z_3), (x_4, y_4, z_4).

Let the required equation of the sphere be

$$(x^2 + y^2 + z^2) + 2ux + 2vy + 2wz + d = 0 \qquad \text{... (i)}$$

Since this sphere passes through four given points, we have

$$(x_1^2 + y_1^2 + z_1^2) + 2ux_1 + 2vy_1 + 2wz_1 + d = 0 \qquad \text{... (ii)}$$

$$(x_2^2 + y_2^2 + z_2^2) + 2ux_2 + 2vy_2 + 2wz_2 + d = 0 \qquad \text{... (iii)}$$

$$(x_3^2 + y_3^2 + z_3^2) + 2ux_3 + 2vy_3 + 2wz_3 + d = 0 \qquad \text{... (iv)}$$

$$(x_4^2 + y_4^2 + z_4^2) + 2ux_4 + 2vy_4 + 2wz_4 + d = 0 \qquad \text{... (v)}$$

Eliminating unknowns u, v, w and d from above five equations, we get

$$\begin{vmatrix} x^2 + y^2 + z^2 & x & y & z & 1 \\ x_1^2 + y_1^2 + z_1^2 & x_1 & y_1 & z_1 & 1 \\ x_2^2 + y_2^2 + z_2^2 & x_2 & y_2 & z_2 & 1 \\ x_3^2 + y_3^2 + z_3^2 & x_3 & y_3 & z_3 & 1 \\ x_4^2 + y_4^2 + z_4^2 & x_4 & y_4 & z_4 & 1 \end{vmatrix} = 0 \qquad \ldots (6)$$

Note : In numerical problems to find the equation of sphere through four points, it is convenient to find centre and radius, which gives from (1), directly the equation of the sphere in centre and radius form.

7.3 TOUCHING SPHERES

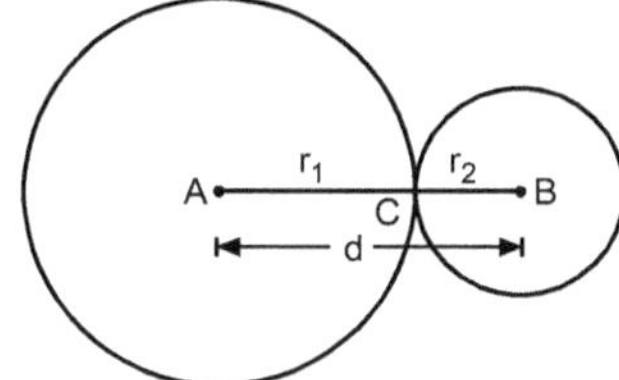

Fig. 7.3

(a) Two spheres touch externally if the distance between their centres is equal to the sum of their radii, i.e. $d = r_1 + r_2$.

The point of contact is the point which divides internally the line, joining the centres in the ratio of radii.

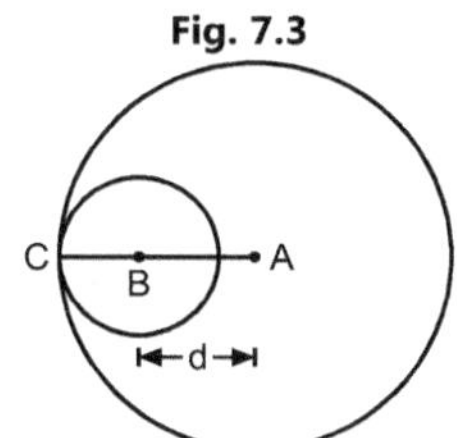

Fig. 7.4

(b) Two spheres touch internally if the distance between their centres is equal to the difference of their radii, i.e. $d = r_1 - r_2$.

The point of contact is the point which divides externally the line joining the centres in the ratio of radii.

7.4 ILLUSTRATIONS ON DIFFERENT FORMS OF EQUATIONS OF SPHERE AND LOCUS PROBLEMS

Ex. 1 : *Find the equation of the sphere whose centre is (–3, 1, 4) and radius is 5. Also write down the centre and radius of the sphere*

$$2x^2 + 2y^2 + 2z^2 - x + 3y - 5z = 10.$$

Sol. : By centre-radius form [article 7.2, (1)], the required equation of the sphere is

$$(x + 3)^2 + (y - 1)^2 + (z - 4)^2 = (5)^2$$

or $\qquad x^2 + y^2 + z^2 + 6x - 2y - 8z + 1 = 0$

For the second part, we note that the equation of the sphere can be written as

$$x^2 + y^2 + z^2 - \frac{1}{2} x + \frac{3}{2} y - \frac{5}{2} z - 5 = 0$$

Comparing with the general equation $x^2 + y^2 + z^2 + 2ux + 2vy + 2wz + d = 0$

we get $\quad u = -\dfrac{1}{4}, \quad v = \dfrac{3}{4}, \quad w = -\dfrac{5}{4}, \quad d = -5$

Hence the required centre is

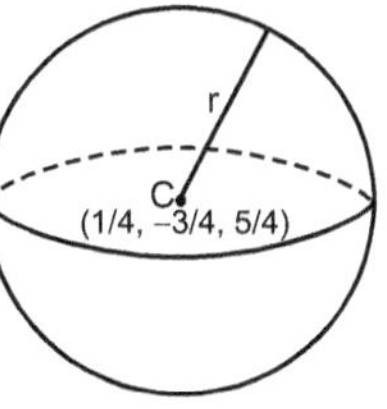
Fig. 7.5

$$\boxed{(-u, -v, -w) = \left(\frac{1}{4}, -\frac{3}{4}, \frac{5}{4}\right)}$$

and the required radius is

$$r = \sqrt{u^2 + v^2 + w^2 - d} = \sqrt{\frac{1}{16} + \frac{9}{16} + \frac{25}{16} + 5}$$

or $\qquad \boxed{r = \dfrac{\sqrt{115}}{4}}$.

Ex. 2 : *Find the equation of the sphere passing through the points (1, 0, –1), (2, 1, 0), (1, 1, –1) and (1, 1, 1).*

Sol. : Let (a, b, c) be the centre of the sphere.

All the points on sphere are equidistant from the centre.

$$\therefore \quad (a-1)^2 + b^2 + (c+1)^2 = (a-2)^2 + (b-1)^2 + c^2$$
$$= (a-1)^2 + (b-1)^2 + (c+1)^2$$
$$= (a-1)^2 + (b-1)^2 + (c-1)^2 \quad \dots (1)$$

This gives us three equations as

$$2a + 2b + 2c - 3 = 0, \quad 2b - 1 = 0, \quad 2b + 4c - 1 = 0$$

Solving these equations, we get $a = 1$, $b = \dfrac{1}{2}$, $c = 0$

Radius of the sphere = Distance of centre (a, b, c) from any one point, say (1, 1, –1) on the sphere

$$= \sqrt{(a-1)^2 + (b-1)^2 + (c+1)^2}$$
$$= \sqrt{\frac{1}{4} + 1} = \frac{\sqrt{5}}{2} \quad \dots (2)$$

From (1) and (2), equation of the sphere is,

$$(x-1)^2 + \left(y - \frac{1}{2}\right)^2 + z^2 = \frac{5}{4}$$

or
$$x^2 - 2x + 1 + y^2 - y + \frac{1}{4} + z^2 = \frac{5}{4}$$

or
$$\boxed{x^2 + y^2 + z^2 - 2x - y = 0}$$

which is the required equation of the sphere.

Ex. 3 : *Find the equation of the sphere through the points (4, –1, 2), (0, –2, 3), (1, 5, –1),   (2, 0, 1).*

Sol. : We can solve this problem by the above method. However, another method of solving this problem is as follows :

Let the required equation of the sphere be

$$x^2 + y^2 + z^2 + 2ux + 2vy + 2wz + d = 0 \quad \dots (1)$$

$\because$ It passes through (4, –1, 2),

$\therefore \qquad\qquad 16 + 1 + 4 + 8u - 2v + 4w + d = 0$

or $\qquad\qquad\qquad 21 + 8u - 2v + 4w + d = 0 \qquad \dots (2)$

Similarly, as the sphere passes through (0, –2, 3), (1, 5, –1), (2, 0, 1)

$\therefore \qquad\qquad\qquad 13 - 4v + 6w + d = 0 \qquad \dots (3)$

and $\qquad\qquad 27 + 2u + 10v - 2w + d = 0 \qquad \dots (4)$

and $\qquad\qquad\quad 5 + 4u + 2w + d = 0 \qquad \dots (5)$

Subtracting (5) from each of (2), (3) and (4) (to eliminate d), we get

$$16 + 4u - 2v + 2w = 0 \ \text{ or } \ 8 + 2u - v + w = 0 \qquad \dots (6)$$
$$8 - 4u - 4v + 4w = 0 \ \text{ or } \ 2 - u - v + w = 0 \qquad \dots (7)$$
$$22 - 2u + 10v - 4w = 0 \ \text{ or } \ 11 - u + 5v - 2w = 0 \qquad \dots (8)$$

Solving (6), (7) and (8), we get $u = -2$, $v = -7$, $w = -11$. Also from (3), $d = 25$.

Substituting these values in (1), the required equation of the sphere is

$$\boxed{x^2 + y^2 + z^2 - 4x - 14y - 22z + 25 = 0.}$$

Ex. 4 :   *Find the equation of the sphere which passes through the points (1, –4, 3), (1, –5, 2), (1, –3, 0) and whose centre lies on the plane x + y + z = 0.*   **(May 2014)**

Sol. : Let the equation of the sphere be

$$x^2 + y^2 + z^2 + 2ux + 2vy + 2wz + d = 0 \quad \dots (1)$$

$\because$ It passes through (1, –4, 3), (1, –5, 2), (1, –3, 0)

$\therefore$ $1 + 16 + 9 + 2u - 8v + 6w + d = 0$ or $26 + 2u - 8v + 6w + d = 0$ $\dots (2)$

and $1 + 25 + 4 + 2u - 10v + 4w + d = 0$ or $30 + 2u - 10v + 4w + d = 0$ $\dots (3)$

and $1 + 9 + 0 + 2u - 6v + d = 0$ or $10 + 2u - 6v + d = 0$ ...(4)

Also, centre of the given sphere $(-u, -v, -w)$ lies on the plane $x + y + z = 0$.

$\therefore$ $-u - v - w = 0$ or $u + v + w = 0$ ... (5)

Solving equation (2), (3), (4) and (5), we get $u = -2$, $v = \dfrac{7}{2}$, $w = -\dfrac{3}{2}$ and $d = 15$.

Substituting these values of u, v, w and d in (1), the required equation of the sphere is

$$\boxed{x^2 + y^2 + z^2 - 4x + 7y - 3z + 15 = 0.}$$

Ex. 5 : *Find the equation of the sphere which touches the co-ordinate axes, whose centre is in the positive octant and has radius 4.*

(Dec. 2006)

Sol. : Let the equation of the sphere be

$$x^2 + y^2 + z^2 + 2ux + 2vy + 2wz + d = 0 \qquad ... (1)$$

Since its radius is 4, we have

$$u^2 + v^2 + w^2 - d = 16 \qquad ... (2)$$

Equations of x-axis are $y = 0$, $z = 0$. Putting in equation (1), we get the points of intersection of the sphere and x-axis as,

$$x^2 + 2ux + d = 0$$

which gives two roots in x, corresponding to the two points of intersection. As x-axis touches the sphere, these two roots must be equal, condition for which is

$$4u^2 = 4d \qquad \therefore u^2 = d$$

Similarly, we shall get $v^2 = d$, $w^2 = d$ as y- and z-axes also touch the sphere.

Substituting in (2), we get

$$2d = 16 \qquad \therefore d = 8$$

$\therefore$ $u^2 = v^2 = w^2 = 8$ or $u = v = w = \pm 2\sqrt{2}$

Since centre of the sphere $(-u, -v, -w)$ is given to be in positive octant, we take

$u = v = w = -2\sqrt{2}$, so that centre is $\left(2\sqrt{2}, 2\sqrt{2}, 2\sqrt{2}\right)$.

Substituting for u, v, w and d in (1), we get the equation of the sphere as

$$\boxed{x^2 + y^2 + z^2 - 4\sqrt{2}\,(x + y + z) + 8 = 0.}$$

Ex. 6 : *Find the equation of the sphere, which passes through the points (1, 0, 0), (0, 2, 0), (0, 0, 3) and has its radius as small as possible.*

Sol. : Let the equation of the sphere be

$$x^2 + y^2 + z^2 + 2ux + 2vy + 2wz + k = 0 \qquad ...(1)$$

Since it passes through the three given points, the co-ordinates of these points must satisfy (1), therefore we have respectively

$$1 + 2u + k = 0, \qquad 4 + 4v + k = 0, \qquad 9 + 6w + k = 0$$

These give $u = -\left(\dfrac{k+1}{2}\right)$, $v = -\left(\dfrac{k+4}{4}\right)$, $w = -\left(\dfrac{k+9}{6}\right)$.

The radius of the sphere (1) is given by,

$$r = \sqrt{u^2 + v^2 + w^2 - k}$$

$$= \sqrt{\frac{(k+1)^2}{4} + \frac{(k+4)^2}{16} + \frac{(k+9)^2}{36} - k}$$

$$= \sqrt{\frac{36(k+1)^2 + 9(k+4)^2 + 4(k+9)^2 - 144k}{144}}$$

$$= \frac{1}{12}\sqrt{49k^2 + 72k + 504}$$

For radius r to be as small as possible, we use the condition $\dfrac{dr}{dk} = 0$.

Thus,
$$\frac{dr}{dk} = \frac{1}{12} \cdot \frac{1\,(98k + 72)}{2\sqrt{49k^2 + 72k + 504}} = 0 \qquad \therefore \quad k = -\frac{36}{49}$$

$$\therefore \quad u = -\left(\frac{-\frac{36}{49} + 1}{2}\right) = -\frac{13}{98}, \quad v = -\left(\frac{-\frac{36}{49} + 4}{4}\right) = -\frac{40}{49}, \quad w = -\left(\frac{-\frac{36}{49} + 9}{6}\right) = -\frac{135}{98}$$

Substituting for u, v, w, k in (1), we get the required equation of the sphere as
$$x^2 + y^2 + z^2 - \frac{13}{49}\,x - \frac{80}{49}\,y - \frac{135}{49}\,z - \frac{36}{49} = 0$$

or
$$\boxed{49\,(x^2 + y^2 + z^2) - 13x - 80y - 135z - 36 = 0}$$

Ex. 7 : *A sphere with its centre in positive octant passes through the origin and cuts the co-ordinate planes xoy, yoz, zox in circles of radii* $\sqrt{10}, \sqrt{2}, \sqrt{10}$ *respectively. Find the equation of the sphere.*

Sol. : Let the sphere passing through the origin (d = 0) be
$$x^2 + y^2 + z^2 + 2ux + 2vy + 2wz = 0 \qquad \dots (1)$$

The equation of xoy plane is z = 0. Putting z = 0 in (1), we have
$$x^2 + y^2 + 2ux + 2vy = 0$$

which is a circle in xoy plane having radius $\sqrt{u^2 + v^2}$, since d = 0.

But the radius is given to be $\sqrt{10}$

$$\therefore \qquad u^2 + v^2 = 10 \qquad \dots (2)$$

Similarly, for circles in yoz plane (i.e. x = 0) and zox plane (i.e. y = 0), we have
$$v^2 + w^2 = 2 \qquad \dots (3)$$
and $\qquad w^2 + u^2 = 10 \qquad \dots (4)$

Adding (2), (3) and (4), we get
$$u^2 + v^2 + w^2 = 11 \qquad \dots (5)$$

From (2) and (5), $w^2 = 1$. Similarly, $u^2 = 9$, $v^2 = 1$

$$\therefore \qquad u = \pm 3, \qquad v = \pm 1, \qquad w = \pm 1.$$

Since the centre (–u, –v, –w) lies in the positive octant, we choose u = –3, v = –1, w = –1.

Hence from (1), required equation of the sphere is
$$\boxed{x^2 + y^2 + z^2 - 6x - 2y - 2z = 0}$$

Ex. 8 : *Find the equation of the sphere circumscribing the tetrahedron, having faces as* $\dfrac{x}{a} + \dfrac{y}{b} = 0,\ \dfrac{y}{b} + \dfrac{z}{c} = 0,\ \dfrac{x}{a} + \dfrac{z}{c} = 0$

and $\dfrac{x}{a} + \dfrac{y}{b} + \dfrac{z}{c} = 1.$

Sol. : Given faces are

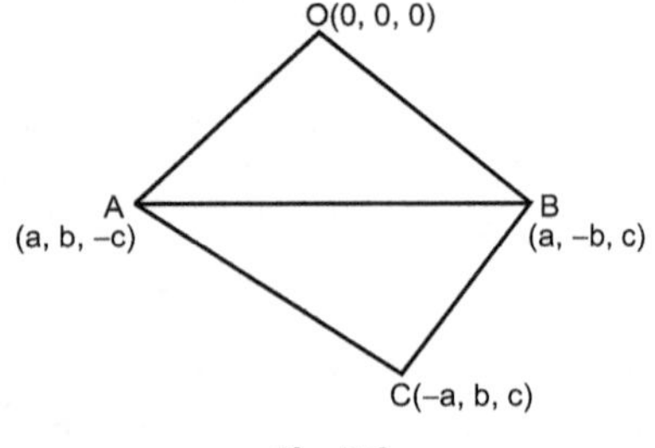

Fig. 7.6

$$\frac{x}{a} + \frac{y}{b} = 0 \qquad \dots (1)$$
$$\frac{y}{b} + \frac{z}{c} = 0 \qquad \dots (2)$$
$$\frac{z}{c} + \frac{x}{a} = 0 \qquad \dots (3)$$
$$\frac{x}{a} + \frac{y}{b} + \frac{z}{c} = 1 \qquad \dots (4)$$

We can get the vertices (O, A, B, C) by solving the equations (1) → (4) of the four faces of the tetrahedron as follows :

(i)   Solve (1), (2), (3) to get the vertex O as O (0, 0, 0)

(ii) Solve (2), (3) and (4) to get point A as A (a, b, –c)

(iii) Solve (2), (1), (4) to get point B as B (a, –b, c)

(iv) Solve (1), (3), (4) to get point C as C (–a, b, c).

Hence vertices of the tetrahedron are $(0, 0, 0)$, $(a, b, - c)$, $(a, - b, c)$ and $(- a, b, c)$.

Let the equation of the required sphere be

$$x^2 + y^2 + z^2 + 2ux + 2vy + 2wz + d = 0 \qquad \dots (5)$$

Since the four points lie on the sphere (5), each point will satisfy the equation. In this process, we get following equations and after solving we get the values of the unknowns u, v, w and d.

Equations are

$$d = 0 \qquad \dots (6)$$

$$a^2 + b^2 + c^2 + 2ua - 2bv + 2cw = 0 \qquad \dots (7)$$

$$a^2 + b^2 + c^2 - 2ua + 2bv + 2cw = 0 \qquad \dots (8)$$

$$a^2 + b^2 + c^2 + 2ua + 2bv - 2cw = 0 \qquad \dots (9)$$

By solving (6) $\to$ (9), we get

$$2u = -\left(\frac{a^2 + b^2 + c^2}{a}\right), \qquad 2v = -\left(\frac{a^2 + b^2 + c^2}{b}\right), \qquad 2w = -\left(\frac{a^2 + b^2 + c^2}{c}\right), \quad d = 0$$

Hence the equation of the sphere will be

$$\boxed{\frac{x^2 + y^2 + z^2}{a^2 + b^2 + c^2} - \frac{x}{a} - \frac{y}{b} - \frac{z}{c} = 0}$$

Ex. 9 : *Find the equation of the sphere inscribed in the tetrahedron having faces $x + y = 0$, $y + z = 0$, $z + x = 0$ and $x + y + z = 1$.*

Sol. : Let (a, b, c) be the centre of the sphere. Since it is equidistant from four faces of the tetrahedron, using formula for length of perpendicular [refer article 7.18 (27)], we have

$$\frac{b + c}{\sqrt{2}} = \frac{c + a}{\sqrt{2}} = \frac{a + b}{\sqrt{2}} = \frac{a + b + c - 1}{\sqrt{3}} \qquad \dots (1)$$

Taking first two ratios, $\qquad b + c = c + a \qquad \therefore \quad a = b \qquad \dots (2)$

Taking second and third ratios, $\quad c + a = a + b \qquad \therefore \quad c = b \qquad \dots (3)$

From (2) and (3), $\qquad a = b = c$

Taking first and fourth ratios, $\dfrac{b + c}{\sqrt{2}} = \dfrac{a + b + c - 1}{\sqrt{3}}$ and using (4), b = a and c = a, we have

$$\frac{2a}{\sqrt{2}} = \frac{3a - 1}{\sqrt{3}} \qquad \therefore \quad a = \frac{3 + \sqrt{6}}{3}$$

The radius r of the sphere = Perpendicular from centre (a, b, c) i.e. (a, a, a) on any one of the face, say $y + z = 0$

$$= \frac{2a}{\sqrt{2}} = \sqrt{2}\, a.$$

Hence, equation of the required sphere in centre-radius form is given by

$$(x - a)^2 + (y - a)^2 + (z - a)^2 = \left(\sqrt{2}\, a\right)^2$$

i.e. $\qquad \boxed{(x^2 + y^2 + z^2) - 2a\, (x + y + z) + a^2 = 0, \text{ where } a = \frac{3 + \sqrt{6}}{3}}$.

Ex. 10 : *Show that the spheres $x^2 + y^2 + z^2 = 25$, $x^2 + y^2 + z^2 - 18x - 24y - 40z + 225 = 0$ touch externally and find their point of contact.*

(Dec. 2018)

Sol. : For the first sphere, centre C_1 is $(0, 0, 0)$ and radius $r_1 = 5$.

For the second sphere, centre C_2 is $(9, 12, 20)$ and radius $r_2 = \sqrt{81 + 144 + 400 - 225} = 20$.

Now distance between the centres is given by;

$$C_1 C_2 = \sqrt{(9 - 0)^2 + (12 - 0)^2 + (20 - 0)^2} = 25 = r_1 + r_2$$

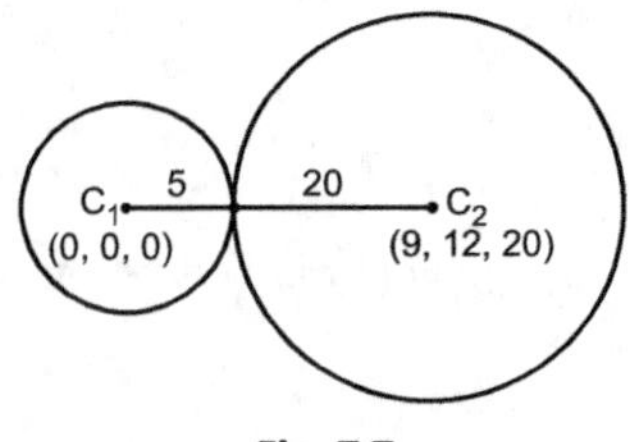

Fig. 7.7

$\therefore$ The two spheres touch externally and point of contact C divides C_1C_2 internally in the ratio 5 : 20 or 1 : 4. Hence the co-ordinates of the point of contact C [refer article 7.7 (5)] are

$$C = \left[\frac{1(9) + 4(0)}{1 + 4} , \frac{1(12) + 4(0)}{1 + 4} , \frac{1(20) + 4(0)}{1 + 4}\right]$$

$$\boxed{C = \left(\frac{9}{5}, \frac{12}{5}, 4\right)} .$$

Ex. 11 : *Find the equation of the sphere which touches the plane* $4x + 3y = 47$ *at the point* (8, 5, 4) *and touches the sphere* $x^2 + y^2 + z^2 = 1$ *internally.*   **(May 2004)**

Sol. :

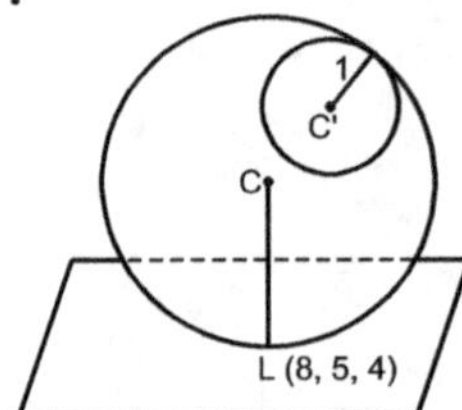

Fig. 7.8

The centre of the required sphere lies on the perpendicular to the plane $4x + 3y = 47$, through the point L (8, 5, 4).

$\therefore$ Equations of CL are $\dfrac{x - 8}{4} = \dfrac{y - 5}{3} = \dfrac{z - 4}{0} = k$ ... (1)

Let co-ordinates of the centre C on this line (1) be (4k + 8, 3k + 5, 4).

$\therefore$ The radius of the sphere (distance between points C and L) is

$$CL = r = \sqrt{(5k + 8 - 8)^2 + (3k + 5 - 5)^2 + (4 - 4)^2} = -5k \qquad \text{(Note)}$$

$\because$ This sphere touches internally the sphere $x^2 + y^2 + z^2 = 1$.

(Centre C' is (0, 0, 0) and radius = 1)

$\therefore$ The radius of the required sphere

$$r = \text{(Distance between centres of the spheres)} + 1$$

$\therefore$ $-5k = \sqrt{(4k + 8)^2 + (3k + 5)^2 + (4)^2} + 1$

or $-5k - 1 = \sqrt{(4k + 8)^2 + (3k + 5)^2 + (4)^2}$

which on squaring gives $k = -\dfrac{26}{21}$.

$\therefore$ Centre of the sphere is $\left(\dfrac{64}{21}, \dfrac{27}{21}, 4\right)$ and the radius is $\dfrac{130}{21}$.

Hence, the required equation of the sphere is

$$\boxed{\left(x - \frac{64}{21}\right)^2 + \left(y - \frac{27}{21}\right)^2 + (z - 4)^2 = \left(\frac{130}{21}\right)^2}$$

LOCUS PROBLEMS

Ex. 12 : *A point moves so that the sum of the squares of its distances from the six faces of a cube is constant. Show that its locus is a sphere.*

Sol. : Take the centre of the cube as origin and planes through the centre parallel to its faces as co-ordinate planes. If 2a is the length of an edge of the cube, then equations of the six faces (three pairs of parallel faces) are x = a, x = –a; y = a, y = –a; z = a, z = –a.

Let $(\bar{x}, \bar{y}, \bar{z})$ be any point on the locus, then sum of squares of distances of point from the six faces is constant = k^2 (say).

$\therefore$ $(\bar{x} - a)^2 + (\bar{x} + a)^2 + (\bar{y} - a)^2 + (\bar{y} + a)^2 + (\bar{z} - a)^2 + (\bar{z} + a)^2 = k^2$

i.e. $2\bar{x}^2 + 2\bar{y}^2 + 2\bar{z}^2 = k^2 - 6a^2$

i.e. $\bar{x}^2 + \bar{y}^2 + \bar{z}^2 = \dfrac{k^2}{2} - 3a^2$

Locus of point $(\bar{x}, \bar{y}, \bar{z})$ is obtained by replacing $(\bar{x}, \bar{y}, \bar{z})$ by (x, y, z) which is

$$\boxed{x^2 + y^2 + z^2 = \frac{k^2}{2} - 3a^2 .}$$

This equation represents a sphere with centre at origin i.e. at the centre of the cube.

Ex. 13 : *A plane passes through a fixed point (a, b, c). Show that the locus of the foot of perpendicular to it from the origin is the sphere* $x^2 + y^2 + z^2 - ax - by - cz = 0$.

Sol. :

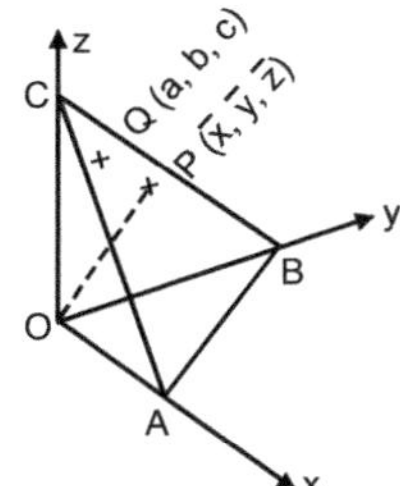

Fig. 7.9

Plane ABC passes through a point Q (a, b, c). Let P $(\bar{x}, \bar{y}, \bar{z})$ be the foot of perpendicular to it from origin (point of locus). Hence the d.r.'s of the normal to the plane will be $(\bar{x} - 0, \bar{y} - 0, \bar{z} - 0)$ i.e. $(\bar{x}, \bar{y}, \bar{z})$ and it passes through (a, b, c). Thus equation of the plane is,

$$\bar{x}(x - a) + \bar{y}(y - b) + \bar{z}(z - c) = 0$$

or $\quad \bar{x}x + \bar{y}y + \bar{z}z - a\bar{x} - b\bar{y} - c\bar{z} = 0$

Hence locus of P $(\bar{x}, \bar{y}, \bar{z})$ can be obtained by generalising $(\bar{x}, \bar{y}, \bar{z})$ i.e. replacing $(\bar{x}, \bar{y}, \bar{z})$ by (x, y, z), we get

$$\boxed{x^2 + y^2 + z^2 - ax - by - cz = 0}$$ which is the equation of the sphere.

Ex. 14 : *A sphere of constant radius r passes through the origin and cuts the axes in A, B, C. Prove that the foot of perpendicular from origin to the plane ABC is given by;*

$$(x^2 + y^2 + z^2)^2 (x^{-2} + y^{-2} + z^{-2}) = 4r^2.$$

Sol. : Let co-ordinates of A, B, C be $(x_1, 0, 0)$, $(0, y_1, 0)$, $(0, 0, z_1)$.

Equation of the sphere passing through origin and meeting axes in A, B, C (intercept form) is

$$x^2 + y^2 + z^2 - x_1 x - y_1 y - z_1 z = 0 \quad \text{[by article 7.2 (4)]}$$

$$\text{Its radius} = \sqrt{\frac{x_1^2}{4} + \frac{y_1^2}{4} + \frac{z_1^2}{4}} = r \text{ (given)}$$

$$\therefore \qquad x_1^2 + y_1^2 + z_1^2 = 4r^2 \qquad \text{... (1)}$$

Now the equation of the plane ABC is $\quad \dfrac{x}{x_1} + \dfrac{y}{y_1} + \dfrac{z}{z_1} = 1$ (intercept form) $\qquad$ ... (2)

Let $(\bar{x}, \bar{y}, \bar{z})$ be the co-ordinates of the foot of perpendicular from origin to the plane ABC (locus point). Hence, d.r.'s of the normal to the plane will be $(\bar{x} - 0, \bar{y} - 0, \bar{z} - 0)$ i.e. $\bar{x}, \bar{y}, \bar{z}$. Thus equation of the plane ABC passing through $(\bar{x}, \bar{y}, \bar{z})$ and having d.r.'s of normal to the plane as $\bar{x}, \bar{y}, \bar{z}$ is

$$\bar{x}(x - \bar{x}) + \bar{y}(y - \bar{y}) + \bar{z}(z - \bar{z}) = 0$$

i.e. $\qquad \bar{x}x + \bar{y}y + \bar{z}z = \bar{x}^2 + \bar{y}^2 + \bar{z}^2$

i.e. $\qquad \dfrac{x}{d/\bar{x}} + \dfrac{y}{d/\bar{y}} + \dfrac{z}{d/\bar{z}} = 1$ where, $d = \bar{x}^2 + \bar{y}^2 + \bar{z}^2 \qquad$ ... (3)

Equations (2) and (3) represent the same plane ABC, we have on comparing,

$$x_1 = \frac{d}{\bar{x}}, \qquad y_1 = \frac{d}{\bar{y}}, \qquad z_1 = \frac{d}{\bar{z}}$$

Substituting these values in (1), we get

$$d^2 \left(\frac{1}{\bar{x}^2} + \frac{1}{\bar{y}^2} + \frac{1}{\bar{z}^2} \right) = 4r^2$$

i.e. $\qquad \left(\bar{x}^2 + \bar{y}^2 + \bar{z}^2 \right)^2 \left(\frac{1}{\bar{x}^2} + \frac{1}{\bar{y}^2} + \frac{1}{\bar{z}^2} \right) = 4r^2$ or $\left(\bar{x}^2 + \bar{y}^2 + \bar{z}^2 \right)^2 \left(\bar{x}^{-2} + \bar{y}^{-2} + \bar{z}^{-2} \right) = 4r^2$

Locus of $(\bar{x}, \bar{y}, \bar{z})$ is obtained by replacing $(\bar{x}, \bar{y}, \bar{z})$ by (x, y, z), which is

$$\boxed{(x^2 + y^2 + z^2)^2 (x^{-2} + y^{-2} + z^{-2}) = 4r^2.}$$

Ex. 15 : *A sphere of constant radius r passes through the origin and meets the co-ordinate axes in A, B, C. Show that the locus of centroid of the triangle ABC is a sphere* $9(x^2 + y^2 + z^2) = 4r^2$. **(Dec. 2004, May 2014)**

Sol. : Let co-ordinates of A, B, C be $(x_1, 0, 0)$, $(0, y_1, 0)$, $(0, 0, z_1)$ respectively, then equation of the sphere passing through origin and meeting axes in A, B, C is

$$x^2 + y^2 + z^2 - x_1 x - y_1 y - z_1 z = 0 \qquad \text{[by article 7.2 (4)]}$$

$$\text{Its radius } = \sqrt{\frac{x_1^2}{4} + \frac{y_1^2}{4} + \frac{z_1^2}{4}} = r \qquad \text{(given)}$$

$$\therefore \qquad x_1^2 + y_1^2 + z_1^2 = 4r^2 \qquad \qquad \text{... (1)}$$

Let $(\bar{x}, \bar{y}, \bar{z})$ be the co-ordinates of the centroid of the triangle ABC (point of locus) then

$$\bar{x} = \frac{x_1 + 0 + 0}{3}; \quad \bar{y} = \frac{0 + y_1 + 0}{3}; \quad \bar{z} = \frac{0 + 0 + z_1}{3}$$

$$\therefore \qquad x_1 = 3\bar{x}, \quad y_1 = 3\bar{y}, \quad z_1 = 3\bar{z}$$

Substituting in (1), we get

$$9\bar{x}^2 + 9\bar{y}^2 + 9\bar{z}^2 = 4r^2$$

Replacing $(\bar{x}, \bar{y}, \bar{z})$ by (x, y, z), required locus is obtained as

$$\boxed{9(x^2 + y^2 + z^2) = 4r^2.}$$

Ex. 16 : *A plane passes through a fixed point (a, b, c) and meets the co-ordinate axes in A, B, C. Show that the locus of the centre of the sphere OABC is* $\dfrac{a}{x} + \dfrac{b}{y} + \dfrac{c}{z} = 2$. **(May 2007, 2005)**

Sol. : Let the plane through (a, b, c) meets the co-ordinate axes in A $(x_1, 0, 0)$, B $(0, y_1, 0)$ and C $(0, 0, z_1)$ respectively, then

Equation of the plane ABC be (intercept form)

$$\frac{x}{x_1} + \frac{y}{y_1} + \frac{z}{z_1} = 1 \qquad \qquad \text{... (1)}$$

Equation of the sphere OABC is given by

$$x^2 + y^2 + z^2 - x_1 x - y_1 y - z_1 z = 0 \qquad \text{[by article 7.2 (4)]}$$

Let $(\bar{x}, \bar{y}, \bar{z})$ be the centre of the sphere (point of locus), then

$$\bar{x} = \frac{x_1}{2}, \quad \bar{y} = \frac{y_1}{2}, \quad \bar{z} = \frac{z_1}{2}$$

$$\therefore \qquad x_1 = 2\bar{x}, \quad y_1 = 2\bar{y}, \quad z_1 = 2\bar{z}$$

Substituting in (1), we get

$$\frac{x}{2\bar{x}} + \frac{y}{2\bar{y}} + \frac{z}{2\bar{z}} = 1 \quad \text{or} \quad \frac{x}{\bar{x}} + \frac{y}{\bar{y}} + \frac{z}{\bar{z}} = 2$$

This plane passes through fixed point (a, b, c).

$$\therefore \qquad \frac{a}{\bar{x}} + \frac{b}{\bar{y}} + \frac{c}{\bar{z}} = 2$$

$\therefore$ Locus of $(\bar{x}, \bar{y}, \bar{z})$ is obtained by replacing $(\bar{x}, \bar{y}, \bar{z})$ by (x, y, z), which is

$$\boxed{\frac{a}{x} + \frac{b}{y} + \frac{c}{z} = 2}$$

Ex. 17 : *Prove that the centres of spheres which touch the lines $y = mx$, $z = c$; $y = -mx$, $z = -c$ lie upon the conicoid $mxy + cz(1 + m^2) = 0$.*

Sol. :

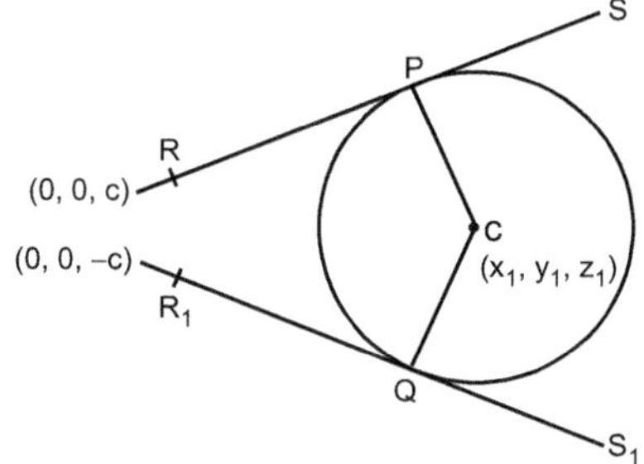

Fig. 7.10

Equations of lines RS, R_1S_1, which touch the sphere are expressed in symmetrical form as,

$$\frac{x}{1} = \frac{y}{m} = \frac{z - c}{0} = k_1$$

$$\frac{x}{1} = \frac{y}{-m} = \frac{z + c}{0} = k_2$$

Let (x_1, y_1, z_1) be the centre of the sphere. Since RS and R_1S_1 touch the sphere, CP = CQ.

$$RP = \text{Projection of CR on RS.} \qquad \text{[by article 7.10 (16)]}$$

$$= \frac{1}{\sqrt{1 + m^2}} [(x_1 - 0) + m(y_1 - 0) + 0(z_1 - c)] = \frac{1}{\sqrt{1 + m^2}} (x_1 + my_1)$$

$$R_1Q = \text{Projection of } CR_1 \text{ on } R_1S_1$$

$$= \frac{1}{\sqrt{1 + m^2}} [(x_1 - 0) - m(y_1 - 0) + 0(z_1 + c)] = \frac{1}{\sqrt{1 + m^2}} (x_1 - my_1)$$

Now,

$$CP^2 = CR^2 - PR^2$$

$$= x_1^2 + y_1^2 + (z_1 - c)^2 - \frac{1}{(1 + m^2)} (x_1 + my_1)^2$$

$$= \frac{1}{(1 + m^2)} [m^2 x_1^2 + y_1^2 + (1 + m^2)(z_1 - c)^2 - 2mx_1 y_1]$$

Similarly,

$$CQ^2 = CR_1^2 - R_1Q^2$$

$$= x_1^2 + y_1^2 + (z_1 + c)^2 - \frac{1}{(1 + m^2)} (x_1 - my_1)^2$$

$$= \frac{1}{(1 + m^2)} [m^2 x_1^2 + y_1^2 + (1 + m^2)(z_1 + c)^2 + 2mx_1 y_1]$$

But

$$CP^2 = CQ^2$$

$$\therefore \quad (1 + m^2)(z_1 - c)^2 - 2m x_1 y_1 = (1 + m^2)(z_1 + c)^2 + 2mx_1 y_1$$

$$\therefore \quad (1 + m^2)(4cz_1) + 4mx_1 y_1 = 0$$

$$\text{or} \quad (1 + m^2) cz_1 + mx_1 y_1 = 0$$

$\therefore$ Locus of (x_1, y_1, z_1) is given by replacing (x_1, y_1, z_1) by (x, y, z).

i.e.

$$\boxed{mxy + cz(1 + m^2) = 0.}$$

Exercise 7.1

1. Find the equation of the sphere whose centre is $(2, -3, 1)$ and radius is 5. **Ans.** $x^2 + y^2 + z^2 - 4x + 6y - 2z - 11 = 0$

2. Find the equation of the sphere with centre $(2, -2, 3)$ and passing through $(7, -3, 5)$.

   Ans. $x^2 + y^2 + z^2 - 4x + 4y - 6z - 13 = 0.$

3. Find the centre and radius of the sphere
   (i) $x^2 + y^2 + z^2 + 6x - 4y + 8z - 10 = 0$
   (ii) $x^2 + y^2 + z^2 + 2x - 4y - 6z + 5 = 0$
   (iii) $2x^2 + 2y^2 + 2z^2 - 2x + 4y + 2z + 3 = 0.$

   Ans. (i) $(-3, 2, -4)$; $\sqrt{39}$ (ii) $(-1, 2, 3)$; 3

   (iii) $\left(\frac{1}{2}, -1, -\frac{1}{2}\right)$; 0. Point sphere.

4. Find the equation of the sphere on the join of $(2, -3, 1)$ and $(1, -2, -1)$ as diameter. **Ans.** $x^2 + y^2 + z^2 - 3x + 5y + 7 = 0.$

5. Find the equation of the sphere which passes through origin and makes equal intercepts of unit length on the axes.

   Ans. $x^2 + y^2 + z^2 - x - y - z = 0.$

6. Find the equation of the sphere passing through the points
   (i) $(0, 0, 0), (0, -1, 1), (-1, 2, 0), (1, 2, 3)$
   (ii) $(1, 2, 3), (0, -2, 4), (4, -4, 2), (3, 1, 4)$

   Ans. (i) $7(x^2 + y^2 + z^2) + 9x - 13y - 27z = 0$

   (ii) $x^2 + y^2 + z^2 - 4x + 2y - 2z - 8 = 0.$

7. Find the equation of the sphere through the four points $(0, 0, 0)$, $(- a, b, c)$, $(a, - b, c)$, $(a, b, - c)$ and determine its radius.

Ans. Equation : $x^2 + y^2 + z^2 - (a^2 + b^2 + c^2) \left(\dfrac{x}{a} + \dfrac{y}{b} + \dfrac{z}{c} \right) = 0$,

$$\text{radius} = \dfrac{a^2 + b^2 + c^2}{2} \sqrt{(a^{-2} + b^{-2} + c^{-2})}$$

8. Obtain the equation of the sphere passing through the points $(3, 0, 2)$, $(-1, 1, 1)$, $(2, -5, 4)$ and having its centre on the plane $2x + 3y + 3z = 6$.

Ans. $x^2 + y^2 + z^2 + 4y - 6z - 1 = 0$.

9. Find the equation of the sphere passing through $(a, 0, 0)$, $(0, b, 0)$, $(0, 0, c)$ and having least possible radius.

[Hint : Let $r^2 = f(d) = \dfrac{1}{4a^2} (a^2 + d)^2 + \dfrac{1}{4b^2} (b^2 + d)^2 + \dfrac{1}{4c^2} (c^2 + d)^2 - d$. For maximum or minimum radius $f'(d) = 0$,

which gives $d = \dfrac{- a^2 b^2 c^2}{a^2 b^2 + b^2 c^2 + c^2 d^2}$ and $f''(d) > 0$. Thus $f(d)$ and therefore r is minimum.]

Ans. $(x^2 + y^2 + z^2) (a^2b^2 + b^2c^2 + c^2a^2) - a^3 (b^2 + c^2) x - b^3 (c^2 + a^2) y - c^3 (a^2 + b^2) z - a^2 b^2 c^2 = 0$.

10. Find the equation of the sphere passing through $(1, 0, 0)$, $(0, 1, 0)$, $(0, 0, 1)$ and having least possible radius.

Ans. $3 (x^2 + y^2 + z^2) - 2 (x + y + z) - 1 = 0$.

11. A sphere with its centre in the positive octant passes through the origin and cuts the co-ordinate planes in circles of radii $a\sqrt{2}$, $b\sqrt{2}$, $c\sqrt{2}$. Find its equation.

Ans. $\dfrac{1}{2} (x^2 + y^2 + z^2) - \left(\sqrt{b^2 + c^2 - a^2} \right) x - \left(\sqrt{c^2 + a^2 - b^2} \right) y - \left(\sqrt{a^2 + b^2 - c^2} \right) z = 0$.

12. Find the equation of the sphere which has its centre in positive quadrant of xoy plane and which cuts the planes $x = 0, y = 0, z = 0$ in circles of radii 3, 4, 5 respectively.

Ans. $x^2 + y^2 + z^2 - 8x - 6y = 0$.

13. Find the equation of the sphere which has its centre at $C (2, 3, -1)$ and touches the line $\dfrac{x + 1}{-5} = \dfrac{y - 8}{3} = \dfrac{z - 4}{4}$.

(May 2016)

[Hint : Any point $P (-5k - 1, 3k + 8, 4k + 4)$ on the given line will be the point of contact if CP is perpendicular to the line and use condition $a_1a_2 + b_1b_2 + c_1c_2 = 0$.]

Ans. $x^2 + y^2 + z^2 - 4x - 6y + 2z + 5 = 0$.

14. Find the equation of the sphere which passes through the points $(2, 1, 1)$ and $(0, 3, 2)$ and has its centre on the line $2x + y + 3z = 0 = x + 2y + 2z$.

Ans. $9 (x^2 + y^2 + z^2) + 28x + 7y - 21z - 96 = 0$.

15. Find the equation of the sphere circumscribing the tetrahedron whose faces are

(i) $x = 0, y = 0,\ z = 0, \dfrac{x}{a} + \dfrac{y}{b} + \dfrac{z}{c} = 1$

(ii) $x = 0, y = 0,\ z = 0,\ ax + by + cz + d = 0$.

Ans. (i) $x^2 + y^2 + z^2 - ax - by - cz = 0$,

(ii) $x^2 + y^2 + z^2 + d \left(\dfrac{x}{a} + \dfrac{y}{b} + \dfrac{z}{c} \right) = 0$.

16. Prove that the two spheres $x^2 + y^2 + z^2 - 2x + 4y - 4z = 0$ and $x^2 + y^2 + z^2 + 10x + 2z + 10 = 0$ touch each other and find the co-ordinates of the point of contact.

(Dec. 2013) Ans. $\left(-\dfrac{11}{7}, -\dfrac{8}{7}, \dfrac{5}{7} \right)$

17. A is the point $(1, 3, 4)$ and B, the point $(1, -2, -1)$. A point P moves so that $3PA = 2PB$. Prove that the locus of P is the sphere $x^2 + y^2 + z^2 - 2x - 14y - 16z + 42 = 0$.

[Hint : Let $P(\bar{x}, \bar{y}, \bar{z})$ be the point on the locus. Given :

$9(PA)^2 = 4(PB)^2$ i.e. $9[(\bar{x} - 1)^2 + (\bar{y} - 3)^2 + (\bar{z} - 4)^2]$

$= 4[(\bar{x} - 1)^2 + (\bar{y} + 2)^2 + (\bar{z} + 1)^2]$. On simplifying or replacing $(\bar{x}, \bar{y}, \bar{z})$ by x, y, z we get locus of $P(\bar{x}, \bar{y}, \bar{z})$.]

18. If 'O' the origin be the centre of the sphere of radius a and A and B be two points on the line with 'O' such that $OA.OB = a^2$. If P be variable point on the sphere, show that $PA : PB = $ constant.

19. A plane passes through the point $(2, 2, 2)$ and cuts the axes in A, B, C. Show that the locus of the centre of the sphere OABC is $x^{-1} + y^{-1} + z^{-1} = 1$.

20. Prove that the locus of a point which moves such that the sum of the squares of its distances from the planes $x + y + z = 0, x - z = 0, x - 2y + z = 0$ is 9, is a sphere of radius 3.

21. A plane passes through a fixed point $(2, 3, 4)$. Find the locus of the foot of the perpendicular from the origin.

Ans. $x^2 + y^2 + z^2 - 2x - 3y - 4z = 0$.

22. A sphere of constant radius 2r passes through the origin and meets the axes in A, B, C. Find the locus of the centroid of (i) tetrahedron OABC, (ii) triangle ABC.

Ans. (i) $x^2 + y^2 + z^2 = r^2$ (ii) $9 (x^2 + y^2 + z^2) = 16r^2$.

7.5 TANGENT PLANE OF THE SPHERE

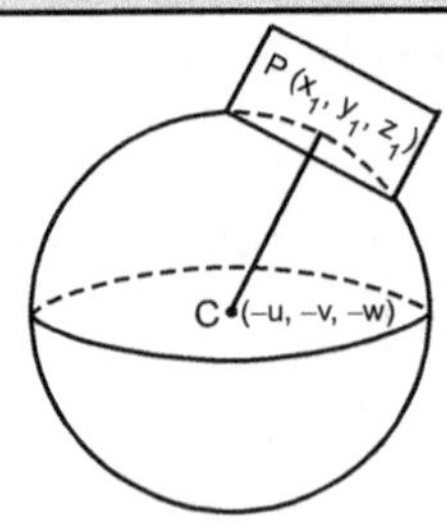

Fig. 7.11

Consider the sphere

$$x^2 + y^2 + z^2 + 2ux + 2vy + 2wz + d = 0 \qquad \ldots (i)$$

To find the equation of the tangent plane at (x_1, y_1, z_1), we use the property that normal to the tangent plane at (x_1, y_1, z_1) passes through the centre.

The d.r.'s of normal CP (Fig. 7.11) are $(x_1 + u, y_1 + v, z_1 + w)$.

The equation of the tangent plane at $P (x_1, y_1, z_1)$ from result (18) of article 7.11, is

$$(x_1 + u) (x - x_1) + (y_1 + v) (y - y_1) + (z_1 + w) (z - z_1) = 0 \qquad \ldots (ii)$$

Rewriting it as

$$xx_1 + yy_1 + zz_1 + ux + vy + wz - ux_1 - vy_1 - wz_1 - x_1^2 - y_1^2 - z_1^2 = 0 \qquad \text{...(iii)}$$

Since point P (x_1, y_1, z_1) lies on the sphere (i), we have

$$x_1^2 + y_1^2 + z_1^2 + 2ux_1 + 2vy_1 + 2wz_1 + d = 0$$

or

$$x_1^2 + y_1^2 + z_1^2 + ux_1 + vy_1 + wz_1 = -ux_1 - vy_1 - wz_1 - d$$

i.e.

$$-x_1^2 - y_1^2 - z_1^2 - ux_1 - vy_1 - wz_1 = ux_1 + vy_1 + wz_1 + d$$

Substituting in (iii), equation of tangent plane takes the form

$$xx_1 + yy_1 + zz_1 + ux + vy + wz + ux_1 + vy_1 + wz_1 + d = 0$$

or

$$\boxed{xx_1 + yy_1 + zz_1 + u(x + x_1) + v(y + y_1) + w(z + z_1) + d = 0} \qquad \text{... (7)}$$

which is the required equation of the tangent plane to the sphere (i).

Note 1 : The equation of tangent plane at any point P (x_1, y_1, z_1) to a sphere is obtained by writing xx_1, yy_1, zz_1 for x^2, y^2, z^2 and $(x + x_1), (y + y_1), (z + z_1)$ for $2x, 2y, 2z$ in the equation (i) of the sphere.

Note 2 : Tangent Plane Property : If a plane touches a sphere, then length of perpendicular from the **centre** of the sphere to the plane must be equal to radius of the sphere.

Note 3 : To find the point of contact, determine the point of intersection of the line perpendicular from centre to the tangent plane.

7.6 SECTION OF A SPHERE BY A PLANE

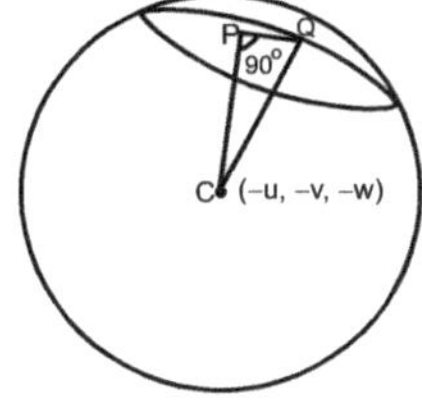

Fig. 7.12

Consider the sphere

$$S \equiv x^2 + y^2 + z^2 + 2ux + 2vy + 2wz + d = 0 \qquad \text{... (i)}$$

which is cut by the plane

$$U \equiv lx + my + nz - p = 0 \qquad \text{... (ii)}$$

We shall now show that the section of the sphere by a plane is a circle. Drop CP, the perpendicular from the centre C $(-u, -v, -w)$ (Fig. 7.12) on the section. P is the foot of the perpendicular.

For a given sphere and a given plane, C and P are fixed points and so also the perpendicular distance CP.

Let Q be any point on the common section, then CQ = Radius of the sphere, which is constant for a given sphere.

In triangle CPQ, 　　　$\angle$ CPQ $= 90°$

$\therefore$ 　　　　　$CQ^2 = CP^2 + PQ^2$

$\therefore$ 　　　　　$PQ = \sqrt{CQ^2 - CP^2} = \sqrt{R^2 - p^2} \qquad \text{... (iii)}$

where, R = CQ is the radius of the sphere and p = CP is the length of the perpendicular from centre C of the sphere (i) on the intersecting plane (ii). Thus PQ is constant for a given sphere and plane and P is a fixed point.

Thus locus of Q is a circle with centre P and radius PQ $= \sqrt{R^2 - p^2}$ which establishes that section of a sphere by a plane is a circle.

Note 1 : Centre of the circle is the foot of perpendicular drawn from the centre of the sphere to the plane. Co-ordinates of centre of the circle is obtained by considering intersection of the line CP, $\dfrac{x + u}{l} = \dfrac{y + v}{m} = \dfrac{z + w}{n} = k$ and plane (ii), and relation (iii) gives radius of circle.

Note 2 : The section of a sphere by a plane through the centre of the sphere is a **great circle**. Its centre and radius are the same as those of the given sphere.

7.7 INTERSECTION OF TWO SPHERES

Let the equations of any two spheres be

$$S_1 \equiv x^2 + y^2 + z^2 + 2u_1x + 2v_1y + 2w_1z + d_1 = 0 \qquad \text{... (i)}$$

$$S_2 \equiv x^2 + y^2 + z^2 + 2u_2x + 2v_2y + 2w_2z + d_2 = 0 \qquad \text{... (ii)}$$

We shall now show that the curve of intersection of two spheres is a circle. Co-ordinates of the points of intersection of (i) and (ii) for which $S_1 = 0$, $S_2 = 0$, also satisfy the equation $S_1 - S_2 = 2(u_1 - u_2)x + 2(v_1 - v_2)y + 2(w_1 - w_2) + d_1 - d_2 = 0$ which is a first degree equation in x, y, z and represents a plane.

Thus the points of intersection of two spheres are same as those of intersection of any of the spheres and the plane $S_1 - S_2 = 0$.

Hence by article 7.6, the curve of intersection of two spheres is a circle.

7.8 EQUATIONS OF A CIRCLE

As we know that two planes together represent line of section of the two planes [see article 7.15 (a)]. Equations of sphere and plane together will represent the circle which is the common section of the sphere and the plane. Thus if $S = 0$ is the equation of a sphere and $U = 0$ that of a plane, the equations

$$\boxed{\begin{aligned} S &= 0 \\ U &= 0 \end{aligned}}$$

... (8)

together represent a circle.

Also circle can be expressed as intersection of two spheres. Thus if $S_1 = 0$, $S_2 = 0$ are equations of spheres, the equations

$$\boxed{\begin{aligned} S_1 &= 0 \\ S_2 &= 0 \end{aligned}}$$

... (9)

together represent a circle.

Note : In three-dimensional geometry, circle will not be represented by a single equation, but by two equations together.

7.9 SPHERE THROUGH A CIRCLE

Consider

$$\left. \begin{aligned} S &\equiv x^2 + y^2 + z^2 + 2ux + 2vy + 2wz + d = 0 \quad \text{... (I)} \\ U &\equiv lx + my + nz - p = 0 \quad \text{... (II)} \end{aligned} \right\}$$

... (i)

In previous section we have seen that (8) represents the circle of intersection of the sphere (I) by the plane (II).

Multiplying (II) by λ and adding to (I), we get

$$\boxed{S + \lambda U = 0}$$

... (10)

i.e. $\quad x^2 + y^2 + z^2 + 2ux + 2vy + 2wz + d + \lambda (lx + my + nz - p) = 0$ ... (ii)

or $\quad x^2 + y^2 + z^2 + x (2u + \lambda l) + y (2u + \lambda m) + z (2w + \lambda n) + d - \lambda p = 0$ ... (iii)

Equation (iii) clearly represents a sphere as coefficients of x^2, y^2, z^2 are equal and terms xy, yz and zx are absent.

Co-ordinates of points which satisfy (I), (II) both i.e. the points lying on the circle (i) clearly satisfy (ii) or (iii). Hence (ii) or (iii) represents sphere passing through the circle given by (i). Corresponding to different values of λ, we get different spheres. Thus equation (ii) or (iii) represents family of spheres passing through the circle (i).

Again if we consider the equations,

$$\left. \begin{aligned} S_1 &\equiv x^2 + y^2 + z^2 + 2u_1x + 2v_1y + 2w_1z + d_1 = 0 \quad \text{... (III)} \\ S_2 &\equiv x^2 + y^2 + z^2 + 2u_2x + 2v_2y + 2w_2z + d_2 = 0 \text{ ... (IV)} \end{aligned} \right\}$$

...(iv)

$$\boxed{S_1 + \lambda S_2 = 0}$$

... (11)

Then (iv) represents the circular section of spheres (III) and (IV).

or $\quad (x^2 + y^2 + z^2) (1 + \lambda) + 2x (u_1 + \lambda u_2) + 2y (v_1 + \lambda v_2) + 2z (w_1 + \lambda w_2) + d_1 + \lambda d_2 = 0$ ... (v)

For $\lambda \neq -1$, (v) clearly represents a sphere passing through the circular section represented by (iv). Different values of λ give different spheres all of which pass through circle (iv), i.e. (v) represents family of spheres passing the circular section (iv).

7.10 ORTHOGONAL SPHERES

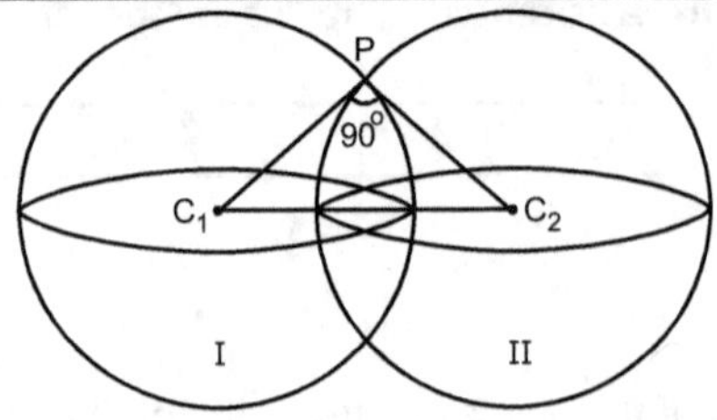

Fig. 7.13

Two spheres are said to be orthogonal if the tangent planes to the two spheres at the points of intersection are at right angles.

It is clear that if the tangent planes at the point of intersection (Fig. 7.13) are at right angles then normals to these tangent planes which pass through the centres C_1, C_2 are also at right angles.

In triangle C_1PC_2, $\angle C_1PC_2 = 90°$

$\therefore$

$$\boxed{C_1C_2^2 = C_1P^2 + C_2P^2}$$

... (12)

If spheres (I) and (II) (Fig. 7.13) are represented by

$$x^2 + y^2 + z^2 + 2u_1x + 2v_1y + 2w_1z + d_1 = 0 \qquad \text{... (i)}$$
$$x^2 + y^2 + 2u_2x + z^2 + 2v_2y + 2w_2z + d_2 = 0 \qquad \text{... (ii)}$$

Then C_1P = radius of sphere (i) $= \sqrt{u_1^2 + v_1^2 + w_1^2 - d_1}$

and C_2P = radius of sphere (ii) $= \sqrt{u_2^2 + v_2^2 + w_2^2 - d_2}$

Also centre of sphere (i) is $(-u_1, -v_1, -w_1)$ and centre of sphere (ii) is $(-u_2, -v_2, -w_2)$.

$$C_1C_2^2 = (u_1 - u_2)^2 + (v_1 - v_2)^2 + (w_1 - w_2)^2$$

Substituting in result (12), we get

$$(u_1 - u_2)^2 + (v_1 - v_2)^2 + (w_1 - w_2)^2 = (u_1^2 + v_1^2 + w_1^2 - d_1) + (u_2^2 + v_2^2 + w_2^2 - d_2)$$

$$\therefore \quad u_1^2 - 2u_1u_2 + u_2^2 + v_1^2 - 2v_1v_2 + v_2^2 + w_1^2 - 2w_1w_2 + w_2^2 = (u_1^2 + v_1^2 + w_1^2 - d_1) + (u_2^2 + v_2^2 + w_2^2 - d_2)$$

which gives $\boxed{2u_1u_2 + 2v_1v_2 + 2w_1w_2 = d_1 + d_2}$ $\qquad$ ... (13)

Result (13) is the analytical condition for the two spheres (i) and (ii) to be orthogonal while result (12) is the geometrical condition for the orthogonality of the two spheres.

7.11 LENGTH OF THE TANGENT

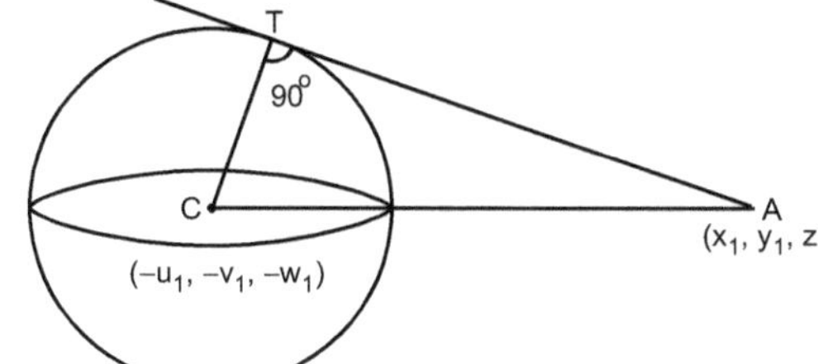

Fig. 7.14

Consider the sphere

$$x^2 + y^2 + z^2 + 2u_1x + 2v_1y + 2w_1z + d_1 = 0 \qquad \text{... (i)}$$

Let $A(x_1, y_1, z_1)$ be any point outside the sphere and AT is the length of the tangent drawn from the point A to the sphere (Fig. 7.14), then CT is perpendicular to AT.

In triangle CTA, $\qquad \angle\ CTA = 90°$

$\therefore \qquad AC^2 = AT^2 + CT^2$

or $\qquad AT^2 = AC^2 - CT^2$

$\therefore \qquad AT^2 = (x_1 + u_1)^2 + (y_1 + v_1)^2 + (z_1 + w_1)^2 - u_1^2 - v_1^2 - w_1^2 + d_1$

or $\qquad \boxed{AT^2 = x_1^2 + y_1^2 + z_1^2 + 2u_1x_1 + 2v_1y_1 + 2w_1z_1 + d_1}$ $\qquad$ ... (14)

which is the formula for the length of the tangent drawn from the point (x_1, y_1, z_1) to the sphere (i).

7.12 RADICAL PLANE

Radical plane of two given spheres is the locus of the point from which the square of the length of tangents to two given spheres are equal.

Consider the equations of the two spheres be

$$S_1 \equiv x^2 + y^2 + z^2 + 2u_1x + 2v_1y + 2w_1z + d_1 = 0 \qquad \text{... (i)}$$
$$S_2 \equiv x^2 + y^2 + z^2 + 2u_2x + 2v_2y + 2w_2z + d_2 = 0 \qquad \text{... (ii)}$$

Let (x, y, z) be any point on the locus then by result (14)

$$x^2 + y^2 + z^2 + 2u_1x + 2v_1y + 2w_1z + d_1 = x^2 + y^2 + z^2 + 2u_2x + 2v_2y + 2w_2z + d_2$$

$\therefore \qquad \boxed{2x(u_1 - u_2) + 2y(v_1 - v_2) + 2z(w_1 - w_2) + d - d_2 = 0}$ $\qquad$ ... (15)

which is the equation of Radical plane of spheres (i) and (ii).

In short, equation of Radical plane is given by

$$\boxed{S_1 - S_2 = 0} \qquad \text{... (16)}$$

Note : The radical plane of two spheres is perpendicular to the line joining their centres.

7.13 ILLUSTRATIONS ON TANGENT PLANE AND CIRCLE

Type I : Problems on Tangent Plane :

Ex. 1 : *Find the equation of tangent plane of the sphere* $3(x^2 + y^2 + z^2) - 2x - 3y - 4z - 22 = 0$ *at the point* (1, 2, 3).

Find also the equation of normal to the sphere at (1, 2, 3).

Sol. : The equation of tangent plane at the point (1, 2, 3) is

$$3x(1) + 3y(2) + 3z(3) - (x + 1) - \frac{3}{2}(y + 2) - 2(z + 3) - 22 = 0 \qquad \text{[refer 7.5 (7)]}$$

or $\qquad 4x + 9y + 14z - 64 = 0$

The d.r.'s of normal to this plane are 4, 9, 14.

∴   Equation of the normal to the sphere at (1, 2, 3) i.e. the line through (1, 2, 3) and perpendicular to the tangent plane are

$$\boxed{\dfrac{x-1}{4} = \dfrac{y-2}{9} = \dfrac{z-3}{14}}.$$

Ex. 2 : *Show that the plane* $4x - 3y + 6z - 35 = 0$ *is tangential to the sphere* $x^2 + y^2 + z^2 - y - 2z - 14 = 0$ *and find the point of contact.*   **(May 2009, 2015, 2017)**

Sol. :

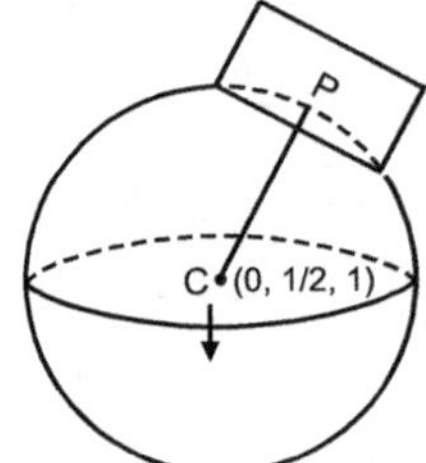

Fig. 7.15

If the plane is tangential then the length of perpendicular from centre of the sphere on the plane is equal to the radius of the sphere.

Given sphere has centre $C\left(0, \dfrac{1}{2}, 1\right)$ and radius, $r = \sqrt{u^2 + v^2 + w^2 - d}$

$$= \sqrt{0 + \frac{1}{4} + 1 + 14} = \frac{\sqrt{61}}{2}$$

Length of perpendicular from $\left(0, \dfrac{1}{2}, 1\right)$ on the plane

$$= \left| \frac{4(0) - 3\left(\frac{1}{2}\right) + 6(1) - 35}{\sqrt{16 + 9 + 36}} \right|$$

$$= \left| \frac{-\frac{61}{2}}{\sqrt{61}} \right| = \frac{\sqrt{61}}{2} \text{ (radius)}$$

Thus the given plane is tangential to the sphere.

To find the point of contact, consider equation of line CP passing through $\left(0, \dfrac{1}{2}, 1\right)$ and whose d.r.'s are 4, −3, 6, as

$$\frac{x-0}{4} = \frac{y - \frac{1}{2}}{-3} = \frac{z-1}{6} = k, \text{ say.}$$

Any point on this line will have coordinates $x = 4k$, $y = -3k + \dfrac{1}{2}$, $z = 6k + 1$. If this is the point of contact, it should satisfy the equation of plane.

∴ $\qquad 4(4k) - 3\left(-3k + \dfrac{1}{2}\right) + 6(6k + 1) - 35 = 0$, this gives $k = \dfrac{1}{2}$.

Hence the point of contact is $\boxed{\left[4\left(\frac{1}{2}\right), -3\left(\frac{1}{2}\right) + \frac{1}{2}, 6\left(\frac{1}{2}\right) + 1\right]}$ i.e. (2, −1, 4).

Ex. 3 : *Show that the plane 2x – 2y + z + 12 = 0 touches the sphere*

$x^2 + y^2 + z^2 – 2x – 4y + 2z – 3 = 0$ *and find the point of contact.*　　　　**(May 06, 16, Dec. 13, 18)**

Sol. :

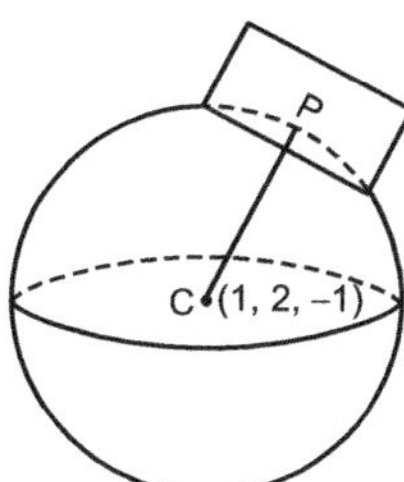

Fig. 7.16

Any plane touches a sphere if perpendicular from centre of the sphere to the plane is equal to radius of the sphere. Centre of the given sphere is C (1, 2, –1) and radius $r = \sqrt{1 + 4 + 1 + 3} = 3$.

Length of perpendicular from centre (1, 2, –1) on the plane is

$$CP = \left| \frac{2\,(1) – 2\,(2) + (–1) + 12}{\sqrt{4 + 4 + 1}} \right| = \frac{9}{3} = 3 \text{ (radius)}$$

Thus, the given plane touches the sphere.

To find the point of contact, consider equations of line CP whose d.r.'s are (2, –2, 1) as,

$$\frac{x – 1}{2} = \frac{y – 2}{– 2} = \frac{z + 1}{1} = k, \text{ say.}$$

∴　(2k + 1, –2k + 2, k – 1) represents general point on this line. If this is the point of contact, it should satisfy the equation of plane.

∴　2 (2k + 1) – 2 (–2k + 2) + (k – 1) + 12 = 0　which gives k = – 1.

Putting value of k in (2k + 1, – 2k + 2, k – 1), we get the point of contact as,

$$\boxed{[2\,(–1) + 1, –2\,(–1) + 2, \ –1 – 1] \text{ i.e. } (–1, 4, –2).}$$

Ex. 4 : *Find the tangent planes to the sphere $x^2 + y^2 + z^2 – 4x + 2y – 6z + 5 = 0$ which are parallel to the plane 2x + 2y – z = 0.*

Sol. : Equation of any plane parallel to given plane is

$$2x + 2y – z + k = 0$$

This will be a tangent plane if its distance from the centre (2, –1, 3) of the given sphere, is equal to its radius which is 3.

∴　$$\frac{2\,(2) + 2\,(–1) – (3) + k}{\sqrt{2^2 + 2^2 + 1}} = \pm\,3, \text{ which gives } k = 10 \text{ or } – 8.$$

Hence, required planes are

$$2x + 2y – z + 10 = 0 \quad \text{and} \quad 2x + 2y – z – 8 = 0.$$

Ex. 5 : *Find the equation of the sphere tangential to the plane x – 2y – 2z = 7 at (3, –1, –1) and passing through the point (1, 1, –3).*

(Dec. 2005, Nov. 2014)

Sol. :

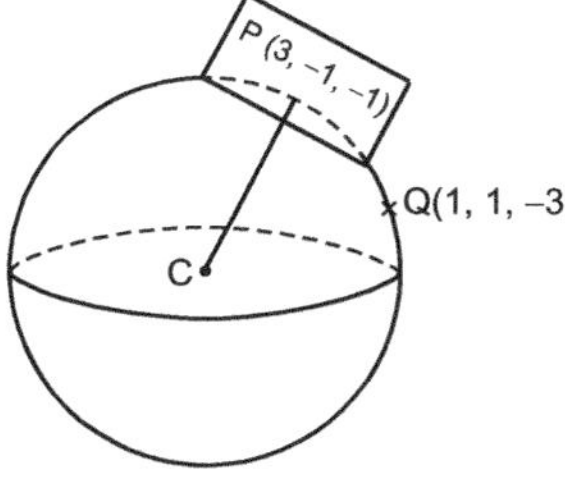

Fig. 7.17

Given plane is tangential to the sphere at (3, –1, –1). Normal to this plane at P must pass through the centre of sphere. Equation of CP can be written as

$$\frac{x – 3}{1} = \frac{y + 1}{– 2} = \frac{z + 1}{– 2} = k \text{ (say)}$$

∴　x = k + 3,

　　y = – 2k – 1,

　　z = – 2k – 1 represent any point on CP

Let co-ordinates of centre C be $(k + 3, -2k - 1, -2k - 1)$

$\therefore$　　　　　　　　　　$CP^2 = k^2 + 4k^2 + 4k^2 = 9k^2$

and　　　　　　　　　　$CQ^2 = (k + 2)^2 + (-2k - 2)^2 + (-2k + 2)^2$

But　　　　　　　　　　$CP^2 = CQ^2$　　　　　　　　　　$(\because CP = CQ = \text{Radius of sphere})$

$\therefore$　　　　　　　　$9k^2 = (k + 2)^2 + 4(k + 1)^2 + 4(k - 1)^2$ which gives $k = -3$

$\therefore$　Centre of sphere is $(0, 5, 5)$ and Radius, $R = \sqrt{9k^2} = 9$.

$\therefore$　Equation of desired sphere is,

$$(x - 0)^2 + (y - 5)^2 + (z - 5)^2 = 81$$

or　　　$\boxed{x^2 + y^2 + z^2 - 10y - 10z - 31 = 0}$.

Ex. 6 : *Find the equations of tangent planes to the sphere $x^2 + y^2 + z^2 = 9$ which pass through the line $x + y = 6$, $x - 2z = 3$.*

Sol. :

Equation of the required plane is of the form $u_1 + \lambda u_2 = 0$ (refer article 7.11 (23)).

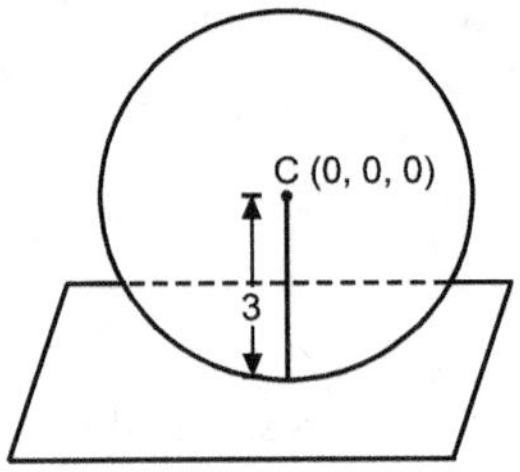

$\therefore$　$(x + y - 6) + \lambda (x - 2z - 3) = 0$

i.e.　$(1 + \lambda) x + y - 2\lambda z - (6 + 3\lambda) = 0$ ... (1)

Centre of the sphere is $C(0, 0, 0)$ and radius of the sphere is 3.

If the plane touches the sphere, perpendicular from centre $(0, 0, 0)$ to the plane is equal to the radius of the sphere.

Fig. 7.18

$\therefore$　$\left| \dfrac{-6 - 3\lambda}{\sqrt{(1 + \lambda)^2 + 1 + 4\lambda^2}} \right| = 3$ which gives $\lambda = 1, -\dfrac{1}{2}$.

Hence, corresponding to the values of $\lambda = 1, -\dfrac{1}{2}$, we get from (1) two tangent planes having equations

$$[(1 + \lambda) x + y - 2\lambda x - (6 + 3\lambda) = 0], \quad \lambda = 1, -\dfrac{1}{2}$$

Hence, required planes are

$\boxed{2x + y - 2z = 9 \quad \text{and} \quad 2x + 2y + 2z = 9}$

Ex. 7 : *Prove that the line $2(x + 1) = 2 - y = z + 3$ touches the sphere $9(x^2 + y^2 + z^2) = 5$ and find the point of contact.*

Sol. : Given line can be written in symmetrical form as,

$$\frac{x + 1}{1} = \frac{y - 2}{-2} = \frac{z + 3}{2} = k$$

Co-ordinates of any point on this line are $(k - 1, -2k + 2, 2k - 3)$. This lies on the sphere if

$$9\left[(k - 1)^2 + (-2k + 2)^2 + (2k - 3)^2\right] = 5$$

i.e.　　$9\left[9k^2 - 22k + 14\right] = 5$

i.e.　　$81k^2 - 198k + 121 = 0$

i.e.　　$(9k - 11)^2 = 0$ which gives $k = \dfrac{11}{9}$ and $\dfrac{11}{9}$.

As the values of k obtained above are equal, the given line intersects the sphere in two coincident points and hence touches the sphere. Thus point of contact is $\boxed{\left(\dfrac{2}{9}, -\dfrac{4}{9}, -\dfrac{5}{9}\right)}$.

Ex. 8 : *Find the equation of the sphere touching three co-ordinate planes and the plane, $2x + y + 2z = 6$.*

Sol. : Let the sphere be,

$$x^2 + y^2 + z^2 + 2ux + 2vy + 2wz + d = 0 \qquad \text{... (1)}$$

Centre of the sphere is $(-u, -v, -w)$. Now, lengths of perpendicular from $(-u, -v, -w)$ to the co-ordinate planes whose equations are $x = 0$, $y = 0$, $z = 0$ and the given plane $2x + y + 2z = 6$ are equal to radius of the sphere.

Using formula for length of perpendicular [refer article 7.13 (27)], we have

$$\frac{-u}{1} = \frac{-v}{1} = \frac{-w}{1} = \frac{-2u - v - 2w - 6}{\sqrt{4 + 1 + 4}} = k, \text{ say radius of the sphere.}$$

$\therefore \qquad u = v = w = -k \qquad$ and $\qquad 2u + v + 2w + 6 = -3k$

which gives $\qquad\qquad -2k - k - 2k + 6 = -3k \qquad\qquad \therefore \quad k = 3$

Also $\qquad\qquad\qquad r = $ radius of the sphere $= \sqrt{u^2 + v^2 + w^2 - d}$

$\therefore \qquad\qquad\qquad r^2 = u^2 + v^2 + w^2 - d \Rightarrow k^2 = k^2 + k^2 + k^2 - d^2$

$\therefore \qquad\qquad\qquad d = 2k^2 = 18$

Substituting the values of $u = v = w = -3$ and $d = 18$ in (1), we get equation of the sphere as

$$\boxed{x^2 + y^2 + z^2 - 6x - 6y - 6z + 18 = 0}$$

Ex. 9 : *Find the equation of the sphere that passes through the points A (1, 0, 0), B (0, 1, 2) and C (1, 1, 1) and touches the line x = 0, y = z.*

Sol. : Let the equation of the sphere be

$$x^2 + y^2 + z^2 + 2ux + 2vy + 2wz + d = 0 \qquad \text{... (1)}$$

Since it passes through the given points A, B, C, we have

$$1 + 2u + d = 0 \qquad \text{... (2)}$$
$$5 + 2v + 4w + d = 0 \qquad \text{... (3)}$$
$$3 + 2u + 2v + 2w + d = 0 \qquad \text{... (4)}$$

Also, since it touches the line $x = 0$, $y = z$, we have, on putting $x = 0$, $y = z$ in (1),

$$2z^2 + 2(v + w)z + d = 0$$

quadratic in z, which must have equal roots.

$$\therefore \qquad\qquad\qquad (v + w)^2 = 2d \qquad \text{... (5)}$$

From (2) and (4), $v + w = -1$ $\therefore$ From (5), $d = \dfrac{1}{2}$.

Substituting for d in (2) and (3), we get $\quad u = -\dfrac{3}{4}$ and $2v + 4w = -\dfrac{11}{2}$.

Solving $v + w = -1$ and $2v + 4w = -\dfrac{11}{2}$, we get $v = \dfrac{3}{4}$ and $w = -\dfrac{7}{4}$.

Hence from (1), the required equation of the sphere is

$$\boxed{x^2 + y^2 + z^2 - \frac{3}{2}x + \frac{3}{2}y - \frac{7}{2}z + \frac{1}{2} = 0}$$

Ex. 10 : *Find the equations of the spheres passing through the points (4, 1, 0), (2, –3, 4), (1, 0, 0) and touching the plane $2x + 2y - z = 11$.*

Sol. : Let the centre be (a, b, c) and radius be r.

$$\therefore \qquad (a - 4)^2 + (b - 1)^2 + c^2 = (a - 2)^2 + (b + 3)^2 + (c - 4)^2$$
$$= (a - 1)^2 + b^2 + c^2$$

This gives us two equations as

$$3a + b = 8 \quad \text{and} \quad a - 3b + 4c = 14$$

$\therefore \qquad b = 8 - 3a$ and $4c = 14 - a + 3b \Rightarrow c = \dfrac{19 - 5a}{2}$

Also, length of perpendicular from centre (a, b, c) to the given plane must be equal to its radius.

$$\left| \frac{2a + 2b - c - 11}{3} \right| = r \text{ (say)}$$

$\therefore \qquad (2a + 2b - c - 11)^2 = 9r^2$

$$= 9 \text{ [distance between (a, b, c) and one point say (1, 0, 0)]}$$
$$= 9 [(a - 1)^2 + b^2 + c^2]$$

Substituting for b and c, we get

$$16a^2 - 99a + 153 = 0 \quad \text{i.e. } (a - 3)(16a - 51) = 0 \text{ giving } a = 3, \frac{51}{16} .$$

Taking $a = 3$, we get $b = -1, c = 2$ and $r^2 = 9$.

$\therefore$ Equation of the sphere is

$$(x - 3)^2 + (y + 1)^2 + (z - 2)^2 = 9$$

or $\qquad x^2 + y^2 + z^2 - 6x + 2y - 4z + 5 = 0$

Similarly on taking $a = \dfrac{51}{16}$, we get $b = 25, c = \dfrac{49}{32}$

Hence, equation of the sphere is

$$\boxed{16 (x^2 + y^2 + z^2) - 102x + 50y - 49z + 86 = 0.}$$

Type II : Problems on Circle :

Ex. 11 : *Find the centre and radius of the circle of intersection of the sphere $x^2 + y^2 + z^2 - 2y - 4z - 11 = 0$ by the plane $x + 2y + 2z = 15$.*　　　　　　　　　　　　　　　　　　　　**(Nov./Dec. 2019, Dec. 2016, May 2017)**

Sol. :

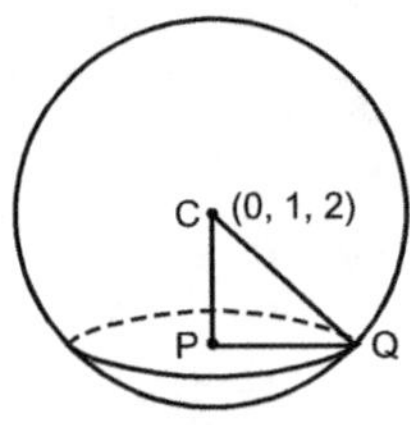

Fig. 7.19

Centre of the circle is the foot of the perpendicular from centre of the sphere C $(0, 1, 2)$ to the plane $x + 2y + 2z = 15$.

The d.r.'s of normal to the given plane are 1, 2, 2.

The equations of line CP through $(0, 1, 2)$ and perpendicular to the plane are

$$\frac{x - 0}{1} = \frac{y - 1}{2} = \frac{z - 2}{2} = k \text{ (say)}$$

Any point P on it is given by $(k, 2k + 1, 2k + 2)$. For foot of perpendicular this must lie on the given plane.

$\therefore \qquad k + 2 (2k + 1) + 2 (2k + 2) = 15$ i.e. $k = 1$.

$\therefore \quad \boxed{\text{Co-ordinates of P, the centre of the circle are } (1, 3, 4).}$

Distance $\qquad\qquad$ CP $= \sqrt{(1 - 0)^2 + (3 - 1)^2 + (4 - 2)^2}$

$\qquad\qquad\qquad\qquad\qquad = \sqrt{1 + 4 + 4} = 3$

Also $\qquad\qquad\qquad$ CQ $=$ radius of the given sphere $= \sqrt{0 + 1 + 4 + 11} = 4$

$\therefore$ Radius of the circle PQ $= \sqrt{CQ^2 - CP^2} = \sqrt{16 - 9}$

$\boxed{\text{Radius of the circle PQ} = \sqrt{7}}$

Note : Distance CP can also be obtained by finding length of the perpendicular from centre C $(0, 1, 2)$ on the given plane, which is

$$\left| \frac{0 + 2 (1) + 2 (2) - 15}{\sqrt{1 + 4 + 4}} \right| = \left| \frac{-9}{3} \right| = 3.$$

Ex. 12 : *Find the centre and radius of the circle $x^2 + y^2 + z^2 - 2x + 4y + 2z - 6 = 0$, $x + 2y + 2z - 4 = 0$*

Find the orthogonal projection of the area of the circle in yoz plane. 　　　　　　　　　　　**(May 2019, Dec. 2017)**

Sol. :

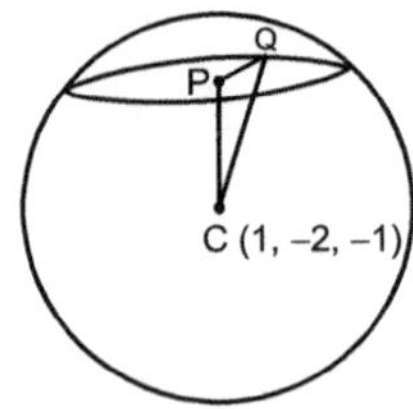

Fig. 7.20

Given plane cuts the sphere in a circle whose centre is P (Fig. 7.20), which is the foot of the perpendicular from the centre C $(1, -2, -1)$.

The d.r.'s of normal to the given plane are 1, 2, 2.

Equation of CP is $\dfrac{x - 1}{1} = \dfrac{y + 2}{2} = \dfrac{z + 1}{2} = k$ (say)

($\because$ d.r.'s of CP are same as d.r.'s of normal to the given plane)

Any point on this line is given by

$$x = k + 1, \qquad y = 2k - 2, \qquad z = 2k - 1 \qquad \text{... (1)}$$

Substituting in equation of plane $(k + 1) + 2(2k - 2) + 2(2k - 1) - 4 = 0$, which gives $k = 1$.

Putting $k = 1$ in (1), we get coordinates of P

i.e. $\boxed{\text{the centre of the circle as } x = 2, \; y = 0, \; z = 1.}$

Distance $CP = \sqrt{(2-1)^2 + (0+2)^2 + (1+1)^2} = \sqrt{1+4+4} = 3$

Also $CQ = $ radius of the given sphere

$$= \sqrt{1+4+1+6} = \sqrt{12}$$

$\therefore$ Radius of the given circle $PQ = \sqrt{CQ^2 - CP^2} = \sqrt{12 - 9} = \sqrt{3}$

$\therefore$ Area of circle $= \pi \left(\sqrt{3}\right)^2 = 3\pi$

Orthogonal projection of area of circle in yoz plane $= 3\pi \cos\theta$, where θ is angle between plane of circle and yoz plane, which is same as angle between normal to the plane of the circle and normal to yoz plane i.e. x-axis.

The d.r.'s of normal to the plane of circle are (1, 2, 2) and d.r.'s of normal to the yoz plane are (1, 0, 0).

$\therefore$ $\cos\theta = \dfrac{(1)\times(1) + (2)\times(0) + (2)\times(0)}{\sqrt{1+4+4}\,\sqrt{1+0+0}} = \dfrac{1}{3}$

$\therefore$ $\boxed{\text{Orthogonal projection} = 3\pi \cos\theta = 3\pi \times \dfrac{1}{3} = \pi}$

Ex. 13 : *Find the equation of the circle which is a section of the sphere* $x^2 + y^2 + z^2 + 6y - 6z - 21 = 0$ *and has its centre at the point* $M(2, -1, 2)$.

Sol. :

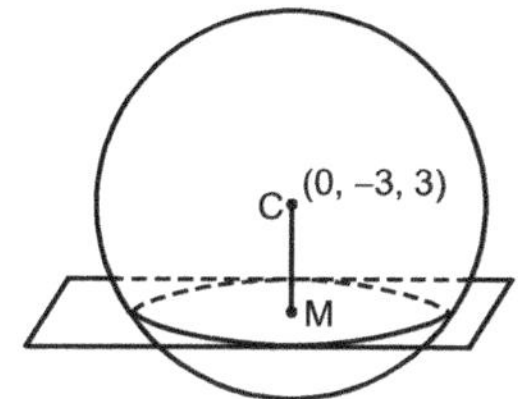

Fig. 7.21

Centre of the given sphere is C (0, −3, 3). CM is perpendicular to the plane of the circle. (Fig. 7.21). Hence the d.r.'s of normal to the plane of circle are $(2, -0, -1 + 3, 2 - 3)$. i.e. $(2, 2, -1)$.

$\therefore$ Equation of the plane passing through the point (2, −1, 2) and having d.r.'s of its normal as (2, 2, −1) is

$$2(x - 2) + 2(y + 1) - 1(z - 2) = 0$$

i.e. $\boxed{2x + 2y - z = 0}$ ... (1)

Equation of the given sphere is

     $\boxed{x^2 + y^2 + z^2 + 6y - 6z - 21 = 0}$ ... (2)

Hence (2) and (1) together represent the equations of circle.

Problems On Sphere Passing Through Circle :

Ex. 14 : *Find the sphere through the circle,* $x^2 + y^2 + z^2 = 4, z = 0$, *meeting the plane* $x + 2y + 2z = 0$ *in a circle of radius 3.*

(May 2007, 2004, 2018)

Sol. :

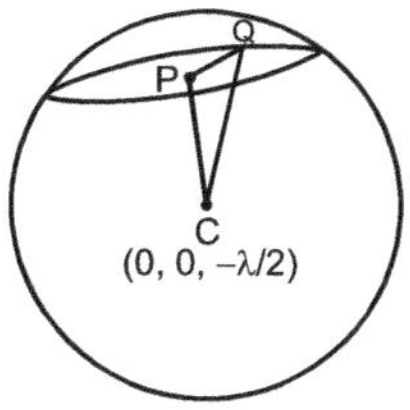

Equation of the sphere through the circle may be taken as

$$x^2 + y^2 + z^2 - 4 + \lambda z = 0 \qquad \text{... (1)}$$

whose centre C is $\left(0, 0, \dfrac{-\lambda}{2}\right)$

and radius $r = \sqrt{\dfrac{\lambda^2}{4} + 4} = \dfrac{\sqrt{\lambda^2 + 16}}{2}$

Fig. 7.22

$\because$ Sphere (1) is met by the plane $x + 2y + 2z = 0$ in a circle of radius 3.

$\therefore \quad PQ = 3, \quad CQ = r = \dfrac{\sqrt{\lambda^2 + 16}}{2}$ and $CP = \dfrac{1 \times 0 + 2 \times 0 + 2\,(-\lambda/2)}{\sqrt{1 + 4 + 4}} = \dfrac{-\lambda}{3}$

But $\qquad\qquad\qquad CQ^2 = CP^2 + PQ^2$ 　　　　　　　(Fig. 7.22)

$\therefore \qquad\qquad \dfrac{\lambda^2 + 16}{4} = \dfrac{\lambda^2}{9} + 9$

$\therefore \quad 5\lambda^2 = 180$ or $\lambda = \pm 6$

Substituting in (1), equation of requisite sphere is

$$\boxed{x^2 + y^2 + z^2 \pm 6z - 4 = 0.}$$

Ex. 15 : *Find the equation of the sphere which passes through the point (3, 1, 2) and meets xoy plane in a circle of radius 3 units with the centre at (1, –2, 0).*　　　　　　**(Dec. 05, 04; May 08, Nov. 15)**

Sol. :

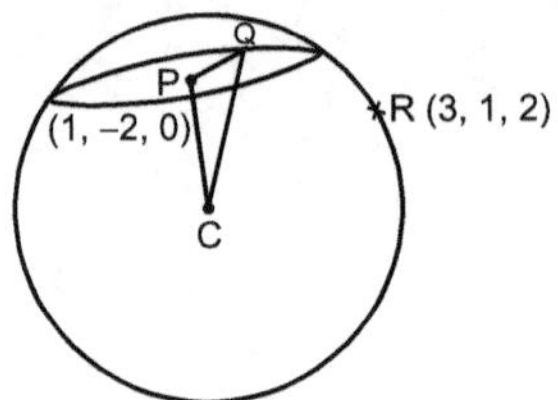

Fig. 7.23

Normal to the xoy plane is z-axis whose d.r.'s are (0, 0, 1). In Fig. 7.23, C is the centre of the sphere and P is the centre of circle. The d.r.'s of CP are (0, 0, 1).

Equation of line CP is

$$\frac{x-1}{0} = \frac{y+2}{0} = \frac{z-0}{1} = k$$

$\therefore \quad x = 1,\ y = -2,\ z = k$ is any point on CP. For some k, let C be the point (1, –2, k) and P is the given point (1, –2, 0).

$\therefore \qquad\qquad CP^2 = k^2$ and $PQ =$ radius of circle $= 3$ 　　　　　(given)

$\qquad\qquad CQ^2 = CP^2 + PQ^2 = k^2 + 9$ (CQ = Radius of sphere)

$\qquad\qquad CR^2 = (1-3)^2 + (-2-1)^2 + (k-2)^2 = 4 + 9 + k^2 - 4k + 4$

$\qquad\qquad\qquad = k^2 - 4k + 17$

$\qquad$ but $\quad CQ^2 = CR^2$ (each is equal to radius of the sphere)

$\therefore \qquad k^2 - 4k + 17 = k^2 + 9$ which gives $4k = 8$ or $k = 2.$

Thus centre of the sphere is (1, –2, 2) and radius of sphere

$$= \sqrt{k^2 + 9} = \sqrt{4 + 9} = \sqrt{13} \qquad\qquad (\because\ CQ^2 = CP^2 + PQ^2)$$

$\therefore \quad$ Equation of required sphere is

$$(x-1)^2 + (y+2)^2 + (z-2)^2 = 13$$

or $\qquad \boxed{x^2 + y^2 + z^2 - 2x + 4y - 4z - 4 = 0}$

Ex. 16 : *Find the equation of the sphere which passes through the point (4, 6, 3) and meets xoy plane in a circle of radius 5 units having centre as (1, 2, 0).*　　　　　　**(May 2005)**

Sol. : Equations of circle in xoy plane are

$$(x-1)^2 + (y-2)^2 = 25,\ z = 0 \text{ (note the step)}$$

i.e. $\qquad x^2 + y^2 - 2x - 4y - 20 = 0,\ z = 0$ 　　　　　　　... (1)

$\therefore \quad$ Equation of the sphere passing through circle (1) is given by

$$(x^2 + y^2 + z^2 - 2x - 4y - 20) + \lambda z = 0 \text{(note the step)} \qquad ... (2)$$

This sphere passes through the point (4, 6, 3)

$\therefore \qquad (16 + 36 + 9 - 8 - 24 - 20) + 3\lambda = 0 \quad \therefore\ \lambda = -3$

Substituting $\lambda = -3$ in (2), the required sphere is

$$\boxed{x^2 + y^2 + z^2 - 2x - 4y - 3z - 20 = 0}$$

Note : This problem can also be solved by method given in Ex. 15.

Ex. 17 : *Find the equation of the sphere which touches the sphere, $x^2 + y^2 + z^2 - x + 3y + 2z - 3 = 0$ at the point (1, 1, −1) and passes through the point (0, 0, 3).*   **(Dec. 2007)**

Sol. : Equation of tangent plane to the given sphere at (1, 1, −1) can be written as (Refer article 7.5 (7))

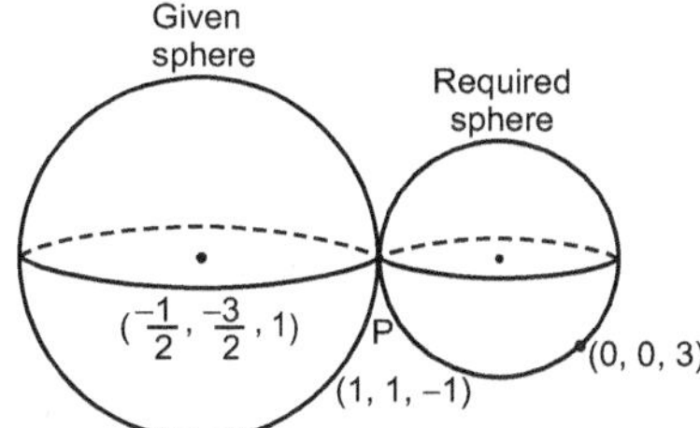

Fig. 7.24

$$x \cdot 1 + y \cdot 1 + z\,(-1) - \frac{1}{2}\,(x + 1) + \frac{3}{2}\,(y + 1) + (z - 1) - 3 = 0$$

i.e. $$2x + 2y - 2z - x - 1 + 3y + 3 + 2z - 2 - 6 = 0$$

or $$x + 5y - 6 = 0$$

$\therefore$
$$\left.\begin{array}{l} x^2 + y^2 + z^2 - x + 3y + 2z - 3 = 0 \\ x + 5y - 6 = 0 \end{array}\right\} \qquad \ldots (1)$$

will represent section of the sphere by a tangent plane which is a point circle.

Taking equation of the sphere through this circle as,

$$x^2 + y^2 + z^2 - x + 3y + 2z - 3 + \lambda\,(x + 5y - 6) = 0 \qquad \ldots (2)$$

Sphere (2) will certainly touch the given sphere at (1, 1, − 1).

If this passes through the point (0, 0, 3), it should satisfy the equation of sphere (2).

$\therefore$ $$9 + 6 - 3 + \lambda\,(-6) = 0 \quad \text{which gives} \quad 6\lambda = 12 \quad \text{or} \quad \lambda = 2$$

Substituting $\lambda = 2$ in (2), we get required sphere as,

$$\boxed{x^2 + y^2 + z^2 + x + 13y + 2z - 15 = 0}$$

Ex. 18 : *Find the equation of the sphere passing through the points (1, 0, 0), (0, 1, 0), (0, 0, 1) and having its radius as small as possible.*

Sol. : First we find the equations of circle passing through the points A (1, 0, 0), B (0, 1, 0) and C (0, 0, 1) as a section of sphere passing through O and intercepts OA = 1, OB = 1, OC = 1 on the axes given by;

$$S \equiv x^2 + y^2 + z^2 - x - y - z = 0 \qquad \text{[see article 7.2 (4)]}$$

and the plane ABC making intercepts 1, 1, 1 on the axes given by;

$$\frac{x}{1} + \frac{y}{1} + \frac{z}{1} = 1 \quad \text{i.e.} \quad x + y + z = 1 \qquad \ldots (2)$$

Equation of any sphere through the circles (1), (2) is

$$S' \equiv (x^2 + y^2 + z^2 - x - y - z) + \lambda\,(x + y + z - 1) = 0 \qquad \ldots (3)$$

i.e. $$S' \equiv x^2 + y^2 + z^2 + (\lambda - 1)\,x + (\lambda - 1)\,y + (\lambda - 1)\,z - \lambda = 0 \qquad \ldots (4)$$

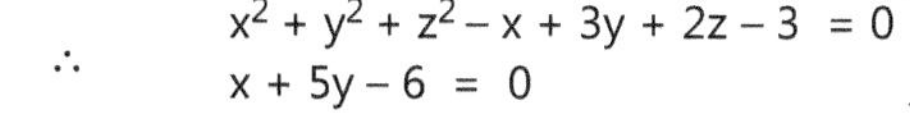

Fig. 7.25

The centre of the sphere (4) is

$$C\left[-\frac{(\lambda - 1)}{2}, \ -\frac{(\lambda - 1)}{2}, \ -\frac{(\lambda - 1)}{2}\right] \qquad \ldots (5)$$

If r is the radius of circle ABC and p is the perpendicular from centre of the sphere to the plane ABC and R is radius of the sphere (4), (Fig. 7.24), then

$$R^2 = p^2 + r^2 \qquad \ldots (6)$$

Since radius r of circle ABC is fixed, R is least if p = 0 i.e. centre C of the sphere (4) must lie on the plane (2).

i.e. $$-\frac{(\lambda - 1)}{2} - \frac{(\lambda - 1)}{2} - \frac{(\lambda - 1)}{2} = 1 \qquad \qquad \therefore \ \lambda = \frac{1}{3}$$

Substituting the value of λ in (3), the required sphere is given by

$$S' \equiv (x^2 + y^2 + z^2 - x - y - z) + \frac{1}{3}(x + y + z - 1) = 0$$

or
$$\boxed{S' \equiv 3(x^2 + y^2 + z^2) - \frac{2}{3}x - \frac{2}{3}y - \frac{2}{3}z - \frac{1}{3} = 0}.$$

Note : This problem can also be solved by the method given in illustrations 7.4, Ex. 6.

Ex. 19 : *Show that the circles,*

$$2(x^2 + y^2 + z^2) + 8x - 13y + 17z - 17 = 0 = 2x + y - 3z + 1$$
$$x^2 + y^2 + z^2 + 3x - 4y + 3z = 0 = x - y + 2z - 4$$

lie on the same sphere. Find its equation. **(Dec. 2006)**

Sol. : Equations of the spheres passing through the given circles can be taken as :

$$2(x^2 + y^2 + z^2) + 8x - 13y + 17z - 17 + \lambda_1(2x + y - 3z + 1) = 0$$

or $\quad 2(x^2 + y^2 + z^2) + x(8 + 2\lambda_1) + y(-13 + \lambda_1) + z(17 - 3\lambda_1) + \lambda_1 - 17 = 0$ ... (1)

and $\quad x^2 + y^2 + z^2 + 3x - 4y + 3z + \lambda_2(x - y + 2z - 4) = 0$

or $\quad x^2 + y^2 + z^2 + x(3 + \lambda_2) + y(-4 - \lambda_2) + z(3 + 2\lambda_2) - 4\lambda_2 = 0$ ... (2)

If the given circles lie on the same sphere, (1) and (2) must be same equations. Comparing (1) and (2), we get

$$\frac{2}{1} = \frac{8 + 2\lambda_1}{3 + \lambda_2} = \frac{-13 + \lambda_1}{-4 - \lambda_2} = \frac{17 - 3\lambda_1}{3 + 2\lambda_2} = \frac{\lambda_1 - 17}{-4\lambda_2}$$

These give four equations in two unknowns which are consistent and give $\lambda_1 = 1$, $\lambda_2 = 2$ putting $\lambda_1 = 1$ in equation (1) or $\lambda_2 = 2$ in equation (2), we get same sphere containing two circles as,

$$\boxed{x^2 + y^2 + z^2 + 5x - 6y + 7z - 8 = 0}$$

Ex. 20 : *A sphere S has points (1, –2, 3), (4, 0, 6) as opposite ends of a diameter. Find the equation of the sphere having the intersection of S with the plane $x + y - 2z + 6 = 0$ as its great circle.* **(May 2008, Nov. 2014)**

Sol. : Equation of the sphere S with (1, –2, 3) and (4, 0, 6) as diametric points is [refer article 7.2 (5)]

$$(x - 1)(x - 4) + (y + 2)(y - 0) + (z - 3)(z - 6) = 0$$

i.e. $\quad x^2 + y^2 + z^2 - 5x + 2y - 9z + 22 = 0$ ... (1)

Equation of the sphere passing through circle is

$$(x^2 + y^2 + z^2 - 5x + 2y - 9z + 22) + \lambda(x + y - 2z + 6) = 0$$

i.e. $\quad x^2 + y^2 + z^2 - (5 - \lambda)x + (2 + \lambda)y - (9 + 2\lambda)z + 22 + 6\lambda = 0$ ... (2)

The centre of the sphere is

$$\left(\frac{5 - \lambda}{2}, -\frac{\lambda + 2}{2}, \frac{9 + 2\lambda}{2}\right)$$

Since the circle of intersection of sphere (1) and the plane

$$x + y - 2z + 6 = 0 \qquad ... (3)$$

is a great circle of the sphere, centre of the sphere must lie on the plane (3).

Hence $\quad \left(\frac{5 - \lambda}{2}\right) - \left(\frac{\lambda + 2}{2}\right) - 2\left(\frac{9 + 2\lambda}{2}\right) + 6 = 0 \qquad \therefore \ \lambda = -\frac{1}{2}$

Substituting for λ in (2), the required sphere is

$$\boxed{2(x^2 + y^2 + z^2) - 11x + 3y - 16z + 38 = 0}$$

Ex. 21 : *Find the equation of the sphere through the circle $x^2 + y^2 = 9$; $z = 0$ and the point (α, β, γ).* **(Dec. 2017)**

Sol. : Here given sphere is $S \equiv x^2 + y^2 - 9 = 0$ and the plane is $U \equiv z = 0$.

But $S + \lambda U \equiv (x^2 + y^2 - 9) + \lambda z = 0$ does not represent sphere, hence we try to represent the same circle as intersection of sphere $x^2 + y^2 + z^2 = 9$ and the plane $z = 0$.

Hence the sphere passing through the given circle is

$$(x^2 + y^2 + z^2 - 9) + \lambda z = 0 \qquad ... (1)$$

Since (1) passes through the point (α, β, γ), we get

$$(\alpha^2 + \beta^2 + \gamma^2 - 9) + \lambda\gamma = 0 \qquad \qquad \ldots (2)$$

Substituting the value of λ from (2) in (1) (i.e. by eliminating λ between (1) and (2)), the required equation of the sphere is given by

$$\boxed{\gamma\,(x^2 + y^2 + z^2 - 9) = (\alpha^2 + \beta^2 + \gamma^2 - 9)\,z}.$$

Ex. 22 : *Find the equation of tangent line in symmetrical form to the circle* $x^2 + y^2 + z^2 + 2x - 4y + 8z - 9 = 0;\ x + y + z = 3$ *at the point (1, 1, 1).*

Sol. : Line joining centre of the circle and the point (1, 1, 1) is perpendicular to the tangent line to the circle. Similarly, normal to the tangent plane at (1, 1, 1) is also perpendicular to the tangent line. Centre of given sphere is $(-1, 2, -4)$.

Equation of the normal to the plane $x + y + z = 3$, passing through $(-1, 2, -4)$ is given by;

$$\frac{x + 1}{1} = \frac{y - 2}{1} = \frac{z + 4}{1} = k.$$

Any point on the line is $(k - 1, k + 2, k - 4)$, putting for x, y, z in the plane $x + y + z = 3$, we get

$$k - 1 + k + 2 + k - 4 = 3 \therefore 3k = 6 \text{ or } k = 2.$$

$\therefore$ Centre of circle P is $(2 - 1, 2 + 2, 2 - 4)$ or $(1, 4, -2)$.

If Q is the point (1, 1, 1), then the d.r.'s of PQ are $(1 - 1, 1 - 4, 1 + 2)$ i.e. $(0, -3, 3)$.

Also, the d.r.'s of normal to the tangent plane at (1, 1, 1) are same as the d.r.'s of line joining centre of the sphere C $(-1, 2, -4)$ and point Q (1, 1, 1). Therefore, the d.r.'s of CQ are $(2, -1, 5)$.

Let (l, m, n) be the d.r.'s of the tangent line at Q (1, 1, 1). Since it is perpendicular to PQ and CQ both, applying condition of perpendicularity (see article 7.9 result 13), we get

$$0 \cdot l - 3m + 3n = 0, \qquad 2l - m + 5n = 0$$

$$\therefore \qquad \frac{l}{-15 + 3} = \frac{m}{6} = \frac{n}{6} \quad \text{or} \quad \frac{l}{2} = \frac{m}{-1} = \frac{n}{-1}$$

$$\therefore \quad \boxed{\text{Equation of tangent line is } \frac{x - 1}{2} = \frac{y - 1}{-1} = \frac{z - 1}{-1}}$$

Ex. 23 : *Find the equations of tangent line to the circle* $x^2 + y^2 + z^2 + 5x - 7y + 2z - 8 = 0;\ 3x - 2y + 4z + 3 = 0$ *at the point* $(-3, 5, 4)$.

Sol. : Tangent line at $(-3, 5, 4)$ is the common section of the plane of the circle $3x - 2y + 4z + 3 = 0$ and the tangent plane to the given sphere at $(-3, 5, 4)$. Equation of tangent plane at $(-3, 5, 4)$ is given by

$$x\,x_1 + y\,y_1 + z\,z_1 + \frac{5}{2}\,(x + x_1) - \frac{7}{2}\,(y + y_1) + (z + z_1) - 8 = 0$$

or $\qquad -3x + 5y + 4z + \dfrac{5}{2}\,(x - 3) - \dfrac{7}{2}\,(y + 5) + (z + 4) - 8 = 0$

or $\qquad x - 3y - 10z + 50 = 0$

Thus the tangent line (in general form) is given by;

$$\boxed{3x - 2y + 4z + 3 = 0;\quad x - 3y - 10z + 50 = 0}.$$

Ex. 24 : *Find the equation of the sphere through the circle,* $x^2 + y^2 + z^2 = 1,\ \ 2x + 3y + 4z = 5$ *and which intersects the sphere* $x^2 + y^2 + z^2 + 3\,(x - y + z) - 56 = 0$ *orthogonally.*   **(Nov./Dec. 2019, May 2009, 2015)**

Sol. : Equation of the sphere through given circle can be taken as,

$$x^2 + y^2 + z^2 - 1 + \lambda\,(2x + 3y + 4z - 5) = 0$$

or $\qquad x^2 + y^2 + z^2 + 2\lambda x + 3\lambda y + 4\lambda z - 1 - 5\lambda = 0 \qquad \qquad \ldots (1)$

If this intersects the sphere, $x^2 + y^2 + z^2 + 3x - 3y + 3z - 56 = 0$ orthogonally, using condition of orthogonality $2u_1 u_2 + 2v_1 v_2 + 2w_1 w_2 = d_1 + d_2$ (refer article 7.10 (13)), we have

$$2\lambda\left(\frac{3}{2}\right) + 3\lambda\left(\frac{-3}{2}\right) + 4\lambda\left(\frac{3}{2}\right) = -1 - 5\lambda - 56$$

$$\therefore \qquad 6\lambda - 9\lambda + 12\lambda = -10\lambda - 114$$

or　　　　　　　　　　　　　$19\lambda = -114$ or $\lambda = -6$

Putting for λ in (1), we get

$$\boxed{x^2 + y^2 + z^2 - 12x - 18y - 24z + 29 = 0}$$

which is the desired equation of the sphere.

Ex. 25 : *Find the equation of the sphere which touches the plane $3x + 2y - z + 2 = 0$ at the point $(1, -2, 1)$ and also cuts orthogonally the sphere $x^2 + y^2 + z^2 - 4x + 6y + 4 = 0$.*

Sol. :　The given plane is

$$3x + 2y - z + 2 = 0 \qquad \qquad \text{... (1)}$$

and the sphere is

$$x^2 + y^2 + z^2 - 4x + 6y + 4 = 0 \qquad \qquad \text{... (2)}$$

$\because$　The required sphere touches the plane (1) at $(1, -2, 1)$.

$\therefore$　Its centre will lie on a line through $(1, -2, 1)$ and perpendicular to the plane. Its equation is

$$\frac{x-1}{3} = \frac{y+2}{2} = \frac{z-1}{-1} = k \text{ (say)}$$

Co-ordinates of centre of the required sphere be $(3k + 1, 2k - 2, -k + 1)$ and radius is the distance of the point from point of contact $(1, -2, 1)$ and is

$$\sqrt{(3k + 1 - 1)^2 + (2k - 2 + 2)^2 + (-k + 1 - 1)^2} = k\sqrt{14}.$$

Centre of the sphere (2) is $(2, -3, 0)$ and radius is 3.

The two spheres cut orthogonally, if

Square of the distance between the centres $=$ Sum of squares of their radii.

$\therefore$　　　$(3k + 1 - 2)^2 + (2k - 2 + 3)^2 + (-k + 1 - 0)^2 = 14k^2 + 9$

or　　　$(3k - 1)^2 + (2k + 1)^2 + (k - 1)^2 = 14k^2 + 9$ which gives $k = -\dfrac{3}{2}$

Hence, the centre is $\left(-\dfrac{7}{2}, -5, \dfrac{5}{2}\right)$ and radius is $\dfrac{3\sqrt{14}}{2}$, and its equation is

$$\left(x + \frac{7}{2}\right)^2 + (y + 5)^2 + \left(z - \frac{5}{2}\right)^2 = \left(\frac{3\sqrt{14}}{2}\right)^2$$

or　　　$\boxed{x^2 + y^2 + z^2 + 7x + 10y - 5z + 12 = 0.}$

MISCELLANEOUS EXAMPLES

Ex. 1 :　*If d is the distance between the two spheres of radii r_1 and r_2, prove that the angle between them is $\cos^{-1}\dfrac{r_1^2 + r_2^2 - d^2}{2\,r_1 r_2}$. Hence find the angle of intersection of the sphere $x^2 + y^2 + z^2 - 2x - 4y - 6z + 10 = 0$ with the sphere, passing through $(1, 2, -3)$ and $(5, 0, 1)$ as extremities of its diameter.*

Sol. : Equation of second sphere [Refer article 7.2 (5)] is

$$(x - 1)(x - 5) + (y - 2)(y - 0) + (z + 3)(z - 1) = 0$$

i.e. $x^2 + y^2 + z^2 - 6x - 2y + 2z + 2 = 0$

The centre C_2 is $(3, 1, -1)$ and radius $r_2^2 = 9 + 1 + 1 - 2 = 9$.　$\therefore$　$r_2 = 3$ 　　　... (1)

For the first sphere, centre C_1 is $(1, 2, 3)$ and radius $r_1^2 = 1 + 4 + 9 - 10 = 4$ $\therefore$ $r_1 = 2$ 　　　... (2)

The distance d between the centres is given by

$$d^2 = (3 - 1)^2 + (1 - 2)^2 + (-1 - 3)^2 = 21 \qquad \qquad \text{... (3)}$$

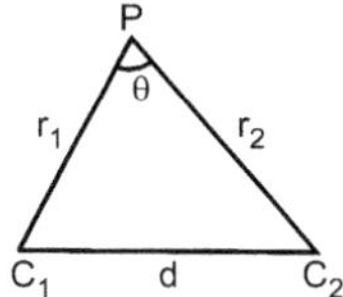

Fig. 7.26

If P is any one of the point of intersection of the spheres then by cosine formula for $\triangle PC_1C_2$, we get

$$\cos\theta = \frac{r_1^2 + r_2^2 - d^2}{2\,r_1 r_2}$$

Hence from (1), (2) and (3), angle of intersection for the given spheres is

$$\cos\theta = \frac{4 + 9 - 21}{2\,(2)\,(3)} = -\frac{8}{12} = -\frac{2}{3}$$

$$\therefore \qquad \theta = \cos^{-1}\left(-\frac{2}{3}\right)$$

Ex. 2 : *Find the equation of the diameter of the sphere $x^2 + y^2 + z^2 = 17$, such that the rotation about it will transfer the point (2, 3, 2) to the point (4, 0, 1) along a great circle of the sphere. Find the angle through which the sphere must be rotated.*

Sol. :

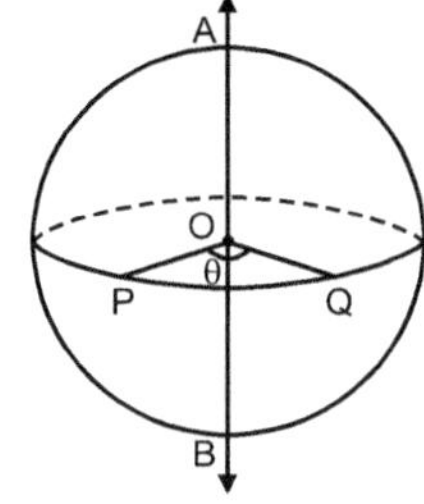

Fig. 7.27

The centre of the sphere is O (0, 0, 0) and given points are P (2, 3, 2), Q (4, 0, 1). Let AB denote the diameter whose equation is required.

Since rotation of the sphere about diameter AB transfers the point P (2, 3, 2) to Q (4, 0, 1) along great circle, therefore angle of rotation is $\angle POQ$ and diameter AB is perpendicular to plane OPQ.

Also, the d.r.'s of OP are 2, 3, 2 and those of OQ are 4, 0, 1. Now, let l, m, n be the d.r.'s of the diameter of rotation AB. Then, since this diameter is perpendicular to both OP and OQ, we have

$$2l + 3m + 2n = 0 \qquad \text{and} \qquad 4l + 0m + 1n = 0$$

Solving these, we get

$$\frac{l}{3-0} = \frac{-m}{2-8} = \frac{n}{0-12} \quad \text{or} \quad \frac{l}{3} = \frac{m}{6} = \frac{n}{-12}$$

i.e.
$$\frac{l}{1} = \frac{m}{2} = \frac{n}{-4}$$

$\therefore$ The equation of the diameter of rotation is

$$\frac{x-0}{1} = \frac{y-0}{2} = \frac{z-0}{-4} \quad \text{i.e.} \quad \frac{x}{1} = \frac{y}{2} = \frac{z}{-4}$$

Next, let
$$\angle POQ = \theta \ (\text{angle of rotation})$$

$\therefore$
$$\cos\theta = \frac{(2)\,(4) + (3)\,(0) + (2)\,(1)}{\sqrt{4 + 9 + 4}\ \sqrt{16 + 0 + 1}} = \frac{10}{17}$$

$\therefore$
$$\boxed{\theta = \cos^{-1}\frac{10}{17}}.$$

Ex. 3 : *The plane ABC whose equation is $\dfrac{x}{a} + \dfrac{y}{b} + \dfrac{z}{b} = 1$ meets the axes in the points A, B, C. Find the equation of circumcircle of the triangle ABC and obtain the coordinates of its centre.*

Sol. : The co-ordinates of the points A, B, C are respectively (a, 0, 0), (0, b, 0), (0, 0, c). The equation of the sphere OABC is

$$x^2 + y^2 + z^2 - ax - by - cz = 0 \qquad \qquad \text{... (1)}$$

and the given plane ABC is

$$\frac{x}{a} + \frac{y}{b} + \frac{z}{c} = 1 \qquad \qquad \text{... (2)}$$

The circumcircle of triangle ABC is the circle which is the section of the sphere (1) by the given plane (2). Thus the equations of the circumcircle are given by

$$x^2 + y^2 + z^2 - ax - by - cz = 0; \quad \frac{x}{a} + \frac{y}{b} + \frac{z}{c} = 1$$

To find the circumcentre of triangle ABC, the perpendicular from centre of the sphere (1) meets the plane ABC in the circumcentre of triangle ABC.

The equation of the line passing plane through the centre $\left(\dfrac{a}{2}, \dfrac{b}{2}, \dfrac{c}{2}\right)$ of the sphere and perpendicular to the given plane are

$$\frac{x - \dfrac{a}{2}}{\dfrac{1}{a}} = \frac{y - \dfrac{b}{2}}{\dfrac{1}{b}} = \frac{z - \dfrac{c}{2}}{\dfrac{1}{c}} = k \text{ (say)}$$

$\therefore$ Any point on this line is $\left(\dfrac{k}{a} + \dfrac{a}{2}, \dfrac{k}{b} + \dfrac{b}{2}, \dfrac{k}{c} + \dfrac{c}{2}\right)$ ... (3)

This point will be the centre of the circle, if it lies on the given plane, i.e. if

$$\frac{1}{a}\left(\frac{k}{a} + \frac{a}{2}\right) + \frac{1}{b}\left(\frac{k}{b} + \frac{b}{2}\right) + \frac{1}{c}\left(\frac{k}{c} + \frac{c}{2}\right) = 1$$

or $k\left(\dfrac{1}{a^2} + \dfrac{1}{b^2} + \dfrac{1}{c^2}\right) = -\dfrac{1}{2}$ i.e. $k = -\dfrac{1}{2\,(a^{-2} + b^{-2} + c^{-2})}$

Putting this value of k in (3), the co-ordinates of the centre as

$$\boxed{\;\frac{a\,(b^{-2} + c^{-2})}{2\,(a^{-2} + b^{-2} + c^{-2})}\,,\; \frac{b - (c^{-2} + a^{-2})}{2\,(a^{-2} + b^{-2} + c^{-2})}\,,\; \frac{c\,(a^{-2} + b^{-2})}{2\,(a^{-2} + b^{-2} + c^{-2})}\;}$$

Ex. 4 : *If r is the radius of the circle $x^2 + y^2 + z^2 + 2ux + 2vy + 2wz + d = 0$ and $lx + my + nz = 0$, prove that*

$$(r^2 + d)\,(l^2 + m^2 + n^2) = (mw - nv)^2 + (nu - lw)^2 + (lv - mu)^2$$

Sol. : The centre of the given sphere is $(-u, -v, -w)$ and radius is

$$R^2 = u^2 + v^2 + w^2 - d.$$

We know that if r is the radius of the circle and p is the perpendicular from the centre $(-u, -v, -w)$ of the sphere to the plane $lx + my + nz = 0$, then

$$r^2 = R^2 - p^2 \qquad\qquad \text{[refer article 7.6 (iii)]}$$

$\therefore$ $r^2 = (u^2 + v^2 + w^2 - d) - \left[\dfrac{-lu - mv - nw}{\sqrt{l^2 + m^2 + n^2}}\right]^2$

$\therefore$ $(r^2 + d)\,(l^2 + m^2 + n^2) = (u^2 + v^2 + w^2)\,(l^2 + m^2 + n^2) - (lu + mv + nw)^2$

$$\boxed{(r^2 + d)\,(l^2 + m^2 + n^2) = (mw - nv)^2 + (nu - lw)^2 + (lv - mu)^2} \qquad \text{(by Lagrange's identity)}$$

Lagrange's identity :

$$\left(l_1^2 + m_1^2 + n_1^2\right)\left(l_2^2 + m_2^2 + n_2^2\right) - (l_1 l_2 + m_1 m_2 + n_1 n_2)^2 = (m_1 n_2 - m_2 n_1)^2 + (n_1 l_2 - n_2 l_1)^2 + (l_1 m_2 - l_2 m_1)^2$$

Note : If l, m, n are the d.c.'s, $l^2 + m^2 + n^2 = 1$, then the given result becomes

$$(r^2 + d) = (mw - nv)^2 + (nu - lw)^2 + (lv - mu)^2$$

Ex. 5 : *If OA, OB, OC are mutually perpendicular lines having direction cosines (l_r, m_r, n_r), $r = 1, 2, 3$. If $OA = a$, $OB = b$, $OC = c$, find the equation of the sphere OABC.*

Sol. : (l_1, m_1, n_1), (l_2, m_2, n_2), (l_3, m_3, n_3) are d.c.'s of three mutually perpendicular lines drawn from origin to meet the sphere in $OA = a$, $OB = b$ and $OC = c$.

$\therefore$ Co-ordinates of A, B, C are respectively $A(l_1 a, m_1 a, n_1 a)$, $B(l_2 b, m_2 b, n_2 b)$, $C(l_3 c, m_3 c, n_3 c)$.

Let the equation of the sphere OABC be

$$x^2 + y^2 + z^2 + 2ux + 2vy + 2wz = 0 \qquad\qquad ... (1)$$

[d = 0 as the sphere passes through the origin]

Substituting the co-ordinates of A in (1), we have

$$a^2\left(l_1^2 + m_1^2 + n_1^2\right) + 2ul_1 a + 2vm_1 a + 2wn_1 a = 0$$

or $a + 2ul_1 + 2vm_1 + 2wn_1 = 0$ $\left[\because l_1^2 + m_1^2 + n_1^2 = 1\right]$ ... (2)

Similarly,

$$b + 2ul_2 + 2vm_2 + 2wn_2 = 0 \qquad \text{... (3)}$$
$$c + 2ul_3 + 2vm_3 + 2wn_3 = 0 \qquad \text{... (4)}$$

To find u, multiply (2), (3), (4) by l_1, l_2, l_3 respectively and add

$$al_1 + bl_2 + cl_3 + 2u\left(l_1^2 + l_2^2 + l_3^2\right) + 2v\,(0) + 2w\,(0) = 0$$

i.e. $\qquad\qquad 2u = -(al_1 + bl_2 + cl_3) \qquad\qquad [\because\ l_1^2 + m_1^2 + n_1^2 = 1, l_1m_1 + l_2m_2 + l_3m_3 = 0 \text{ etc.}]$

Similarly, $\qquad\qquad 2v = -(am_1 + bm_2 + cm_3)$

$$2w = -(an_1 + bn_2 + cn_3)$$

Substituting in (1), the required equation of the sphere OABC is

$$\boxed{x^2 + y^2 + z^2 - (al_1 + bl_2 + cl_3)\,x - (am_1 + bm_2 + cm_3)\,y - (an_1 + bn_2 + cn_3)\,z = 0}$$

Ex. 6 : *If three mutually perpendicular lines whose direction cosines are* (l_r, m_r, n_r), $r = 1, 2, 3$ *are drawn from origin 'O' to meet the sphere* $x^2 + y^2 + z^2 = a^2$ *in A, B, C, find the equation of the plane ABC and also radius of circle ABC.*

Sol. : (l_1, m_1, n_1), (l_2, m_2, n_2), (l_3, m_3, n_3) are d.c.'s of three lines drawn from origin to meet the sphere in A, B, C.

∴ Coordinates of A, B, C are respectively $A(l_1a, m_1a, n_1a)$, $B(l_2a, m_2a, n_2a)$, $C(l_3a, m_3a, n_3a)$.

Let the equation of the plane ABC be,

$$a_1x + b_1y + c_1z = 1 \qquad \text{... (1)}$$

Substituting coordinates of A, B, C in (1), we have

$$a_1l_1a + b_1m_1a + c_1n_1a = 1 \qquad \text{... (2)}$$
$$a_1l_2a + b_1m_2a + c_1n_2a = 1 \qquad \text{... (3)}$$
$$a_1l_3a + b_1m_3a + c_1n_3a = 1 \qquad \text{... (4)}$$

Multiplying (2), (3), (4) by l_1, l_2, l_3 and using conditions of perpendicularity,

$\left(l_1^2 + l_2^2 + l_3^2 = 1,\ l_1m_1 + l_2m_2 + l_3m_3 = 0 \text{ etc.}\right)$ and adding, we get $a_1a = l_1 + l_2 + l_3$ or $a_1 = \dfrac{1}{a}\,(l_1 + l_2 + l_3)$.

Similarly multiplying (2), (3), (4) by m_1, m_2, m_3 and then by n_1, n_2, n_3 and adding, we get

$$b_1 = \frac{1}{a}\,(m_1 + m_2 + m_3), \quad c_1 = \frac{1}{a}\,(n_1 + n_2 + n_3)$$

Substituting for a_1, b_1, c_1 in (1), plane of ABC is obtained as,

$$\boxed{(l_1 + l_2 + l_3)\,x + (m_1 + m_2 + m_3)\,y + (n_1 + n_2 + n_3)\,z = a}\ .$$

If p is length of perpendicular from $(0, 0, 0)$, centre of the sphere on the plane ABC, then

$$p^2 = \frac{a^2}{(l_1 + l_2 + l_3)^2 + (m_1 + m_2 + m_3)^2 + (n_1 + n_2 + n_3)^2}$$

∴ $\qquad p^2 = \dfrac{a^2}{3}$ (using $\sum l_1^2 = \sum l_2^2 = \sum l_3^2 = 1$, $\sum l_1l_2 = \sum l_2l_3 = \sum l_3l_1 = 0$)

Radius of sphere $= a$. If $r =$ radius of circle, then $r^2 = a^2 - p^2$ [Refer article 7.6 (iii)]

∴ $\qquad r^2 = a^2 - \dfrac{a^2}{3} = \dfrac{2a^2}{3}$

∴ $\qquad \boxed{r = \sqrt{\dfrac{2}{3}}\,a}$

EXERCISE 7.2

1. Find the equation of tangent plane and normal to the sphere $x^2 + y^2 + z^2 - 4x + 2y = 4$ at the point $(4, -2, 2)$.

   Ans. $2x - y + 2z = 14;\ \dfrac{x - 4}{2} = \dfrac{y + 2}{-1} = \dfrac{z - 2}{2}$.

2. Find the sphere, extremities of whose diameter are $(1, 2, 2)$, $(-3, 1, 4)$. Find the tangent planes at these points.

   Ans. Equation : $x^2 + y^2 + z^2 + 2x - 3y - 6z + 7 = 0$, Tangent planes are : $4x + y - 2z - 2 = 0$ and $4x + y - 2z + 19 = 0$.

3. Prove that plane $x + y + z = 1$ touches the sphere

   $3\,(x^2 + y^2 + z^2) - 30x + 12y - 18z + 89 = 0$ and find the co-ordinates of the point of contact. **Ans.** $\left(\dfrac{10}{3}, -\dfrac{11}{3}, \dfrac{4}{3}\right)$.

4. Prove that the sphere $x^2 + y^2 + z^2 + 2x - 4y - 2z - 3 = 0$ touch the plane $2x - 2y + z + 16 = 0$ and find the point of contact. **(May 11) Ans.** $(3, 4, -2)$.

5. Find the equation of the sphere which passes through the point $(-1, 0, 0)$ and touches the plane $2x - y - 2z - 4 = 0$ at the point $(1, 2, -2)$.   **(Dec. 2011) Ans.** $x^2 + y^2 + z^2 + 2x - 6y + 1 = 0$.

6. Find the equations of the tangent planes to the sphere
   (i) $x^2 + y^2 + z^2 + 2x - 4y + 6z - 7 = 0$ which passes through the line $6x - 3y - 23 = 0 = 3z + 2$
   (ii) $x^2 + y^2 + z^2 + 6x - 2z + 1 = 0$ which passes through the line

   $$3(16 - x) = 3z = 2y + 30 \ (\text{or } x + z = 16 \text{ and } 2y - 3z + 30 = 0)$$ **Ans.** (i) $6x - 3y + 12z - 15 = 0$ and $6x - 3y - \dfrac{3}{2}z - 24 = 0$

   (ii) $2x + 2y - z - 2 = 0$ and $x + 2y - 2z + 14 = 0$ (for $k = 1$ and $\dfrac{1}{2}$)

7. Find the tangent planes to the sphere $x^2 + y^2 + z^2 - 2x + 4y - 6z + 5 = 0$ perpendicular to y-axis. Find the point of contact.
   [**Hint :** Equation of any plane $\perp$ to y-axis is $y = k$ or $y - k = 0$. The $\perp$ distance from centre $(1, -2, 3)$ to the plane $y - k = 0$ is $|-2 - k| = 3$ (radius). $\therefore$ $k = 1$ or -5. For point of contact, put $y = 1$ and $y = -5$ in the equation of sphere and then find point of contact.]   **Ans.** Planes are $y = 1$ and $y = -5$. Points of contact are $(1, 1, 3)$ and $(1, -5, 3)$.

8. Find the tangent planes to the sphere with $x = y = z$ as diameter and touching the planes $x + y + z = 0$, $x + y + z = 8$.
   [**Hint :** Let the centre be (k, k, k). Now apply tangent plane property.]   **Ans.** $3(x^2 + y^2 + z^2) + 8(x + y + z) = 0$.

9. Find the equation of tangent planes at the points where the line
   $$\frac{x + 3}{4} = \frac{y + 4}{3} = \frac{z - 8}{-5}$$ intersect the sphere $x^2 + y^2 + z^2 + 2x - 10y = 23$.

   [**Hint :** Any point on the given line is $x = 4k - 3$, $y = 3k - 4$, $z = -5k + 8$, substitute in equation of sphere to find $\lambda = 1, 2$ and thus the points of intersection $(1, -1, 3)$ and $(5, 2, -2)$ and then find tangent planes.]
                    Ans. $6x - 3y - 2z - 28 = 0$ and $2x - 6y + 3z - 17 = 0$.

10. Find the equations of the two spheres in positive octant which
    (i) touches three co-ordinate planes and the plane $x + y + z = 1$.
    (ii) touches three co-ordinate planes and the plane $x + 2y + 2z = 8$.   **Ans.** (i) radius a, centre (a, a, a), where $a = \dfrac{1}{6}\left(3 \pm \sqrt{3}\right)$

    (ii) $x^2 + y^2 + z^2 - 2x - 2y - 2z + 2 = 0$ and $x^2 + y^2 + z^2 - 8x - 8y - 8z + 32 = 0$.

11. Find the equation of tangent plane of the sphere $x^2 + y^2 + z^2 = a^2$ at (r, θ, ϕ), these being polar co-ordinates.
    [**Hint :** $x_1 = r \sin\theta \cos\phi$, $y_1 = r \sin\theta \sin\phi$, $z_1 = r \cos\theta$, where $r = a$]   **Ans.** $x \cos\theta \sin\phi + y \sin\theta \sin\phi + z \cos\phi = a$.

12. Show that, the condition that the plane $lx + my + nz = p$ should touch the sphere $x^2 + y^2 + z^2 + 2ux + 2vy + 2wz + d = 0$ is
    $$(ul + vm + wn + p)^2 = (l^2 + m^2 + n^2)(u^2 + v^2 + w^2 - d).$$

13. If any tangent plane to the sphere $x^2 + y^2 + z^2 = r^2$ makes intercepts a, b, c on the axes, then show that $\dfrac{1}{a^2} + \dfrac{1}{b^2} + \dfrac{1}{c^2} = \dfrac{1}{r^2}$.

    [**Hint :** Equation of tangent plane at (x_1, y_1, z_1) is $xx_1 + yy_1 + zz_1 = r^2$ and equation of the plane $\dfrac{x_1}{a} + \dfrac{y_1}{b} + \dfrac{z_1}{c} = 1$.

    $\therefore$ $ax_1 = r^2$, $by_1 = r^2$ and $cz_1 = r^2$ and also point (x_1, y_1, z_1) lies on the sphere.]

14. Determine the centre and radius of the circle
    (i) $x^2 + y^2 + z^2 = 16$; $2x + y + 2z = 9$
    (ii) $x^2 + y^2 + z^2 + 2x - 2y - 4z - 19 = 0$; $x + 2y + 2z + 7 = 0$.   **(Dec. 2007)**
    Also find the orthogonal projection of the area of the circle in xoy plane.
    [**Hint :** (ii) Centre of the given sphere $(-1, 1, 2)$ and Radius $= 5$. Perpendicular distance from $C(-1, 1, 2)$ to the plane

    $CC_1 = \left|\dfrac{-1 + 2 + 4 + 7}{\sqrt{1 + 4 + 4}}\right| = 4$. Therefore radius of circle $= \sqrt{25 - 16} = 3$. Equation of CC_1 is $\dfrac{x + 1}{1} = \dfrac{y - 1}{2} = \dfrac{z - 2}{2} = k$ and

    $k = -\dfrac{4}{3}$.]   **Ans.** (i) Centre $(2, 1, 2)$, Radius $= \sqrt{7}$ and orthogonal projection $= \dfrac{14\pi}{3}$.

    (ii) Centre $\left(-\dfrac{7}{3}, -\dfrac{5}{3}, -\dfrac{2}{3}\right)$, Radius $= 3$ and orthogonal projection $= 6\pi$.

15. Find the centre and radius of the section of the sphere $x^2 + y^2 + z^2 = 9$ by the plane $2x + 3y + 4z = 5$ and the area of the orthogonal projection of this section on the yoz plane.
                    Ans. Centre $\left(\dfrac{10}{29}, \dfrac{15}{29}, \dfrac{20}{29}\right)$, Radius $= \sqrt{\dfrac{236}{29}}$ and Orthogonal projection $= \dfrac{472\pi}{29\sqrt{29}}$.

16. Find the equation of the circle which is a section of the sphere
$x^2 + y^2 + z^2 + 4x - 6y + 2z - 10 = 0$ and has its centre at the point M $(1, 2, 2)$.   **Ans.** The given sphere and $3x - y + 3z - 7 = 0$.

17. Obtain the equations of the circle lying on the sphere
(i) $x^2 + y^2 + z^2 - 2x + 4y - 6z - 43 = 0$ and having its centre $(2, 1, -4)$.
(ii) $x^2 + y^2 + z^2 - 2x + 4y - 6z + 3 = 0$ and having its centre at $(2, 3, -4)$.

Ans. (i) The given sphere and $x + 3y - 7z - 33 = 0$ (ii) The given sphere and $x + 5y - 7z - 45 = 0$.

18. Find the equation of the sphere   **(May 2011, Nov. 2015)**
(i) passing through the circle $x^2 + y^2 + z^2 = 9$; $2x + 3y + 4z = 5$ and point $(1, 2, 3)$.
[Hint : Equation of the sphere through circle is $x^2 + y^2 + z^2 - 9 + \lambda (2x + 3y + 4z - 5) = 0$. It passes through $(1, 2, 3)$, therefore,
$\lambda = -1/3$.]   **Ans.** $3 (x^2 + y^2 + z^2) - 2x - 3y - 4z - 22 = 0$.

(ii) passing through the circle $x^2 + y^2 + z^2 + 5x - 4y + 6z - 11 = 0$;
$x^2 + y^2 + z^2 - 3x + 8y - 10z - 3 = 0$ and the point $(1, -2, 3)$.
[Hint : Plane of circle $u = s_1 - s_2 = 0$. Equation of the sphere is $s_1 + \lambda u = 0$.]   **Ans.** $2 (x^2 + y^2 + z^2) + 7y - 8z - 12 = 0$.

(iii) for which the circle $x^2 + y^2 + z^2 + 7y - 2z + 2 = 0$; $2x + 3y + 4z = 8$ is a great circle.
[Hint : Centre $\left(-\lambda, \dfrac{7 + 3\lambda}{-2}, \dfrac{4k - 2}{-2}\right)$ and $\lambda = -1$.]   **Ans.** $x^2 + y^2 + z^2 - 2x + 4y - 6z + 10 = 0$.

(iv) which touches the sphere $4 (x^2 + y^2 + z^2) + 10x - 25y - 2z = 0$ at the point $(1, 2, -2)$ and passes through the point $(-1, 0, 0)$.   **(May 2003)**
[Hint : The equation of tangent plane at the point $(1, 2, -2)$ is $u = 2x - y - 2z - 4 = 0$. Thus $s = 0$, $u = 0$ represents a point circle. The sphere through the point circle is $s + \lambda u = 0$. This sphere passes through $(-1, 0, 0)$, which gives $\lambda = -1/4$.
Substituting the value of $\lambda = -\dfrac{1}{4}$ in $s + \lambda u = 0$.]   **Ans.** $x^2 + y^2 + z^2 + 2x - 6y + 1 = 0$.

(v) passing through the circle $x^2 + y^2 + z^2 = 5$, $x + 2y + 3z = 3$ and touch the plane $4x + 3y = 15$.
Ans. $x^2 + y^2 + z^2 + 2x + 4y + 6z - 11 = 0$.
$x^2 + y^2 + z^2 - 4x - 8y - 12z - 13 = 0$.

(vi) passing through the circle $x^2 + y^2 + z^2 - 4x - y - 3z + 12 = 0$,
$2x + 3y - 7x = 10$ and touch the plane $x - 2y + 2z = 1$.   **Ans.** $x^2 + y^2 + z^2 - 2x + 2y - 4z + 2 = 0$
$x^2 + y^2 + z^2 - 6x - 4y + 10z + 22 = 0$

(vii) which passes through the circle $x^2 + y^2 + z^2 - 2x + 2z - 2 = 0$ $y = 0$ and touches the plane $y - z - 7 = 0$.   **(Dec. 2009)**
[Hint : Sphere through the circle is $s + \lambda u = 0$, whose centre is $\left(1, -\dfrac{\lambda}{2}, -1\right)$ and radius $= \sqrt{4 + \dfrac{\lambda^2}{4}} = \left|\dfrac{-(\lambda/2 + 6)}{\sqrt{2}}\right|$,
perpendicular from centre to the plane. This gives $\lambda = -4, 28$.]   **Ans.** $x^2 + y^2 + z^2 - 2x - 4y + 2z - 2 = 0$
$x^2 + y^2 + z^2 - 2x + 28y + 2z - 2 = 0$

19. Find the centre and radius of circle passing through the points $(1, 0, 0)$, $(0, 2, 0)$, $(0, 0, 3)$.

Ans. Radius $= \dfrac{5\sqrt{26}}{14}$ and Centre $\left(\dfrac{13}{98}, \dfrac{40}{49}, \dfrac{123}{98}\right)$.

20. Prove that :
(i) the plane $x + 2y - z = 4$ cuts the sphere $x^2 + y^2 + z^2 - x + z - 2 = 0$ in circle of radius unity and find the equation of the sphere which has this circle for one of its great circle.
[Hint : The equation of the sphere through the given circle is $s + \lambda u = 0$. Radius of this sphere is
$r^2 = \left(\dfrac{\lambda - 1}{2}\right)^2 + \lambda^2 + \left(\dfrac{1 - \lambda}{2}\right)^2 + (2 + 4\lambda)$. For the circle to be great, we must have $r^2 = 1$, which gives $\lambda = -1, -1$. On
substituting the value of $\lambda = -1$.]
(ii) the sphere $x^2 + y^2 + z^2 + 3x + 5y - 7z - 4 = 0$ intersects the sphere
$x^2 + y^2 + z^2 + 2x + 4y - 8z - 3 = 0$ in great circle of the second sphere.

21. Prove that the following circles lie on the same sphere and find the equations :
(i) $x^2 + y^2 + z^2 - y + 2z = 0$; $x - y + z = 2$
and $x^2 + y^2 + z^2 + x - 3y + z - 5 = 0$; $2x - y + 4z = 1$ **Ans.** $x^2 + y^2 + z^2 + 3x - 4y + 5z - 6 = 0$.
(ii) $x^2 + y^2 + z^2 - 2x + 3y + 4z - 5 = 0$; $5y + 6z + 1 = 0$
and $x^2 + y^2 + z^2 + 3x - 4y + 5z - 6 = 0$; $x + 2y - 7z = 0$ **Ans.** $x^2 + y^2 + z^2 - 2x - 2y - 2z - 6 = 0$.

22. Find the equation of the sphere through   **(May 2019)**

 (i) the circle $x^2 + y^2 = 4$, $z = 4$ & cutting the sphere $x^2 + y^2 + z^2 + 10y - 4z - 8 = 0$ orthogonally.

 (ii) the circle $x^2 + y^2 + z^2 - 2x + 3y - 4z + 6 = 0$, $3x - 4y + 5z - 15 = 0$ and cutting the sphere

  $x^2 + y^2 + z^2 + 2x + 4y - 6z + 11 = 0$ orthogonally.   **(Dec. 2011, May 2010)**

 (iii) the points (a, 0, 0), (0, b, 0), (0, 0, c) so as to cut the sphere $x^2 + y^2 + z^2 - 2ax - 2by - 2cz = 0$ orthogonally.

Ans. (i) $x^2 + y^2 + z^2 + 6z - 4 = 0$ (ii) $5(x^2 + y^2 + z^2) - 13x + 19y - 25z + 45 = 0$

(iii) $x^2 + y^2 + z^2 - ax - by - cz + \dfrac{a^2 + b^2 + c^2}{2}\left(\dfrac{x}{a} + \dfrac{y}{b} + \dfrac{z}{c} - 1\right) = 0$

23. Prove that the equation of the sphere which cuts orthogonally each of the spheres

 $x^2 + y^2 + z^2 = a^2 + b^2 + c^2$, $x^2 + y^2 + z^2 + 2ax = a^2$

 $x^2 + y^2 + z^2 + 2by = b^2$, $x^2 + y^2 + z^2 + 2cz = c^2$

 is $x^2 + y^2 + z^2 + \dfrac{b^2 + c^2}{a}x + \dfrac{c^2 + a^2}{b}y + \dfrac{a^2 + b^2}{c}z + a^2 + b^2 + c^2 = 0.$

24. Prove that every sphere through the circle $x^2 + y^2 - 2ax + r^2 = 0$, $z = 0$ cuts orthogonally every sphere through circle $x^2 + z^2 = r^2$, $y = 0$.

 [Hint : Any sphere through first circle $x^2 + y^2 + z^2 - 2ax + r^2 + \lambda_1 z = 0$ and any sphere through second circle is $x^2 + y^2 + z^2 - r^2 + \lambda_2 y = 0$ and show that condition of orthogonality holds.]

25. Two spheres of radii r_1 and r_2 cut orthogonally. Prove that the radius of common circle is $\dfrac{r_1 r_2}{\sqrt{r_1^2 + r_2^2}}$.

 [Hint : Let common circle be $x^2 + y^2 + z^2 = a^2$; $z = 0$, then equations of two spheres through the circle are

 $x^2 + y^2 + z^2 - a^2 + \lambda_1 z = 0$ and $x^2 + y^2 + z^2 - a^2 + \lambda_2 z = 0$ where $r_1^2 = \dfrac{\lambda_1^2}{4} + a^2$ and $r_2^2 = \dfrac{\lambda_2^2}{4} + a^2$. Now condition of orthogonality will give the required radius.]

26. Find the equation of the diameter of the sphere $x^2 + y^2 + z^2 = 29$ such that rotation about it will transfer the point A (4, –3, 2) to the point B (5, 0, –2) along a great circle of the sphere. Find the angle through which the sphere must be rotated.

Ans. $\dfrac{x}{2} = \dfrac{y}{6} = \dfrac{z}{5}$, $\theta = \cos^{-1}\dfrac{16}{29}.$

27. A sphere $x^2 + y^2 + z^2 + 2ux + 2vy + 2wz + d = 0$ cuts the sphere

 $x^2 + y^2 + z^2 + 2u'x + 2v'y + 2w'z + d = 0$ in a great circle of the second sphere, then show that

 $2(u'^2 + v'^2 + w'^2) - d' = 2(uu' + vv' + ww') - d.$

28. POP$'$ is a variable diameter of the ellipse $\dfrac{x^2}{a^2} + \dfrac{y^2}{b^2} = 1$, $z = 0$. A circle is described in a plane PP$'$zz$'$ on PP$'$ as diameter. Prove

 that as PP$'$ varies the circle generates the surface $(x^2 + y^2 + z^2)\left(\dfrac{x^2}{a^2} + \dfrac{y^2}{b^2}\right) = x^2 + y^2.$

 [Hint : If P $(x_1, y_1, 0)$ and P$'$ $(-x_1, -y_1, 0)$ are diametric points, then $\dfrac{x_1^2}{a^2} + \dfrac{y_1^2}{b^2} = 1$. Equation of sphere on PP$'$ as diameter is

 $x^2 + y^2 + z^2 = x_1^2 + y_1^2$. Plane through z-axis (x = 0, y = 0), containing P is $\dfrac{x}{x_1} - \dfrac{y}{y_1} = 0$ and since P$'$ satisfy it, therefore it

 represents plane PP$'$zz$'$. Eliminating x_1, y_1 between these equations, we get $x_1 = kx$, $y_1 = ky$. Substituting and eliminating k^2, we get locus of circle as required surface.]

29. A variable plane passing through the point A(a, b, c) cuts the sphere $x^2 + y^2 + z^2 = r^2$. Find the locus of the centre of section of the sphere by this plane.

 [Hint : Let centre of the section be P $(\bar{x}, \bar{y}, \bar{z})$. Since join of P to the centre (0, 0, 0) of the sphere is normal to the section,

 $\therefore$ equation of plane of section passing through A (a, b, c) is $\bar{x}(x - a) + \bar{y}(y - b) + \bar{z}(z - c) = 0]$

Ans. Locus of centre of the section is $x^2 + y^2 + z^2 - ax - by - cz = 0.$

□□□

CHAPTER-8

THE CONE AND THE CYLINDER

THE CONE

8.1 DEFINITION

A cone is a surface generated by a straight line which passes through a fixed point and satisfies one more condition for example, it intersects a given curve or touches a given surface.

The fixed point is called the **Vertex** and the given curve (or surface) the **Guiding Curve** of the cone. Any straight line lying on the surface of the cone is called its **generator**.

Note : Cones whose equations are of the second degree in x, y, z are called the **Quadratic Cones**. In what follows, we shall be concerned only with quadratic cones.

8.2 CONE WITH VERTEX AT THE ORIGIN

The Equation of a Cone, whose Vertex is at the Origin, is Homogeneous in x, y, z.

Let us suppose that the general equation of second degree in x, y and z represents a cone whose vertex is at origin be

$$ax^2 + by^2 + cz^2 + 2fyz + 2gzx + 2hxy + 2ux + 2vy + 2wz + d = 0 \qquad \text{... (i)}$$

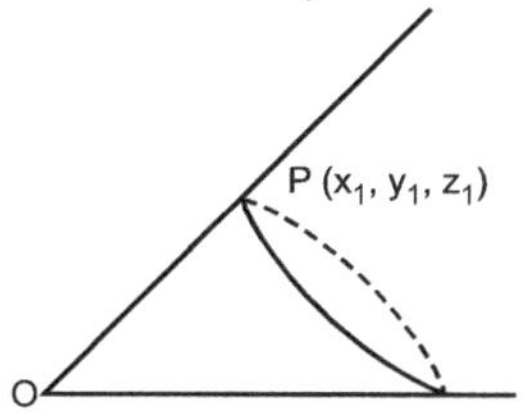

Let $P(x_1, y_1, z_1)$ be any point on the cone (Fig. 8.1). Then the equation of OP is

$$\frac{x}{x_1} = \frac{y}{y_1} = \frac{z}{z_1} = k \text{ (say)} \qquad \text{... (ii)}$$

Any point $Q(kx_1, ky_1, kz_1)$ on this generator must lie on the cone for all values of k.

Fig. 8.1

$$\therefore k^2 (ax_1^2 + by_1^2 + cz_1^2 + 2fy_1z_1 + 2gz_1x_1 + 2hx_1y_1) + 2k (ux_1 + vy_1 + wz_1) + d = 0 \qquad \text{... (iii)}$$

which should be satisfied for all values of k, i.e. (iii) should be an identity and as such we have the following conditions :

$$ax_1^2 + by_1^2 + cz_1^2 + 2fy_1z_1 + 2gz_1x_1 + 2hx_1y_1 = 0 \qquad \text{... (iv)}$$

$$ux_1 + vy_1 + wz_1 = 0 \qquad \text{... (v)}$$

$$d = 0 \qquad \text{... (vi)}$$

From condition (v), it is clear that u = v = w = 0, otherwise, any point $P(x_1, y_1, z_1)$ on the conical surface satisfies the first degree equation ux + vy + wz = 0 representing a plane which is a contradiction.

Putting u = v = w = d = 0 in (i), the required equation of the cone reduces to the form

$$\boxed{ax^2 + by^2 + cz^2 + 2fyz + 2gzx + 2hxy = 0} \qquad \text{... (1)}$$

which is a homogeneous equation.

Conversely, Every Homogeneous Equation of Second Degree in x, y, z Represents a Cone whose Vertex is at the Origin.

Let the homogeneous equation of second degree in x, y, z be

$$ax^2 + by^2 + cz^2 + 2fyz + 2gzx + 2hxy = 0 \qquad \text{... (vii)}$$

If the co-ordinates of any point $P(x_1, y_1, z_1)$ satisfy the equation (vii), then since it is homogeneous, the coordinates of point $Q(kx_1, ky_1, kz_1)$ must also satisfy the equation (vii) for all values of k and hence lies on locus represented by (vii). But the point $Q(kx_1, ky_1, kz_1)$ is any point on the line joining the origin O to the point P. Therefore, line OP lies entirely on the locus of (vii).

Thus, the surface is generated by lines through the origin and hence by definition, it is a cone with vertex at origin.

Remark : Rule : Since the equation of the cone with vertex at the origin is homogeneous then to find equation of cone for

(1)   if one of the equations of the guiding curve be of first degree, then make the other equation homogeneous with its help,

(2)   if neither of the equations of guiding curve be of first degree then make each of them homogeneous by multiplying with proper powers of variable t and eliminate t between the two.

Cor : If the line $\dfrac{x}{l} = \dfrac{y}{m} = \dfrac{z}{n}$ is a generator of the cone (whose vertex is at the origin)

$ax^2 + by^2 + cz^2 + 2fyz + 2gzx + 2hxy = 0$, then direction cosines (or direction ratios) l, m, n satisfy the equation of the cone.

Any point on the generator is (lk, mk, nk). Since it lies on the given cone,

$$\boxed{al^2 + bm^2 + cn^2 + 2fmn + 2gnl + 2hlm = 0}$$ ... (2)

Conversely, if quantities l, m, n satisfy the relation

$al^2 + bm^2 + cn^2 + 2fmn + 2gnl + 2hlm = 0$, then the line with d.c.'s l, m, n is a generator of the cone

$ax^2 + by^2 + cz^2 + 2fyz + 2gzx + 2hxy = 0$.

8.3 QUADRATIC CONE THROUGH THE AXES

The General Equation of a Cone of Second Degree Passing through the three Coordinate Axes is fyz + gzx + hxy = 0.

Since the co-ordinate axes intersect at origin and the cone passes through the axes, the origin must be vertex of the cone. Hence its general equation is given by

$$ax^2 + by^2 + cz^2 + 2fyz + 2gzx + 2hxy = 0$$ ... (i)

Since x-axis is the generator of the cone, its d.c.'s 1, 0, 0 must satisfy the equation (i)

$\therefore$ $a = 0$.

Similarly, since y-axis and z-axis are generators with d.c.'s 0, 1, 0 and 0, 0, 1,

$\therefore$ $b = 0$ and $c = 0$.

Hence from (i), the required equation of the cone is

$$\boxed{fyz + gzx + hxy = 0}$$ ... (3)

8.4 CONE WITH A GIVEN VERTEX AND GIVEN PLANE CURVE AS THE GUIDING CURVE

To find the equation of a cone whose vertex is the point (α, β, γ) and base (guiding curve) the conic,

$ax^2 + 2hxy + by^2 + 2gx + 2fy + c = 0, \ z = 0$

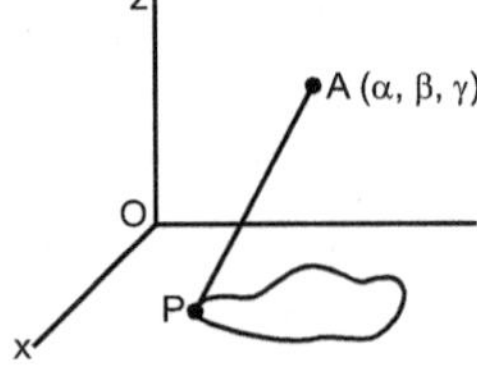

Fig. 8.2

The equation of any line whose d.r.'s are l, m, n and passing through the vertex A (α, β, γ) (Fig. 8.2) is given by;

$$\frac{x - \alpha}{l} = \frac{y - \beta}{m} = \frac{z - \gamma}{n}$$ ... (ii)

This line meets the plane $z = 0$ at the point $P\left(\alpha - \dfrac{l\gamma}{n}, \ \beta - \dfrac{m\gamma}{n}, \ 0\right)$

This point lies on the given conic, if

$$a\left(\alpha - \frac{l\gamma}{n}\right)^2 + 2h\left(\alpha - \frac{l\gamma}{n}\right)\left(\beta - \frac{m\gamma}{n}\right) + b\left(\beta - \frac{m\gamma}{n}\right)^2 + 2g\left(\alpha - \frac{l\gamma}{n}\right) + 2f\left(\beta - \frac{m\gamma}{n}\right) + c = 0$$ ...(iii)

This is the condition for the line (ii) to intersect the conic (i).

To find the locus of the line (ii), eliminate l, m, n between (ii) and (iii), i.e. by putting values of $\dfrac{l}{n} = \dfrac{x - \alpha}{z - \gamma}$ and $\dfrac{m}{n} = \dfrac{y - \beta}{z - \gamma}$ from (ii) in (iii), we have

$$a\left(\alpha - \frac{x - \alpha}{z - \gamma}\cdot\gamma\right)^2 + 2h\left(\alpha - \frac{x - \alpha}{z - \gamma}\cdot\gamma\right)\left(\beta - \frac{y - \beta}{z - \gamma}\cdot\gamma\right) + b\left(\beta - \frac{y - \beta}{z - \gamma}\cdot\gamma\right)^2 + 2g\left(\alpha - \frac{x - \alpha}{z - \gamma}\cdot\gamma\right) + 2f\left(\beta - \frac{y - \beta}{z - \gamma}\cdot\gamma\right) + c = 0$$

Multiplying throughout by $(z - \gamma)^2$ and simplifying, we get

$$a\,(\alpha z - \gamma x)^2 + 2h\,(\alpha z - \gamma x)(\beta z - \gamma y) + b\,(\beta z - \gamma y)^2 + 2g\,(\alpha z - \gamma x)(z - \gamma) + 2f\,(\beta z - \gamma y)(z - \gamma) + c\,(z - \gamma)^2 = 0$$ ... (4)

which is the required equation of the cone.

8.5 GENERAL SECOND DEGREE EQUATION

Condition for the General Equation of Second Degree to Represent a Cone and to find the Co-ordinates of the Vertex :

Let general equation of second degree of cone with vertex (α, β, γ) be

$$F (x, y, z) \equiv ax^2 + by^2 + cz^2 + 2fyz + 2gzx + 2hxy + 2ux + 2vy + 2wz + d = 0 \qquad \text{... (i)}$$

Shifting the origin to the point (α, β, γ) by transformations $x = X + \alpha$, $y = Y + \beta$, $z = Z + \gamma$, equation (i) reduces to

$$a (X + \alpha)^2 + b (Y + \beta)^2 + c (Z + \gamma)^2 + 2f (Y + \beta) (Z + \gamma) + 2g (Z + \gamma) (X + \alpha) + 2h (X + \alpha) (Y + \beta) + 2u (X + \alpha) + 2v (Y + \beta)$$
$$+ 2w (Z + \gamma) + d = 0$$

or $\quad aX^2 + bY^2 + cZ^2 + 2fYZ + 2gZX + 2hXY + 2 (a\alpha + h\beta + g\gamma + u) X + 2 (h\alpha + b\beta + f\gamma + v) Y$

$$+ 2 (g\alpha + f\beta + c\gamma + w) Z + (a\alpha^2 + b\beta^2 + c\gamma^2 + 2f\beta\gamma + 2g\gamma\alpha + 2h\alpha\beta + 2u\alpha + 2v\beta + 2w\gamma + d) = 0 \qquad \text{... (ii)}$$

Since equation (ii) represents a cone with vertex at origin, it must be a second degree homogeneous in X, Y, Z. Therefore the coefficients of X, Y, Z and the absolute term should be zero and as such we have following conditions :

$$a\alpha + h\beta + g\gamma + u = 0 \qquad \text{... (iii)}$$
$$h\alpha + b\beta + f\gamma + v = 0 \qquad \text{... (iv)}$$
$$g\alpha + f\beta + c\gamma + w = 0 \qquad \text{... (v)}$$

and $\quad a\alpha^2 + b\beta^2 + c\gamma^2 + 2f\beta\gamma + 2g\gamma\alpha + 2h\alpha\beta + 2u\alpha + 2v\beta + 2w\gamma + d = 0 \qquad \text{... (vi)}$

Condition (vi) can also be written as

$\alpha \qquad (a\alpha + h\beta + g\gamma + u) + \beta (h\alpha + b\beta + f\gamma + v) + \gamma (g\alpha + f\beta + c\gamma + w) + (u\alpha + v\beta + w\gamma + d) = 0$

Using relations (iii), (iv) and (v), we get

$$u\alpha + v\beta + w\gamma + d = 0 \qquad \text{... (vii)}$$

The required condition is obtained by eliminating α, β, γ from equations (iii), (iv), (v) and (vii), which is

$$\begin{vmatrix} a & h & g & u \\ h & b & f & v \\ g & f & c & w \\ u & v & w & d \end{vmatrix} = 0 \qquad \text{... (5)}$$

Vertex is obtained from (iii), (iv) and (v).

However, in practice, the equations (iii), (iv), (v) and (vii) can be obtained as follows :

Make the equation of cone homogeneous by multiplying with proper power of a variable t, equation of the cone (i) becomes

$$F (x, y, z, t) = ax^2 + by^2 + cz^2 + 2fyz + 2gzx + 2hxy + 2uxt + 2vyt + 2wzt + dt^2 = 0$$

Differentiating the above equation partially with respect to x, y, z, t and then putting t = 1, we get

$$\frac{\partial F}{\partial x} = 0, \quad \frac{\partial F}{\partial y} = 0, \quad \frac{\partial F}{\partial z} = 0 \quad \text{and} \quad \frac{\partial F}{\partial t} = 0.$$

These equations are, infact, the equations (iii), (iv), (v) and (vii) respectively. Solving the first three equations for x, y, z we get vertex and if these co-ordinates satisfy the fourth equation, then the equation (i) represents the cone.

8.6 RIGHT CIRCULAR CONE

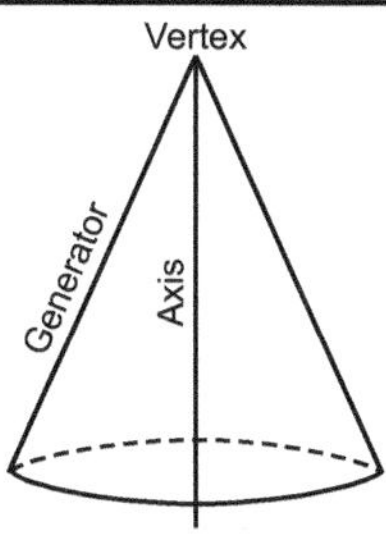

Definition : A right circular cone is a surface generated by a straight line which passes through a fixed point and makes a constant angle with a fixed straight line through the vertex.

The fixed point is called **Vertex**, the fixed line the **Axis** of the cone and the angle is known as the **Semi-vertical Angle** of the cone.

Remark : The section of right circular cone by any plane perpendicular to its axis is circle.

8.7 EQUATION OF RIGHT CIRCULAR CONE

The Equation Of Right Circular Cone whose Vertex is at (α, β, γ), Semi-vertical Angle α and Axis the line $\dfrac{x - \alpha}{l} = \dfrac{y - \beta}{m} = \dfrac{z - \gamma}{n}$.

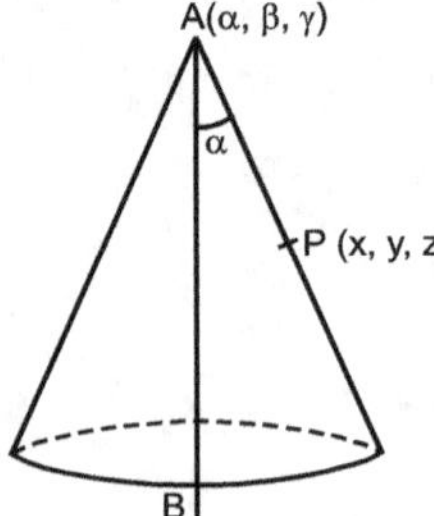

Let P(x, y, z) be any point on the cone and AB, the axis of the cone whose d.r.'s are l, m, n and which passes through the vertex $A(\alpha, \beta, \gamma)$ (Fig. 8.4). The direction ratios of AP are $x - \alpha$, $y - \beta$, $z - \gamma$.

Now the angle between the lines AP and AB is α.

Fig. 8.4

$$\therefore \quad \cos \alpha = \frac{l\,(x - \alpha) + m\,(y - \beta) + n\,(z - \gamma)}{\sqrt{(l^2 + m^2 + n^2)}\,\sqrt{(x - \alpha)^2 + (y - \beta)^2 + (z - \gamma)^2}}$$

or

$$\boxed{[l\,(x - \alpha) + m\,(y - \beta) + n\,(z - \gamma)]^2 = (l^2 + m^2 + n^2)\,[(x - \alpha)^2 + (y - \beta)^2 + (z - \gamma)^2]\cos^2 \alpha} \qquad \text{... (6)}$$

which is the required equation of right circular cone.

Particular Cases :

(A) **Equation of Right Circular Cone with Vertex at Origin :**

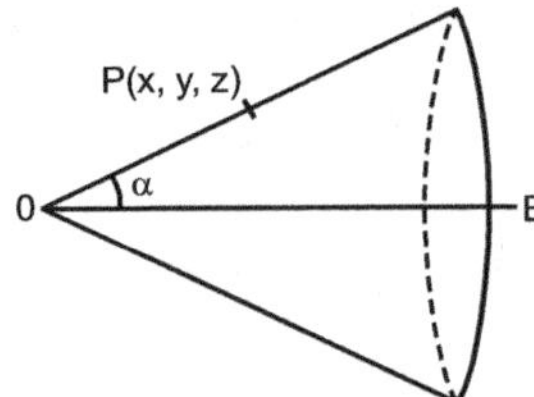

Putting $\alpha = \beta = \gamma = 0$ in (6), we get the required equation of cone as

$$\boxed{(lx + my + nz)^2 = (l^2 + m^2 + n^2)\,(x^2 + y^2 + z^2)\cos^2 \alpha} \qquad \text{... (7)}$$

Fig. 8.5

(B) **Equation of Right Circular Cone with Vertex at Origin and Axis along z-axis :**

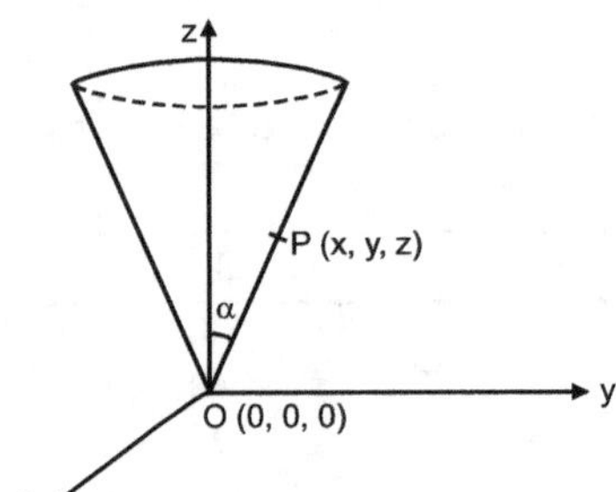

Putting $l = 0$, m = 0, n = 1 in (7), we get the required equation of cone as

$$z^2 = (x^2 + y^2 + z^2)\cos^2 \alpha$$

or $\quad x^2 + y^2 = z^2\,(\sec^2 \alpha - 1)$

$\therefore \quad \boxed{x^2 + y^2 = z^2 \tan^2 \alpha} \qquad \text{...(8)}$

Fig. 8.6

8.8 ENVELOPING CONE

Definition : The locus of the tangent lines from a given point to a given surface is a cone and is called the enveloping cone of the surface with the given point at its vertex.

8.9 EQUATION OF THE ENVELOPING CONE

To find the Equation of the Enveloping Cone of the Sphere $x^2 + y^2 + z^2 = a^2$ with Vertex at the Point (α, β, γ).

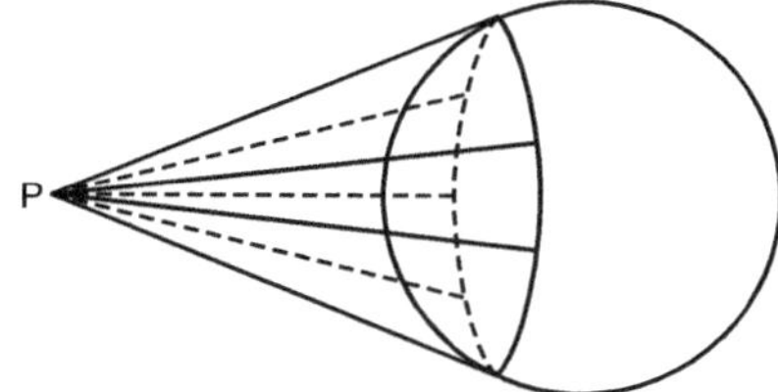

Fig. 8.7

Any line through $P(\alpha, \beta, \gamma)$ is given by,

$$\frac{x - \alpha}{l} = \frac{y - \beta}{m} = \frac{z - \gamma}{n} = k \text{ (say)} \qquad \ldots \text{(i)}$$

Any point on the line (i) is $(\alpha + lk, \beta + mk, \gamma + nk)$. This point will lie on the given surface (sphere), if

$$(\alpha + lk)^2 + (\beta + mk)^2 + (\gamma + nk)^2 = a^2$$

or $k^2 (l^2 + m^2 + n^2) + 2k (\alpha l + \beta m + \gamma n) + \alpha^2 + \beta^2 + \gamma^2 - a^2 = 0 \qquad \ldots \text{(ii)}$

The line (i) will touch the given sphere if this quadratic has equal roots, i.e. if

$$(\alpha l + \beta m + \gamma n)^2 = (l^2 + m^2 + n^2) (\alpha^2 + \beta^2 + \gamma^2 - a^2) \qquad \ldots \text{(iii)}$$

which is the condition for line (i) to touch the given sphere.

The locus of the tangent line (i) is obtained by eliminating l, m, n between (i) and (iii). Thus,

$$\boxed{[\alpha (x - \alpha) + \beta (y - \beta) + \gamma (z - \gamma)]^2 = [(x - \alpha)^2 + (y - \beta)^2 + (z - \gamma)^2] (\alpha^2 + \beta^2 + \gamma^2 - a^2)} \qquad \ldots \text{(9)}$$

which is the required equation of enveloping cone.

If we use the notations

$$S \equiv x^2 + y^2 + z^2 - a^2$$
$$S_1 \equiv \alpha^2 + \beta^2 + \gamma^2 - a^2$$
$$T \equiv \alpha x + \beta y + \gamma z - a^2$$

then the equation (9) can also be written as

$$(T - S_1)^2 = (S + S_1 - 2T) S_1$$

or $\qquad \boxed{SS_1 = T^2} \qquad \ldots \text{(10)}$

Note : If the given sphere is $S \equiv x^2 + y^2 + z^2 + 2ux + 2vy + 2wz + d = 0$, then enveloping cone will still be found to be

$$SS_1 = T^2$$

where, $\qquad S_1 \equiv \alpha^2 + \beta^2 + \gamma^2 + 2u\alpha + 2v\beta + 2w\gamma + d$

$$T \equiv x\alpha + y\beta + z\gamma + u (x + \alpha) + v (y + \beta) + w (z + \gamma) + d$$

8.10 ILLUSTRATIONS

Type I : Problems on Cone with Vertex at Origin :

Ex. 1 : *Find the equation of the cone whose vertex is at the origin and which passes through the curve of intersection of the plane $lx + my + nz = p$ and the surface $ax^2 + by^2 + cz^2 = 1$.*

Sol. : [**Note :** We make second degree equation homogeneous with the help of first equation.]

The equation of the surface is $ax^2 + by^2 + cz^2 = 1$. $\qquad \ldots \text{(1)}$

The equation of the plane can be written as $\dfrac{lx + my + nz}{p} = 1$ $\qquad \ldots \text{(2)}$

Since the equation of the cone with vertex at the origin is a homogeneous equation of second degree in x, y and z, it is obtained by making equation (1) homogeneous with the help of (2).

Now, (1) can be written as

$$ax^2 + by^2 + cz^2 = (1)^2$$

$\therefore \qquad ax^2 + by^2 + cz^2 = \left(\dfrac{lx + my + nz}{p}\right)^2$

$\therefore \qquad \boxed{p^2 (ax^2 + by^2 + cz^2) = (lx + my + nz)^2}$

which is the required equation of the cone.

Ex. 2 : *Find the equation of the cone with vertex at the origin and which passes through the curve of intersection of spheres*

$$x^2 + y^2 + z^2 - x - 1 = 0; \qquad x^2 + y^2 + z^2 + y - 2 = 0.$$

Sol. : The equations of the guiding curve are

$$x^2 + y^2 + z^2 - x - 1 \ = \ 0 \qquad \text{... (1)}$$

and

$$x^2 + y^2 + z^2 + y - 2 \ = \ 0 \qquad \text{... (2)}$$

which is equivalent to

$$x + y \ = \ 1 \qquad \text{... (3)}$$

($\because$ $S_1 - S_2 = 0$, represents the plane of circle as intersection of two spheres.)

Equation of the cone is obtained by making (1) or (2) homogeneous with the help of (3).

Now, (1) can be written as

$$x^2 + y^2 + z^2 - x\,(1) - (1)^2 \ = \ 0$$

$\therefore \qquad x^2 + y^2 + z^2 - x\,(x + y) - (x + y)^2 \ = \ 0$

or $\qquad x^2 + y^2 + z^2 - x^2 - xy - x^2 - 2xy - y^2 \ = \ 0$

or $\qquad \boxed{z^2 - 3xy - x^2 = 0 \qquad \text{or} \qquad x^2 + 3xy - z^2 = 0}$

which is the required equation of the cone.

Ex. 3 : *Find the equation of the cone whose vertex is the origin and base the circle $x = a$, $y^2 + z^2 = b^2$ and show that the section of the cone by a plane parallel to the plane XOY is hyperbola.*

Sol. : The equation of the cone, on making the equation $y^2 + z^2 = b^2$ homogeneous with the help of $x = a$ $\left(\text{i.e. } \dfrac{x}{a} = 1 \right)$ is

$$y^2 + z^2 \ = \ b^2 \left(\frac{x}{a} \right)^2 \qquad \text{or} \qquad a^2 \,(y^2 + z^2) \ = \ b^2\, x^2.$$

Now any plane parallel to XOY plane is $z = c$. Hence the section of this cone by the plane $z = c$ is the curve

$$a^2 \,(y^2 + c^2) \ = \ b^2\, x^2$$

or $\qquad b^2\, x^2 - a^2\, y^2 \ = \ a^2\, c^2$

or $\qquad \boxed{\dfrac{x^2}{\left(\dfrac{a^2 c^2}{b^2} \right)} - \dfrac{y^2}{c^2} = 1}$

which clearly represents a hyperbola in the plane $z = c$.

Ex. 4 : *The plane $\dfrac{x}{a} + \dfrac{y}{b} + \dfrac{z}{c} = 1$ meets the co-ordinate axes in A, B, C. Prove that the equation of the cone generated by lines drawn from O to meet the circle ABC is*

$$yz \left(\frac{b}{c} + \frac{c}{b} \right) + zx \left(\frac{c}{a} + \frac{a}{c} \right) + xy \left(\frac{a}{b} + \frac{b}{a} \right) \ = \ 0.$$

Sol. : The plane $\dfrac{x}{a} + \dfrac{y}{b} + \dfrac{z}{c} = 1$ $\qquad \text{... (1)}$

meets the axes in A $(a, 0, 0)$, B $(0, b, 0)$, C $(0, 0, c)$.

The equation of the sphere OABC is

$$x^2 + y^2 + z^2 - ax - by - cz \ = \ 0 \qquad \text{... (2)}$$

(1) and (2) together represent circle ABC.

Hence, the equation of the cone generated by lines drawn through O to meet the circle ABC is obtained by making equation (2) homogeneous with the help of (1), i.e.

$$x^2 + y^2 + z^2 - (ax + by + cz) \left(\frac{x}{a} + \frac{y}{b} + \frac{z}{c} \right) \ = \ 0$$

or $\qquad \boxed{yz \left(\dfrac{b}{c} + \dfrac{c}{b} \right) + zx \left(\dfrac{c}{a} + \dfrac{a}{c} \right) + xy \left(\dfrac{a}{b} + \dfrac{b}{a} \right) \ = \ 0}$

Ex. 5 : *Show that the lines drawn through the point (α, β, γ) whose direction cosines satisfy $al^2 + bm^2 + cn^2 = 0$ generates the cone $a\,(x - \alpha)^2 + b\,(y - \beta)^2 + c\,(z - \gamma)^2 = 0$.*

Sol. : Any line through (α, β, γ) is given by;

$$\frac{x - \alpha}{l} = \frac{y - \beta}{m} = \frac{z - \gamma}{n} = k \text{ (say)} \qquad \ldots (1)$$

Its direction cosines satisfy the relation

$$al^2 + bm^2 + cn^2 = 0 \qquad \ldots (2)$$

Eliminating l, m, n between (1) and (2), we get

$$\boxed{a\,(x - \alpha)^2 + b\,(y - \beta)^2 + c\,(z - \gamma)^2 = 0}$$

which is the required equation of the cone.

Ex. 6 : *Find the equation of the cone with vertex at the origin and direction cosines of its generators satisfying the relation $3l^2 - 4m^2 + 5n^2 = 0$.*

Sol. : Any line through origin $(0, 0, 0)$ is given by;

$$\frac{x}{l} = \frac{y}{m} = \frac{z}{n} \qquad \ldots (1)$$

Given that, the d.c.'s of generator of the cone satisfy relation

$$3l^2 - 4m^2 + 5n^2 = 0 \qquad \ldots (2)$$

Eliminating l, m, n between (1) and (2), we get,

$$\boxed{3x^2 + 4y^2 + 5z^2 = 0}$$

which is the required equation of the cone.

Ex. 7 : *Show that a cone of second degree can be found to pass through two sets of rectangular axes through the same origin.*

OR

Show that a cone can be found to contain any two sets of three mutually perpendicular concurrent lines as generators.

Sol. : Let the equation of the cone of second degree having vertex at origin be

$$ax^2 + by^2 + cz^2 + 2fyz + 2gzx + 2hxy = 0 \qquad \ldots (1)$$

[**Note :** On dividing (1) by 'a', it is found that (1) contains five constants and as such a cone can be made to pass through five lines.]

Let the co-ordinate axes be one set of rectangular axes through which the cone passes. Then the equation of the cone passing through the co-ordinate axes is

$$fyz + gzx + hxy = 0 \qquad \ldots (2)$$

Let the second sets of rectangular axes through the same origin be OP, OQ, OR having direction cosines $l_1, m_1, n_1; l_2, m_2, n_2; l_3, m_3, n_3$. Then, since lines are mutually perpendicular, we have

$$m_1 n_1 + m_2 n_2 + m_3 n_3 = 0 \qquad \ldots (3)$$

$$n_1 l_1 + n_2 l_2 + n_3 l_3 = 0 \qquad \ldots (4)$$

$$l_1 m_1 + l_2 m_2 + l_3 m_3 = 0 \qquad \ldots (5)$$

Let the cone (2) passes through OP and OQ (i.e. it contains two of the three lines of second set), then

$$f\, m_1 n_1 + g\, n_1 l_1 + h\, l_1 m_1 = 0 \qquad \ldots (6)$$

$$f\, m_2 n_2 + g\, n_2 l_2 + h\, l_2 m_2 = 0 \qquad \ldots (7)$$

On adding (6) and (7), we get

$$f\,(m_1 n_1 + m_2 n_2) + g\,(n_1 l_1 + n_2 l_2) + h\,(l_1 m_1 + l_2 m_2) = 0$$

By using (3), (4) and (5), it becomes

$$f\, m_3 n_3 + g\, n_3 l_3 + h\, l_3 m_3 = 0.$$

This shows that the cone (2) also passes through OR.

Ex. 8 : *Find the equation of the quadratic cone which passes through the three co-ordinate axes and the three mutually perpendicular lines*

$$\frac{x}{1} = \frac{y}{-2} = \frac{z}{3}, \ \frac{x}{1} = \frac{y}{-1} = \frac{z}{-1}, \ \frac{x}{5} = \frac{y}{4} = \frac{z}{1}.$$

Sol. : The general equation of the cone passing through the coordinate axes is

$$fyz + gzx + hxy = 0 \qquad \text{... (1)}$$

Since, it passes through the line $\frac{x}{1} = \frac{y}{-2} = \frac{z}{3}$, the d.r.'s of this line must satisfy (1).

$\therefore \qquad f(-2)(3) + g(3)(1) + h(1)(-2) = 0$

or $\qquad 6f - 3g + 2h = 0 \qquad \text{... (2)}$

Similarly, as (1) passes through $\frac{x}{1} = \frac{y}{-1} = \frac{z}{-1}$, the d.r.'s of this line also satisfy (1).

$\therefore \qquad f(-1)(-1) + g(-1)(1) + h(1)(-1) = 0$

or $\qquad f - g - h = 0 \qquad \text{... (3)}$

Solving (2) and (3), we have

$$\frac{f}{3+2} = \frac{-g}{-6-2} = \frac{h}{-6+3}$$

or $\qquad \dfrac{f}{5} = \dfrac{g}{8} = \dfrac{h}{-3} \qquad \text{... (4)}$

Eliminating f, g, h from (1) and (4), we have

$$\boxed{5yz + 8zx - 3xy = 0} \qquad \text{... (5)}$$

Also, the d.r.'s of third generator $\frac{x}{5} = \frac{y}{4} = \frac{z}{1}$ satisfy the equation (5).

Hence (5) represents the required equation of the cone.

Type II : Problems on Cone with a given Vertex and given Guiding Curve :

Ex. 9 : *Find the equation of the cone whose vertex is (1, 1, 1) and base the circle $x^2 + y^2 = 4$, $z = 2$.* **(Dec. 2006, May 2005)**

Sol. : Let the equation of the line (generator) passing through (1, 1, 1) be

$$\frac{x-1}{l} = \frac{y-1}{m} = \frac{z-1}{n} \qquad \text{... (1)}$$

This line meets the plane z = 2, where

$$\frac{x-1}{l} = \frac{y-1}{m} = \frac{2-1}{n} \quad \text{or} \quad \frac{x-1}{l} = \frac{y-1}{m} = \frac{1}{n}$$

$\therefore \qquad x = 1 + \dfrac{l}{n}, \qquad y = 1 + \dfrac{m}{n}, \qquad z = 2$

This point lies on the circle $x^2 + y^2 = 4$, if

$$\left(1 + \frac{l}{n}\right)^2 + \left(1 + \frac{m}{n}\right)^2 = 4 \qquad \text{... (2)}$$

To find locus of line (1), eliminate l, m, n between (1) and (2), by writing

$$\frac{x-1}{z-1} = \frac{l}{n} \ \text{ and } \ \frac{y-1}{z-1} = \frac{m}{n} \ \text{ from (1) in (2).}$$

Hence the required equation of the cone is

$$\left(1 + \frac{x-1}{z-1}\right)^2 + \left(1 + \frac{y-1}{z-1}\right)^2 = 4$$

or $\qquad (x + z - 2)^2 + (y + z - 2)^2 = 4(z-1)^2$

or $\qquad \boxed{x^2 + y^2 - 2z^2 + 2zx + 2zy - 4x - 4y + 4 = 0}$

Ex. 10 : *Find the equation of the cone whose vertex is at the point (1, 1, 3) and which passes through the ellipse $4x^2 + z^2 = 1$, $y = 4$.*

(Dec. 2010, 2004; May 2007)

Sol. : Let the equation of the line (generator) passing through (1, 1, 3) be

$$\frac{x-1}{l} = \frac{y-1}{m} = \frac{z-3}{n} \qquad \qquad \dots (1)$$

This line meets the plane $y = 4$, where

$$\frac{x-1}{l} = \frac{4-1}{m} = \frac{z-3}{n} \quad \text{or} \quad \frac{x-1}{l} = \frac{3}{m} = \frac{z-3}{n}$$

$$\therefore \qquad x = 1 + \frac{3l}{m}, \qquad y = 4, \qquad z = 3 + \frac{3n}{m}$$

This point lies on the ellipse $4x^2 + z^2 = 1$, if

$$4\left(1 + \frac{3l}{m}\right)^2 + \left(3 + \frac{3n}{m}\right)^2 = 1 \qquad \qquad \dots (2)$$

To find locus of line (1), eliminate l, m, n between (1) and (2), by writing

$$\frac{l}{m} = \frac{x-1}{y-1} \quad \text{and} \quad \frac{n}{m} = \frac{z-3}{y-1} \text{ from (1) in (2).}$$

Hence the required equation of the cone is

$$4\left[1 + 3\left(\frac{x-1}{y-1}\right)\right]^2 + \left[3 + 3\left(\frac{z-3}{y-1}\right)\right]^2 = 1$$

or　　　$4(3x + y - 4)^2 + (3y + 3z - 12)^2 = (y-1)^2$

or　　　$\boxed{12x^2 + 4y^2 + 3z^2 + 6yz + 8xy - 32x - 34y - 24z + 69 = 0}$

Ex. 11 : *Find the equation of the cone with vertex (5, 4, 3) and with $3x^2 + 2y^2 = 6$, $y + z = 0$ as base.*

Sol. : Let equation of any line (generator) through (5, 4, 3) is

$$\frac{x-5}{l} = \frac{y-4}{m} = \frac{z-3}{n} \qquad \qquad \dots (1)$$

This line meets the plane $y + z = 0$, where

$$\frac{-z-4}{m} = \frac{z-3}{n} \quad \text{or} \quad z = \frac{3m - 4n}{m + n}, \text{ so that } y = -z = \frac{4n - 3m}{m + n}$$

Thus $\dfrac{x-5}{l} = \dfrac{y-4}{m}$, giving $x = 5 + l\left(\dfrac{y-4}{m}\right) = 5 - \dfrac{7l}{m+n}$.

Hence (1) meets the plane $y + z = 0$ in point

$$\left(5 - \frac{7l}{m+n}, \frac{4n - 3m}{m+n}, \frac{3m - 4n}{m+n}\right)$$

which lies on the given conic if

$$3\left(5 - \frac{7l}{m+n}\right)^2 + 2\left(\frac{4n - 3m}{m+n}\right)^2 = 6$$

i.e.　　　$3(5m + 5n - 7l)^2 + 2(4n - 3m)^2 = 6(m + n)^2$

or　　　$147l^2 + 17m^2 + 101n^2 - 210lm + 9mn - 210lm = 0 \qquad \dots (2)$

Eliminating l, m, n between (1) and (2), we get locus of line (1) as;

$$147(x-5)^2 + 87(y-4)^2 + 101(x-3)^2 - 210(x-5)(y-4)$$
$$+ 90(y-4)(z-3) - 210(x-5)(x-3) = 0$$

or　　　$\boxed{147x^2 + 87y^2 + 101z^2 - 210xy + 90yz - 210zx + 84y + 84z - 294 = 0}$

which is the required equation of the cone.

Ex. 12 : *Obtain the equation of the cone with vertex at (1, 2, 3) and the guiding curve given by $x^2 - 2y^2 + z^2 = 4$, $x - y + z = 3$.*

Sol. : Equation of any line in the direction l, m, n and passing through (1, 2, 3) is

$$\frac{x-1}{l} = \frac{y-2}{m} = \frac{z-3}{n} = k \text{ (say)} \qquad \qquad \dots (1)$$

Any point on this line is $(lk + 1, mk + 2, nk + 3)$.

This point lies on the given guiding curve (circle) if for the some value of k

$$(lk + 1)^2 - 2(mk + 2)^2 + (nk + 3)^2 = 4$$

and $(lk + 1) - (mk + 2) + (nk + 3) = 3$

i.e. $(l^2 - 2m^2 + n^2)\, k^2 + (2l - 8m + 6n)\, k - 2 = 0$ ... (2)

and $(l - m + n)\, k = 1$ ... (3)

From (3), we get

$$k = \frac{1}{l - m + n}$$

Substituting in (2), we have

$$(l^2 - 2m^2 + n^2)\left(\frac{1}{l - m + n}\right)^2 + (2l - 8m + 6n)\left(\frac{1}{l - m + n}\right) - 2 = 0$$

i.e. $l^2 + 4m^2 + 5n^2 - 10mn + 4nl - 6lm = 0$ ... (4)

which is the condition that l, m, n must satisfy, so that the line (1) may be the generator of the required cone.

Eliminating l, m, n from (1) and (4), we get locus of the line (1), i.e. the equation of the required cone. Thus we have

$$(x - 1)^2 + 4\,(y - 2)^2 + 5\,(z - 3)^2 - 10\,(y - 2)\,(z - 3) + 4\,(z - 3)\,(x - 1) - 6\,(x - 1)\,(y - 2) = 0$$

or

$$\boxed{x^2 + 4y^2 + 5z^2 - 10yz + 4zx - 6xy - 2x + 20y - 14z + 2 = 0}$$

which is the required cone.

Ex. 13 : *The section of a cone whose guiding curve is the ellipse* $\dfrac{x^2}{a^2} + \dfrac{y^2}{b^2} = 1$, $z = 0$ *by the plane* $x = 0$, *is a rectangular hyperbola.*

Prove that the locus of the vertex is the surface $\dfrac{x^2}{a^2} + \dfrac{y^2 + z^2}{b^2} = 1$

Sol. : Let (α, β, γ) be the vertex of the cone. Then the equations of the line (generator) are

$$\frac{x - \alpha}{l} = \frac{y - \beta}{m} = \frac{z - \gamma}{n}$$ ... (1)

This line meets the plane $z = 0$ at $\left(\alpha - \dfrac{l\gamma}{n},\ \beta - \dfrac{m\gamma}{n},\ 0\right)$ which will lie on the given ellipse if

$$\frac{1}{a^2}\left(\alpha - \frac{l\gamma}{n}\right)^2 + \frac{1}{b^2}\left(\beta - \frac{m\gamma}{n}\right)^2 = 1$$ ... (2)

Eliminating l, m, n from (1) and (2) by writing $\dfrac{x - \alpha}{z - \gamma} = \dfrac{l}{n}$ and $\dfrac{y - \beta}{z - \gamma} = \dfrac{m}{n}$ from (1) in (2).

Hence the equation of the required cone is

$$\frac{1}{a^2}\left(\alpha - \frac{x - \alpha}{z - \gamma}\cdot\gamma\right)^2 + \frac{1}{b^2}\left(\beta - \frac{y - \beta}{z - \gamma}\cdot\gamma\right)^2 = 1$$

or $\dfrac{1}{a^2}\,(\alpha z - \gamma x)^2 + \dfrac{1}{b^2}\,(\beta z - \gamma y)^2 = (z - \gamma)^2$ ... (3)

The section of this cone (3) by the plane $x = 0$ is the curve given by

$$\frac{\alpha^2 z^2}{a^2} + \frac{1}{b^2}\,(\beta z - \gamma y)^2 = (z - \gamma)^2$$

or $\dfrac{\alpha^2 z^2}{a^2} + \dfrac{1}{b^2}\,(\beta^2 z^2 + \gamma^2 y^2 - 2\beta\gamma yz) - (z^2 + \gamma^2 - 2z\gamma) = 0$

or $\dfrac{\gamma^2 y^2}{b^2} + \left(\dfrac{\alpha^2}{a^2} + \dfrac{\beta^2}{b^2} - 1\right) z^2 - \dfrac{2\beta\gamma yz}{b^2} + 2z\gamma - \gamma^2 = 0$

This section will represent a rectangular hyperbola in yz-plane if

    coeff. of y^2 + coeff. of $z^2 = 0$

i.e. $\dfrac{\gamma^2}{b^2} + \left(\dfrac{\alpha^2}{a^2} + \dfrac{\beta^2}{b^2} - 1\right) = 0$

Hence the locus of (α, β, γ) is the surface

$$\boxed{\frac{x^2}{a^2} + \frac{y^2 + z^2}{b^2} = 1.}$$

Ex. 14 : *Prove that the equation*

$$x^2 - 2y^2 + 3z^2 - 4xy + 5yz - 6zx + 8x - 19y - 2z - 20 = 0$$

represents a cone. Find the co-ordinates of its vertex.   *(May 98)*

Sol. : Making the equation homogeneous by introducing a new variable t, we get

$$F(x, y, z, t) \equiv x^2 - 2y^2 + 3z^2 - 4xy + 5yz - 6zx + 8xt - 19yt - 2zt - 20t^2 = 0.$$

Now equating to zero partial derivatives of F, we get

$$\frac{\partial F}{\partial x} = 2x - 4y - 6z + 8t = 0 \qquad \qquad \text{... (1)}$$

$$\frac{\partial F}{\partial y} = -4y - 4x + 5z - 19t = 0 \qquad \qquad \text{... (2)}$$

$$\frac{\partial F}{\partial z} = 6z + 5y - 6x - 2t = 0 \qquad \qquad \text{... (3)}$$

$$\frac{\partial F}{\partial t} = 8x - 19y - 2z - 40t = 0 \qquad \qquad \text{... (4)}$$

Taking t = 1 in (1), (2), (3) and (4), we get

$$2x - 4y - 6z + 8 = 0 \qquad \qquad \text{... (5)}$$

$$-4x - 4y + 5z - 19 = 0 \qquad \qquad \text{... (6)}$$

$$-6x + 5y + 6z - 2 = 0 \qquad \qquad \text{... (7)}$$

$$8x - 19y - 2z - 40 = 0 \qquad \qquad \text{... (8)}$$

Solving equations (5), (6), (7), we get

$$\boxed{x = 1, \quad y = -2, \quad z = 3}.$$

These values of x, y, z i.e. co-ordinates (1, – 2, 3) also satisfy equation (8). Hence, the given equation represents a cone whose vertex is at the point (1, – 2, 3).

Ex. 15 : *Prove that the equation*

$$2x^2 + 2y^2 + 7z^2 - 10yz - 10zx + 2x + 2y + 26z - 17 = 0$$

represents a cone whose vertex is (2, 2, 1).

Sol. : Making the equation homogeneous by introducing a new variable t, we get

$$F(x, y, z, t) \equiv 2x^2 + 2y^2 + 7z^2 - 10yz - 10zx + 2xt + 2yt + 26zt - 17t^2 = 0$$

Now equating to zero partial derivatives of F, we get

$$\frac{\partial F}{\partial x} = 4x - 10z + 2t = 0 \qquad \text{i.e.} \qquad 2x - 5z + t = 0 \qquad \text{... (1)}$$

$$\frac{\partial F}{\partial y} = 4y - 10z + 2t = 0 \qquad \text{i.e.} \qquad 2y - 5z + t = 0 \qquad \text{... (2)}$$

$$\frac{\partial F}{\partial z} = 14z - 10y - 10x + 26t = 0 \qquad \text{i.e.} \qquad 5x + 5y - 7z - 13t = 0 \qquad \text{... (3)}$$

$$\frac{\partial F}{\partial t} = 2x + 2y + 26z - 34t = 0 \qquad \text{i.e.} \qquad x + y + 13z - 17t = 0 \qquad \text{...(4)}$$

Taking t = 1 and solving equations (1), (2) and (3), we get

$$x = 2, \qquad y = 2, \qquad z = 1$$

These values of x, y, z i.e. co-ordinates (2, 2, 1) also satisfy equation (4). Hence, the given equation represents a cone whose vertex is at the point (2, 2, 1).

Type III : Problems on Right Circular Cone and Enveloping Cone :

Ex. 16 : *Find the equation of the cone with vertex at (1, 2, – 3), semi-vertical angle* $\cos^{-1}\left(\dfrac{1}{\sqrt{3}}\right)$ *and the line* $\dfrac{x-1}{1} = \dfrac{y-2}{2} = \dfrac{z+1}{-1}$ *as*

axis of the cone.   *(Dec. 05, 13, 16, Nov. 14, May 16)*

Sol. : Axis of the cone is $\dfrac{x-1}{1} = \dfrac{y-2}{2} = \dfrac{z+1}{-1}$, so that its d.r.'s are 1, 2, – 1.

The vertex of the cone is A (1, 2, –3) and let P (x, y, z) be any point on the surface of the cone, then d.r.'s of the line joining vertex A (1, 2, –3) to the point P (x, y, z) are (x – 1), (y – 2), (z + 3).

If α is the angle made by AP with the axis of the cone, then

$$\cos \alpha = \frac{(1)\,(x-1) + (2)\,(y-2) + (-1)\,(z+3)}{\sqrt{(1+4+1)}\,\sqrt{(x-1)^2 + (y-2)^2 + (z+3)^2}}$$

But, given that the semi-vertical angle, $\alpha = \cos^{-1}\left(\dfrac{1}{\sqrt{3}}\right)$.

$\therefore \qquad \dfrac{(x-1) + 2\,(y-2) - (z+3)}{\sqrt{6}\,\sqrt{(x-1)^2 + (y-2)^2 + (z+3)^2}} = \dfrac{1}{\sqrt{3}}$

On squaring, we have

$$\frac{1}{6}\,[(x-1) + 2\,(y-2) - (z+3)]^2 = \frac{1}{3}\,[(x-1)^2 + (y-2)^2 + (z+3)^2]$$

Hence the required equation of the cone is

$$\boxed{x^2 - 2y^2 + z^2 + 4yz + 2zx - 4xy + 12x - 4y - 4z - 36 = 0.}$$

Ex. 17 : *Find the equation of the right circular cone whose vertex is at the origin, whose axis is the line $\dfrac{x}{1} = \dfrac{y}{2} = \dfrac{z}{3}$, and which has a semi-vertical angle of 30°.*

(May 07, 13, Nov. 15)

Sol. :

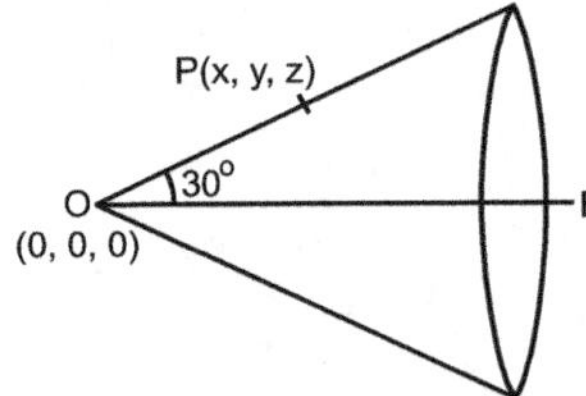

Fig. 8.8

The axis say OB (Fig. 8.8) of the cone is

$\dfrac{x}{1} = \dfrac{y}{2} = \dfrac{z}{3}$, so that its d.r.'s are 1, 2, 3.

Let $P(x, y, z)$ be any point on the cone, then d.r.'s of OP are x, y, z and $\angle$ POB $= 30°$ (semi-vertical angle).

$\therefore \qquad \cos 30° = \dfrac{x\,(1) + y\,(2) + (z)\,(3)}{\sqrt{(x^2 + y^2 + z^2)}\,\sqrt{1 + 4 + 9}}$

or $\qquad \dfrac{\sqrt{3}}{2} = \dfrac{x + 2y + 3z}{\sqrt{(x^2 + y^2 + z^2)}\,\sqrt{14}}$

On squaring, we have

$$\frac{3}{4}\,(x^2 + y^2 + z^2) = \frac{1}{14}\,(x + 2y + 3z)^2$$

or $\qquad 21\,(x^2 + y^2 + z^2) = 2\,(x + 2y + 3z)^2$

or $\qquad \boxed{19x^2 + 13y^2 + 3z^2 - 8xy - 24yz - 12zx = 0}$

which is the required equation of the cone.

Ex. 18 : *Find the equation of the right circular cone which passes through the point (1, 1, 2), has its axis at the line $6x = -3y = 4z$ and vertex at origin.*

(May 05, 04, 17; Dec. 06, Nov. 2014)

Sol. : The axis of the cone is $\dfrac{x}{2} = \dfrac{y}{-4} = \dfrac{z}{3}$, so that its direction ratios are $2, -4, 3$. The direction ratios of the line joining the vertex O $(0, 0, 0)$ of the cone to the given point A $(1, 1, 2)$ are 1, 1, 2.

Now, α, the semi-vertical angle of the cone is the angle between the axis and the line OA.

$\therefore \qquad \cos \alpha = \dfrac{(2)\,(1) + (-4)\,(1) + (3)\,(2)}{\sqrt{(4 + 16 + 9)}\,\sqrt{(1 + 1 + 4)}} = \dfrac{4}{\sqrt{174}}$

Let P (x, y, z) be any point on the cone, then the direction ratios of generator OP are x, y, z. Since OP must make an angle with the axis equal to the semi-vertical angle α of the cone, we have

$$\cos \alpha = \dfrac{(x)\,(2) + (y)\,(-4) + (z)\,(3)}{\sqrt{(x^2 + y^2 + z^2)}\,\sqrt{(4 + 16 + 9)}}$$

Thus $\qquad \dfrac{2x - 4y + 3z}{\sqrt{(x^2 + y^2 + z^2)}\,\sqrt{29}} = \dfrac{4}{\sqrt{174}}$

Squaring we have

$$3 (2x - 4y + 3z)^2 = 8 (x^2 + y^2 + z^2)$$

or $\boxed{4x^2 + 40y^2 + 19z^2 - 48xy - 72yz + 36zx = 0}$

as the required equation of the cone.

Ex. 19 : *Obtain the equation of a right circular cone which passes through the point (2, 1, 3) with vertex at (1, 1, 2) and axis parallel to*
the line $\dfrac{x - 2}{2} = \dfrac{y - 1}{-4} = \dfrac{z + 2}{3}.$ **(Dec. 07, 13, 16, 17)**

Sol. :

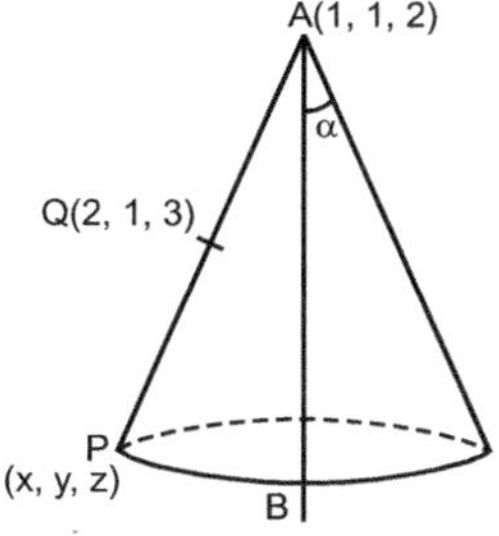

Fig. 8.9

Here the semi-vertical angle of the cone is not given, hence first we determine it as an angle between AB parallel to the given line which passes through the vertex A (1, 1, 2) and the given point Q(2, 1, 3) on the cone (Fig. 8.9).

∴   The d.r.'s of AB, which is the axis of the cone, are 2, – 4, 3 and the d.r.'s of AQ, which is the generator of the cone, are 1, 0, 1.

If α is the semi-vertical angle of the cone, then

$$\cos \alpha = \frac{(2)\,(1) + (-4)\,(0) + (3)\,(1)}{\sqrt{4 + 16 + 9}\,\sqrt{1 + 0 + 1}} = +\frac{5}{\sqrt{58}}$$

Let P(x, y, z) be any point on the cone whose vertex is A (1, 1, 2), then the direction ratios of generator AP are $(x - 1)$, $(y - 1)$, $(z - 2)$. Now α, the semi-vertical angle of the cone is the angle between axis AB and line AP.

∴ $$\frac{(2)\,(x - 1) + (-4)\,(y - 1) + 3\,(z - 2)}{\sqrt{4 + 16 + 9}\,\sqrt{(x - 1)^2 + (y - 1)^2 + (z - 2)^2}} = +\frac{5}{\sqrt{58}}$$

On squaring, we have

$$\frac{1}{29} [2\,(x - 1) - 4\,(y - 1) + 3\,(z - 2)]^2 = \frac{25}{58} [(x - 1)^2 + (y - 1)^2 + (z - 2)^2]$$

or $\boxed{17x^2 - 7y^2 + 7z^2 + 48yz - 24zx + 32xy - 18x - 114y - 52z + 118 = 0}$

which is the required equation of the cone.

Ex. 20 : *The axis of a right circular cone whose vertex is origin O, makes equal angles with the co-ordinate axes and the cone passes*
through the line drawn from O with direction cosines proportional to 1, –2, 2. Find the equation of the cone.   **(May 10, 15)**

Sol. : Let the axis of the cone be $\dfrac{x}{1} = \dfrac{y}{1} = \dfrac{z}{1}$ as it is equally inclined to the axes of co-ordinates.

Any generator of the cone is $\dfrac{x}{1} = \dfrac{y}{-2} = \dfrac{z}{2}$ as its d.r.'s are given to be 1, – 2, 2.

∴ $$\cos \alpha = \frac{1 - 2 + 2}{\sqrt{(1 + 1 + 1)}\,\sqrt{1 + 4 + 4}} = \frac{1}{3\sqrt{3}}$$ ... (1)

If P (x, y, z) be any point on the cone, then d.r.'s of OP are x, y, z and it is inclined at an angle α to axis of the cone.

∴ $$\cos \alpha = \frac{(x)\,(1) + (y)\,(1) + (z)\,(1)}{\sqrt{(x^2 + y^2 + z^2)}\,\sqrt{(1 + 1 + 1)}} = \frac{1}{3\sqrt{3}}$$ [(by (1)]

∴ $$9 (x + y + z)^2 = (x^2 + y^2 + z^2)$$

or $\boxed{4 (x^2 + y^2 + z^2) + 9 (xy + yz + zx) = 0}$

Ex. 21 : *Lines are drawn from the origin with direction cosines proportional to (1, 2, 2), (2, 3, 6), (3, 4, 12). Find the direction cosines of*
the axis of right circular cone through them, and prove that semi-vertical angle of the cone is $\cos^{-1} \dfrac{1}{\sqrt{3}}$ *. Also find the equation of the right*
circular cone.   **(May 14)**

Sol. : Direction cosines of the given lines are

$$\frac{1}{3},\ \frac{2}{3},\ \frac{2}{3}\ ;\qquad \frac{2}{7},\ \frac{3}{7},\ \frac{6}{7}\ ;\qquad \frac{3}{13},\ \frac{4}{13},\ \frac{12}{13}.$$

Let l, m, n be the direction cosines of the axis of the right circular cone through the given lines, and α the semi-vertical angle of the cone, then

$$\cos \alpha = l\left(\frac{1}{3}\right) + m\left(\frac{2}{3}\right) + n\left(\frac{2}{3}\right) = l\left(\frac{2}{7}\right) + m\left(\frac{3}{7}\right) + n\left(\frac{6}{7}\right) = l\left(\frac{3}{13}\right) + m\left(\frac{4}{13}\right) + n\left(\frac{12}{13}\right).$$

$$\therefore \quad \frac{l + 2m + 2n}{3} = \frac{2l + 3m + 6n}{7} = \frac{3l + 4m + 12n}{13}$$

The first two members give $\qquad l + 5m - 4n = 0$ ... (1)

The first and the third members give $\quad 2l + 7m - 5n = 0$ ... (2)

Solving (1) and (2), we have

$$\frac{l}{-25 + 28} = \frac{-m}{-5 + 8} = \frac{n}{7 - 10}$$

or

$$\frac{l}{-1} = \frac{m}{1} = \frac{n}{1} = \frac{\sqrt{l^2 + m^2 + n^2}}{\sqrt{1 + 1 + 1}} = \frac{1}{\sqrt{3}}$$

$$\therefore \quad l = -\frac{1}{\sqrt{3}}, \quad m = \frac{1}{\sqrt{3}}, \quad n = \frac{1}{\sqrt{3}}$$

which are the required direction cosines of the axis.

Also

$$\cos \alpha = \frac{l + 2m + 2n}{3} = \frac{-\dfrac{1}{\sqrt{3}} + \dfrac{2}{\sqrt{3}} + \dfrac{2}{\sqrt{3}}}{3} = \frac{1}{\sqrt{3}}$$

$$\therefore \quad \alpha = \cos^{-1}\left(\frac{1}{\sqrt{3}}\right)$$

Now, let $P(x, y, z)$ be any point on the cone. The direction ratios of OP are x, y, z and those of the axis are $-1, 1, 1$. Since OP makes angle α with axis,

$$\therefore \quad \cos \alpha = \frac{x(-1) + y(1) + z(1)}{\sqrt{x^2 + y^2 + z^2}\sqrt{1 + 1 + 1}}$$

or

$$\frac{1}{\sqrt{3}} = \frac{-x + y + z}{\sqrt{x^2 + y^2 + z^2}\,\sqrt{3}}$$

or

$$\boxed{(x^2 + y^2 + z^2) = (-x + y + z)^2 \quad \text{or} \quad xy - yz + zx = 0}$$

which is the required equation of the right circular cone.

Ex. 22 : *Find the equation of the right circular cone generated by the straight lines drawn from the origin to cut the circle through the three points $(1, 2, 2)$, $(2, 1, -2)$ and $(2, -2, 1)$.*

Sol. :

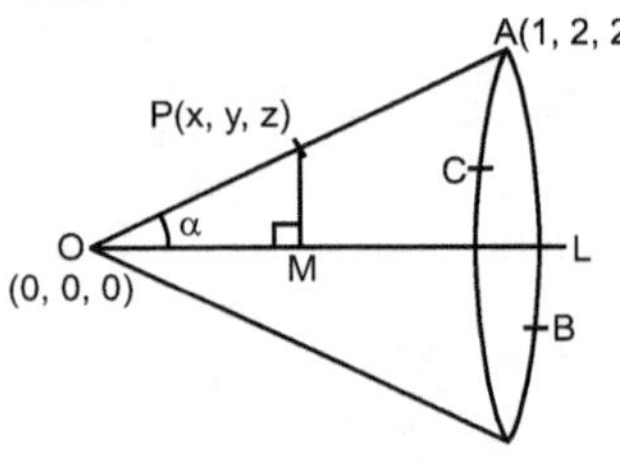

Fig. 8.10

Let A $(1, 2, 2)$, B $(2, 1, -2)$ and C $(2, -2, 1)$ be the given points and l, m, n be the d.r.'s of the axis OL. If α is the semi-vertical angle, then OA, OB, OC make the same angle α with the axis OL (Fig. 8.10).

The d.r.'s of OA, OB, OC are

$$1 - 0, 2 - 0, 2 - 0; \quad 2 - 0, 1 - 0, -2 - 0; \quad 2 - 0, -2 - 0, 1 - 0.$$

i.e. $\qquad 1, 2, 2; \qquad 2, 1, -2; \qquad 2, -2, 1$ respectively.

$$\therefore \quad \cos \alpha = \cos \angle AOL = \frac{l(1) + m(2) + n(2)}{\sqrt{1 + 4 + 4}} = \frac{l + 2m + 2n}{3} \qquad \text{... (1)}$$

$$\cos \alpha = \cos \angle BOL = \frac{l(2) + m(1) + n(-2)}{\sqrt{4 + 1 + 4}} = \frac{2l + m - 2n}{3} \qquad \text{... (2)}$$

$$\cos \alpha = \cos \angle COL = \frac{l(2) + m(-2) + n(1)}{\sqrt{4 + 4 + 1}} = \frac{2l - 2m + n}{3} \qquad \text{... (3)}$$

From (1) and (2), $l - m - 4n = 0$ ... (4)

From (2) and (3), $0l + m - n = 0$ ... (5)

Solving (4) and (5), we have

$$\frac{l}{1+4} = \frac{-m}{-1-0} = \frac{n}{1-0}$$

or

$$\frac{l}{5} = \frac{m}{1} = \frac{n}{1} = \frac{\sqrt{l^2 + m^2 + n^2}}{\sqrt{25 + 1 + 1}} = \frac{1}{\sqrt{27}}$$

∴

$$l = \frac{5}{\sqrt{27}}, \quad m = \frac{1}{\sqrt{27}}, \quad n = \frac{1}{\sqrt{27}}$$

∴   From (1), $$\cos \alpha = \frac{5 + 2 + 2}{3\sqrt{27}} = \frac{1}{\sqrt{3}}$$ ... (6)

Now let P (x, y, z) be any point on the cone. The d.r.'s of OP are x, y, z and those of the axis are 5, 1, 1. Since OP makes angle α with axis,

∴

$$\cos \alpha = \frac{x(5) + y(1) + z(1)}{\sqrt{x^2 + y^2 + z^2}\,\sqrt{25 + 1 + 1}}$$

or

$$\frac{1}{\sqrt{3}} = \frac{5x + y + z}{\sqrt{x^2 + y^2 + z^2}\,\sqrt{27}}$$

or $9(x^2 + y^2 + z^2) = (5x + y + z)^2$

or $\boxed{8x^2 - 4y^2 - 4z^2 + 5xy + 5zx + yz = 0}$

which is the required equation of the cone.

Ex. 23 : *Find the equation of the cone with vertex at the origin and generators touching the sphere $x^2 + y^2 + z^2 - 2x + 4z = 1$.*

Sol. : The equation of the given sphere is $x^2 + y^2 + z^2 - 2x + 4z = 1$ and the vertex is $(0, 0, 0)$.

Equation of enveloping cone is obtained by applying formula

$SS_1 = T^2$.   [Refer article 8.9 (10)].

Here $S = x^2 + y^2 + z^2 - 2x + 4z - 1$ and $x_1 = 0, \ y_1 = 0, \ z_1 = 0$

$S_1 = x_1^2 + y_1^2 + z_1^2 - 2x_1 + 4z_1 - 1 = -1$

and $T = xx_1 + yy_1 + zz_1 - (x + x_1) + 2(y + y_1) - 1 = -x + 2y - 1$

∴   Equation of the required enveloping cone is

$(x^2 + y^2 + z^2 - 2x + 4z - 1)(-1) = (-x + 2z - 1)^2$

which reduces to $\boxed{2x^2 + y^2 + 5z^2 - 4xz = 0}$

Ex. 24 : *Show that the plane $z = 0$ cuts the enveloping cone of the sphere $x^2 + y^2 + z^2 = 11$ which has its vertex at $(2, 4, 1)$ in a rectangular hyperbola.*

Sol. : The equation of the given sphere is $x^2 + y^2 + z^2 - 11 = 0$ and the vertex is $(2, 4, 1)$.

Here $S = x^2 + y^2 + z^2 - 11$ and $x_1 = 2, \ y_1 = 4, z_1 = 1$

$S_1 = x_1^2 + y_1^2 + z_1^2 - 11 = 4 + 16 + 1 - 11 = 10$

and $T = xx_1 + yy_1 + zz_1 - 11 = 2x + 4y + z - 11.$

∴   Equation of the enveloping cone using formula $SS_1 = T^2$ is

$(x^2 + y^2 + z^2 - 11)(10) = (2x + 4y + z - 11)^2$

The plane $z = 0$ cuts it in the conic

$10(x^2 + y^2 - 11) = (2x + 4y - 11)^2$ ... (1)

This represents a rectangular hyperbola in XOY plane,

if $\boxed{\text{coeff. of } x^2 + \text{coeff. of } y^2 = 0}$

i.e. if $(10 - 4) + (10 - 16) = 0$, which is true.

Hence the result.

Type IV : Problems on Angle Between the Lines in which a Plane through the Vertex Cuts a Cone :

Ex. 25 : *Find the angle between the lines of section of the plane $3x + y + 5z = 0$ and cone $6yz - 2zx + 5xy = 0$.*

Sol. : Let $\dfrac{x}{l} = \dfrac{y}{m} = \dfrac{z}{n}$ be the line of section (one of the generators) of the cone $6yz - 2zx + 5xy = 0$ by the plane $3x + y + 5z = 0$,

so that

$$6\,mm - 2nl + 5lm = 0 \ \text{ and } \ 3l + m + 5n = 0 \qquad \qquad \text{... (1)}$$

Eliminating m from these equations, we get

$$-6n\,(3l + 5n) - 2nl - 5l\,(3l + 5n) = 0$$

or $15l^2 + 45ln + 30n^2 = 0 \quad \therefore \ (l + n)\,(l + 2n) = 0.$

When $l + 2n = 0$ and from (1), $3l + m + 5n = 0$, then $\dfrac{l}{1} = \dfrac{m}{2} = \dfrac{n}{-1}$

When $l + 2n = 0$ and from (1), $3l + m + 5n = 0$, then $\dfrac{l}{-2} = \dfrac{m}{1} = \dfrac{n}{1}$

Thus the d.r.'s of two lines are $1, 2, -1$ and $-2, 1, 1$.

If θ is the angle between two lines, then

$$\cos\theta \ = \ \frac{(1)\,(-2) + (2)\,(1) + (-1)\,(1)}{\sqrt{(1 + 4 + 1)}\,\sqrt{(4 + 1 + 1)}} \ = \ -\frac{1}{6} = \frac{1}{6} \ \text{ (numerically)}$$

$\therefore$ $\boxed{\theta \ = \ \cos^{-1}\left(\dfrac{1}{6}\right).}$

Hence the result.

Ex. 26 : *Prove that the plane $ax + by + cz = 0$ cuts the cone $yz + zx + xy = 0$ in perpendicular lines if $\dfrac{1}{a} + \dfrac{1}{b} + \dfrac{1}{c} = 0$.*

Sol. : Let $\dfrac{x}{l} = \dfrac{y}{m} = \dfrac{z}{n}$ be the line of section (one of the generators) of the cone $yz + zx + xy = 0$ by the plane $ax + by + cz = 0$.

$\therefore$ $mn + nl + lm = 0$ and $al + bm + cn = 0$

Eliminating n between these equations, we get

$$\left(-\frac{al + bm}{c}\right)(m + l) + lm = 0 \qquad\qquad \left(\because n = -\frac{al + bm}{c}\right)$$

or $al^2 + (a + b - c)\,lm + bm^2 = 0$

or $a\left(\dfrac{l}{m}\right)^2 + (a + b - c)\left(\dfrac{l}{m}\right) + b = 0$

Let roots of quadratic in $\left(\dfrac{l}{m}\right)$ be $\dfrac{l_1}{m_1}$ and $\dfrac{l_2}{m_2}$,

$\therefore$ products of the roots $= \dfrac{l_1}{m_1} \cdot \dfrac{l_2}{m_2} = \dfrac{b}{a}$ or $a\,l_1 l_2 = b\,m_1 m_2$

or $\dfrac{l_1 l_2}{\dfrac{1}{a}} = \dfrac{m_1 m_2}{\dfrac{1}{b}} = \dfrac{n_1 n_2}{\dfrac{1}{c}}$ (by symmetry)

Now two lines will be at right angles, if $l_1 l_2 + m_1 m_2 + n_1 n_2 = 0$ or if $\boxed{\dfrac{1}{a} + \dfrac{1}{b} + \dfrac{1}{c} = 0}$, which is the required condition.

MISCELLANEOUS EXAMPLES

Ex. 1 : *Prove that the equation of the cone generated by rotating the line $\dfrac{x}{l} = \dfrac{y}{m} = \dfrac{z}{n}$ about the line $\dfrac{x}{a} = \dfrac{y}{b} = \dfrac{z}{c}$ as axis is given by*

$$(l^2 + m^2 + n^2)\,(ax + by + cz)^2 = (al + bm + cn)^2\,(x^2 + y^2 + z^2)$$

Sol. : Axis of the cone is $\dfrac{x}{a} = \dfrac{y}{b} = \dfrac{z}{c}$ ... (1)

Any generator of the cone is $\dfrac{x}{l} = \dfrac{y}{m} = \dfrac{z}{n}$ ... (2)

$\therefore$ $\cos\alpha \ = \ \dfrac{al + bm + cn}{\sqrt{(a^2 + b^2 + c^2)}\,\sqrt{(l^2 + m^2 + n^2)}}$ ... (3)

If P (x, y, z) is any point on the cone, then d.r.'s of generator OP are x, y, z and it is inclined to axis (1) at an angle α given by (3).

$$\therefore \quad \cos \alpha = \frac{ax + by + cz}{\sqrt{(a^2 + b^2 + c^2)} \sqrt{(x^2 + y^2 + z^2)}} = \frac{al + bm + cn}{\sqrt{a^2 + b^2 + c^2} \sqrt{l^2 + m^2 + n^2}} \qquad \text{[by 3]}$$

Squaring, we get the required equation of the cone as

$$\boxed{(l^2 + m^2 + n^2) \ (ax + by + cz)^2 = (al + bm + cn)^2 \ (x^2 + y^2 + z^2)}$$

Ex. 2 : *Find the equation of right circular cone generated when the straight line 2y + 3z = – 6, x = 0 revolves about z-axis.*

(Dec. 2005, May 2009)

Sol. :

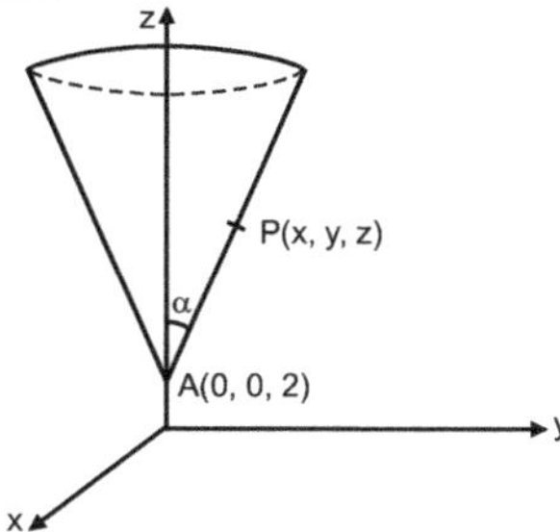

Fig. 8.11

Since the given line revolves about z-axis, cone generated is a right circular cone whose vertex is the point of intersection of the line $2y + 3z = 6$, $x = 0$ and z-axis (Fig. 8.11) i.e. x = 0, y = 0.

Equation of the given line can be written as $2y = -3 (z - 2)$, x = 0 or $\dfrac{x}{0} = \dfrac{y}{1/2} = \dfrac{z - 2}{-1/3}$

$\therefore$ The vertex is A (0, 0, 2) and generator of the cone is given by

$$\frac{x}{0} = \frac{y}{3} = \frac{z - 2}{- 2} \qquad \qquad \text{...(1)}$$

Hence d.r.'s of the generator are 0, 3, – 2 and that of axis (z-axis) are 0, 0, 1 and the semi-vertical angle α is, therefore, given by

$$\cos \alpha = \frac{(0) (0) + (3) (0) + (-2) (1)}{\sqrt{0 + 9 + 4} \sqrt{0 + 0 + 1}} = -\frac{2}{\sqrt{13}} \qquad \text{... (2)}$$

Let P(x, y, z) be any point on the cone, so that the d.r.'s of AP are x, y, z – 2. Since AP makes an angle α with Az, the axis of the cone, then we have

$$\cos \alpha = \frac{(x) (0) + (y) (0) + (z - 2) (1)}{\sqrt{x^2 + y^2 + (z - 2)^2} \sqrt{0 + 0 + 1}}$$

Thus from (2) and (3), we have

$$\frac{(z - 2)^2}{x^2 + y^2 + (z - 2)^2} = \frac{4}{13} \qquad \text{... (3)}$$

or $\boxed{4x^2 + 4y^2 - 9z^2 + 36z - 36 = 0}$

which is the required equation of the cone.

Ex. 3 : *Find the equation of right circular cone which has its vertex at the point (0, 0, 10) and whose intersection with the plane XOY is a circle of diameter 10.*

(Nov./Dec. 2019, May 2006, 2014, Dec. 2018)

Sol. :

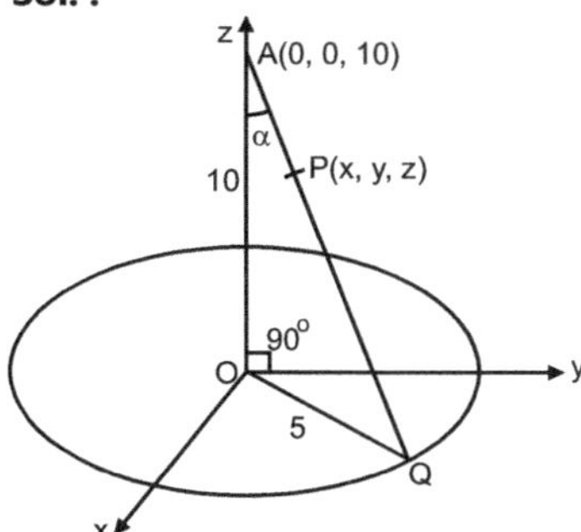

Fig. 8.12

Since the intersection of the cone with the plane XOY is a circle, the plane XOY must be perpendicular to the axis of the cone (Fig. 8.12). Therefore the axis of the cone is the line through the vertex A(0, 0, 10) perpendicular to the plane XOY. i.e. z-axis. The distance of (0, 0, 10) from the plane XOY is 10, which may be taken as height of the cone.

Let Q be any point on the circle, then OQ = radius of the circle = 5.

If α is the semi-vertical angle of the cone, then $\tan \alpha = \dfrac{5}{10} = \dfrac{1}{2}$ or $\cos \alpha = \dfrac{2}{\sqrt{5}}$.

If P (x, y, z) be any point on the cone, then d.r.'s of AP are x, y, z – 10. Also d.r.'s of z-axis are 0, 0, 1.

$$\therefore \qquad \cos \alpha = \frac{x (0) + y (0) + (z - 10) (1)}{\sqrt{x^2 + y^2 + (z - 10)^2} \sqrt{(0 + 0 + 1)}}$$

or $$\frac{2}{\sqrt{5}} = \frac{z - 10}{\sqrt{x^2 + y^2 + (z - 10)^2}}$$

or $4 [x^2 + y^2 + (z - 10)^2] = 5 (z - 10)^2$

or $\boxed{4x^2 + 4y^2 - z^2 + 20z - 100 = 0}$

which is the required equation of the cone.

Ex. 4 : *Find the semi-vertical angle and the equation of the right circular cone having its vertex at the origin and passing through the circle $x^2 + z^2 = 25$, $y = 4$.*

(*Dec. 2009, May 2011*)

Sol. :

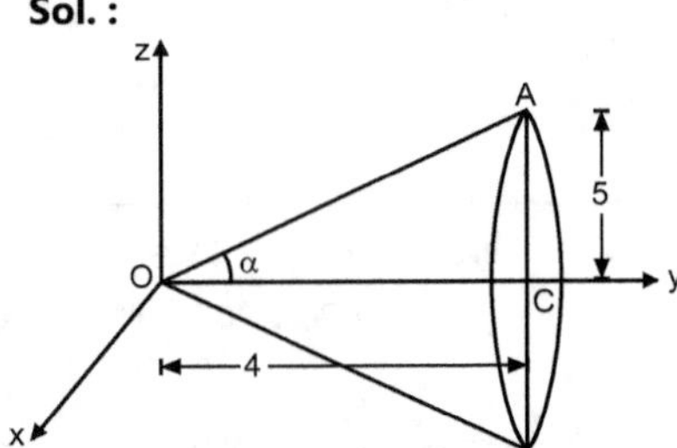

Fig. 8.13

Here the axis of the cone must be the line through vertex (origin) perpendicular to the plane of circle $x^2 + z^2 = 25$, $y = 5$ (Fig. 8.13). Therefore the axis of the cone is y-axis.

If α is the semi-vertical angle of the cone, then

$$\tan \alpha = \frac{CA}{OC} = \frac{5}{4} \quad \therefore \alpha = \tan^{-1}\left(\frac{5}{4}\right).$$

To find equation of the cone, make $x^2 + z^2 = 25$ homogeneous with the help of $y = 4$ or $\frac{y}{4} = 1$.

Thus $\qquad x^2 + z^2 = 25\left(\frac{y}{4}\right)^2$

or $\qquad \boxed{16(x^2 + z^2) = 25 y^2}$

which is the required equation of the cone.

EXERCISE 8.1

1. Find the equation of the cones with vertex at the origin and which pass through the curves given by;

   (i) $x^2 + y^2 = 4$, $z = 2$

   (ii) $ax^2 + by^2 = 2z$, $lx + my + nz = p$

   (iii) $x^2 + y^2 + z^2 = 9$, $x + y + z = 1$

   (iv) $\frac{x^2}{4} + \frac{y^2}{9} + \frac{z^2}{1} = 1$, $x + y + z = 1$

   (v) $x^2 + y^2 + z^2 + 2x - y + 3z - 1 = 0$,

   $\qquad x^2 + y^2 + z^2 + x + 2z - 5 = 0$.

   (vi) $\frac{x^2}{a^2} + \frac{z^2}{c^2} = 1$, $y = b$

   Ans. (i) $x^2 + y^2 - z^2 = 0$ (ii) $p(ax^2 + by^2) = 2z(lx + my + nz)$
   (iii) $4(x^2 + y^2 + z^2) + 9(xy + yz + zx) = 0$ (iv) $27x^2 + 32y^2 + 72(yz + zx + xy) = 0$
   (v) $3x^2 + 11y^2 + 3z^2 + 14xy + 18yz - 30zx = 0$ (vi) $\frac{x^2}{a^2} - \frac{y^2}{b^2} + \frac{z^2}{c^2} = 0$.

2. A variable plane is parallel to the given plane $\frac{x}{a} + \frac{y}{b} + \frac{z}{c} = 0$, and meets the co-ordinate axes in A, B, C. Find the equation of the cone whose vertex is the origin and guiding curve is the circle ABC.

   [**Hint :** Let the plane parallel to given plane is $\frac{x}{a} + \frac{y}{b} + \frac{z}{c} = k$ or $\frac{x}{ak} + \frac{y}{bk} + \frac{z}{ck} = 1$ meets the axes at points A(ka, 0, 0), B(0, kb, 0), C(0, 0, kc). The equation of sphere OABC is $x^2 + y^2 + z^2 - k(ax + by + cz)(1) = 0$. Make this equation homogeneous with the help of equation of plane.]
   Ans. $yz\left(\frac{b}{c} + \frac{c}{b}\right) + zx\left(\frac{c}{a} + \frac{a}{c}\right) + xy\left(\frac{a}{b} + \frac{b}{a}\right) = 0.$

3. Prove that the line $\frac{x}{l} = \frac{y}{m} = \frac{z}{n}$, where $2l^2 + 3m^2 - 5n^2 = 0$, is a generator of the cone $2x^2 + 3y^2 - 5z^2 = 0$.

   [**Hint :** Eliminate l, m, n between given line equation and $2l^2 + 3m^2 - 5n^2 = 0$.]

4. Find the equation of the cone which passes through the three co-ordinate axes as well as the two lines $\frac{x}{1} = \frac{y}{-2} = \frac{z}{3}$,

   $\frac{x}{3} = \frac{y}{-1} = \frac{z}{1}$

   [**Hint :** The d.r.'s of lines (1, –2, 3) and (3, –1, 1) satisfy equation of cone through co-ordinate axes $fyz + gzx + hxy = 0$. Solve $-6f + 3g - 2h = 0$, $-f + 3g - 3h = 0$ for f, g and h.]
   Ans. $3yz + 16zx + 15xy = 0$

5. Find the equation of the cone which passes through the three coordinate axes and the three mutually perpendicular lines

   $\frac{x}{2} = \frac{y}{1} = \frac{z}{-1}$, $\frac{x}{1} = \frac{y}{3} = \frac{z}{5}$, $\frac{x}{8} = \frac{y}{-11} = \frac{z}{5}$.

   (*May 2006*)

   [**Hint :** The d.r.'s of lines (2, 1, –1) and (1, 3, 5) satisfy $fyz + gzx + hxy = 0$. Solve $-f - 2g + 2h = 0$, $15f + 8g + 3h = 0$ for f, g and h. Also, d.r.'s (8, –11, 5) of third line satisfy the equation of the cone.]
   Ans. $16yz - 33zx - 25xy = 0$.

6. Show that the equation of the cone which contains the three co-ordinate axes and the lines through the origin having direction cosines l_1, m_1, n_1 and l_2, m_2, n_2 is $l_1 l_2 (m_1 n_2 - m_2 n_1) yz + m_1 m_2 (n_1 l_2 - n_2 l_1) zx + n_1 n_2 (l_1 m_2 - l_2 m_1) xy = 0$.

[**Hint :** Equation of cone through axes is $fyz + gzx + hxy = 0$. It contains lines through origin with given d.c.'s.
$\therefore$ $fm_1n_1 + gn_1l_1 + hl_1m_1 = 0$, $fm_2n_2 + gn_2l_2 + hl_2m_2 = 0$.]

On putting for f, g, h in first equation from last two equations, we get required equation of the cone.

[**Note :** Hence a cone of second degree can be found to pass through any five concurrent lines.]

7. Find the equation of the cone whose

   (i) vertex is (α, β, γ) and guiding curve is the parabola $y^2 = 4ax$, $z = 0$.   **Ans.** $(\gamma y - \beta z)^2 = 4a (z - \gamma) (\alpha z - \gamma x)$.

   (ii) vertex is $(1, 1, 0)$ and guiding curve is $z^2 + x^2 = 4$, $y = 0$.   **Ans.** $x^2 - 3y^2 + z^2 - 2xy + 8y - 4 = 0$.

   (iii) vertex is at the point $(0, 0, 5)$ and guiding curve is $x^2 + 2xy + 3y^2 = 1$, $z = 3$.   **Ans.** $4x^2 + 8xy + 12y^2 - z^2 + 10z - 25 = 0$.

   (iv) vertex is at the point $(3, 1, 2)$ and whose generating line passing through the ellipse $2x^2 + 3y^2 = 1$, $z = 0$.
        Ans. $2x^2 + 3y^2 + 5z^2 - 3yz - 3zx + z - 1 = 0$.

   (v) vertex is $(3, 4, 5)$ and with $3y^2 + 4z^2 = 16$ and $z + 2x = 0$ as base.
        Ans. $48x^2 + 33y^2 + 16z^2 - 48xy - 24yz - 32zx - 64x + 32z - 176 = 0$.

   (vi) vertex is $(1, 2, 3)$ and guiding curve is the circle $x^2 + y^2 + z^2 = 4$, $x + y + z = 1$.   **(May 04, Dec. 07)**
        Ans. $5x^2 + 3y^2 + z^2 - 2xy - 6yz - 4zx + 6x + 8y + 10z = 26$.

8. Prove that the equation of the cone whose vertex is the origin and base is the curve $z = k$, $f(x, y) = 0$ is $f\left(\dfrac{xk}{z}, \dfrac{yk}{z}\right) = 0$.

   [**Hint :** Let $\dfrac{x}{l} = \dfrac{y}{m} = \dfrac{z}{n}$ be a generator. It meets the base $z = k$, $f(x, y) = 0$. Thus, $x = \dfrac{lk}{n}$, $y = \dfrac{mk}{n}$. Now equation of the cone
   is obtained by putting for x, y in $f(x, y) = 0$ and eliminating l, m, n from the two equations.]

9. The vertex of a cone is (a, b, c) and yz-plane cuts it in the curve $F(y, z) = 0$, $x = 0$. Show that the zx plane cuts it in the curve
   $y = 0$, $F\left(\dfrac{bx}{x - a}, \dfrac{cx - az}{x - a}\right) = 0$.

   [**Hint :** Let $\dfrac{x - a}{l} = \dfrac{y - b}{m} = \dfrac{z - c}{n}$ be a generator. It meets the plane $x = 0$ in point $\left(0, b - \dfrac{m}{l}, c - \dfrac{n}{l}\right)$. Now find the equation of
   the cone and put $y = 0$.]

10. Find the equation of the cone whose vertex is the point $P(0, 0, 1)$ and whose directing curve is the ellipse $\dfrac{x^2}{25} + \dfrac{y^2}{9} = 1$, $z = 3$.
    Obtain the section of the derived cone by plane $y = 0$.

    [**Hint :** Let $\dfrac{x - 0}{l} = \dfrac{y - 0}{m} = \dfrac{z - 1}{n}$ be a generator. It meets the plane $z = 3$ in point $\left(\dfrac{2l}{n}, \dfrac{2m}{n}, 3\right)$. This point lies on the given
    ellipse, $\dfrac{4}{25}\left(\dfrac{l}{n}\right)^2 + \dfrac{4}{9}\left(\dfrac{m}{n}\right)^2 = 1$. Required equation of cone is obtained by eliminating l, m, n by using $\dfrac{l}{n} = \dfrac{x}{z - 1}$ and
    $\dfrac{m}{n} = \dfrac{y}{z - 1}$.]   **Ans.** $36x^2 + 100y^2 = 225 (z - 1)^2$; $2x = \pm 5 (z - 1)$, $y = 0$.

11. Prove that the equations :
    (i) $x^2 + 2y^2 + z^2 - 4yz - 6zx - 2x + 8y - 2z + 9 = 0$ represents a cone whose vertex is $(1, -2, 0)$.
    (ii) $4x^2 - y^2 + 2z^2 - 3yz + 2xy + 12x - 21y + 6z + 4 = 0$ represents a cone whose vertex is $(-1, -2, -3)$.
    (iii) $2y^2 - 8yz - 4zx - 8xy + 6x - 4y - 2z + 5 = 0$ represents a cone whose vertex is $\left(-\dfrac{7}{6}, \dfrac{1}{3}, \dfrac{5}{6}\right)$.

12. Prove that the equation $ax^2 + by^2 + cz^2 + 2ux + 2vy + 2wz + d = 0$ represents a cone if $\dfrac{u^2}{a} + \dfrac{v^2}{b} + \dfrac{w^2}{c} = d$.

    [**Hint :** $x = -\dfrac{u}{a}$, $y = -\dfrac{v}{b}$ and $z = -\dfrac{w}{c}$ and $F_t = 0$, gives $2 (ux + vy + wz + dt) = 0$.]

13. Find the equation of the right circular cone whose

    (i) vertex is $(1, -1, 2)$, axis is the line $\dfrac{x - 1}{2} = \dfrac{y + 1}{1} = \dfrac{z - 2}{-2}$ and semi-vertical angle $45°$.   **(May 11, 09, 15, Dec. 17)**

    [**Hint :** The d.r.'s of axis is $2, 1, -2$. The vertex $A(1, -1, 2)$. Let $P(x, y, z)$ be any point on the surface of the cone, then d.r.'s of AP
    are $(x - 1)$, $(y + 1)$, $(z - 2)$. Therefore, $\cos 45° = \dfrac{1}{\sqrt{2}} = \dfrac{2(x - 1) + (y + 1) - 2(z - 2)}{3\sqrt{(x - 1)^2 + (y + 1)^2 + (z - 2)^2}}$. Simplifying, we get the equation of the
    cone.]

         Ans. $x^2 + 7y^2 + z^2 - 8xy + 8yz + 16zx - 42x + 6y - 12z + 36 = 0$.

(ii) vertex is (1, 2, 3), axis has direction ratios 2, –1, 4 and semi-vertical angle 60°. **(May 2017)**

Ans. $5x^2 + 17y^2 – 43z^2 + 16xy + 32yz – 64zx + 150x – 180y + 258z – 282 = 0.$

(iii) vertex is (1, –1, 1). The axis parallel to $x = -\dfrac{y}{2} = -z$ and one of its generators has direction cosines proportional to (2, 2, 1).

(Dec. 11, Nov. 15) Ans. $3y^2 + 4yz – 2zx – 4xy – 2x + 6y + 6z + 1 = 0.$

(iv) vertex at (0, 0, 0), semi-vertical angle $\dfrac{\pi}{2}$ and axis along the line $x = -2y = z.$ **(May 2019)**

Ans. $x^2 + 7y^2 + z^2 + 8xy + 8yz – 16zx = 0.$

(v) vertex is at the point (1, 0, 1) which passes through the point (1, 1, 1) and axis of the cone is equally inclined to the co-ordinate axes.

[**Hint :** The axes has d.r.'s 1, 1, 1.]   **Ans.** $xy + yz + zx – x – 2y – z + 1 = 0.$

(vi) passing through (2, –2, 1) with vertex at origin and axis parallel to the line $\dfrac{x-2}{5} = \dfrac{y-1}{1} = \dfrac{z+2}{1}$

(Nov./Dec. 2019, Dec. 2011, May 2016)

[**Hint :** The d.r.'s of axis are 5, 1, 1, $\cos \alpha = \dfrac{(2)\,(5) + (-2)\,(1) + (1)\,(1)}{\sqrt{4 + 4 + 1}\ \sqrt{25 + 1 + 1}} = \dfrac{1}{\sqrt{3}}$, d.r.'s of generator are $x - 0,\ y - 0,\ z - 0$. Thus

equation of cone is $\dfrac{5x + y + z}{\sqrt{x^2 + y^2 + z^2}\ \sqrt{27}} = \dfrac{1}{\sqrt{3}}$ or $(5x + y + z)^2 = 9\,(x^2 + y^2 + z^2)]$

Ans. $16x^2 – 8y^2 – 8z^2 + 10xy + 10xz + 2yz = 0.$

14. Prove that $x^2 – y^2 + z^2 – 2x + 4y + 6z + 6 = 0$ represents a right circular cone whose vertex is the point (1, 2, – 3), whose axis is parallel to OY and whose semi-vertical angle is 45°.

15. Show that the right circular cone which contains the three co-ordinate axes as generator is $yz + zx + xy = 0.$

[**Hint :** Axis of the cone equally inclined to axes is $\dfrac{x}{1} = \dfrac{y}{1} = \dfrac{z}{1}$.]

16. If α is the semi-vertical angle of the right circular cone which passes through OX, OY and the line $x = y = z$, show that $\cos \alpha = \left(9 – 4\sqrt{3}\right)^{-1/2}$.

[**Hint :** Let axis of the cone be $\dfrac{x}{l} = \dfrac{y}{m} = \dfrac{z}{n}$. Vertex is (0, 0, 0), line OX is $\dfrac{x}{1} = \dfrac{y}{0} = \dfrac{z}{0}$, line OY is $\dfrac{x}{0} = \dfrac{y}{1} = \dfrac{z}{0}$ and given line is

$\dfrac{x}{1} = \dfrac{y}{1} = \dfrac{z}{1}$.]

17. Find the equation of the right circular cone generated by rotating

(i) the line $\dfrac{x}{1} = \dfrac{y}{2} = \dfrac{z}{3}$ about the line $\dfrac{x}{-1} = y = \dfrac{z}{2}$. **(Dec. 2009)**   **Ans.** $5x^2 + 5y^2 – z^2 + 4xy – 8yz + 8zx = 0.$

(ii) the line $2x + 3y = 6,\ z = 0$ about the y-axis.   **Ans.** $4x^2 – 9y^2 + 4z^2 + 36y – 36 = 0.$

18. Find the equation of the cone of revolution with vertex at origin, axis the y-axis and semi-vertical angle 30°.   **(Dec. 2018)**

Ans. $3\,(x^2 + z^2) = y^2.$

19. The vertex of the right circular cone is (0, 0, 0). Its generator intersects the circle through the points (1, 2, 2), (1, –2, 2), (2, –1, –2). Find the equation of the cone.   **Ans.** $12x^2 – 4y^2 – 3z^2 + 8xz = 0.$

20. Prove that $\sqrt{fx} + \sqrt{gy} = \sqrt{hz} = 0$ represents a cone touching the co-ordinate planes.

[**Hint :** On squaring, we get $fx + gy – hz = -2\sqrt{fgxy}$. Again squaring, we get second degree homogeneous equation representing cone. Putting $z = 0$, we get $(fx – gy)^2 = 0$. Therefore, the cone touches plane $z = 0$ along the generator $fx – gy = 0.$

21. Show that the lines $x = 2y = 8z,\ x = y = 2z,\ 4x = 7y = 7z$ lie on a circular cone of semi-vertical angle, $\cos^{-1}\left(\dfrac{11}{\sqrt{126}}\right).]$

22. Find the semi-vertical angle, the axis and the equation of the cone with vertex at the origin and passing through the straight lines $\dfrac{x}{3} = \dfrac{y}{6} = \dfrac{z}{-2},\ \dfrac{x}{2} = \dfrac{y}{2} = \dfrac{z}{-1}$ and $\dfrac{x}{-1} = \dfrac{y}{2} = \dfrac{z}{2}.$ **Ans.** $\cos^{-1}\dfrac{1}{\sqrt{3}};\ x = y = z;\ xy + yz + zx = 0.$

23. Find the equation of the cone of tangents drawn from the point P (2, 0, 0) to the sphere $x^2 + y^2 + z^2 = 1.$
[**Hint :** Use equation of enveloping cone as $SS_1 = T^2$.]   **Ans.** $x^2 – 3y^2 – 3z^2 – 4x + 4 = 0.$

24. Find the equation of the cone with vertex at (1, 1, 1) and generators touching the sphere $x^2 + y^2 + z^2 – 2x + 4z = 1.$
[**Hint :** Use equation of enveloping cone as $SS_1 = T^2$.]   **Ans.** $4x^2 + 3y^2 – 5z^2 – 6yz – 8x + 16z – 4 = 0.$

25. Find the enveloping cone of the sphere $x^2 + y^2 + z^2 + 2x – 2y = 2$ with its vertex at (1, 1, 1).

Ans. $3x^2 – y^2 + 4zx – 10x + 2y – 4z + 6 = 0.$

26. Find the equation of the line in which the plane $2x + y - z = 0$ cuts the cone $4x^2 - y^2 + 3z^2 = 0$.

Ans. $\dfrac{x}{-1} = \dfrac{y}{2} = \dfrac{z}{0}$ and $\dfrac{x}{-1} = \dfrac{y}{4} = \dfrac{z}{2}$.

27. Find the condition that the plane $ux + vy + wz = 0$ may cut the cone $ax^2 + by^2 + cz^2 = 0$ in perpendicular lines.

Ans. $(b + c)\, u^2 + (c + a)\, v^2 + (a + b)\, w^2 = 0$.

28. Find the equation of the cone which passes through the three co-ordinate axes and the lines $\dfrac{x}{1} = \dfrac{y}{2} = \dfrac{z}{3}$ and $\dfrac{x}{3} = \dfrac{y}{1} = \dfrac{z}{-4}$.

[**Hint :** The general equation of the cone passing through the coordinate axes is $fyz + gzx + hxy = 0$ or $Fyz + Gzx + xy = 0$,

where $F = \dfrac{f}{h}$ and $G = \dfrac{g}{h}$. Also d.r.'s of given lines satisfy the cone equations : $6F + 3G + 2 = 0$ and $-4F - 12G + 3 = 0$.

Solving these, $F = \dfrac{-11}{20}$ and $G = \dfrac{26}{60}$.]   **Ans.** $33yz - 26xz - 60xy = 0$

THE CYLINDER

8.11 DEFINITION

A cylinder is a surface generated by a straight line which is always parallel to a fixed straight line and satisfies one more condition for example, it intersects a given curve or touches a given surface.

The fixed straight line is called the **Axis** and the given curve (or surface) the **Guiding Curve** of the cylinder. The moving straight line is called the **Generator** of the cylinder.

8.12 EQUATION OF A CYLINDER

To find the equation of the cylinder whose generators are parallel to the line $\dfrac{x}{l} = \dfrac{y}{m} = \dfrac{z}{n}$ ... (i)

and intersecting the curve (conic)

$$ax^2 + 2hxy + by^2 + 2gx + 2fy + c = 0, \quad z = 0 \qquad \text{... (ii)}$$

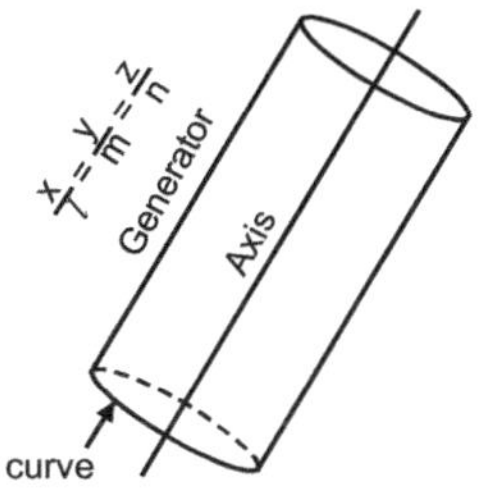

Let $P(x_1, y_1, z_1)$ be any point on the cylinder (Fig. 8.14). Then the equation of a generator through this point and parallel to the line (i) is

$$\frac{x - x_1}{l} = \frac{y - y_1}{m} = \frac{z - z_1}{n} \qquad \text{... (iii)}$$

This line meets the plane $z = 0$ at the point $\left(x_1 - \dfrac{lz_1}{n}, y_1 - \dfrac{mz_1}{n}, 0 \right)$

The generator (iii) will intersect the given guiding curve (ii), if the point

$$\left(x_1 - \frac{lz_1}{n}, y_1 - \frac{mz_1}{n}, 0 \right)$$

Fig. 8.14

lies on (ii) i.e.

$$a \left(x_1 - \frac{lz_1}{n} \right)^2 + 2h \left(x_1 - \frac{lz_1}{n} \right) \left(y_1 - \frac{mz_1}{n} \right) + b \left(y_1 - \frac{mz_1}{n} \right)^2 + 2g \left(x_1 - \frac{lz_1}{n} \right) + 2f \left(y_1 - \frac{mz_1}{n} \right) + c = 0$$

$\therefore$ The locus of $P\,(x_1, y_1, z_1)$ is

$$a \left(x - \frac{lz}{n} \right)^2 + 2h \left(x - \frac{lz}{n} \right) \left(y - \frac{mz}{n} \right) + b \left(y - \frac{mz}{n} \right)^2 + 2g \left(x - \frac{lz}{n} \right) + 2f \left(y - \frac{mz}{n} \right) + c = 0$$

or $\boxed{a\,(nx - lz)^2 + 2h\,(nx - lz)\,(ny - mz) + b\,(ny - mz)^2 + 2gn\,(nx - lz) + 2fn\,(ny - mz) + cn^2 = 0}$ ... (11)

which is the required equation of the cylinder.

Important Particular Case :

If the generators are parallel to the z-axis, then $l = 0$, $m = 0$ and $n = 1$.

$\therefore$ The general equation (11) of the cylinder reduces to

$$ax^2 + 2hxy + by^2 + 2gx + 2fy + c = 0 \qquad \text{... (iv)}$$

Note : The z-co-ordinate does not appear in the equation of cylinder whose generators are parallel to z-axis. Therefore the equation of the cylinder whose generators are parallel to z-axis and intersect the curve

$$f(x, y, z) = 0 \qquad \phi\,(x, y, z) = 0$$

is obtained by eliminating z between the two equations of the curve.

8.13 RIGHT CIRCULAR CYLINDER

Definition : A right circular cylinder is a surface generated by a straight line which intersects a fixed circle (guiding curve) and is perpendicular to the plane of the circle.

The normal to the plane of the guiding circle through its centre is called the **Axis** of the cylinder and radius of the circle is the **Radius** of the cylinder.

Section of a right circular cylinder by plane perpendicular to the axis is called normal section, which is a circle.

Remark : The length of the perpendicular from any point on right circular cylinder to its axis is equal to the radius of the cylinder.

8.14 EQUATION OF A RIGHT CIRCULAR CYLINDER

To Find the Equation of the Right Circular Cylinder whose Radius is r and Axis is the Line $\dfrac{x-\alpha}{l} = \dfrac{y-\beta}{m} = \dfrac{z-\gamma}{n}$.

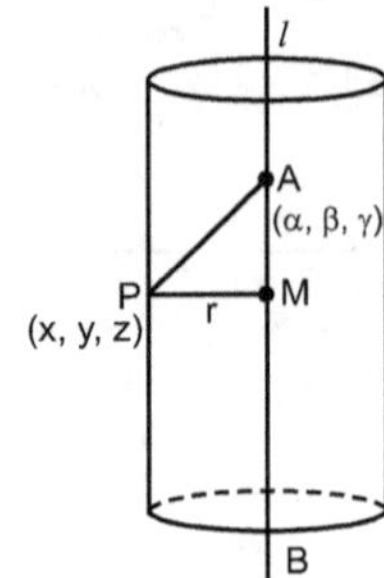

Let P (x, y, z) be any point on the cylinder. A (α, β, γ) be a fixed point on the axis AB (Fig. 8.15).

Then   PM = Radius of the cylinder = r (given).

and   AP = Distance between points A and P.

$\qquad\qquad = \sqrt{(x-\alpha)^2 + (y-\beta)^2 + (z-\gamma)^2}$

$\quad$ AM = Projection of AP on the axis AB   [Refer article 7.10.1 (16)]

$\qquad\qquad = \dfrac{l\,(x-\alpha) + m\,(y-\beta) + n\,(z-\gamma)}{\sqrt{l^2 + m^2 + n^2}}$

Note : If the d.r.'s of the axis are l, m, n, then the d.c.'s of the axis are $\dfrac{l}{\sqrt{\Sigma l^2}}, \dfrac{m}{\sqrt{\Sigma l^2}}, \dfrac{n}{\sqrt{\Sigma l^2}}$.

Fig. 8.15

Now   from $\triangle$ PMA, $AP^2 = AM^2 + PM^2$ or $AP^2 - AM^2 = PM^2$

$\therefore \qquad (x-\alpha)^2 + (y-\beta)^2 + (z-\gamma)^2 - \left\{\dfrac{l\,(x-\alpha) + m\,(y-\beta) + n\,(z-\gamma)}{\sqrt{l^2 + m^2 + n^2}}\right\}^2 = r^2$ ... (12)

which is the required equation of the right circular cylinder.

Particular Cases :

(A) Equation of a Right Circular Cylinder whose Axis is $\dfrac{x}{l} = \dfrac{y}{m} = \dfrac{z}{n}$

Putting $\alpha = 0$, $\beta = 0$, $\gamma = 0$ in (12), we get the required equation of the cylinder as

$$(x^2 + y^2 + z^2) - \left\{\dfrac{lx + my + nz}{\sqrt{l^2 + m^2 + n^2}}\right\}^2 = r^2 \qquad ... (13)$$

(B) Equation of a Right Circular Cylinder whose Axis is z-axis

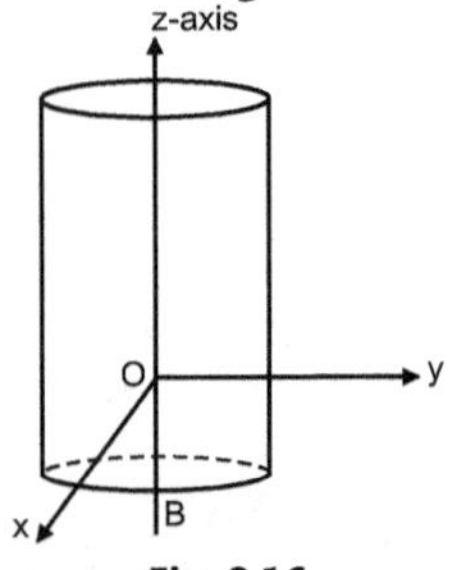

Putting $l = 0$, $m = 0$, $n = 1$ in (13),

We get the required equation of the cylinder as

$$x^2 + y^2 = r^2 \qquad ... (14)$$

Fig. 8.16

8.15 ENVELOPING CYLINDER

Definition : The locus of the tangent lines to a given surface and parallel to a given line is a cylinder and is called the enveloping cylinder of the surface.

8.16 EQUATION OF ENVELOPING CYLINDER

To Find the Equation of the Enveloping Cylinder of the Sphere $x^2 + y^2 + z^2 = a^2$, whose Generators are Parallel to the Line $\dfrac{x}{l} = \dfrac{y}{m} = \dfrac{z}{n}$.

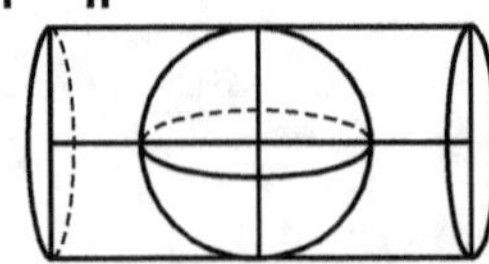

The given sphere is

$\qquad\qquad x^2 + y^2 + z^2 = a^2$ ... (i)

and the line is $\dfrac{x}{l} = \dfrac{y}{m} = \dfrac{z}{n}$ ... (ii)

Fig. 8.17

Let P (α, β, γ) be any point on the cylinder (locus), then any line through (α, β, γ) parallel to line (ii) is

$$\frac{x - \alpha}{l} = \frac{y - \beta}{m} = \frac{z - \gamma}{n} = k \text{ (say)}. \qquad \text{... (iii)}$$

Any point on the line (iii) is $(\alpha + lk, \ \beta + mk, \ \gamma + nk)$.

This point will lie on the sphere (i), if

$$(\alpha + lk)^2 + (\beta + mk)^2 + (\gamma + nk)^2 = a^2$$

or $\qquad k^2 (l^2 + m^2 + n^2) + 2k (\alpha l + \beta m + \gamma n) + \alpha^2 + \beta^2 + \gamma^2 - a^2 = 0 \qquad$ ... (iv)

The line (iii) will touch the given sphere if the quadratic equation (iv) has equal roots, i.e. if

$$(\alpha l + \beta m + \gamma n)^2 = (l^2 + m^2 + n^2)(\alpha^2 + \beta^2 + \gamma^2 - a^2)$$

Hence the locus of P(α, β, γ) is obtained by replacing α, β, γ by x, y, z.

$$\boxed{(lx + my + nz)^2 = (l^2 + m^2 + n^2)(x^2 + y^2 + z^2 - a^2)} \qquad \text{... (15)}$$

8.17 ILLUSTRATIONS

Type I : Problems on cylinder whose generator and guiding curve are given :

Ex. 1 : *Find the equation of the cylinder whose generators are parallel to z-axis and intersect the curve* $ax^2 + by^2 = 2z$, $lx + my + nz = p$. **(Dec. 2010)**

Sol. : The given guiding curve is

$$ax^2 + by^2 = 2z \qquad \text{... (1)}$$
$$lx + my + nz = p \qquad \text{... (2)}$$

Since the equation of the cylinder whose axis is z-axis, is the relation between x and y only, it is obtained by eliminating z from the given equations (1) and (2).

Now from (1), we have $\qquad z = \dfrac{ax^2 + by^2}{2}$

Putting in (2), we get

$$lx + my + n\left(\frac{ax^2 + by^2}{2}\right) = p$$

or $\qquad 2lm + 2my + n(ax^2 + by^2) = 2p$

or $\qquad \boxed{n(ax^2 + by^2) + 2lx + 2my - 2p = 0}$

which is the required equation of the cylinder.

Ex. 2 : *Find the equation of the cylinder whose generators are parallel to x-axis and whose guiding curve is* $ax^2 + by^2 + cz^2 = 1$, $lx + my + nz = p$ **(May 2005)**

Sol. : The given guiding curve is

$$ax^2 + by^2 + cz^2 = 1 \qquad \text{... (1)}$$
$$lx + my + nz = p \qquad \text{... (2)}$$

Since generators of the cylinder are parallel to x-axis, the equation of the cylinder will be free from x. Thus we eliminate x from equations (1) and (2) to get the equation of the cylinder.

Now from (2), we have $x = -\dfrac{my + nz - p}{l}$

Putting in (1), we get

$$a\left(-\frac{my + nz - p}{l}\right)^2 + by^2 + cz^2 = 1$$

or $\qquad a(my + nz - p)^2 + bl^2 y^2 + cl^2 z^2 = l^2$

or $\qquad a(m^2 y^2 + n^2 z^2 + p^2 - 2mpy - 2npz + 2mnyz) + bl^2 y^2 + cl^2 z^2 - l^2 = 0$

or $\qquad \boxed{(am^2 + bl^2) y^2 + (an^2 + cl^2) z^2 + 2amnyz - 2ampy - 2anpz + ap^2 - l^2 = 0}$

which is the required equation of the cylinder.

Ex. 3 : *Find the equation of the cylinder whose generators are parallel to the line* $\dfrac{x}{1} = \dfrac{y}{-2} = \dfrac{z}{3}$ *and guiding curve is the ellipse* $x^2 + 2y^2 = 1$, $z = 3$. **(Dec. 2006)**

Sol. : Let (x_1, y_1, z_1) be any point on the cylinder, then the equations of the generator through this point and parallel to the given line are

$$\frac{x - x_1}{1} = \frac{y - y_1}{-2} = \frac{z - z_1}{3}$$

This generator meets the plane $z = 3$,

where
$$\frac{x - x_1}{1} = \frac{y - y_1}{-2} = \frac{3 - z_1}{3}$$

$\therefore$
$$x = x_1 + \frac{3 - z_1}{3} = \frac{3(x_1 + 1) - z_1}{3}$$

$$y = y_1 - \frac{2(3 - z_1)}{3} = \frac{3(y_1 - 2) + 2z_1}{3}$$

$$z = z_1 + \frac{3(3 - z_1)}{3} = \frac{3(z_1 + 3) - 3z_1}{3}$$

If this point lies on the guiding curve, then

$$\left[\frac{3(x_1 + 1) - z_1}{3}\right]^2 + 2\left[\frac{3(y_1 - 2) + 2z_1}{3}\right]^2 = 1$$

or
$$3\left(x_1^2 + 2y_1^2 + z_1^2\right) + 8y_1z_1 - 2z_1x_1 + 6x_1 - 24y_1 + 24 = 0$$

Hence the locus of the point (x_1, y_1, z_1) is obtained by writing x, y, z for x_1, y_1, z_1 in the above equation, we get

$$\boxed{3(x^2 + 2y^2 + z^2) + 8yz - 2zx + 6x - 24y + 24 = 0}$$

which is the required equation of the cylinder.

Remark : We can also obtain the above result by writing the equation of generator as $\dfrac{x - x_1}{1} = \dfrac{y - y_1}{-2} = \dfrac{z - z_1}{3} = k$. Any point on the generator is $(x_1 + k, y_1 - 2k, z_1 + 3k)$. If this point lies on the guiding curve, then $(x_1 + k)^2 + 2(y_1 - 2k)^2 = 1$ and $z_1 + 3k = 3$. Thus,

$k = \dfrac{z - z_1}{3}$ and $\left[x_1 + \left(\dfrac{z - z_1}{3}\right)\right]^2 + 2\left[y_1 - 2\left(\dfrac{z - z_1}{3}\right)\right]^2 = 1$, which is the same equation.

Ex. 4 : *Find the equation of the cylinder whose generators are parallel to the line* $\dfrac{x}{3} = \dfrac{y}{1} = \dfrac{z}{\sqrt{6}}$ *and whose guiding curve is*

$x^2 + y^2 = 25,\ z = 0.$ **(Dec. 2004, May 2006)**

Sol. : Let (x_1, y_1, z_1) be any point on the generator which is parallel to the given line, then the equations of this generator are

$$\frac{x - x_1}{3} = \frac{y - y_1}{1} = \frac{z - z_1}{\sqrt{6}}$$

This generator meets the plane $z = 0$,

where
$$\frac{x - x_1}{3} = \frac{y - y_1}{1} = \frac{0 - z_1}{\sqrt{6}}$$

$\therefore$
$$x = x_1 - \frac{3z_1}{\sqrt{6}},\quad y = y_1 - \frac{z_1}{\sqrt{6}},\quad z = 0$$

If this point lies on the guiding curve, then

$$\left(x_1 - \frac{3z_1}{\sqrt{6}}\right)^2 + \left(y_1 - \frac{z_1}{\sqrt{6}}\right)^2 = 25$$

or
$$\left(\sqrt{6}\,x_1 - 3z_1\right)^2 + \left(\sqrt{6}\,y_1 - z_1\right)^2 = 150$$

Hence the locus of (x_1, y_1, z_1) is

$$\left(\sqrt{6}\,x - 3z\right)^2 + \left(\sqrt{6}\,y - z\right)^2 = 150$$

or
$$6x^2 + 9z^2 - 6\sqrt{6}\,xz + 6y^2 + z^2 - 2\sqrt{6}\,yz = 150$$

or
$$\boxed{6x^2 + 6y^2 + 10z^2 - 2\sqrt{6}\,yz - 6\sqrt{6}\,xz - 150 = 0}$$

which is the required equation of the cylinder.

Ex. 5 : *Find the equation of the cylinder whose generating lines have direction cosines l, m, n and which pass through circumference of the fixed circle* $x^2 + z^2 = a^2$ *in the ZOX plane.*

Sol. : Let (x_1, y_1, z_1) be any point on the cylinder, then the equation of the generator through this point in the direction l, m, n is

$$\frac{x - x_1}{l} = \frac{y - y_1}{m} = \frac{z - z_1}{n} \qquad \text{... (1)}$$

Given guiding circle is $x^2 + z^2 = a^2$, $y = 0$ ... (2)

The generator (1) meets the plane $y = 0$,

where $\dfrac{x - x_1}{l} = \dfrac{0 - y_1}{m} = \dfrac{z - z_1}{n}$

$\therefore$ $x = x_1 - \dfrac{l}{m}\, y_1, \quad y = 0, \quad z = z_1 - \dfrac{n}{m}\, y_1$

If this point lies on circle (2), then

$$\left(x_1 - \frac{l}{m}\, y_1\right)^2 + \left(z_1 - \frac{n}{m}\, y_1\right)^2 = a^2$$

or $(mx_1 - ly_1)^2 + (mz_1 - ny_1)^2 = a^2 m^2$

Hence the locus of (x_1, y_1, z_1) is obtained by replacing x_1, y_1, z_1 by x, y, z

$\therefore$ $\boxed{(mx - ly)^2 + (mz - ny)^2 = a^2 m^2}$

which is the required equation of the cylinder.

Ex. 6 : *Find the cylinder with generator parallel to* $\dfrac{x}{1} = \dfrac{y}{1} = \dfrac{z}{1}$ *and with guiding curve*

$x^2 + 2y^2 + 6xy - 2z + 8 = 0; \quad x - 2y + 3 = 0$ **(May 2007)**

Sol. : Let (x_1, y_1, z_1) be any point on the cylinder, then the equation of the generator through this point and parallel to given line is

$\dfrac{x - x_1}{1} = \dfrac{y - y_1}{1} = \dfrac{z - z_1}{1} = k$ (say)   ... (1)

$\therefore$ The co-ordinates of any point on this line are

$x = x_1 + k, \quad y = y_1 + k, \quad z = z_1 + k$

The generator meets the plane $x - 2y + 3 = 0$ if

$(x_1 + k) - 2(y_1 + k) + 3 = 0 \quad$ or $\quad k = x_1 - 2y_1 + 3$

$\therefore$ $x = x_1 + (x_1 - 2y_1 + 3) = 2x_1 - 2y_1 + 3$

      $y = y_1 + (x_1 - 2y_1 + 3) = x_1 - y_1 + 3$

      $z = z_1 + (x_1 - 2y_1 + 3) = x_1 - 2y_1 + z_1 + 3$

Substituting these values of x, y, z in $x^2 + 2y^2 + 6xy - 2z + 8 = 0$, we get

$(2x_1 - 2y_1 + 3)^2 + 2(x_1 - y_1 + 3)^2 + 6(2x_1 - 2y_1 + 3)(x_1 - y_1 + 3) - 2(x_1 - 2y_1 + z_1 + 3) + 8 = 0$

Hence the locus of (x_1, y_1, z_1) is

$(2x - 2y + 3)^2 + 2(x - y + 3)^2 + 6(2x - 2y + 3)(x - y + 3) - 2(x - 2y + z + 3) + 8 = 0$

or $\boxed{18x^2 + 18y^2 - 36xy + 76x - 74y + 83 = 0}$

which is the required equation of the cylinder.

Type II : Problems on Right Circular Cylinder and Enveloping Cylinder :

Ex. 7 : *Find the equation of the right circular cylinder of radius 2 whose axis is the line*

$\dfrac{x - 1}{2} = \dfrac{y - 2}{1} = \dfrac{z - 3}{2}$.   *(Dec. 2010, 2005, 2013, 2018; May 2011, 2007, 2018)*

Sol. :

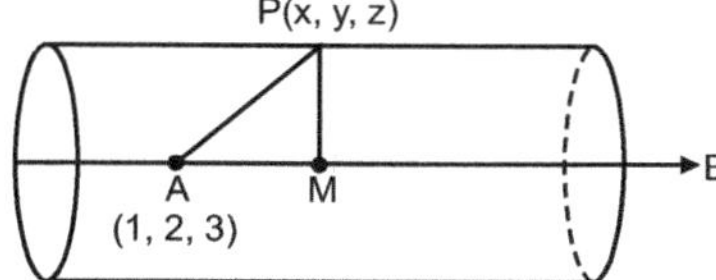

Fig. 8.18

Let P(x, y, z) be any point on the cylinder, A(1, 2, 3) be a fixed point on the axis AB (Fig. 8.18) whose equation is

$\dfrac{x - 1}{2} = \dfrac{y - 2}{1} = \dfrac{z - 3}{2}$ ... (1)

The d.r.'s of AB are 2, 1, 2 $\therefore$ Its actual d.c.'s are $\dfrac{2}{3}, \dfrac{1}{3}, \dfrac{2}{3}$. Join AP and draw PM perpendicular on AB, so that

     PM = Radius of the cylinder = 2 (given).

and   AP = Distance between points A and P.

         $= \sqrt{(x - 1)^2 + (y - 2)^2 + (z - 3)^2}$

Also $\qquad$ AM = Projection of AP on the axis AB.

$$= \frac{2}{3}(x-1) + \frac{1}{3}(y-2) + \frac{2}{3}(z-3) \qquad \text{[Ref. article 7.10.1 (16)]}$$

$$= \frac{2x + y + 2z - 10}{3}$$

Now, $\qquad AP^2 = AM^2 + MP^2 \qquad$ or $\qquad AP^2 - AM^2 = PM^2$

$\therefore \qquad [(x-1)^2 + (y-2)^2 + (z-3)^2] - \left(\dfrac{2x+y+2z-10}{3}\right)^2 = 4$

or $\qquad 9[(x-1)^2 + (y-2)^2 + (z-3)^2] - (2x+y+2z-10)^2 = 36$

or $\qquad \boxed{5x^2 + 8y^2 + 5z^2 - 4yz - 8zx - 4xy + 22x - 16y - 14z - 10 = 0}$

which is the required equation of the cylinder.

Ex. 8 : *Find the equation of right circular cylinder of radius 2 whose axis passes through (1, 2, 3) and has direction cosines proportional to 2, – 3, 6.*

(Nov./Dec. 2019, May 06, 04, 13, 15, 17; Dec. 18)

Sol. : $\qquad\qquad\qquad\qquad$ Given that the axis of the cylinder passes through A(1, 2, 3) and its d.r.'s are 2, – 3, 6.

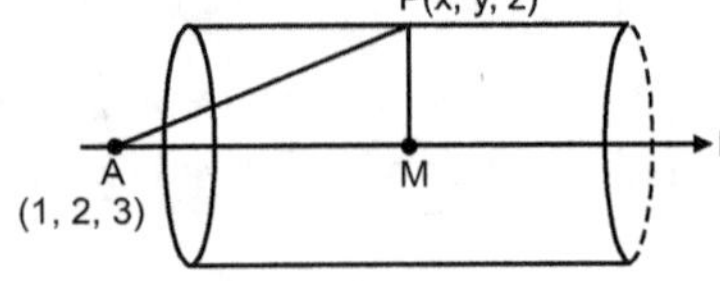

Fig. 8.19

$\therefore$ Its actual d.c.'s are $\dfrac{2}{7}, -\dfrac{3}{7}, \dfrac{6}{7}$.

Let P(x, y, z) be any point on the cylinder (Fig. 8.19) & PM be perpendicular from P on the axis AB, then

$\qquad$ PM = Radius of the cylinder = 2 $\qquad\qquad$ (given)

and $\qquad$ AP = Distance between points A and P = $\sqrt{(x-1)^2 + (y-2)^2 + (z-3)^2}$

Also $\qquad$ AM = Projection of AP on the axis AB

$$= \frac{2}{7}(x-1) - \frac{3}{7}(y-2) + \frac{6}{7}(z-3) \qquad \text{[Ref. article 6.10.1 (16)]}$$

$$= \frac{2x - 3y + 6z - 14}{7}$$

Now $\qquad AP^2 = AM^2 + MP^2 \qquad$ or $\qquad AP^2 - AM^2 = PM^2$

$\therefore \qquad [(x-1)^2 + (y-2)^2 + (z-3)^2] - \left(\dfrac{2x-3y+6z-14}{7}\right)^2 = 4$

or $\qquad 49[(x-1)^2 + (y-2)^2 + (z-3)^2] - (2x-3y+6z-14)^2 = 196$

or $\qquad \boxed{45x^2 + 40y^2 + 13z^2 + 12xy + 36yz - 24zx - 42x - 280y - 126z + 294 = 0}$

which is the required equation of the cylinder.

Ex. 9 : *Show that the foot of the perpendicular from the point (α, β, γ) on the line $x = y = -z$ is $\left(\dfrac{\alpha + \beta - \gamma}{3}, \dfrac{\alpha + \beta - \gamma}{3}, -\dfrac{\alpha + \beta - \gamma}{3}\right)$*

Deduce the equation of a right circular cylinder of radius a having its axis along the given line.

Sol. : $\qquad\qquad\qquad$ Given line is

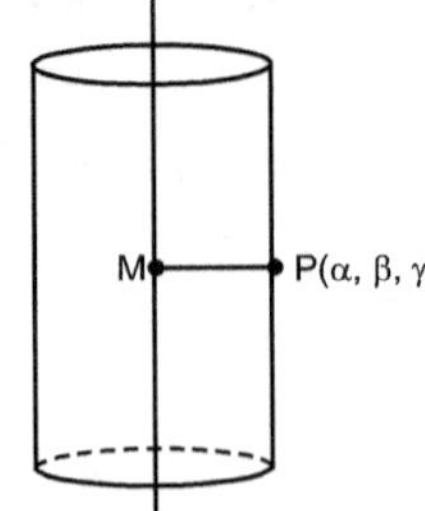

Fig. 8.20

$$\frac{x}{1} = \frac{y}{1} = \frac{z}{-1} = k \ (\text{say}) \qquad\qquad \text{... (1)}$$

Any point on (1) is given by M (k, k, –k). Let P(α, β, γ) represent any point on the cylinder (Fig. 8.20), then d.r.'s of PM are ($\alpha - k$), ($\beta - k$), ($\gamma + k$).

Since PQ is perpendicular to the line (1),

$\therefore \qquad (1)(\alpha - k) + (1)(\beta - k) + (-1)(\gamma + k) = 0 \ $ or $\ k = \dfrac{\alpha + \beta - \gamma}{3}$

Thus foot M of the perpendicular from point P (α, β, γ) on the line (1) is

$$\left(\frac{\alpha + \beta - \gamma}{3}, \ \frac{\alpha + \beta - \gamma}{3}, \ -\frac{\alpha + \beta - \gamma}{3} \right)$$

Now if PM = a = radius of the cylinder, then P always lies on the required right circular cylinder.

$$\therefore \qquad \left(\alpha - \frac{\alpha + \beta - \gamma}{3} \right)^2 + \left(\beta - \frac{\alpha + \beta - \gamma}{3} \right)^2 + \left(\gamma + \frac{\alpha + \beta - \gamma}{3} \right)^2 \ = a^2$$

or $\qquad (2\alpha - \beta + \gamma)^2 + (2\beta - \alpha + \gamma)^2 + (2\gamma + \alpha + \beta)^2 \ = \ 9a^2$

or $\qquad 2\,(\alpha^2 + \beta^2 + \gamma^2) - 2\alpha\beta + 2\beta\gamma + 2\gamma\alpha \ = \ 3a^2$

Hence locus of P(α, β, γ) is obtained by replacing α, β, γ by x, y, z.

$$\therefore \qquad \boxed{2\,(x^2 + y^2 + z^2) - 2xy + 2yz + 2zx = 3a^2}$$

which is the required equation of the right circular cylinder.

Ex. 10 : *Find the equation of the right circular cylinder whose guiding curve is*

$$x^2 + y^2 + z^2 \ = \ 9, \quad x - y + z \ = \ 3. \hspace{4cm} \textbf{(May 10, 05; Dec. 06, 16)}$$

Sol. : The axis of the cylinder is the line through the centre L of the given circle (which also passes through centre O of the given sphere) and perpendicular to the plane of the circle (Fig. 8.21).

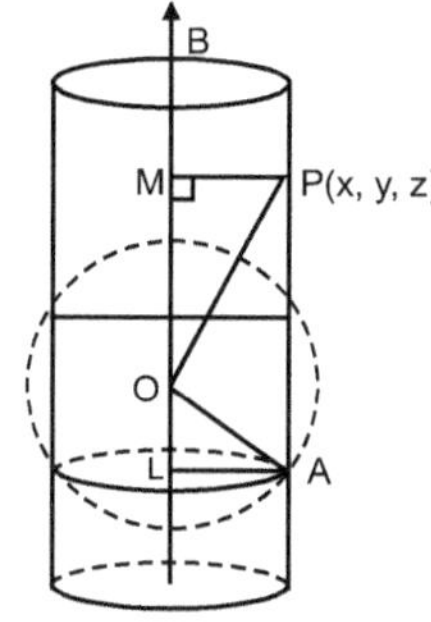

Fig. 8.21

The given sphere is

$$x^2 + y^2 + z^2 = 9 \hspace{5cm} \ldots (1)$$

and the plane is

$$x - y + z \ = \ 3 \hspace{5cm} \ldots (2)$$

The centre of the sphere is O (0, 0, 0) and radius = 3, numerically. The equations of the line through the centre O (0, 0, 0) of the sphere and perpendicular to the plane (2) are

$$\frac{x}{1} \ = \ \frac{y}{-1} \ = \ \frac{z}{1} \hspace{5cm} \ldots (3)$$

which represents the axis of the cylinder whose d.r.'s are 1, –1, 1. Therefore, its actual d.c.'s are $\dfrac{1}{\sqrt{3}}, \dfrac{-1}{\sqrt{3}}, \dfrac{1}{\sqrt{3}}$.

Also $\qquad\qquad$ OL $\ = \ $ Perpendicular distance from O (0, 0, 0) on the plane (2)

$$= \ \left| \frac{0 - 0 + 0 - 3}{\sqrt{1 + 1 + 1}} \right| \ = \ \sqrt{3} \ , \quad \text{numerically}$$

$$\text{LA} \ = \ \text{Radius of the circle} = \ \sqrt{\text{OA}^2 - \text{OL}^2} \ = \sqrt{9 - 3}$$

$$= \ \sqrt{6} \hspace{5cm} \ldots (4)$$

$\therefore \qquad$ Radius of the cylinder $\ = \ \sqrt{6}$.

Let P (x, y, z) be any point on the cylinder, join OP and draw PM perpendicular on axis OB, so that

$$\text{PM} \ = \ \text{Radius of the cylinder} = \sqrt{6}$$

and $\qquad\qquad$ OP $\ = \ $ Distance between points O and P

$$= \ \sqrt{x^2 + y^2 + z^2}$$

and $\qquad\qquad$ OM $\ = \ $ Projection of OP on the axis OB

$$= \ \frac{1}{\sqrt{3}} \ (x - 0) \ - \frac{1}{\sqrt{3}} \ (y - 0) \ + \frac{1}{\sqrt{3}} \ (z - 0)$$

$$= \ \frac{x - y + z}{\sqrt{3}}$$

Now $\qquad\qquad$ OP^2 $\ = \ $ OM^2 + MP^2 $\qquad$ or $\qquad$ OP^2 – OM^2 $\ = \ $ MP^2

$\therefore \qquad 3\,(x^2 + y^2 + z^2) - (x - y + z)^2 = 18$

or $\qquad \boxed{x^2 + y^2 + z^2 \ + xy + yz - zx = 9}$

which is the required equation of the sphere.

Ex. 11 : *Find the equation of the right circular cylinder described on the circle through (a, 0, 0), (0, a, 0), (0, 0, a).*   **(May 2004)**

Sol. :

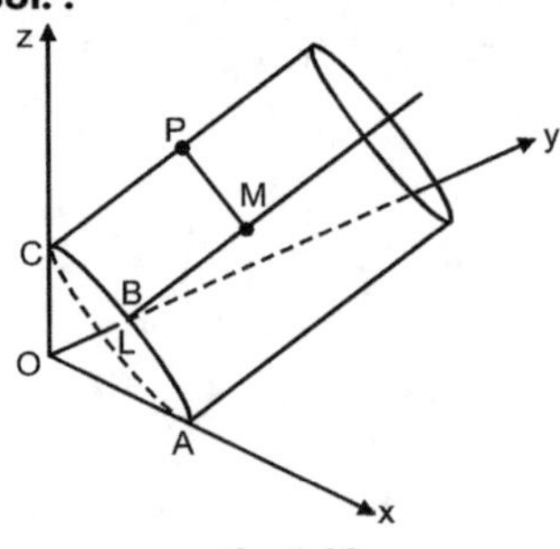
Fig. 8.22

The radius of the guiding circle through A(a, 0, 0), B(0, a, 0), C(0, 0, a), which is the normal section, is the radius of the required cylinder and normal to the plane of circle through its centre is the axis of the cylinder.

Now equation of the sphere passing through the origin and intercepts a, a, a on the axes is

$$x^2 + y^2 + z^2 - ax - ay - az = 0 \qquad \text{... (1)}$$

and the equation of the plane in intercept form is

$$\frac{x}{a} + \frac{y}{a} + \frac{z}{a} = 1 \quad \text{or} \quad x + y + z = a \qquad \text{... (2)}$$

Section of the sphere (1) by plane (2) is circle, hence (1) and (2) together represent equation of circle.

The equation of perpendicular from the centre $\left(\dfrac{a}{2}, \dfrac{a}{2}, \dfrac{a}{2}\right)$ of the sphere to the plane (2) is

$$\frac{x - \frac{a}{2}}{1} = \frac{y - \frac{a}{2}}{1} = \frac{z - \frac{a}{2}}{1} \qquad \text{... (3)}$$

which represents the axis of the sphere.

The centre of the circle ABC is the point of intersection of the axis (3) and the plane (2) and is given by;

$$L\left(\frac{a}{3}, \frac{a}{3}, \frac{a}{3}\right) \qquad \text{... (4)}$$

and the radius of the circle ABC and therefore cylinder is

$$AL^2 = \left(\frac{a}{3} - a\right)^2 + \left(\frac{a}{3} - 0\right)^2 + \left(\frac{a}{3} - 0\right)^2$$

$$= \frac{2}{3} a^2 \qquad \text{... (5)}$$

Let P (x, y, z) be any point on the cylinder, draw PM perpendicular from P on the axis of the cylinder, then

$$PL^2 = LM^2 + MP^2 \quad \text{or} \quad PL^2 - LM^2 = PM^2$$

$$\therefore \quad \left(x - \frac{a}{3}\right)^2 + \left(y - \frac{a}{3}\right)^2 + \left(z - \frac{a}{3}\right)^2 - \left[\frac{1}{\sqrt{3}}\left(x - \frac{a}{3}\right) + \frac{1}{\sqrt{3}}\left(y - \frac{a}{3}\right) + \frac{1}{\sqrt{3}}\left(z - \frac{a}{3}\right)\right]^2 = \text{(radius of cylinder)}^2$$

Note : Here LM = Projection of LP on the axis of the cylinder whose d.c.'s are $\dfrac{1}{\sqrt{3}}, \dfrac{1}{\sqrt{3}}, \dfrac{1}{\sqrt{3}}$.

$$\therefore \quad 3\left[(3x - a)^2 + (3y - a)^2 + (3z - a)^2\right] - \left[(3x - a) + (3y - a) + (3z - a)\right]^2 = 18a^2$$

or
$$\boxed{x^2 + y^2 + z^2 - xy - yz - zx = a^2}$$

which is the required equation of the cylinder.

Ex. 12 : *Find the equation of the enveloping cylinder of the sphere $x^2 + y^2 + z^2 - 2x + 4y = 1$ having its generators parallel to the line $x = y = z$.*

Sol. : The given sphere is

$$x^2 + y^2 + z^2 - 2x + 4y = 1 \qquad \text{... (1)}$$

and the given line is

$$\frac{x}{1} = \frac{y}{1} = \frac{z}{1} \qquad \text{... (2)}$$

Centre of the sphere (1) is (1, –2, 0) and radius $= \sqrt{u^2 + v^2 + w^2 - d} = \sqrt{1 + 4 + 0 + 1} = \sqrt{6}$.

Let (α, β, γ) be any point on the cylinder (locus), then any line through this point and parallel to the line (2) is

$$\frac{x - \alpha}{1} = \frac{y - \beta}{1} = \frac{z - \gamma}{1} = k \text{ (say)} \qquad \text{... (3)}$$

Any point on this line is $(\alpha + k, \beta + k, \gamma + k)$.

This point will lie on the sphere (1), if

$$(\alpha + k)^2 + (\beta + k)^2 + (\gamma + k)^2 - 2(\alpha + k) + 4(\beta + k) = 1$$

or
$$3k^2 + 2k(\alpha + \beta + \gamma - 1) + (\alpha^2 + \beta^2 + \gamma^2 - 2\alpha + 4\beta - 1) = 0 \qquad \text{... (4)}$$

The line (3) will touch the given sphere if this quadratic in k has equal roots i.e. if

$$4 (\alpha + \beta + \gamma - 1)^2 - 12 (\alpha^2 + \beta^2 + \gamma^2 - 2\alpha + 4\beta - 1) = 0$$

$$\left.\begin{array}{l}\text{For equal roots}\\\text{of quadratic,}\\ b^2 - 4ac = 0.\end{array}\right.$$

or $3 (\alpha^2 + \beta^2 + \gamma^2 - 2\alpha + 4\beta - 1) = (\alpha + \beta + \gamma - 1)^2$

Hence the locus of (α, β, γ) is obtained by replacing α, β, γ by x, y, z, we get

$$3 (x^2 + y^2 + z^2 - 2x + 4y - 1) = (x + y + z - 1)^2$$

or $\boxed{x^2 + y^2 + z^2 - xy - yz - zx - 4x + 5y - z - 2 = 0}$

which is the required equation of enveloping cylinder.

EXERCISE 8.2

1. Find the equations of the cylinder :
   (i) whose generators are parallel to z-axis and which passes through the curve of intersection of $x^2 + y^2 + z^2 = 1$ and $x + y + z = 1$.
   (ii) whose generators are parallel to z-axis and the guiding curve is $ax^2 + by^2 = 2z$; $lx + my + nz = p$.
     Ans. (i) $x^2 + y^2 + xy - x - y = 0$, (ii) $n (ax^2 + by^2) + 2lx + 2my - 2p = 0$.

2. Find the equations of the cylinder with generators parallel to OX, OY, OZ respectively, which pass through the curve of intersection of the surfaces represented by $x^2 + y^2 + 2z^2 = 12$ and $x - y + z = 1$.
   [**Hint :** Eliminate x, y, z in turn between the given equations.] **Ans.** (i) $2y^2 + 3z^2 - 2yz + 2y - 2z - 11 = 0$
     (ii) $2x^2 + 3z^2 + 2zx - 2x - 2z - 11 = 0$ (iii) $3x^2 + 3y^2 - 4xy - 4x + 4y - 10 = 0$

3. Find the equation of the cylinder whose generators are parallel to the line $\dfrac{x}{1} = \dfrac{y}{2} = \dfrac{z}{3}$ and pass through the curve $x^2 + y^2 = 16$, $z = 0$.

   [**Hint :** Let (x_1, y_1, z_1) be any point on the generator. The equation of generator is $\dfrac{x - x_1}{1} = \dfrac{y - y_1}{2} = \dfrac{z - z_1}{3} = k$. Any point $(x_1 + k, y_1 + 2k, z_1 + 3k)$ lying on the guiding curve requires $(x_1 + k)^2 + (y_1 + 2k)^2 = 16$ and $z_1 + 3k = 0$. $\therefore k = -\dfrac{z_1}{3}$ and hence the locus is $(3x_1 - z_1)^2 + (3y_1 - 2z_1)^2 = 144$. Replacing x_1, y_1, z_1 by x, y, z we get the required equation.]
     Ans. $9x^2 + 9y^2 + 5z^2 - 6zx - 12yz - 144 = 0$.

4. Find the equation of the cylinder whose generators are parallel to the line $\dfrac{x}{-1} = \dfrac{y}{2} = \dfrac{z}{3}$ and whose guiding curve is $x^2 + y^2 = 9$, $z = 1$. **(Dec. 2007)**
   [**Hint :** Let (x_1, y_1, z_1) be any point on the generator. The equation of generator through this point and parallel to given line is $\dfrac{x - x_1}{-1} = \dfrac{y - y_1}{2} = \dfrac{z - z_1}{3} = k$. Any point $(x_1 - k, y_1 + 2k, z_1 + 3k)$ lying on the guiding curve requires $(x_1 - k)^2 + (y_1 + 2k)^2 = 9$ and $z_1 + 3k = 1$. Thus, $k = \dfrac{1 - z_1}{3}$ and hence locus is $(3x_1 - 1 + z_1)^2 + (3y_1 + 2 - 2z_1)^2 = 81$. Replace x_1, y_1, z_1 by x, y, z. Thus $(3x + z - 1)^2 + (3y - 2z + 2)^2 = 81$.] **Ans.** $9x^2 + 9y^2 + 5z^2 + 6xz - 12yz - 6x + 12y - 10z - 76 = 0$.

5. Find the equation of the cylinder whose generators are parallel to the line $6x = -3y = 2z$ and the guiding curve is $x^2 + \dfrac{y^2}{4} = \dfrac{1}{4}$, $z = 0$. **Ans.** $36x^2 + 9y^2 + 17z^2 + 6yz - 48zx - 9 = 0$.

6. Find the equation of the cylinder whose generators are parallel to $6x = -3y = 2z$ and the guiding curve is $x^2 + y^2 - 1 = 0$, $z - 3 = 0$. **Ans.** $9 (x^2 + y^2) + 5z^2 + 12yz - 6zx + 18x - 36y = 0$.

7. Obtain the equation of a right circular cylinder of radius 5 and axis is $\dfrac{x - 2}{2} = \dfrac{y - 3}{1} = \dfrac{z + 1}{1}$. **(Dec. 11, 13, 17, May 15)**
     Ans. $2x^2 + 5y^2 + 5z^2 - 4xy - 4xz - 2yz - 24y + 24z - 102 = 0$

8. Find the equation of the right circular cylinder of radius 3 whose axis is the line $\dfrac{x - 1}{2} = \dfrac{y - 3}{2} = \dfrac{z - 5}{-1}$.
     (Dec. 2009, 2018; May 2009, 2014, 2017 Nov. 2014)
     Ans. $5x^2 + 5y^2 + 8z^2 + 4yz + 4zx - 8xy - 6x - 42y - 96z + 225 = 0$.

9. Find the equation of the right circular cylinder of radius 2 whose axis is the line $\dfrac{x - 1}{2} = \dfrac{y - 2}{1} = \dfrac{z - 3}{2}$. **(May 2009, Nov. 2015)** **Ans.** $5x^2 + 8y^2 + 5z^2 - 4yz - 8zx - 4xy + 22x - 16y - 14z - 10 = 0$.

10. Find the equation of the right circular cylinder of radius 2 whose axis passes through A $(1, -2, 4)$ and has d.c.'s proportional to 2, 3, 6.

[**Hint :** The d.r.'s of the axis are 2/7, 3/7, 6/7; PM = 2; $AP^2 = (x - 1)^2 + (y + 2)^2 + (z - 4)^2$; $AM = \frac{2}{7}(x - 1) + \frac{3}{7}(y + 2) + \frac{6}{7}(z - 4)$.

Equation of required cylinder is obtained from $AP^2 - AM^2 = PM^2$.]

Ans. $45x^2 + 40y^2 + 13z^2 - 36yz - 24zx - 12xy - 18x - 316y - 152z + 433 = 0$

11. Find the equation of the right circular cylinder whose axis is

$\dfrac{x-2}{2} = \dfrac{y-1}{1} = \dfrac{z}{3}$ and which passes through the point $(0, 0, 3)$. **(May 2008, Nov. 2014, Dec. 2016, 2017)**

[**Hint :** A(2, 1, 0), Radius of cylinder is PM = r, d.r.'s of axis are 2, 1, 3. $AP^2 - AM^2 = PM^2$,

$(x - 2)^2 + (y - 1)^2 + (z - 0)^2 - \left[\dfrac{2(x - 2) + (y - 1) + 3z}{\sqrt{14}}\right]^2 = r^2$, it passes through (0, 0, 3), therefore $r^2 = \dfrac{90}{7}$.]

Ans. $10x^2 + 13y^2 + 5z^2 - 4xy - 6yz - 12zx - 36x - 18y + 30z - 135 = 0.$

12. The axis of right circular cylinder of radius 2 is $\dfrac{x-1}{2} = \dfrac{y}{3} = \dfrac{z-3}{1}$.

Show that its equation is $10x^2 + 5y^2 + 13z^2 - 12xy - 6yz - 4xz - 8x + 30y - 74z + 59 = 0.$

13. Find the equation of the right circular cylinder which passes through the section of the sphere $x^2 + y^2 + z^2 = 25$ made by the plane $x + 2y + 2z = 0$. **(May 11, 08)**

[**Hint :** A(0, 0, 0), Radius of cylinder is PM = 5, d.r.'s of axis are 1, 2, 2. $\therefore AP^2 - AM^2 = PM^2$,

$x^2 + y^2 + z^2 - \left[\dfrac{x + 2y + 2z}{3}\right]^2 = 25$]

Ans. $8x^2 + 5y^2 + 5z^2 - 4xy - 4xz - 8yz = 225.$

14. The radius of normal section of a right circular cylinder is 2 units, the axis lies along the straight line $\dfrac{x-1}{2} = \dfrac{y+3}{-1} = \dfrac{z-2}{5}$.

Find its equation. **(Dec. 2004)**

[**Hint :** $\therefore AP^2 - AM^2 = PM^2$, $(x - 1)^2 + (y + 3)^2 + (z - 2)^2 - \left[\dfrac{2x - y + 5z - 15}{\sqrt{30}}\right]^2 = 4.$]

Ans. $26x^2 + 29y^2 + 5z^2 + 4xy + 10yz - 20zx + 150y + 30z + 75 = 0.$

15. Find the equation of the right cylinder cylinder of radius a, whose axis passes through the origin and makes equal angles with the co-ordinate axes. **(May 2014) Ans.** $2(x^2 + y^2 + z^2 - xy - yz - zx) = 3a^2.$

16. Show that the equation of the right circular cylinder described on the circle through three points $(1, 0, 0)$, $(0, 1, 0)$, $(0, 0, 1)$ as the guiding curve is $x^2 + y^2 + z^2 - xy - yz - zx = 1.$

17. Find the equation of the cylinder whose directing curve is $x^2 + z^2 - 4x - 2z + 4 = 0$, $y = 0$ and whose axis contains the point $(0, 3, 0)$. Find also the area of the section of the cylinder by a plane parallel to xz plane.

[**Hint :** The guiding curve is a circle in the xoz plane. The centre of the circle is (2, 0, 1). The axis of the cylinder is the line joining (0, 3, 0) and (2, 0, 1) and the d.r.'s of axis are 2, –3, 1.] **Ans.** $9x^2 + 5y^2 + 9z^2 + 12xy + 6yz - 36x - 30y - 18z + 36 = 0; \pi.$

18. Find the equation of the right circular cylinder whose axis is $x = 2y = -z$ and radius is 4. Prove that the area of the section of this cylinder by the plane $z = 0$ is 24π. **(Nov./Dec. 2019, Dec. 2011, 2007; May 2013)**

Ans. $5x^2 + 8y^2 + 5z^2 + 4yz + 8zx - 4xy - 144 = 0.$

[**Hint :** Section by plane z = 0, is the conic $5x^2 + 8y^2 - 4xy - 144 = 0$. The equation can be written as $Ax^2 + 2Hxy + By^2 = 1$, where $A = \dfrac{5}{144}$, $H = -\dfrac{1}{72}$, $B = \dfrac{1}{18}$. Since squares of the semi-axes are the roots of $\dfrac{1}{r^4} - (A + B)\dfrac{1}{r^2} + (AB - H^2) = 0$

$\therefore \dfrac{1}{r^4} - \dfrac{13}{144}\dfrac{1}{r^2} + \dfrac{1}{576} = 0$ or $r^4 - 52r^2 + 576 = 0$ giving r = 6, 4. Conic is ellipse, whose area = $\pi ab = \pi \cdot 6 \cdot 4 = 24\pi$.]

19. Show that the cylinder with generators parallel to the z-axis and the guiding curve $3x^2 + 2y^2 + z^2 = 18$, $y - z = 3$ is a right circular cylinder. Find the area of the section of the cylinder by xoy plane. **Ans.** $4\pi.$

20. Find the equation of the enveloping cylinder of the sphere $x^2 + y^2 + z^2 - 2y - 4z - 11 = 0$ having its generators parallel to the line $x = -2y = 2z$. **Ans.** $5(x^2 + y^2 + z^2) + 2(xy + yz - zx + x - 7y - 11z) - 67 = 0.$

❑❑❑

CHAPTER-9

MULTIPLE INTEGRALS : DOUBLE AND TRIPLE INTEGRATIONS

9.1 INTRODUCTION

In earlier chapters on integration, we have discussed topics like, Reduction formula, Beta and Gamma functions, DUIS, Rectification etc. In these discussions, we came across integrals involving one variable. Applications of integration related to problems like Areas, Volumes, Centre of gravity etc. require use of more than one variables. In two-dimensional problems like finding areas bounded by plane curves, moment of inertia or centre of gravity of plane laminas, we require two variables for integration, as areas in question are to be covered in two movements, one parallel to x-axis and other parallel to y-axis. Similarly in three-dimensional problems like determination of volumes, mass of solids, we shall require three variables for integration. For such types of problems, the concept of multiple integration plays an important role. Basically, it is the extension of the idea of single integration with one variable used so far. In evaluation of multiple integration problems, we have to formulate the problem first and then decide, with respect to which variable, we should carry out integration first, so that actual integration process is simplified. These and other related problems will be taken-up for discussion in this chapter.

9.2 DOUBLE INTEGRATION

Representation of Area as a Double Integral : Consider the region bounded by the curve y = f(x), the ordinates x = a, x = b and the x-axis. This area ABCD can be considered as the limit of the sum of an infinite number of inscribed rectangles like PMRS, where P(x, y) and Q(x + δx, y + δy) are two adjacent points on the curve y = f (x), and the expression for the area ABCD is,

$$A = \lim_{\delta x \to 0} \sum_{x=a}^{x=b} y \cdot \delta x$$

which is expressed in integral notation as

$$A = \int_a^b y \, dx \quad \text{or} \quad \int_a^b f(x) \, dx$$

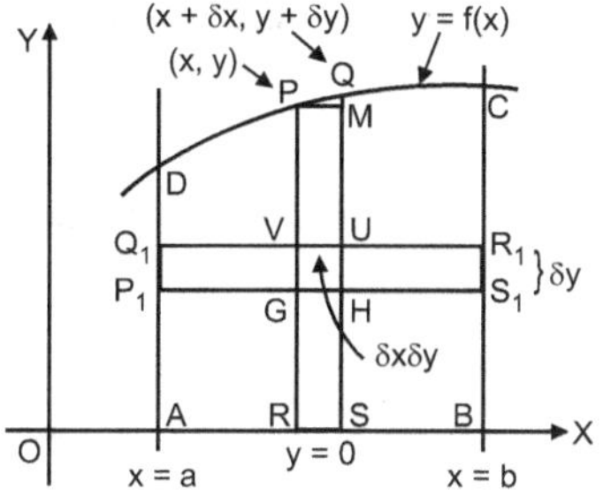

Fig. 9.1

Let us now consider the horizontal strip $P_1Q_1R_1S_1$ of width δy, which intersects the vertical strip PQRS in an element GHUV of area = δx δy. Thus we have, δA = δx δy

Now, we can first consider the summation in a vertical direction and area PQRS can be considered as the limit of a sum of an infinite number of elements like GHUV situated along the strip PQRS. Therefore, area A_1 of the strip PQRS can be expressed as,

$$A_1 = \lim_{\delta y \to 0} \{\Sigma \, \delta y\} \, \delta x$$

or,
$$A_1 = \left\{ \int_0^{f(x)} dy \right\} \delta x$$

[replacing summation sign by integral sign, which is from y = 0 i.e. x-axis to y = f(x) i.e. the curve]

Next we consider the summation in horizontal direction and consider the sum of all area elements like PQRS situated over the area ABCD, or in other words, we move the strip PQRS along x-axis covering the whole area ABCD and express area ABCD as,

$$A = \lim_{\delta x \to 0} \sum \left\{ \delta x \int_0^{f(x)} dy \right\}$$

or, replacing the summation sign by integral sign,

$$A = \int_{x=a}^{x=b} dx \int_{y=0}^{f(x)} dy \quad \text{Or} \quad A = \int_a^b \int_0^{f(x)} dx \, dy$$

Thus, the area ABCD is expressed in terms of double integral in which inner integration is carried out with respect to y treating x as constant and outer integration is then carried out with respect to x.

Question may arise as to why the area ABCD which can be computed by evaluating a single integral should be expressed in terms of double integral. Answer to this question is, double integral can handle more complicated regions than the region ABCD with simple boundaries. Then, with the help of double integral, we can compute many physical quantities like Mass, Moment of Inertia, Centre of Gravity, Centre of Pressure etc.

Consider for example, the problem of determination of mass of area ABCD, where the density is variable and is a function of (x, y), say f(x, y).

It means density varies from point to point in the region. Now, rectangular element GHUV can be considered as so small that we can treat density as constant for the region δA = GHUV. Considering δM as the mass of element δA, we can write,

$$\delta M = \rho \cdot \delta A, \qquad \text{[where } \rho = f(x, y) \text{ is density]}.$$
$$= f(x, y)\, \delta x\, \delta y, \qquad [\delta A = \delta x\, \delta y]$$

Then, considering the summation first in a vertical direction and then in a horizontal direction, we get mass of area ABCD as,

$$M = \int_a^b \int_0^{f(x)} f(x, y)\, dx\, dy$$

Depending upon what f(x, y) represents, the double integral $\int_a^b \int_0^{f(x)} f(x, y)\, dx\, dy$ will represent different physical quantities in relation to are

Irrespective of what physical quantity is represented by $\iint f(x, y)\, dx\, dy$, we shall consider evaluation of double integral $\iint f(x, y)\, dx\, dy$ over different regions.

We note that $\iint_R dx\, dy$ represents the area of the region R and $\iint_R f(x, y)\, dx\, dy$ represents some physical quantity related to the area of region R.

9.3 PROPERTIES OF THE DOUBLE INTEGRAL

1. $\displaystyle\iint_R k\, f(x, y)\, dA = k \iint_R f(x, y)\, dA$ where, k is free from x and y i.e. constant.

2. $\displaystyle\iint_R [\, f(x, y) \pm g(x, y)\,]\, dA = \iint_R f(x, y)\, dA \pm \iint_R g(x, y)\, dA$

3. If $R = R_1 \cup R_2$ and $R_1 \cap R_2 = \phi$ then,

$$\iint_R f(x, y)\, dA = \iint_{R_1} f(x, y)\, dA + \iint_{R_2} f(x, y)\, dA.$$

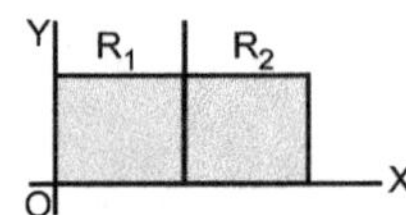

Fig. 9.2

This result holds when R is the union of two non-overlapping regions R_1 and R_2 as shown in Fig. 9.2.

Evaluation of Double Integrals : Double integrals over a region R may be evaluated by two successive integrations as follows :

1. **Suppose that R can be described by x = a, x = b, y = $f_1(x)$ and y = $f_2(x)$, then,**

$$I = \iint_R f(x, y)\, dy\, dx = \int_a^b \left\{ \int_{f_1(x)}^{f_2(x)} f(x, y)\, dy \right\} dx \qquad \text{...(I)}$$

We first integrate the inner integral w.r.t. y keeping x as a constant between the limits $y = f_1(x)$, $y = f_2(x)$ and the resulting expression, say g(x), be integrated w.r.t. x between the limits x = a, x = b. We then obtain the value of the double integral in (I).

2.   **Suppose that R can be described by y = c, y = d, x = f_1(y) and x = f_2(y), then,**

$$I = \iint_R f(x, y)\ dx\ dy = \int_c^d \left\{ \int_{f_1(y)}^{f_2(y)} f(x, y)\ dx \right\} dy \qquad \ldots (II)$$

We first integrate the inner integral w.r.t. x keeping y as a constant between the limits x = f_1 (y), x = f_2 (y) and the resulting expression, say h (y), be integrated w.r.t. y between the limits y = c, y = d. We then obtain the value of the double integral in (II).

3.   **Suppose that R can be described by x = a, x = b, y = c and y = d (a rectangle), then**

$$I = \int_a^b \left\{ \int_c^d f(x, y)\ dy \right\} dx \qquad \text{or} \qquad I = \int_c^d \left\{ \int_a^b f(x, y)\ dx \right\} dy$$

Note :   When a, b, c, d are constants then, the order of integration must be clearly specified.

4.   **Suppose R is same as mentioned in 3 above and the integrand is separable i.e. f(x, y) = u (x) · v (y), then**

$$I = \int_{x=a}^{b} \int_{y=c}^{d} f(x, y)\ dy\ dx = \int_{x=a}^{b} u(x)\ dx \cdot \int_{y=c}^{d} v(y)\ dy$$

9.4 DETERMINING THE LIMITS OF INTEGRATION

To evaluate $\iint_R f(x, y)\ dx\ dy$ over the region R given by x = 0,   y = 0,   x + y = a.

Method – I : Integrating first w.r.t. y and then w.r.t. x

We take the following steps.

Step 1 : Draw the region R given by x = 0, y = 0, x + y = a. Here R is the triangle OAB.

Step 2 : We have I = $\int \left\{ \int f(x, y)\ dy \right\}$ dx. Since, we are integrating first w.r.t. y, *always imagine a vertical strip LK anywhere in the region R*. Refer Fig. 9.3 (a).

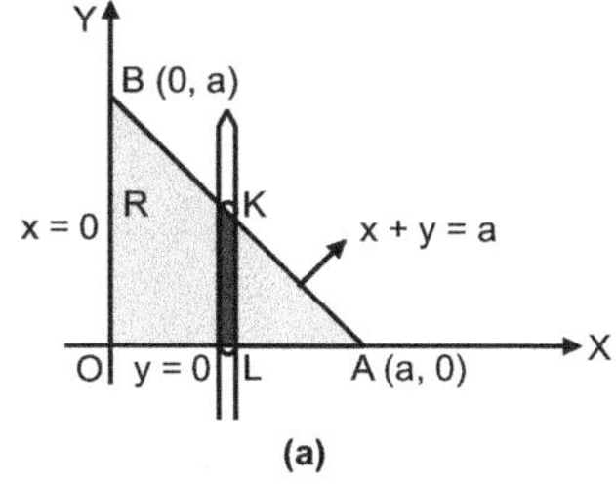

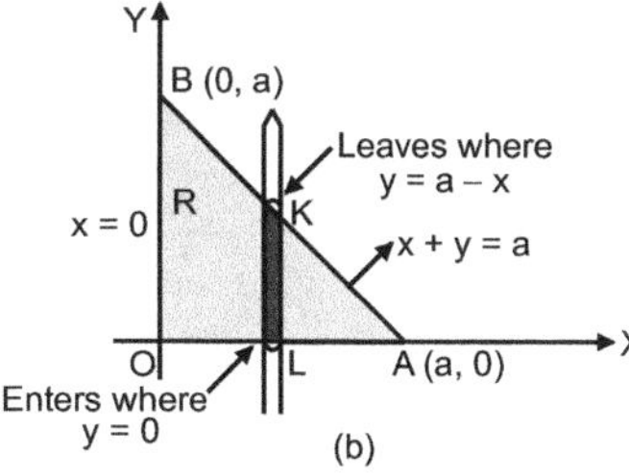

Fig. 9.3

Step 3 : To find the limits for y : Since, the lower end of strip enters the region R where y = 0 and upper end of strip leaves the region R where y = a − x [Refer Fig. 9.3 (b)].

Step 4 : To find the limits for x : Move the strip in horizontal direction from left to right covering the entire region R. Therefore, x varies from x = 0 to x = a.

Complete limits are, $I = \int_0^a \left\{ \int_0^{a-x} f(x, y)\ dy \right\} dx$

Method – II : Integrating first w.r.t. x and then w.r.t. y

We take the following steps :

Step 1 : Draw the region R given by x = 0, y = 0, x + y = a. Here R is the triangle OAB.

Step 2 : We have, $I = \int \left\{ \int f(x, y) \, dx \right\} dy$. Since, we are integrating first w.r.t. x, *always imagine a horizontal strip LK anywhere in the region R* [Refer Fig. 9.4 (a)].

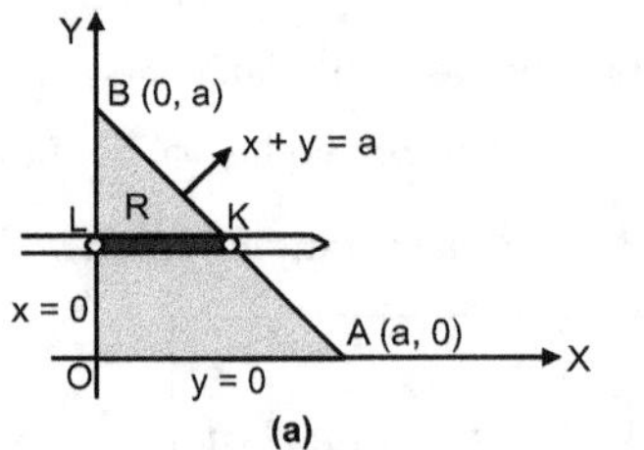

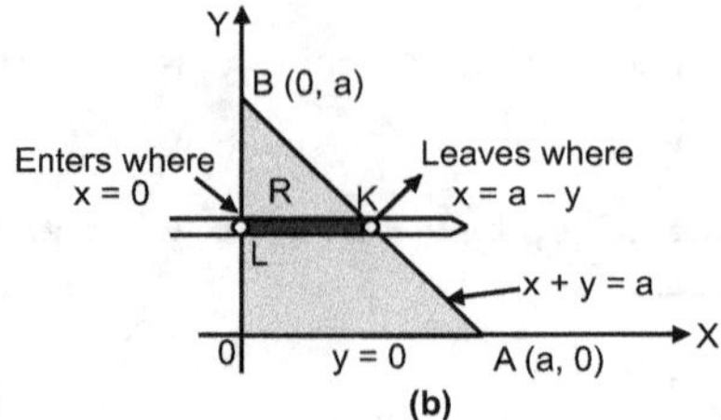

Fig. 9.4

Step 3 : To find the limits for x : Since, the left end of strip enters the region R where x = 0, and right end of strip leaves the region R where x = a – y [Refer Fig. 9.4 (b)].

Step 4 : To find the limits for y : Move the strip in vertical direction from bottom to top covering the entire region R. Therefore y varies from y = 0 to y = a.

Complete limits are $\qquad I = \int_0^a \left\{ \int_0^{a-y} f(x, y) \, dx \right\} dy$

Note : Please refer article 9.7 (Page 9.75) Dirichlet's Theorem before examples.

Problems on Double Integration are Mainly Divided into the following Types :

Type 1 – **Problems on Direct Evaluation of Double Integral.**

Type 2 – **Problems on Integrals when the Limits are not Provided.**

Type 3 – **Problems on Change of Order of Integration.**

Type 1 : Illustrations on Direct Evaluation of Double Integral.

Ex. 1 : *Evaluate* $\int_0^1 \int_0^y xy \, dx \, dy$.

Sol. : Since, limits of inner integral are functions of y, integrate first, w.r.t. x keeping y constant.

$$I = \int_0^1 \left\{ \int_0^y xy \, dx \right\} dy = \int_0^1 \left[y \cdot \frac{x^2}{2} \right]_0^y dy = \int_0^1 \frac{y^3}{2} dy = \frac{1}{2} \left[\frac{y^4}{4} \right]_0^1$$

$$\boxed{I = \frac{1}{8}}$$

Ex. 2 : *Evaluate* $\int_0^1 \int_0^{1-x} (x + y) \, dy \, dx$.

Sol. : Since, limits of inner integral are functions of x, integrate first w.r.t. y keeping x constant.

$$I = \int_0^1 \left\{ \int_0^{1-x} (x + y) \, dy \right\} dx = \int_0^1 \left[xy + \frac{y^2}{2} \right]_0^{1-x} dx \qquad \text{(Note this step carefully)}$$

$$= \int_0^1 \left[x(1-x) + \frac{(1-x)^2}{2} \right] dx = \left[\frac{x^2}{2} - \frac{x^3}{3} + \frac{(1-x)^3}{2(-3)} \right]_0^1$$

$$\boxed{I = \frac{1}{3}}$$

Ex. 3 : *Evaluate* $\displaystyle\int_0^1 \int_0^1 \frac{dx\,dy}{(1+x^2)(1+y^2)}$.

Sol. : Here we note that limits of both the integrals are constants, and variables can be separated, hence the double integral is a product of two single integrals.

$\therefore \qquad I = \displaystyle\int_0^1 \frac{dx}{1+x^2} \int_0^1 \frac{dy}{1+y^2} = [\tan^{-1} x]_0^1 \cdot [\tan^{-1} y]_0^1 = \frac{\pi}{4} \cdot \frac{\pi}{4} = \frac{\pi^2}{16}$ \qquad (Note this step carefully)

$$\boxed{I = \frac{\pi^2}{16}}$$

Ex. 4 : *Evaluate* $\displaystyle\int_0^1 \int_0^{\sqrt{\frac{1}{2}(1-y^2)}} \frac{dx\,dy}{\sqrt{1-x^2-y^2}}$.

Sol. : Let,

$$I = \int_0^1 dy \int_0^{\sqrt{\frac{1}{2}(1-y^2)}} \frac{dx}{\sqrt{1-y^2-x^2}} = \int_0^1 I_1 \, dy, \text{ say.}$$

Step 1 : Where

$$I_1 = \int_0^{\frac{1}{\sqrt{2}}\sqrt{(1-y^2)}} \frac{dx}{\sqrt{1-y^2-x^2}} \qquad (\text{put } 1-y^2 = a^2)$$

$$= \int_0^{\frac{a}{\sqrt{2}}} \frac{dx}{\sqrt{a^2-x^2}} = \left[\sin^{-1}\frac{x}{a}\right]_0^{\frac{a}{\sqrt{2}}} = \left[\sin^{-1}\frac{1}{\sqrt{2}} - 0\right] = \frac{\pi}{4}$$

Step 2 : Hence,

$$I = \int_0^1 \left(\frac{\pi}{4}\right) dy = \frac{\pi}{4} [y]_0^1 = \frac{\pi}{4}$$

$$\boxed{I = \frac{\pi}{4}}$$

Ex. 5 : *Evaluate* $\displaystyle\int_{-1}^1 \int_0^{1-x} x^{1/3}\, y^{-1/2}\, (1-x-y)^{1/2}\; dx\,dy$.

Sol. : Let,

$$I = \int_{-1}^1 x^{1/3}\, dx \int_0^{1-x} y^{-1/2}(1-x-y)^{1/2}\, dy = \int_{-1}^1 x^{1/3} I_1\, dx, \text{ say.}$$

Step 1 : Where

$$I_1 = \int_0^{1-x} y^{-1/2}(1-x-y)^{1/2}\, dy \qquad \text{Put } 1-x = a$$

$$= \int_0^a y^{-1/2}(a-y)^{1/2}\, dy \qquad \text{Put } y = at$$

y	0	a
t	0	1

$$I_1 = \int_0^1 a^{-1/2}\, t^{-1/2}(a-at)^{1/2} \cdot a\, dt = a\int_0^1 t^{-1/2}(1-t)^{1/2}\, dt$$

$$= a\,B\left(\frac{1}{2}, \frac{3}{2}\right) = a\,\frac{\overline{|1/2}\;\;\overline{|3/2}}{\overline{|2}} = \frac{a\sqrt{\pi}\,\frac{1}{2}\sqrt{\pi}}{1} = \frac{\pi a}{2}$$

Step 2 : Hence,

$$I = \int_{-1}^{1} x^{1/3} \cdot \frac{\pi a}{2} \cdot dx = \frac{\pi}{2} \int_{-1}^{1} x^{1/3} (1-x)\, dx \qquad (\because\ a = 1-x)$$

$$= \frac{\pi}{2} \int_{-1}^{1} (x^{1/3} - x^{4/3})\, dx = \frac{\pi}{2}\left[\frac{3}{4} x^{4/3} - \frac{3}{7} x^{7/3}\right]_{-1}^{1} = \frac{3\pi}{2}\left[\left(\frac{1}{4} - \frac{1}{7}\right) - \left(\frac{1}{4} + \frac{1}{7}\right)\right]$$

$$\boxed{I = -\frac{3\pi}{7}}$$

Ex. 6 : *Show that* $\displaystyle\int_{0}^{\infty}\int_{0}^{\infty} e^{-x^2(1+y^2)}\, x\, dx\, dy = \frac{\pi}{4}$. **(Dec. 2004)**

Proof : Here all the limits of integrals are constants and we note that integration first w.r.t. x is easy.

Let,

$$I = \int_{0}^{\infty}\left\{\int_{0}^{\infty} e^{-x^2(1+y^2)}\, x\, dx\right\} dy = \int_{0}^{\infty} I_1\, dy, \ \text{say.}$$

Step 1 : Where

$$I_1 = \int_{0}^{\infty} e^{-x^2(1+y^2)} \cdot x\, dx \qquad \text{(Treat } (1+y^2) \text{ as constant, say a)}$$

$\therefore$

$$I_1 = \int_{0}^{\infty} e^{-ax^2} \cdot x\, dx = -\frac{1}{2a}\int_{0}^{\infty} e^{-ax^2}(-2ax)\, dx$$

$$I_1 = -\frac{1}{2a}\left[e^{-ax^2}\right]_{0}^{\infty} = -\frac{1}{2a}\left[e^{-\infty} - e^{0}\right] = -\frac{1}{2a}[0-1] = \frac{1}{2a} = \frac{1}{2(1+y^2)}$$

Step 2 : Hence,

$$I = \int_{0}^{\infty} \frac{1}{2(1+y^2)}\, dy = \frac{1}{2}\left[\tan^{-1} y\right]_{0}^{\infty} = \frac{1}{2}\left[\tan^{-1}\infty - \tan^{-1} 0\right] = \frac{1}{2}\left[\frac{\pi}{2} - 0\right] \qquad \text{... Proved.}$$

$$\boxed{I = \frac{\pi}{4}}$$

EXERCISE 9.1

Show that

1. $\displaystyle\int_{0}^{1}\int_{x^2}^{2-x} y\, dy\, dx = \frac{16}{15}$

2. $\displaystyle\int_{-2}^{1}\int_{x^2}^{2-x} y\, dx\, dy = \frac{36}{5}$

3. $\displaystyle\int_{0}^{1}\int_{x^2}^{x} xy\,(x+y)\, dx\, dy = \frac{3}{56}$

4. $\displaystyle\int_{0}^{1}\int_{-\sqrt{y}}^{-y^2} xy\, dx\, dy = -\frac{1}{12}$

5. $\displaystyle\int_{0}^{1}\int_{0}^{\sqrt{1+x^2}} \frac{dy\, dx}{1+x^2+y^2} = \frac{\pi}{4}\log(1+\sqrt{2})$ **(May 2019)**

6. $\displaystyle\int_{0}^{a\sqrt{3}}\int_{0}^{\sqrt{x^2+a^2}} \frac{x\, dy\, dx}{y^2+x^2+a^2} = \frac{\pi a}{4}$

7. $\displaystyle\int_{0}^{1}\int_{0}^{1} \frac{dy\, dx}{\sqrt{(1-x^2)(1-y^2)}} = \frac{\pi^2}{4}$

8. $\displaystyle\int_{0}^{1}\int_{0}^{x^2} e^{y/x}\, dy\, dx = \frac{1}{2}$

9. $\displaystyle\int_{0}^{1}\int_{0}^{\sqrt{1-x^2}} x^2\, y\, dy\, dx = \frac{1}{15}$

10. $\displaystyle\int_{0}^{1}\int_{x}^{\sqrt{x}} (x^2+y^2)\, dy\, dx = \frac{3}{35}$

11. $\displaystyle\int_0^{\pi/2} \int_0^{3(1-\cos t)} x^2 \sin t \, dt \, dx = \frac{9}{4}$

12. $\displaystyle\int_0^1 \int_0^y xy \, e^{-x^2} \, dx \, dy = \frac{1}{4e}$

13. $\displaystyle\int_0^5 \int_0^{x^2} x \, (x^2 + y^2) \, dx \, dy = 18880 \, \frac{5}{24}$

14. $\displaystyle\int_0^{\pi/2} dx \int_{\pi/2}^{\pi} \cos (x + y) \, dy = -2$

15. $\displaystyle\int_0^1 \int_{x^2}^{2-x} xy \, dy \, dx = \frac{3}{8}$

16. $\displaystyle\int_0^1 \int_{x^2}^{2-x^2} xy \, dy \, dx = \frac{1}{2}$

17. $\displaystyle\int_0^1 \int_0^{\sqrt{1-x^2}} 4 \, xy \, e^{x^2} \, dy \, dx = e - 2$

18. $\displaystyle\int_0^1 \int_{2x^2}^{2\sqrt{x}} xy^2 \, dy \, dx = \int_0^2 \int_{\frac{y^2}{4}}^{\sqrt{\frac{y}{2}}} xy^2 \, dx \, dy$ **Ans.** $\dfrac{3}{7}$

19. $\displaystyle\int_0^1 \int_0^{1-x} xy \sqrt{1-x-y} \, dy \, dx = \frac{16}{945}$ (Refer Solved Ex. 5)

20. $\displaystyle\int_0^\infty dx \int_0^1 e^{-x^a \cdot y} \, dy = \frac{\left|\frac{1}{a}\right.}{a-1}$ $(a > 1)$.

[**Hint :** $I = \displaystyle\int_0^1 dy \int_0^\infty e^{-x^a y} \, dx$

Put $x^a \cdot y = t$, $x = \left(\dfrac{t}{y}\right)^{1/a}$ (use Gamma function)]

21. $\displaystyle\int_0^a \int_0^{\sqrt{a^2-y^2}} \sqrt{a^2 - x^2 - y^2} \, dx \, dy = \frac{\pi a^3}{6}$

[**Hint :** Use $\displaystyle\int \sqrt{b^2 - x^2} \, dx = \frac{x}{2} \sqrt{b^2 - x^2} + \frac{b^2}{2} \sin^{-1} \frac{x}{b}$, $b^2 = a^2 - y^2$.]

Type 2 : Illustrations on Integrals when the Limits are Not Provided.

If the limits of integral are not provided but the region of integration is given, then we first find the points of intersections of the curves and draw the given region.

Note :

1. Vertical strip (or horizontal strip) is to be chosen, in order to integrate w.r. to y (or x) first, in such a way that the evaluation of inner integral becomes possible and simple.

2. For determination of limits, refer article 9.4.

Ex. 1 : *Evaluate* $\displaystyle\iint x^2 \, y^2 \, dx \, dy$ *over the positive quadrant of circle* $x^2 + y^2 = 1$.

Sol. : Region of integration is shaded as shown in Fig. 9.5. i.e. R is OAB.

Let us evaluate this integral first w.r.t. y and then w.r.t. x.

Therefore, to substitute the limits in double integral, consider a vertical strip in the region, the lower end of which lies on x-axis for which y = 0 and upper end touches the circle $x^2 + y^2 = 1$ for which $y = \sqrt{1 - x^2}$. Thus y varies along the vertical strip from y = 0 to $y = \sqrt{1 - x^2}$. Then horizontally move this strip to cover the whole positive quadrant of the circle, for which x varies from x = 0 to x = 1.

Thus, $I = \displaystyle\iint x^2 \, y^2 \, dx \, dy$

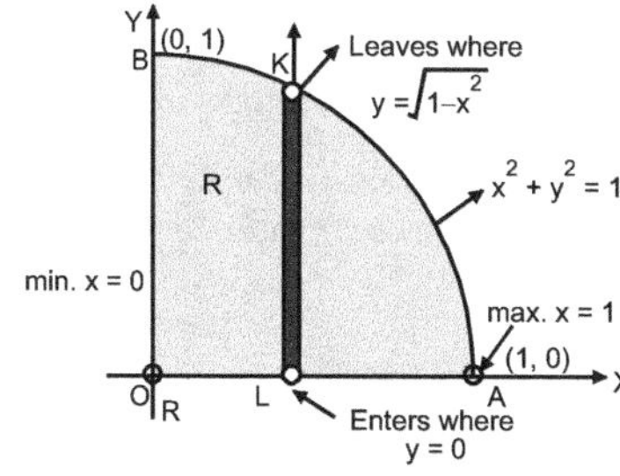

Fig. 9.5

over the positive quadrant of circle can be written, after substitution of limits, as :

$$I = \int_0^1 \int_0^{\sqrt{1-x^2}} x^2 y^2 \, dx \, dy = \int_0^1 \left\{ \int_0^{\sqrt{1-x^2}} x^2 y^2 \, dy \right\} dx$$

$$= \int_0^1 x^2 \left\{ \int_0^{\sqrt{1-x^2}} y^2 \, dy \right\} dx = \int_0^1 x^2 \cdot I_1 \, dx, \text{ say.}$$

Step 1 : Where

$$I_1 = \int_0^{\sqrt{1-x^2}} y^2 \, dy = \left[\frac{y^3}{3} \right]_0^{\sqrt{1-x^2}} = \frac{(1-x^2)^{3/2}}{3}$$

Step 2 : Hence, $I = \int_0^1 x^2 \cdot \dfrac{(1-x^2)^{3/2}}{3} \, dx$. Put $x = \sin\theta$; $dx = \cos\theta \, d\theta$;

x	0	1
θ	0	π/2

$$\therefore \qquad I = \frac{1}{3} \int_0^{\pi/2} \sin^2\theta \, (1 - \sin^2\theta)^{3/2} \, \cos\theta \, d\theta$$

$$= \frac{1}{3} \int_0^{\pi/2} \sin^2\theta \, \cos^4\theta \, d\theta = \frac{1}{3} \frac{(1) \cdot (3 \cdot 1)}{6 \cdot 4 \cdot 2} \cdot \frac{\pi}{2}$$

$$\boxed{I = \frac{\pi}{96}}$$

Note that in evaluation of this integral, we could have considered first integration in a horizontal direction by considering horizontal strip, the end limits of which would be $x = 0$ to $x = \sqrt{1-y^2}$ and then moving the strip in a vertical direction, covering whole region the variation in y being from $y = 0$ to $y = 1$.

Then,

$$I = \int_0^1 \int_0^{\sqrt{1-y^2}} x^2 y^2 \, dx \, dy$$

Here the integration is to be carried out w.r.t. x first and then w.r.t. y. The actual result will obviously be the same as $\dfrac{\pi}{96}$.

In this problem, $f(x, y) = x^2 y^2$ is symmetric w.r.t. x and y i.e. interchange of x, y does not alter the function, hence the order of integration is immaterial. But in many problems, we have to first decide the order of integration for evaluation of integral to be possible, easy, convenient and then proceed further. This is illustrated in examples solved below.

Ex. 2 : *Evaluate* $I = \iint \dfrac{1}{x^4 + y^2} \, dx \, dy$ *over the region* $y \geq x^2$, $x \geq 1$. **(May 2016)**

Sol. : Here $y = x^2$ represents a parabola with y-axis as axis of symmetry and the point of intersection of line $x = 1$ and parabola is $x = 1, y = 1$ i.e. $(1, 1)$ as shown in Fig. 9.6.

The region is spread to the right of the line $x = 1$ and above the parabola $y = x^2$ extending upto infinity.

In evaluation of this integral if we decide to integrate w.r.t. x first we would be confronted with x^4 term in the denominator.

$$\int \frac{1}{x^4 + y^2} \, dx$$

with y constant is very complicated, therefore we integrate w.r.t. y first which is very simple as integral is directly in standard form.

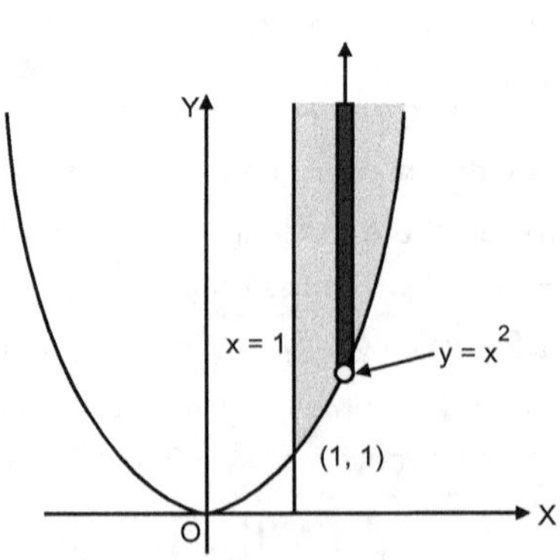

Fig. 9.6

Considering a vertical strip, the end limits of which are from $y = x^2$ to ∞ and then in moving the strip to cover the whole region, x will vary from $x = 1$ to $x = \infty$.

$$I = \int_1^\infty \int_{y=x^2}^\infty \frac{dx\,dy}{x^4 + y^2}$$

$$I = \int_1^\infty \frac{1}{x^2} \left[\tan^{-1} \frac{y}{x^2} \right]_{x^2}^\infty dx \qquad\qquad \left[\because \int \frac{dy}{a^2 + y^2} = \frac{1}{a} \tan^{-1}\frac{y}{a} \right]$$

$$= \int_1^\infty \frac{1}{x^2} \left[\tan^{-1} \infty - \tan^{-1} \frac{x^2}{x^2} \right] dx = \int_1^\infty x^{-2} \left[\frac{\pi}{2} - \frac{\pi}{4} \right] dx = \frac{\pi}{4} \left[\frac{x^{-1}}{-1} \right]_1^\infty = \frac{\pi}{4} \left[-\frac{1}{x} \right]_1^\infty = \frac{\pi}{4} [0 + 1]$$

$$\boxed{I = \frac{\pi}{4}}$$

Ex. 3 : *Evaluate* $\displaystyle\iint_R \sqrt{xy\,(1 - x - y)}\ dx\,dy$, *where R is the area bounded by* $x = 0$, $y = 0$ *and* $x + y = 1$. **(May 2011)**

Sol. : Let, $\quad I = \displaystyle\iint_R \sqrt{xy\,(1 - x - y)}\ dx\,dy$

Here we first evaluate w.r.t. y then w.r.t. x. The region of integration is as shown shaded in Fig. 9.7. Here we imagine a vertical strip in region R.

$\therefore \qquad I = \displaystyle\int \sqrt{x}\, dx \int \sqrt{y\,(1 - x - y)}\ dy$

$$= \int_0^1 \sqrt{x}\ dx \int_0^{1-x} \sqrt{y\,(1 - x - y)}\ dy = \int_0^1 \sqrt{x}\, I_1\, dx, \text{ (say)}$$

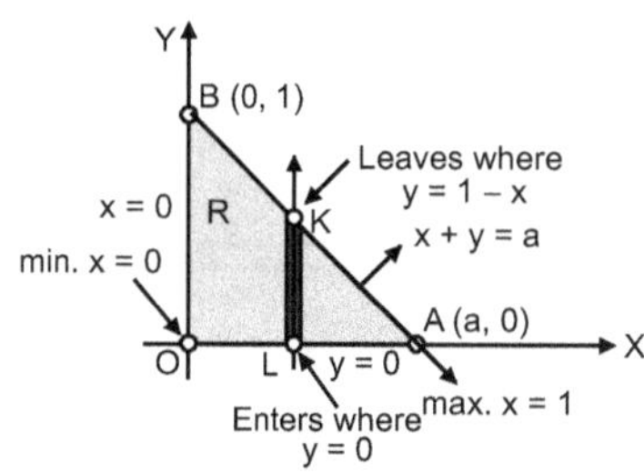

Fig. 9.7

Step 1 : Where $\quad I_1 = \displaystyle\int_0^{1-x} \sqrt{y\,(1 - x - y)}\ dy,$ (Assume $1 - x = a$)

$$= \int_0^a y^{1/2}\,(a - y)^{1/2}\, dy \qquad \text{Put } y = at, \quad dy = a\, dt$$

y	0	a
t	0	1

$$= \int_0^1 a^{1/2}\, t^{1/2}\,(a - at)^{1/2} \cdot a\, dt = a^2 \int_0^1 t^{1/2}\,(1 - t)^{1/2}\, dt$$

$$= a^2\, B\left(\frac{3}{2}, \frac{3}{2}\right) = a^2 \frac{\overline{|3/2}\ \overline{|3/2}}{\overline{|3}} = \frac{a^2 \cdot \left(\frac{1}{2}\sqrt{\pi}\right)^2}{2} = \frac{\pi a^2}{8}$$

Step 2 : Hence, $\quad I = \displaystyle\int_0^1 \sqrt{x} \cdot dx \left(\frac{\pi a^2}{8}\right) = \frac{\pi}{8} \int_0^1 x^{1/2}\,(1 - x)^2\, dx.$ $(\because a = 1 - x)$

$$= \frac{\pi}{8}\, B\left(\frac{3}{2}, 3\right) = \frac{\pi}{8} \frac{\overline{|3/2}\ \overline{|3}}{\overline{|9/2}} = \frac{\pi}{8} \cdot \frac{\overline{|3/2} \cdot 2}{\frac{7}{2} \cdot \frac{5}{2} \cdot \frac{3}{2} \overline{|3/2}}$$

$$\boxed{I = \frac{2\pi}{105}}$$

Ex. 4 : *Evaluate* $\iint\limits_{A} x^{m-1} y^{n-1}\, dx\, dy$, *where A is bounded by* $x + y = h,\ x = 0,\ y = 0.$

Sol. : $I = \iint\limits_{A} x^{m-1} y^{n-1}\, dx\, dy$

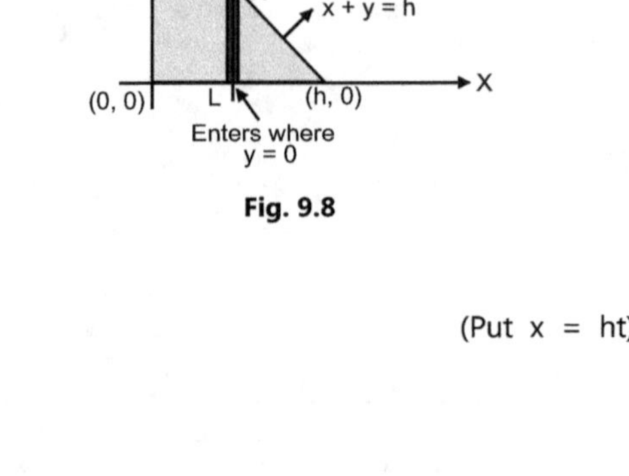

$\therefore$ $I = \int\limits_{0}^{h} x^{m-1}\, dx \int\limits_{0}^{h-x} y^{n-1}\, dy$ (Refer Fig. 9.8)

Step 1 : $I_1 = \int\limits_{0}^{h-x} y^{n-1}\, dy = \left[\dfrac{y^n}{n}\right]_{0}^{h-x} = \dfrac{(h-x)^n}{n}$

Step 2 : $I = \int\limits_{0}^{h} x^{m-1} \dfrac{(h-x)^n}{n}\, dx = \dfrac{1}{n}\int\limits_{0}^{h} x^{m-1}(h-x)^n\, dx$ (Put $x = ht$)

$$= \frac{1}{n}\int\limits_{0}^{1} h^{m-1} t^{m-1}(h - ht)\cdot h\, dt = \frac{h^{m+n}}{n}\int\limits_{0}^{1} t^{m-1}(1-t)^n\, dt.$$

$$= \frac{h^{m+n}}{n}\, B(m,\, n+1) = \frac{h^{m+n}}{n}\, \frac{\overline{|m}\ \overline{|n+1}}{\overline{|m+n+1}} = \frac{h^{m+n}}{n}\, \frac{\overline{|m}\ n\overline{|n}}{(m+n)\overline{|m+n}}$$

$$\boxed{\ I = \frac{h^{m+n}}{m+n}\, \frac{\overline{|m}\ \overline{|n}}{\overline{|m+n}}\ }$$

Ex. 5 : *Evaluate* $\iint y\, dx\, dy$ *over the area bounded by* $y = x^2$ *and* $x + y = 2.$ **(May 2008)**

Sol. : The points of intersection of parabola and straight line are given by,

$$x + x^2 - 2 = 0 \ \text{ or } \ (x+2)(x-1) = 1 \ \therefore \ x = -2 \ \text{ and } \ x = 1.$$

For $x = -2,\ y = 2 - x = 4$ and for $x = 1,\ y = 2 - x = 1.$

Thus, $A(1,\,1),\ B(-2,\,4)$ are the points of intersection as shown in Fig. 9.9. The region of integration is shown shaded in the figure. To evaluate the integral, we first consider a vertical strip, the end limits of which are $y = x^2$ and $y = 2 - x$. Then move the strip in horizontal direction to cover the region of integration, for which x varies from $x = -2$ to $x = 1.$

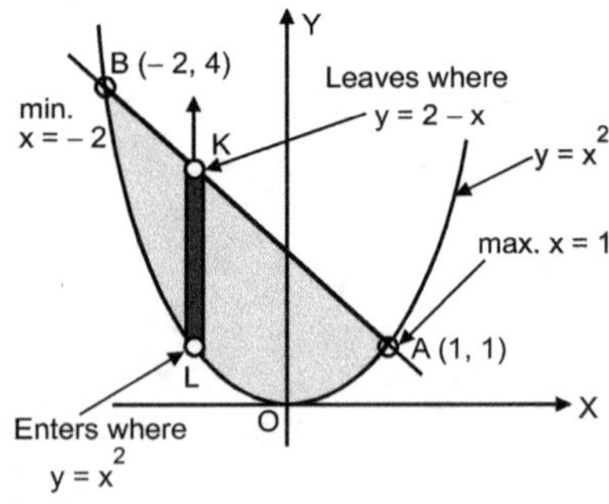

$$I = \int\limits_{-2}^{1}\int\limits_{x^2}^{2-x} y\, dx\, dy$$

Integrating w.r.t. y treating x as constant

$$= \int\limits_{-2}^{1}\left[\frac{y^2}{2}\right]_{x^2}^{2-x} dx = \frac{1}{2}\int\limits_{-2}^{1}\{(2-x)^2 - x^4\}\, dx = \frac{1}{2}\int\limits_{-2}^{1}\{4 - 4x + x^2 - x^4\}\, dx$$

$$= \frac{1}{2}\left[4x - 4\frac{x^2}{2} + \frac{x^3}{3} - \frac{x^5}{5}\right]_{-2}^{1}$$

$$= \frac{1}{2}\left[4 \times 1 - 4 \times (-2) - 2\,(1)^2 + 2\,(-2)^2 + \frac{1}{3}\,(1)^3 - \frac{1}{3}\,(-2)^3 - \frac{1}{5}\,(1)^5 + \frac{1}{5}\,(-2)^5\right]$$

$$= \frac{1}{2}\left[4 + 8 - 2 + 8 + \frac{1}{3} + \frac{8}{3} - \frac{1}{5} - \frac{32}{5}\right] = \frac{1}{2}\left[21 - \frac{33}{5}\right] = \frac{1}{2}\left[\frac{105 - 33}{5}\right]$$

$$\boxed{I = \frac{36}{5}}$$

In the above problem, if the integration is to be carried out w.r.t. x first, we have to consider a horizontal strip where its right end shifts the curves (across y = 1) in the vertical movement. Hence, the region is splitted up into two parts A_1, A_2 with the help of demarking line y = 1. Here A_1 is OAP ; A_2 is OBP. (Refer Fig. 9.10)

Over the region A_1, the ends of the horizontal strip touch parabolic arcs for which x varies from $-\sqrt{y}$ to $\sqrt{y}$ and in movement of the strip, vertically, y varies from 0 to 1. In the region A_2, the end limits of horizontal strip are $x = -\sqrt{y}$ and $x = 2 - y$, y varies between y = 1 and y = 4.

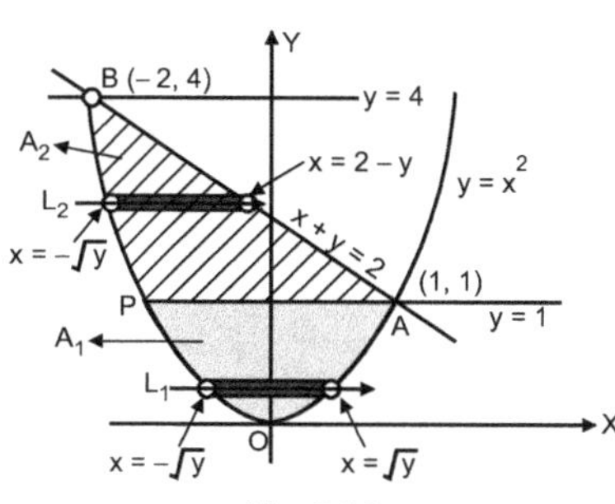

Fig. 9.10

Thus,
$$I = \iint\limits_{A_1} y \, dx \, dy + \iint\limits_{A_2} y \, dx \, dy = \int_0^1 \int_{-\sqrt{y}}^{\sqrt{y}} y \, dy \, dx + \int_1^4 \int_{-\sqrt{y}}^{2-y} y \, dy \, dx$$

After evaluation, the sum of integrals should give value of I as $\frac{36}{5}$ as obtained earlier, verification of this result is left to the students as an exercise.

EXERCISE 9.2

Evaluate the following Integrals :

1. $\iint\limits_R y \, dx \, dy$, where R is $y = x^2$, $x + y = 2$, $x = 0$ in first quadrant.   **Ans. :** $\dfrac{16}{15}$

2. $\iint\limits_R y \, dx \, dy$, where R is $y = x^2$, $y = x$.   **Ans. :** $\dfrac{1}{15}$

3. $\iint\limits_R xy \, (x + y) \, dx \, dy$, where R is $y = x^2$, $y = x$.   **Ans. :** $\dfrac{3}{56}$

4. $\iint\limits_R xy \, dx \, dy$, where R is $x^2 = y$, $y^2 = -x$.   **Ans. :** $-\dfrac{1}{12}$

5. $\iint\limits_R xy \left(\dfrac{x^2}{a^2} + \dfrac{y^2}{b^2}\right)^{n/2} dx \, dy$, over positive quadrant of $\dfrac{x^2}{a^2} + \dfrac{y^2}{b^2} = 1$.   **Ans. :** $\dfrac{a^2 b^2}{2\,(n + 4)}$

6. $\iint\limits_R x^{m-1} \, y^{n-1} \, dx \, dy$, over positive quadrant of $\dfrac{x^2}{a^2} + \dfrac{y^2}{b^2} = 1$.   **Ans. :** $\dfrac{a^m \, b^n}{2\,(m + n)} \cdot \dfrac{\overline{|m/2}\;\overline{|n/2}}{\overline{|(m + n)/2}}$

7. $\iint\limits_R \dfrac{xy \, dx \, dy}{\sqrt{1 - y^2}}$, over positive quadrant of $x^2 + y^2 = 1$.   **Ans. :** $\dfrac{1}{6}$

8. $\iint\limits_R x^p \, y^q \, dx \, dy$, over positive quadrant of $x^2 + y^2 = a^2$.   **Ans. :** $\dfrac{a^{p+q+2}}{2\,(p + q + 2)} \cdot \dfrac{\overline{|(p + 1)/2}\;\overline{|(q + 1)/2}}{\overline{|(p + q + 2)/2}}$

9. $\iint\limits_R x^3 \, y \, dx \, dy$, over positive quadrant of $\dfrac{x^2}{a^2} + \dfrac{y^2}{b^2} = 1$.   **Ans. :** $\dfrac{b^2 \, a^4}{24}$

10. $\iint\limits_R e^{ax + by} \, dx \, dy$, where R is $x = 0$, $y = 0$, $ax + by = 1$.   **Ans. :** $\dfrac{1}{ab}$

11. $\iint\limits_{R} (5 - 2x - y)\, dx\, dy$, where R is $y = 0$, $x + 2y = 3$, $x = y^2$. **Ans. :** $\dfrac{217}{60}$

12. $\iint\limits_{R} (x^2 - y^2)\, dx\, dy$, over the triangle with vertices $(0, 1)$, $(1, 1)$ and $(1, 2)$. **(May 2007)** **Ans. :** $\left(-\dfrac{2}{3}\right)$

13. $\iint\limits_{R} xy\sqrt{1 - x - y}\, dx\, dy$, over the region $x \geq 0$, $y \geq 0$ and $x + y \leq 1$. **Ans. :** $\dfrac{16}{945}$

14. $\iint\limits_{R} xy\, dx\, dy$, over R bounded by x-axis, $x = 2a$ and the curve $x^2 = 4\,ay$. **Ans. :** $\dfrac{a^4}{3}$

15. $\iint\limits_{R} xy^2\, dx\, dy$, where R is the region bounded by $y = x^2$, $y = 0$, $x = 1$. **Ans. :** $\dfrac{1}{24}$

16. $\iint\limits_{R} xy\,(x + y)\, dx\, dy$, over the region R bounded by $y^2 = x$, $x^2 = y$. **Ans. :** $\dfrac{3}{28}$

17. $\iint\limits_{R} xy\, dx\, dy$, where R is the region bounded by $\dfrac{x}{a} + \dfrac{y}{b} = 1$, $x = 0$, $y = 0$. **Ans. :** $\dfrac{a^2 b^2}{24}$

18. $\iint\limits_{R} x^2\, y\, dx\, dy$, where R is the region bounded by x-axis, $y = x$, $y = 2 - x$. **Ans. :** $\dfrac{11}{30}$

19. $\iint\limits_{R} xy\, dx\, dy$, over the region R bounded by $y^2 = 4x$, $y = 2x - 4$. **Ans. :** $\dfrac{45}{2}$

20. $\iint\limits_{R} e^{y^2}\, dx\, dy$, over the triangle whose vertices are $(0, 0)$, $(2, 1)$, $(0, 1)$. **Ans. :** $e - 1$

Type 3 : Illustrations on Change of Order of Integration :

It may happen that the integrand $f(x, y)$ in the double integral $\int \left\{ \int f(x, y)\, dy \right\} dx$ is difficult, or even impossible to integrate w.r.t. y first, but can be easily integrated w.r.t. x first. In such an event, it becomes necessary to reverse (i.e. change) the order of integration in the double integral i.e. to work it out in the form $\int \left\{ \int f(x, y)\, dx \right\} dy$.

Similarly, in $I = \int \left\{ \int f(x, y)\, dx \right\} dy$, if it is difficult to integrate first w.r.t. x, then, by changing the order of integration, we get,

$$I = \int \left\{ \int f(x, y)\, dy \right\} dx.$$

How to Change the Order of Integration ?

Method - I : Let the given integral be $I = \displaystyle\int_{x=a}^{x=b} \left\{ \int_{y = f_1(x)}^{y = f_2(x)} f(x, y)\, dy \right\} dx$

Step 1 : Given region R is bounded by

x	a	b
y	$f_1(x)$	$f_2(x)$

and find points of intersections. Sketch the region of integration R and check whether it is correct or wrong by drawing *a vertical strip*.

Step 2 : Now to reverse the order, draw a *horizontal strip* cutting through region R. Write down the given integral I with order of integration reversed as

$$I = \int \left\{ \int f(x, y)\, dx \right\} dy$$

Step 3 : Find the new limits : Limits for x : A function of y, say $x = g_1(y)$ to $x = g_2(y)$

Limits for y : constants, say $y = c$ to $y = d$.

∴ By changing the order of integration,

$$I = \int_c^d \left\{ \int_{g_1(y)}^{g_2(y)} f(x, y)\, dx \right\} dy$$

Method-II : Let the given integral be $I = \int_{y=c}^{y=d} \left\{ \int_{x=g_1(y)}^{x=g_2(y)} f(x, y)\, dx \right\} dy$

Step 1 : Given region R is bounded by

y	c	d
x	$g_1(y)$	$g_2(y)$

and find points of intersections.

Sketch the region of integration R and check whether it is correct or wrong by drawing *a horizontal strip*.

Step 2 : Now to reverse the order, draw a *vertical strip* cutting through region R. Write down the given integral I with order of integration reversed as

$$I = \int \left\{ \int f(x, y)\, dy \right\} dx$$

Step 3 : Find the new limits : Limits for y : A function of x say $y = f_1(x)$ to $y = f_2(x)$

Limits for x : always constants say $x = a$ to $x = b$.

∴ By changing the order of integration,

$$I = \int_a^b \left\{ \int_{f_1(x)}^{f_2(x)} f(x, y)\, dy \right\} dx$$

Ex. 1 : *Evaluate* $\displaystyle \int_0^1 \int_0^{\sqrt{1-x^2}} \frac{dx\, dy}{(1 + e^y)\sqrt{1 - x^2 - y^2}}$ *(Dec. 2006)*

Sol. : Let, $\displaystyle I = \int_0^1 dx \int_0^{\sqrt{1-x^2}} \frac{dy}{(1 + e^y)\sqrt{1 - x^2 - y^2}}$.

Here integral first w.r.t. y is difficult to solve, therefore, we should change the order of integration. The given region is bounded by $x = 0$, $x = 1$, $y = 0$ and $y = +\sqrt{1^2 - x^2}$, represents a positive quadrant of a circle $x^2 + y^2 = a^2$.

For changing the order of integration, imagine a horizontal strip LK cutting through region R as shown in Fig. 9.11.

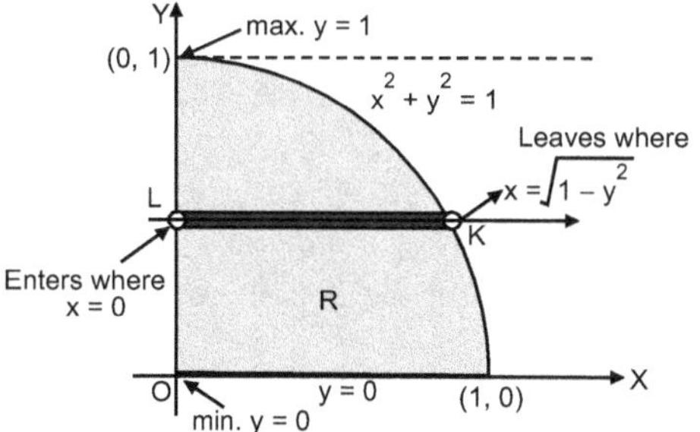

Fig. 9.11

Step 1 : Order is reversed.

$$I = \int \frac{dy}{1 + e^y} \int \frac{dx}{\sqrt{1 - y^2 - x^2}}$$

Step 2 : New limits for integration,

$$I = \int_0^1 \frac{dy}{1 + e^y} \int_0^{\sqrt{1-y^2}} \frac{dx}{\sqrt{1 - y^2 - x^2}}$$

Step 3 : Let
$$I_1 = \int_0^{\sqrt{1-y^2}} \frac{dx}{\sqrt{1 - y^2 - x^2}} \quad \text{Let } 1 - y^2 = b^2$$

$$= \int_0^b \frac{dx}{\sqrt{b^2 - x^2}} = \left[\sin^{-1}\frac{x}{b}\right]_0^b = \frac{\pi}{2} - 0 = \frac{\pi}{2}$$

Step 4 :
$$I = \int_0^1 \frac{dy}{1 + e^y} \cdot \frac{\pi}{2} = \frac{\pi}{2} \int_0^1 \frac{1 + e^y - e^y}{1 + e^y} \, dy$$

$$= \frac{\pi}{2} \int_0^1 \left(1 - \frac{e^y}{1 + e^y}\right) dy = \frac{\pi}{2} \left[y - \log (1 + e^y)\right]_0^1$$

$$= \frac{\pi}{2} \left[1 - \log (1 + e) + \log 2\right] = \frac{\pi}{2} \left[\log e + \log 2 - \log (1 + e)\right]$$

$$\boxed{I = \frac{\pi}{2} \log \left(\frac{2e}{1 + e}\right)}$$

Ex. 2 : *Show that* $\displaystyle \int_0^a \int_{\frac{y^2}{a}}^{y} \frac{y \, dy \, dx}{(a - x)\sqrt{ax - y^2}} = \frac{\pi a}{2}$ *(May 2013)*

Proof :
$$I = \int_0^a y \cdot \left\{ \int_{\frac{y^2}{a}}^{y} \frac{dx}{(a - x)\sqrt{ax - y^2}} \right\} dy$$

Here inner integral w.r.t. x is difficult to solve, therefore, we should change the order of integration. The given region is bounded by $y = 0$, $y = a$, $x = \dfrac{y^2}{a}$ i.e. $y^2 = ax$ and $x = y$.

When $y = 0$, $x = 0$. Also $y = a$ then $x = a$. Therefore, $(0, 0)$ and (a, a) are the points of *intersection*. The region is shown shaded. (Refer Fig. 9.12)

Now, new order of integration is first w.r.t. y, then w.r.t. x, for which we will imagine a vertical strip cutting through R. Limits of y : $y = x$ to $y = \sqrt{ax}$; Limits of $x : x = 0$ to $x = a$.

∴   With the order of integration reversed,

$$I = \int_0^a \frac{1}{a - x} \left\{ \int_{x}^{\sqrt{ax}} \frac{y \, dy}{\sqrt{ax - y^2}} \right\} dx$$

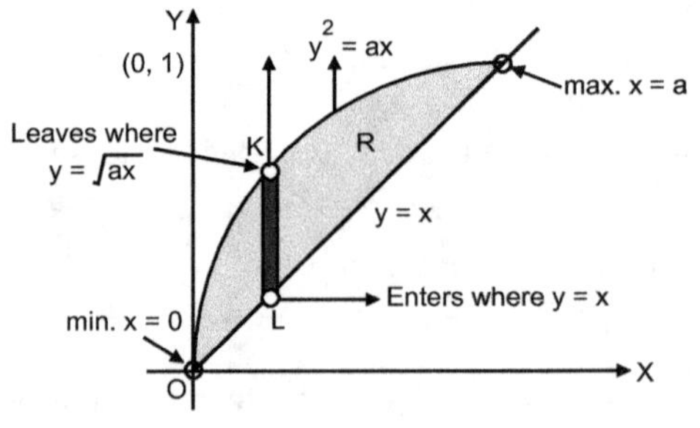

Fig. 9.12

Step 1 : Let,

$$I_1 = \int_x^{\sqrt{ax}} \frac{y\,dy}{\sqrt{ax - y^2}} = -\frac{1}{2} \int_x^{\sqrt{ax}} \frac{-2y\,dy}{\sqrt{ax - y^2}}$$ (Note the adjustment)

$$= -\frac{1}{2}\left[2\sqrt{ax - y^2}\right]_x^{\sqrt{ax}}$$ $\left(\because \int \frac{f'}{\sqrt{f}} = 2\sqrt{f}\right)$

$$= -\left[0 - \sqrt{ax - x^2}\right] = \sqrt{x}\,\sqrt{a - x}$$

Step 2 : Hence,

$$I = \int_0^a \frac{1}{a - x}\,\sqrt{x}\,\sqrt{a - x}\,dx = \int_0^a x^{1/2}(a - x)^{-1/2}\,dx. \quad \text{Put } x = at;\ dx = a\,dt$$

x	0	a
t	0	1

$$I = \int_0^1 a^{1/2}\,t^{1/2}\,a^{-1/2}(1 - t)^{-1/2}\,a\,dt = a\int_0^1 t^{1/2}(1 - t)^{-1/2}\,dt.$$

$$= a \cdot B\left(\frac{3}{2}, \frac{1}{2}\right) = a\,\frac{\left\lfloor\frac{3}{2}\right. \left\lfloor\frac{1}{2}\right.}{\left\lfloor 2\right.}$$

$$\boxed{I = \frac{a \cdot \frac{1}{2}\sqrt{\pi}\sqrt{\pi}}{1!} = \frac{\pi a}{2}}$$

Ex. 3 : *Show that* $\displaystyle \int_0^1 \int_1^{\sqrt{2 - y^2}} \frac{y\,dx\,dy}{\sqrt{(2 - x^2)\,(1 - x^2 y^2)}} = 1 - \frac{\pi}{4}$

Proof : Let,

$$I = \int_0^1 y\left\{\int_1^{\sqrt{2 - y^2}} \frac{dx}{\sqrt{(2 - x^2)(1 - x^2 y^2)}}\right\} dy.$$

Here inner integral w.r.t. x is difficult to solve, therefore we should change the order of integration. The given region is bounded by $y = 0$, $y = 1$ and $x = 1$, $x = \sqrt{2 - y^2}$ i.e. $x^2 + y^2 = 2$. When $y = 0$, $x = \sqrt{2}$ and when $y = 1$, $x = 1$. Therefore, $(\sqrt{2},\ 0)$, $(1,\ 1)$ are the points of intersection. The region R is shown shaded in Fig. 9.13.

Now, new order of integration is first w.r.t. y, then w.r.t. x, for which we will imagine a vertical strip cutting through R.

Limits of y : $y = 0$ to $y = \sqrt{2 - x^2}$;   Limits of x : $x = 1$ to $x = \sqrt{2}$.

∴ With the order of integration reversed,

$$I = \int_1^{\sqrt{2}} \frac{dx}{\sqrt{2 - x^2}} \int_0^{\sqrt{2 - x^2}} \frac{y\,dy}{\sqrt{1 - x^2 y^2}}$$

Step 1 : $$I_1 = \int_0^{\sqrt{2 - x^2}} \frac{y\,dy}{\sqrt{1 - x^2 y^2}} = -\frac{1}{2x^2} \int_0^{\sqrt{2 - x^2}} \frac{-2x^2\,y\,dy}{\sqrt{1 - x^2 y^2}}$$

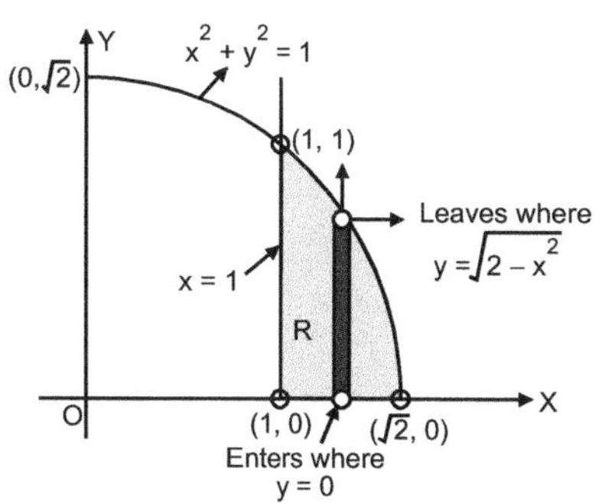

Fig. 9.13

(Note the adjustment as x^2 is constant)

$$I_1 = -\frac{1}{2x^2}\left[2\sqrt{1-x^2y^2}\right]_0^{\sqrt{2-x^2}} \qquad \left(\because \int \frac{f'}{\sqrt{f}} = 2\sqrt{f}\right)$$

$$= -\frac{1}{x^2}\left[\sqrt{1-x^2(2-x^2)}-1\right] = \frac{1}{x^2}\left[1-\sqrt{1-2x^2+x^4}\right]$$

$$= \frac{1}{x^2}\left[1-\sqrt{(x^2-1)^2}\right] = \frac{1}{x^2}(1-x^2+1) = \frac{2-x^2}{x^2}$$

Step 2 :

$$I = \int_1^{\sqrt{2}} \frac{1}{\sqrt{2-x^2}}\left(\frac{2-x^2}{x^2}\right)\cdot dx = \int_1^{\sqrt{2}} \frac{\sqrt{2-x^2}}{x^2}\cdot dx$$

Put

$$x^2 = 2\sin^2\theta$$

x	1	$\sqrt{2}$
θ	$\pi/4$	$\pi/2$

$\therefore$

$$I = \int_{\pi/4}^{\pi/2} \frac{\sqrt{2-2\sin^2\theta}}{2\sin^2\theta}\cdot\sqrt{2}\cos\theta\,d\theta = \int_{\pi/4}^{\pi/2} \frac{\cos^2\theta}{\sin^2\theta}\,d\theta$$

$$= \int_{\pi/4}^{\pi/2} (\text{cosec}^2\theta - 1)\,d\theta = \left[-\cot\theta - \theta\right]_{\pi/4}^{\pi/2} = \left(0-\frac{\pi}{2}\right) - \left(-1-\frac{\pi}{4}\right)$$

$\therefore$

$$\boxed{I = 1 - \frac{\pi}{4}}$$

Ex. 4 : *Show that* $\displaystyle \int_0^a \int_0^y \frac{x\,dx\,dy}{\sqrt{(a^2-x^2)(a-y)(y-x)}} = \pi a.$

Proof : Let,

$$I = \int_0^a \frac{dy}{\sqrt{a-y}} \int_0^y \frac{x\,dx}{\sqrt{(a^2-x^2)(y-x)}}$$

Here first integral w.r.t. x is difficult to solve, hence we should change the order of integration.

Given region R is y = 0 to y = a ; x = 0 to x = y.

$\therefore$ (0, 0), (a, a) are the points of intersection (Refer Fig. 9.14)

Now, with the order of integration reversed, we integrate first w.r.t. y and then w.r.t. x, for which we will imagine a vertical strip cutting through R.

Limits of y : y = x to y = a ; Limits of x : x = 0 to x = a.

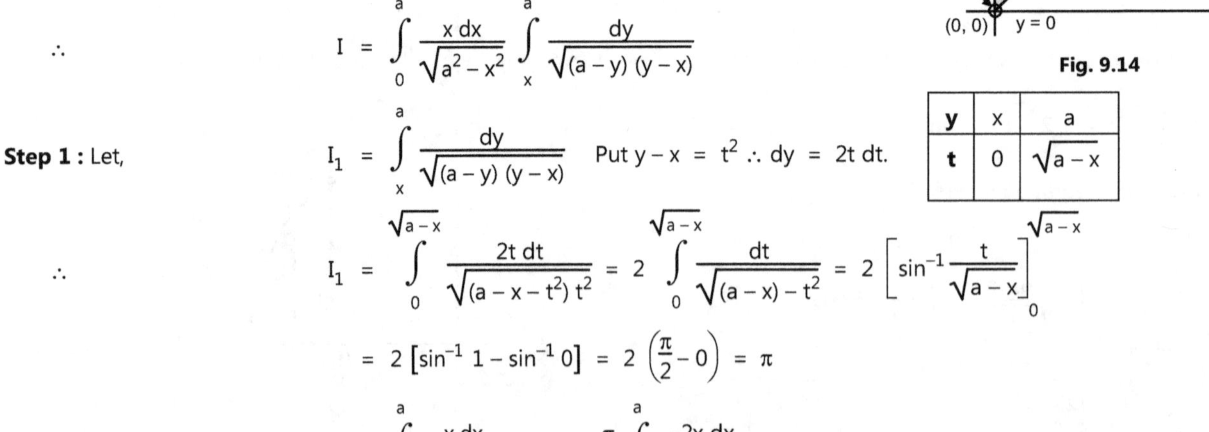

$\therefore$

$$I = \int_0^a \frac{x\,dx}{\sqrt{a^2-x^2}} \int_x^a \frac{dy}{\sqrt{(a-y)(y-x)}}$$

Step 1 : Let,

$$I_1 = \int_x^a \frac{dy}{\sqrt{(a-y)(y-x)}} \qquad \text{Put } y-x = t^2 \therefore dy = 2t\,dt.$$

y	x	a
t	0	$\sqrt{a-x}$

$\therefore$

$$I_1 = \int_0^{\sqrt{a-x}} \frac{2t\,dt}{\sqrt{(a-x-t^2)\,t^2}} = 2\int_0^{\sqrt{a-x}} \frac{dt}{\sqrt{(a-x)-t^2}} = 2\left[\sin^{-1}\frac{t}{\sqrt{a-x}}\right]_0^{\sqrt{a-x}}$$

$$= 2\left[\sin^{-1}1 - \sin^{-1}0\right] = 2\left(\frac{\pi}{2}-0\right) = \pi$$

Step 2 : Hence,

$$I = \int_0^a \frac{x\,dx}{\sqrt{a^2-x^2}}\,(\pi) = -\frac{\pi}{2}\int_0^a \frac{-2x\,dx}{\sqrt{a^2-x^2}} \qquad \text{(Note this step)}$$

$$I = -\frac{\pi}{2}\left[2\sqrt{a^2 - x^2}\right]_0^a = -\pi\,[0 - a] \qquad \left(\because \int \frac{f'}{\sqrt{f}} = 2\sqrt{f}\right)$$

$$\boxed{I = \pi a}$$

...... Hence proved.

Ex. 5 : *Show that* $\displaystyle\int_0^1 \int_x^{1/x} \frac{y\,dx\,dy}{(1 + xy)^2\,(1 + y^2)} = \frac{\pi - 1}{4}$ **(Dec. 2011)**

Proof : Let, $\displaystyle I = \int_0^1 dx \int_x^{1/x} \frac{y\,dy}{(1 + xy)^2\,(1 + y^2)}$

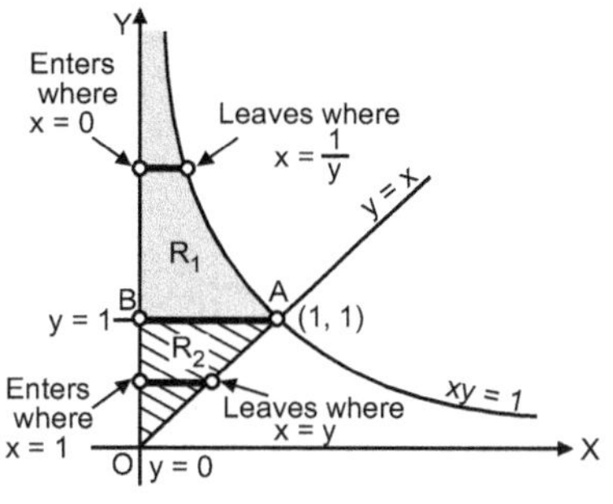

Here, integral first w.r.t. y is difficult to evaluate, therefore we should change the order of integration.

To evaluate first w.r.t. x and then w.r.t. y, we have to consider a horizontal strip where ends shift the curves in the vertical movement. Hence, the region is splitted up into two parts R_1, R_2 with the help of demarking line y = 1. (Refer Fig. 9.15)

Fig. 9.15

Given region is x = 0, x = 1, y = x, $y = \dfrac{1}{x}$ i.e. xy = 1.

R_1 is OAB ; R_2 is ABC. Consider two horizontal strips in R_1 and R_2 respectively.

Limits for R_1 : y = 0 to y = 1 ; x = 0 to x = y

Limits for R_2 : y = 1 to y = ∞ ; x = 0 to $x = \dfrac{1}{y}$.

Thus, $\displaystyle I = \iint_{R_1} \frac{y\,dy\,dx}{(1 + xy)^2\,(1 + y^2)} + \iint_{R_2} \frac{y\,dy\,dx}{(1 + xy)^2\,(1 + y^2)} = I_1 + I_2.$

$$I_1 = \int_0^1 \frac{y\,dy}{1 + y^2} \int_0^y \frac{dx}{(1 + xy)^2} = \int_0^1 \frac{y\,dy}{1 + y^2}\left[\frac{-1}{y\,(1 + xy)}\right]_0^y$$

$$= \int_0^1 \frac{dy}{1 + y^2}\left[-\frac{1}{1 + y^2} + \frac{1}{1}\right] = \int_0^1 -\frac{dy}{(1 + y^2)^2} + \int_0^1 \frac{dy}{1 + y^2}$$

In the first integral, put y = tan θ, dy = $\sec^2 θ\,dθ$

y	0	1
θ	0	π/4

$$I_1 = -\int_0^{\pi/4} \frac{\sec^2 θ\,dθ}{\sec^2 θ} + \left[\tan^{-1} y\right]_0^1 = -\int_0^{\pi/4} \frac{1 + \cos 2θ}{2}\,dθ + \frac{\pi}{4}$$

$$= -\frac{1}{2}\left[\frac{\pi}{4} + \frac{\sin 2θ}{2}\Bigg|_0^{\pi/4}\right] + \frac{\pi}{4} = \frac{\pi}{8} - \frac{1}{4}$$

$$I_2 = \int_1^\infty \frac{y\,dy}{1 + y^2} \int_0^{1/y} \frac{dx}{(1 + xy)^2} = \int_1^\infty \frac{y\,dy}{1 + y^2}\left[-\frac{1}{y\,(1 + xy)}\right]_0^{1/y}$$

$$= \int_1^\infty -\frac{dy}{1 + y^2}\left[\frac{1}{2} - 1\right] = \frac{1}{2}\int_1^\infty \frac{dy}{1 + y^2} = \frac{1}{2}\left[\tan^{-1} y\right]_1^\infty = \frac{1}{2}\left[\frac{\pi}{2} - \frac{\pi}{4}\right] = \frac{\pi}{8}$$

$\therefore$ $\displaystyle I = \frac{\pi}{8} - \frac{1}{4} + \frac{\pi}{8}$

$$\boxed{I = \frac{\pi - 1}{4}}$$

...... Hence proved.

Ex. 6 : *Evaluate* $\displaystyle\int_0^{\pi/2}\int_0^{y}\cos 2y\sqrt{1-a^2\sin^2 x}\ dy\,dx.$ **(May 2009, Dec. 2013)**

Sol. : Here it is difficult to integrate w.r.t. x first. Changing the order of integration. Refer Fig. 9.16 (by considering vertical strip).

New limits are $y = x$ to $y = \dfrac{\pi}{2}$ and $x = 0$ to $x = \dfrac{\pi}{2}$.

$$I = \int_0^{\pi/2}\int_x^{\pi/2}\cos 2y\sqrt{1-a^2\sin^2 x}\ dx\,dy$$

$$I = \int_0^{\pi/2}\sqrt{1-a^2\sin^2 x}\left[\frac{\sin 2y}{2}\right]_x^{\pi/2} dx = -\frac{1}{2}\int_0^{\pi/2}\sqrt{1-a^2\sin^2 x}\ \sin 2x\,dx$$

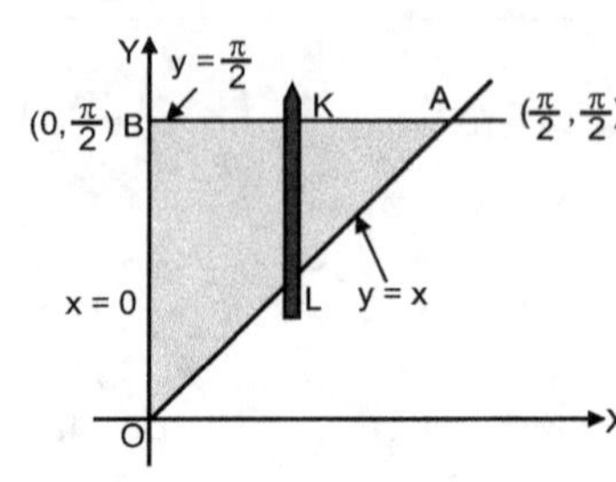

Fig. 9.16

Put $\sin^2 x = t$ $\therefore$ $2\sin x\cos x\,dx = dt$ Or $\sin 2x\,dx = dt$

When $x = 0,\ t = 0,\ x = \dfrac{\pi}{2},\ t = 1.$

$$= -\frac{1}{2}\int_0^{1}\sqrt{1-a^2 t}\ dt = -\frac{1}{2}\left[\frac{(1-a^2 t)^{3/2}}{(-a^2)\cdot\frac{3}{2}}\right]_0^1 = \frac{1}{3a^2}\left[(1-a^2\cdot t)^{3/2}\right]_0^1$$

$$\boxed{I = \frac{1}{3a^2}\left[(1-a^2)^{3/2}-1\right]}$$

Ex. 7 : *Evaluate* $\displaystyle\int_0^{1}\int_{1-\sqrt{1-y}}^{1+\sqrt{1-y}}\frac{dx\,dy}{(x^2-2x+y-3)^2}.$

Sol. : Region of integration is bounded by

$$x = 1-\sqrt{1-y} \qquad \text{i.e.} \qquad (x-1)^2 = 1-y$$

to

$$x = 1+\sqrt{1-y} \qquad \text{i.e.} \qquad (x-1)^2 = 1-y$$

and

$$y = 0 \text{ to } y = 1$$

{Shifting the origin to (1, 1), it becomes $(x + 1 - 1)^2 = 1 - (y + 1)$ or $x^2 = -y$ which is a downward parabola with vertex at (1, 1).}

Curve passes through (0, 0) and (2, 0). Other boundaries are $y = 0, y = 1.$

Region of integration is as shown in Fig. 9.17. It is convenient to integrate w.r.t. y first.

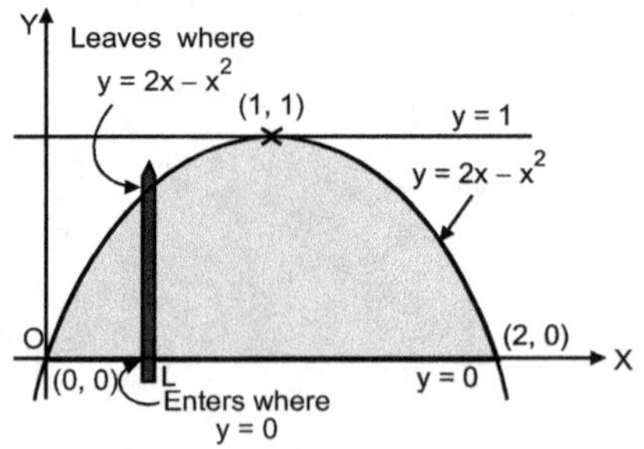

Fig. 9.17

$$I = \int_0^{2}\int_0^{2x-x^2}\left[\frac{dy\cdot dx}{(x^2-2x+y-3)^2}\right] = \int_0^{2}\left[-\frac{1}{(x^2-2x+y-3)}\right]_0^{2x-x^2} dx$$

$$= \int_0^{2}\left[-\frac{1}{(x^2-2x+2x-x^2-3)} + \frac{1}{(x^2-2x+0-3)}\right] dx$$

$$= \int_0^{2}\left[\frac{1}{3} + \frac{1}{(x-3)(x+1)}\right] dx = \int_0^{2}\left[\frac{1}{3} + \frac{1}{4}\frac{1}{(x-3)} - \frac{1}{4}\frac{1}{(x+1)}\right] dx$$

$$= \int_0^2 \left[\frac{1}{3} - \frac{1}{4} \frac{1}{(3-x)} - \frac{1}{4} \frac{1}{(x+1)} \right] dx \qquad \text{[Note this step } (\because 0 < x < 2)]$$

$$= \left[\frac{1}{3} x + \frac{1}{4} \log(3-x) - \frac{1}{4} \log(x+1) \right]_0^2$$

$$= \left[\frac{2}{3} - \frac{1}{4} \log 3 - \frac{1}{4} \log 3 \right]$$

$$\boxed{I = \frac{2}{3} - \frac{1}{2} \log 3}$$

Ex. 8 : *Evaluate* $\int_0^1 dx \int_1^\infty e^{-y} y^x \log y \, dy.$ **(Nov. 2015)**

Sol. : $I = \int_0^1 dx \int_1^\infty e^{-y} y^x \log y \cdot dy.$

Here, it is advantageous to integrate w.r.t. x first. *Since both the limits are constants, we can just interchange the order of integration, without plotting the region, as* :

$$I = \int_1^\infty e^{-y} \log y \, dy \int_0^1 y^x \, dx \qquad \text{(Integrate w.r.t. x, y is a constant)}$$

$$= \int_1^\infty e^{-y} \log y \left[\frac{y^x}{\log y} \right]_0^1 dy = \int_1^\infty e^{-y} \log y \frac{(y-1)}{\log y} \, dy$$

$$= \int_1^\infty e^{-y} (y-1) \, dy, \qquad \text{Let } ye^{-y} = u \quad \therefore (y-1) e^{-y} dy = -du$$

y	1	∞
u	1/e	0

$$= \int_{1/e}^0 (-du) = [-u]_{1/e}^0$$

$$\boxed{I = \frac{1}{e}}$$

Ex. 9 : *Show that* $\int_0^a \int_0^{a-\sqrt{a^2-y^2}} \frac{xy \log(x+a)}{(x-a)^2} \, dx \, dy = \frac{a^2}{8} (2 \log a + 1).$ **(May 2007, 2004)**

Proof : Here the RoI (Region of integration) R is bounded by lines $y = 0$, $y = a$, $x = 0$ and the curve $x = a - \sqrt{a^2 - y^2}$.

$\therefore$ $x - a = -\sqrt{a^2 - y^2} < 0$ and $(x-a)^2 + y^2 = a^2$

i.e. $x = a - \sqrt{a^2 - y^2}$ represents a semi-circle on the negative or left side of the line $x - a = 0$. The circle is centred at $(a, 0)$ and has radius 'a'. x changes (imagine horizontal strip) from $x = 0$ to the circle and the strip moves from $y = 0$ (bottom) to $y = a$ (top). The RoI – R is as shaded in Fig. 9.18. As it is convenient to first integrate w.r. to y, taking a vertical strip in R, to change the order of integration, y changes from the circle $y = +\sqrt{a^2 - (x-a)^2} = +\sqrt{2ax - x^2}$ to $y = a$, and x changes from $x = 0$ to $x = a$. Hence, after changing the order, given double integral becomes :

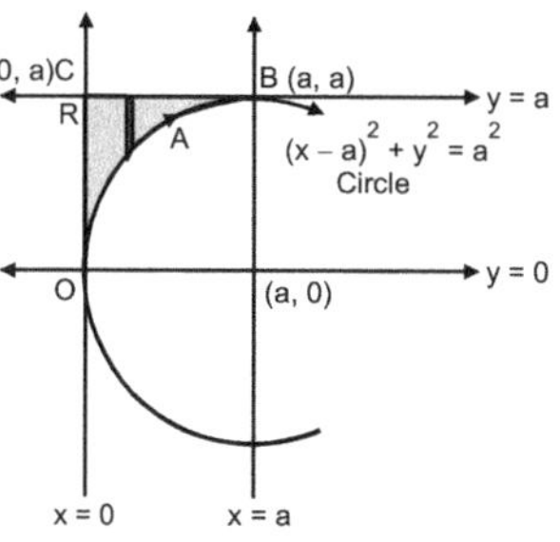

Fig. 9.18

$$I = \int\limits_{x=0}^{a} \int\limits_{y=\sqrt{2ax-x^2}}^{a} \frac{xy \log(x+a)}{(x-a)^2}\, dx\, dy = \int\limits_{x=0}^{a} \frac{x \log(x+a)}{(x-a)^2} \left\{\frac{y^2}{2}\right\}_{\sqrt{2ax-x^2}}^{a} dx$$

$$= \frac{1}{2} \int\limits_{x=0}^{a} x \log(x+a)\, dx = \frac{1}{2} \left\{ \left[\log(x+a)\cdot\frac{x^2}{2}\right]_0^a - \int\limits_0^a \frac{1}{x+a}\cdot\frac{x^2}{2}\, dx \right\}$$

$$= \frac{a^2}{4} \log 2a - \frac{1}{4} \int\limits_0^a \frac{x^2}{x+a}\, dx \qquad\qquad \left[\text{Use } \frac{x^2}{x+a} = \frac{x^2 - a^2 + a^2}{x+a} = x - a + \frac{a^2}{x+a} \right]$$

$$\therefore\quad I = \frac{a^2}{4} \log 2a - \frac{1}{4} \int\limits_0^a \left[x - a + a^2\cdot\frac{1}{x+a} \right] dx = \frac{a^2}{4} \log 2a - \frac{1}{4} \left[\frac{x^2}{2} - ax + a^2 \log(x+a) \right]_0^a \quad \text{and simplify,}$$

$$= \frac{a^2}{8} + \frac{a^2}{4} \log a$$

$$\boxed{I = \frac{a^2}{8}\,(1 + 2\log a)} \qquad\qquad\qquad\qquad\qquad \text{... Proved.}$$

Ex. 10 : *Change the order of integration in double integral* $\displaystyle\int\limits_{0}^{a} \int\limits_{\sqrt{a^2-y^2}}^{y+a} f(x,y)\, dx\, dy.$ **(May 2018)**

Sol. : Let, $\displaystyle I = \int\limits_{0}^{a} \left\{ \int\limits_{\sqrt{a^2-y^2}}^{y+a} f(x,y)\, dx \right\} dy .$

Region of integration is bounded by $y = 0$, $y = a$, $x = \sqrt{a^2-y^2}$ i.e. $x^2 + y^2 = a^2$ and $x = y + a$ i.e. $x - y = a$.

Points of intersection are $(0, a)$, $(2a, a)$ $(a, 0)$ and the region R is shaded in Fig. 9.19. To change the order of integration, we have to integrate first w.r.t. y and then w.r.t. x.

i.e. $\displaystyle I = \int \left\{ \int f(x,y)\, dy \right\} dx$

$\therefore$ We have to imagine a vertical strip and to note that : the lower end of vertical strip changes its y-value from circle (i.e. $\sqrt{a^2-x^2}$) to straight line (i.e. $x-a$) across $x = a$.

$\therefore$ We divide the region R into two parts R_1, R_2 by the line AC i.e. $x = a$.

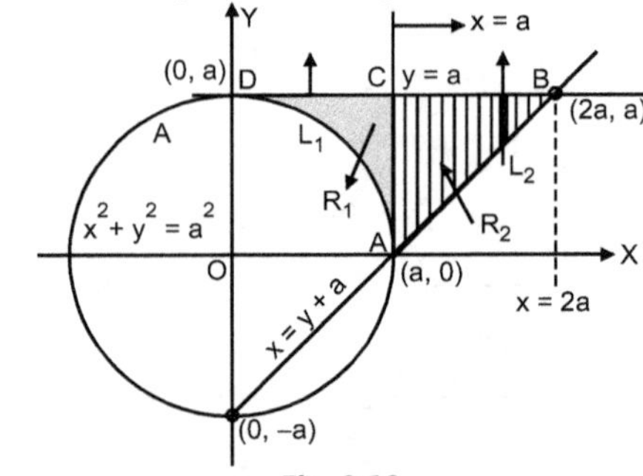

Let R_1 is ACD and R_2 is ABC.

$$\therefore\quad I = \iint\limits_{ABCD} f(x,y)\, dy\, dx = \iint\limits_{ACD} f(x,y)\, dy\, dx + \iint\limits_{ABC} f(x,y)\, dy\, dx$$

Now, to find the new limits for R_1 and R_2, we will imagine two vertical strips in R_1, R_2 respectively. By changing the order of integration, given integral takes the form :

$$\boxed{I = \int\limits_{0}^{a} \left\{ \int\limits_{\sqrt{a^2-x^2}}^{a} f(x,y)\, dy \right\} dx + \int\limits_{a}^{2a} \left\{ \int\limits_{x-a}^{a} f(x,y)\, dy \right\} dx}$$

Ex. 11 : *Change the order of integration in the double integration* $\displaystyle\int\limits_{0}^{5} \int\limits_{2-x}^{2+x} f(x,y)\, dy\, dx.$ **(May 2015)**

Sol. : Let, $\displaystyle I = \int\limits_{0}^{5} \left\{ \int\limits_{2-x}^{2+x} f(x,y)\, dy \right\} dx.$

Region of integration is bounded by $x = 0$, $x = 5$, $y = 2 - x$ i.e. $x + y = 2$; $y = 2 + x$. Points of intersection are $(0, 2)$, $(5, -3)$, $(5, 7)$ and the region R is shaded in Fig. 9.20. To change the order of integration, we have to integrate first w.r.t. x and then w.r.t. y i.e.

$$I = \int \left\{ \int f(x, y)\, dx \right\} dy$$

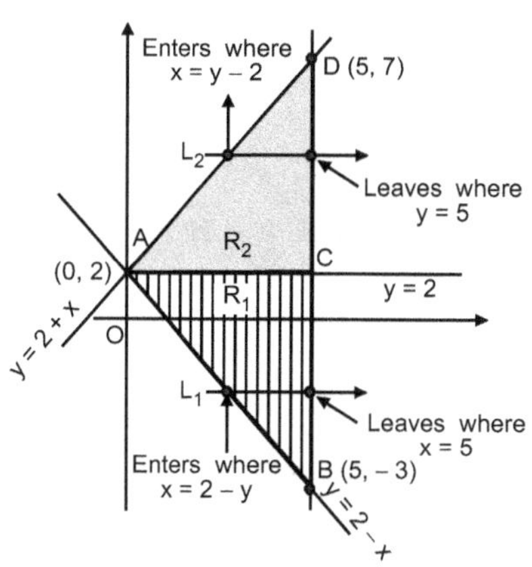

Fig. 9.20

$\therefore$ We have to imagine a horizontal strip and also we note that at the point A$(0, 2)$, the left end of horizontal strip changes its x-value from the straight line $x = 2 - y$ to $x = y - 2$. Therefore, we divide the region into two parts R_1, R_2 by the line AC i.e. $y = 2$.

Let R_1 is ABC and R_2 is ACD.

$$I = \iint_{ABCD} f(x, y)\, dx\, dy = \iint_{ABC} f(x, y)\, dx\, dy + \iint_{ACD} f(x, y)\, dx\, dy.$$

Now to find the new limits for R_1, R_2, we will imagine two horizontal strips, in regions R_1 and R_2 respectively.

By changing the order of integration, given integral becomes

$$I = \int_{-3}^{2} \left\{ \int_{2-y}^{5} f(x, y)\, dx \right\} dy + \int_{2}^{7} \left\{ \int_{y-2}^{5} f(x, y)\, dx \right\} dy$$

Ex. 12 : *Change the order of integration in the double integral,* $\displaystyle \int_{0}^{a} \int_{-a + \sqrt{a^2 - y^2}}^{a + \sqrt{a^2 - y^2}} f(x, y)\, dx\, dy$

Sol. : Region is bounded by $y = 0$, $y = a$

$$
\begin{array}{lll}
x = -a + \sqrt{a^2 - y^2} & \text{and} & x = a + \sqrt{a^2 - y^2} \\
x + a = \sqrt{a^2 - y^2} & & x - a = \sqrt{a^2 - y^2} \\
(x + a)^2 + (y - 0)^2 = a^2 & & (x - a)^2 + (y - 0)^2 = a^2
\end{array}
$$

Circle : Centre $(-a, 0)$, $r = a$. Circle : Centre $(a, 0)$, $r = a$.

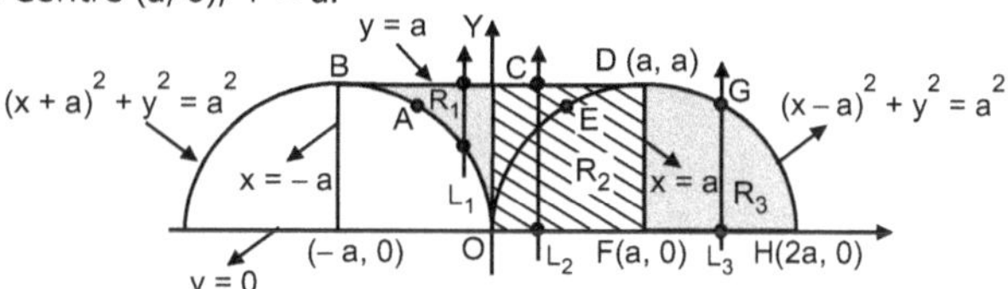

Fig. 9.21

For changing the order of integration, we first integrate w.r.t. y and then, w.r.t. x.

We note that $x = -a + \sqrt{a^2 - y^2}$ represents the arc OAB and $x = a + \sqrt{a^2 - y^2}$ represents the arc DGH.

Hence, the region of integration is OABCDGHFO (Refer Fig. 9.21). We divide the region into three parts by drawing the lines $x = 0$ and $x = a$ (i.e. OC and FD).

$\therefore$ R_1 is OABC ; R_2 is OCDF and R_3 is DGHF.

For R_1 : min. $x = -a$, max. $x = 0$; $y = \sqrt{a^2 - (x + a)^2}$ to $y = a$

For R_2 : $x = 0$ to $x = a$; $y = 0$ to $y = a$

For R_3 : min. $x = a$, max. $x = 2a$; $y = 0$ to $y = \sqrt{a^2 - (x - a)^2}$

$\therefore$

$$I = \iint_{R_1} f(x, y)\, dy\, dx + \iint_{R_2} f(x, y)\, dy\, dx + \iint_{R_3} f(x, y)\, dy\, dx$$

$$\boxed{\; I = \int_{-a}^{0} \left\{ \int_{\sqrt{a^2 - (x + a)^2}}^{a} f(x, y)\, dy \right\} dx + \int_{0}^{a} \left\{ \int_{0}^{a} f(x, y)\, dy \right\} dx + \int_{a}^{2a} \left\{ \int_{0}^{\sqrt{a^2 - (x - a)^2}} f(x, y)\, dy \right\} dx \;}$$

Ex. 13 : *Combine into a single term integral* $I = \int\limits_0^a \int\limits_0^y f(x, y)\, dy\, dx + \int\limits_a^\infty \int\limits_0^{a^2/y} f(x, y)\, dy\, dx.$

Sol. : Region of integration for first integral on R.H.S. is bounded by the lines x = 0, x = y, y = 0, y = a it is region A_1 shown shaded in Fig. 9.22.

Region for second integral on R.H.S. is bounded by $x = 0$, $x = \dfrac{a^2}{y}$ i.e. $xy = a^2$

y = a and y = ∞. It is region A_2 extending upto infinity in vertical direction as shown in the figure.

In the combined region A_1, A_2 if we consider a vertical line, its lower end lies on the line y = x and upper end touches hyperbola $xy = a^2$. Thus, along the vertical strip, y varies from y = x to y = $\dfrac{a^2}{x}$.

If this strip is moved horizontally to cover the whole region, x shows variation from x = 0 to x = a. Thus, by changing the order of integration, the sum of two given integrals can be combined into a single integral as,

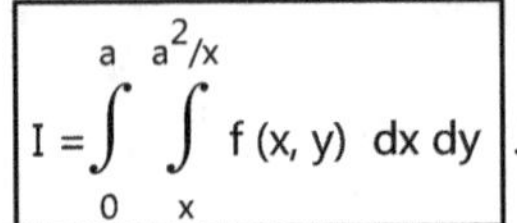

$$I = \int\limits_0^a \int\limits_x^{a^2/x} f(x, y)\ dx\, dy\ .$$

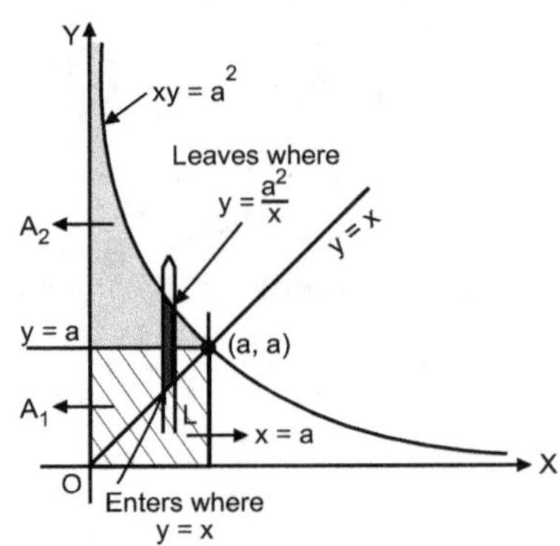

Fig. 9.22

EXERCISE 9.3

I.   Problems on evaluation by using change of order of integration.

Prove that (Ex. 1 to 11),

1. $\displaystyle\int\limits_0^1 \int\limits_0^{\sqrt{1-x^2}} \frac{y\, dy\, dx}{(1 + y^2)\, \sqrt{1 - x^2 - y^2}} = \frac{\pi}{4}\ \log 2$

2. $\displaystyle\int\limits_0^1 \int\limits_0^{\sqrt{1-y^2}} \frac{\cos^{-1} x\, dx\, dy}{\sqrt{(1 - x^2 - y^2)\,(1 - x^2)}} = \frac{\pi^3}{16}$

(Nov./Dec. 2019, May 11, Dec. 17)

3. $\displaystyle\int\limits_0^\infty \int\limits_0^x x\, e^{-x^2/y}\, dy\, dx = \frac{1}{2}$ **(Dec. 2016)**

4. $\displaystyle\int\limits_0^1 \int\limits_{4y}^4 e^{x^2}\, dx\, dy = \frac{e^{16} - 1}{8}$

5. $\displaystyle\int\limits_0^\infty \int\limits_y^\infty \frac{e^{-x}}{x}\, dx\, dy = 1$

6. $\displaystyle\int\limits_0^a \int\limits_0^x x\, \sqrt{(a^2 - y^2)\,(x^2 - y^2)}\ dy\, dx = \frac{8a^5}{45}$

7. $\displaystyle\int\limits_0^\infty \int\limits_x^\infty \frac{e^{-y}}{y}\, dx\, dy = 1$ **(Dec. 2007, Nov. 2014)**

8. $\displaystyle\int\limits_0^\infty dx \int\limits_0^1 \frac{dy}{1 + yx^2} = \pi$

9. $\displaystyle\int\limits_0^a \int\limits_0^x \frac{\tan^{-1}\frac{y}{a}\, dy\, dx}{(a^2 + y^2)\, \sqrt{(a - x)\,(x - y)}} = \frac{\pi^3}{32\, a}$

10. $\displaystyle\int\limits_0^a \int\limits_0^x \frac{\sin y\, dy\, dx}{(5 \cos y - 4)\, \sqrt{(a - x)\,(x - y)}} = -\frac{\pi}{5}\ \log (5 \cos a - 4)$

11. $\displaystyle\int\limits_0^a \int\limits_0^x \frac{dy\, dx}{(y + a)\, \sqrt{(a - x)\,(x - y)}} = \pi \log 2$

12. $\displaystyle\int\limits_0^a \int\limits_0^y \frac{dx\, dy}{\sqrt{(a^2 + x^2)\,(a - y)\,(y - x)}} = \pi \log (1 + \sqrt{2}\,)$

13. Change the order of integration and evaluate $\displaystyle\int_0^2 \int_{2-\sqrt{4-y^2}}^{2+\sqrt{4-y^2}} dx\ dy$ **Hint :** By changing the order, $I = \displaystyle\int_0^4 \int_0^{\sqrt{4-(x-2)^2}} dx \cdot dy$ **Ans. :** 2π

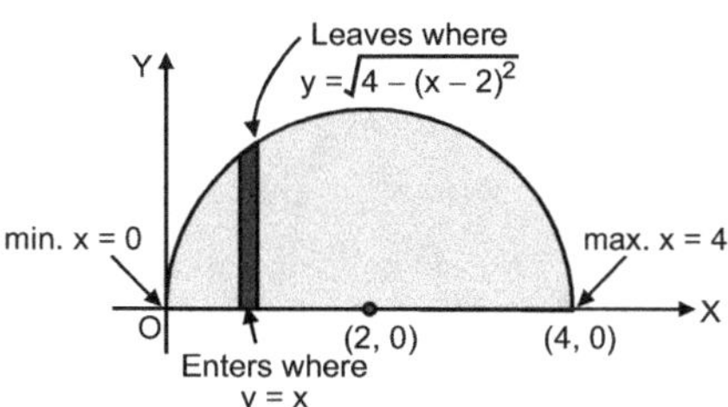

Fig. 9.23

II. Problems on Change of Order of Integration.

Change the order of integration of following double integrals :

1. $\displaystyle\int_1^2 \int_y^{y^2} f(x, y)\ dx\ dy$

2. $\displaystyle\int_0^1 \int_x^{\frac{1}{x}} f(x, y)\ dy\ dx$

3. $\displaystyle\int_{-a}^a \int_0^{\frac{y^2}{a}} f(x, y)\ dx\ dy$

4. $\displaystyle\int_0^{a/\sqrt{2}} \int_y^{\sqrt{a^2-y^2}} f(x, y)\ dx\ dy$

5. $\displaystyle\int_1^2 \int_0^{\frac{a^2}{x}} f(x, y)\ dy\ dx$

6. $\displaystyle\int_0^{a\cos\alpha} \int_{x\tan\alpha}^{\sqrt{a^2-x^2}} f(x, y)\ dy\ dx$

7. $\displaystyle\int_0^{4a} \int_{x^2/4a}^{2\sqrt{ax}} f(x, y)\ dy\ dx$

8. $\displaystyle\int_0^1 \int_{x^2}^{\sqrt{2-x^2}} f(x, y)\ dy\ dx$

9. $\displaystyle\int_{-1}^2 \int_{x^2}^{x+2} f(x, y)\ dy\ dx$

10. $\displaystyle\int_0^a \int_{\frac{x^2}{a}}^{2a-x} f(x, y)\ dy\ dx$

11. $\displaystyle\int_{-2}^1 \int_{x^2}^{2-x} f(x, y)\ dy\ dx$

12. $\displaystyle\int_0^3 \int_{\frac{y^2}{9}}^{\sqrt{10-y^2}} f(x, y)\ dx\ dy$

13. $\displaystyle\int_0^a \int_{-a+\sqrt{a^2-x^2}}^{a+\sqrt{a^2-x^2}} f(x, y)\ dy\ dx.$

14. $\displaystyle\int_0^{2a} \int_{\sqrt{2ax-x^2}}^{\sqrt{2ax}} f(x, y)\ dy\ dx$

15. $\displaystyle\int_1^2 \int_{1-\sqrt{2x-x^2}}^{1+\sqrt{2x-x^2}} f(x, y)\ dy\ dx$

16. $\displaystyle\int_{-a/2}^0 \int_{\sqrt{-2ax-x^2}}^{\sqrt{a^2-x^2}} f(x, y)\ dy\ dx$

17. $\displaystyle\int_0^a \int_{\sqrt{2ay-y^2}}^{a+\sqrt{a^2-y^2}} f(x, y)\ dx\ dy$

18. $\displaystyle\int_0^1 \int_{1-\sqrt{1-y}}^{1+\sqrt{1-y}} f(x, y)\ dx\ dy$

19. $\displaystyle\int_{-2a}^0 \int_{2a-\sqrt{4a^2-y^2}}^{a+\frac{y^2}{4a}} f(x, y)\ dy\ dx$

20. $\displaystyle\int_0^b \int_{a-\frac{a}{b}\sqrt{b^2-y^2}}^{a\frac{y^2}{b^2}} f(x, y)\ dx\ dy$

21. $\displaystyle\int_0^a \int_{\sqrt{a^2-x^2}}^{x+2a} f(x, y)\ dy\ dx$

22. $\displaystyle\int_0^a \int_{mx}^{nx} f(x, y)\ dy\ dx,\ n > m$

III. Express the following Double Integrals as Single Term Integrals.

1. $\displaystyle\int_0^1 \int_{-\sqrt{y}}^{\sqrt{y}} dx\, dy + \int_1^2 \int_{-1}^1 dx\, dy$ and evaluate.

   Ans. : $\dfrac{10}{3}$

2. $\displaystyle\int_0^1 \int_0^y (x^2 + y^2)\, dx\, dy + \int_1^2 \int_0^{2-y} (x^2 + y^2)\, dx\, dy$ and evaluate.

   (Dec. 10, 08, 05) **Ans. :** $\dfrac{4}{3}$

3. $\displaystyle\int_{-3}^2 \int_{2-y}^5 f(x, y)\, dx\, dy + \int_2^7 \int_{y-2}^5 f(x, y)\, dx\, dy$

   Ans. : $\displaystyle\int_0^5 \int_{2-x}^{2+x} f(x, y)\, dy\, dx.$

4. $\displaystyle\int_1^2 \int_{\sqrt{x}}^x f(x, y)\, dy\, dx + \int_2^4 \int_{\sqrt{x}}^2 f(x, y)\, dy\, dx.$

   Ans. : $\displaystyle\int_1^2 \int_y^{y^2} f(x, y)\, dx\, dy$

5. $\displaystyle\int_0^1 \int_0^y f(x, y)\, dx\, dy + \int_1^\infty \int_0^{\frac{1}{y}} f(x, y)\, dx\, dy$

   Ans. : $\displaystyle\int_0^1 \int_x^{\frac{1}{x}} f(x, y)\, dy\, dx$

9.5 TRANSFORMATION OF CARTESIAN DOUBLE INTEGRAL INTO POLAR DOUBLE INTEGRAL

Consider the area OAB bounded by polar curve $r = f(\theta)$, and lines $\theta = \alpha$ and $\theta = \beta$ as shown in Fig. 9.24.

Let $P(r, \theta)$, $Q(r + \delta r, \theta + \delta\theta)$ be two adjacent points on curve AB. Join OP and OQ. Draw two consecutive arcs of circle of radii r and $r + \delta r$ cutting the elementary area GHUV. Arc GV $= r\,\delta\theta$, GH $= \delta r$.

$$\delta A = A_{GHUV} = r\,\delta\theta\,\delta r$$

Considering summation along OP (along direction of r), we get an elementary sector area OPQ known as wedge.

$$A_{OPQ} = \lim_{\delta r \to 0} \left\{ \sum_{r=0}^{f(\theta)} r\,\delta r \right\}.\,\delta\theta = \left\{ \int_0^{f(\theta)} r\,dr \right\}.\,\delta\theta \qquad \text{... [Replacing summation sign by integral sign]}$$

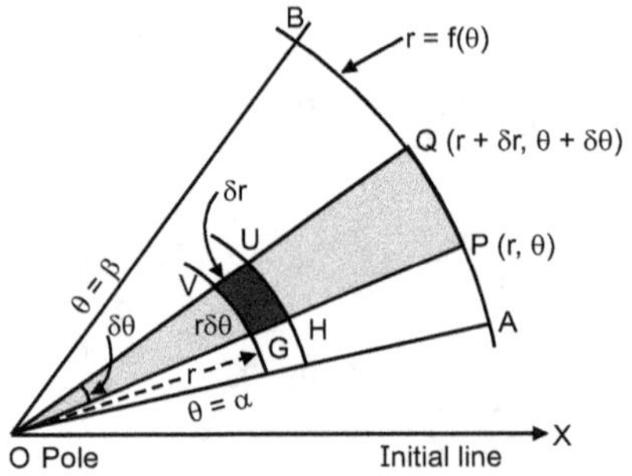

Fig. 9.24

Next we can consider area OAB as consisting of infinite number of elementary areas like OPQ. Considering the summation of all such elementary areas, we get :

$$A_{OAB} = \lim_{\delta\theta \to 0} \sum_{\theta=\alpha}^{\beta} \left\{ \int_0^{f(\theta)} r\,dr \right\}.\,\delta\theta \qquad \text{... [Summation by variation of } \theta\text{]}$$

$$= \int_{\theta=\alpha}^{\beta} d\theta \int_0^{f(\theta)} r\,dr \qquad \text{...} \begin{bmatrix} \text{Replacing summation sign} \\ \text{by integral sign as before} \end{bmatrix}$$

Or

$$= \int_\alpha^\beta \int_0^{f(\theta)} r\, d\theta\, dr$$

Here the inner integration is w.r.t. r, treating θ as constant and outer integration is w.r.t. θ. Order of integration can also be changed by considering summation in the direction of θ increasing first and then w.r.t. r. In most of the problems on polar double integration, *it is quite convenient to integrate w.r.t. r first and then w.r.t. θ.*

Just as $\displaystyle\int_{\alpha}^{\beta}\int_{0}^{f(\theta)} r\, d\theta\, dr$ represents the area OAB, $\displaystyle\int_{\alpha}^{\beta}\int_{0}^{f(\theta)} F(r, \theta)\, r\, d\theta\, dr$ represents some physical quantity related to the area OAB.

If F (r, θ) is density function,

$\displaystyle\int_{\alpha}^{\beta}\int_{0}^{f(\theta)} F(r, \theta)\, r\, d\theta\, dr$ represents mass of lamina OAB. Similarly, it can represent any other physical quantity depending upon nature of F (r, θ).

In many cases, it is quite advantageous to convert $\displaystyle\int\int f(x, y)\, dx\, dy$ into polar form by transformations $x = r\cos\theta$, $y = r\sin\theta$.

Function f (x, y) gets converted to F(r, θ), equations of bounding cartesian curves are expressed in polar form. The area element dx dy is replaced by corresponding area element r dθ dr.

[One may recall Jacobians from MI paper.

$$dx\, dy = |J|\, dr\, d\theta = \left|\frac{\partial (x, y)}{\partial (r, \theta)}\right|\, dr\, d\theta = r\, dr\, d\theta, \text{ where } x = r\cos\theta, \ y = r\sin\theta]$$

Limits in polar double integral $\displaystyle\int\int F(r, \theta)\, r\, d\theta\, dr$ are substituted as explained in following examples :

1. For a complete circle $x^2 + y^2 = a^2$

$$I = \int_{0}^{2\pi}\left\{\int_{0}^{a} f(r, \theta)\, r\, dr\right\} d\theta$$

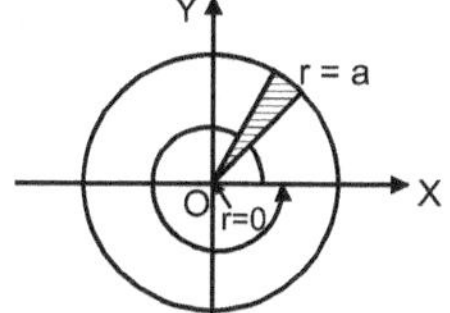

Fig. 9.25 (a)

For a semi-circle $x^2 + y^2 = a^2,\ y \geq 0$

$$I = \int_{0}^{\pi}\left\{\int_{0}^{a} f(r, \theta)\, r\, dr\right\} d\theta$$

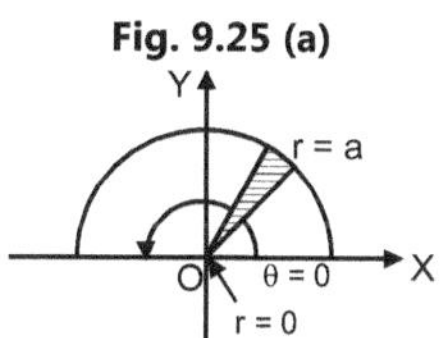

Fig. 9.25 (b)

For a positive quadrant of a circle

$$I = \int_{0}^{\pi/2}\left\{\int_{0}^{a} f(r, \theta)\, r\, dr\right\} d\theta$$

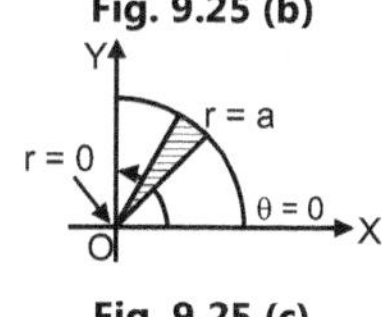

Fig. 9.25 (c)

Note : In all these three cases, polar equation of circle $x^2 + y^2 = a^2$ becomes r = a.

2. For a circle $x^2 + y^2 = 2ax$ i.e. $x^2 - 2ax + a^2 + y^2 = a^2$

    OR $(x - a)^2 + (y - 0)^2 = a^2$ having centre (a, 0), r = a

We write polar equation for $x^2 + y^2 = 2ax$ as $r^2 = 2ar\cos\theta$ or $r = 2a\cos\theta$

$$I = \int_{-\pi/2}^{\pi/2}\left\{\int_{0}^{2a\cos\theta} f(r, \theta)\, r\, dr\right\} d\theta$$

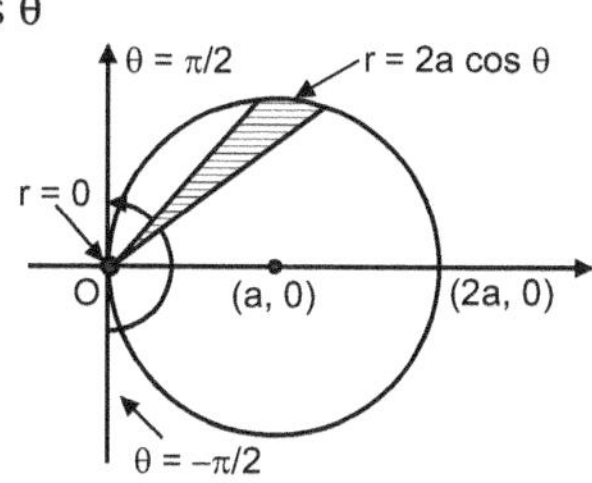

Fig. 9.26 (a)

For upper half of the circle

$x^2 + y^2 = 2ax$

$$I = \int_0^{\pi/2} \left\{ \int_0^{2a\cos\theta} f(r, \theta)\, r\, dr \right\} d\theta$$

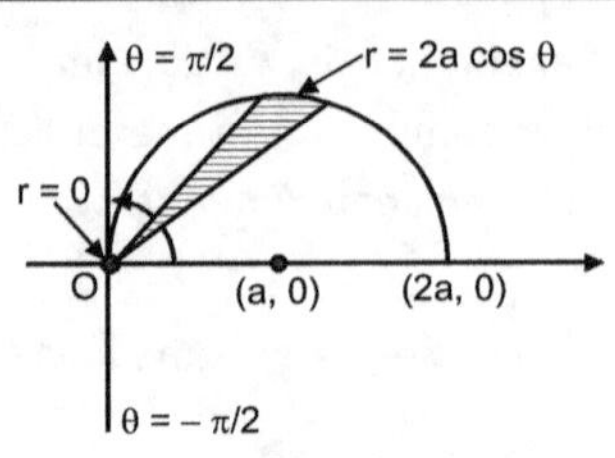

Fig. 9.26 (b)

3. For a circle $x^2 + y^2 = 2ay$ i.e. $x^2 + y^2 - 2ay + a^2 = a^2$

    i.e. $(x - 0)^2 + (y - a)^2 = a^2$ having centre $(0, a)$, $r = a$

    Polar equation of $x^2 + y^2 = 2ay$ is $r = 2a\sin\theta$

    $$I = \int_0^{\pi} \left\{ \int_0^{2a\sin\theta} f(r, \theta)\, r\, dr \right\} d\theta$$

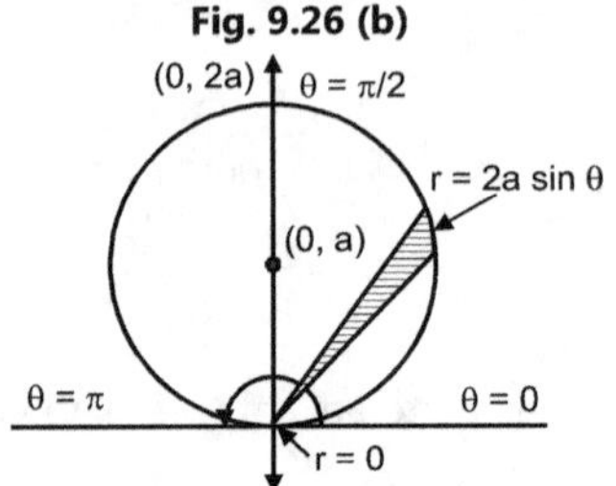

Fig. 9.27

4. For a cardioide $r = a(1 + \cos\theta)$

    $$I = \int_0^{2\pi} \left\{ \int_0^{a(1+\cos\theta)} f(r, \theta)\, r\, dr \right\} d\theta$$

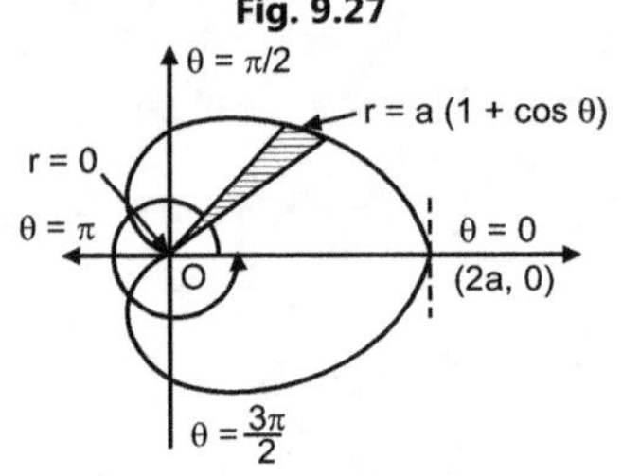

Fig. 9.28 (a)

For a cardioide $r = a(1 - \cos\theta)$

$$I = \int_0^{2\pi} \left\{ \int_0^{a(1-\cos\theta)} f(r, \theta)\, r\, dr \right\} d\theta$$

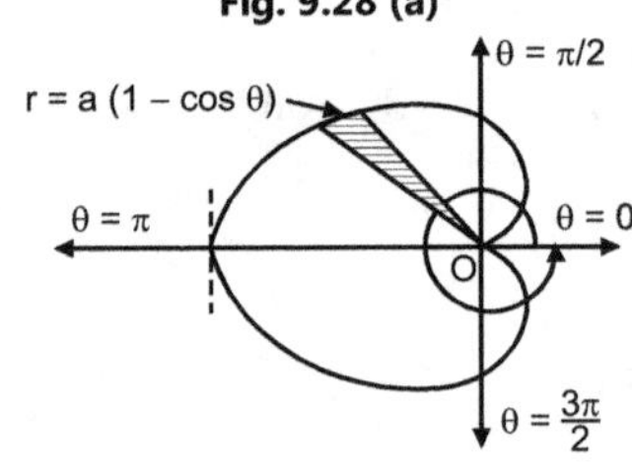

Fig. 9.28 (b)

5. For $r^2 = a^2 \cos 2\theta$ (Bernoullie's Lemniscate)

    $$I = 2 \int_{-\pi/4}^{\pi/4} \left\{ \int_0^{a\sqrt{\cos 2\theta}} f(r, \theta)\, r\, dr \right\} d\theta$$

    (for whole curve i.e. two loops)

    $$I = \int_{-\pi/4}^{\pi/4} \left\{ \int_0^{a\sqrt{\cos 2\theta}} f(r, \theta)\, r\, dr \right\} d\theta \quad \text{(for one loop)}$$

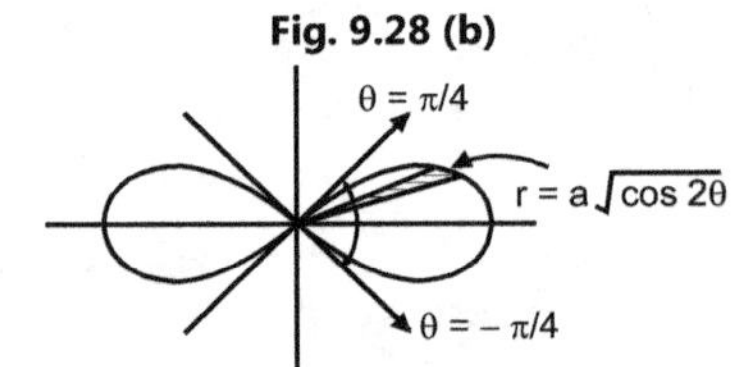

Fig. 9.29

6. For a ellipse $\dfrac{x^2}{a^2} + \dfrac{y^2}{b^2} = 1$; Use $x = ar\cos\theta$, $y = br\sin\theta$, $dx\, dy = ab\, r\, dr\, d\theta$

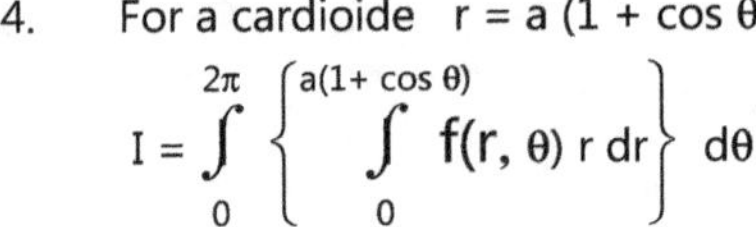
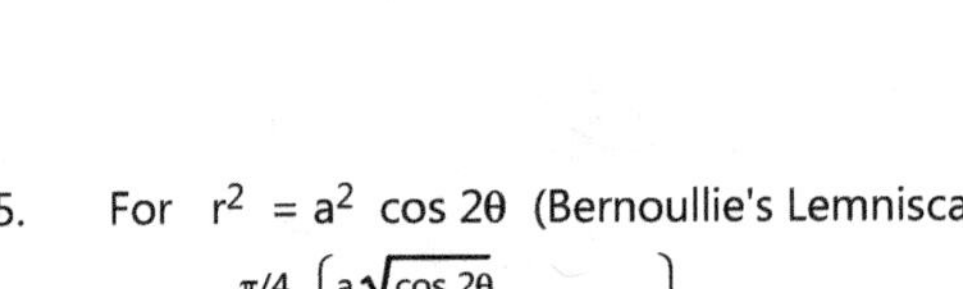
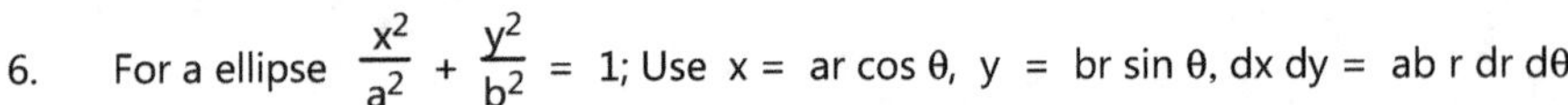
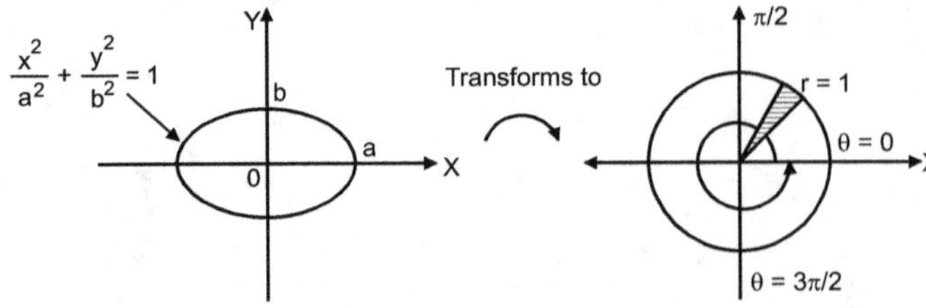

Fig. 9.30 (a)

$$I = \int_0^{2\pi} \left\{ \int_0^1 f(r, \theta)\, ab \;\; r \; dr \right\} d\theta \quad \text{(for complete ellipse)}$$

For a positive quadrant of an ellipse,

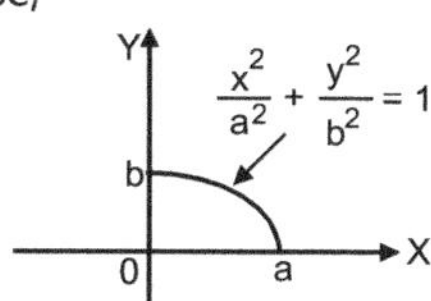
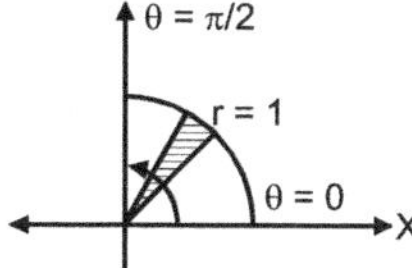

Fig. 9.30 (b)

$$I = \int_0^{\pi/2} \left\{ \int_0^1 f(r, \theta)\, ab \;\; r \; dr \right\} d\theta$$

7. For a triangle $y = 0$, $x = a$, $x = y$

$$I = \int_0^{\pi/4} \left\{ \int_0^{a \sec \theta} f(r, \theta)\, r \; dr \right\} d\theta$$

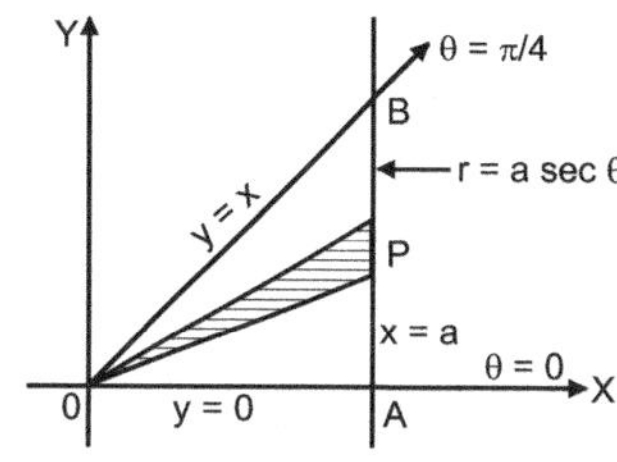

Fig. 9.31

Problems on double integration by transforming to polar form are mainly divided in the following types.

Type I : Direct Evaluation.

Type II : Evaluation by Finding Limits of Integration.

Type I : Illustrations on Direct Evaluation of Double Integral by using Polar Co-ordinates.

Ex. 1 : *Evaluate* $\displaystyle \int_0^a \int_0^{\sqrt{a^2 - x^2}} \sin\left\{ \frac{\pi}{a^2} (a^2 - x^2 - y^2) \right\} dx\, dy.$ *(Nov. 2015)*

Sol. : Region of integration is bounded by $y = 0$, $y = \sqrt{a^2 - x^2}$.
Or $y^2 = a^2 - x^2$ i.e. $x^2 + y^2 = a^2$, $x = 0$, $x = a$ i.e. positive quadrant of circle OAB as shaded in the adjoining figure. Equation of circle in polar coordinates is $r = a$. Along radius vector OP, r varies between 0 to a and when this radius OP is rotated to cover the whole region, θ varies between $\theta = 0$ to $\theta = \dfrac{\pi}{2}$.

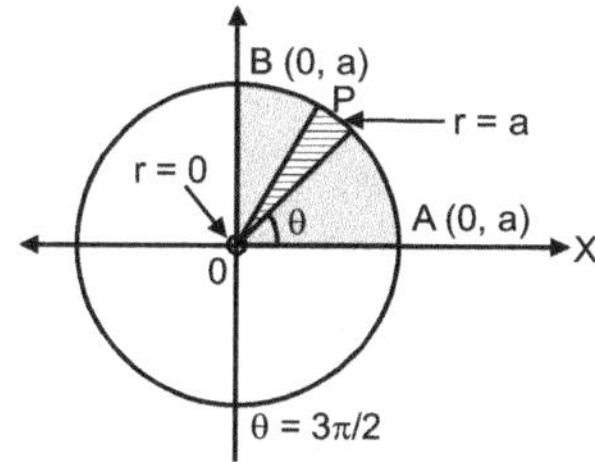

Fig. 9.32

$$I = \int_{\theta = 0}^{\pi/2} \int_{r = 0}^{a} \sin\left\{ \frac{\pi}{a^2} (a^2 - r^2) \right\} r\, d\theta\, dr$$

$$= \frac{1}{2} \int_0^{\pi/2} d\theta \int_0^{a} \sin\left\{ \frac{\pi}{a^2} (a^2 - r^2) \right\} d\,(r^2) \quad [2r\, dr = d(r^2)]$$

$$= \frac{1}{2} \int_0^{\pi/2} \left[\frac{-\cos\left\{ \frac{\pi}{a^2} (a^2 - r^2) \right\}}{-\frac{\pi}{a^2}} \right]_0^{a} d\theta = \frac{1}{2} \int_0^{\pi/2} \frac{a^2}{\pi} \{\cos 0 - \cos \pi\}\, d\theta$$

$$= \frac{1}{2} \frac{a^2}{\pi} \int_0^{\pi/2} \{1 - (-1)\}\, d\theta = \frac{a^2}{\pi} [\theta]_0^{\pi/2} = \frac{a^2}{\pi} \cdot \frac{\pi}{2}$$

$$\boxed{I = \frac{a^2}{2}}$$

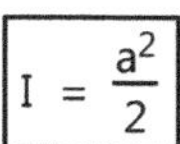

Ex. 2 : *Evaluate* $\displaystyle\int_0^a \int_{\sqrt{ax-x^2}}^{\sqrt{a^2-x^2}} \frac{xy}{x^2+y^2}\, e^{-(x^2+y^2)}\, dx\, dy.$ *(Dec. 2005)*

Sol. : Region of integration is bounded by

$$y = \sqrt{ax-x^2} \quad \text{i.e.} \quad y^2 = ax - x^2$$

Or $\qquad x^2 + y^2 = ax \quad \text{Or} \quad \left(x - \dfrac{a}{2}\right)^2 + y^2 = \dfrac{a^2}{4}.$ [Polar equation $r = a\cos\theta$]

It is a circle with centre $\left(\dfrac{a}{2},\ 0\right)$ and radius $= \dfrac{a}{2}$

$$y = \sqrt{a^2-x^2} \quad \text{Or} \quad y^2 = a^2 - x^2 \quad \text{i.e.} \quad x^2 + y^2 = a^2$$

[$r = a$], $x = 0$, $x = a$. Region is shown shaded in Fig. 9.33.

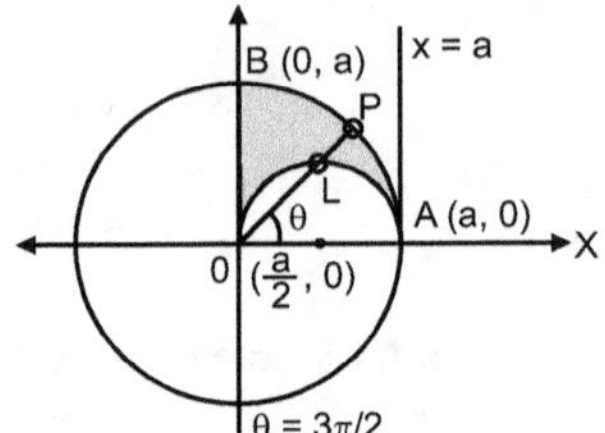

Fig. 9.33

To obtain limits, consider a radius vector OLP passing through the region, the end limits of which in the region are, L : r = a cos θ to P : r = a. For the entire region, θ varies from 0 to π/2. Substituting x = r cos θ, y = r sin θ, given integral takes the form

$$I = \int_{\theta=0}^{\pi/2} \int_{r=a\cos\theta}^{a} \frac{r\cos\theta\ r\sin\theta}{r^2}\, e^{-r^2} \cdot r\, d\theta\, dr$$

$$= \int_0^{\pi/2} \int_{a\cos\theta}^{a} \sin\theta\cos\theta\ e^{-r^2} r\, dr\, d\theta = \frac{1}{2}\int_0^{\pi/2} \int_{a\cos\theta}^{a} \sin\theta\cos\theta\, e^{-r^2} d(r^2)\, d\theta$$

$$= \frac{1}{2}\int_0^{\pi/2} \sin\theta\cos\theta \left[\frac{e^{-r^2}}{-1}\right]_{a\cos\theta}^{a} d\theta$$

$$= -\frac{1}{2}\int_0^{\pi/2} \sin\theta\ \cos\theta\ (e^{-a^2})\, d\theta + \frac{1}{2}\int_0^{\pi/2} \sin\theta\cos\theta\ e^{-a^2\cos^2\theta}\, d\theta = I_1 + I_2$$

$$I_1 = -\frac{1}{2}\int_0^{\pi/2} \frac{\sin 2\theta}{2}\, e^{-a^2}\, d\theta = -\frac{1}{2}\, e^{-a^2}\left(-\frac{\cos 2\theta}{4}\right)_0^{\pi/2} = \frac{1}{8}\, e^{-a^2}[\cos\pi - \cos 0] = -\frac{1}{4}\, e^{-a^2}$$

$$I_2 = \frac{1}{2}\int_0^{\pi/2} e^{-a^2\cos^2\theta}\, \sin\theta\cos\theta\, d\theta \qquad\qquad \text{Put} -\cos^2\theta = t$$
$$\therefore\ 2\cos\theta\sin\theta\ d\theta = dt$$

$$= \frac{1}{4}\int_{-1}^{0} e^{a^2 t}\, dt = \frac{1}{4}\left[\frac{e^{a^2 t}}{a^2}\right]_{-1}^{0} \qquad\qquad \text{when } \theta = 0,\ t = -1\,;\ \theta = \frac{\pi}{2},\ t = 0$$

$$= \frac{1}{4a^2} - \frac{1}{4a^2}\, e^{-a^2}$$

$$I = I_1 + I_2 = -\frac{1}{4}\, e^{-a^2} + \frac{1}{4a^2} - \frac{1}{4a^2}\, e^{-a^2}$$

$$\boxed{I = \frac{1}{4a^2}\left[1 - (a^2 + 1)\, e^{-a^2}\right]}$$

Ex. 3 : *Evaluate* $\displaystyle\int_0^a \int_{2\sqrt{ax}}^{\sqrt{5ax-x^2}} \frac{\sqrt{x^2+y^2}}{y^2}\, dx\, dy.$

Sol. : Region is bounded by $y = 2\sqrt{ax}$ i.e. $y^2 = 4ax$, a parabola whose polar equation is;

$$r^2\sin^2\theta = 4ar\cos\theta \quad \text{i.e.} \quad r = \frac{4a\cos\theta}{\sin^2\theta}$$

Circle $y = \sqrt{5ax - x^2}$ i.e. $y^2 = 5ax - x^2$

or $\qquad x^2 + y^2 = 5ax \quad$ or $\quad \left(x - \dfrac{5a}{2}\right)^2 + y^2 = \dfrac{25\,a^2}{4}$

Polar equation of circle is $r^2 = 5ar \cos\theta$ or $r = 5a \cos\theta$. Points of intersection of parabola and circle are given by solving $y^2 = 4ax$ and $y^2 = 5ax - x^2$.

$\therefore\ 4ax = 5ax - x^2$ that gives $x = 0$ and $x = a$. $x = 0$ gives $y = 0$, $x = a$ gives $y = 2a$.

Thus circle and parabola intersect at the points $(0, 0)$ and $(a, 2a)$.

Lines $x = 0$ and $x = a$ are the other boundaries. Region is shown shaded in Fig. 9.34 (a). Considering a radius vector in the region, r varies from $\dfrac{4\,a\cos\theta}{\sin^2\theta}$ to $r = 5a\cos\theta$ and θ varies from

$\tan^{-1} 2$ (Line OA makes an angle $\tan^{-1} 2$ with x-axis) to $\dfrac{\pi}{2}$.

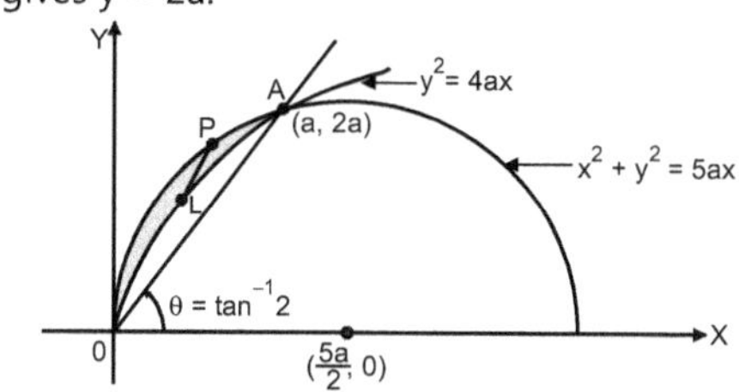

Fig. 9.34 (a)

$$\left(\because\ \tan\theta = \frac{y}{x} = \frac{2a}{a} \text{ along OA, } \therefore \tan\theta = 2 \ \therefore\ \theta = \tan^{-1} 2\right)$$

$$I = \int_{\tan^{-1} 2}^{\pi/2} \int_{\frac{4a\cos\theta}{\sin^2\theta}}^{5a\cos\theta} \frac{r}{r^2 \sin^2\theta}\, r\, dr\, d\theta = \int_{\tan^{-1} 2}^{\pi/2} [r]_{4a\cos\theta/\sin^2\theta}^{5a\cos\theta}\, \frac{d\theta}{\sin^2\theta}$$

$$= \int_{\tan^{-1} 2}^{\pi/2} \left\{5a\cos\theta - \frac{4a\cos\theta}{\sin^2\theta}\right\} \frac{1}{\sin^2\theta}\, d\theta$$

$$= \int_{\tan^{-1} 2}^{\pi/2} [5a\ \text{cosec}\,\theta\ \cot\theta - 4a\ \text{cosec}^2\theta\ \text{cosec}\,\theta\ \cot\theta]\, d\theta$$

$$= \int_{\tan^{-1} 2}^{\pi/2} 5a\ \text{cosec}\ \theta\ \cot\theta\, d\theta + 4a \int_{\tan^{-1} 2}^{\pi/2} \text{cosec}^2\theta\ d\,(\text{cosec}\,\theta)$$

$$= [-5\,a\ \text{cosec}\,\theta]_{\tan^{-1} 2}^{\pi/2} + 4a\left[\frac{\text{cosec}^3\theta}{3}\right]_{\tan^{-1} 2}^{\pi/2}$$

$$= -5a + 5a\frac{\sqrt{5}}{2} + \frac{4a}{3} - \frac{4a}{3}\frac{5\sqrt{5}}{8}$$

$$\left[\text{From the triangle AOB, } \tan\theta = \frac{2}{1},\ \sin\theta = \frac{2}{\sqrt{5}},\ \text{cosec}\,\theta = \frac{\sqrt{5}}{2}\right]$$

$$= \frac{-11\,a}{3} + \frac{5\sqrt{5}\,a}{2}\left(1 - \frac{1}{3}\right)$$

$$= \frac{-11a}{3} + \frac{5\sqrt{5}\,a}{3}$$

$$\boxed{I = \frac{a}{3}\left(5\sqrt{5} - 11\right)}$$

A (1, 2)

Fig. 9.34 (b)

Ex. 4 : *Sketch the area of double integration and evaluate :*

$$\int_{0}^{a\sqrt{2}} \int_{y}^{\sqrt{a^2 - y^2}} \log_e (x^2 + y^2)\ dx\, dy. \qquad\qquad \textbf{(May 2016, Dec. 2018)}$$

Sol. : Region of integration is bounded by $x = y$, $x = \sqrt{a^2 - y^2}$ i.e. $x^2 + y^2 = a^2$ $(r = a)$ and $y = 0$, $y = \dfrac{a}{\sqrt{2}}$. It is shown in Fig. 9.35.

To evaluate the integral, it is simpler to transform to polars. Considering a radius vector in the region, r varies from 0 to a and θ varies from 0 to $\frac{\pi}{4}$.

$$I = \int_{\theta=0}^{\pi/4} \int_{r=0}^{a} \log_e (r^2)\ r\ d\theta\ dr$$

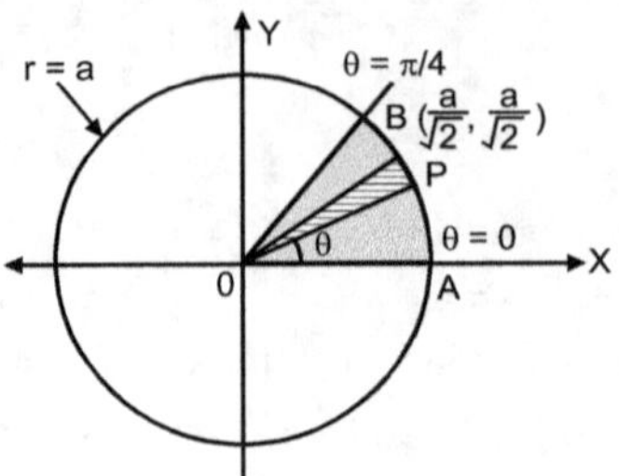

Fig. 9.35

Consider inner integral

$$\int_0^a \log r^2 \cdot r\ dr = 2 \int_0^a r \log r\ dr$$ Integrating by parts,

$$= 2\left[\left\{\log r \cdot \frac{r^2}{2}\right\}_0^a - \int_0^a \frac{1}{r}\cdot\frac{r^2}{2}\ dr\right]$$

$$= a^2\ \log a - \left\{\frac{r^2}{2}\right\}_0^a$$ $$\left[\begin{array}{c}\text{Lim}\\ r\to 0\end{array} \log r \cdot \frac{r^2}{2} = 0\right]$$

$$= a^2\ \log a - \frac{a^2}{2}$$

$$\therefore\qquad I = \int_0^{\pi/4}\left(a^2\ \log a - \frac{a^2}{2}\right) d\theta$$

$$\boxed{I = \frac{\pi}{4}a^2\left(\log a - \frac{1}{2}\right)}$$

Ex. 5 : Show that $\displaystyle\int_0^a \int_0^{\sqrt{a^2-x^2}} e^{-x^2-y^2}\ dx\ dy = \frac{\pi}{4}\left(1 - e^{-a^2}\right)$ **(May 2015)**

Sol. : Transforming to polars,

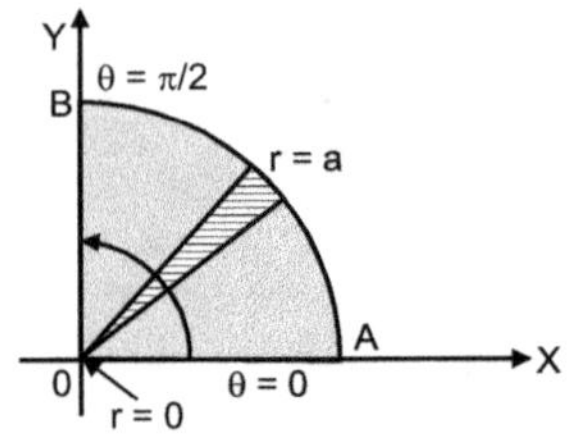

$$I = \int_0^{\pi/2}\int_0^a e^{-r^2} r\ dr\ d\theta = \int_0^{\pi/2} d\theta \left(-\frac{1}{2}\right)\int_0^a e^{-r^2}\ (-2\ r\ dr)$$

$$= -\frac{1}{2}\ (\pi/2)\left[e^{-r^2}\right]_0^a$$

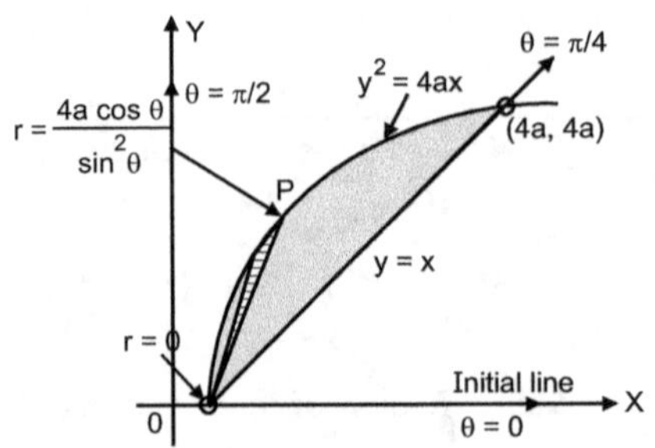

Fig. 9.36

$$\boxed{I = \frac{\pi}{4}[1 - e^{-a^2}]}$$

Ex. 6 : Evaluate $\displaystyle\int_0^{4a}\int_{y^2/4a}^{y} \frac{x^2-y^2}{x^2+y^2}\ dx\ dy.$ **(May 2009)**

Sol. : $$I = \int_0^{4a}\left\{\int_{y^2/4a}^{y} \frac{x^2-y^2}{x^2+y^2}\ dx\right\} dy$$

Fig. 9.37

Since the limits of inner integral are functions of y i.e. $x = \dfrac{y^2}{4a}$ i.e. $y^2 = 4ax$, $x = y$ and the limits of outer integral are $y = 0$, $y = 4a$.

The points of intersection of $y = x$, $y^2 = 4ax$ is $x^2 = 4ax$ i.e. $x(x - 4a) = 0$ i.e. $x = 0$, $x = 4a$. $\therefore$ $(0, 0)$ $(4a, 4a)$ are the points of intersection.

Region of integration is as shown in Fig. 9.37.

It is difficult to evaluate the integral in cartesian co-ordinates but becomes simple in polar co-ordinates.

Put $x = r\cos\theta$, $y = r\sin\theta$, $dx\,dy = r\,dr\,d\theta$, $x^2 + y^2 = r^2$,

$$I = \iint \left(\frac{r^2\cos^2\theta - r^2\sin^2\theta}{r^2}\right) r\,dr\,d\theta$$

$$= \int_{\pi/4}^{\pi/2} (\cos^2\theta - \sin^2\theta)\,d\theta \int_0^{4a\cos\theta/\sin^2\theta} r\,dr$$

$$= \int_{\pi/4}^{\pi/2} (\cos^2\theta - \sin^2\theta)\,d\theta \left[\frac{r^2}{2}\right]_0^{\frac{4a\cos\theta}{\sin^2\theta}}$$

$$= 8a^2 \int_{\pi/4}^{\pi/2} (\cos^2\theta - \sin^2\theta)\frac{\cos^2\theta}{\sin^4\theta}\,d\theta = 8a^2 \int_{\pi/4}^{\pi/2} (\cot^4\theta - \cot^2\theta)\,d\theta$$

$$= 8a^2 \int_{\pi/4}^{\pi/2} \cot^2\theta(\cot^2\theta - 1)\,d\theta = 8a^2 \int_{\pi/4}^{\pi/2} \cot^2\theta[\csc^2\theta - 1 - 1]\,d\theta$$

$$= 8a^2 \int_{\pi/4}^{\pi/2} \cot^2\theta\,\csc^2\theta\,d\theta - 16a^2 \int_{\pi/4}^{\pi/2} (\csc^2\theta - 1)\,d\theta$$

$$= 8a^2 \left[-\frac{\cot^3\theta}{3}\right]_{\pi/4}^{\pi/2} - 16a^2 \left[-\cot\theta - \theta\right]_{\pi/4}^{\pi/2}$$

$$= \frac{8a^2}{3}[0 + 1] + 16a^2\left[0 + \frac{\pi}{2} - 1 - \frac{\pi}{4}\right] = \frac{8a^2}{3} + 16a^2\,[\pi/4 - 1]$$

$$= 8a^2\left(\frac{\pi}{2} - \frac{5}{3}\right)$$

Ex. 7 : *Show that* $\displaystyle\int_0^a \int_y^{a+\sqrt{a^2-y^2}} \frac{dx\,dy}{(4a^2 + x^2 + y^2)^2} = \frac{1}{8a^2}\left(\frac{\pi}{4} - \frac{1}{\sqrt{2}}\tan^{-1}\frac{1}{\sqrt{2}}\right)$

Proof : Let, $\displaystyle I = \int_0^a \left\{\int_y^{a+\sqrt{a^2-y^2}} \frac{dx}{(4a^2 + x^2 + y^2)^2}\right\} dy$

Region of integration is bounded by $y = 0$, $y = a$, $x = y$, $x = a + \sqrt{a^2 - y^2}$

i.e. $x - a = \sqrt{a^2 - y^2}$ OR $(x - a)^2 + (y - 0)^2 = a^2$ OR $x^2 + y^2 = 2ax$.

Point of intersection of circle $x^2 + y^2 = 2ax$ and the line $y = x$ is (a, a). By using polar co-ordinates $x = r\cos\theta$, $y = r\sin\theta$, $x^2 + y^2 = r^2$, $dx\,dy = r\,dr\,d\theta$

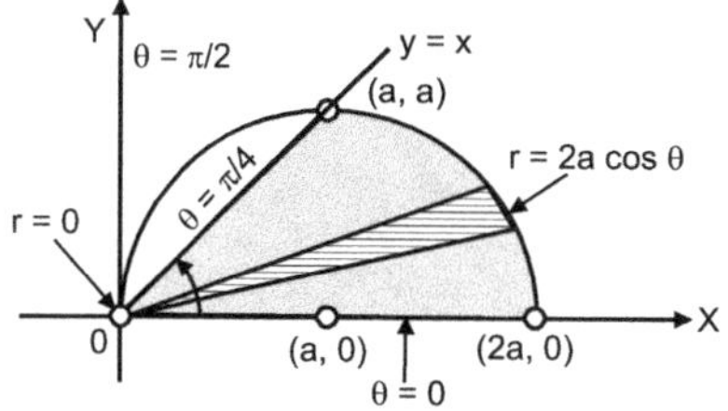

Fig. 9.38

$$I = \int_0^{\pi/4} \int_0^{2a\cos\theta} \frac{r\,dr\,d\theta}{(4a^2 + r^2)^2} = \frac{1}{2} \int_0^{\pi/4} d\theta \int_0^{2a\cos\theta} \frac{2r\,dr}{(4a^2 + r^2)^2}$$

$$= \frac{1}{2} \int_0^{\pi/4} \left[\frac{-1}{4a^2 + r^2} \right]_0^{2a\cos\theta} d\theta = \frac{1}{2} \int_0^{\pi/4} \left(\frac{-d\theta}{4a^2 + 4a^2\cos^2\theta} + \frac{d\theta}{4a^2} \right)$$

$$= \frac{1}{8a^2} \left(\frac{\pi}{4} \right) - \frac{1}{8a^2} \int_0^{\pi/4} \frac{1}{1 + \cos^2\theta}\,d\theta, \qquad \text{divide numerator and denominator by } \cos^2\theta$$

$$= \frac{1}{8a^2} \left(\frac{\pi}{4} \right) - \frac{1}{8a^2} \int_0^{\pi/4} \frac{\sec^2\theta\,d\theta}{\sec^2\theta + 1} = \frac{1}{8a^2} \left(\frac{\pi}{4} \right) - \frac{1}{8a^2} \int_0^{\pi/4} \frac{\sec^2\theta\,d\theta}{2 + \tan^2\theta} \qquad (\text{Put } \tan\theta = t)$$

$$= \frac{1}{8a^2} \left(\frac{\pi}{4} \right) - \frac{1}{8a^2} \int_0^1 \frac{dt}{2 + t^2} = \frac{1}{8a^2} \left(\frac{\pi}{4} \right) - \frac{1}{8a^2} \left[\frac{1}{\sqrt{2}} \tan^{-1} \frac{t}{\sqrt{2}} \right]_0^1$$

$$= \frac{1}{8a^2} \left(\frac{\pi}{4} \right) - \frac{1}{8a^2} \left(\frac{1}{\sqrt{2}} \tan^{-1} \frac{1}{\sqrt{2}} \right)$$

$$\boxed{I = \frac{1}{8a^2} \left[\frac{\pi}{4} - \frac{1}{\sqrt{2}} \tan^{-1} \frac{1}{\sqrt{2}} \right]}$$

Ex. 8 : *Evaluate* $\displaystyle \int_0^2 \int_{1-\sqrt{2x-x^2}}^{1+\sqrt{2x-x^2}} \frac{dx\,dy}{(x^2 + y^2)^2}$

Sol. :
$$I = \int_0^2 \left\{ \int_{1-\sqrt{2x-x^2}}^{1+\sqrt{2x-x^2}} \frac{dy}{(x^2 + y^2)^2} \right\} dx$$

The region of integration R is bounded by $x = 0$, $x = 2$; $y = 1 \pm \sqrt{2x - x^2}$ i.e. $y - 1 = \pm\sqrt{2x - x^2}$ or $(y-1)^2 + x^2 - 2x + 1 = 1$ or $(x-1)^2 + (y-1)^2 = (1)^2$, a circle with centre (1, 1) and $r = 1$. (Refer Fig. 9.39). By using polar co-ordinates $x = r\cos\theta$, $y = r\sin\theta$, $x^2 + y^2 = r^2$, $dx\,dy = r\,dr\,d\theta$, equation of circle becomes $(r\cos\theta - 1)^2 + (r\sin\theta - 1)^2 = 1$

$$r^2 - 2(\sin\theta + \cos\theta)r + 1 = 0$$

$$r = \frac{2(\sin\theta + \cos\theta) \pm \sqrt{4(\sin\theta + \cos\theta)^2 - 4}}{2} = \sin\theta + \cos\theta \pm \sqrt{\sin 2\theta}$$

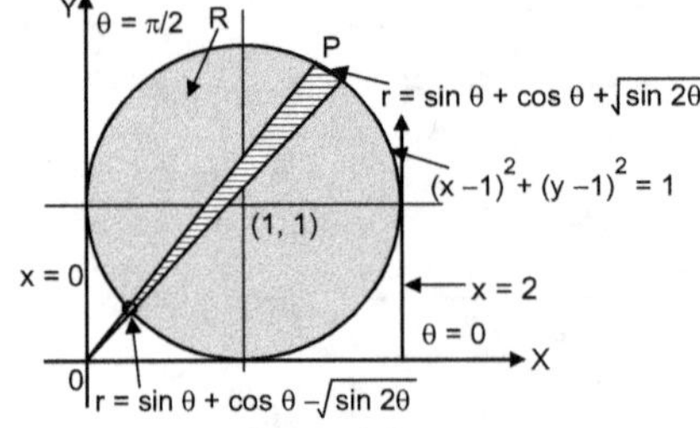

Fig. 9.39

$$I = \iint_R \frac{dx\,dy}{(x^2 + y^2)^2} = \int_0^{2\pi} \int_{\sin\theta + \cos\theta - \sqrt{\sin 2\theta}}^{\sin\theta + \cos\theta + \sqrt{\sin 2\theta}} \frac{r\,dr\,d\theta}{(r^2)^2}$$

$$= \int_0^{\pi/2} d\theta \int_{A-B}^{A+B} \frac{dr}{r^3}, \quad \text{where } A = \sin\theta + \cos\theta,\ B = \sqrt{\sin 2\theta}$$

$$= \int_0^{\pi/2} d\theta \left[-\frac{1}{2r^2}\right]_{A-B}^{A+B} = \frac{1}{2} \int_0^{\pi/2} \left[\frac{1}{(A-B)^2} - \frac{1}{(A+B)^2}\right] d\theta$$

$$= \frac{1}{2} \int_0^{\pi/2} \frac{4\,AB}{(A^2 - B^2)^2}\, d\theta = 2 \int_0^{\pi/2} (\sin\theta + \cos\theta)\, \sqrt{\sin 2\theta}\, d\theta$$

$$= 2\sqrt{2} \int_0^{\pi/2} \sin^{3/2}\theta \, \cos^{\frac{1}{2}}\theta\, d\theta + 2\sqrt{2} \int_0^{\pi/2} \sin^{\frac{1}{2}}\theta \cos^{\frac{3}{2}}\theta\, d\theta$$

$$= 2\sqrt{2}\,(2) \left[\frac{1}{2} \frac{\overline{|5/4}\; \overline{|3/4}}{\overline{|2}}\right] = 2\sqrt{2}\, \frac{1}{4}\, \overline{\left|\frac{1}{4}\right.}\; \overline{\left|\frac{3}{4}\right.}$$

$$= \frac{1}{\sqrt{2}}\, \overline{\left|\frac{1}{4}\right.}\; \overline{\left|1-\frac{1}{4}\right.} = \frac{1}{\sqrt{2}}\, \frac{\pi}{\sin\frac{\pi}{4}} = \frac{1}{\sqrt{2}}\,\left(\sqrt{2}\,\pi\right) = \pi$$

$$\boxed{I = \pi}$$

Type II : Illustrations on Integrals when the limits are not provided : (Use Polar Co-ordinates)

Ex. 9 : *Evaluate* $\displaystyle\iint_A \frac{dx\,dy}{(1 + x^2 + y^2)^2}$ *where A is one loop of the lemniscate* $(x^2 + y^2)^2 = x^2 - y^2$.

Sol. : Equation of the curve in polar form is $r^4 = r^2 (\cos^2\theta - \sin^2\theta)$ or $r^2 = \cos 2\theta$, whose shape is as shown in the figure (one loop is shown shaded). (Refer Fig. 9.40).

Transforming to polars, given integral becomes :

$$I = \int_{-\pi/4}^{\pi/4} \int_0^{\sqrt{\cos 2\theta}} \frac{r\, d\theta\, dr}{(1 + r^2)^2}$$

$$I = \frac{1}{2} \int_{-\pi/4}^{\pi/4} d\theta \int_0^{\sqrt{\cos 2\theta}} \frac{d(r^2)}{(1 + r^2)^2}$$

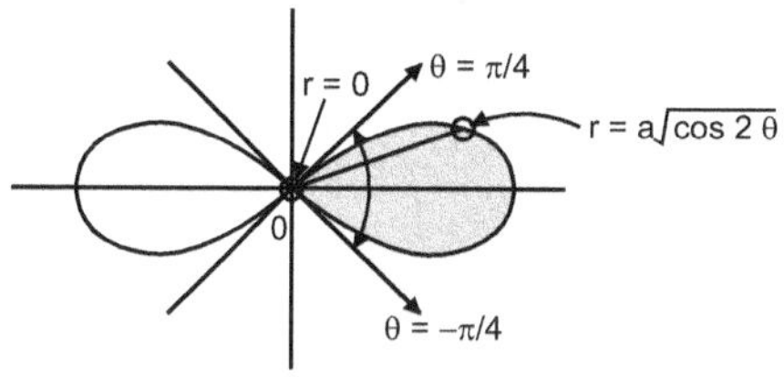

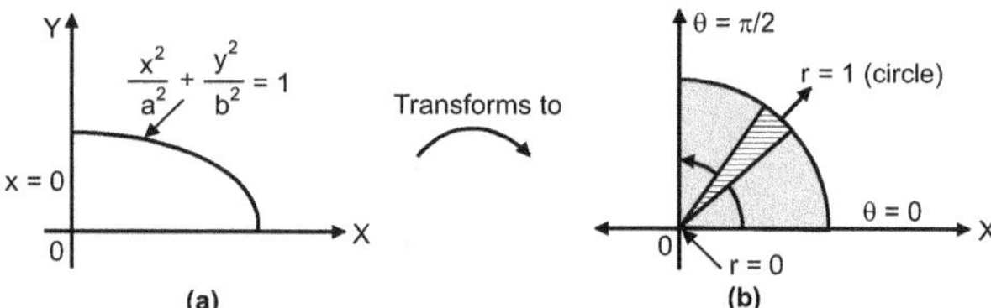

Fig. 9.40

$$= \frac{1}{2} \int_{-\pi/4}^{\pi/4} \left[-\frac{1}{(1 + r^2)}\right]_0^{\sqrt{\cos 2\theta}} d\theta$$

$$= \frac{1}{2} \int_{-\pi/4}^{\pi/4} \left(1 - \frac{1}{1 + \cos 2\theta}\right) d\theta = \frac{1}{2}\, 2 \int_0^{\pi/4} \left[1 - \frac{1}{(1 + \cos 2\theta)}\right] d\theta = \int_0^{\pi/4} \left[1 - \frac{1}{2\cos^2\theta}\right] d\theta$$

$$= \left[\theta - \frac{1}{2} \tan\theta\right]_0^{\pi/4}$$

$$\boxed{I = \frac{\pi}{4} - \frac{1}{2}}$$

Ex. 10 : *Evaluate* $\displaystyle\iint_R xy \left(\frac{x^2}{a^2} + \frac{y^2}{b^2}\right)^{n/2} dx\,dy$, *where R is positive quadrant of* $\dfrac{x^2}{a^2} + \dfrac{y^2}{b^2} = 1$.

Sol. : Given region is positive quadrant of ellipse $\dfrac{x^2}{a^2} + \dfrac{y^2}{b^2} = 1$ [Refer Fig. 9.41 (a)].

Fig. 9.41

By using **elliptical polar co-ordinates** $x = ar \cos \theta$, $y = br \sin \theta$, $dx\, dy = ab\, r\, dr\, d\theta$, $\dfrac{x^2}{a^2} + \dfrac{y^2}{b^2} = r^2$, ellipse gets transformed into

$r^2 = 1$ i.e. $r = 1$, a circle [Refer Fig. 9.41 (b)].

$$I = \iint ar \cos \theta\; br \sin \theta\; (r^2)^{n/2}\; ab\, r\, dr\, d\theta = a^2\, b^2 \int_0^{\pi/2} \cos \theta\; \sin \theta\; d\theta \int_0^1 r^{n+3}\, dr = a^2\, b^2 \left(\frac{1}{2}\right) \left[\frac{r^{n+4}}{n+4}\right]_0^1$$

$$\boxed{I = \frac{a^2\, b^2}{2\,(n+4)}}$$

Ex. 11 : *Evaluate* $\displaystyle\iint_R \frac{(x^2 + y^2)^2}{x^2\, y^2}\, dx\, dy$, *where R is common to* $x^2 + y^2 = ax$, $x^2 + y^2 = by$, $a > b > 0$.

Sol. : $\qquad\qquad I = \displaystyle\iint \frac{(x^2 + y^2)^2}{x^2\, y^2}\, dx\, dy$

We will convert to polar form by using $x = r \cos \theta$, $y = r \sin \theta$, $x^2 + y^2 = r^2$, $dx\, dy = r\, dr\, d\theta$.

The region of integration common to $x^2 + y^2 = ax$, $x^2 + y^2 = by$ is shown shaded in Fig. 9.42. Equation of line OB is $ax = by$.

$\therefore \qquad\qquad \dfrac{y}{x} = \dfrac{a}{b} = \tan \theta \qquad \therefore\; \theta = \tan^{-1} \dfrac{a}{b}$

Observe that an upper end of the wedge in R changes the curve across the line OB. Hence divide R into R_1 and R_2 across OB, as shown. Let OP and OQ be the two radial strips (wedges) in R_1 and R_2 respectively.

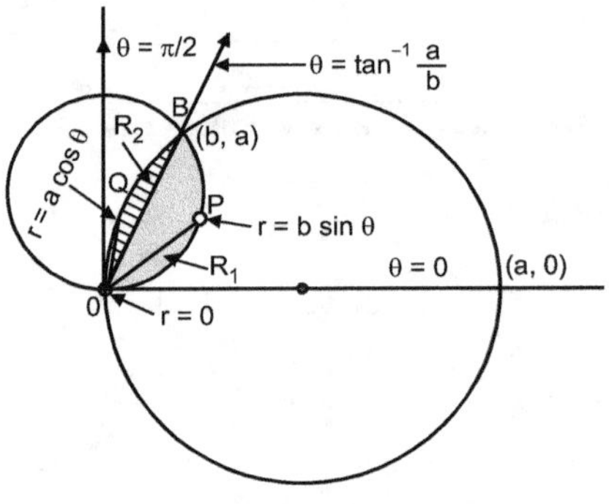

Fig. 9.42

Limits for R_1 $\qquad:\; \theta = 0$ to $\theta = \tan^{-1} \dfrac{a}{b}$; $r = 0$ to $r = b \sin \theta$

Limits for R_2 $\qquad:\; \theta = \tan^{-1} \dfrac{a}{b}$ to $\theta = \dfrac{\pi}{2}$; $r = 0$ to $r = a \cos \theta$

$$I = \iint_{R_1} \frac{(r^2)^2}{r^2 \cos^2 \theta\; r^2 \sin^2 \theta}\, r\, dr\, d\theta + \iint_{R_2} \frac{(r^2)^2}{r^2 \cos^2 \theta\; r^2 \sin^2 \theta}\, r\, dr\, d\theta$$

$$= \int_0^{\tan^{-1}\frac{a}{b}} \frac{1}{\sin^2 \theta \cos^2 \theta}\, d\theta \int_0^{b \sin \theta} r\, dr + \int_{\tan^{-1}\frac{a}{b}}^{\pi/2} \frac{1}{\sin^2 \theta \cos^2 \theta}\, d\theta \int_0^{a \cos \theta} r\, dr$$

$$= \int_0^{\tan^{-1}\frac{a}{b}} \frac{d\theta}{\sin^2 \theta \cos^2 \theta} \left(\frac{b^2 \sin^2 \theta}{2}\right) + \int_{\tan^{-1}\frac{a}{b}}^{\pi/2} \frac{d\theta}{\sin^2 \theta \cos^2 \theta} \left(\frac{a^2 \cos^2 \theta}{2}\right)$$

$$= \frac{b^2}{2} \int_0^{\tan^{-1}\frac{a}{b}} \sec^2 \theta\; d\theta + \frac{a^2}{2} \int_{\tan^{-1}\frac{a}{b}}^{\pi/2} \operatorname{cosec}^2 \theta\; d\theta$$

$$= \frac{b^2}{2} \left[\tan \theta\right]_0^{\tan^{-1}\frac{a}{b}} + \frac{a^2}{2} \left[-\cot \theta\right]_{\tan^{-1}\frac{a}{b}}^{\pi/2}$$

$$= \frac{b^2}{2}\left(\frac{a}{b}\right) + \frac{a^2}{2}\left[-\frac{1}{\tan\theta}\right]_{\tan^{-1}\frac{a}{b}}^{\pi/2}$$

$$= \frac{ab}{2} - \frac{a^2}{2}\left[0 - \frac{1}{a/b}\right] = \frac{ab}{2} + \frac{ab}{2} = ab$$

Ex. 12 : *Evaluate* $\displaystyle\iint_R \sqrt{\dfrac{a^2 b^2 - b^2 x^2 - a^2 y^2}{a^2 b^2 + b^2 x^2 + a^2 y^2}}\; dx\; dy$ *over the ellipse* $\dfrac{x^2}{a^2} + \dfrac{y^2}{b^2} = 1.$

Sol. : By using elliptical polar co-ordinates $x = ar\cos\theta$, $y = br\sin\theta$, $dx\,dy = ab\,r\,dr\,d\theta$, the ellipse gets transformed into $r^2 = 1$ i.e. $r = 1$, a circle.

We have
$$a^2 b^2 - b^2 x^2 - a^2 y^2 = a^2 b^2 - b^2 a^2 r^2 \cos^2\theta - a^2 b^2 r^2 \sin^2\theta$$
$$= a^2 b^2 - a^2 b^2 r^2 = a^2 b^2 (1 - r^2)$$

Similarly,
$$a^2 b^2 + b^2 x^2 + a^2 y^2 = a^2 b^2 (1 + r^2)$$

$$I = \iint \sqrt{\frac{a^2 b^2 (1 - r^2)}{a^2 b^2 (1 + r^2)}}\; ab\,r\,dr\,d\theta = ab \int_0^{2\pi} d\theta \int_0^1 \sqrt{\frac{1 - r^2}{1 + r^2}}\; r\,dr$$

$$= ab\,(2\pi) \int_0^1 \frac{1 - r^2}{\sqrt{1 - r^4}}\; r\,dr, \quad \text{Put } r^2 = \sin t, \quad r\,dr = \frac{\cos t}{2}\,dt$$

r	0	1
t	0	$\pi/2$

$$= 2\pi\,ab \int_0^{\pi/2} \frac{1 - \sin t}{\cos t} \cdot \frac{1}{2}\cos t\;dt = \pi\,ab\,[t + \cos t]_0^{\pi/2} = \pi\,ab\left[\frac{\pi}{2} - 1\right]$$

Ex. 13 : *Express the following as a single term integral and evaluate :*

$$\int_0^{a\sqrt{2}} \int_0^x \cos\{k\,(x^2 + y^2)\}\; dx\,dy + \int_{a\sqrt{2}}^a \int_0^{\sqrt{a^2 - x^2}} \cos\{k\,(x^2 + y^2)\}\; dx\,dy$$

Sol. : Let regions of integration for both the integrals be R_1 and R_2 respectively i.e.

$$R_1 : \quad y = 0 \text{ to } y = x \quad \text{and} \quad R_2 : y = 0 \text{ to } y = \sqrt{a^2 - x^2} \text{ i.e. } x^2 + y^2 = a^2.$$

$$x = 0 \text{ to } x = \frac{a}{\sqrt{2}} \qquad\qquad x = \frac{a}{\sqrt{2}} \text{ to } x = a$$

The region is shown in Fig. 9.43. OMB is region R_1 for first integral and MBA is R_2 for second integral. To combine the two integrals, we consider combined region OAB. Considering a horizontal strip and end limits of which are $x = y$ and $x = \sqrt{a^2 - y^2}$, moving the strip vertically y varies from 0 to $\dfrac{a}{\sqrt{2}}$. Combined integral can be written as

$$I = \int_0^{a/\sqrt{2}} \int_y^{\sqrt{a^2 - y^2}} \cos\{k\,(x^2 + y^2)\}\; dx\,dy$$

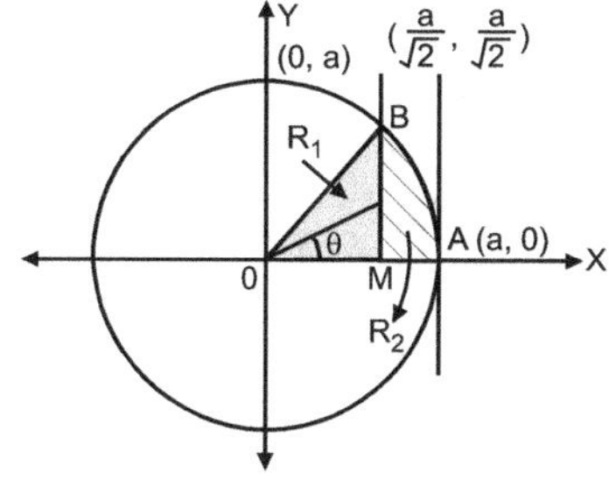

Fig. 9.43

To evaluate the integral, we transform to polars. Considering a wedge in the region R, r varies from 0 to a and to cover the complete region, θ varies from 0 to $\dfrac{\pi}{4}$.

$$I = \int_{\theta=0}^{\pi/4} \int_{r=0}^{a} \cos\{kr^2\} \, r \, d\theta \, dr = \frac{1}{2} \int_{0}^{\pi/4} d\theta \int_{0}^{a} \cos\{kr^2\} \, d(r^2), \; [d(r^2) = 2r \, dr]$$

$$= \frac{1}{2} \int_{0}^{\pi/4} \left\{\frac{\sin(kr^2)}{k}\right\}_{0}^{a} \, d\theta = \frac{1}{2k} \sin(ka^2) \, [\theta]_{0}^{\pi/4}$$

$$\boxed{I = \frac{\pi}{8k} \sin(ka^2)}$$

Ex. 14 : *Prove that* $\quad B(m, n) = \dfrac{\overline{|m} \; \overline{|n}}{\overline{|m + n}}$

Sol. : $\qquad\qquad B(m, n) = 2 \displaystyle\int_{0}^{\pi/2} \sin^{2m-1}\theta \; \cos^{2n-1}\theta \, d\theta$

$$\overline{|m} = 2 \int_{0}^{\infty} e^{-x^2} x^{2m-1} \, dx \qquad \overline{|n} = 2 \int_{0}^{\infty} e^{-y^2} y^{2n-1} \, dy$$

$$\overline{|m} \; \overline{|n} = 4 \int_{0}^{\infty} \int_{0}^{\infty} e^{-(x^2+y^2)} x^{2m-1} \, y^{2n-1} \, dx \, dy$$

Transforming to polars, the region of integration being entire positive quadrant in xy plane.

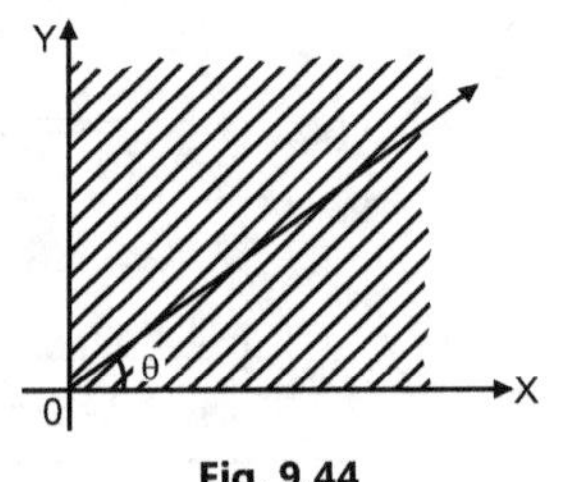

$$\overline{|m} \; \overline{|n} = 4 \int_{\theta=0}^{\pi/2} \int_{r=0}^{\infty} e^{-r^2} \, r^{2m-1} \cos^{2m-1}\theta \; \infty \; r^{2n-1} \sin^{2n-1}\theta \; r \, dr \, d\theta$$

$$= 4 \int_{0}^{\pi/2} \int_{0}^{\infty} e^{-r^2} r^{2(m+n)-1} \cos^{2m-1}\theta \; \sin^{2n-1}\theta \, dr \, d\theta$$

Observe that all the limits are constant and integrand is separable.

∴ This can be written as product of two single integrals as :

$$\overline{|m} \; \overline{|n} = 2 \int_{0}^{\pi/2} \cos^{2m-1}\theta \, \sin^{2n-1}\theta \, d\theta \; . \; 2 \int_{0}^{\infty} e^{-r^2} r^{2(m+n)-1} \, dr = B(m, n) \times \overline{|m + n}$$

∴ $\qquad\qquad \boxed{B(m, n) = \dfrac{\overline{|m} \; \overline{|n}}{\overline{|m + n}}}$ which is the required result.

EXERCISE 9.4

I.   Problems on Evaluation by Transforming to Polar Form :

Prove that :

1. $\displaystyle\int_{0}^{1} \int_{0}^{\sqrt{1-x^2}} x^2 y^2 \, dy \, dx = \frac{\pi}{96}$

2. $\displaystyle\int_{0}^{1} \int_{x}^{\sqrt{2-x^2}} \frac{x}{\sqrt{x^2+y^2}} \, dy \, dx = 1 - \frac{1}{\sqrt{2}}$

3. $\displaystyle\int_{0}^{a} \int_{\sqrt{ax-x^2}}^{\sqrt{a^2-x^2}} \frac{dy \, dx}{\sqrt{a^2-x^2-y^2}} = a$

4. $\displaystyle\int_{0}^{1} 4x \, e^{-x^2} \, dx \int_{0}^{\sqrt{x-x^2}} \frac{y \, e^{-y^2}}{x^2+y^2} \, dy = \frac{1}{e}$ **(Dec. 2011)**

5. $\displaystyle\int_{0}^{a} \int_{0}^{\sqrt{a^2-x^2}} xy \, dx \, dy = \frac{a^4}{8}$

6. $\displaystyle\int_{0}^{a} \int_{0}^{\sqrt{a^2-x^2}} \sqrt{a^2-x^2-y^2} \, dy \, dx = \frac{\pi a^3}{6}$

7. $\displaystyle\int_0^1 \int_{x^2}^{x} \frac{1}{\sqrt{x^2 + y^2}}\, dy\ dx = \sqrt{2} - 1$

8. $\displaystyle\int_0^1 \int_{x}^{\sqrt{2x-x^2}} (x^2 + y^2)\, dy\ dx = \frac{3\pi}{8} - 1$

9. $\displaystyle\int_0^2 \int_0^{\sqrt{2x-x^2}} \frac{x\, dy\, dx}{\sqrt{x^2 + y^2}} = \frac{4}{3}$

10. $\displaystyle\int_{-\infty}^{\infty} dx \int_{-\infty}^{\infty} \frac{dy}{(1 + x^2 + y^2)^{3/2}} = 4\pi$

11. $\displaystyle\int_0^a \int_y^a \frac{x^2\, dx\, dy}{(x^2 + y^2)^{3/2}} = \frac{a}{\sqrt{2}}$

12. $\displaystyle\int_0^a \int_0^{\sqrt{ax-x^2}} (a^2 - x^2 - y^2)^{3/2}\, dy\ dx = \frac{a^5}{5}\left(\frac{\pi}{2} - \frac{8}{15}\right)$

II. Illustrations on Integrals when the Limits are not Provided : (Transform to Polar Form)

Evaluate :

1. $\displaystyle\iint_R x^2 y^2\, dy\ dx$ over positive quadrant of $x^2 + y^2 = 1$. **(May 18) Ans. :** $\dfrac{\pi}{96}$

2. $\displaystyle\iint_R \frac{x^2 y^2}{x^2 + y^2}\, dx\, dy$, where R is annulus between $x^2 + y^2 = 4$, $x^2 + y^2 = 9$. **Ans. :** $\dfrac{65\pi}{16}$ **(Dec. 07, May 05, 17)**

3. $\displaystyle\iint_R \sqrt{\frac{1 - x^2 - y^2}{1 + x^2 + y^2}}\, dx\ dy$, over positive quadrant of $x^2 + y^2 = 1$ **Ans. :** $\dfrac{\pi}{8}(\pi - 2)$

4. $\displaystyle\iint_R y^2\, dx\, dy$ where R is outside $x^2 + y^2 = ax$ and inside $x^2 + y^2 = 2ax$. **Ans. :** $\dfrac{15\,\pi\,a^4}{64}$

5. $\displaystyle\iint_R \frac{r\, dr\, d\theta}{\sqrt{a^2 + r^2}}$ over one loop of $r^2 = a^2 \cos 2\theta$. **Ans. :** $2a\left(1 - \dfrac{\pi}{4}\right)$

6. $\displaystyle\iint_R \sin(x^2 + y^2)\, dx\, dy$, where R is $x^2 + y^2 = a^2$, $x \geq 0$ **(Nov. 2014) Ans. :** $\dfrac{\pi}{2}(1 - \cos a^2)$

7. $\displaystyle\iint_R \frac{dx\, dy}{(1 + x^2 + y^2)^{3/2}}$, where R is $x = 0$, $x = 1$, $y = 0$, $y = 1$. **Ans. :** $\dfrac{\pi}{6}$

8. $\displaystyle\iint_R \frac{dx\, dy}{\sqrt{xy}}$, where R is $x^2 + y^2 - x = 0$, $y = 0$, $y > 0$. **Ans. :** $\dfrac{\pi}{\sqrt{2}}$

$$\left[\textbf{Hint} : \text{Use } \int_0^{\pi/2} \sin^p \theta\ \cos^q \theta\ d\theta = \frac{1}{2} \frac{\overline{|(p + 1)/2}\ \ \overline{|(q + 1)/2}}{\overline{|(p + q + 2)/2}}\right]$$

9. $\displaystyle\iint_R r\, e^{-r^2/a^2} \sin\theta\, \cos\theta\, dr\, d\theta$, over the upper half of the circle $r = 2a \cos\theta$. **Ans. :** $\dfrac{a^2}{16}(3 + e^{-4})$

10. $\displaystyle\iint_R dx\, dy$, over the cardioide $r = 1 + \cos\theta$. **Ans. :** $\dfrac{3\pi}{2}$

11. Express as a single integral and evaluate : $\displaystyle\int_0^{a\sqrt{2}} \int_0^{x} x\ dy\ dx + \int_{a\sqrt{2}}^{a} \int_0^{\sqrt{a^2-x^2}} x\ dy\ dx$ (Refer Solved Ex. 13) **Ans. :** $\dfrac{a^3}{3\sqrt{2}}$

9.6 TRIPLE INTEGRATION

Triple integration is an extension of concept of double integration to third dimension. Basically, it can be expressed in terms of volume of a body or it can represent any physical quantity such as Mass, Moment of inertia, Centre of gravity related to three-dimensional configuration.

Let us consider a specific problem of determination of volume of a tetrahedron bounded by the planes $x = 0$, $y = 0$, $z = 0$ i.e. three co-ordinate planes and a plane $x + y + z = 1$. We shall express this volume in terms of a triple integral and then find it by evaluation of it. Tetrahedron is as shown in Fig. 9.45. Consider two strips one parallel to x-axis and other parallel to y-axis of widths δx and δy, intersecting an area GHUV $= \delta x\ \delta y$. With area GHUV as base, erect a rectangular parallelopiped which will cut the surface ABC i.e. the plane $x + y + z = 1$. Now, draw two planes parallel to xy plane, one at a distance z and other at a distance $z + \delta z$ which will cut a volume element $\delta V = \delta x\ \delta y\ \delta z$ from the rectangular parallelopiped. Now, in the first summation, the volume of rectangular parallelopiped can be obtained by expressing it as a sum of infinite number of volume elements like δV. Thus the volume of rectangular parallelopiped erected on element $\delta x\ \delta y$ as base, say V_1 is written as :

$$V_1 = \lim_{\delta z \to 0} \Sigma\ \delta x\ \delta y\ \delta z$$

Replacing summation sign by integral sign,

$$V_1 = \left\{ \int_0^{1-x-y} dz \right\} \delta x\ \delta y$$

Next consider the summation of all such volumes like V_1, situated along a strip parallel to y-axis. This volume V_2 is expressed as :

$$V_2 = \lim_{\delta y \to 0} \Sigma \left\{ \int_0^{1-x-y} dz \right\} \delta y \cdot \delta x = \left\{ \int_0^{1-x} dy \int_0^{1-x-y} dz \right\} \delta x$$

Note here that the strip parallel to y-axis has its ends on the straight lines $y = 0$ (i.e. x-axis) and $x + y = 1$ (i.e. on the line AB).

Then in the third summation, the strip parallel to y-axis is moved along x-axis from 0 to A in which case x varies from $x = 0$ to $x = 1$. Thus the entire volume V is expressed as :

$$V = \lim_{\delta x \to 0} \Sigma \left\{ dx \int_0^{1-x} dy \int_0^{1-x-y} dz \right\}$$

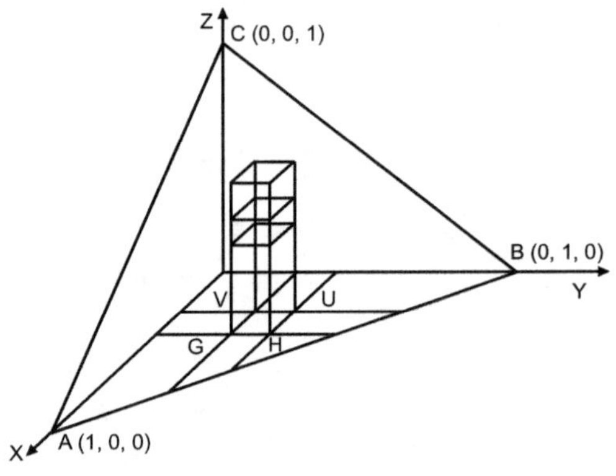

Fig. 9.45

Or replacing the summation sign by integral sign, the volume V of the tetrahedron is expressed as :

$$V = \int_{x=0}^{1} dx \int_{y=0}^{1-x} dy \int_{z=0}^{1-x-y} dz \quad \text{i.e.} \quad V = \int_{x=0}^{1} \int_{y=0}^{1-x} \int_{z=0}^{1-x-y} dz\ dy\ dx$$

In this triple integration, the innermost integration is w.r.t. z, treating x, y as constants, second integration is w.r.t. y, treating x as constant and third or outer integration is w.r.t. x.

Let us find V by evaluating this triple integral.

$$V = \int_0^1 \int_0^{1-x} [z]_0^{1-x-y}\ dx\ dy = \int_0^1 \int_0^{1-x} (1-x-y)\ dx\ dy$$

$$= \int_0^1 \left[\frac{(1-x-y)^2}{-2} \right]_0^{1-x} dx = \int_0^1 \frac{(1-x)^2}{2}\ dx = \left[\frac{(1-x)^3}{-6} \right]_0^1 = \frac{1}{6}$$

If the problem had been the determination of mass of solid tetrahedron where density ρ at any point is given by $\rho = f(x, y, z)$

We could have started with mass of volume element δV as

$$\delta m = \rho \, \delta x \, \delta y \, \delta z = f(x, y, z) \, \delta x \, \delta y \, \delta z$$

Then three summations as above would give mass M of tetrahedron as

$$M = \int_0^1 \int_0^{1-x} \int_0^{1-x-y} f(x, y, z) \, dx \, dy \, dz$$

Depending upon nature of $f(x, y, z)$ $\int \int \int f(x, y, z) \, dx \, dy \, dz$ will represent corresponding physical quantity related to the volume of body under consideration.

Note : Please Refer Article 9.7 (Page 9.75) Dirichlet's Theorem before examples.

1. **Use of notation :**

$$\int_a^b \int_{f_1(x)}^{f_2(x)} \int_{\phi_1(x,y)}^{\phi_2(x,y)} f(x, y, z) \, dz \, dy \, dx = \int_a^b \left\{ \int_{f_1(x)}^{f_2(x)} \left[\int_{\phi_1(x,y)}^{\phi_2(x,y)} f(x, y, z) \, dz \right] dy \right\} dx$$

Here solve inner integral first w.r.t. z then w.r.t. y and outermost integral finally w.r.t. x.

Note : The order of integration depends upon the distribution of limits.

2. **Use of spherical polar co-ordinates :**

Spherical polar co-ordinates are given by :

$$x = r \sin\theta \cos\phi, \ y = r\sin\theta \sin\phi, \ z = r\cos\theta, \ x^2 + y^2 + z^2 = r^2, \ \frac{\partial(x, y, z)}{\partial(r, \theta, \phi)} = r^2 \sin\theta \ \therefore \ dx \, dy \, dz = r^2 \sin\theta \, dr \, d\theta \, d\phi$$

$$\therefore \ I = \int \int \int_V f(r, \theta, \phi) \, r^2 \sin\theta \, dr \, d\theta \, d\phi$$

Standard limits :

(i)   For complete sphere $x^2 + y^2 + z^2 = a^2$, $\theta \to 0$ to π, $\phi \to 0$ to 2π, $r \to 0$ to a

(ii)   For hemisphere $x^2 + y^2 + z^2 = a^2$, $z > 0 : \theta \to 0$ to $\pi/2$, $\phi \to 0$ to 2π, $r \to 0$ to a

(iii) For positive octant of a sphere $x^2 + y^2 + z^2 = a^2$, $\theta \to 0$ to $\pi/2$, $\phi \to 0$ to $\pi/2$, $r \to 0$ to a

(iv)   Also for ellipsoid $\dfrac{x^2}{a^2} + \dfrac{y^2}{b^2} + \dfrac{z^2}{c^2} = 1$, we use

$$x = ar \sin\theta \cos\phi, \ y = br\sin\theta \sin\phi, \ z = cr\cos\theta, \ \frac{x^2}{a^2} + \frac{y^2}{b^2} + \frac{z^2}{c^2} = r^2, \ dx \, dy \, dz = abcr^2 \sin\theta \, dr \, d\theta \, d\phi$$

$$\therefore \qquad I = \int \int \int_V f(r, \theta, \phi) \, abc \, r^2 \sin\theta \, dr \, d\theta \, d\phi$$

Standard limits for ellipsoid : $\theta \to 0$ to π, $\phi \to 0$ to 2π, $r \to 0$ to 1

3. **Use of cylindrical polar co-ordinates :** Cylindrical polar co-ordinates are given by :

$$x = \rho \cos\phi, \ y = \rho \sin\phi, \ z = z, \ x^2 + y^2 = \rho^2, \ dx \, dy \, dz = \rho \, d\rho \, d\phi \, dz, \ \frac{\partial(x, y, z)}{\partial(\rho, \phi, z)} = \rho$$

$$\therefore \qquad I = \int \int \int_V f(\rho, \phi, z) \, \rho \, d\rho \, d\phi \, dz$$

Standard limits : For a cylinder $x^2 + y^2 = a^2$, $z = 0$, $z = h$, $\rho \to 0$ to a, $\phi \to 0$ to 2π, $z \to 0$ to h

Problems on triple integration are mainly divided in the following types :

Type I : Problems on Direct Evaluation of Triple Integral.

Type II : Problems on Triple Integral when the Limits are not provided.

Type I : Problems on Direct Evaluation of Triple Integral

Ex. 1 : *Evaluate* $\displaystyle\int_0^2 \int_0^x \int_0^{2x+2y} e^{x+y+z}\ dx\ dy\ dz.$ *(May 2015)*

Sol. : Let,

$$I = \int_0^2 \int_0^x \left[e^{x+y+z}\right]_0^{2x+2y} dx\ dy \qquad \left\{\begin{array}{l}\text{integrated w.r.t. z treating}\\ \text{x and y constants}\end{array}\right.$$

$$= \int_0^2 \int_0^x \{e^{x+y+2x+2y} - e^{x+y}\}\ dx\ dy = \int_0^2 \left\{\frac{e^{3x+3y}}{3} - \frac{e^{x+y}}{1}\right\}_0^x dx$$

$$= \int_0^2 \left\{\frac{e^{3x+3x}}{3} - \frac{e^{3x}}{3} - \frac{e^{x+x}}{1} + \frac{e^x}{1}\right\} dx$$

$$= \int_0^2 \left\{\frac{e^{6x}}{3} - \frac{e^{3x}}{3} - \frac{e^{2x}}{1} + \frac{e^x}{1}\right\} dx = \left[\frac{e^{6x}}{18} - \frac{e^{3x}}{9} - \frac{e^{2x}}{2} + \frac{e^x}{1}\right]_0^2$$

$$= \frac{e^{12}}{18} - \frac{1}{18} - \frac{e^6}{9} + \frac{1}{9} - \frac{e^4}{2} + \frac{1}{2} + \frac{e^2}{1} - 1$$

$$= \frac{1}{18}\ (e^{12} - 2e^6 - 9e^4 + 18e^2 - 1 + 2 + 9 - 18)$$

$$\boxed{I = \frac{1}{18}\ (e^{12} - 2e^6 - 9e^4 + 18e^2 - 8)}$$

Ex. 2 : *Evaluate* $\displaystyle\int_{-1}^1 dz \int_0^z dx \int_{x-z}^{x+z} (x+y+z)\ dy.$ *(Dec. 2017)*

Sol. :

$$I = \int_{-1}^1 dz \int_0^z dx \left[(x+z)\ y + \frac{y^2}{2}\right]_{x-z}^{x+z} \qquad \left\{\begin{array}{l}\text{integrated w.r.t. y treating}\\ \text{x and z constants}\end{array}\right.$$

$$= \int_{-1}^1 dz \int_0^z \left[(x+z)^2 + \frac{(x+z)^2}{2} - (x^2-z^2) - \frac{(x-z)^2}{2}\right] dx$$

$$= \int_{-1}^1 dz \left[\frac{3}{2}\frac{(x+z)^3}{3} - \frac{x^3}{3} + z^2 x - \frac{(x-z)^3}{6}\right]_0^z$$

$$= \int_{-1}^1 \left(4z^3 - \frac{z^3}{3} + z^3 - \frac{z^3}{2} + \frac{z^3}{6}\right) dz = \int_{-1}^1 \frac{13\ z^3}{3}\ dz \qquad \boxed{I = 0} \qquad \text{(odd function of z)}$$

Ex. 3 : *Evaluate* $\displaystyle\int_0^\pi 2\ d\theta \int_0^{a(1+\cos\theta)} r\ dr \int_0^{h\left[1-\frac{r}{a(1+\cos\theta)}\right]} dz.$

Sol. :

$$I = \int_0^\pi 2\ d\theta \int_0^{a(1+\cos\theta)} r\ dr\ h\left(1 - \frac{r}{a(1+\cos\theta)}\right) = 2h \int_0^\pi d\theta \int_0^{a(1+\cos\theta)} \left[r\ dr - \frac{r^2\ dr}{a\ (1+\cos\theta)}\right]$$

$$= 2h \int_0^\pi d\theta \left[\frac{r^2}{2} - \frac{r^3}{3a\ (1+\cos\theta)}\right]_0^{a(1+\cos\theta)}$$

$$= 2h \int_0^\pi \left[\frac{a^2\ (1+\cos\theta)^2}{2} - \frac{a^2\ (1+\cos\theta)^2}{3}\right] d\theta = \frac{ha^2}{3} \int_0^\pi (1+\cos\theta)^2\ d\theta$$

$$= \frac{ha^2}{3} \int_0^\pi \left(2 \cos^2 \frac{\theta}{2}\right)^2 d\theta = \frac{4\,ha^2}{3} \int_0^{\pi/2} \cos^4 t \,.\, 2\, dt \ \left(\text{by putting } \frac{\theta}{2} = t\right)$$

$$= \frac{8ha^2}{3} \left(\frac{3}{4} \frac{1}{2} \frac{\pi}{2}\right)$$

$$\boxed{I = \frac{h\,\pi\,a^2}{2}}$$

Ex. 4 : *Evaluate* $\displaystyle \int_0^\infty dx \int_0^\infty)dy \int_0^\infty \frac{dz}{(1 + x^2 + y^2 + z^2)^2}$. *(Nov. 2015)*

Sol. : Here x, y, z all vary from 0 to ∞, indicating that region of integration is entire positive octant in space.

Transforming to spherical polars, r varies from 0 to ∞, θ and ϕ both vary from 0 to $\frac{\pi}{2}$.

$$I = \int_{\phi=0}^{\pi/2} \int_{\theta=0}^{\pi/2} \int_{r=0}^{\infty} \frac{r^2 \sin\theta \, dr \, d\theta \, d\phi}{(1 + r^2)^2}$$

Consider inner integral $\displaystyle I_1 = \int_0^\infty \frac{r^2}{(1 + r^2)^2} \, dr$ Put $r = \tan t$ $\therefore$ $dr = \sec^2 t \, dt$

When $r = 0, t = 0, r = \infty, t = \frac{\pi}{2}$

$$I_1 = \int_0^{\pi/2} \frac{\tan^2 t}{\sec^4 t} \,.\, \sec^2 t \ dt = \int_0^{\pi/2} \sin^2 t \ dt = \frac{1}{2} \,.\, \frac{\pi}{2} = \frac{\pi}{4}$$

$$I = \int_0^{\pi/2} d\phi \int_0^{\pi/2} \frac{\pi}{4} \sin\theta \, d\theta = \frac{\pi}{4} \int_0^{\pi/2} (-\cos\theta)\Big|_0^{\pi/2} \, d\phi = \frac{\pi}{4} \,[\phi]_0^{\pi/2} = \frac{\pi}{4} \,.\, \frac{\pi}{2}$$

$$\boxed{I = \frac{\pi^2}{8}}$$

Note : This 'I' could have been evaluated by the product of three single integrals w.r.t. r, θ and ϕ separately, simultaneously.

Ex. 5 : *Evaluate* $\displaystyle \int_0^a \int_0^\alpha \int_0^\beta (x^2 + y^2 + z^2) \ dx \, dy \, dz,$ *where a is constant,*

$\alpha = \sqrt{a^2 - z^2}$ *and* $\beta = \sqrt{a^2 - y^2 - z^2}$

Sol. : It is clear that innermost integration is w.r.t. x where limits of x are from $x = 0$ to $x = \sqrt{a^2 - y^2 - z^2}$ i.e. $x^2 + y^2 + z^2 = a^2$.

Second integration is w.r.t. y, the limits of y being from $y = 0$ to $y = \sqrt{a^2 - z^2}$ i.e. $y^2 + z^2 = a^2$.

Third integration is w.r.t. z, which varies from 0 to a.

Region of integration is positive octant of sphere. Transforming to spherical polar system,

$$I = \int_{\phi=0}^{\pi/2} \int_{\theta=0}^{\pi/2} \int_{r=0}^{a} r^2 \,.\, r^2 \sin\theta \, dr \, d\theta \, d\phi$$

$$= \int_0^{\pi/2} d\phi \int_0^{\pi/2} \left[\frac{r^5}{5}\right]_0^a \sin\theta \, d\theta = \frac{a^5}{5} \int_0^{\pi/2} d\phi \int_0^{\pi/2} \sin\theta \, d\theta$$

$$= \frac{a^5}{5} \,.\, \frac{\pi}{2} \,.\, 1$$

$$\boxed{I = \frac{\pi\, a^5}{10}}$$

Type II : Illustrations on Triple Integration when the Limits are not provided.

Ex. 1 : Integrate $\iiint x^2 y z \; dx \, dy \, dz$ throughout the volume bounded by the planes $x = 0, y = 0, z = 0, \dfrac{x}{a} + \dfrac{y}{b} + \dfrac{z}{c} = 1.$

(See also Ex. 1 on page 9.76)

(Dec. 2011, 2004)

Sol. : To simplify the problem, we put $x = au, \; y = bv, z = cw$.

So that $dx = a \, du, \; dy = b \, dv, \; dz = c \, dw$

Given integral transforms to :

$$I = a^2 \, bc \iiint u^2 \, v \, w \; abc \; du \; dv \; dw$$

where, volume is bounded by $u = 0, v = 0, w = 0$ and plane $u + v + w = 1$.

$$I = a^3 \, b^2 \, c^2 \int_{u=0}^{1} \int_{v=0}^{1-u} \int_{w=0}^{1-u-v} u^2 \, v \, w \; du \, dv \, dw$$

$$I = a^3 \, b^2 \, c^2 \int_{0}^{1} \int_{0}^{1-u} u^2 \, v \left[\frac{w^2}{2} \right]_{0}^{1-u-v} du \; dv$$

$$= \frac{a^3 \, b^2 \, c^2}{2} \int_{0}^{1} \int_{0}^{1-u} u^2 \, v \, (1 - u - v)^2 \; du \, dv$$

$$= \frac{a^3 \, b^2 \, c^2}{2} \int_{0}^{1} \int_{0}^{1-u} u^2 \, v \, \{(1 - u)^2 - 2 (1 - u) \, v + v^2\} \; du \; dv$$

$$= \frac{a^3 \, b^2 \, c^2}{2} \int_{0}^{1} u^2 \left[(1 - u)^2 \, \frac{v^2}{2} - 2 (1 - u) \, \frac{v^3}{3} + \frac{v^4}{4} \right]_{0}^{1-u} du$$

$$= \frac{a^3 \, b^2 \, c^2}{2} \int_{0}^{1} u^2 \left[\frac{(1 - u)^4}{2} - \frac{2}{3} (1 - u)^4 + \frac{1}{4} (1 - u)^4 \right] du$$

$$= \frac{a^3 \, b^2 \, c^2}{2} \int_{0}^{1} u^2 \, \frac{(1 - u)^4}{12} \; du = \frac{a^3 \, b^2 \, c^2}{24} \int_{0}^{1} u^2 \, (1 - u)^4 \, du$$

$$= \frac{a^3 \, b^2 \, c^2}{24} \, B \, (3, 5) = \frac{a^3 \, b^2 \, c^2}{24} \, \frac{\lfloor 3 \; \lfloor 5}{\lfloor 8}$$

$$= \frac{a^3 \, b^2 \, c^2}{24} \left(\frac{2 ! \, 4 !}{7 !} \right)$$

$$\boxed{I = \frac{a^3 \; b^2 \, c^2}{2520}}$$

Ex. 2 : Evaluate $\iiint \dfrac{dx \, dy \, dz}{\sqrt{1 - x^2 - y^2 - z^2}}$ taken throughout the volume of the sphere $x^2 + y^2 + z^2 = 1$ in the positive octant.

(Nov./Dec. 2019, May 2011, 2008, 2018, 2016; Dec. 2013, 2018)

Sol. : Transforming to spherical polar system, $x = r \sin \theta \, \cos \phi, \; y = r \sin \theta \sin \phi, \quad z = r \cos \theta$ giving $x^2 + y^2 + z^2 = r^2$.

For positive octant of the sphere, θ and ϕ both vary from 0 to $\dfrac{\pi}{2}$. The volume element $dx \, dy \, dz$ gets transformed to $r^2 \sin \theta \, dr \, d\theta \, d\phi$.

Converting to sperical polar, we get

$$I = \int_{\phi=0}^{\pi/2} \int_{\theta=0}^{\pi/2} \int_{r=0}^{1} \frac{1}{\sqrt{1-r^2}} \; r^2 \sin\theta \; dr \; d\theta \; d\phi$$

Consider

$$I_1 = \int_{0}^{1} \frac{1}{\sqrt{1-r^2}} \; r^2 \; dr.$$

Putting $r = \sin t$, $dr = \cos t \; dt$, when $r = 0, t = 0$; $r = 1, t = \pi/2$

$$I_1 = \int_{0}^{\pi/2} \sin^2 t \; dt = \frac{1}{2} \cdot \frac{\pi}{2} = \frac{\pi}{4}$$

$$\therefore \qquad I = \int_{0}^{\pi/2} d\phi \int_{0}^{\pi/2} \sin\theta \; d\theta \; \frac{\pi}{4} = \frac{\pi}{2} \, (1) \, \frac{\pi}{4} = \frac{\pi^2}{8}$$

Ex. 3 : *Evaluate in terms of Gamma function* $\iiint x^{a-1} \; y^{b-1} \; z^{c-1} \; dx \; dy \; dz$ *taken throughout the volume of the tetrahedron given by* $x \geq 0, y \geq 0, z \geq 0, x + y + z \leq 1.$

Sol. :

$$I = \int_{x=0}^{1} \int_{y=0}^{1-x} \int_{z=0}^{1-x-y} x^{a-1} \; y^{b-1} \; z^{c-1} \; dx \; dy \; dz$$

$$= \int_{0}^{1} \int_{0}^{1-x} x^{a-1} \; y^{b-1} \; \left\{ \frac{z^c}{c} \right\}_{0}^{1-x-y} dx \; dy = \frac{1}{c} \int_{0}^{1} \int_{0}^{1-x} x^{a-1} \; y^{b-1} (1-x-y)^c \; dx \; dy$$

$$= \frac{1}{c} \int_{0}^{1} x^{a-1} \; dx \int_{0}^{1-x} y^{b-1} \; (1-x-y)^c \; dy \qquad\qquad (\text{Let } A = 1-x)$$

$$= \frac{1}{c} \int_{0}^{1} x^{a-1} \; dx \int_{0}^{A} y^{b-1} \; (A-y)^c \, dy, \quad \text{Put } y = At, \, dy = A \, dt$$

y	0	A
t	0	1

$$I = \frac{1}{c} \int_{0}^{1} x^{a-1} \; dx \int_{0}^{1} A^{b-1} \; t^{b-1} \; (A - At)^c \cdot A \, dt$$

$$= \frac{1}{c} \int_{0}^{1} x^{a-1} \; dx \; A^{b+c} \int_{0}^{1} t^{b-1} \; (1-t)^c \cdot dt$$

$$= \frac{1}{c} \int_{0}^{1} x^{a-1} \; (1-x)^{b+c} \; dx \int_{0}^{1} t^{b-1} \; (1-t)^c \cdot dt$$

$$= \frac{1}{c} \, B \, (a, b+c+1) \; B \, (b, c+1) = \frac{1}{c} \frac{\overline{|a} \; \overline{|b+c+1}}{\overline{|a+b+c+1}} \frac{\overline{|b} \; \overline{|c+1}}{\overline{|b+c+1}}$$

$$= \frac{1}{c} \frac{\overline{|a} \; \overline{|b} \; c\overline{|c}}{(a+b+c) \, \overline{|a+b+c}} = \frac{\overline{|a} \; \overline{|b} \; \overline{|c}}{(a+b+c) \, \overline{|a+b+c}}$$

$$\boxed{I = \frac{\overline{|a} \; \overline{|b} \; \overline{|c}}{\overline{|a+b+c+1}}}$$

This example provides an idea for proving Dirichlet's Theorem (refer article 9.7, page 9.48) for the case of three variables in particular and n variables in general.

Ex. 4 : *Evaluate* $\iiint\limits_{V} \sqrt{1 - \dfrac{x^2}{a^2} - \dfrac{y^2}{b^2} - \dfrac{z^2}{c^2}}\; dx\, dy\, dz$ *throughout the volume of ellipsoid* $\dfrac{x^2}{a^2} + \dfrac{y^2}{b^2} + \dfrac{z^2}{c^2} = 1.$ **(May 2007, 2015)**

Sol. : By using $x = ar \sin\theta \cos\phi,\ y = br \sin\theta \sin\phi,\ z = cr \cos\theta,\ \dfrac{x^2}{a^2} + \dfrac{y^2}{b^2} + \dfrac{z^2}{c^2} = r^2,$

$dx\, dy\, dz = abc\, r^2 \sin\theta\; d\theta\; d\phi\; dr$ ellipsoid gets transformed into a unit sphere $r = 1.$

$$I = \iiint \sqrt{1 - r^2}\; abc\, r^2\; \sin\theta\, d\theta\; d\phi\; dr$$

$$= abc \int_0^\pi \sin\theta\, d\theta \int_0^{2\pi} d\phi \int_0^1 \sqrt{1 - r^2}\, r^2\, dr$$

$$= abc\, (2)\, (2\pi) \int_0^{\pi/2} \cos t \cdot \sin^2 t\; \cos t\; dt \qquad\qquad \text{(by putting } r = \sin t\text{)}$$

$$= 4\pi\, abc\, \frac{1 \cdot 1}{4 \cdot 2}\, \frac{\pi}{2} \qquad \boxed{I = \frac{\pi^2\, abc}{4}}$$

Ex. 5 : *Evaluate* $\iiint\limits_{V} \sqrt{x^2 + y^2}\; dx\, dy\, dz,$ *where V is* $x^2 + y^2 = z^2, z > 0$ *and* $z = 0,\ z = 1.$ **(May 2013, Dec. 2016)**

Sol. : $z^2 = x^2 + y^2$ is a right circular cone. Transforming to cylindrical polar co-ordinates $x = \rho \cos\phi,\ y = \rho \sin\phi,\ z = z,$ $x^2 + y^2 = \rho^2,\ dx\, dy\, dz = \rho\, d\rho\; d\phi\; dz.$

Limits : $\rho \to 0$ to 1 ; $\phi \to 0$ to 2π ; $z \to \rho$ to 1

$$I = \iiint \sqrt{\rho^2}\; \rho\; d\rho\; d\phi\, dz = \int_0^1 \rho^2\, d\rho \int_0^{2\pi} d\phi \int_\rho^1 dz$$

$$= \int_0^1 \rho^2\, d\rho\, (2\pi)\, (1 - \rho) = 2\pi \left[\frac{\rho^3}{3} - \frac{\rho^4}{4}\right]_0^1 = 2\pi \left(\frac{1}{3} - \frac{1}{4}\right)$$

$$\boxed{I = \frac{\pi}{6}}$$

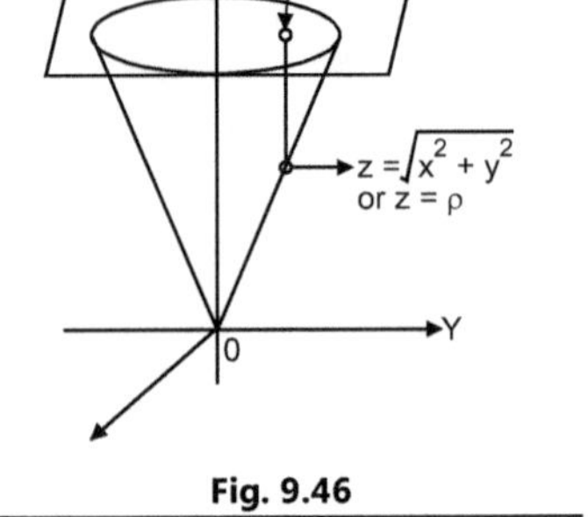

Fig. 9.46

Ex. 6 : *Evaluate* $\iiint (x^2 y^2 + y^2 z^2 + z^2 x^2)\; dx\, dy\, dz$ *throughout the volume of the sphere* $x^2 + y^2 + z^2 = a^2.$ **(May 05, 17, Nov. 15)**

Sol. : $I = \iiint (x^2 y^2 + y^2 z^2 + z^2 x^2)\; dx\, dy\, dz.$

Here V is volume of the sphere. By using spherical polar co-ordinates, $x = r \sin\theta \cos\phi,\ y = r \sin\theta \sin\phi,\ z = r \cos\theta\ dx\, dy\, dz = r^2 \sin\theta\, d\theta\, d\phi\; dr.$

Limits : $\theta \to 0$ to π ; $\phi \to 0$ to 2π ; $r \to 0$ to a

$$I = \iiint [r^4 \sin^4\theta \sin^2\phi\; \cos^2\phi + r^4 \sin^2\theta \cos^2\theta \sin^2\phi + r^4 \sin^2\theta\; \cos^2\theta \cos^2\phi]\, r^2 \sin\theta\, d\theta\, d\phi\, dr$$

$$= \int_0^\pi \sin\theta\, d\theta \int_0^{2\pi} [\sin^4\theta\; \sin^2\phi\; \cos^2\phi + \sin^2\theta \cos^2\theta]\, d\phi \int_0^a r^6\, dr$$

$$= \int_0^\pi \sin\theta\, d\theta \left[\sin^4\theta\; 4 \int_0^{\pi/2} \sin^2\phi\; \cos^2\phi\; d\phi + \sin^2\theta\; \cos^2\theta\, (2\pi)\right] \left(\frac{a^7}{7}\right)$$

$$= \frac{a^7}{7} \left\{\int_0^\pi 4 \sin^5\theta\; \left(\frac{1 \cdot 1}{4 \cdot 2}\, \frac{\pi}{2}\right) d\theta + 2\pi \int_0^\pi \sin^3\theta \cos^2\theta\, d\theta\right\}$$

$$= \frac{a^7}{7} \left\{ \frac{\pi}{4} \cdot 2 \int_0^{\pi/2} \sin^5 \theta \, d\theta + 2\pi \cdot 2 \int_0^{\pi/2} \sin^3 \theta \cos^2 \theta \, d\theta \right\} = \frac{a^7}{7} \left\{ \frac{\pi}{2} \cdot \frac{4}{5} \cdot \frac{2}{3} + 4\pi \left(\frac{2 \cdot 1}{5 \cdot 3 \cdot 1} \right) \right\}$$

$$\boxed{I = \frac{4a^7 \pi}{35}}$$

Ex. 7 : Evaluate $\iiint z^2 \, dx \, dy \, dz$ over the volume common to sphere $x^2 + y^2 + z^2 = a^2$ and cylinder $x^2 + y^2 = ax$.

Sol. :

$$I = \iiint z^2 \, dx \, dy \, dz = \int \int dx \, dy \int_{-\sqrt{a^2 - x^2 - y^2}}^{\sqrt{a^2 - x^2 - y^2}} z^2 \, dz$$

$$= \iint dx \, dy \left[\frac{z^3}{3} \right]_{-\sqrt{a^2 - x^2 - y^2}}^{\sqrt{a^2 - x^2 - y^2}} = \frac{1}{3} \iint \left[(a^2 - x^2 - y^2)^{3/2} + (a^2 - x^2 - y^2)^{3/2} \right] dx \, dy$$

$$= \frac{2}{3} \int\int_R (a^2 - x^2 - y^2)^{3/2} \, dx \, dy$$

where, R is $x^2 + y^2 = ax$ (Refer Fig. 9.47) and transforming to polars,

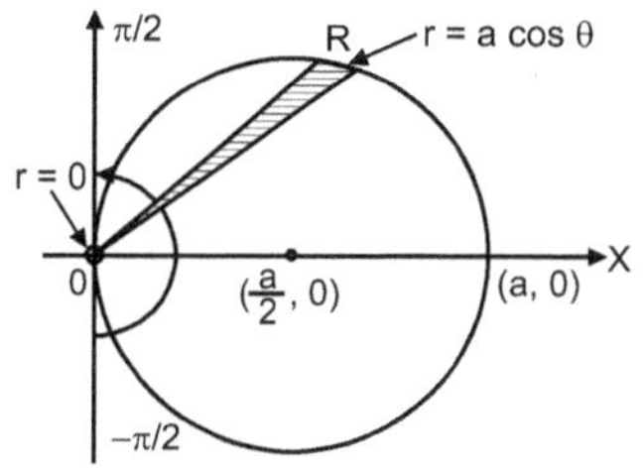

$$I = \frac{2}{3} \int_{-\frac{\pi}{2}}^{\pi/2} \int_0^{a\cos\theta} (a^2 - r^2)^{3/2} \, r \, dr \, d\theta$$

$$= \frac{-1}{3} \int_{-\frac{\pi}{2}}^{\pi/2} d\theta \int_0^{a\cos\theta} (a^2 - r^2)^{3/2} \, (-2r \, dr)$$

$$= \frac{-1}{3} \int_{-\frac{\pi}{2}}^{\pi/2} \left[\frac{2}{5} (a^2 - r^2)^{5/2} \right]_0^{a\cos\theta} d\theta$$

$$= \frac{-2}{15} \cdot 2 \int_0^{\pi/2} (a^5 \sin^5 \theta - a^5) \, d\theta \qquad \left(\because \sin\theta = \pm \sqrt{1 - \cos^2\theta} \right)$$

$$= \frac{4}{15} a^5 \int_0^{\pi/2} (1 - \sin^5 \theta) \, d\theta = \frac{4}{15} a^5 \left[\frac{\pi}{2} - \frac{4}{5} \cdot \frac{2}{3} \right]$$

$$\boxed{I = \frac{4a^5}{15} \left(\frac{\pi}{2} - \frac{8}{15} \right)}$$

Fig. 9.47

Ex. 8 : Evaluate $\iiint z^2 \, dx \, dy \, dz$ over the volume bounded by surfaces $x^2 + y^2 = a^2$, $x^2 + y^2 = z$ and $z = 0$.

Sol. :

$$I = \iiint z^2 \, dx \, dy \, dz = \iint dx \, dy \int_0^{x^2 + y^2} z^2 \, dz = \iint \left[\frac{z^3}{3} \right]_0^{x^2 + y^2} dx \, dy$$

$$= \frac{1}{3} \int\int_R (x^2 + y^2)^3 \, dx \, dy$$

Transforming to polar (Refer Fig. 9.48)

$$I = \frac{1}{3} \iint (r^2)^3 \, r \, dr \, d\theta$$

$$= \frac{1}{3} \int_0^{2\pi} d\theta \int_0^a r^7 \, dr = \frac{1}{3} (2\pi) \frac{a^8}{8}$$

$$\boxed{I = \frac{\pi a^8}{12}}$$

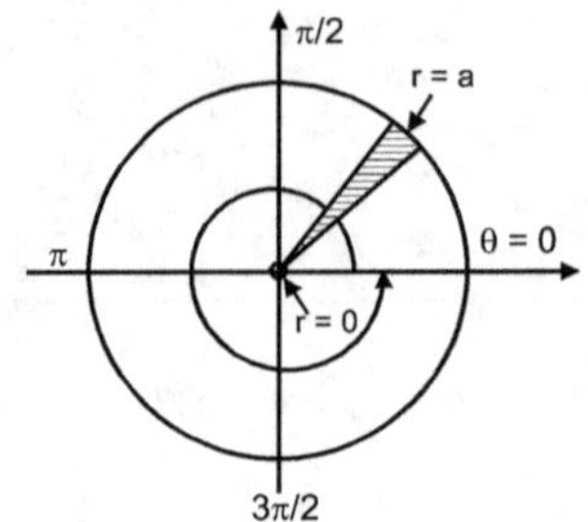

Fig. 9.48

Ex. 9 : *Evaluate* $\iiint \dfrac{dx \, dy \, dz}{(1 + x + y + z)^3}$ *over the volume of tetrahedron bounded by* $x = 0, y = 0, z = 0$ *and* $x + y + z = 1.$

Sol. :

$$I = \int_0^1 dx \int_0^{1-x} dy \int_0^{1-x-y} \frac{dz}{(1 + x + y + z)^3}$$

$$= \int_0^1 dx \int_0^{1-\pi} \left[\frac{(1 + x + y + z)^{-2}}{-2} \right]_0^{1-x-y}$$

$$= \frac{1}{2} \int_0^1 dx \int_0^{1-x} \left[\frac{1}{(1 + x + y)^2} - \frac{1}{4} \right] dy$$

$$= \frac{1}{2} \int_0^1 dx \left[\frac{-1}{1 + x + y} - \frac{y}{4} \right]_0^{1-x}$$

$$= \frac{1}{2} \int_0^1 \left[\frac{1}{1 + x} - \frac{1}{2} - \frac{(1 - x)}{4} \right] dx$$

$$= \frac{1}{2} \left[\log (1 + x) - \frac{1}{2} x + \frac{(1 - x)^2}{8} \right]_0^1$$

$$= \frac{1}{2} \left[\log 2 - \frac{1}{2} - \frac{1}{8} \right]$$

$$\boxed{I = \frac{1}{2} \left[\log 2 - \frac{5}{8} \right]}$$

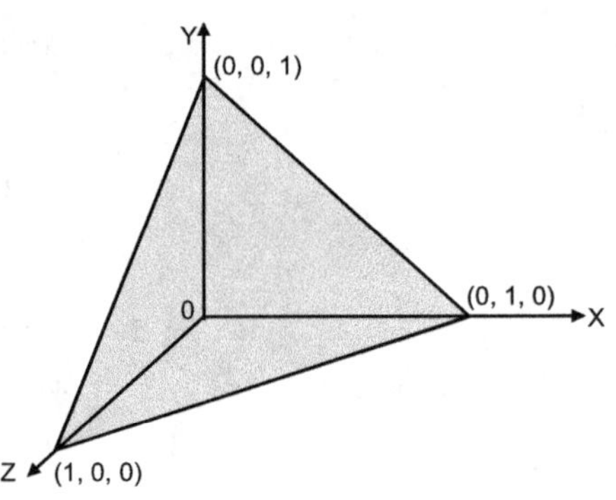

Fig. 9.49

Ex. 10 : *Evaluate* $\displaystyle\iiint_V \dfrac{z^2}{x^2 + y^2 + z^2} \, dx \, dy \, dz,$ *where, V is volume bounded by* $x^2 + y^2 + z^2 = z.$ **(May 2004)**

Sol. : For upper half of the sphere, $I = \displaystyle\iiint \dfrac{z^2}{x^2 + y^2 + z^2} \, dx \, dy \, dz,$

$x = r \sin \theta \cos \theta,$ $y = r \sin \theta \sin \phi, z = r \cos \theta$

$x^2 + y^2 + z^2 = r^2.$ $dx \, dy \, dz = r^2 \sin \theta \, d\theta \, d\phi \, dr$

$x^2 + y^2 + z^2 = z \Rightarrow r^2 = r \cos \theta$ OR $r = \cos \theta$

$$= 4 \int_0^{\pi/2} \int_0^{\pi/2} \int_0^{\cos \theta} \frac{r^2 \cos^2 \theta \, r^2 \sin \theta \, d\theta \, d\phi \, dr}{r^2} = 4 \int_0^{\pi/2} \cos^2 \theta \sin \theta \, d\theta \int_0^{\pi/2} d\phi \int_0^{\cos \theta} r^2 \, dr$$

$$= 4 \int_0^{\pi/2} \cos^2 \theta \, \sin \theta \, d\theta \left(\frac{\pi}{2} \right) \left(\frac{\cos^3 \theta}{3} \right) = 2\pi \int_0^{\pi/2} \cos^5 \theta \, \sin \theta \, d\theta = 2\pi \left(\frac{1}{6} \right)$$

$$\boxed{I = \frac{\pi}{3}}$$

9.7 DIRICHLET'S THEOREM

The theorem states that

$$\iiint \cdots\cdots \int x_1^{l_1-1} \; x_2^{l_2-1} \; x_3^{l_3-1} \;\cdots\cdots\; x_n^{l_n-1} \; dx_1 \cdot dx_2 \cdot dx_3 \;\cdots\cdots\; dx_n = \frac{\overline{|l_1} \; \overline{|l_2} \; \overline{|l_3} \;\cdots\; \overline{|l_n}}{\overline{|1 + l_1 + l_2 + l_3 + \cdots\cdots + l_n}}$$

where, the integral is extended to all positive values of the variables subjected to the condition $x_1 + x_2 + x_3 + \cdots\cdots + x_n \leq 1$. (This theorem could be proved with the idea used in Ex. 3, page 9.47).

For two variables :

$$\iint x^{m-1} \; y^{n-1} \, dx \; dy \;=\; \frac{\overline{|m} \; \overline{|n}}{\overline{|1 + m + n}} \; , \text{ where } x + y \leq 1$$

For three variables :

$$\iiint x^{l-1} \; y^{m-1} \, z^{n-1} \, dx \, dy \, dz = \frac{\overline{|l} \; \overline{|m} \; \overline{|n}}{\overline{|1 + l + m + n}} \; , \text{ where } x + y + z \leq 1$$

Particular case :

(i)
$$\iint x^{m-1} \; y^{n-1} \; dx \; dy \;=\; \frac{\overline{|m} \; \overline{|n}}{\overline{|1 + m + n}} \; h^{m+n} \; \text{ if } \; x + y \leq h$$

(ii)
$$\iiint x^{l-1} \; y^{m-1} \; z^{n-1} \, dx \, dy \, dz = \frac{\overline{|l} \; \overline{|m} \; \overline{|n}}{\overline{|1 + l + m + n}} \; h^{l+m+n} \; \text{ if } \; x + y + z < h$$

Ex. 1 : Evaluate $\iiint x^2 \, y \, z \; dx \, dy \, dz$ throughout the volume bounded by the plane

$x = 0, \; y = 0, z = 0, \; \dfrac{x}{a} + \dfrac{y}{b} + \dfrac{z}{c} = 1.$ **(Dec. 2004)**

Sol. : $I = \iiint x^2 \, y \, z \; dx \, dy \, dz$ Put $x = au, dx = a \, du, \; y = bv, dy = b \, dv,$

                                                      $z = cw, dz = c \, dw$

$$= \int \int \int a^2 \cdot u^2 \; b \, v \, c \, w \; abc \; du \, dv \, dw$$

$$= a^3 \, b^2 \, c^2 \iiint u^2 \, v \, w \; du \, dv \, dw, \; \text{ where } u + v + w \leq 1$$

$$= a^3 \, b^2 \, c^2 \iiint u^{3-1} \; v^{2-1} \; w^{2-1} \; du \, dv \, dw$$

$$= a^3 \, b^2 \, c^2 \; \frac{\overline{|3} \; \overline{|2} \; \overline{|2}}{\overline{|1 + 3 + 2 + 2}} \; \text{ by using Dirichlet's theorem}$$

$$= a^3 \, b^2 \, c^2 \; \frac{2! \; 1!}{7!}$$

$$\boxed{I = \frac{a^3 \, b^2 \, c^2}{25 \; 20}}$$

Ex. 2 : Evaluate $\iiint xyz \; dx \, dy \, dz$ taken throughout the ellipsoid $\dfrac{x^2}{a^2} + \dfrac{y^2}{b^2} + \dfrac{z^2}{c^2} \leq 1.$

Sol. : $\dfrac{x^2}{a^2} = u$ $x \, dx = \dfrac{a^2}{2} \, du \; ; \; \dfrac{y^2}{b^2} = v \; \therefore \; y \, dy = \dfrac{b^2}{2} \, dv \; ; \; \dfrac{z^2}{c^2} = w \; \therefore \; z \, dz = \dfrac{c^2}{2} \, dw$

Let us evaluate the given integral for positive octant first.

$$I = \iiint \frac{a^2}{2} \frac{b^2}{2} \frac{c^2}{2} u^{1-1} v^{1-1} w^{1-1} \, du \, dv \, dw, \text{ where } u + v + w \le 1$$

$$= \frac{a^2 b^2 c^2}{8} \frac{\overline{|1} \ \overline{|1} \ \overline{|1}}{\overline{|1+1+1+1}} = \frac{a^2 b^2 c^2}{8} \frac{1}{3!} = \frac{a^2 b^2 c^2}{48}$$

Hence for the whole ellipsoid, the given integral is $8 \times \dfrac{a^2 b^2 c^2}{48}$

$$\boxed{I = \frac{a^2 b^2 c^2}{6}}$$

Ex. 3 : *Evaluate* $\iiint (x^2 y^2 + y^2 z^2 + z^2 x^2) \, dx \, dy \, dz$ *throughout the volume of the sphere* $x^2 + y^2 + z^2 = a^2$.   **(Nov. 2015)**

Sol. : $I = \iiint (x^2 y^2 + y^2 z^2 + z^2 x^2) \, dx \, dy \, dz$

Put

$$\frac{x^2}{a^2} = u, \quad x = a \, u^{1/2}, \quad dx = \frac{a}{2} u^{-1/2} \, du$$

$$\frac{y^2}{a^2} = v, \quad y = a \, v^{1/2}, \quad dy = \frac{a}{2} v^{-1/2} \, dv$$

$$\frac{z^2}{a^2} = w, \quad z = a \, w^{1/2}, \quad dz = \frac{a}{2} w^{-1/2} \, dw$$

where $u + v + w \le 1$

By considering 8 times positive octant of a sphere,

$$I = 8 \iiint (a^4 uv + a^4 vw + a^4 uw) \frac{a^3}{8} u^{-1/2} v^{-1/2} w^{-1/2} (du \, dv \, dw)$$

$$= a^7 \iiint (u^{1/2} v^{1/2} w^{-1/2} + u^{-1/2} v^{1/2} w^{1/2} + u^{1/2} v^{-1/2} w^{1/2}) \, du \, dv \, dw$$

$$= a^7 \, 3 \left(\frac{\overline{|3/2} \ \overline{|3/2} \ \overline{|1/2}}{\overline{|1 + 3/2 + 3/2 + 1/2}} \right) \qquad \text{(By using Dirichlet's theorem)}$$

$$= 3a^7 \frac{\frac{1}{2}\sqrt{\pi} \ \frac{1}{2}\sqrt{\pi} \ \sqrt{\pi}}{\frac{7}{2} \frac{5}{2} \frac{3}{2} \frac{1}{2}\sqrt{\pi}}$$

$$\boxed{I = \frac{4 a^7 \pi}{35}}$$

EXERCISE 9.5

I.   Problems on Direct Evaluation of Triple Integral :

Prove that :

1. $\displaystyle \int_0^1 \int_0^{\sqrt{1-x^2}} \int_0^{\sqrt{1-x^2-y^2}} \frac{dz \, dy \, dx}{\sqrt{1-x^2-y^2-z^2}} = \frac{\pi^2}{8}$

2. $\displaystyle \int_0^{\pi/2} \int_0^{a \sin \theta} \int_0^{\frac{a^2-r^2}{a}} r \, d\theta \, dr \, dz = \frac{5 \pi a^3}{64}$

3. $\displaystyle \int_{-2}^{2} \int_{-\sqrt{\frac{4-x^2}{2}}}^{\sqrt{\frac{4-x^2}{2}}} \int_{x^2+3y^2}^{8-x^2-y^2} dz\ dy\ dx = 8\sqrt{2}\,\pi$

4. $\displaystyle \int_{0}^{a} \int_{0}^{a-x} \int_{0}^{a-x-y} x^2\ dz\ dy\ dx = \frac{a^5}{60}$

5. $\displaystyle \int_{0}^{a} \int_{0}^{a-x} \int_{0}^{a-x-y} (x^2+y^2+z^2)\ dz\ dy\ dx = \frac{a^5}{20}$

6. $\displaystyle \int_{0}^{a} \int_{-\sqrt{a^2-x^2}}^{\sqrt{a^2-x^2}} \int_{0}^{mx} x^2\ dx\ dy\ dz = \frac{4}{45}\,m^3\,a^5$

7. $\displaystyle \int_{0}^{\log 2} \int_{0}^{x} \int_{0}^{x+y} e^{x+y+z}\ dx\ dy\ dz = \frac{5}{8}$ **(May 2019)**

8. $\displaystyle \int_{0}^{1} dx \int_{0}^{2} dy \int_{1}^{2} x^2\,y\,z\ dz = 1$

9. $\displaystyle \int_{0}^{1} dy \int_{y^2}^{1} dx \int_{0}^{1-x} x\ dz = \frac{4}{35}$ **(Dec. 2010)**

10. $\displaystyle \int_{0}^{1} \int_{0}^{1-x} \int_{0}^{x+y} e^z\ dz\ dy\ dx = \frac{1}{2}$

11. $\displaystyle \int_{0}^{4} \int_{0}^{2\sqrt{z}} \int_{0}^{\sqrt{4z-x^2}} dy\ dx\ dz = 8\pi$

12. $\displaystyle \int_{0}^{a} \int_{0}^{x} \int_{0}^{\sqrt{x+y}} z\ dz\ dy\ dx = \frac{a^3}{4}$

13. $\displaystyle \int_{0}^{a} \int_{0}^{\sqrt{a^2-x^2}} \int_{0}^{\sqrt{a^2-x^2-y^2}} xyz\ dz\ dy\ dx = \frac{a^6}{48}$

14. $\displaystyle \int_{0}^{2a} \int_{-\sqrt{2ax-x^2}}^{\sqrt{2ax-x^2}} \int_{0}^{\sqrt{4a^2-x^2-y^2}} dx\ dy\ dz = \frac{8}{3}\pi a^3$

15. $\displaystyle \int_{0}^{\pi/2} \int_{0}^{a\cos\theta} \int_{0}^{\sqrt{a^2-r^2}} r\ dz\ dr\ d\theta = \frac{a^3}{3}\left(\frac{\pi}{2} - \frac{2}{3}\right)$

16. $\displaystyle \int_{0}^{1} dx \int_{0}^{1} dy \int_{\sqrt{x^2+y^2}}^{2} xyz\ dz = \frac{3}{8}$

17. $\displaystyle \int_{1}^{2} dz \int_{2}^{3} dy \int_{1}^{3} (x^2\,y+z)\ dx = \frac{74}{3}$

18. $\displaystyle \int_{1}^{2} dx \int_{2}^{3} dy \int_{1}^{3} (x^2\,y+z)\ dz = \frac{47}{3}$

19. $\displaystyle \int_{1}^{2} dx \int_{2}^{3} dy \int_{0}^{3} (x^2+y^3+z)\ dz = \frac{241}{4}$

20. $\displaystyle \int_{1}^{2} dx \int_{2}^{3} dy \int_{0}^{3} x^2\,y^3\,z\ dz = \frac{195}{8}$

21. $\displaystyle \int_{0}^{a} \int_{0}^{\sqrt{a^2-x^2}} \int_{0}^{\sqrt{a^2-y^2}} y\ dx\ dy\ dz = \frac{a^4}{4}$

22. $\displaystyle \int_{0}^{2a} dx \int_{0}^{x} dy \int_{y}^{x} xyz\ dz = \frac{4}{3}$

23. $\displaystyle \int_{1}^{e} dy \int_{1}^{\log y} dx \int_{1}^{e^x} \log z\ dz = \frac{e^2-8e+13}{4}$

24. $\displaystyle \int_{0}^{a} \int_{0}^{a} \int_{0}^{a} (yz+zx+xy)\ dx\ dy\ dz = \frac{3}{4}\,a^5$

25. $\displaystyle \int_{0}^{a} \int_{0}^{\sqrt{a^2-x^2}} \int_{0}^{\sqrt{a^2-x^2-y^2}} (x^2+y^2+z^2)\ dx\ dy\ dz = \frac{\pi a^5}{10}$

26. $\displaystyle \int_{0}^{2} \int_{0}^{x} \int_{0}^{x+y} e^{x+y+z}\ dx\ dy\ dz = \frac{e^8-6e^4+8e^2-3}{8}$

II. Problems on Triple Integration when the Limits are not Provided.

Evaluate :

1. $\iiint\limits_{V} \dfrac{dx\,dy\,dz}{(x^2 + y^2 + z^2)^{3/2}}$, where V is annulus between sphere $x^2 + y^2 + z^2 = a^2$, $x^2 + y^2 + z^2 = b^2$, $a > b > 0$.   **Ans. :** $4\pi \log \dfrac{a}{b}$

2. $\iiint\limits_{V} x^{l-1}\, y^{m-1}\, z^{n-1}\ dx\,dy\,dz$ taken throughout the volume of $x = 0,\ y = 0,\ z = 0,\ \dfrac{x}{a} + \dfrac{y}{b} + \dfrac{z}{c} = 1.$

$$\textbf{Ans. : } a^l\ b^m\ c^n\ \dfrac{\overline{\lvert l}\ \ \overline{\lvert m}\ \ \overline{\lvert n}}{\overline{\lvert 1 + l + m + n}}$$

3. $\iiint\limits_{V} \dfrac{z^2\,dx\,dy\,dz}{(x^2 + y^2 + z^2)}$, where V is the volume of sphere $x^2 + y^2 + z^2 = 2.$ **Ans. :** $\dfrac{8\sqrt{2}\,\pi}{9}$

4. $\iiint\limits_{V} (x^2 + y^2)\,dx\,dy\,dz$, where V is $x^2 + y^2 = 2z$ and plane $z = 2.$ **Ans. :** $\dfrac{16\,\pi}{3}$

5. $\iiint xyz\ dx\,dy\,dz$ over the tetrahedron formed by $x = 0,\ y = 0,\ z = 0,\ x + y + z = a.$ **Ans. :** $\dfrac{a^5}{720}$

6. $\iiint x^2\ dx\,dy\,dz$ over the tetrahedron formed by $x = 0,\ y = 0,\ z = 0,\ \dfrac{x}{a} + \dfrac{y}{b} + \dfrac{z}{c} = 1.$ **Ans. :** $\dfrac{a^3bc}{60}$

7. $\iiint z\,dx\,dy\,dz$ over the hemisphere $x^2 + y^2 + z^2 = a^2,\ z \geq 0$ **Ans. :** $\dfrac{\pi\,a^4}{16}$

8. $\iiint xyz\ dx\,dy\,dz$ over the first octant of the sphere $x^2 + y^2 + z^2 = a^2.$ **Ans. :** $\dfrac{a^6}{48}$

9. $\iiint (x + y + z)\,dx\,dy\,dz$ over the positive octant of the sphere $x^2 + y^2 + z^2 = a^2.$ **Ans. :** $\dfrac{3\pi\,a^4}{16}$

10. $\iiint (x + y + z)\,dx\,dy\,dz$ over the tetrahedron formed by $x = 0,\ y = 0,\ z = 0,\ x + y + z = 1.$ **Ans. :** $\dfrac{1}{8}$

❑❑❑

CHAPTER-10

AREA, VOLUME, MEAN AND ROOT MEAN SQUARE VALUES

10.1 INTRODUCTION

In this chapter, we shall discuss some important applications of double integrals which occur quite often in science and engineering. These include problems involving area, volume, mass etc. Formulae for these in terms of multiple integrals are developed and their use in examples on these topics is illustrated.

10.2 REPRESENTATION OF AREA AS A DOUBLE INTEGRAL

TYPE I **Area Enclosed by Plane Curves Expressed in Cartesian Co-ordinates**

Consider the area enclosed by the curves $y = f_1(x)$ and $y = f_2(x)$ and the ordinates $x = a$, $x = b$ $(a < b)$. Refer Fig. 10.1.

Divide this area into vertical strips of width δx. If $P(x, y)$, $R(x + \delta x, y + \delta y)$ are two neighbouring points, then the area of small rectangle PQRS = $\delta x\, \delta y$.

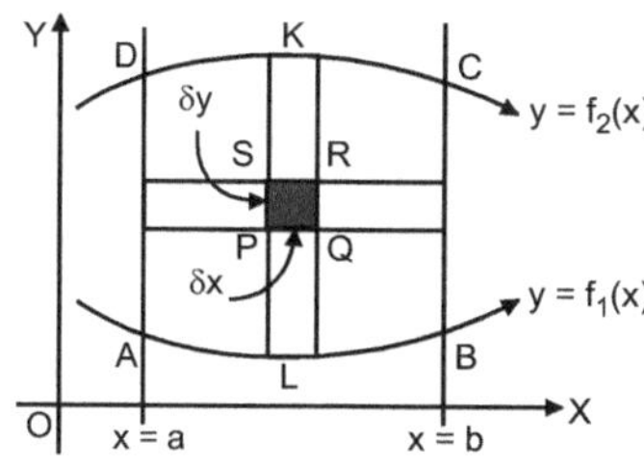

$$\therefore \quad \text{Area of strip KL} = \underset{\delta y \to 0}{\text{Lim}} \left(\Sigma\, \delta x\, \delta y \right)$$

Since, for all rectangles in this strip, δx is the same and y varies from $y = f_1(x)$ to $y = f_2(x)$.

Fig. 10.1

$$\therefore \quad \text{Area of the vertical strip } KL = \delta x\, \underset{\delta y \to 0}{\text{Lim}} \sum_{f_1(x)}^{f_2(x)} \delta y = \delta x \int_{f_1(x)}^{f_2(x)} dy \qquad \text{(replacing summation sign by integral sign)}$$

Now adding up all such strips from $x = a$ to $x = b$, we get the required area

$$\text{ABCD} = \underset{\delta x \to 0}{\text{Lim}} \left(\sum_{a}^{b} \delta x \right) \int_{f_1(x)}^{f_2(x)} dy$$

$$\boxed{\text{Area} = \int_{a}^{b} dx \int_{f_1(x)}^{f_2(x)} dy} \qquad \text{(replacing summation sign by integral sign)}$$

Thus, the area ABCD is expressed in terms of double integral as,

$$\text{Area} = \int_{a}^{b} \left\{ \int_{f_1(x)}^{f_2(x)} dy \right\} dx.$$

Here inner integration is carried out w.r.t. y treating x as constant and outer integration is then carried out w.r.t. 'x'.

Similarly, dividing the area EFGH (Refer Fig. 10.2) into horizontal strips of width δy, we get the area

$$\text{EFGH} = \int_{c}^{d} \left\{ \int_{g_1(y)}^{g_2(y)} dx \right\} dy$$

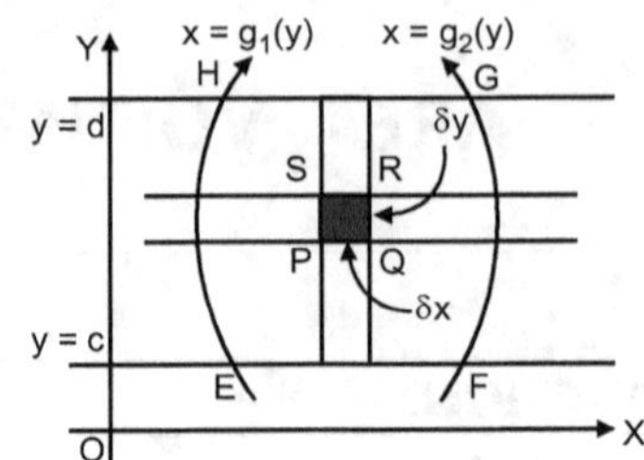

$$\boxed{\text{Area} = \int_{c}^{d} dy \int_{g_1(y)}^{g_2(y)} dx}$$

Fig. 10.2

Note :

1. The area A included by the curve $y = f(x)$, the x-axis, and the ordinates $x = a$ and $x = b$ is given by;

$$\boxed{A = \int_{a}^{b} y\, dx = \int_{a}^{b} f(x)\, dx}$$

[Refer Fig. 10.3 (a)]

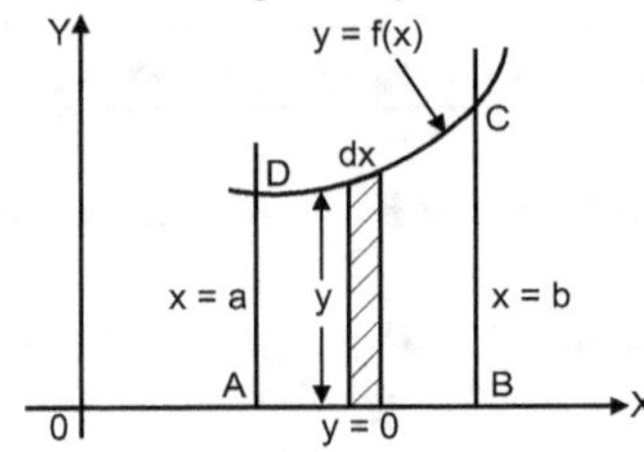

Fig. 10.3 (a)

2. The area A included by the curve $x = f(y)$, the y-axis and the abscissa $y = c$ and $y = d$ is given by

$$\boxed{A = \int_{c}^{d} x\, dy = \int_{c}^{d} f(y)\, dy}$$

[Refer Fig. 10.3 (b)]

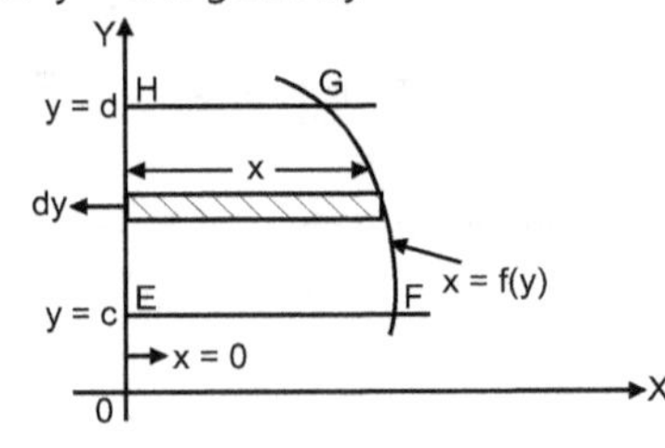

Fig. 10.3 (b)

3. Double integration is generally not suitable for parametric equations of curves. Instead we employ the following formula for area as :

$$\boxed{\begin{aligned} \text{Area} &= \int y\, dx \text{ if x-axis is the boundary of the area} \\ \text{Area} &= \int x\, dy \text{ if y-axis is the boundary of the area} \end{aligned}}$$

4. Whenever we are finding the area of a plane region always consider *the symmetry*.

5. Area is always to be considered as positive.

10.3 TYPE II : AREA ENCLOSED BY PLANE CURVES EXPRESSED IN POLAR CO-ORDINATES

Consider the area enclosed by the polar curves $r = f_1(\theta)$ and $r = f_2(\theta)$ and the line $\theta = \alpha$, $\theta = \beta$ ($\alpha < \beta$). Refer Fig. 10.4.

Let $P(r, \theta)$, $Q(r + \delta r, \theta + \delta \theta)$ be two neighbouring points. Mark circular areas of radii r and $(r + \delta r)$ meeting OQ in R and OP (produced) in S.

Since arc $PR = r\,\delta\theta$ and arc $PS = \delta r$

∴ Area of the curvilinear rectangle, PQRS is approximately $=$ (PR) (PS) $=$ $(r\,\delta\theta)\,(\delta r)$. If the whole area is divided into such curvilinear rectangles, the sum $\sum\sum r\,\delta\theta\,\delta r$, taken for all these rectangles, gives in the limit the area A.

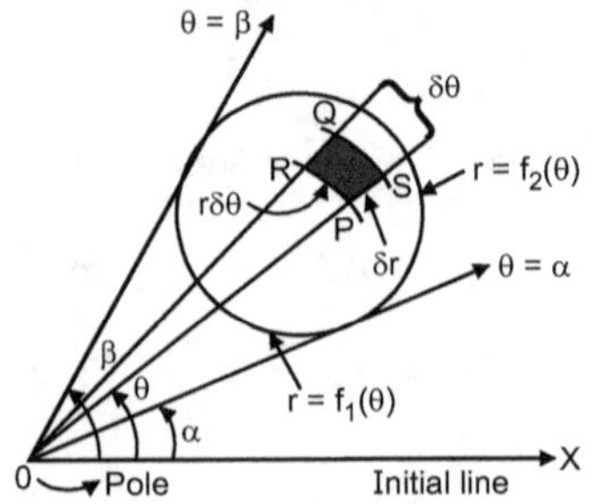

Fig. 10.4

Hence,

$$A = \lim_{\substack{\delta r \to 0 \\ \delta\theta \to 0}} \sum_{\alpha}^{\beta} \sum_{f_1(\theta)}^{f_2(\theta)} r \, \delta\theta \, \delta r$$

$$\boxed{\text{Area} = \int_{\alpha}^{\beta} \left\{ \int_{f_1(\theta)}^{f_2(\theta)} r \, dr \right\} d\theta}$$

Note :

1. The area (Refer Fig. 10.5) bounded by the curve $r = f(\theta)$ and the lines $\theta = \alpha$ and $\theta = \beta$ is

$$\boxed{\text{Area} = \frac{1}{2} \int_{\alpha}^{\beta} r^2 \, d\theta}$$

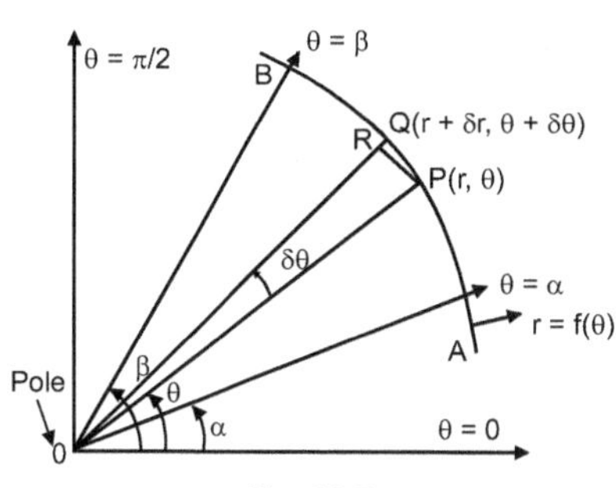

Fig. 10.5

2. For the equation of the curve in implicit form, if the loop does not lie on the x or y-axis, then it is inclined to them. In case of *inclined loop*, we change the equation to polar co-ordinates with $x = r \cos\theta$, $y = r \sin\theta$.

3. Always consider the symmetry.

10.4 ILLUSTRATIONS ON AREA ENCLOSED BY PLANE CURVES

Ex. 1 : Find the area between the curves $y^2 = 4x$ and $2x - 3y + 4 = 0$.   *(May 2016)*

Sol. : The points of intersection of parabola $y^2 = 4x$ and the straight line $2x - 3y + 4 = 0$ is obtained by solving the two equations

i.e. $\dfrac{y^2}{2} = 2x$

$\therefore \quad \dfrac{y^2}{2} - 3y + 4 = 0.$

or $\quad y^2 - 6y + 8 = 0$ or $(y - 4)(y - 2) = 0$

When $y = 4$, $x = 4$ and when $y = 2$, $x = 1$. (1, 2) and (4, 4) are the points of intersection. The required area is shown shaded in Fig. 10.6. For straight line put $y = 0$, $x = -2$ and put $x = 0$, $y = \dfrac{4}{3}$.

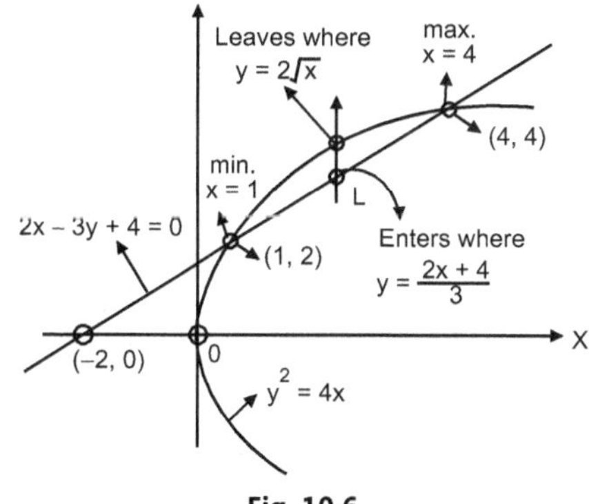

Fig. 10.6

Note that there is no symmetry.

$\therefore \quad$ Required area is $A = \int \left\{ \int dy \right\} dx$.

Imagine vertical line L cutting through region R.

Limits for y : $y = \dfrac{2x + 4}{3}$ to $y = 2\sqrt{x}$; Limits for x : $x = 1$ to $x = 4$.

$$A = \int_{1}^{4} \left\{ \int_{\frac{2x+4}{3}}^{2\sqrt{x}} dy \right\} dx = \int_{1}^{4} [y]_{\frac{2x+4}{3}}^{2\sqrt{x}} dx$$

$$= \int_{1}^{4} \left[2\sqrt{x} - \left(\frac{2x+4}{3}\right) \right] dx = \left[2 \cdot \frac{x^{3/2}}{3/2} - \frac{2}{3} \cdot \frac{x^2}{2} - \frac{4}{3} x \right]_{1}^{4} = \left(\frac{32}{3} - \frac{16}{3} - \frac{16}{3}\right) - \left(\frac{4}{3} - \frac{1}{3} - \frac{4}{3}\right) = \frac{1}{3}$$

Note : We can also use $A = \int \left\{ \int dx \right\} dy$, then imagine horizontal line and

$$A = \int_{2}^{4} \left\{ \int_{y^2/4}^{3y - 4/2} dx \right\} dy, \text{ we will get the same answer } A = \frac{1}{3}$$

Ex. 2 : *Find the area between the curve* $\dfrac{y+8}{x} = x-2$ *and x-axis.*

Sol. : $\dfrac{y+8}{x} = x-2$ i.e. $y+8 = x^2-2x$ or $y = x^2-2x-8$ represents a parabolic shape. $\therefore\ y = (x-4)(x+2)$

It's intersection with x-axis is (4, 0), (–2, 0) and with y-axis is (0, –8). Hence required area lies below x-axis as shown shaded in Fig. 10.7. Since the equation of the parabola is expressed directly in terms of x (i.e. $y = x^2-2x-8$), hence we will integrate first w.r.t. y and then w.r.t. x.

$\therefore$ Imagine a vertical line L cutting through region R.

Limits for y : $y = x^2-2x-8$ to $y = 0$

Limits for x : $x = -2$ to $x = 4$.

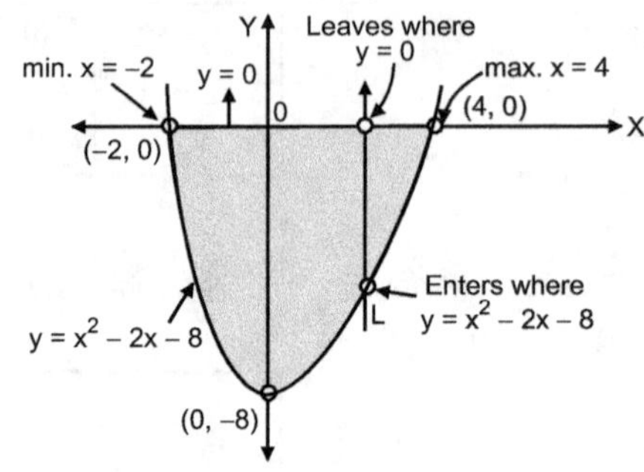

Fig. 10.7

$$\text{Area} = \int\left\{\int dy\right\}dx = \int_{-2}^{4}\left\{\int_{x^2-2x-8}^{0} dy\right\}dx$$

$$= \int_{-2}^{4}[0-(x^2-2x-8)]\,dx = \int_{-2}^{4}(-x^2+2x+8)\,dx$$

$$= \left[\frac{-x^3}{3}+\frac{2x^2}{2}+8x\right]_{-2}^{4} = \left(\frac{-64}{3}+16+32-\frac{8}{3}-4+16\right)$$

$$= \frac{108}{3}$$

$$\boxed{\text{Area} = 36}$$

Ex. 3 : *Find by double integration the area included between the curves $y = 3x^2-x-3$ and $y = -2x^2+4x+7$.*

Sol. : $y = 3x^2-x-3$ and $y = -2x^2+4x+7$ represent parabolic shape. We will solve the two equations.

i.e. $3x^2-x-3 = -2x^2+4x+7$ or $5x^2-5x-10 = 0$ or $x^2-x-2 = 0$

$\therefore$ $(x-2)(x+1) = 0$

$\therefore$ when $x = -1$, $y = 1$; when $x = 2$, $y = 7$. Hence, (–1, 1), (2, 7) are the points of intersection (Refer Fig. 10.8).

For $y = 3x^2-x-3$:

x	0	1	2
y	–3	–1	7

and

For $y = -2x^2+4x+7$:

x	0	1	2
y	7	9	7

$\therefore$ Required area $= \int\left\{\int dy\right\}dx = \int_{-1}^{2}\left\{\int_{3x^2-x-3}^{-2x^2+4x+7} dy\right\}dx$

$$= \int_{-1}^{2}(-2x^2+4x+7-3x^2+x+3)\,dx$$

$$= \int_{-1}^{2}(-5x^2+5x+10)\,dx$$

$$= \left[\frac{-5x^3}{3}+\frac{5x^2}{2}+10x\right]_{-1}^{2} = \frac{45}{2}$$

$$\boxed{\text{Area} = \frac{45}{2}}$$

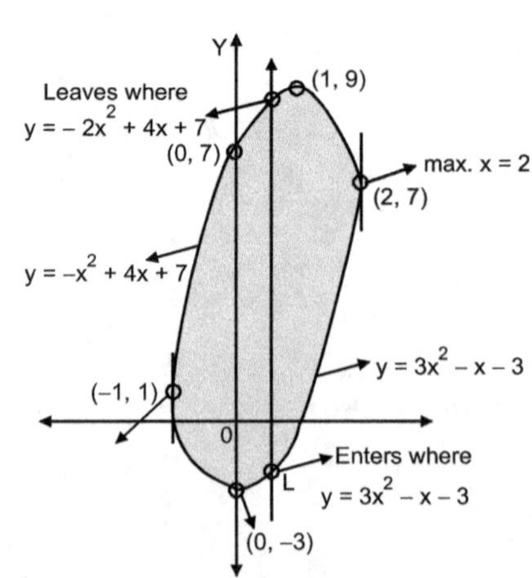

Fig. 10.8

Ex. 4 : *Find the total area included between the two branches of the curve* $y^2 (4 - x)(x - 2) = x^2$ *and the two asymptotes.*

Sol. :
$$y^2 = \frac{x^2}{(4 - x)(x - 2)}$$

It is symmetrical about x-axis. Intersection with the co-ordinate axis is only at (0, 0). x = 2, x = 4 are the asymptotes to the given curve, when x = 3, y = ± 3. A rough sketch of the curve is as shown in Fig. 10.9. Because of the symmetry,

$$\text{Required area} = 2 \int_2^4 \left\{ \int dy \right\} dx = 2 \int_2^4 \int_0^{\frac{x}{\sqrt{(4-x)(x-2)}}} dy\, dx = 2 \int_2^4 \frac{x}{\sqrt{(4-x)(x-2)}} dx$$

Put $x - 2 = 2\sin^2\theta$, $dx = 4\sin\theta\cos\theta\, d\theta$

x	2	4
θ	0	π/2

$$\text{Area} = 2 \int_0^{\pi/2} \frac{(2 + 2\sin^2\theta)\, 4\sin\theta\cos\theta\, d\theta}{\sqrt{(2 - 2\sin^2\theta)\, 2\sin^2\theta}}$$

$$= 8 \int_0^{\pi/2} (1 + \sin^2\theta)\, d\theta = 8 \left(\frac{\pi}{2} + \frac{1}{2}\frac{\pi}{2} \right)$$

$$\boxed{\text{Area} = 6\pi}$$

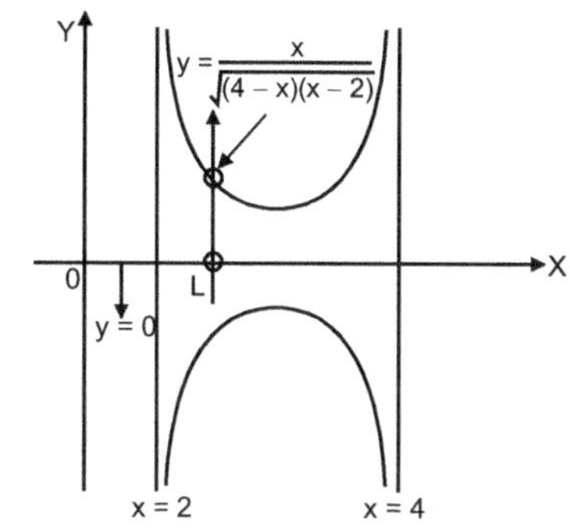

Fig. 10.9

Ex. 5 : *Find by double integration the area between the curve* $y^2 x = 4a^2 (2a - x)$ *and its asymptote.*   **(Dec. 2004, May 2013)**

Sol. :
$$y^2 = \frac{4a^2 (2a - x)}{x}$$

It is symmetrical about x-axis and does not pass through (0, 0). Also x = 0 i.e. y-axis is the asymptote to the curve and intersects x-axis at (2a, 0). For x < 0, x > 2a, curve does not exist and for x = a, y = ± 2a. This famous curve is known as "Witch of Agnesi" and it is shown shaded in Fig. 10.10. Because of the symmetry,

$$\text{Required area} = 2 \int_0^{2a} \left\{ \int dy \right\} dx = 2 \int_0^{2a} \int_0^{2a\sqrt{2a-x}/\sqrt{x}} dy\, dx$$

$$\text{Area} = 2 \int_0^{2a} 2a \cdot \sqrt{\frac{2a - x}{x}}\, dx$$

$$= 4a \int_0^{2a} x^{-1/2} (2a - x)^{1/2}\, dx \qquad \text{Put } x = 2at, \ dx = 2a\, dt$$

x	0	2a
t	0	1

$$\text{Area} = 4a \int_0^1 (2at)^{-1/2} (2a)^{1/2} (1 - t)^{1/2}\, 2a\, dt$$

$$= 8a^2 \int_0^1 t^{-1/2} (1 - t)^{1/2}\, dt = 8a^2\, B\left(\frac{1}{2}, \frac{3}{2} \right)$$

$$= 8a^2\, \frac{\underline{|1/2}\ \underline{|3/2}}{\underline{|2}} = 8a^2 \sqrt{\pi}\, \frac{1}{2}\, \sqrt{\pi}$$

$$\boxed{\text{Area} = 4\pi a^2}$$

Fig. 10.10

Ex. 6 : *Find the area bounded by the curve* $x(x^2 + y^2) = a(y^2 - x^2)$ *and its asymptote. Also find the area of the loop of the curve.*

Sol. : | **Part I** | $x(x^2 + y^2) = a(y^2 - x^2)$ i.e. $ay^2 - xy^2 = x^3 + ax^2$

or $y^2(a - x) = x^2(a + x)$ i.e. $y^2 = \dfrac{x^2(a + x)}{a - x}$

It is symmetrical about x-axis and passes through (0, 0). Also $y = \pm x$ are the tangents at (0, 0). It intersects x-axis at (0, 0) (–a, 0). Here $x = a$ is the asymptote to the given curve. The required area between the curve and its asymptote is shown shaded in Fig. 10.11. Because of symmetry,

$$\text{Area} = 2\int_0^a \left\{\int dy\right\} dx = 2\int_0^a \left\{\int_0^{x\sqrt{\frac{a+x}{a-x}}} dy\right\} dx$$

$$= \int_0^a \frac{x\sqrt{a+x}}{\sqrt{a-x}} dx = 2\int_0^a \frac{x(a+x)}{\sqrt{a^2-x^2}} dx$$

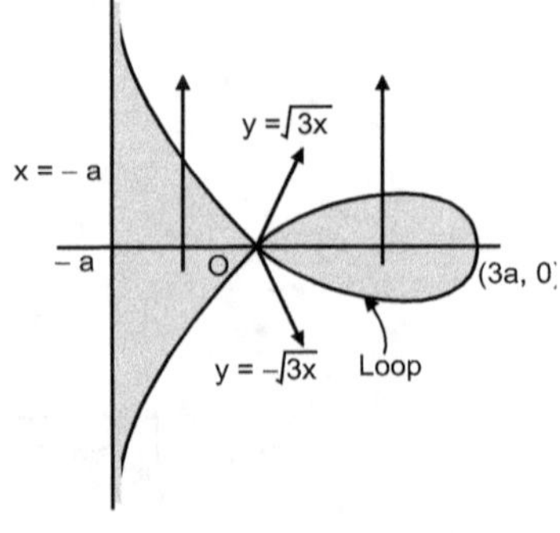

Fig. 10.11

$$= 2\int_0^{\pi/2} \frac{a\sin\theta\,(a + a\sin\theta)\,a\cos\theta\,d\theta}{a\cos\theta}$$

(by putting $x = a\sin\theta$)

$$= 2a^2\int_0^{\pi/2} (\sin\theta + \sin^2\theta)\,d\theta = 2a^2\left[1 + \frac{1}{2}\cdot\frac{\pi}{2}\right]$$

$$\boxed{\text{Area} = 2a^2\left[1 + \frac{\pi}{4}\right]}$$

| **Part II** | To find the area of the loop, we have,

$$\text{Area} = 2\int \left\{\int dy\right\} dx = 2\int_{-a}^0 \left\{\int_0^{x\sqrt{\frac{a+x}{a-x}}} dy\right\} dx$$

$$= 2\int_{-a}^0 \frac{x\sqrt{a+x}}{\sqrt{a-x}} dx = 2\int_{-a}^0 \frac{x(a+x)}{\sqrt{a^2-x^2}} dx$$

Put $x = -a\sin\theta$

$$= 2\int_{\pi/2}^0 \frac{(-a\sin\theta)\,(a - a\sin\theta)\,(-a\cos\theta\,d\theta)}{a\cos\theta} = -2a^2\int_0^{\pi/2}(\sin\theta - \sin^2\theta)\,d\theta = -2a^2\left[1 - \frac{\pi}{4}\right]$$

$$\boxed{\text{Area} = 2a^2\left(1 - \frac{\pi}{4}\right)}$$

(numerically)

Ex. 7 : *Show that the area of the loop of the curve* $y^2(a + x) = x^2(3a - x)$ *is equal to the area between the curve and its asymptote.*

Sol. : $y^2 = \dfrac{x^2(3a - x)}{a + x}$. It is symmetrical about x-axis, passes through origin and $y = \pm\sqrt{3}\,x$ are tangents at origin. It intersects x-axis at (3a, 0) (0, 0). Also $x = -a$ is an asymptote to the curve.

The required area is shown shaded in Fig. 10.12.

| **Part I** | $\text{Area of the loop} = 2\displaystyle\int_0^{3a} \int_0^{x\sqrt{\frac{3a-x}{a+x}}} dy\,dx$

$$\text{Area} = 2\int_0^{3a} x\sqrt{\frac{3a-x}{a+x}}\,dx$$

Fig. 10.12

The substitution employed here is obtained by the following rule useful for integration of expressions of the type under consideration.

Rule : Half the difference between a + x and 3a − x is $\frac{1}{2}$ [(a + x) − (3a − x)] i.e. x − a, and half their sum is $\frac{1}{2}$ [(a + x) + (3a − x)] i.e. 2a.

We therefore put $x - a = 2a \sin\theta$ (or $x - a = 2a \cos\theta$) ∴ $dx = 2a \cos\theta \, d\theta$

x	0	3a
θ	−π/6	π/2

$$\text{Area of the loop} \;=\; 2 \int_{-\pi/6}^{\pi/2} \frac{(a + 2a\sin\theta)\sqrt{3a - a - 2a\sin\theta} \cdot 2a\cos\theta \, d\theta}{\sqrt{a + a + 2a\sin\theta}}$$

$$= 4a^2 \int_{-\pi/6}^{\pi/2} \frac{(1 + 2\sin\theta)\sqrt{1 - \sin\theta}}{\sqrt{1 + \sin\theta}} \cos\theta \, d\theta$$

$$= 4a^2 \int_{-\pi/6}^{\pi/2} \frac{(1 + 2\sin\theta)(1 - \sin\theta)}{\sqrt{1 - \sin^2\theta}} \cos\theta \, d\theta = 4a^2 \int_{-\pi/6}^{\pi/2} (1 + \sin\theta - 2\sin^2\theta) \, d\theta$$

$$= 4a^2 \int_{-\pi/6}^{\pi/2} (1 + \sin\theta - 1 + \cos 2\theta) \, d\theta = 4a^2 \int_{-\pi/6}^{\pi/2} (\sin\theta + \cos 2\theta) \, d\theta$$

$$= 4a^2 \left[-\cos\theta + \frac{\sin 2\theta}{2} \right]_{-\pi/6}^{\pi/2} = 4a^2 \left[0 + \cos(-\pi/6) - \frac{1}{2}\sin\left(\frac{-\pi}{3}\right) \right]$$

$$= 4a^2 \left[\frac{\sqrt{3}}{2} - \frac{1}{2}\left(-\frac{\sqrt{3}}{2} \right) \right]$$

$$\boxed{\text{Area of the loop} \;=\; 3\sqrt{3}\, a^2} \qquad\qquad \text{... (i)}$$

Part II Area between the curve and its asymptote $= 2 \int_{-a}^{0} \int_{0}^{x\sqrt{\dfrac{3a - x}{a + x}}} dy \, dx.$

$$\text{Area} \;=\; 2 \int_{-a}^{0} \frac{x\sqrt{3a - x}}{\sqrt{a + x}} \, dx \quad \text{as in the part (i) Put } x - a = 2a\sin\theta$$

x	−a	0
θ	−π/2	−π/6

$$= 2 \int_{-\pi/2}^{-\pi/6} \frac{(a + 2a\sin\theta)\sqrt{2a(1 - \sin\theta)} \, 2a\cos\theta \, d\theta}{\sqrt{2a(1 + \sin\theta)}}$$

$$= 4a^2 \int_{-\pi/2}^{-\pi/6} (\sin\theta + \cos 2\theta) \, d\theta \qquad\qquad \text{(As simplified in part I)}$$

$$= 4a^2 \left[-\cos\theta + \frac{\sin 2\theta}{2} \right]_{-\pi/2}^{-\pi/6} = 4a^2 \left[-\cos\left(\frac{-\pi}{6}\right) + \frac{1}{2}\sin\left(\frac{-\pi}{3}\right) - 0 \right]$$

$$= 4a^2 \left[-\cos \pi/6 - \frac{1}{2}\sin \pi/3 \right]$$

$$= -4a^2 \left[\frac{\sqrt{3}}{2} + \frac{1}{2}\frac{\sqrt{3}}{2} \right]$$

$$= -4a^2 \left(\frac{3\sqrt{3}}{4} \right) = -3\sqrt{3}\, a^2 = 3\sqrt{3}\, a^2 \text{ (numerically)} \qquad\qquad \text{... (ii)}$$

From (i) and (ii), it is clear that $\boxed{\text{Area between the curve and its asymptote} \;=\; \text{Area of the loop of the curve.}}$

Ex. 8 : *Show that the area of the curve $a^2 x^2 = y^3 (2a - y)$ is πa^2.*

Sol. : $\qquad\qquad a^2 x^2 = y^3 (2a - y)$

It is symmetrical about y-axis, passes through origin and there exists a cusp at origin. It intersects y-axis at (0, 0) (0, 2a). Also $\left(\dfrac{dx}{dy}\right)_{(0,\,2a)} = \infty \Rightarrow$ tangent at (0, 2a) is parallel to x-axis. The required area is shown shaded in Fig. 10.13. Since the loop is on y-axis,

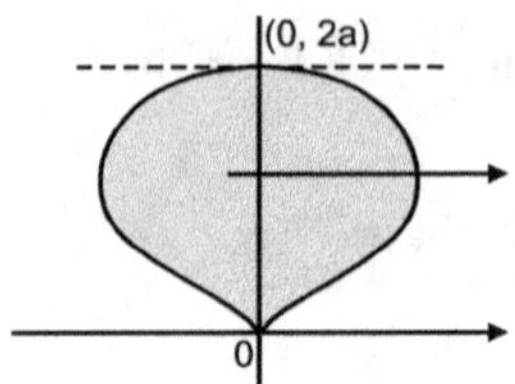

Fig. 10.13

$$\text{Area} = 2 \int x \, dy = 2 \int_0^{2a} \frac{y^{3/2} (2a - y)^{1/2}}{a} \, dy$$

$$= \frac{2}{a} \int_0^{\pi/2} (2a \sin^2 \theta)^{3/2} [2a(1 - \sin^2 \theta)]^{1/2} \cdot 4a \sin \theta \cos \theta \, d\theta \qquad \text{(putting } y = 2a \sin^2 \theta\text{)}$$

$$= 32a^2 \int_0^{\pi/2} \sin^4 \theta \cos^2 \theta \, d\theta = 32a^2 \left[\frac{3 \cdot 1 \cdot 1}{6 \cdot 4 \cdot 2}\right] \frac{\pi}{2}$$

$$\boxed{\text{Area} = \pi a^2}$$

Ex. 9 : *Find the area of the curve $x^4 - 3ax^3 + a^2 (2x^2 + y^2) = 0$.*

Sol. : $x^4 - 3ax^3 + a^2 (2x^2 + y^2) = 0 \quad$ or $\quad a^2 y^2 = 3ax^3 - x^4 - 2a^2 x^2$

i.e. $\qquad\qquad a^2 y^2 = x^2 (3ax - x^2 - 2a^2) \quad$ or $\quad a^2 y^2 = x^2 (2a - x) (x - a)$

It is symmetrical about x-axis, passes through (0, 0). It intersects x-axis at (0, 0), (a, 0), (2a, 0). No asymptote. For 0 < x < a, x > 2a, curve does not exist. The required area is shown shaded in Fig. 10.14.

$$\text{Area} = 2 \int y \, dx$$

$$\text{Area} = 2 \int_a^{2a} \frac{x \sqrt{(x - a)(2a - x)}}{a} \, dx$$

$$= \frac{2}{a} \int_a^{2a} x \sqrt{(x - a)(2a - x)} \, dx$$

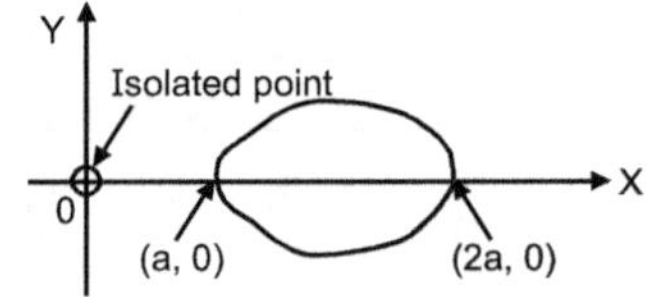

Fig. 10.14

Important Note : $\boxed{\text{For } I = \int_\alpha^\beta f(x - \alpha, \beta - x) \, dx \text{ always put } x = \alpha \cos^2 \theta + \beta \sin^2 \theta}$

Here, $\alpha = a, \quad \beta = 2a \quad \therefore$ We put $x = a \cos^2 \theta + 2a \sin^2 \theta$

$\therefore \qquad x - a = a \cos^2 \theta + 2a \sin^2 \theta - a = 2a \sin^2 \theta - a \sin^2 \theta = a \sin^2 \theta$

$$dx = 2a \sin \theta \cos \theta \, d\theta$$

x	a	2a
θ	0	π/2

and $\qquad\qquad 2a - x = 2a - a - a \sin^2 \theta = a \cos^2 \theta$

$$\therefore \qquad \text{Area} = \frac{2}{a} \int_0^{\pi/2} (a + a \sin^2 \theta) \sqrt{a \sin^2 \theta \cdot a \cos^2 \theta} \; (2a \sin \theta \cos \theta \, d\theta)$$

$$= 4a^2 \left[\int_0^{\pi/2} \sin^2 \theta \cos^2 \theta \, d\theta + \int_0^{\pi/2} \sin^4 \theta \cos^2 \theta \, d\theta \right]$$

$$= 4a^2 \left[\frac{1 \cdot 1}{4 \cdot 2} \frac{\pi}{2}\right] + \frac{3 \cdot 1 \cdot 1}{6 \cdot 4 \cdot 2} \frac{\pi}{2}$$

$$\boxed{\text{Area} = \frac{3\pi a^2}{8}}$$

Ex. 10 : *Find the total area included between the two cardioides* $r = a(1 + \cos\theta)$ *and* $r = a(1 - \cos\theta)$.

Sol. : Required area is as shown shaded in Fig. 10.15.

$$\text{Area} = 4 \int_0^{\pi/2} \left\{ \int_0^{a(1-\cos\theta)} r\, dr \right\} d\theta = 4 \int_0^{\pi/2} \frac{a^2(1-\cos\theta)^2}{2}\, d\theta$$

$$= 2a^2 \int_0^{\pi/2} (1 - 2\cos\theta + \cos^2\theta)\, d\theta = 2a^2 \left[\frac{\pi}{2} - 2(1) + \frac{1}{2}\frac{\pi}{2} \right]$$

$$\boxed{\text{Area} = 2a^2 \left[\frac{3\pi}{4} - 2 \right]}$$

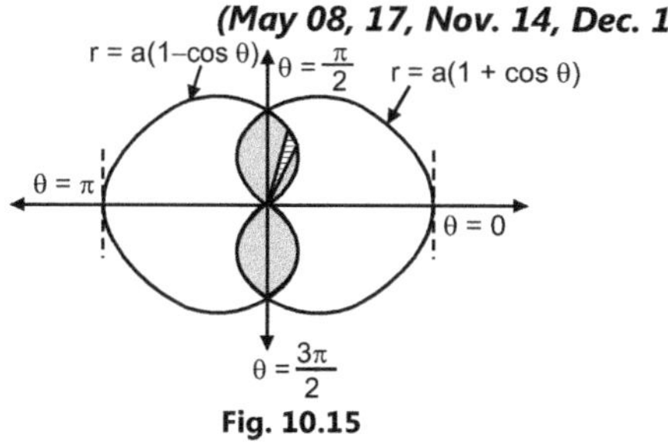

Ex. 11 : *Find the area common to the circles* $x^2 + y^2 = a^2$ *and* $x^2 + y^2 = 2ax$.   **(May 2011)**

Sol. : Required area is shown shaded in Fig. 10.16. Equation of both the circles in polar co-ordinates is $r = a$, $r = 2a\cos\theta$. Solving area $a = 2a\cos\theta$, $\cos\theta = \dfrac{1}{2}$, $\theta = \dfrac{\pi}{3}$.

$$\text{Area} = 2 \int_0^{\pi/3} \left\{ \int_0^{a} r\, dr \right\} d\theta + 2 \int_{\pi/3}^{\pi/2} \left\{ \int_0^{2a\cos\theta} r\, dr \right\} d\theta$$

$$= 2 \int_0^{\pi/3} \frac{a^2}{2}\, d\theta + 2 \int_{\pi/3}^{\pi/2} \frac{4a^2\cos^2\theta}{2}\, d\theta$$

$$= \frac{\pi a^2}{3} + 2a^2 \int_{\pi/3}^{\pi/2} (1 + \cos 2\theta)\, d\theta$$

$$= \frac{\pi a^2}{3} + 2a^2 \left((\pi/2 - \pi/3) + \left[\frac{\sin 2\theta}{2} \right]_{\pi/3}^{\pi/2} \right)$$

$$= \frac{\pi a^2}{3} + 2a^2 \left(\frac{\pi}{6} + 0 - \frac{1}{2}\frac{\sqrt{3}}{2} \right) = \frac{\pi a^2}{3} + 2a^2 \left(\frac{\pi}{6} - \frac{\sqrt{3}}{4} \right)$$

$$\boxed{\text{Area} = \frac{2\pi a^2}{3} - \frac{\sqrt{3}\, a^2}{2}}$$

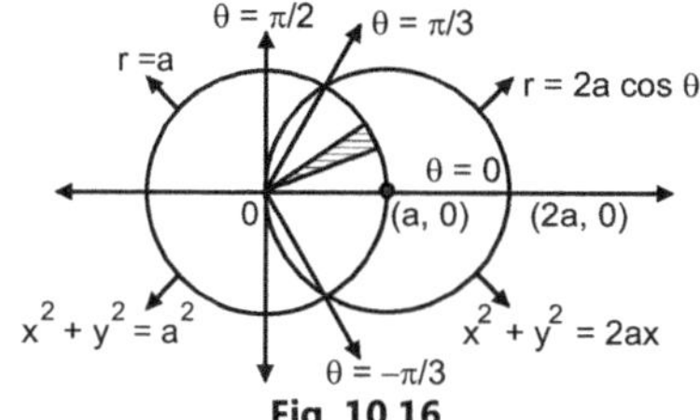

Ex. 12 : *Find by double integration the area inside the circle* $r = a\sin\theta$ *and outside the cardioide* $r = a(1 - \cos\theta)$.

(Nov./Dec. 2019, May 2007, 2006; Dec. 2008, 2017)

Sol. : The required area is as shown shaded in Fig. 10.17.

$$\text{Area} = \int_0^{\pi/2} \left\{ \int_{a(1-\cos\theta)}^{a\sin\theta} r\, dr \right\} d\theta$$

$$= \frac{1}{2} \int_0^{\pi/2} [a^2\sin^2\theta - a^2(1-\cos\theta)^2]\, d\theta$$

$$= \frac{a^2}{2} \int_0^{\pi/2} [\sin^2\theta - 1 + 2\cos\theta - \cos^2\theta]\, d\theta$$

$$= \frac{a^2}{2} \left[\frac{1}{2}\frac{\pi}{2} - \frac{\pi}{2} + 2(1) - \frac{1}{2}\frac{\pi}{2} \right] = a^2 \left(1 - \frac{\pi}{4} \right)$$

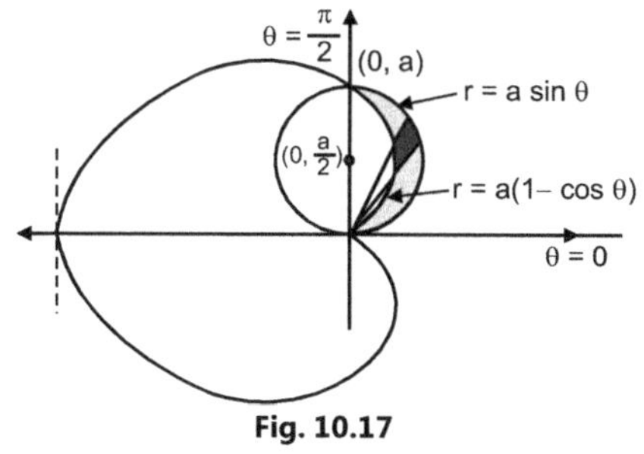

Ex. 13 : *Find the area inside the cardioide* $r = 2a(1 + \cos\theta)$ *and outside the parabola* $r = \dfrac{2a}{1 + \cos\theta}$.

Sol. : Here $r = 2a(1 + \cos\theta)$ is a cardioide and $r = \dfrac{2a}{1 + \cos\theta}$ is a parabola.

θ	0	$\pi/2$	π	$3\pi/2$	2π
$r = 2a(1 + \cos\theta)$	4a	2a	0	2a	4a
$r = \dfrac{2a}{1 + \cos\theta}$	a	2a	∞	2a	a

The required region of integration is as shown shaded in Fig. 10.18.

$$\text{Area} = 2 \int_0^{\pi/2} \left\{ \int_{\frac{2a}{1+\cos\theta}}^{2a(1+\cos\theta)} r\, dr \right\} d\theta$$

$$= 2 \int_0^{\pi/2} \frac{1}{2} \left(4a^2 (1+\cos\theta)^2 - \frac{4a^2}{(1+\cos\theta)^2} \right) d\theta$$

$$= 4a^2 \int_0^{\pi/2} (1 + 2\cos\theta + \cos^2\theta)\, d\theta - 4a^2 \int_0^{\pi/2} \frac{d\theta}{4\cos^4\frac{\theta}{2}}$$

$$= 4a^2 \left[\frac{\pi}{2} + 2(1) + \frac{1}{2}\frac{\pi}{2} \right] - a^2 \int_0^{\pi/4} \sec^4 t \; 2\, dt \qquad \left(\because \; \frac{\theta}{2} = t \right)$$

$$= 4a^2 \left[\frac{3\pi}{4} + 2 \right] - 2a^2 \int_0^{\pi/4} (1 + \tan^2 t)\sec^2 t\, dt \qquad (\because \; \tan t = u)$$

$$= 4a^2 \left[\frac{3\pi}{4} + 2 \right] - 2a^2 \int_0^1 (1 + u^2)\, du = 4a^2 \left[\frac{3\pi}{4} + 2 \right] - 2a^2 \left[1 + \frac{1}{3} \right]$$

$$= 4a^2 \left[\frac{3\pi}{4} + 2 - \frac{2}{3} \right]$$

$$\boxed{\text{Area} = 4a^2 \left[\frac{3\pi}{4} + \frac{4}{3} \right]}$$

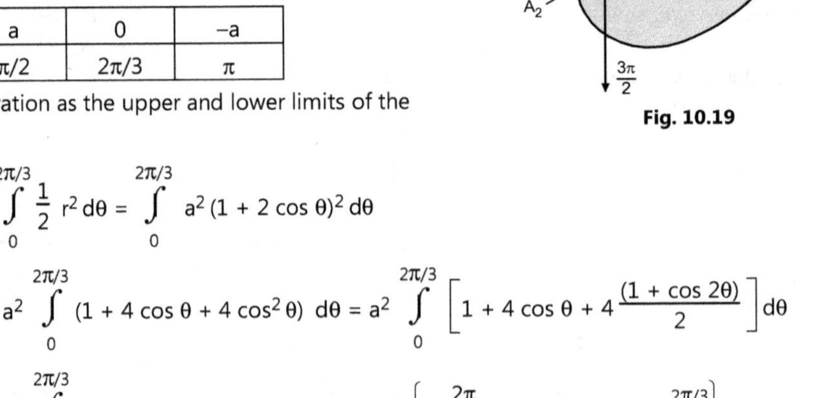

Fig. 10.18

Ex. 14 : *Find the area of the region included between the two loops of the curve* $r = a(1 + 2\cos\theta)$.

Sol. : Given $r = a(1 + 2\cos\theta)$. It is symmetrical about the initial line. Put

$r = 0$, $\cos\theta = -\dfrac{1}{2}$, $\theta = \dfrac{2\pi}{3}$.

It passes through pole. For $2\pi/3 < \theta < \pi$,
r is negative, there exists a inner loop and because of symmetry the curve is drawn as shown in Fig. 10.19.

r	3a	a	0	−a
θ	0	π/2	2π/3	π

Here we cannot use double integration as the upper and lower limits of the inner integral will be same.

$$\text{Area of outer loop} = A_1 = 2 \int_0^{2\pi/3} \frac{1}{2} r^2 d\theta = \int_0^{2\pi/3} a^2 (1 + 2\cos\theta)^2 d\theta$$

$$A_1 = a^2 \int_0^{2\pi/3} (1 + 4\cos\theta + 4\cos^2\theta)\, d\theta = a^2 \int_0^{2\pi/3} \left[1 + 4\cos\theta + 4\frac{(1+\cos 2\theta)}{2} \right] d\theta$$

$$= a^2 \int_0^{2\pi/3} (3 + 4\cos\theta + 2\cos 2\theta)\, d\theta = a^2 \left\{ 3 \cdot \frac{2\pi}{3} + [4\sin\theta + \sin 2\theta]_0^{2\pi/3} \right\}$$

$$= a^2 \left\{ 2\pi + \frac{4\sqrt{3}}{2} - \frac{\sqrt{3}}{2} \right\} = a^2 \left(2\pi + \frac{3\sqrt{3}}{2} \right)$$

$$\text{Area of inner loop,} \quad A_2 = 2 \int_{2\pi/3}^{\pi} \frac{1}{2} r^2 d\theta = \int_{2\pi/3}^{\pi} a^2 (1 + 2\cos\theta)^2 d\theta$$

The figure for Fig. 10.19 shows:

Fig. 10.19

$$A_2 = a^2 [3\theta + 4 \sin \theta + \sin 2\theta]_{2\pi/3}^{\pi}$$

$$= a^2 \left[\pi - 4 \frac{\sqrt{3}}{2} + \frac{\sqrt{3}}{2} \right] = a^2 \left[\pi - \frac{3\sqrt{3}}{2} \right]$$

$$\text{Required area} = A_1 - A_2 = a^2 \left[2\pi + \frac{3\sqrt{3}}{2} - \pi + \frac{3\sqrt{3}}{2} \right]$$

$$\boxed{\text{Required area} = a^2 \left(\pi + 3\sqrt{3} \right)}$$

Ex. 15 : *Find the area between the curve $r = a (\sec \theta + \cos \theta)$ and the asymptote $r = a \sec \theta$.*

Sol. : Here $r = a \sec \theta$ i.e. $r = \dfrac{a}{\cos \theta}$ or $r \cos \theta = a$ i.e. $x = a$

$$r = a (\sec \theta + \cos \theta)$$

i.e.

$$\sqrt{x^2 + y^2} = a \left(\frac{\sqrt{x^2 + y^2}}{x} + \frac{x}{\sqrt{x^2 + y^2}} \right)$$

or

$$x (x^2 + y^2) = a (x^2 + y^2 + x^2)$$

or

$$y^2 = \frac{x^2 (2a - x)}{x - a}$$

It is symmetrical about x-axis and passes through origin (isolated point). $x = a$ is the asymptote and it intersects x-axis at $(2a, 0)$, $(0, 0)$. For $0 < x < a$, $x > 2a$ curve does not exist. The required area is as shown shaded in Fig. 10.20.

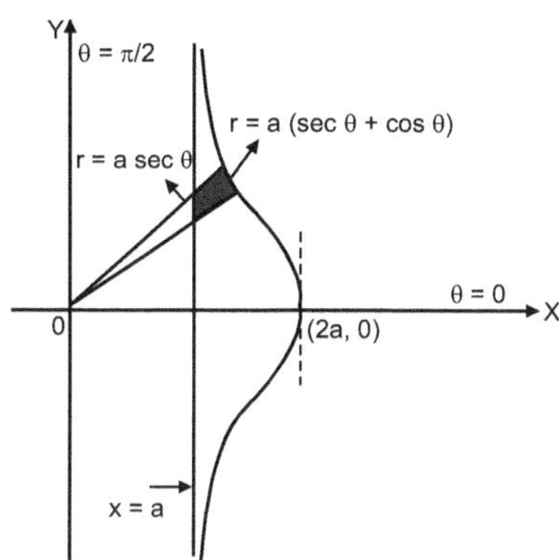

Fig. 10.20

$$\text{Area} = 2 \int_{0}^{\pi/2} \int_{a \sec \theta}^{a (\sec \theta + \cos \theta)} r \, dr \, d\theta$$

$$= \int_{0}^{\pi/2} [a^2 (\sec^2 \theta + 2 \sec \theta \cos \theta + \cos^2 \theta) - a^2 \sec^2 \theta] \, d\theta$$

$$= a^2 \int_{0}^{\pi/2} (2 + \cos^2 \theta) \, d\theta = a^2 \left[\pi + \frac{\pi}{4} \right]$$

$$\boxed{\text{Area} = \frac{5\pi}{4} a^2}$$

Ex. 16 : *Prove that the area of the loop of the Foulium of DesCarte's $x^3 + y^3 = 3axy$ is three times the area of one loop of $r^2 = a^2 \cos 2\theta$, Bernoullie's lemniscate.*

Sol. : For $x^3 + y^3 = 3axy$, it is symmetrical about $y = x$, passes through $(0, 0)$, x and y-axis are tangents at $(0, 0)$ and there exists a node at $(0, 0)$. It intersects $y = x$ at $\left(\dfrac{3a}{2}, \dfrac{3a}{2} \right)$ and $x + y + a = 0$ is the asymptote to the given curve. The curve does not exist in the third quadrant. The curve $x^2 + y^3 = 3axy$ is known as Foulium of DesCarte's and is shown shaded in Fig. 10.21 (a).

For $r^2 = a^2 \cos 2\theta$, it is symmetrical about $\theta = 0$, pole, $\theta = \dfrac{\pi}{2}$. Maximum value of $r = a$ and minimum value of $r = 0$. $\theta = \dfrac{\pi}{4}, \dfrac{3\pi}{4}, \dfrac{5\pi}{4}, \dfrac{7\pi}{4}$ are tangents at $(0, 0)$. No asymptote. The curve $r^2 = a^2 \cos 2\theta$ is known as Bernoullie's lemniscate and is shown shaded in Fig. 10.21 (b).

Part I : Area of Loop of $x^3 + y^3 = 3axy$:

In case of inclined loop, we change the equation to polar form

$$r^3 (\cos^3 \theta + \sin^3 \theta) = 3ar^2 \sin \theta \cos \theta \implies r = \frac{3a \sin \theta \cos \theta}{\cos^3 \theta + \sin^3 \theta}$$

$$\text{Area of the loop} = \frac{1}{2} \int_{0}^{\pi/2} r^2 \, d\theta = \frac{1}{2} \int_{0}^{\pi/2} \frac{9a^2 \sin^2 \theta \cos^2 \theta}{(\cos^3 \theta + \sin^3 \theta)^2} \, d\theta = \frac{9a^2}{2} \int_{0}^{\pi/2} \frac{\tan^2 \theta \cdot \sec^2 \theta}{(1 + \tan^3 \theta)^2} \, d\theta$$

(dividing by $\cos^6 \theta$ and put $1 + \tan^3 \theta = t$)

$$= \frac{3a^2}{2} \int_1^\infty \frac{dt}{t^2} = \frac{3a^2}{2} \left[-\frac{1}{t} \right]_1^\infty = \frac{3a^2}{2} [0 + 1] = \frac{3a^2}{2} \qquad \text{... (i)}$$

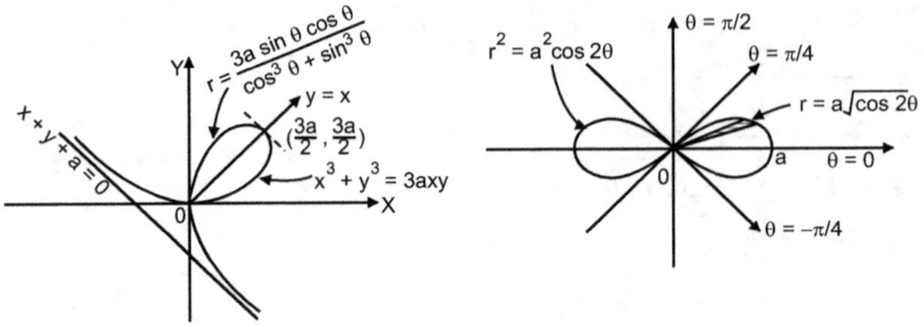

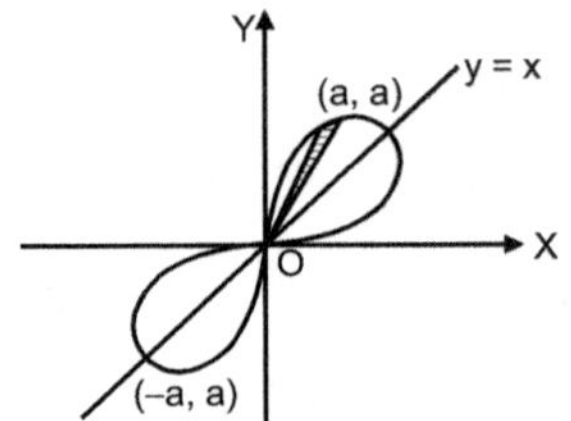

(a)   (b)

Fig. 10.21

Part II : Area of One Loop of $r^2 = a^2 \cos 2\theta$:

$$\text{Area of the loop} = 2 \int_0^{\pi/4} \frac{1}{2} r^2 \, d\theta = \int_0^{\pi/4} a^2 \cos 2\theta \, d\theta = a^2 \left. \frac{\sin 2\theta}{2} \right|_0^{\pi/4} = \frac{a^2}{2}(1) = \frac{a^2}{2} \qquad \text{... (ii)}$$

From (i) and (ii), we note that :

> Area of loop of $x^3 + y^3 = 3axy = 3$ times area of one loop of $r^2 = a^2 \cos 2\theta$.

Ex. 17 : *Find by double integration the area of the loop of curve $x^4 + y^4 = 2a^2 xy$.*

Sol. : It is symmetrical about $y = x$ and in opposite quadrant. It passes through $(0, 0)$, x and y-axis are tangents at $(0, 0)$ and there exists a node at $(0, 0)$. It intersects the line $y = x$ at (a, a) $(-a, -a)$. No asymptote. It does not exist in 2nd and 4th quadrants. A rough sketch of the curve is as shown in Fig. 10.22.

For the inclined loop, we convert the equation of curve to polar form

$$r^4 (\cos^4 \theta + \sin^4 \theta) = 2a^2 r^2 \sin \theta \cos \theta \Rightarrow r^2 = \frac{2a^2 \sin \theta \cos \theta}{\cos^4 \theta + \sin^4 \theta}$$

$$\text{Area} = 2 \int_0^{\pi/2} \frac{r^2}{2} \, d\theta$$

$$\text{Area} = \int_0^{\pi/2} \frac{2a^2 \sin \theta \cos \theta}{\cos^4 \theta + \sin^4 \theta} \, d\theta$$

$$= 2a^2 \int_0^{\pi/2} \frac{\tan \theta \sec^2 \theta \, d\theta}{1 + (\tan^2 \theta)^2} \qquad (\tan^2 \theta = t)$$

$$= a^2 \int_0^\infty \frac{dt}{1 + t^2} = a^2 [\tan^{-1} t]_0^\infty = a^2 [\pi/2]$$

> $$\text{Area} = \frac{\pi a^2}{2}$$

Fig. 10.22

Ex. 18 : *Show that the area of a loop of the curve $x^5 + y^5 = 5a x^2 y^2$ is $\dfrac{5a^2}{2}$.*

Sol. : It is symmetrical about $y = x$. It passes through $(0, 0)$ and x and y-axis are tangents at origin. There exists a node at $(0, 0)$. It intersects $y = x$ at $\left(\dfrac{5a}{2}, \dfrac{5a}{2}\right)$ and $x + y = a$ is the asymptote. The curve does not exist in 3rd quadrant. The required area of the loop is shown shaded in Fig. 10.23.

For an inclined loop, we convert the equation of curve to polar form

$$r^5 (\cos^5 \theta + \sin^5 \theta) = 5ar^4 \sin^2 \theta \cos^2 \theta \quad \text{or} \quad r = \frac{5a \sin^2 \theta \cos^2 \theta}{\cos^5 \theta + \sin^5 \theta}$$

$$\text{Area of the loop} \ = \ \int_0^{\pi/2} \frac{1}{2}\, r^2\, d\theta$$

$$\text{Area} \ = \ \frac{1}{2} \int_0^{\pi/2} \frac{25a^2 \sin^4\theta \cos^4\theta\, d\theta}{(\cos^5\theta + \sin^5\theta)^2} \quad \text{(dividing by } \cos^{10}\theta)$$

$$= \ \frac{25a^2}{2} \int_0^{\pi/2} \frac{\tan^4\theta \sec^2\theta}{(1 + \tan^5\theta)^2}\, d\theta, \quad (1 + \tan^5\theta = t)$$

$$= \ \frac{5a^2}{2} \int_1^{\infty} \frac{dt}{t^2} \ = \ \frac{5a^2}{2} \left[-\frac{1}{t} \right]_1^{\infty}$$

$$\boxed{\text{Area} \ = \ \frac{5a^2}{2}}$$

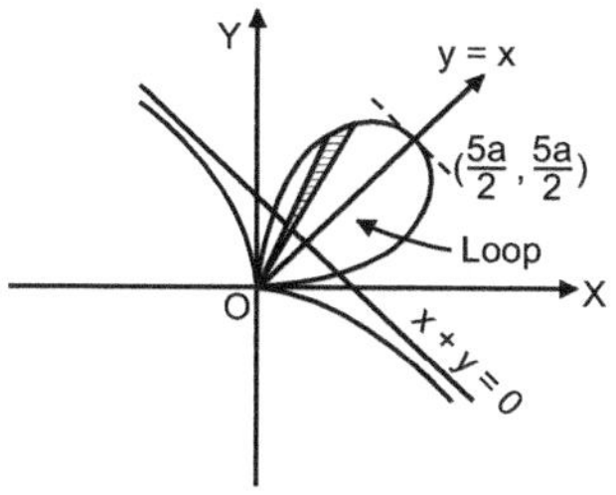

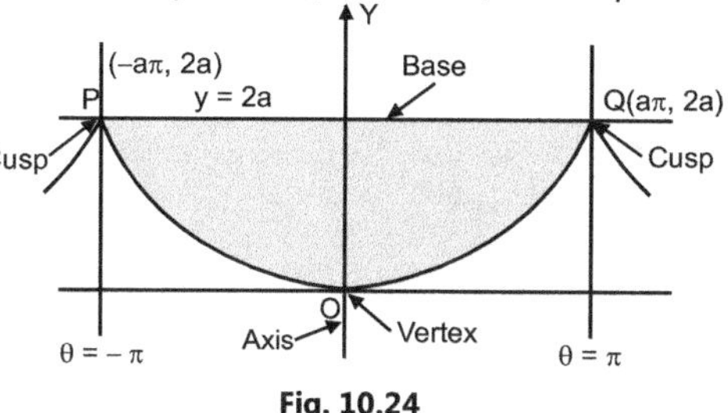

Fig. 10.23

Ex. 19 : *In the cycloid x = a (θ + sin θ), y = a (1 − cos θ), find the area between its base and portion of the curve from cusp to cusp.*

Sol. : It is symmetrical about y-axis and passes through (0, 0), maximum value of $y = 2a$, minimum value of $y = 0$ when $\theta = \pi$, $x = a\pi$, $y = 2a$ and when $\theta = -\pi$, $x = -a\pi$, $y = 2a$. There exists a cusp at P($-a\pi$, 2a) and Q(aπ, 2a). $\dfrac{dy}{dx} = 0$ when $\theta = 0$, $\dfrac{dy}{dx} = \infty$ when $\theta = \pm\pi$. The straight line PQ is called base of the cycloid, O is vertex and OY is axis of the cycloid. Curve does not exist for $y < 0$. The required area is as shown shaded in Fig. 10.24.

Fig. 10.24

$$\text{Required area} \ = \ 2 \int x\, dy \ = \ 2 \int_0^{\pi} x \frac{dy}{d\theta}\, d\theta$$

$$\text{Area} \ = \ 2 \int_0^{\pi} a\,(\theta + \sin\theta)\, a \sin\theta\, d\theta$$

$$= \ 2a^2 \int_0^{\pi} (\theta \sin\theta + \sin^2\theta)\, d\theta$$

$$= \ 2a^2 \left[[-\theta \cos\theta + \sin\theta]_0^{\pi} + 2 \cdot \frac{1}{2} \cdot \frac{\pi}{2} \right] = 2a^2 \left[\pi + \frac{\pi}{2} \right]$$

$$\boxed{\text{Area} \ = \ 3\pi a}$$

Ex. 20 : *Find the total area of the Astroid $x^{2/3} + y^{2/3} = a^{2/3}$.*

Sol. : The parametric equations of Astroid are $x = a \cos^3\theta$, $y = a \sin^3\theta$. It is symmetrical about both axes and does not pass through (0, 0). Here x and y cannot exceed a. $\dfrac{dy}{dx} = 0$ when $\theta = 0$, $\pi \cdot \dfrac{dy}{dx} = \infty$ when $\theta = \dfrac{\pi}{2}$, $\dfrac{3\pi}{2}$.

Hence the required area is as shown shaded in Fig. 10.25.

$$\text{Area} \ = \ 4 \int x\, dy \ = \ 4 \int_0^{\pi/2} x \frac{dy}{d\theta}\, d\theta$$

$$= \ 4 \int_0^{\pi/2} a \cos^3\theta \cdot 3a \sin^2\theta \cos\theta\, d\theta$$

$$= \ 12a^2 \int_0^{\pi/2} \sin^2\theta \cos^4\theta\, d\theta \ = \ 12a^2\, \frac{1 \cdot 3 \cdot 1}{6 \cdot 4 \cdot 2}\, \frac{\pi}{2}$$

$$\boxed{\text{Area} \ = \ \frac{3\pi a^2}{8}}$$

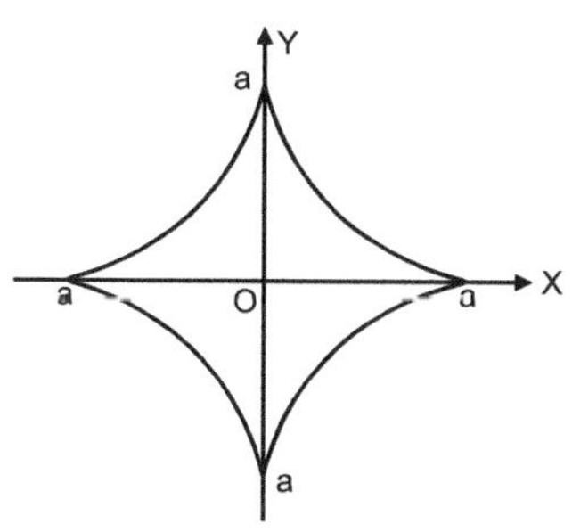

Fig. 10.25

Ex. 21 : Show that the area of a loop of the curve $r = a \cos n\theta$ is $\dfrac{\pi a^2}{4n}$ and state the total area in cases n is odd, n is even. Also show that the area contained between the circle $r = a$ and the curve $r = a \cos 5\theta$ is equal to three-fourth of the area of the circle.

Sol. : Part I : For $r = a \cos n\theta$, put $r = 0$, $\cos n\theta = 0$

$$n\theta = -\frac{\pi}{2}, \frac{\pi}{2}, \frac{3\pi}{2}, \frac{5\pi}{2} \ldots\ldots \text{ or } \theta = -\frac{\pi}{2n}, \frac{\pi}{2n}, \frac{3\pi}{2n}, \frac{5\pi}{2n} \ldots\ldots\ldots$$

First loop lies between $-\dfrac{\pi}{2n}$ to $\dfrac{\pi}{2n}$

$$\text{Area of a loop} = \int_{-\pi/2n}^{\pi/2n} \frac{r^2}{2} d\theta = \frac{1}{2} \int_{-\pi/2n}^{\pi/2n} a^2 \cos^2 n\theta \, d\theta$$

$$= \frac{a^2}{2} \cdot 2 \int_0^{\pi/2n} \left(\frac{1 + \cos 2n\theta}{2}\right) d\theta = \frac{a^2}{2} \left[\frac{\pi}{2n} + \left(\frac{\sin 2n\theta}{2n}\right)_0^{\pi/2n}\right]$$

$$= \frac{\pi a^2}{4n}$$

Case (i) : If n is odd, $r = a \cos n\theta$ contains n equal loops.

$$\therefore \qquad \text{Total area} = n \int_{-\pi/2n}^{\pi/2n} \frac{r^2}{2} d\theta = n \cdot \frac{\pi a^2}{4n} = \frac{\pi a^2}{4}$$

Case (ii) : If n is even, $r = a \cos n\theta$ contains (2n) equal loops.

$$\therefore \qquad \text{Total area} = 2n \int_{-\pi/2n}^{\pi/2n} \frac{r^2}{2} d\theta = 2n \left(\frac{\pi a^2}{4n}\right) = \frac{\pi a^2}{2}$$

Hence for $r = a \cos n\theta$

$$\boxed{\begin{aligned} &\text{if n is odd, Total area} = \frac{\pi a^2}{4} \\ &\text{if n is even, Total area} = \frac{\pi a^2}{2} \end{aligned}}$$

Part II : $r = a \cos 5\theta$ contains 5 equal loops and the entire curve lies within a circle of radius $r = a$. Put $r = 0$, $\cos 5\theta = 0$.

$$\therefore \qquad 5\theta = -\frac{\pi}{2}, \frac{\pi}{2}, \frac{3\pi}{2}, \frac{5\pi}{2}, \frac{7\pi}{2}, \frac{9\pi}{2}, \frac{11\pi}{2}, \frac{13\pi}{2}, \frac{15\pi}{2}, \frac{17\pi}{2}, \ldots\ldots$$

$$\therefore \qquad \theta = -\frac{\pi}{10}, \frac{\pi}{10}, \frac{3\pi}{10}, \frac{\pi}{2}, \frac{7\pi}{10}, \frac{9\pi}{10}, \frac{11\pi}{10}, \frac{13\pi}{10}, \frac{3\pi}{2}, \frac{17\pi}{10}$$

First loop lies between $-\dfrac{\pi}{10}$ to $\dfrac{\pi}{10}$. Place the loops between alternate division. Required area is as shown shaded in Fig. 10.26.

Area of the curve $r = a \cos 5\theta$ is

$$\text{Area} = 5 \int_{-\pi/10}^{\pi/10} \frac{r^2}{2} d\theta = \frac{5}{2} \int_{-\pi/10}^{\pi/10} a^2 \cos^2 5\theta \, d\theta$$

$$= 5a^2 \int_0^{\pi/10} \cos^2 5\theta \, d\theta \quad (5\theta = t)$$

$$= 5a^2 \int_0^{\pi/2} \cos^2 t \cdot \frac{dt}{5} = a^2 \cdot \frac{1}{2} \cdot \frac{\pi}{2} = \frac{\pi a^2}{4}$$

Area of circle, $\qquad r = a$ is πa^2.

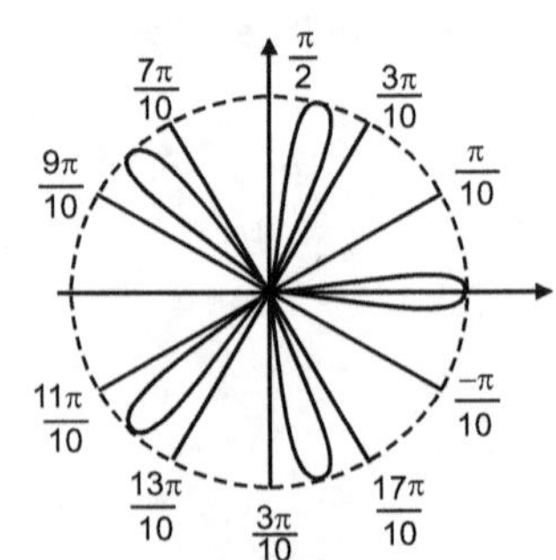

Fig. 10.26

Hence the required area $= \pi a^2 - \dfrac{\pi a^2}{4} = \dfrac{3}{4} \pi a^2$ = three-fourth of the area of the circle

Ex. 22 : *Show that the area bounded by the spiral $r = a\,e^{k\theta}$ and two radii is proportional to the difference of the squares of these radii.*

Sol. : Let r_1 and r_2 be the two radii corresponding to the values θ_1 and θ_2 respectively of the vectorial angle θ. The shape of the curve is shown in Fig. 10.27.

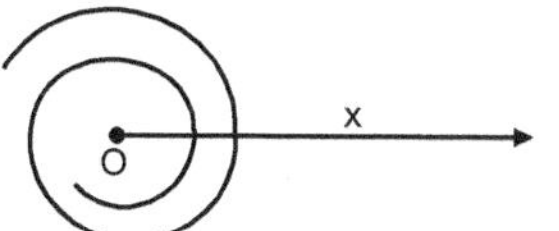

Fig. 10.27

$$\text{The required area} = \int_{\theta_1}^{\theta_2} \int_0^r r\,dr\,d\theta \quad \text{where} \quad \begin{array}{l} r_1 = a\,e^{k\theta_1} \\ r_2 = a\,e^{k\theta_2} \end{array}$$

$$\text{Area} = \frac{1}{2}\int_{\theta_1}^{\theta_2} a^2 e^{2k\theta}\,d\theta = \frac{a^2}{2}\left[\frac{e^{2k\theta}}{2k}\right]_{\theta_1}^{\theta_2} = \frac{a^2}{4k}\left[e^{2k\theta_2} - e^{2k\theta_1}\right] = \frac{a^2}{4k}\left[\frac{r_2^2}{a^2} - \frac{r_1^2}{a^2}\right]$$

$$\boxed{\text{Area} = \frac{1}{4k}(r_2^2 - r_1^2) \text{ which is proportional to } r_2^2 - r_1^2}$$

EXERCISE 10.1

I.   Find the Area Bounded by :

1. $y^2 = 4x$ and $2x - y - 4 = 0$ **Ans.** 9

2. $y^2 = 4ax$ and $x^2 = 4ay$. **Ans.** $\frac{16}{3}a^2$

3. $y^2 = x$ and $x^2 = -8y$ **Ans.** $\frac{8}{3}$

4. $y = x(4 - x)$ and x-axis **Ans.** $\frac{32}{3}$

5. $y^2 = 4x$ and $y = x - 8$. **Ans.** 72

6. $y^2 = 12x$ and $y = 3(x - 5)$ **Ans.** $56\frac{8}{9}$

7. $y = x^2 - 6x + 3$ and $y = 2x - 9$ **Ans.** $\frac{32}{3}$

8. $x^2 = 4y$ and $x - 2y + 4 = 0$ **Ans.** 9

9. $9xy = 4$ and $2x + y = 2$. **Ans.** $\frac{1}{3} - \frac{4}{9}\log 2$

10. $y^2 = 4a(x + a)$ and $y^2 = 4b(b - x)$ **Ans.** $\frac{8}{3}(a + b)\sqrt{ab}$

11. $xy^2 = a^2(a - x)$, and $(a - x)y^2 = a^2 x$ **Ans.** $(\pi - 2)a^2$

12. $\frac{x^2}{a^2} + \frac{y^2}{b^2} = 1$ and $\frac{x^2}{b^2} + \frac{y^2}{a^2} = 1$.

13. $3y^2 = 25x$ and $5x^2 = 9y$. **Ans.** 5

Hint : Change to polars **Ans.** $4ab\,\tan^{-1}\frac{a}{b}$

14. $y = ax^2$ and $y = 1 - \frac{x^2}{a}$, $a > 0$, is $\frac{4}{3}\sqrt{\frac{a}{a^2 + 1}}$

15. $x^{2m} + y^{2n} = 1$ in first quadrant. **Ans.** $\dfrac{\left\lceil\dfrac{1}{2m}\right.\left\lceil\dfrac{1}{2n}\right.}{4mn\left\lceil\dfrac{1}{2m} + \dfrac{1}{2n} + 1\right.}$

II.   Find the Area between the Curve and its Asymptote :

16. $y^2(4 - x) = x(x - 2)^2$ **Ans.** $2(\pi + 4)$

17. $y^2(a + x) = (a - x)^3$ **Ans.** $3\pi a^2$

18. $x(x^2 + y^2) = a(x^2 - y^2)$ **(Dec. 2011)** **Ans.** $\frac{a^2}{2}(\pi + 4)$

19. $x^2(x^2 + y^2) = a^2(y^2 - x^2)$ **Ans.** $(\pi + 2)a^2$

20. $x^2 y^2 = a^2(y^2 - x^2)$ **Ans.** $4a^2$

21. $(a - x)y^2 = x^3$ **Ans.** $\frac{3\pi a^2}{4}$

22. $x(x^2 + y^2) = a(y^2 - x^2)$ **Ans.** $a^2\left(2 + \frac{\pi}{2}\right)$

23. $xy^2 = a^2(a - x)$ **Ans.** πa^2

24. $y(x^2 + 4a^2) = 8a^3$ **Ans.** $4\pi a^2$

25. $x^2(1 - y)y = 1$ **Ans.** 2π

III. Find the Area of Loop of the Curve :

26. $x(x^2 + y^2) = a(x^2 - y^2)$ **Ans.** $a^2\left(2 - \dfrac{\pi}{2}\right)$

27. $y^2(4 - x) = x(x - 2)^2$ **Ans.** $2(4 - \pi)$

28. $x^4 - 2a^2 xy + a^2 y^2 = 0$ (Convert to polar)   **Ans.** $\dfrac{2a^2}{3}$

29. $r(\cos\theta + \sin\theta) = 2a\sin\theta\cos\theta$ **Ans.** $\dfrac{a^2}{4}(4 - \pi)$

30. $x(x^2 + y^2) = a(y^2 - x^2)$ **Ans.** $a^2\left(2 - \dfrac{\pi}{2}\right)$

31. $3ay^2 = x(x - a)^2$ **Ans.** $\dfrac{8a^2}{15\sqrt{3}}$

32. $a^4 y^2 = x^5(2a - x)$ **(May 2009)**   **Ans.** $\dfrac{5\pi a^2}{4}$

33. $(x + y)(x^2 + y^2) = 2axy$ **Ans.** $a^2\left(1 - \dfrac{\pi}{4}\right)$

34. $y^2 = x^2(4 - x^2)$ **Ans.** $\dfrac{16}{3}$

35. $xy^2 + (x + a)^2(x + 2a) = 0$ **Ans.** $\dfrac{a^2}{2}(4 - \pi)$

36. $ay^2 = (x - a)(x - 5a)^2$ **Ans.** $\dfrac{128}{15}a^2$

IV. Find the Total Area of the Curve :

37. $x^2(x^2 + y^2) = a^2(x^2 - y^2)$ **Ans.** $a^2(\pi - 2)$

38. $x^6 + y^6 = a^2 x^2 y^2$ **Ans.** $\dfrac{\pi a^2}{3}$

39. $(x^2 + y^2)^2 = a^2(x^2 - y^2)$ **Ans.** a^2

40. $r = a(1 + \cos\theta)$ **(May 2005)**   **Ans.** $\dfrac{3\pi a^2}{2}$

41. $r = a(1 - \cos\theta)$ **Ans.** $\dfrac{3\pi a^2}{2}$

42. $r = a\sin 3\theta$ **Ans.** $\dfrac{\pi a^2}{4}$

43. $r = a\cos 3\theta$ **Ans.** $\dfrac{\pi a^2}{4}$

44. $r = a\sin 2\theta$ **Ans.** $\dfrac{\pi a^2}{2}$

45. $r = a\cos 2\theta$ **Ans.** $\dfrac{\pi a^2}{2}$

46. $r = a + b\cos\theta \ (a > b)$ **Ans.** $\left(a^2 + \dfrac{b^2}{2}\right)\pi$

47. $\left(\dfrac{x}{a}\right)^4 + \left(\dfrac{y}{b}\right)^{10} = 1$ **Ans.** $ab\, B\left(\dfrac{11}{10}, \dfrac{1}{4}\right)$

48. $a^4 y^2 = x^4(a^2 - x^2)$ **Ans.** $\dfrac{\pi a^2}{4}$

V. 49. Show that area common to the circles $x^2 + y^2 - 4y = 0$, $x^2 + y^2 - 4x - 4y + 4 = 0$ is $4\left(\dfrac{2\pi}{3} - \dfrac{\sqrt{3}}{2}\right)$.

50. Show that the area outside the circle $x^2 + y^2 = a^2$ and inside the circle $x^2 + y^2 = 2ax$ is $2\displaystyle\int_{0}^{\pi/3} \int_{a}^{2a\cos\theta} r\, dr\, d\theta$ and evaluate it.

Ans. $a^2\left(\dfrac{\pi}{3} + \dfrac{\sqrt{3}}{2}\right)$

51. Show that the area outside $x^2 + y^2 = a^2$ and inside $r = a(1 + \cos\theta)$ is $\dfrac{a^2}{4}(8 + \pi)$.   **(Dec. 2009, 2007)**

52. Prove that the area of the *crescent* bounded by the circles $r = a\sqrt{2}$ and $r = 2a\cos\theta$ is a^2.

53. Prove that the area of the astroid $\left(\dfrac{x}{a}\right)^{2/3} + \left(\dfrac{y}{b}\right)^{2/3} = 1$ is $\dfrac{3}{8}\pi ab$.

54. Find the area enclosed by one arch of the cycloid $x = a(\theta - \sin\theta)$, $y = a(1 - \cos\theta)$ and its base.   **Ans.** $3\pi a^2$

55. Find the area included between two loops of the curve $r = a(\sqrt{2}\cos\theta - 1)$.   **Ans.** $a^2(\pi + 3)$

56. Find the area included between two loops of the curve $r = a(2\cos\theta + \sqrt{3})$.   **Ans.** $\dfrac{a^2}{3}\left(10\pi + 9\sqrt{3}\right)$.

57. Find the area enclosed by the curve $a^4 y^2 + b^2 x^4 = a^2 b^2 x^2$.   **Ans.** $\dfrac{4ab}{3}$

58. Obtain the area in the first quadrant bounded by the curve $b^4 y^2 = (a^2 - x^2)^3$ and the co-ordinate axes.   **Ans.** $\dfrac{3\pi a^4}{16\, b^2}$

59. Find the area enclosed by the curve $y^2 = (x - a)(b - x)$, $0 < a < b$.   **Ans.** $\dfrac{\pi}{4}(a - b)^2$

10.5 MASS OF A LAMINA

If the surface density ρ of a plane lamina is a function of the position of a point of the lamina, then the mass of an elementary area dA is $\rho\, dA$ and the total mass of the lamina is $\int \rho\, dA$.

In Cartesian co-ordinates, if $\rho = f(x, y)$, the mass of the lamina, $M = \iint f(x, y)\, dx\, dy$.

In polar co-ordinates, if $\rho = F(r, \theta)$, the mass of the lamina, $M = \iint F(r, \theta)\, r\, dr\, d\theta$.

Both the integrals being taken over area of the lamina.

10.6 ILLUSTRATIONS ON APPLICATIONS TO MASS

Ex. 1 : *A lamina is bounded by the curves $y = x^2 - 3x$ and $y = 2x$. If the density at any point is given by λxy, find the mass of the lamina.*

Sol. : The density at any point $P(x, y)$ of the lamina is $\rho = \lambda xy$ and the mass of the elementary lamina at P is $\rho\, dx\, dy$ i.e. $\lambda\, xy\, dx\, dy$.

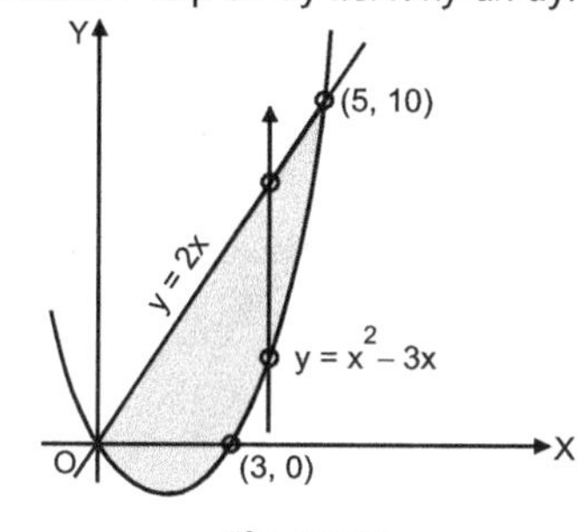

$$\text{Mass of the lamina} = \iint_R \rho \cdot dA = \iint \lambda\, xy\, dx\, dy = \lambda \int_0^5 \left\{ \int_{x^2-3x}^{2x} xy\, dy \right\} dx$$

$$= \lambda \int_0^5 \left[\frac{xy^2}{2} \right]_{x^2-3x}^{2x} dx = \frac{\lambda}{2} \int_0^5 [4x^3 - x(x^2 - 3x)^2]\, dx$$

$$= \frac{\lambda}{2} \int_0^5 (-x^5 + 6x^4 - 5x^3)\, dx = \frac{\lambda}{2} \left[-\frac{x^6}{6} + \frac{6x^5}{5} - \frac{5x^4}{4} \right]_0^5$$

$$\boxed{\text{Mass of the lamina} = \frac{4375}{24}\lambda}$$

Fig. 10.28

Ex. 2 : *The density at any point of a non-uniform circular lamina of radius a varies as its distance from a fixed point on the circumference of the circle. Find the mass of the lamina.*

Sol. : Take the fixed point in the circumference of the circle as the origin and the diameter through it as x-axis, and using the polar co-ordinates, the polar equation of the circle is $r = 2a \cos \theta$. The density at any point $P(r, \theta)$ of the lamina is $\rho = \lambda r$ and the mass of the elementary lamina at P is $(\lambda r)\, r\, dr\, d\theta$.

Integrating this over the area of the circle, we have

$$\begin{array}{l}\text{Mass of} \\ \text{the lamina}\end{array} = \lambda - \int_{-\pi/2}^{\pi/2} d\theta \int_0^{2a\cos\theta} r^2\, dr$$

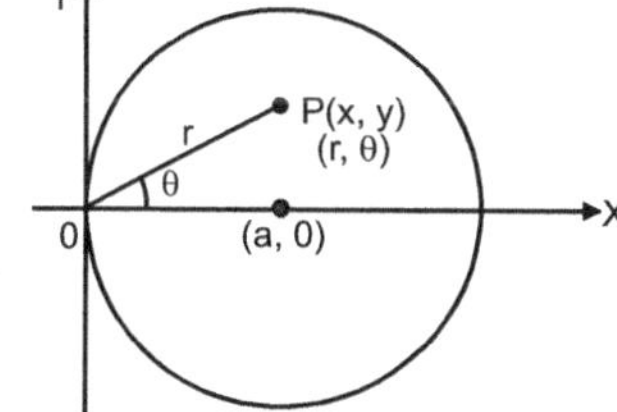

Fig. 10.29

$$= \frac{\lambda}{3} \int_{-\pi/2}^{\pi/2} 8a^3 \cos^3 \theta\, d\theta = \frac{8\lambda a^3}{3}\, 2 \int_0^{\pi/2} \cos^3 \theta\, d\theta$$

$$\boxed{\text{Mass of the lamina} = \frac{16\lambda a^3}{3}\left(\frac{2}{3}\right) = \frac{32\lambda a^3}{9}}$$

EXERCISE 10.2

1. Find the mass distributed over the area bounded by the curve $16y^2 = x^3$ and the line $2y = x$, assuming that the density at a point of the area varies as the distance of the point from x-axis.

    Ans. $\dfrac{2}{3}\lambda$

2. If the density at any point on a circular lamina is k times the square of its distance from a fixed point in its circumference, find its mass.

    Ans. $\dfrac{3}{2}\pi\lambda a^4$

3. A lamina in the form of a parabolic segment of mass M, height h and base 2k has density at a point given by $\lambda\, pq^3$ per unit area, where p and q are distances from the base and the axis respectively. Determine the value of λ.

[**Hint :** $y^2 = 4ax$ be parabola, $q = y$, $p = h - x$ $\rho = \lambda\,(h - x)\,y^3$,

$$M = 2\int_0^h \int_0^{2\sqrt{ax}} \lambda\,(h - x)\,y^3\,dx\,dy$$

$\therefore\ M = \dfrac{2}{3}\,\lambda\,a^2\,h^4$, Q (h, k) lies on $y^2 = 4ax$ $\therefore\ a = \dfrac{k^2}{4h}$]

     Ans. $\dfrac{24\,M}{h^2\,k^4}$.

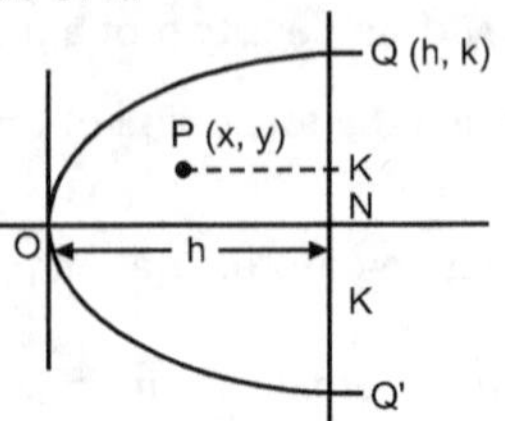

Fig. 10.30

4. The density at any point of a non-uniform circular lamina of radius unity is k times its distance from a given diameter. Find the total mass of the lamina.

     Ans. $\dfrac{4k}{3}$

5. The density at any point of a cardioide $r = a\,(1 + \cos\theta)$ varies as the square of the distance from the axis of symmetry of the cardioide. Find the mass of the lamina.

     Ans. $\dfrac{21}{32}\,\pi\,ka^4$

6. In a lamina in the form of an ellipse $\dfrac{x^2}{a^2} + \dfrac{y^2}{b^2} = 1$, the density at any point varies as the product of the distances from the axes of the ellipse. Find the mass of the lamina.

     Ans. $\dfrac{ka^2\,b^2}{2}$

7. If mass per unit area varies as the square of the ordinate of the point, find the mass of the lamina bounded by the cycloid $x = a\,(\theta + \sin\theta)$, $y = a\,(1 - \cos\theta)$ the ordinates from the two cusps and the tangent at the vertex.

     Ans. $\dfrac{5\,\lambda\,\pi\,a^4}{12}$

8. Find the mass of a lamina bounded by the curves $y^2 = ax$ and $x^2 = ay$, when the mass per unit area varies as the square of the distance of the point from the origin.

     Ans. $\dfrac{6}{35}\cdot\lambda\,a^4$.

10.7 VOLUMES OF SOLIDS

To express the volume of a solid as a triple integral, we note that the volume of an elementary cuboid is dx dy dz; and so the volume of the solid is given by;

$$\boxed{\text{Volume} = \iiint\limits_{V} dx\,dy\,dz}$$

where, the limits of integration w.r.t. z (if we integrate first w.r.t. z) are z_1 and z_2 obtained from its equations to the top and bottom of the given surface and then the double integration is w.r.t. x and y is performed over the area of projection of the given solid on the XOY plane.

If $\rho = f\,(x, y, z)$ is the density of the solid at the point P (x, y, z), then the mass of the solid is

$$\boxed{\text{Mass} = \iiint\limits_{V} \rho\,dx\,dy\,dz = \iiint\limits_{V} f(x, y, z)\,dx\,dy\,dz}$$

with appropriate limits of integrations.

In spherical polar system,

$$\boxed{V = \iiint r^2 \sin\theta\,dr\,d\theta\,d\phi}$$

In cylindrical polar system,

$$\boxed{V = \iiint \rho\,d\rho\,d\phi\,dz}$$

10.8 ILLUSTRATIONS ON VOLUMES OF SOLIDS

Ex. 1 : *Prove that the volume bounded by the cylinders $y^2 = x$, $x^2 = y$ and the planes $z = 0$, $x + y + z = 2$ is $\dfrac{11}{30}$.*   **(May 2008)**

Sol. :

$$\text{Volume} = V = \iiint dx\, dy\, dz = \iint dx\, dy \int_0^{2-x-y} dz.$$

$$V = \iint_R (2 - x - y)\, dx\, dy,$$

where, R is the region in XOY plane as shown in Fig. 10.31.

$$V = \int_0^1 \int_{x^2}^{\sqrt{x}} (2 - x - y)\, dy\, dx$$

$$= \int_0^1 \left[2y - xy - \frac{y^2}{2} \right]_{x^2}^{\sqrt{x}} dx$$

$$= \int_0^1 \left[\left(2\sqrt{x} - x^{3/2} - \frac{x}{2} \right) - \left(2x^2 - x^3 - \frac{x^4}{2} \right) \right] dx$$

$$= \left[2\frac{x^{3/2}}{3/2} - \frac{x^{5/2}}{5/2} - \frac{x^2}{4} - \frac{2x^3}{3} + \frac{x^4}{4} + \frac{x^5}{10} \right]_0^1 = \frac{4}{3} - \frac{2}{5} - \frac{1}{4} - \frac{2}{3} + \frac{1}{4} + \frac{1}{10}$$

$$\boxed{V = \frac{11}{30}}$$

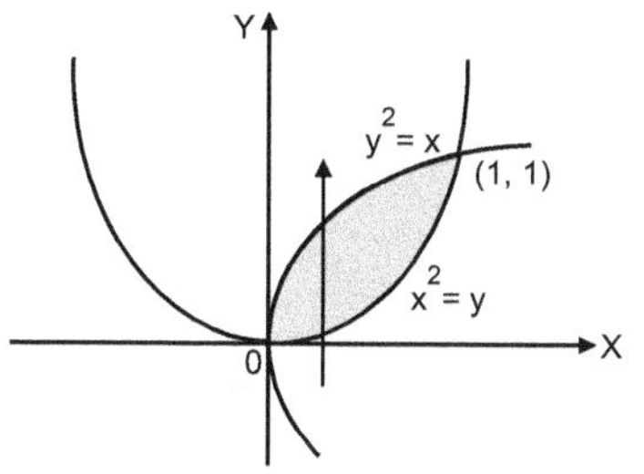

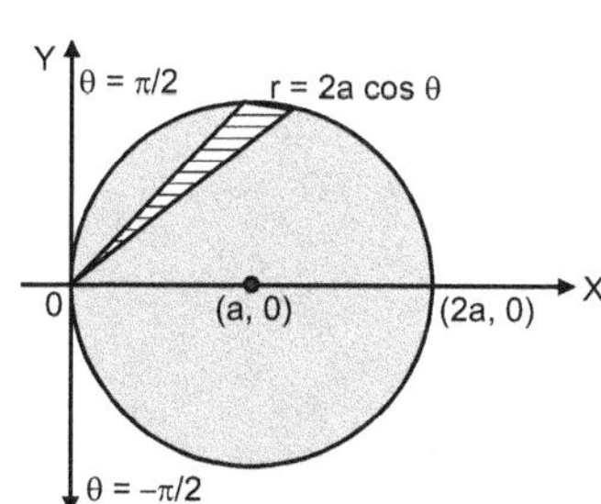

Fig. 10.31

Ex. 2 : *Prove that the volume enclosed between the cylinders $x^2 + y^2 = 2ax$ and $z^2 = 2ax$ is $\dfrac{128\, a^3}{15}$.*   **(Dec. 2008, 2004)**

Sol. : The required volume is $V = 2\iiint_0^{\sqrt{2ax}} dx\, dy\, dz = 2\iint \sqrt{2ax}\, dx\, dy$

Transforming to polars,

$$V = 4 \int_0^{\pi/2} \int_0^{2a\cos\theta} \sqrt{2ar\cos\theta}\; r\, dr\, d\theta$$

$$= 4\sqrt{2a} \int_0^{\pi/2} \sqrt{\cos\theta}\, d\theta \int_0^{2a\cos\theta} r^{3/2}\, dr$$

$$= 4\sqrt{2a} \int_0^{\pi/2} \sqrt{\cos\theta}\, d\theta \left[\frac{2}{5} r^{5/2} \right]_0^{2a\cos\theta}$$

$$= \frac{8}{5}\sqrt{2a} \int_0^{\pi/2} \sqrt{\cos\theta}\; (2a)^{5/2}\, \cos^{5/2}\theta\, d\theta$$

$$= \frac{64}{5} a^3 \int_0^{\pi/2} \cos^3\theta\, d\theta = \frac{64}{5} a^3 \left(\frac{2}{3} \right)$$

$$\boxed{V = \frac{128\, a^3}{15}}$$

Fig. 10.32

Ex. 3 : *Find the volume cut off from the paraboloid* $x^2 + \dfrac{y^2}{4} + z = 1$ *by the plane z = 0.*   **(Nov./Dec. 2019, May 2017)**

Sol. : Required volume is $V = 4 \displaystyle\iiint_0^{1-x^2-\frac{y^2}{4}} dx\, dy\, dz$

$$V = 4 \iint_R \left(1 - x^2 - \frac{y^2}{4}\right) dx\, dy, \quad \text{where R is } x^2 + \frac{y^2}{4} = 1, \text{ ellipse.}$$

By using elliptical polar co-ordinates $x = r\cos\theta$, $y = 2r\sin\theta$, $x^2 + \dfrac{y^2}{4} = r^2$, $dx\, dy = 2r\, dr\, d\theta$, $0 \le \theta \le \dfrac{\pi}{2}$, $0 \le r \le 1$

$$V = 4 \int_0^{\pi/2} \int_0^1 (1 - r^2)\, 2r\, dr\, d\theta = 8 \int_0^{\pi/2} d\theta \int_0^1 (r - r^3)\, dr$$

$$= 8 \left(\frac{\pi}{2}\right)\left(\frac{r^2}{2} - \frac{r^4}{4}\right)_0^1 = 4\pi \left(\frac{1}{2} - \frac{1}{4}\right) = \pi$$

$$\boxed{V = \pi}$$

Ex. 4 : *Find the volume of the region enclosed by the cone* $z = \sqrt{x^2 + y^2}$ *and paraboloid* $z = x^2 + y^2$.   **(Dec. 2009, 2017, May 2004)**

Sol. : $V = \displaystyle\iint \int_{x^2 + y^2}^{\sqrt{x^2 + y^2}} dx\, dy\, dz$

$$V = \iint_R \left[\sqrt{x^2 + y^2} - (x^2 + y^2)\right] dx\, dy$$

where R is $\sqrt{x^2 + y^2} = x^2 + y^2$ or $x^2 + y^2 = 1$

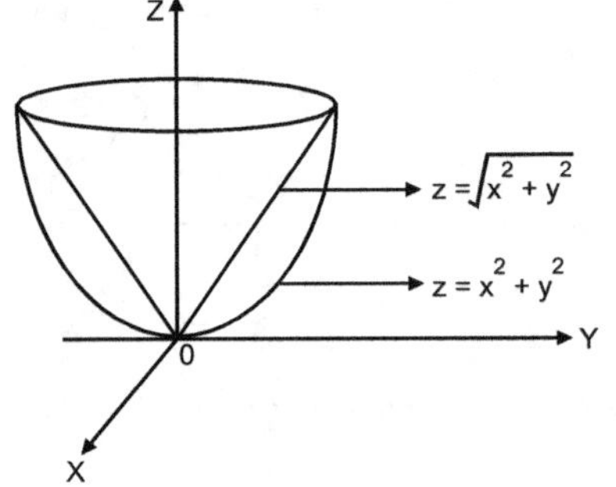

Fig. 10.33

Transforming to polar coordinates,

$$V = 4 \int_0^{\pi/2} \int_0^1 (r - r^2)\, r\, dr\, d\theta = 4\left(\frac{\pi}{2}\right)\left(\frac{r^3}{3} - \frac{r^4}{4}\right)_0^1 = 2\pi\left(\frac{1}{12}\right)$$

$$\boxed{V = \frac{\pi}{6}}$$

Ex. 5 : *Find the volume bounded by the paraboloid* $x^2 + y^2 = az$, *the cylinder* $x^2 + y^2 = 2ay$ *and the plane z = 0.*

Sol. : $V = \displaystyle\iint \int_0^{\frac{x^2+y^2}{a}} dx\, dy\, dz = \iint_R \frac{x^2 + y^2}{a}\, dx\, dy$

$$V = \frac{1}{a} \iint_R (x^2 + y^2)\, dx\, dy$$

$$= \frac{2}{a} \int_0^{\pi/2} \int_0^{2a\sin\theta} r^2 \cdot r\, dr\, d\theta$$

$$= \frac{2}{a} \int_0^{\pi/2} \frac{16a^4 \sin^4\theta}{4}\, d\theta = 8a^3 \left(\frac{3}{4} \cdot \frac{1}{2} \cdot \frac{\pi}{2}\right)$$

$$\boxed{V = \frac{3\pi a^3}{2}}$$

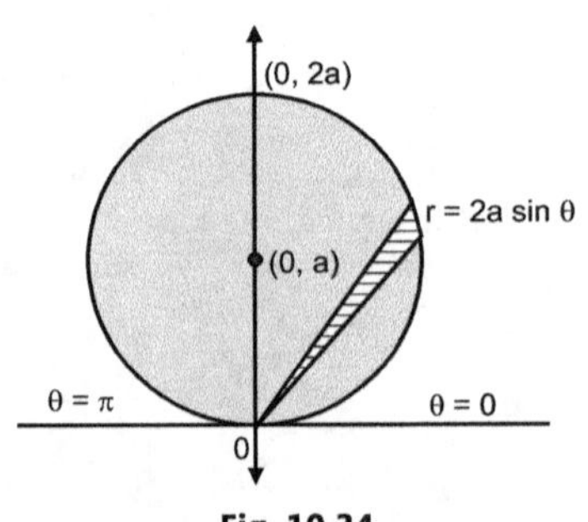

Fig. 10.34

Ex. 6 : *Find the volume of the cylinder $x^2 + y^2 = 2ax$, intercepted between the paraboloid $x^2 + y^2 = 2az$ and the XY-plane.*

Sol. :
$$V = \iint \int_0^{\frac{x^2+y^2}{2a}} dx\,dy\,dz = \iint \frac{x^2+y^2}{2a}\,dx\,dy \qquad\text{(May 2005, 2013)}$$

$$V = \frac{1}{2a} \iint_R (x^2 + y^2)\,dx\,dy$$

where R is $x^2 + y^2 = 2ax$

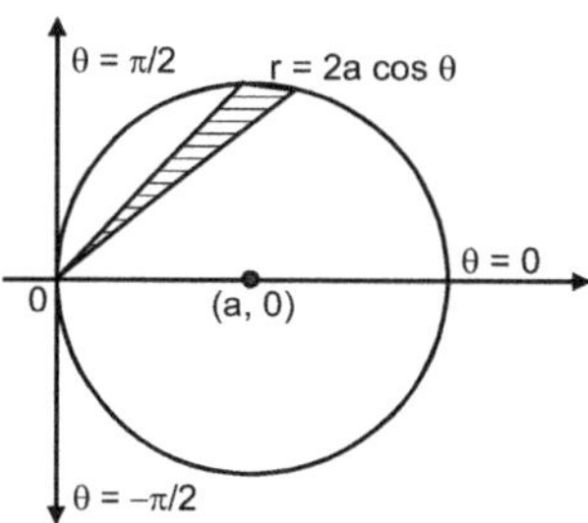

Fig. 10.35

$$= \frac{1}{2a} \cdot 2 \int_0^{\pi/2} \int_0^{2a\cos\theta} r^2\, r\,dr\,d\theta$$

$$= \frac{1}{a} \int_0^{\pi/2} \frac{16a^4\cos^4\theta}{4}\,d\theta$$

$$= 4a^3 \left(\frac{3}{4}\cdot\frac{1}{2}\cdot\frac{\pi}{2}\right)$$

$$\boxed{V = \frac{3\pi a^3}{4}}$$

Ex. 7 : *Find the volume of the solid cut off from the sphere $x^2 + y^2 + z^2 = a^2$ by the cylinder $x^2 + y^2 = ax$.* **(May 2009)**

Sol. : By symmetry $V = 4\iiint dx\,dy\,dz$. Now using cylindrical polar co-ordinates $x = \rho\cos\phi$, $y = \rho\sin\phi$, $z = z$, $x^2 + y^2 = \rho^2$, $dx\,dy\,dz = \rho\,d\rho\,d\phi\,dz$. We note that z varies from z = 0 to $z = \sqrt{a^2-\rho^2}$, ρ varies from $\rho = 0$ to $\rho = a\cos\phi$, finally ϕ varies from 0 to $\frac{\pi}{2}$.

$$V = 4 \int_0^{\pi/2} \int_0^{a\cos\phi} \int_0^{\sqrt{a^2-\rho^2}} \rho\,d\rho\,d\phi\,dz$$

$$= 4 \int_0^{\pi/2} d\phi \int_0^{a\cos\phi} \rho\,d\rho\,\left(\sqrt{a^2-\rho^2}\right) = -2 \int_0^{\pi/2} d\phi \left[\frac{2}{3}(a^2-\rho^2)^{3/2}\right]_0^{a\cos\phi}$$

$$= \frac{4}{3} \int_0^{\pi/2} (a^3 - a^3\sin^3\phi)\,d\phi$$

$$\boxed{V = \frac{4}{3}a^3\left(\frac{\pi}{2} - \frac{2}{3}\right)}$$

Ex. 8 : *A cylindrical hole of radius b is bored symmetrical through a sphere of radius a. Find the volume of the remaining solid.*

Sol. : Let the equation of the sphere be $x^2 + y^2 + z^2 = a^2$.

$$V = 2\iint \int_0^{\sqrt{a^2-x^2-y^2}} dx\,dy\,dz = 2 \iint_R \sqrt{a^2-x^2-y^2}\,dx\,dy$$

where R is the region in XOY plane bounded by the circles r = a, r = b (a > b) in polars. Transforming to polars,

$$V = 2 \int_0^{2\pi} \int_0^{a} \sqrt{a^2-r^2}\, r\,dr\,d\theta = -\int_0^{2\pi} d\theta \left[\frac{2}{3}(a^2-r^2)^{3/2}\right]_b^{a}$$

$$= \frac{2}{3}(2\pi)(a^2-b^2)^{3/2}$$

$$\boxed{V = \frac{4\pi}{3}(a^2-b^2)^{3/2}}$$

Ex. 9 : *Find the volume cut from the sphere $x^2 + y^2 + z^2 = a^2$ by the cone $x^2 + y^2 = z^2$.*

Sol. :

$$V = \iint \int_{\sqrt{x^2 + y^2}}^{\sqrt{a^2 - x^2 - y^2}} dx\, dy\, dz$$

$$V = \iint_R \left[\sqrt{a^2 - x^2 - y^2} - \sqrt{x^2 + y^2} \right] dx\, dy, \text{ where R is } x^2 + y^2 = \frac{a^2}{2}$$

$$= \int_0^{2\pi} \int_0^{a/\sqrt{2}} \left(\sqrt{a^2 - r^2} - r \right) r\, dr\, d\theta$$

$$= 2\pi \left[-\frac{1}{2} \cdot \frac{2}{3} (a^2 - r^2)^{3/2} - \frac{r^3}{3} \right]_0^{a/\sqrt{2}}$$

$$\boxed{V = \frac{2\pi a^3}{3} \left(1 - \frac{1}{\sqrt{2}} \right)}$$

Ex. 10 : *Find the volume of the region bounded by $z = x^2 + y^2$ and $z = 2x$.*

Sol. :

$$V = \iint \int_{x^2 + y^2}^{2x} dx\, dy\, dz$$

$$= \iint_R [2x - (x^2 + y^2)]\, dx\, dy \text{ where R is } x^2 + y^2 = 2x$$

$$= 2 \int_0^{\pi/2} \int_0^{2\cos\theta} (2r\cos\theta - r^2)\, r\, dr\, d\theta = 2 \int_0^{\pi/2} \left[2\cos\theta \frac{r^3}{3} - \frac{r^4}{4} \right]_0^{2\cos\theta} d\theta$$

$$= 2 \int_0^{\pi/2} \left[\frac{16\cos^4\theta}{3} - 4\cos^4\theta \right] d\theta$$

$$= \frac{8}{3} \int_0^{\pi/2} \cos^4\theta\, d\theta = \frac{8}{3} \left(\frac{3}{4} \frac{1}{2} \frac{\pi}{2} \right)$$

$$\boxed{V = \frac{\pi}{2}}$$

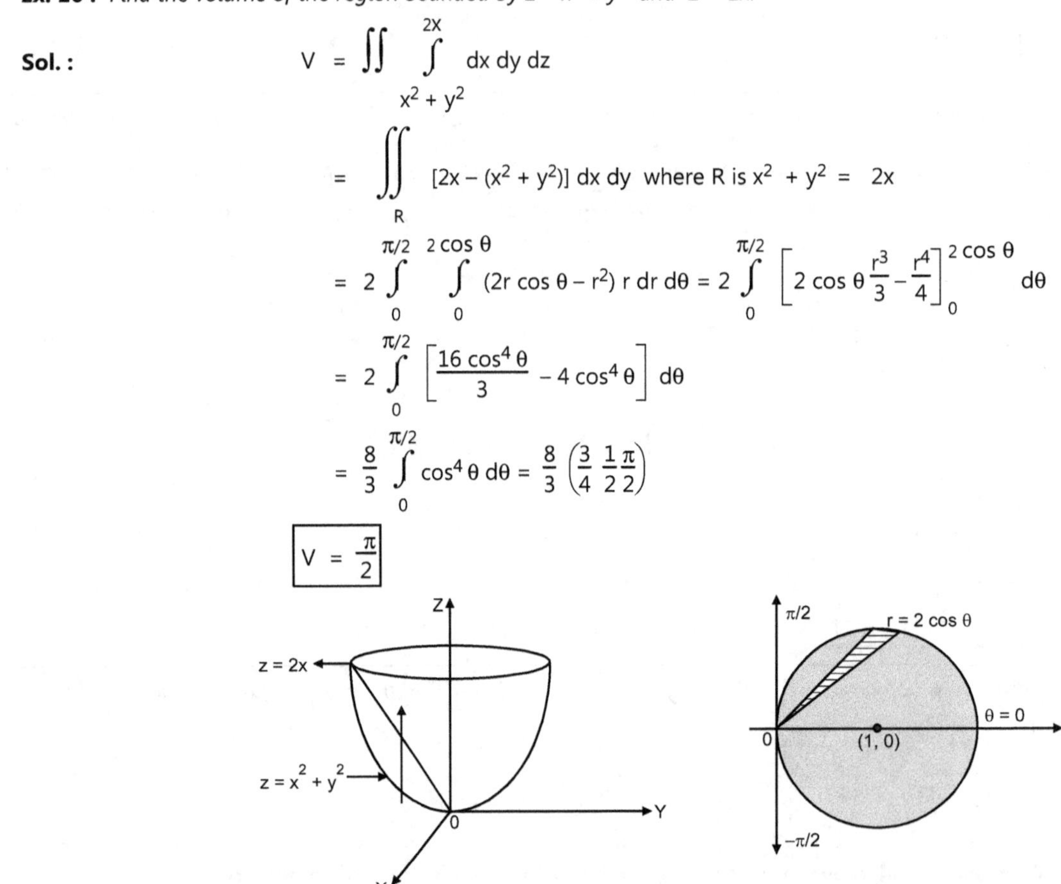

Fig. 10.36

Ex. 11 : *Find volume common to the cylinders $x^2 + y^2 = a^2$, $x^2 + z^2 = a^2$.*   *(Dec. 07, 16, May 16)*

OR

Two equal right circular cylinders of radius a have their axes intersecting at right angles. Prove that the volume common to both is $\frac{16}{3} a^3$.

Sol. : We may take the equations of the cylinders as $x^2 + y^2 = a^2$, $x^2 + z^2 = a^2$. Since their axes are along z-axis and y-axis respectively, which intersect at right angles,

$$V = \int_{x=-a}^{x=a} \int_{y=-\sqrt{a^2-x^2}}^{y=\sqrt{a^2-x^2}} \int_{z=-\sqrt{a^2-x^2}}^{z=\sqrt{a^2-x^2}} dx\, dy\, dz$$

$$= 8 \int_0^a \int_0^{\sqrt{a^2-x^2}} \sqrt{a^2-x^2}\, dx\, dy = 8 \int_0^a (a^2 - x^2)\, dx$$

$$= 8 \left[a^2 x - \frac{x^3}{3} \right]_0^a$$

$$\boxed{V = \frac{16a^3}{3}}$$

Ex. 12 : *Find the volume of region bounded by paraboloid $x^2 + y^2 = 2z$ and cylinder $x^2 + y^2 = 4$.*   **(Dec. 2018)**

Sol. : By using cylindrical polar co-ordinates, $x = \rho \cos\phi$, $y = \rho \sin\phi$, $z = z$, $x^2 + y^2 = \rho^2$, $dx\, dy\, dz = \rho\, d\rho\ d\phi\, dz$

$$V = 4 \iiint dx\, dy\, dz$$

$$= 4 \iiint \rho\, d\rho\, d\phi\, dz = 4 \int_0^{\pi/2} d\phi \int_0^2 \rho\, d\rho \int_0^{\rho^2/2} dz$$

$$= 4 \frac{\pi}{2} \int_0^2 \rho \left(\frac{\rho^2}{2} \right) d\rho = \pi \left[\frac{\rho^4}{4} \right]_0^2$$

$$\boxed{V = 4\pi}$$

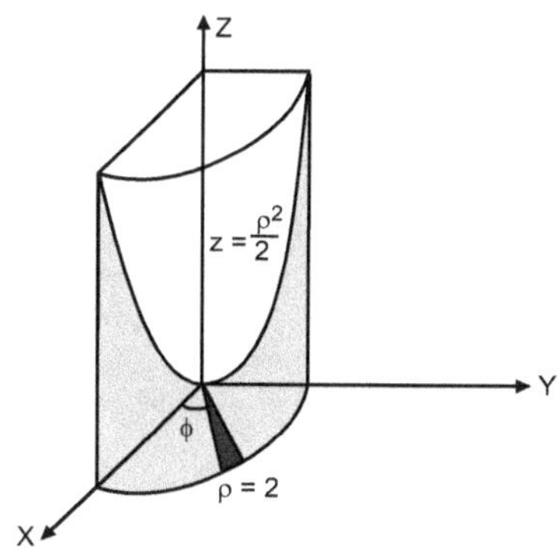

Fig. 10.37

Ex. 13 : *Find the volume of the tetrahedron bounded by the co-ordinate planes and the plane $\dfrac{x}{a} + \dfrac{y}{b} + \dfrac{z}{c} = 1$.* **(May 14, 18, Nov. 14)**

Sol. :
$$V = \iiint dx\, dy\, dz$$

Put $x = au$, $dx = a\, du$, $y = bv$, $dy = b\, dv$, $z = cw$, $dz = c\, dw$

$$V = \iiint abc\, du\, dv\, dw$$

$$= abc \iiint u^{1-1} v^{1-1} w^{1-1}\, du\, dv\, dw, \qquad \text{where, } u + v + w \le 1 \text{ (Note this step)}$$

$$= abc\, \frac{\overline{|1}\ \overline{|1}\ \overline{|1}}{\overline{|1+1+1+1}} = \frac{abc}{3!} \qquad \text{(by Dirichlet's theorem)}$$

$$\boxed{V = \frac{abc}{6}}$$

Ex. 14 : *If the density at a point varies as the square of the distance of the point from XOY plane, find the mass of the volume common to the sphere $x^2 + y^2 + z^2 = a^2$ and cylinder $x^2 + y^2 = ax$.*

Sol. : Distance of any point (x, y, z) from XOY plane $= z$

$\rho \propto z^2$, $\rho = \lambda z^2$.

$\therefore$ The required mass is

$$M = \iiint \rho \, dx \, dy \, dz \quad \iiint \lambda z^2 \, dx \, dy \, dz$$

Put
$x = \rho \cos \phi$
$y = \rho \sin \phi$
$z = z$
$dx \, dy \, dz = \rho \, d\rho \, d\phi \, dz$

$$= 4\lambda \int_{\phi=0}^{\pi/2} \int_{\rho=0}^{a \cos \phi} \int_{z=0}^{\sqrt{a^2 - \rho^2}} z^2 \rho \, d\rho \, d\phi \, dz$$

$$= 4\lambda \int_{0}^{\pi/2} d\phi \int_{0}^{a \cos \phi} \rho \, d\rho \, \frac{(a^2 - \rho^2)^{3/2}}{3}$$

$$= -\frac{2\lambda}{3} \int_{0}^{\pi/2} d\phi \left[\frac{2}{5} (a^2 - \rho^2)^{5/2} \right]_{0}^{a \cos \phi}$$

$$= \frac{4\lambda}{15} \int_{0}^{\pi/2} (a^5 - a^5 \sin^5 \phi) \, d\phi = \frac{4\lambda \, a^5}{15} \left[\frac{\pi}{2} - \frac{4}{5}\frac{2}{3} \right]$$

$$\boxed{M = \frac{4\lambda \, a^5}{15} \left(\frac{\pi}{2} - \frac{8}{15} \right)}$$

Ex. 15 : *Find the mass of the octant of the ellipsoid* $\dfrac{x^2}{a^2} + \dfrac{y^2}{b^2} + \dfrac{z^2}{c^2} = 1$, *the density at any point (x, y, z) being k xyz.*

Sol. : Here $\qquad\qquad \rho = k \, xyz$

$\therefore$ Required mass is $\qquad M = \iiint \rho \, dx \, dy \, dz = \iiint k \, xyz \, dx \, dy \, dz \quad$ Put $x = ar \sin \theta \cos \phi$, $y = br \sin \theta \sin \phi$, $z = cr \cos \theta$

$$dx \, dy \, dz = abc \, r^2 \sin \theta \cdot d\theta \, d\phi \, dr$$

$$= k \iiint a \, b \, c \, r^3 \sin^2 \theta \cos \theta \sin \phi \cos \phi \, abc \, r^2 \sin \theta \, d\theta \, d\phi \, dr$$

$$= k \, a^2 b^2 c^2 \int_{0}^{\pi/2} \sin^3 \theta \cos \theta \, d\theta \int_{0}^{\pi/2} \sin \phi \cos \phi \, d\phi \int_{0}^{1} r^5 \, dr$$

$$= k \, a^2 b^2 c^2 \left(\frac{1}{4} \right) \left(\frac{1}{2} \right) \left(\frac{1}{6} \right)$$

$$\boxed{M = \frac{k \, a^2 b^2 c^2}{48}}$$

EXERCISE 10.3

1. Find the volume enclosed between the two surfaces $z = x^2 + 3y^2$ and $z = 8 - x^2 - y^2$.

   [**Hint :** $V = \iint \int_{x^2 + 3y^2}^{8 - x^2 - y^2} dx \, dy \, dz$ and further $x = 2r \cos \theta$, $y = \sqrt{2} \, r \sin \theta$, $dx \, dy = 2\sqrt{2} \, r \, dr \, d\theta$.]   **Ans.** $8\pi \sqrt{2}$

2. Find the volume of the region bounded below by the xy plane and above by the paraboloid $z = x^2 + y^2$, and lying inside the cylinder $x^2 + y^2 = 4$.

   [**Hint :** $V = 4 \iint \int_{0}^{x^2 + y^2} dx \, dy \, dz$ and further transform to polars.]   **Ans.** 8π.

3. Find the volume of the tetrahedron bounded by the plane $\dfrac{x}{2} + \dfrac{y}{3} + \dfrac{z}{4} = 1$ and the co-ordinate planes.   **Ans.** 4

4. Find the volume of an ellipsoid of semi-axes a, b, c.

   [**Hint :** Ellipsoid is $\dfrac{x^2}{a^2} + \dfrac{y^2}{b^2} + \dfrac{z^2}{c^2} = 1$.]   **Ans.** $\dfrac{4}{3} \pi \, abc$

5. Find the volume bounded by the cylinder $x^2 + y^2 = 4$ and the planes $y + z = 4$ and $z = 0$. **(Dec. 2006) Ans.** 16π

6. A triangular prism is formed by planes $ay = bx$, $y = 0$, $x = a$. Prove that volume of prism between planes $z = 0$ to $z = c + xy$ is $\dfrac{ab}{8}(4c + ab)$.

7. Find the volume between paraboloides $z = 4 - x^2 - \dfrac{y^2}{4}$ and $z = 3x^2 + \dfrac{y^2}{4}$. **Ans.** $4\sqrt{2}\,\pi$.

8. A right circular cylinder $x^2 + y^2 = ax$ is formed by planes $z = 0$ to $z = a$. Find the volume of portion of cylinder which is inside the cone $x^2 + y^2 = z^2$.

 [**Hint :** $V = \displaystyle\iint \int_{\sqrt{x^2 + y^2}}^{a} dz\, dy\, dx$ and then transform to polar.] **Ans.** $\dfrac{a^3}{36}(9\pi - 16)$

9. Find the volume bounded by sphere $x^2 + y^2 + z^2 = a^2$ and cylinder $x^2 + y^2 = ay$. **Ans.** $\dfrac{2}{9}a^3(3\pi - 4)$

10. Find the volume in first octant bounded by cylinder $x^2 + x^2 = 2$ and $z = x + y$, $y = x, z = 0, x = 0$. **Ans.** $\dfrac{2\sqrt{2}}{3}$

11. Find the volume of paraboloid of revolution $x^2 + y^2 = 4z$ cut off by the plane $z = 4$. **(May 2019, Dec. 2005)**

 [**Hint :** $V = \displaystyle\iint \int_{\frac{x^2 + y^2}{4}}^{4} dz\, dy\, dx$ and then transform to polars.] **Ans.** 32π

12. Find the volume for the region enclosed by sphere $x^2 + y^2 + z^2 = a^2$ and the cone $z^2 \sin^2 \alpha = (x^2 + y^2) \cos^2 \alpha$, where α is constant and $0 < \alpha < \pi$.

 [**Hint :** Transform to spherical polar co-ordinates $V = \displaystyle\int_{0}^{2\pi} d\phi \int_{0}^{\alpha} d\theta \int_{0}^{a} r^2 \sin\theta\, dr\, d\theta\, d\phi$] **Ans.** $\dfrac{2\pi a^3}{3}(1 - \cos\alpha)$

13. Find the volume of the region bounded by $z = 4 - x^2 - y^2$ and the xy-plane. **(May 2006) Ans.** 8π

14. Find the volume bounded by the sphere $x^2 + y^2 + z^2 = 4$ and the paraboloid $x^2 + y^2 = 3z$. **(May 2010)**

 [**Hint :** $V = \displaystyle\iint \int_{\frac{x^2 + y^2}{3}}^{\sqrt{4 - x^2 - y^2}} dz\, dy\, dx$ and evaluate double integral along $x^2 + y^2 = 3$ which is intersection of two surfaces.] **Ans.** $\dfrac{19\pi}{6}$

15. Find the volume enclosed between the elliptic paraboloid $\dfrac{x^2}{p} + \dfrac{y^2}{q} = 2z$, the cylinder $x^2 + y^2 = a^2$ and the plane $z = 0$.

                      Ans. $\dfrac{\pi}{8}a^4\left(\dfrac{1}{p} + \dfrac{1}{q}\right)$

16. Find the volume included between the surfaces $y^2 = x$, $z = 0$, $x + y + z = 2$. **Ans.** $\dfrac{81}{20}$

17. If the density at any point of the solid octant of the ellipsoid $\dfrac{x^2}{a^2} + \dfrac{y^2}{b^2} + \dfrac{z^2}{c^2} = 1$ vary as $x^2 y^2 z^2$, find the mass of the solid.

                      Ans. $\dfrac{k\,\pi\,a^3\,b^3\,c^3}{1890}$

18. If the density at any point (x, y, z) being $\lambda x^2 y\, z$, find the mass of the volume bounded by the planes $x = 0$, $y = 0$, $z = 0$,

 $\dfrac{x}{a} + \dfrac{y}{b} + \dfrac{c}{z} = 1$.   **Ans.** $\dfrac{\lambda\,a^3\,b^2\,c^2}{2520}$

10.9 MEAN AND R.M.S. VALUES

1.   Mean Value : The mean value of the ordinate y, of a function y = f(x), over the range x = a to x = b is the limit of mean value of the equidistant ordinates y_1, y_2, ... y_n, as n tends to infinity. Thus, if y_m is the mean value,

$$y_m = \lim_{n \to \infty} \frac{y_1 + y_2 + y_3 + ... + y_n}{n} = \lim_{n \to \infty} \frac{y_1\,\delta x + y_2\,\delta x + ... + y_n\,\delta x}{n\,\delta x}$$

Now,
$$n\,\delta x = (b - a) = \int_a^b dx \text{ and } \lim_{n \to \infty} (y_1\,\delta x + y_2\,\delta x + ... + y_n\,\delta x)$$

$$= \lim_{\substack{n \to \infty \\ \delta x \to 0}} \sum_{x = a}^{x = b} y \cdot \delta x = \int_a^b y\,dx$$

$$\therefore \quad y_m = \frac{\int_a^b y\,dx}{\int_a^b dx} = \frac{\int_a^b f(x)\,dx}{\int_a^b dx}$$

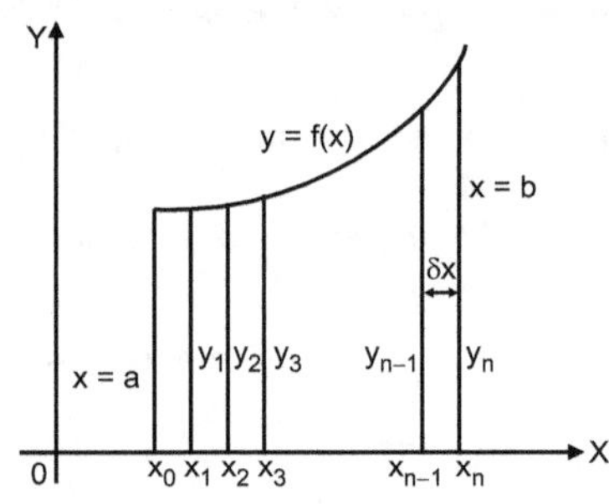

Fig. 10.38

2.   Mean Square Value of y = f(x) over (a, b) is defined as :

$$\text{Mean square value of } y = \frac{\int_a^b y^2\,dx}{\int_a^b dx} = \frac{\int_a^b [f(x)]^2\,dx}{\int_a^b dx}$$

The analysis can be extended to the mean values of the functions of two variables, say z = f (x, y) over an area A is,

3.   Mean Value of Z = f(x, y) :

$$\text{Mean value of } z = f(x, y) \text{ over an area A } = Z_m = \frac{\iint_A f(x, y)\,dx\,dy}{\iint_A dx\,dy}$$

Similarly,

4.

$$\text{Mean value of } u = f(x, y, z) \text{ over a region of volume V } = u_m = \frac{\iiint f(x, y, z)\,dx\,dy\,dz}{\iiint dx\,dy\,dz}$$

5. Root Mean Square Value (R.M.S. Value) : This term is usually applied to periodic functions. If y is a periodic function of x, of period p, the root mean square value of y is the square root of the mean value of y^2 over the range x = c to x = c + p, c being any constant.

∴ 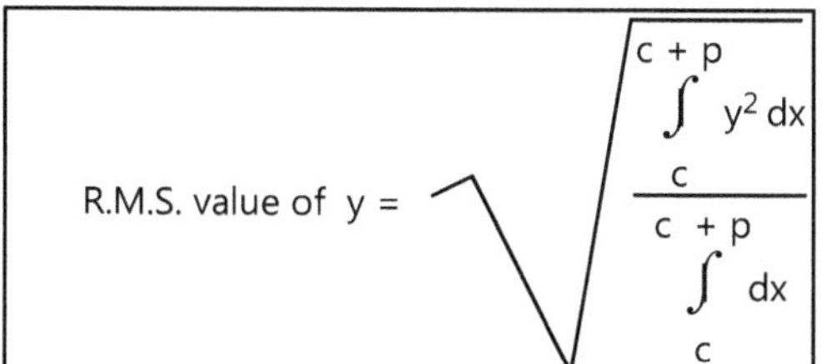

R.M.S. value of $y = \sqrt{\dfrac{\int_{c}^{c+p} y^2\, dx}{\int_{c}^{c+p} dx}}$

Note : The R.M.S. value is very useful in calculations of A.C. electrical circuits.

10.10 ILLUSTRATIONS ON MEAN AND R.M.S. VALUES

Ex. 1 : *Find the mean height of portion of parabola y = (x – 2) (3 – x) which lies above x-axis.*

Sol. : Mean height means mean of y. The parabola y = (x – 2) (3 – x) is shown in Fig. 10.39.

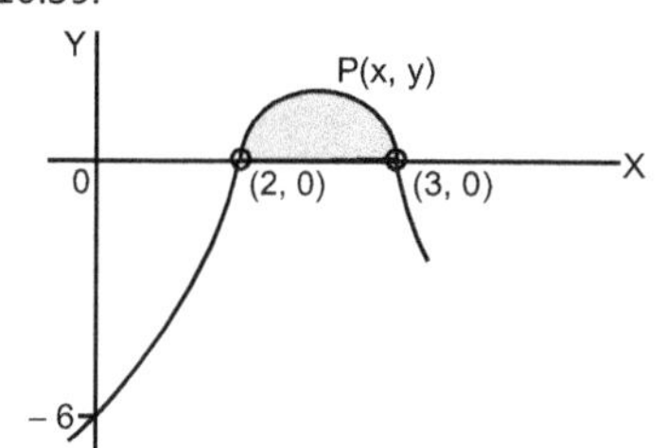

Fig. 10.39

Mean value of $y = y_m = \dfrac{\int_{2}^{3} y\, dx}{\int_{2}^{3} dx}$

$$y_m = \frac{\int_{2}^{3} (x-2)(3-x)\, dx}{(3-2)}$$

$$= \int_{2}^{3} (-x^2 + 5x - 6)\, dx = \left[-\frac{x^3}{3} + \frac{5x^2}{2} - 6x \right]_{2}^{3}$$

$$\boxed{y_m = \frac{1}{6}}$$

Ex. 2 : *A rod of length l is divided into two parts at random. Find average of sum of squares of these parts. Also find mean value of rectangle contained by these two segments.*

Sol. : Length of rod OA $= l$. Let OB = x, ∴ BA = $l - x$. Required two parts be x, $l - x$.

(Refer Fig. 10.40)

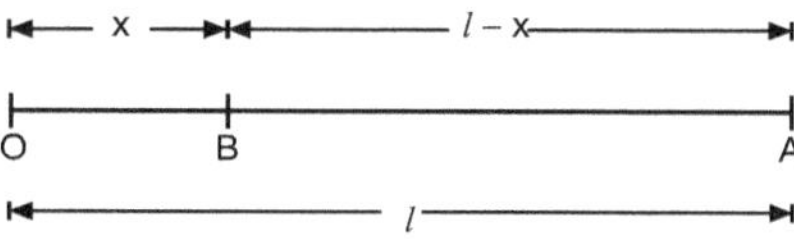

Fig. 10.40

Sum of squares of these parts = $x^2 + (l - x)^2$

$$\text{Average value} = \frac{\int_{0}^{l} [x^2 + (l-x)^2]\, dx}{\int_{0}^{l} dx}$$

$$= \frac{1}{l} \left[\frac{x^3}{3} + \frac{(l-x)^3}{-3} \right]_{0}^{l} = \frac{1}{l} \left[\frac{l^3}{3} + \frac{l^3}{3} \right] = \frac{2l^2}{3}$$

Rectangle contained by x, $l - x$ is $x(l - x)$

$$\text{Mean value } = \frac{\displaystyle\int_0^l x(l-x)\,dx}{l} = \frac{1}{l}\left[\frac{lx^2}{2} - \frac{x^3}{3}\right]_0^l$$

$$\boxed{\text{Mean value} = \frac{l^2}{6}}$$

Ex. 3 : *If* $V = L\dfrac{dI}{dt} + RI$, *where t is time in seconds, show that if* $I = I_0 \sin (pt)$ *where* I_0 *and p are constants, then the mean value of power P is* $\dfrac{1}{2} I_0^2 R$.

Sol. : We have power $P = V.I.$, given that $I = I_0 \sin pt$.

$$\frac{dI}{dt} = p\,I_0 \cos pt.$$

Also given
$$V = L\frac{dI}{dt} + RI$$
$$V = Lp\,I_0 \cos pt + R\,I_0 \sin pt$$
$$P = VI = I_0^2\,[Lp \cos pt + R \sin pt] \sin pt$$

As P = 0 when I = 0 i.e. $\sin pt = 0$ $\therefore$ $pt = 0,\ \pi,\ \ldots\ldots$

$\therefore$ $t = 0$ and $t = \pi/p$,

$$\text{M.V. of power P } = \frac{\displaystyle\int_0^{\pi/p} I_0^2\left[\frac{Lp}{2}\sin 2pt + R\left(\frac{1-\cos 2pt}{2}\right)\right]dt}{\displaystyle\int_0^{\pi/p} dt}$$

$$= \frac{I_0^2\,p}{\pi}\left[-\frac{Lp}{2}\frac{\cos 2pt}{2p} + \frac{R}{2}\left(t - \frac{\sin 2pt}{2p}\right)\right]_0^{\pi/p} = \frac{I_0^2\,p}{\pi}\left[\left(-\frac{L}{4} + \frac{R\pi}{2p}\right) - \left(-\frac{L}{4}\right)\right]$$

$$\boxed{\text{M.V. of power P} = \frac{I_0^2\,R}{2}}$$

Ex. 4 : *Find the mean value of the product "i.e" for a complete period, when* $i = I \sin\left(\dfrac{2\pi t}{T} + \alpha\right)$ *and* $e = E \sin\left(\dfrac{2\pi t}{T} + \beta\right)$.

Sol. : Let
$$f(t) = i.e = I.E.\ \sin(\theta + \alpha)\sin(\theta + \beta),\ \text{where } \theta = \frac{2\pi t}{T},\ dt = \frac{T}{2\pi}\,d\theta$$
$$f(t + T) = f(t) \Rightarrow f(t) \text{ is periodic function with period T}$$

$$\text{M.V. of "i.e" } = \frac{\displaystyle\int_0^T i.e\ dt}{\displaystyle\int_0^T dt} = \frac{1}{T}\int_0^{2\pi} IE \sin(\theta + \alpha)\sin(\theta + \beta)\frac{T}{2\pi}\,d\theta$$

$$= \frac{IE}{2\pi}\frac{1}{2}\int_0^{2\pi}[\cos(\alpha - \beta) - \cos(2\theta + \alpha + \beta)]\,d\theta = \frac{IE}{4\pi}[2\pi \cos(\alpha - \beta)]$$

$$\boxed{\text{M.V. of "i.e"} = \frac{IE}{2}\cos(\alpha - \beta)}$$

Ex. 5 : *Find the root mean square value of an electric circuit given by;*   **(Dec. 2005)**

$$I = I_0 + I_1 \sin\left(\frac{2\pi t}{T} + \alpha_1\right) + I_2 \sin\left(\frac{4\pi t}{T} + \alpha_2\right).$$

Sol. : It is easily noted that for $I = I_0 + I_1 \sin\left(\dfrac{2\pi t}{T} + \alpha_1\right) + I_2 \sin\left(\dfrac{4\pi t}{T} + \alpha_2\right)$, the periodic time is T

$$\therefore \quad I_{r.m.s.} = \sqrt{\frac{\int_0^T I^2\, dt}{\int_0^T dt}} = \sqrt{\frac{1}{T}\int_0^T I^2\, dt}$$

Consider

$$\int_0^T I^2\, dt = \int_0^T \left\{ I_0^2 + I_1^2 \sin^2\left(\frac{2\pi t}{T} + \alpha_1\right) + I_2^2 \sin^2\left(\frac{4\pi t}{T} + \alpha_2\right) \right.$$

$$\left. + 2\, I_0 I_1 \sin\left(\frac{2\pi t}{T} + \alpha_1\right) + 2 \cdot I_0 \cdot I_2 \sin\left(\frac{4\pi t}{T} + \alpha_2\right) + 2\, I_1 I_2 \sin\left(\frac{2\pi t}{T} + \alpha_1\right) \sin\left(\frac{4\pi t}{T} + \alpha_2\right) \right\}\, dt$$

$$\int_0^T I_0^2\, dt = I_0^2\, T;$$

$$\int_0^T I_1^2 \sin^2\left(\frac{2\pi t}{T} + \alpha_1\right) dt = \frac{I_1^2}{2} \int_0^T \left[1 - \cos\left(\frac{4\pi t}{T} + 2\alpha_1\right)\right] dt = \frac{I_1^2}{2}\, [T - 0] = \frac{I_1^2\, T}{2}$$

$$\int_0^T I_2^2 \sin^2\left(\frac{4\pi t}{T} + \alpha_2\right) dt = \frac{I_2^2\, T}{2}$$

$$\int_0^T \sin\left(\frac{2\pi t}{T} + \alpha_1\right) dt = 0, \quad \int_0^T \sin\left(\frac{4\pi t}{T} + \alpha_2\right) dt = 0$$

$$\int_0^T \sin\left(\frac{2\pi t}{T} + \alpha_1\right) \sin\left(\frac{4\pi t}{T} + \alpha_2\right) dt = \frac{1}{2} \int_0^T \left[\cos\left(\frac{2\pi t}{T} + \alpha_2 - \alpha_1\right) - \cos\left(\frac{6\pi t}{T} + \alpha_2 + \alpha_1\right)\right] dt = 0$$

$$\frac{1}{T} \int_0^T I^2\, dt = \frac{1}{T}\left[I_0^2\, T + I_1^2 \frac{T}{2} + I_2^2 \frac{T}{2}\right] = I_0^2 + \frac{I_1^2}{2} + \frac{I_2^2}{2}$$

$$\boxed{I_{r.m.s.} = \sqrt{I_0^2 + \frac{I_1^2}{2} + \frac{I_2^2}{2}}}$$

Ex. 6 : *Find the R.M.S. value of a sin pt + b cos qt + c sin rt + d cos st + ... p, q, r, s ... being integers.*

Sol. : Given function is a periodic function with period 2π.

$$\text{Mean square value} = \frac{\int_0^{2\pi} [a \sin pt + b \cos qt + c \sin rt + d \cos st + ...]^2\, dt}{\int_0^{2\pi} dt}$$

$$= \frac{1}{2\pi} \int_0^{2\pi} \{a^2 \sin^2 pt + b^2 \cos^2 qt + c^2 \sin^2 rt + d^2 \cos^2 st + ...$$

$$+ 2ab \sin pt \cos qt + 2\, ac \sin pt \sin rt + ...\}\, dt$$

$$\int_0^{2\pi} \sin^2 pt\, dt = \int_0^{2\pi} \frac{1 - \cos 2pt}{2}\, dt = \frac{2\pi}{2} - 0 = \pi \text{ and } \int_0^{2\pi} \cos^2 qt\, dt = \pi$$

$$\int_0^{2\pi} \sin pt \cos qt\, dt = 0 \text{ and so on.}$$

$$\text{Mean square value} = \frac{1}{2\pi} [a^2\pi + b^2\pi + c^2\pi + d^2\pi + ...] = \frac{1}{2}(a^2 + b^2 + c^2 + d^2 + ...)$$

$$\boxed{\text{R.M.S. value} = \sqrt{\frac{a^2 + b^2 + c^2 + d^2 + ...}{2}}}$$

Ex. 7 : *The law of density ρ of a sphere of radius a is $\rho = \rho_0 \dfrac{\sin (kr)}{nr}$, where 'r' is the distance from the centre, ρ_0, k and n are constants. Find the average density.*

(May 2008)

Sol. : Average density $= \dfrac{\text{Total mass}}{\text{Volume}} = \dfrac{\text{Total mass}}{\dfrac{4}{3}\pi a^3}$

To find the total mass, divide the sphere into an infinite number of thin spherical shells concentric with the sphere and consider a typical shell of radius r and thickness dr. (Refer Fig. 10.41).

$$\text{Its volume} = 4\pi r^2\, dr$$

$$\text{The mass of this shell} = dm = 4\pi r^2\, dr \cdot \rho$$

$$dm = 4\pi r^2\, dr \frac{\rho_0 \sin kr}{nr} = \frac{4\pi\rho_0}{n}(r \sin kr)\, dr$$

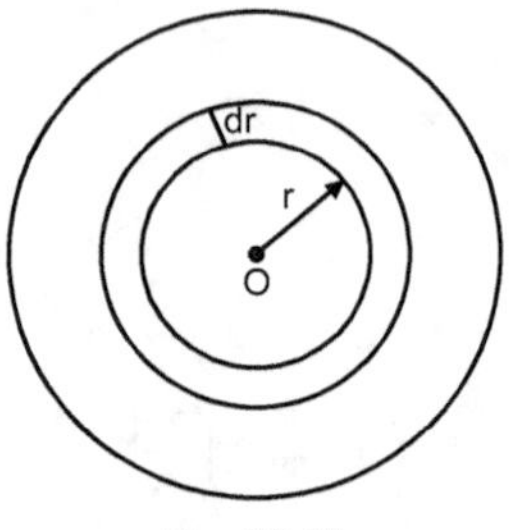

Since r varies from 0 to a,

$$\text{Total mass} = \int_0^a dm = \int_0^a \frac{4\pi\,\rho_0}{n}(r \sin kr)\, dr$$

Fig. 10.41

$$= \frac{4\pi\,\rho_0}{n}\left[r\left(\frac{-\cos kr}{k}\right) - (1)\left(\frac{-\sin kr}{k^2}\right) \right]_0^a = \frac{4\pi\,\rho_0}{n}\left[\frac{-a\cos ka}{k} + \frac{\sin ka}{k^2}\right] = \frac{4\pi\,\rho_0}{nk^2}[\sin ak - ak \cos ak]$$

$$\text{Average density} = \frac{\dfrac{4\pi\,\rho_0}{nk^2}(\sin ak - ak \cos ak)}{(4\pi a^3)/3}$$

$$\boxed{\text{Average density} = \frac{3\,\rho_0}{nk^2 a^3}(\sin ak - ak \cos ak)}$$

Ex. 8 : *Find the M.V. and R.M.S. value of the ordinate of the cycloid $x = a(\theta + \sin \theta)$, $y = a(1 - \cos \theta)$ over the range $\theta = -\pi$ to $\theta = \pi$.*

(May 2005)

Sol. : Let P(x, y) be any point on the cycloid. Its ordinate is y.

$$y_m = \frac{\displaystyle\int y\, dx}{\displaystyle\int dx} = \frac{\displaystyle\int_{-\pi}^{\pi} a(1 - \cos \theta)\, a(1 + \cos \theta)\, d\theta}{\displaystyle\int_{-\pi}^{\pi} a(1 + \cos \theta)\, d\theta}$$

$$y_m = \frac{a\displaystyle\int_{-\pi}^{\pi} \sin^2 \theta\, d\theta}{[\theta + \sin \theta]_{-\pi}^{\pi}} = \frac{a}{2\pi}\, 2\int_0^{\pi} \sin^2 \theta\, d\theta = \frac{a}{\pi}\, 2\int_0^{\pi/2} \sin^2\theta\, d\theta = \frac{2a}{\pi}\, \frac{1}{2}\, \frac{\pi}{2}$$

$$\boxed{y_m = \frac{a}{2}}$$

$$\text{R.M.S. value} = \frac{\displaystyle\int y^2\, dx}{\displaystyle\int dx} = \frac{1}{2\pi a}\int_{-\pi}^{\pi} a^2(1 - \cos \theta)^2\, a(1 + \cos \theta)\, d\theta$$

$$= \frac{a^2}{2\pi}\, 2\int_0^{\pi} 4\sin^4\frac{\theta}{2}\, 2\cos^2\frac{\theta}{2}\, d\theta \qquad \ldots\left(\because \frac{\theta}{2} = t\right)$$

$$= \frac{8a^2}{\pi}\int_0^{\pi/2} \sin^4 t \cos^2 t\, 2\, dt = \frac{16a^2}{\pi}\, \frac{3 \cdot 1 \cdot 1}{6 \cdot 4 \cdot 2}\, \frac{\pi}{2} = \frac{a^2}{2}$$

$$\boxed{\text{RMS value} = \frac{a^2}{2}}$$

$\therefore$ R.M.S. value $= \dfrac{a}{\sqrt{2}}$

Ex. 9 : *If a point moves with uniform acceleration, prove that the mean square of velocities at infinitely small intervals of time is* $\frac{1}{3}\left(v_0^2 + v_1 v_0 + v_1^2\right)$, *where v_0 and v_1 are the initial and final velocities.*

Sol. : Acceleration $= \dfrac{dv}{dt} = f$ (say) $\qquad\qquad \therefore\ dt = \dfrac{1}{f}\,dv$

$$\text{M.S. value} = \frac{\int v^2\,dt}{\int dt} = \frac{\displaystyle\int_{v_0}^{v_1} v^2\,\frac{1}{f}\,dv}{\displaystyle\int_{v_0}^{v_1}\frac{1}{f}\,dv} = \frac{\left[\dfrac{v^3}{3}\right]_{v_0}^{v_1}}{v_1 - v_0} = \frac{1}{3}\frac{\left(v_1^3 - v_0^3\right)}{(v_1 - v_0)} = \frac{1}{3}\frac{(v_1 - v_0)\left(v_1^2 + v_1 v_0 + v_0^2\right)}{v_1 - v_0}$$

$$\boxed{\ \text{M.S. value} = \frac{1}{3}\left(v_0^2 + v_1 v_0 + v_1^2\right)\ }$$

Ex. 10 : *If a point moves with constant acceleration, show that the space average of velocity over any distance is* $\dfrac{2}{3}\left(\dfrac{v_1^2 + v_1 v_0 + v_0^2}{v_1 + v_0}\right)$, *where v_0 and v_1 are the initial and final velocities.*

Sol. : $v\dfrac{dv}{ds} = $ Acceleration $= f$ say, $ds = \dfrac{1}{f}\,v\,dv$

$$\therefore\ \text{Space average of velocity over any distance} = \frac{\int v\,ds}{\int ds} = \frac{\int v\,\frac{v}{f}\,dv}{\int \frac{v}{f}\,dv} = \frac{\displaystyle\int_{v_0}^{v_1} v^2\,dv}{\displaystyle\int_{v_0}^{v_1} v\,dv} = \frac{(v_1^3 - v_0^3)/3}{(v_1^2 - v_0^2)/2} = \frac{2}{3}\frac{(v_1 - v_0)(v_1^2 + v_1 v_0 + v_0^2)}{(v_1 - v_0)(v_1 + v_0)}$$

$$\boxed{\ \text{Space average of velocity over any distance} = \frac{2}{3}\left(\frac{v_1^2 + v_1 v_0 + v_0^2}{v_1 + v_0}\right)\ }$$

Ex. 11 : *Find R.M.S. value of distances from origin of points within $x^2 + y^2 - 2ax = 0$ cut off by $x = a$ where $x \geq a$.*

Sol. : Given region is ABCDA shown in Fig. 10.42. $x = a \Rightarrow r\cos\theta = a$ or $r = a\sec\theta$ $\therefore$ $x^2 + y^2 = 2ax \Rightarrow r = 2a\cos\theta$. The distance of any point P (r, θ) from origin O being r, then the

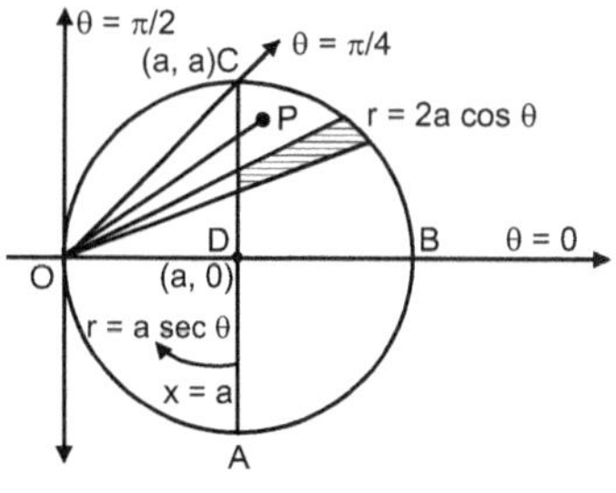

$$\text{M.S. value} = \frac{\iint r^2\,r\,dr\,d\theta}{\iint r\,dr\,d\theta} = \frac{2\displaystyle\int_0^{\pi/4} d\theta \int_{a\sec\theta}^{2a\cos\theta} r^3\,dr}{\dfrac{\pi a^2}{2}}$$

$\theta = \pi/2$, (a, a)C, $\theta = \pi/4$, P, $r = 2a\cos\theta$, D $(a, 0)$, B, $\theta = 0$, O, $r = a\sec\theta$, $x = a$, A

Fig. 10.42

$$= \frac{4}{\pi a^2}\int_0^{\pi/4} \frac{16a^4\cos^4\theta - a^4\sec^4\theta}{4}\,d\theta$$

$$= \frac{a^2}{\pi}\left\{\int_0^{\pi/4} 16\,\frac{(1 + \cos 2\theta)^2}{4}\,d\theta - \int_0^{\pi/4}\sec^2\theta\,(1 + \tan^2\theta)\,d\theta\right\}$$

$$= \frac{a^2}{\pi}\left\{4\int_0^{\pi/2}(1 + 2\cos t + \cos^2 t)\,\frac{dt}{2} - \int_0^1 (1 + u^2)\,du\right\} \qquad \begin{cases} 2\theta = t \\ \tan\theta = u \end{cases}$$

$$= \frac{a^2}{\pi}\left\{2\left(\frac{\pi}{2} + 2 + \frac{1}{2}\frac{\pi}{2}\right) - \left(1 + \frac{1}{3}\right)\right\} = a^2\left(\frac{3}{2} + \frac{8}{3\pi}\right) \qquad \boxed{\ \text{R.M.S. value} = a\sqrt{\frac{3}{2} + \frac{8}{3\pi}}\ }$$

Ex. 12 : *Prove that the mean distance of points within a circular area of radius a from a fixed point on the circumference is $\dfrac{32a}{9\pi}$.*

Sol. : Take the fixed point O on the circumference as the pole and the diameter through O as the initial line. Then the polar equation of the circle is r = 2a cos θ. Refer Fig. 10.43. By symmetry we may consider only the upper semi-circle for the required mean distance. The distance of any point P(r, θ) of the region from O being r, the required M.V. of the distance is

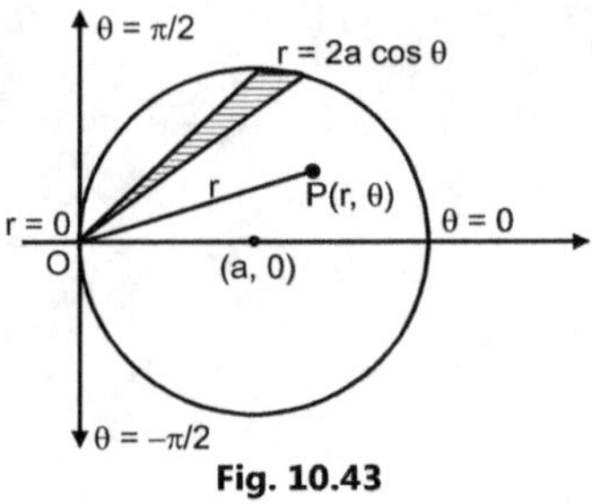

Fig. 10.43

$$\text{M.V.} \;=\; \frac{\displaystyle\iint r\,r\,dr\,d\theta}{\displaystyle\iint r\,dr\,d\theta} \;=\; \frac{\displaystyle\int_0^{\pi/2}\int_0^{2a\cos\theta} r^2\,dr\,d\theta}{\dfrac{\pi a^2}{2}}$$

$$=\; \frac{2}{\pi a^2}\int_0^{\pi/2}\frac{8a^3\cos^3\theta}{3}\,d\theta \;=\; \frac{16a}{3\pi}\cdot\frac{2}{3} \qquad \boxed{\text{M.V.} \;=\; \frac{32a}{9\pi}}$$

Ex. 13 : *A rod of length a is divided into three parts at random. Find the M.V. of*
(i) the sum of the squares of these parts, (ii) their product.   **(Dec. 2008, May 2006, 2016)**

Sol. : If OC be the rod, OC = a. We first divide it at A into two random parts x, a − x so that x varies from 0 to a. We now divide the portion AC at random into parts AB = y and BC = a − x − y so that y varies from 0 to (a − x).

The three parts so obtained viz. x, y, a − x − y are shown in Fig. 10.44.

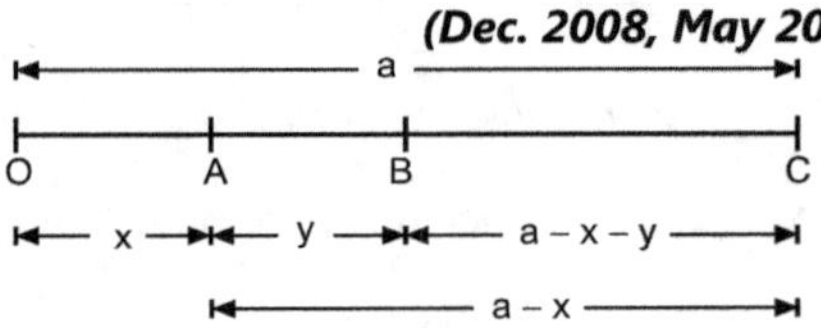

Fig. 10.44

Part (i) : The required M.V.
$$=\; \frac{\displaystyle\int_0^a\int_0^{a-x}[x^2+y^2+(a-x-y)^2]\,dx\,dy}{\displaystyle\int_0^a\int_0^{a-x} dx\,dy} \;=\; \frac{\displaystyle\int_0^a\left[x^2y+\frac{y^3}{3}+\frac{(a-x-y)^3}{-3}\right]_0^{a-x}dx}{\displaystyle\int_0^a (a-x)\,dx}$$

$$=\; \frac{\displaystyle\int_0^a\left[x^2(a-x)+\frac{2}{3}(a-x)^3\right]dx}{\left[ax-\dfrac{x^2}{2}\right]_0^a} \;=\; \frac{\left[a\dfrac{x^3}{3}-\dfrac{x^4}{4}+\dfrac{2}{3}\dfrac{(a-x)^4}{-4}\right]_0^a}{\dfrac{a^2}{2}} \;=\; \frac{a^4/4}{a^2/2} \;=\; \frac{a^2}{2}$$

Part (ii) :
$$\text{M.V.} \;=\; \frac{\displaystyle\int_0^a\int_0^{a-x} xy\,(a-x-y)\,dy\,dx}{\displaystyle\int_0^a\int_0^{a-x} dy\,dx} \;=\; \frac{a^5/120}{a^2/2} \qquad\qquad \text{(details left to student)}$$

$$\boxed{\text{M.V.} \;=\; \frac{a^3}{60}}$$

Ex. 14 : *The surface density of an electrified circular disc of radius a at the distance r from the centre is $\dfrac{K}{\sqrt{a^2-r^2}}$. Find the mean density and ratio of mean density to density at the centre.*

Sol. :
$$\text{Mean density} \;=\; \frac{\text{Total mass}}{\text{Total area}} \;=\; \frac{\displaystyle\iint \rho\,r\,dr\,d\theta}{\displaystyle\iint r\,dr\,d\theta}$$

$$=\; \frac{\displaystyle\int_0^{2\pi}\int_0^{a}\frac{k}{\sqrt{a^2-r^2}}\,r\,dr\,d\theta}{\pi a^2}$$

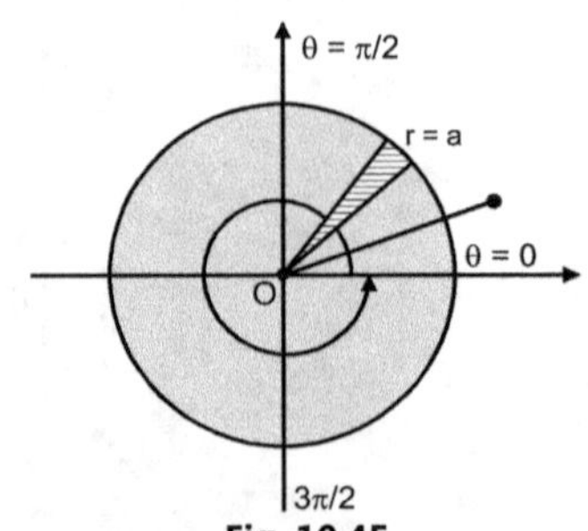

Fig. 10.45

$$= \frac{-k}{2\pi a^2}\,(2\pi)\,\left(2\sqrt{a^2 - r^2}\right)_0^a = \frac{2k}{a^2}\,(a)$$

$$\boxed{\text{Mean density } = \frac{2k}{a}}$$

The density at the centre ($r = 0$) is $\rho_O = \dfrac{k}{\sqrt{a^2 - 0}} = \dfrac{k}{a}$

$\therefore$ $\boxed{\text{Ratio } = \dfrac{\text{M.D.}}{\rho_O} = \dfrac{2k/a}{k/a} = \dfrac{2}{1} \text{ i.e. } 2:1}$

Ex. 15 : *Find the M.V. of $x^2\,y^2\,z^2$ over the positive octant of the ellipsoid $\dfrac{x^2}{a^2} + \dfrac{y^2}{b^2} + \dfrac{z^2}{c^2} = 1$.*

Sol. : $\text{M.V.} = \dfrac{\iiint x^2\,y^2\,z^2\,dx\,dy\,dz}{\iiint dx\,dy\,dz}$ $\because \iiint dx\,dy\,dz = \dfrac{4}{3}\pi\,abc.$ **(Dec. 2004)**

Put $x = ar\sin\theta\cos\phi$, $y = br\sin\theta\sin\phi$, $z = cr\cos\theta$, $dx\,dy\,dz = abc\,r^2\sin\theta\,d\theta\,d\phi\,dr$

$$\text{M.V.} = \frac{\displaystyle\int_0^{\pi/2}\int_0^{\pi/2}\int_0^1 a^2\,b^2\,c^2\,r^6\sin^4\theta\cos^2\theta\,\sin^2\phi\,\cos^2\phi\;abc\,r^2\sin\theta\,d\theta\,d\phi\,dr}{\dfrac{1}{8}\cdot\dfrac{4}{3}\,\pi\,abc}$$

$$= \frac{6}{\pi}\,a^2\,b^2\,c^2\int_0^{\pi/2}\sin^5\theta\cos^2\theta\,d\theta\int_0^{\pi/2}\sin^2\phi\cos^2\phi\,d\phi\int_0^1 r^8\,dr = \frac{6a^2\,b^2\,c^2}{\pi}\,\frac{4\cdot2\cdot1}{7\cdot5\cdot3\cdot1}\,\frac{1\cdot1}{4\cdot2}\,\frac{\pi}{2}\,\frac{1}{9}$$

$$\boxed{\text{M.V.} = \frac{a^2\,b^2\,c^2}{315}}$$

$$\boxed{\text{EXERCISE 10.4}}$$

1. Find the average density of the sphere of radius a whose density at a distance r from the   centre of the sphere is $\rho = \rho_O\left(1 + k\,\dfrac{r^3}{a^3}\right)$.   **Ans.** $\rho_O\left(1 + \dfrac{k}{2}\right)$ (Refer Solved Ex. 7)

2. A point moves from rest with uniform acceleration. Prove that in any interval from the start, the space average of velocity is $\dfrac{4}{3}$ of the time average.

3. Find the r.m.s. value of the expression a sin pt + b cos qt over the interval 0 to 2π.   **Ans.** $\sqrt{\dfrac{a^2 + b^2}{2}}$

4. Find the M.V. of $e^{-x^2 - y^2}$ over $x^2 + y^2 = 1$. **(Dec. 2010, 2006)**   **Ans.** $\dfrac{e - 1}{e}$

5. Find M.V. of distances from one corner of a square of side a to any point in the square.   **Ans.** $\dfrac{a}{3}\left[\sqrt{2} + \log\left(1 + \sqrt{2}\right)\right]$

6. Find M.V. of xyz over positive octant of ellipsoid $\dfrac{x^2}{a^2} + \dfrac{y^2}{b^2} + \dfrac{z^2}{c^2} = 1$.   **Ans.** $\dfrac{abc}{8\pi}$

7. Find M.V. of rectangle contained by the abscissa and ordinate of any point in first quadrant of ellipse $\dfrac{x^2}{a^2} + \dfrac{y^2}{b^2} = 1$.   **Ans.** $\dfrac{ab}{2\pi}$

8. A particle moves from rest with uniform acceleration. Prove that in any interval of time from starting point, the space average of velocity is $\dfrac{2}{3}$ v, where v is final velocity.

$$\left[\textbf{Hint :} \text{ Space average velocity } = \frac{\int v\,ds}{\int ds} = \frac{\displaystyle\int_0^v v^2\,dv}{\displaystyle\int_0^v v\,dv} = \frac{2}{3}\,v\right]$$

9. A particle moves from rest with uniform acceleration. Prove that, at any interval of time from starting point, the time average of velocity is $\dfrac{v}{2}$.

[**Hint :** Average velocity $= \dfrac{\int v\,dt}{\int dt} = \dfrac{\displaystyle\int_0^v v\,dv}{\displaystyle\int_0^v dv} = \dfrac{v}{2}$]

10. Find the M.S. value and R.M.S. value of the distance from a point within a cardioide $r = a\,(1 + \cos\theta)$ from vertex.

Ans. $\dfrac{35a^2}{24}$, $\dfrac{a}{2}\sqrt{\dfrac{35}{6}}$

11. If the density at a distance r from centre of sphere is given by $\rho_0\left(1 - \dfrac{kr^2}{a^2}\right)$, prove that the average density is $\rho_0\left(1 - \dfrac{3k}{5}\right)$, where a is radius of sphere.

[**Hint :** Average density $= \dfrac{\int \rho\,dv}{\int dv}$ and use $v = \dfrac{4}{3}\pi r^3,\ \ dv = 4\pi r^2\,dr.$]

12. If in a spherical mass whose density ρ is a function of r where, r is distance from the centre. If D denotes mean density of matter of the mass included within a concentric sphere of radius r, prove that $\rho = D + \dfrac{r}{3}\cdot\dfrac{dD}{dr}$.

Hint : $D = \dfrac{\int \rho\,dv}{\int dv},\ \ \dfrac{4}{3}\pi r^3 D = \int \rho\,4\pi r^2\,dr,\ \ \dfrac{r^3}{3}D = \int \rho\,r^2\,dr.$ Now differentiate w.r.t. r

13. Show that the mean distance of points on a spherical surface of radius 'a' from a point P which is at a distance 'c' from centre is $c + \dfrac{a^2}{3c}$ or $a + \dfrac{c^2}{3a}$ according the point is outside or inside the sphere.

Hint : $OP = c,\ \ PL = x,\ \ OL = c - x,$ Mean of AP $= \dfrac{\int AP\,dx}{\int dx}$

$AP^2 = AL^2 + PL^2 = OA^2 - OL^2 + PL^2$

$AP = (a^2 - c^2 + 2cx)^{1/2}$, when A is on PO $\ x = c - a$

and when A is on PO produced, $x = c + a$

Case (i) : P is outside sphere, $c > a,$

Case (ii) : P is inside sphere, $a > c.$

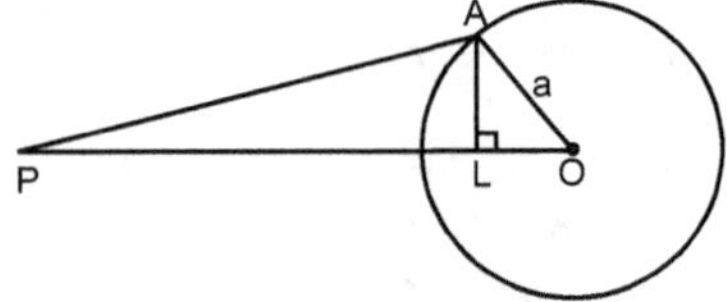

Fig. 10.46

14. Find the mean value of reciprocals of distances of points on a circular area of radius 'a' from a fixed point on its circumference.

[**Hint :** Take $r = 2a\cos\theta$]

Ans. $\dfrac{4}{\pi a}$

15. Do the Ex. 14 if the distance is from the centre.

Hint : Take $x^2 + y^2 = a^2$ **Ans.** $\dfrac{2}{a}$

16. Find M.V. of xy over the area of loop of $x^3 + y^3 = 3axy.$

Ans. $\dfrac{3a^2}{4}$

17. Find the M.V. of the ordinates of a semi-circle of radius a.

Ans. $\dfrac{\pi a}{4}$

❑❑❑

CHAPTER-11

CENTRE OF GRAVITY AND MOMENT OF INERTIA

11.1 DEFINITIONS

1. **Centre of Mass :** The centre of mass of a body is the point through which the resultant mass acts.

2. **Centre of Gravity :** The centre of gravity (C.G.) of a body is the point through which the resultant weight acts. Since, weight is proportional to mass, the centre of mass is the same point as the centre of gravity.

11.2 THE CENTRE OF MASS OF n POINT MASSES

A small boy on one side of a seesaw (which we regard as weightless) can balance a bigger boy on the other side.

For example, the two boys in Fig. 11.1 (a) balance.

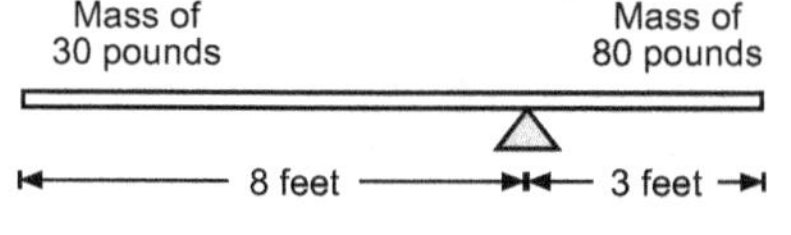

Fig. 11.1 (a)

The small mass with the long lever arm balances the large mass with the small lever arm. Each boy contributes the same tendency to turn but in opposite directions.

This tendency is called the *moment*. $\boxed{\text{Moment = (Mass)} \cdot \text{(Lever arm)}}$ where the lever arm can be positive or negative.

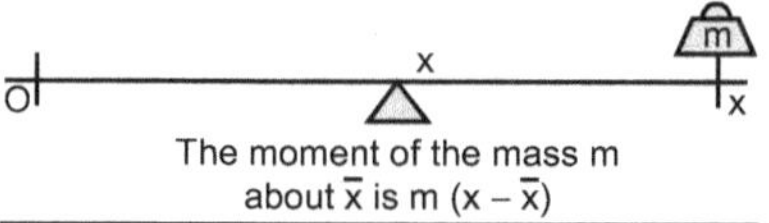

Fig. 11.1 (b)

If a mass m is located on a line at co-ordinate x, define its moment about the point having co-ordinate $\bar{x}$ as the product $m (x - \bar{x})$. Refer Fig. 11.1 (b).

Now, consider several point masses $m_1, m_2, m_3, \ldots\ldots m_n$. If mass m_i is located at x_i, with i = 1, 2, ..., n, then $\displaystyle\sum_{i=1}^{n} m_i (x_i - \bar{x})$ is the total moment of all the masses about the point $\bar{x}$.

The balancing point of masses $m_1, m_2, m_3, \ldots m_n$, located respectively at $x_1, x_2, x_3, \ldots x_n$ on x-axis, is found by solving the equation $\displaystyle\sum_{i=1}^{n} m_i (x_i - \bar{x}) = 0$ for the number $\bar{x}$.

$$\therefore \quad \sum_{i=1}^{n} m_i x_i = \bar{x} \sum_{i=1}^{n} m_i$$

$$\Rightarrow \quad \boxed{\bar{x} = \frac{\displaystyle\sum_{i=1}^{n} m_i x_i}{\displaystyle\sum_{i=1}^{n} m_i}} \qquad \ldots (1)$$

$\therefore$ *The number $\bar{x}$ given by equation (1) is called the centre of mass of the system of masses.*

It is the point about which all the masses balance. The centre of the mass is found by dividing the total moment about O by the total mass.

If point masses are situated in a plane, then

$$\bar{x} = \frac{\displaystyle\sum_{i=1}^{n} m_i x_i}{\displaystyle\sum_{i=1}^{n} m_i} \quad ; \qquad \bar{y} = \frac{\displaystyle\sum_{i=1}^{n} m_i y_i}{\displaystyle\sum_{i=1}^{n} m_i}$$

If point masses are situated in a space, then

$$\bar{x} = \frac{\sum\limits_{i=1}^{n} m_i\, x_i}{\sum\limits_{i=1}^{n} m_i} \quad ; \qquad \bar{y} = \frac{\sum\limits_{i=1}^{n} m_i\, y_i}{\sum\limits_{i=1}^{n} m_i} \quad ; \qquad \bar{z} = \frac{\sum\limits_{i=1}^{n} m_i\, z_i}{\sum\limits_{i=1}^{n} m_i}$$

Instead of discrete masses, if the mass distribution is continuous (i.e. if the body is rigid body) then the summations will be replaced by the corresponding integrals and we have,

$$\bar{x} = \frac{\int x\, dm}{\int dm} \quad ; \qquad \bar{y} = \frac{\int y\, dm}{\int dm} \quad ; \qquad \bar{z} = \frac{\int z\, dm}{\int dm}$$

where, dm is an element of the distributed mass of the body, $(\bar{x},\bar{y},\bar{z})$ may be considered as C.G. of mass distribution.

11.3 C.G. OF AN ARC OF A CURVE

Let the mass be distributed in the form of the wire "CD" in the plane XOY. Let ρ be the density per unit length. Take an elementary arc "ds" of the curve at the point P (x, y) whose equation is $y = f(x)$. Refer Fig. 11.2.

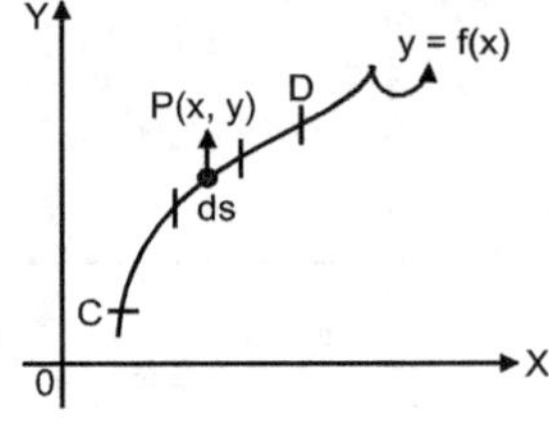

$\quad \therefore\quad$ Mass $=$ Density (length)

$\quad \therefore\quad$ The mass of this element is dm $= (\rho)\,(ds)$ and is situated at (x, y). Hence, C.G. $\left(\bar{x},\bar{y}\right)$ of arc CD will be given by :

Fig. 11.2

$$\boxed{\bar{x} = \frac{\int x\, dm}{\int dm} = \frac{\int x\,\rho\, ds}{\int \rho\, ds}} \quad ; \qquad \boxed{\bar{y} = \frac{\int y\, dm}{\int dm} = \frac{\int y\,\rho\, ds}{\int \rho\, ds}}$$

If ρ is constant, we have

$$\boxed{\bar{x} = \frac{\int x\, ds}{\int ds}} \quad ; \qquad \boxed{\bar{y} = \frac{\int y\, ds}{\int ds}}$$

Note :

1.	For $y = f(x)$ replace ds $= \sqrt{1 + \left(\dfrac{dy}{dx}\right)^2}\cdot dx$
2.	For $x = f(y)$ replace ds $= \sqrt{1 + \left(\dfrac{dx}{dy}\right)^2}\cdot dy$
3.	For $x = f_1(t),\ y = f_2(t)$ replace ds $= \sqrt{\left(\dfrac{dx}{dt}\right)^2 + \left(\dfrac{dy}{dt}\right)^2}\cdot dt$
4.	For $r = f(\theta)$ replace ds $= \sqrt{r^2 + \left(\dfrac{dr}{d\theta}\right)^2}\cdot d\theta$
5.	For $\theta = f(r)$ replace ds $= \sqrt{1 + r^2\left(\dfrac{d\theta}{dr}\right)^2}\cdot dr$

(For the detail proof, refer chapter 6)

11.4 C.G. OF A PLANE AREA OR LAMINA

Let C be the bounding curve of a plane lamina and ρ be the mass per unit area.

∵   Mass = (Density) × (Area)

∴   The mass of an elementary area dA at P(x, y) is dm = ρ(dA).

Let $(\bar{x}, \bar{y})$ be the co-ordinates of C.G. of the lamina (Refer Fig. 11.3), then,

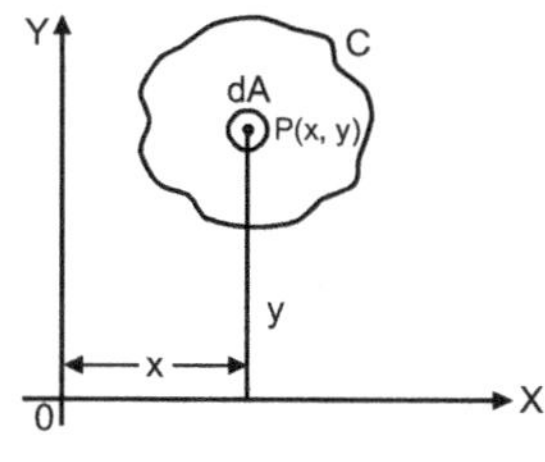

Fig. 11.3

$$\bar{x} = \frac{\int x\, dm}{\int dm} = \frac{\int x \rho\, dA}{\int \rho\, dA}; \quad \bar{y} = \frac{\int y\, dm}{\int dm} = \frac{\int y \rho\, dA}{\int \rho\, dA}$$

or in Double integral form, Cartesian System (dA = dx dy)

∴

$$\bar{x} = \frac{\iint x \rho\, dx\, dy}{\iint \rho\, dx\, dy}; \quad \bar{y} = \frac{\iint y \rho\, dx\, dy}{\iint \rho\, dx\, dy}$$

If ρ is a constant;

∴

$$\bar{x} = \frac{\iint x\, dx\, dy}{\iint dx\, dy}; \quad \bar{y} = \frac{\iint y\, dx\, dy}{\iint dx\, dy}$$

If the curve is given in polar co-ordinates,

dA = r dr dθ; x = r cos θ; y = r sin θ; then

$$\bar{x} = \frac{\iint r \cos\theta \cdot \rho \cdot r\, dr\, d\theta}{\iint \rho\, r\, dr\, d\theta}; \quad \bar{y} = \frac{\iint r \sin\theta\, \rho\, r\, dr\, d\theta}{\iint \rho\, r\, dr\, d\theta}$$

If ρ is a constant,

∴

$$\bar{x} = \frac{\iint r \cos\theta\, r\, dr\, d\theta}{\iint r\, dr\, d\theta}; \quad \bar{y} = \frac{\iint r \sin\theta\, r\, dr\, d\theta}{\iint r\, dr\, d\theta}$$

Note :

The following results, for the C.G. of a plane lamina in special cases, which involve only single integrals and are also useful, are stated without proof. They can be established easily by the principles used in the above cases.

(a) For the co-ordinates of the C.G. of a plane lamina bounded by the curve y = f(x), the x-axis and the ordinates x = a and x = b, (Refer Fig. 11.4) consider strip of width dx.

If ρ is mass per unit area, dm = $\rho \cdot y \cdot dx$.

C.G. of strip is at its midpoint $\left(x, \dfrac{y}{2}\right)$.

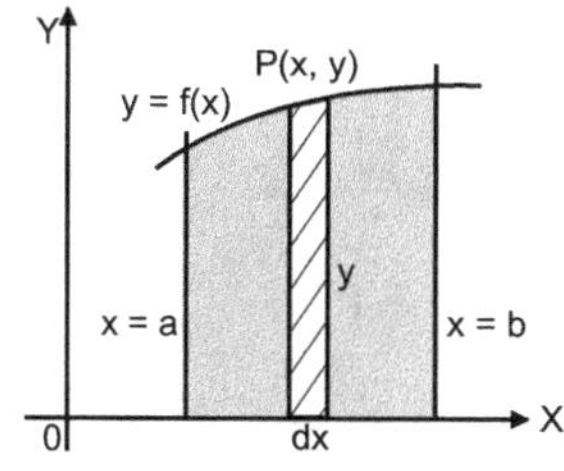

Fig. 11.4

$$\bar{x} = \frac{\int x \rho\, y\, dx}{\int \rho\, y\, dx}; \quad \bar{y} = \frac{\int \frac{y}{2} \rho\, y\, dx}{\int \rho\, y\, dx} = \frac{\frac{1}{2}\int \rho\, y^2\, dx}{\int \rho\, y\, dx}$$

(b) The co-ordinates of the C.G. of a sectorial element OAB, bounded by the curve $r = f(\theta)$ and the radii vectors $\theta = \alpha$, $\theta = \beta$ (Refer Fig. 11.5). Let ρ be mass per unit area. Area of $\triangle OPQ = \dfrac{1}{2}\, r^2\, d\theta$.

$$\therefore \quad dm = (\rho)\left(\frac{1}{2} r^2\, d\theta\right)$$

Its C.G. is on median through O at a point $\left(\dfrac{2}{3} r \cos\theta,\ \dfrac{2}{3} r \sin\theta\right)$

$$\bar{x} = \frac{\dfrac{2}{3}\displaystyle\int_{\alpha}^{\beta} r^3 \cos\theta\, d\theta}{\displaystyle\int_{\alpha}^{\beta} r^2\, d\theta} \quad ; \quad \bar{y} = \frac{\dfrac{2}{3}\displaystyle\int_{\alpha}^{\beta} r^3 \sin\theta\, d\theta}{\displaystyle\int_{\alpha}^{\beta} r^2\, d\theta}$$

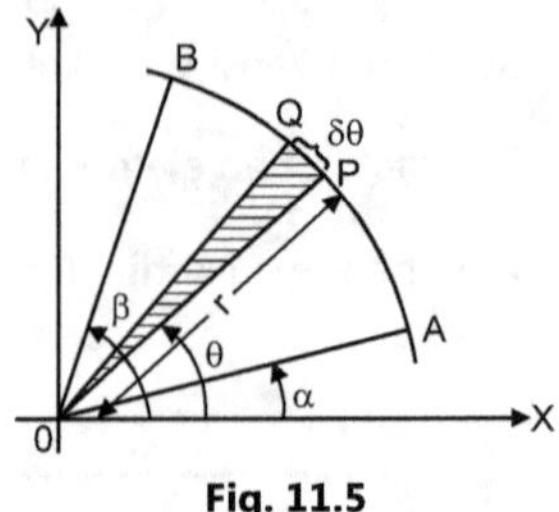

Fig. 11.5

11.5 C.G. OF A SOLID

Here we take an elementary volume dx dy dz, at the point P(x, y, z), so that $dm = (\rho)\,(dx\,dy\,dz)$ if ρ is the mass per unit volume, we have

$$\bar{x} = \frac{\iiint x\,\rho\,dx\,dy\,dz}{\iiint \rho\,dx\,dy\,dz} \quad ; \quad \bar{y} = \frac{\iiint y\,\rho\,dx\,dy\,dz}{\iiint \rho\,dx\,dy\,dz} \quad ; \quad \bar{z} = \frac{\iiint z\,\rho\,dx\,dy\,dz}{\iiint \rho\,dx\,dy\,dz}$$

Conversion to spherical polar or cylindrical systems can be easily made.

11.6 ILLUSTRATIONS ON C.G. OF AN ARC OF A CURVE

Ex. 1 : *If 's' is the length of the arc of the catenary* $y = c \cosh\dfrac{x}{c}$ *from vertex to (x, y), prove that C.G. of the arc is given by,*

$$\bar{x} = x - \frac{c\,(y-c)}{s}, \quad \bar{y} = \frac{y}{2} + \frac{cx}{2s}$$

Sol. : We have, $\quad \bar{x} = \dfrac{\displaystyle\int x\, ds}{\displaystyle\int ds},\ \bar{y} = \dfrac{\displaystyle\int y\, ds}{\displaystyle\int ds}$

Now, $\quad \dfrac{dy}{dx} = \sinh\dfrac{x}{c}$, so that

$$ds = \sqrt{1 + \left(\frac{dy}{dx}\right)^2}\; dx = \sqrt{1 + \sinh^2\frac{x}{c}}\; dx = \cosh\left(\frac{x}{c}\right) dx$$

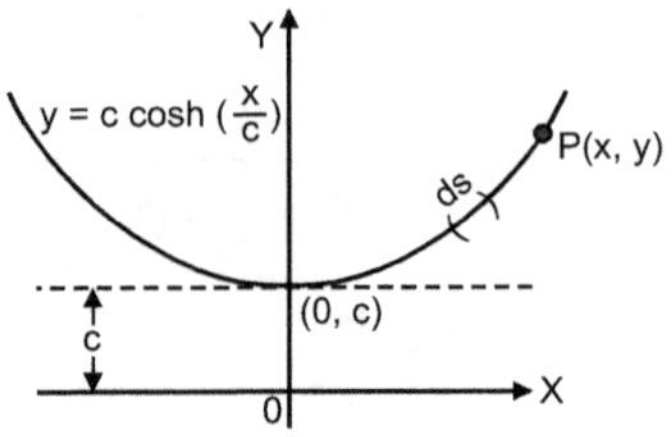

Fig. 11.6

$$\therefore \quad \int x\, ds = \int_0^x x \cosh\frac{x}{c}\, dx \text{ (Integrate by parts)}$$

$$= \left[x \cdot c\, \sinh\frac{x}{c} - 1 \cdot c^2 \cosh\frac{x}{c}\right]_0^x$$

$$= cx \sinh\frac{x}{c} - c^2 \cosh\frac{x}{c} + c^2 \qquad\qquad \dots \text{(i)}$$

$$\int y\, ds = \int_0^x c \cosh\frac{x}{c} \cdot \cosh\frac{x}{c}\, dx$$

$$= \frac{c}{2}\int_0^x \left[1 + \cosh\left(\frac{2x}{c}\right)\right] dx = \frac{c}{2}\left[x + \frac{c}{2}\sinh\frac{2x}{c}\right]_0^x$$

$$= \frac{c}{2}\left(x + c \sinh\frac{x}{c}\cosh\frac{x}{c}\right) \qquad\qquad \dots \text{(ii)}$$

Also, $\quad \displaystyle\int ds = \int_0^x \cosh\frac{x}{c}\, dx = c \sinh\frac{x}{c} \Rightarrow s = c \sinh\frac{x}{c} \qquad\qquad \dots \text{(iii)}$

From (i), (ii), (iii),

$$\int x \, ds = xs - cy + c^2; \quad \int y \, ds = \frac{c}{2}\left(x + s \cdot \frac{y}{c}\right) = \frac{cx}{2} + \frac{sy}{2}$$

$$\therefore \quad \bar{x} = \frac{xs - cy + c^2}{s} = x - \frac{c(y - c)}{s}, \quad \bar{y} = \frac{cx/2 + sy/2}{s} = \frac{y}{2} + \frac{cx}{2s}$$

$$\boxed{\bar{x} = -\frac{c(y - c)}{s} \; ; \; \bar{y} = \frac{y}{2} + \frac{cx}{2s}}$$

Ex. 2 : *Find the C.G. of the arc from the cusp to cusp of the cycloid $x = a(\theta - \sin\theta)$, $y = a(1 - \cos\theta)$.*

Sol. : The cycloid is as shown in Fig. 11.7. It is symmetrical about the line $x = a\pi$. $\therefore \bar{x} = a\pi$

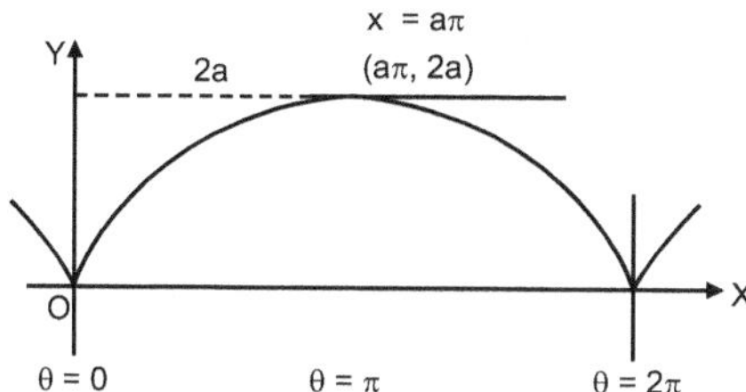

Fig. 11.7

Now,
$$\bar{y} = \frac{\int y \, ds}{\int ds}$$

$$ds = \sqrt{\left(\frac{dx}{d\theta}\right)^2 + \left(\frac{dy}{d\theta}\right)^2} \, d\theta$$

$$ds = \sqrt{a^2(1 - \cos\theta)^2 + a^2 \sin^2\theta} \, d\theta$$

$$= 2a \sin\frac{\theta}{2} \, d\theta$$

$$\int y \, ds = \int_0^{2\pi} a(1 - \cos\theta) \cdot 2a \sin\frac{\theta}{2} \, d\theta = 4a^2 \int_0^{2\pi} \sin^3\frac{\theta}{2} \, d\theta \qquad \left(\because \frac{\theta}{2} = t\right)$$

$$= 8a^2 \int_0^{\pi} \sin^3 t \, dt = 16a^2 \int_0^{\pi/2} \sin^3 t \, dt = 16a^2 \frac{2}{3} = \frac{32a^2}{3}$$

$$\int ds = \int_0^{2\pi} 2a \sin\frac{\theta}{2} \, d\theta = 2a \left[-2\cos\frac{\theta}{2}\right]_0^{2\pi} = 4a[1 + 1] = 8a$$

$$\bar{y} = \frac{32a^2/3}{8a} = \frac{4a}{3}$$

Required co-ordinates of C.G. are $\left(a\pi, \dfrac{4a}{3}\right)$

Ex. 3 : *Show that the centroid of a wire bent into the form of a cardioide $r = a(1 + \cos\theta)$ and with line density $k \sec\dfrac{\theta}{2}$, k being constant, is on the axis of the cardioide at a distance $\dfrac{a}{2}$ from the cusp.*

Sol. : Here as the line density is not constant, we will use $\bar{x} = \dfrac{\int x \rho \, ds}{\int \rho \, ds}$

Because of symmetry, C.G. lies on x-axis. $\therefore \bar{y} = 0$. Also $\rho = k \sec\dfrac{\theta}{2}$.

$$ds = \sqrt{r^2 + \left(\frac{dr}{d\theta}\right)^2} \, d\theta = \sqrt{a^2(1 + \cos\theta)^2 + a^2 \sin^2\theta} \, d\theta$$

$$= a\sqrt{1 + 2\cos\theta + \cos^2\theta + \sin^2\theta} \, d\theta$$

$$= a\sqrt{2(1 + \cos\theta)} \, d\theta$$

$$ds = 2a \cos\frac{\theta}{2} \, d\theta$$

Also replace $x = r\cos\theta = a(1 + \cos\theta) \cdot \cos\theta$

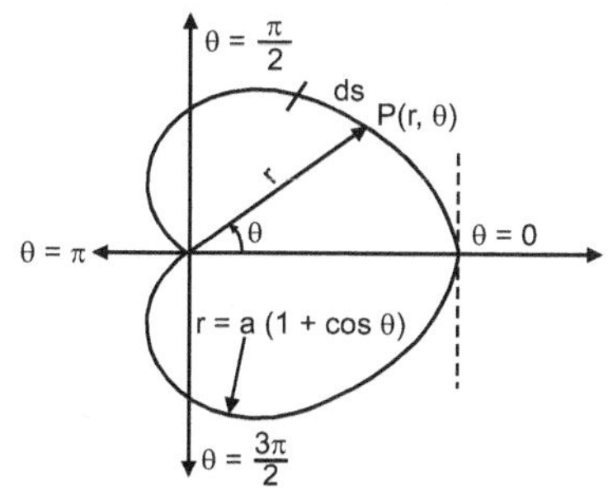

Fig. 11.8

$$\int x \rho \, ds = 2 \int_0^{\pi} a(1 + \cos\theta)\cos\theta \, k \sec\frac{\theta}{2} \cdot 2a \cos\frac{\theta}{2} \, d\theta$$

$$= 4a^2 k \int_0^{\pi} (\cos\theta + \cos^2\theta) \, d\theta$$

$$\int x\rho\,ds = 4a^2 k\left[0 + 2\int_0^{\pi/2}\cos^2\theta\,d\theta\right] = 8a^2 k\left(\frac{1}{2}\frac{\pi}{2}\right) = 2\pi a^2 k$$

Also

$$\int \rho\,ds = 2\int_0^{\pi} k\sec\frac{\theta}{2}\cdot 2a\cos\frac{\theta}{2}\,d\theta = 4ak\int_0^{\pi} d\theta = 4ak\pi$$

∴

$$\boxed{\overline{x} = \frac{2\pi a^2 k}{4\pi\,ak} = \frac{a}{2}}$$

Thus, the C.G. of the wire is on the axis of the cardioide and at a distance $\dfrac{a}{2}$ from the origin, which is its cusp.

Ex. 4 : *Find the C.G. of arc in the positive quadrant of* $x^{2/3} + y^{2/3} = a^{2/3}$.

Sol. : For the arc ABC, because of symmetry about y = x, we have $\overline{x} = \overline{y}$.

$$\overline{x} = \frac{\int x\,ds}{\int ds}\quad (\rho = \text{constant})$$

Parametric equations of the astroid are $x = a\cos^3\theta$, $y = a\sin^3\theta$

$$ds = \sqrt{\left(\frac{dx}{d\theta}\right)^2 + \left(\frac{dy}{d\theta}\right)^2}\,d\theta = \sqrt{(-3a\cos^2\theta\sin\theta)^2 + (3a\sin^2\theta\cos\theta)^2}\,d\theta$$

$$ds = 3a\sin\theta\cos\theta\,d\theta$$

In the positive quadrant, θ varies from 0 to $\dfrac{\pi}{2}$.

$$\int x\,ds = \int_0^{\pi/2} a\cos^3\theta\cdot 3a\sin\theta\cos\theta\,d\theta$$

$$= 3a^2\int_0^{\pi/2}\cos^4\theta\sin\theta\,d\theta = \frac{3a^2}{5}$$

$$\int ds = \int_0^{\pi/2} 3a\sin\theta\cos\theta\,d\theta = 3a\int_0^{\pi/2}\sin\theta\cos\theta\,d\theta = \frac{3a}{2}$$

$$\overline{x} = \frac{3a^2/5}{3a/2} = \frac{2a}{5}\quad \therefore\ \overline{y} = \frac{2a}{5}$$

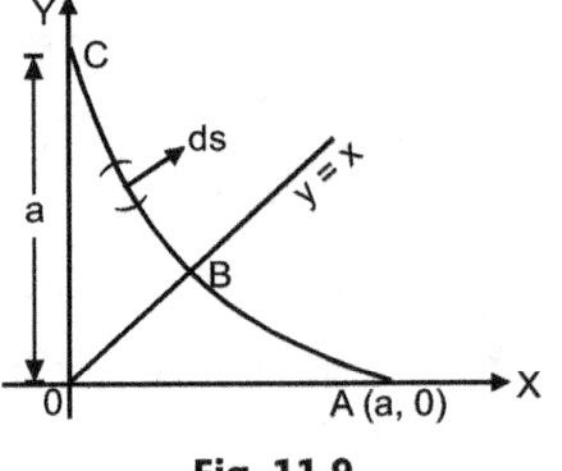

Fig. 11.9

∴ C.G. of arc is $\boxed{\left(\dfrac{2a}{5},\dfrac{2a}{5}\right)}$.

Ex. 5 : *If the density at any point of the curve* $x = a\,(\theta + \sin\theta)$, $y = a\,(1 - \cos\theta)$ *varies as its distance from the x-axis, find the distance of its C.G. of arc from the x-axis.*

Sol. : Required distance of its C.G. from the x-axis is $\overline{y}$. Also $\rho \propto y$ ∴ $\rho = ky$, $dm = \rho\,ds$ ∴ $dm = ky\,ds$.

$$\overline{y} = \frac{\int y\,dm}{\int dm} = \frac{\int y\,k\,y\,ds}{\int k\,y\,ds} = \frac{\int y^2\,ds}{\int y\,ds}$$

$$ds = \sqrt{\left(\frac{dx}{d\theta}\right)^2 + \left(\frac{dy}{d\theta}\right)^2}\,d\theta = \sqrt{(a\,(1 + \cos\theta))^2 + (a\sin\theta)^2}\,d\theta = 2a\cos\frac{\theta}{2}\,d\theta$$

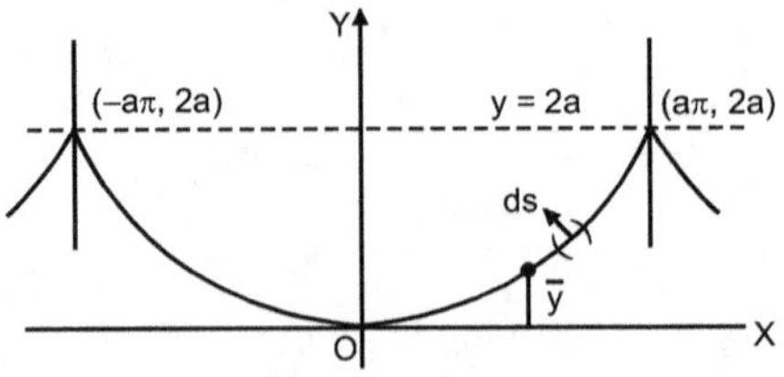

Fig. 11.10

$$\int y^2\, ds = 2 \int_0^\pi a^2 (1 - \cos\theta)^2\, 2a \cos\frac{\theta}{2}\, d\theta = 4a^3 \int_0^\pi 4 \sin^4\frac{\theta}{2} \cos\frac{\theta}{2}\, d\theta$$

$$= 16a^3\, 2 \int_0^{\pi/2} \sin^4 t \cos t\, dt = 32a^3 \left(\frac{1}{5}\right) = \frac{32a^3}{5}$$

$$\int y\, ds = 2 \int_0^\pi a (1 - \cos\theta)\, 2a \cos\frac{\theta}{2}\, d\theta = 4a^2 \int_0^\pi 2 \sin^2\frac{\theta}{2} \cos\frac{\theta}{2}\, d\theta$$

$$= 8a^2\, 2 \int_0^{\pi/2} \sin^2 t \cos t\, dt = \frac{16a^2}{3}$$

$$\boxed{\overline{y} = \frac{32a^3/5}{16a^2/3} = \frac{6a}{5}}$$

Ex. 6 : *Find the C.G. of the arc of a uniform sector of a circle of radius 'a' angle at centre being 2α. Deduce the result for semi-circle.* **(Dec. 2008)**

Sol. : Let the equation of the circle be $x^2 + y^2 = a^2$. Parametric equations of the circle are $x = a \cos\theta$, $y = a \sin\theta$. We choose x-axis bisecting the central angle of sector. Because of symmetry, C.G. of arc AB lies on x-axis. $\therefore \ \overline{y} = 0$.

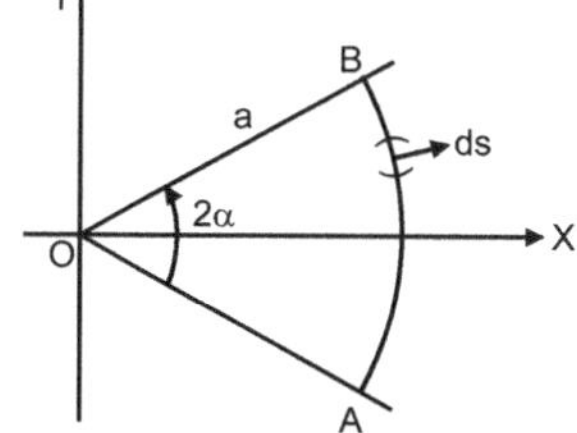

Fig. 11.11

$$\overline{x} = \frac{\int x\, \rho\, ds}{\int \rho\, ds} = \frac{\int x\, ds}{\int ds} \quad (\rho = \text{constant}) \quad \because\ s = a\theta \quad \therefore\ ds = a\, d\theta$$

$$\int x\, ds = 2 \int_0^\alpha a \cos\theta\, a\, d\theta = 2a^2 \sin\alpha; \qquad \int ds = 2 \int_0^\alpha a\, d\theta = 2a\alpha$$

$$\therefore \qquad \overline{x} = \frac{2a^2 \sin\alpha}{2a\alpha} = \frac{a \sin\alpha}{\alpha}$$

We note that for a semi-circle $\alpha = \dfrac{\pi}{2}$

$$\therefore \qquad \boxed{\overline{x} = \frac{a \cdot \sin \pi/2}{\pi/2} = \frac{2a}{\pi}}.$$

Ex. 7 : *Find the C.G. of arc of the curve $x = a\,(\theta + \sin\theta)$, $y = a\,(1 - \cos\theta)$ in the positive quadrant.*

Sol. : We have, $\quad \overline{x} = \dfrac{\int x\, ds}{\int ds}, \quad \overline{y} = \dfrac{\int y\, ds}{\int ds}$

Now, $\quad ds = \sqrt{\left(\dfrac{dx}{d\theta}\right)^2 + \left(\dfrac{dy}{d\theta}\right)^2}\, d\theta$

$$= \sqrt{a^2 (1 + \cos\theta)^2 + a^2 \sin^2\theta}\; d\theta$$

$$= a\sqrt{1 + 2\cos\theta + \cos^2\theta + \sin^2\theta}\; d\theta$$

$$= a\sqrt{2(1 + \cos\theta)}\; d\theta$$

$$= 2a \cos\frac{\theta}{2}\; d\theta$$

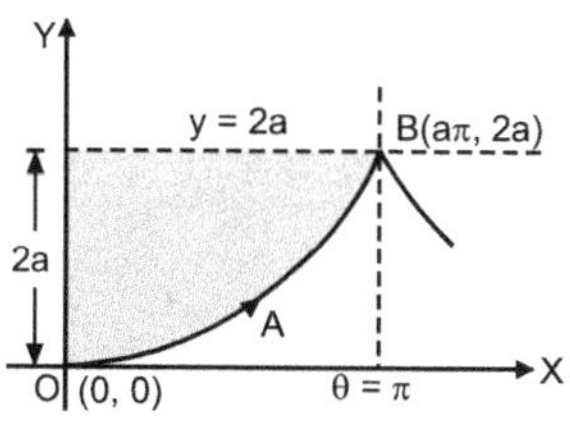

Y
y = 2a
B(aπ, 2a)
2a
A
O (0, 0)
θ = π
X

Fig. 11.12

$$\therefore \quad \int x\, ds = \int_0^\pi a\,(\theta + \sin\theta)\, 2a \cos\frac{\theta}{2}\, d\theta = 2a^2 \int_0^\pi \left(\theta \cos\frac{\theta}{2} + 2 \cos^2\frac{\theta}{2} \sin\frac{\theta}{2}\right) d\theta$$

$$= 2a^2 \left[\theta \cdot 2 \sin\frac{\theta}{2} - \int 1.2 \sin\frac{\theta}{2}\, d\theta - 4 \frac{\cos^3 \theta/2}{3}\right]_0^\pi = 2a^2 \left[2\theta \sin\frac{\theta}{2} + 4 \cos\frac{\theta}{2} - \frac{4}{3} \cos^3\frac{\theta}{2}\right]_0^\pi$$

$$\int x \, ds \;=\; 2a^2 \left(2\pi - \frac{8}{3}\right) \;;\; \int y \, ds \;=\; \int_0^{\pi} a(1-\cos\theta)\cdot 2a\cos\frac{\theta}{2} \, d\theta$$

$$=\; 2a^2 \int_0^{\pi} 2\sin^2\frac{\theta}{2}\,\cos\frac{\theta}{2}\,d\theta \;=\; 4a^2\cdot\left[2\,\frac{\sin^3\theta/2}{3}\right]_0^{\pi} \;=\; \frac{8a^2}{3}$$

$$\int ds \;=\; \int_0^{\pi} 2a\cos\frac{\theta}{2}\,d\theta \;=\; 2a\left[2\sin\frac{\theta}{2}\right]_0^{\pi} \;=\; 4a$$

$$\therefore \quad \overline{x} \;=\; \frac{2a^2\left(2\pi - \frac{8}{3}\right)}{4a} \;=\; a\left(\pi - \frac{4}{3}\right) \;;\; \overline{y} \;=\; \frac{8a^2/3}{4a} \;=\; \frac{2a}{3}$$

Hence, the C.G. of the arc is $\left[a\left(\pi - \frac{4}{3}\right),\; \frac{2a}{3}\right]$.

11.7 ILLUSTRATIONS ON C.G. OF AN AREA

Ex. 8 : *A lamina bounded by the parabolas $y^2 = 4x$ and $2x^2 = y$ has a variable density ρ given by $\rho = 1 + xy$. Prove that $\overline{y} = 2\overline{x}$.*

Sol. : The points of intersection of two parabolas are (0, 0) (1, 2), density $\rho = 1 + xy$ (given).

$$\therefore \quad \overline{x} \;=\; \frac{\displaystyle\iint x\,\rho\,dx\,dy}{\displaystyle\iint \rho\,dx\,dy} \;;\; \overline{y} \;=\; \frac{\displaystyle\iint y\,\rho\,dx\,dy}{\displaystyle\iint \rho\,dx\,dy}$$

$$\iint x\,\rho\,dx\,dy \;=\; \iint x\,(1+xy)\,dx\,dy \;=\; \int_0^1 x\,dx \int_{2x^2}^{2\sqrt{x}} (1+xy)\,dy$$

$$=\; \int_0^1 x\,dx\cdot\left[y + \frac{xy^2}{2}\right]_{2x^2}^{2\sqrt{x}} \;=\; \int_0^1 x\left[2\sqrt{x} - 2x^2 + 2x^2 - 2x^5\right]dx$$

$$=\; \left(2\cdot\frac{2x^{5/2}}{5} - 2\,\frac{x^7}{7}\right)\bigg|_0^1 \;=\; \frac{18}{35} \qquad\qquad \ldots \text{(i)}$$

$$\iint y\,\rho\,dx\,dy \;=\; \iint y\,(1+xy)\,dx\,dy$$

$$=\; \int_0^1 dx \int_{2x^2}^{2\sqrt{x}} (y + xy^2)\,dy \;=\; \int_0^1 dx\left[\frac{y^2}{2} + \frac{xy^3}{3}\right]_{2x^2}^{2\sqrt{x}}$$

$$=\; \int_0^1 \left(2x - 2x^4 + \frac{8}{3}x^{5/2} - \frac{8}{3}x^7\right)dx$$

$$=\; \left[x^2 - \frac{2}{5}x^5 + \frac{16}{21}x^{7/2} - \frac{x^8}{3}\right]_0^1$$

$$=\; \frac{36}{35} \qquad\qquad \ldots \text{(ii)}$$

$$\iint \rho\,dx\,dy \;=\; \iint (1+xy)\,dx\,dy \;=\; \int_0^1 dx \int_{2x^2}^{2\sqrt{x}} (1+xy)\,dy \;=\; \int_0^1 \left[y + \frac{xy^2}{2}\right]_{2x^2}^{2\sqrt{x}} dx$$

$$=\; \int_0^1 \left(2\sqrt{x} - 2x^2 + 2x^2 - 2x^5\right)dx \;=\; \left[\frac{4}{3}x^{3/2} - \frac{x^6}{3}\right]_0^1 \;=\; 1 \qquad\qquad \ldots \text{(iii)}$$

From result (i), (ii) and (iii), $\overline{x} = \dfrac{18}{35}$; $\overline{y} = \dfrac{36}{35}$ $\quad\therefore\quad$ $\overline{y} = 2\,\overline{x}$.

Ex. 9 : *Find the co-ordinates of the C.G. of the area in the first quadrant, bounded by the curve* $\left(\dfrac{x}{a}\right)^{2/3} + \left(\dfrac{y}{b}\right)^{2/3} = 1$, *the density being given by* $\rho = \lambda\, xy$, *where* λ *is a constant.*

Sol. : Here, $\qquad\qquad \rho = \lambda\, xy$

$\therefore \qquad \bar{x} = \dfrac{\iint x \rho\, dx\, dy}{\iint \rho\, dx\, dy} \;;\; \bar{y} = \dfrac{\iint y \rho\, dx\, dy}{\iint \rho\, dx\, dy}$

$\iint x \rho\, dx\, dy = \iint \lambda\, x^2 y\, dx\, dy$

$\qquad\qquad = \lambda \displaystyle\int_0^a x^2\, dx \int_0^y y\, dy$ (Note down the limits for x and y)

$\qquad\qquad = \lambda \displaystyle\int_0^a x^2 \cdot \dfrac{y^2}{2} \cdot \dfrac{dx}{d\theta} \cdot d\theta$ $\qquad\qquad$ (Putting $x = a \cos^3\theta$, $y = b \sin^3\theta$)

$\qquad\qquad = \dfrac{\lambda}{2} \displaystyle\int_0^{\pi/2} a^2 \cos^6\theta \cdot b^2 \sin^6\theta \cdot 3a \cos^2\theta \sin\theta \cdot d\theta$

$\qquad\qquad = \dfrac{3\lambda\, a^3\, b^2}{2} \displaystyle\int_0^{\pi/2} \sin^7\theta \cos^8\theta\, d\theta$

$\qquad\qquad = \dfrac{3\lambda\, a^3\, b^2}{2} \left[\dfrac{6 \cdot 4 \cdot 2 \cdot 7 \cdot 5 \cdot 3 \cdot 1}{15 \cdot 13 \cdot 11 \cdot 9 \cdot 7 \cdot 5 \cdot 3 \cdot 1}\right] = \dfrac{8\lambda\, a^3\, b^2}{15 \cdot 13 \cdot 11}$

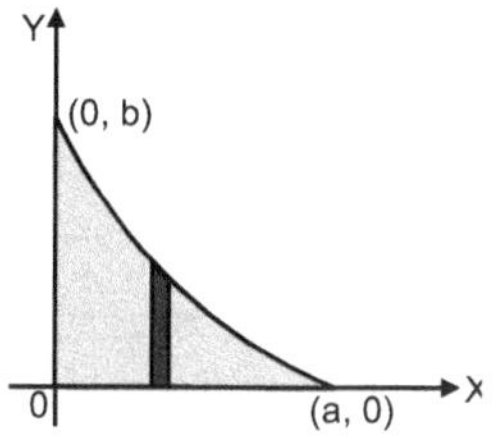

$\iint y \rho\, dx\, dy = \iint \lambda\, xy^2\, dx\, dy = \lambda \displaystyle\int_0^a x\, dx \int_0^y y^2\, dy$

$\qquad\qquad = \dfrac{\lambda}{3} \displaystyle\int_0^a x \cdot y^3 \dfrac{dx}{d\theta}\, d\theta = \dfrac{\lambda}{3} \displaystyle\int_0^{\pi/2} a \cos^3\theta \cdot b^3 \sin^9\theta \cdot (3a \cos^2\theta \sin\theta) \cdot d\theta$

$\qquad\qquad = \lambda\, a^2\, b^3 \displaystyle\int_0^{\pi/2} \sin^{10}\theta \cos^5\theta\, d\theta$

$\qquad\qquad = \lambda\, a^2\, b^3\, \dfrac{9 \cdot 7 \cdot 5 \cdot 3 \cdot 1 \cdot 4 \cdot 2}{15 \cdot 13 \cdot 11 \cdot 9 \cdot 7 \cdot 5 \cdot 3 \cdot 1}$

$\qquad\qquad = \dfrac{8\lambda\, a^2\, b^3}{15 \cdot 13 \cdot 11}$

$\iint \rho\, dx\, dy = \lambda \displaystyle\int_0^a x\, dx \int_0^y y\, dy = \dfrac{\lambda}{2} \displaystyle\int_0^a xy^2 \dfrac{dx}{d\theta}\, d\theta$

$\qquad\qquad = \dfrac{\lambda}{2} \displaystyle\int_0^{\pi/2} a \cos^3\theta \cdot b^2 \sin^6\theta \cdot 3a \cos^2\theta \sin\theta\, d\theta$

$\qquad\qquad = \dfrac{3\lambda\, a^2\, b^2}{2} \displaystyle\int_0^{\pi/2} \sin^7\theta \cdot \cos^5\theta\, d\theta = \dfrac{3\lambda\, a^2\, b^2}{2}\left(\dfrac{6 \cdot 4 \cdot 2 \cdot 4 \cdot 2}{12 \cdot 10 \cdot 8 \cdot 6 \cdot 4 \cdot 2}\right) = \dfrac{\lambda\, a^2\, b^2}{80}$

$\qquad\qquad \bar{x} = \dfrac{8\lambda\, a^3\, b^2}{15 \cdot 13 \cdot 11} \dfrac{80}{\lambda\, a^2\, b^2} = \dfrac{128\, a}{429}$

$\qquad\qquad \bar{y} = \dfrac{8\lambda\, a^2\, b^3}{15 \cdot 13 \cdot 11} \dfrac{80}{\lambda\, a^2\, b^2} = \dfrac{128\, b}{429}$

$\therefore \quad \boxed{\text{Co-ordinates of C.G. are } \left(\dfrac{128\, a}{429},\ \dfrac{128\, b}{429}\right)}$.

Fig. 11.14

Ex. 10 : *Find the centroid of the loop of the curve* $r^2 = a^2 \cos 2\theta$.   *(May 2008, 15, Dec. 2013, 2018)*

Sol. : C.G. lies on x-axis $\therefore \bar{y} = 0$ and $\bar{x} = \dfrac{\dfrac{2}{3} \displaystyle\int_{-\pi/4}^{\pi/4} r^3 \cos\theta \, d\theta}{\displaystyle\int_{-\pi/4}^{\pi/4} r^2 \, d\theta} = \dfrac{N}{D}$

(Refer formula (b) given in special note after 11.4)

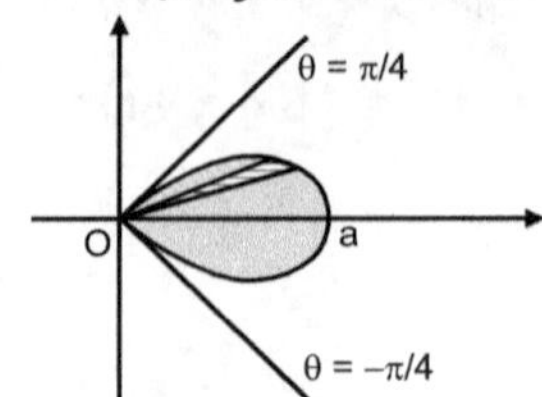

Fig. 11.15

$$N = \frac{2}{3} \int_{-\pi/4}^{\pi/4} \left(a\sqrt{\cos 2\theta}\right)^3 \cos\theta \, d\theta = \frac{4a^3}{3} \int_{0}^{\pi/4} (1 - 2\sin^2\theta)^{3/2} \cos\theta \, d\theta$$

$\because$ $2\sin^2\theta = \sin^2 t, \ \sqrt{2}\sin\theta = \sin t; \ \cos\theta \, d\theta = \dfrac{1}{\sqrt{2}}\cos t \, dt$

θ	0	$\pi/4$
t	0	$\pi/2$

$$= \frac{4a^3}{3} \int_{0}^{\pi/2} (1 - \sin^2 t)^{3/2} \cdot \frac{1}{\sqrt{2}} \cos t \, dt$$

$$= \frac{4a^3}{3\sqrt{2}} \int_{0}^{\pi/2} \cos^4 t \, dt$$

$$= \frac{4a^3}{3\sqrt{2}} \cdot \frac{3}{4}\frac{1}{2}\frac{\pi}{2} = \frac{\pi a^3}{4\sqrt{2}}$$

$$D = \int_{-\pi/4}^{\pi/4} a^2 \cos 2\theta \, d\theta = 2a^2 \int_{0}^{\pi/4} \cos 2\theta \, d\theta = 2a^2 \frac{1}{2} = a^2$$

$$\boxed{\bar{x} = \frac{\pi a^3}{4\sqrt{2}\, a^2} = \frac{\pi a}{4\sqrt{2}}}$$

$\therefore$ C.G. is $\left(\dfrac{\pi a}{4\sqrt{2}}, 0\right)$

Ex. 11 : *The density at a point on a circular lamina varies as the distance from a point O on the circumference. Show that the centre of gravity divides the diameter through 'O' in the ratio 3 : 2.*

Sol. : Take the fixed point O on the circumference as the pole, the diameter OA as the initial line and 'a' the radius of the circle.

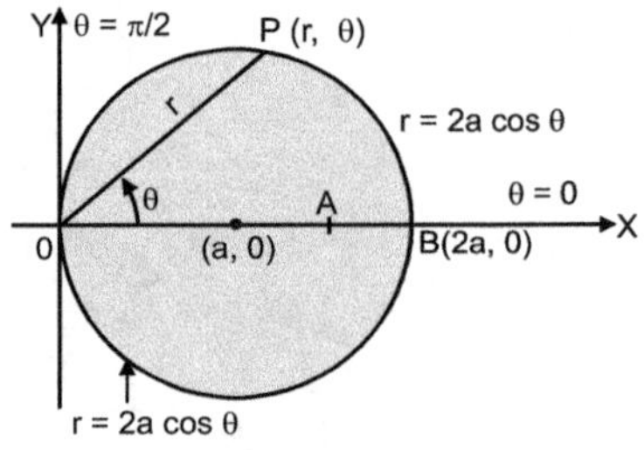

Then the polar equation of the circle is $r = 2a\cos\theta$. By symmetry, the C.G. $(\bar{x}, \bar{y})$ of the lamina lies on the initial line, so that $\bar{y} = 0$. $\rho \propto r$. $\therefore \rho = \lambda r$.

Fig. 11.16

$$\bar{x} = \frac{\displaystyle\iint x\rho \, dx \, dy}{\displaystyle\iint \rho \, dx \, dy} = \frac{\displaystyle\iint r\cos\theta \, \lambda r \, r \, dr \, d\theta}{\displaystyle\iint \lambda r \, r \, dr \, d\theta}$$

$$= \frac{2\lambda \displaystyle\int_{0}^{\pi/2} \cos\theta \, d\theta \displaystyle\int_{0}^{2a\cos\theta} r^3 \, dr}{2\lambda \displaystyle\int_{0}^{\pi/2} d\theta \displaystyle\int_{0}^{2a\cos\theta} r^2 \, dr}$$

$$= \frac{\displaystyle\int_{0}^{\pi/2} \frac{\cos\theta \cdot 16a^4 \cos^4\theta}{4} \, d\theta}{\displaystyle\int_{0}^{\pi/2} \frac{8a^3 \cos^3\theta}{3} \, d\theta} = \frac{3}{2}a \frac{\displaystyle\int_{0}^{\pi/2} \cos^5\theta \, d\theta}{\displaystyle\int_{0}^{\pi/2} \cos^3\theta \, d\theta} = \frac{3a}{2} \left(\frac{\frac{4}{5}\cdot\frac{2}{3}}{\frac{2}{3}}\right) = \frac{6\,a}{5}$$

$\therefore$ C.G. is A $\left(\dfrac{6a}{5}, 0\right)$ as shown in Fig. 11.16.

We have
$$OB = 2a, \quad OA = \dfrac{6a}{5} \; ;$$
$$AB = OB - OA = 2a - \dfrac{6a}{5} = \dfrac{4a}{5}$$
$$\boxed{OA : AB = \dfrac{6a}{5} : \dfrac{4a}{5} = 3:2}.$$

Ex. 12 : *Find the centre of mass of a homogeneous lamina having the shape of the region bounded by the parabola $y = 2 - 3x^2$ and the line $3x + 2y - 1 = 0$.*

Sol. : The line intersects the parabola at P $\left(-\dfrac{1}{2}, \dfrac{5}{4}\right)$, Q $(1, -1)$.

$$\bar{x} = \frac{\iint x \, dx \, dy}{\iint dx \, dy}, \qquad \bar{y} = \frac{\iint y \, dx \, dy}{\iint dx \, dy}$$

$$\iint x \, dx \, dy = \int_{-1/2}^{1} x \, dx \int_{\frac{1-3x}{2}}^{2-3x^2} dy = \int_{-1/2}^{1} x \left[2 - 3x^2 - \frac{1}{2} + \frac{3x}{2}\right] dx$$

$$= \left[\frac{3x^2}{4} - \frac{3x^4}{4} + \frac{x^3}{2}\right]_{-1/2}^{1} = \frac{27}{64}$$

$$\iint dx \, dy = \int_{-1/2}^{1} dx \int_{\frac{1-3x}{2}}^{2-3x^2} dy = \int_{-1/2}^{1} \left(2 - 3x^2 - \frac{1}{2} + \frac{3x}{2}\right) dx$$

$$= \left[\frac{3}{2}x - x^3 + \frac{3}{4}x^2\right]_{-1/2}^{1} = \frac{27}{16}$$

$$\iint y \, dx \, dy = \int_{-1/2}^{1} dx \int_{\frac{1-3x}{2}}^{2-3x^2} y \, dy = \frac{1}{2} \int_{-1/2}^{1} [y^2]_{\frac{1-3x}{2}}^{2-3x^2} dx$$

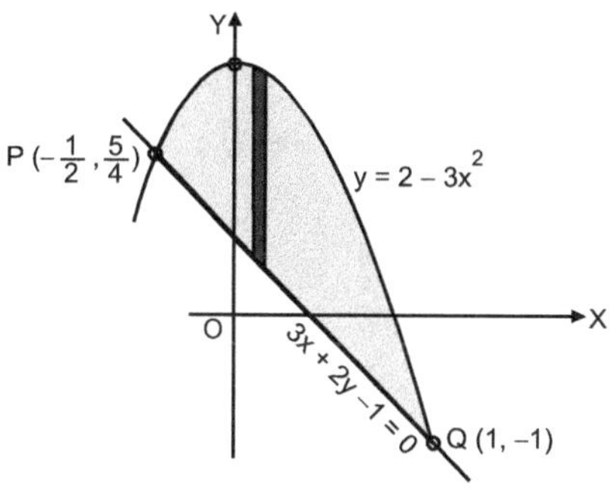

Fig. 11.17

$$= \frac{1}{2} \int_{-1/2}^{1} \left[(2 - 3x^2)^2 - \left(\frac{1-3x}{2}\right)^2\right] dx = \frac{27}{20} \text{ (after simplification)}$$

$$\bar{x} = \frac{27}{64} \times \frac{16}{27} = \frac{1}{4}$$

$$\bar{y} = \frac{27}{20} \times \frac{16}{27} = \frac{4}{5}$$

$\boxed{\text{Hence } \left(\dfrac{1}{4}, \dfrac{4}{5}\right) \text{ is C.G. of the area}}$.

Ex. 13 : *A plate is in the form of a quadrant of an ellipse and is of small but varying thickness, the thickness at any point being proportional to the product of the distances of the point from the axes. Find the C.G. of the plate.*

Sol. : $\rho \propto x \cdot y \implies \rho = \lambda xy,$

$$\bar{x} = \frac{\iint x \rho \, dx \, dy}{\iint \rho \, dx \, dy}, \qquad \bar{y} = \frac{\iint y \rho \, dx \, dy}{\iint \rho \, dx \, dy}$$

For ellipse,
$$\frac{x^2}{a^2} + \frac{y^2}{b^2} = 1$$

We will use $\quad x = ar\cos\theta, \; y = br\sin\theta, \; dx\,dy = ab\,r\,dr\,d\theta$

Equation of ellipse reduces to $r^2 = 1$ i.e. $r = 1$

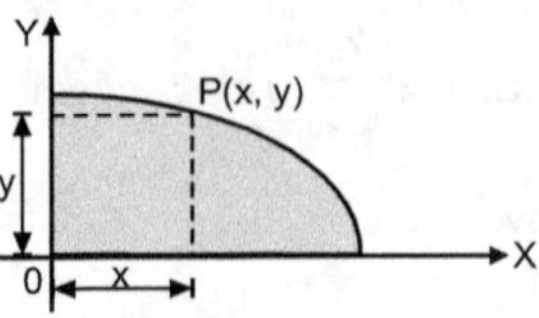
Fig. 11.18 (a)

$$\iint x\,\rho\,dx\,dy = \iint (ar\cos\theta)^2\,\lambda \cdot br\sin\theta\,ab\,r\,dr\,d\theta$$

$$= \lambda\,a^3 b^2 \int_0^{\pi/2}\cos^2\theta\sin\theta\,d\theta \int_0^1 r^4\,dr = \lambda\,a^3 b^2\,\frac{1}{3}\cdot\frac{1}{5} = \frac{\lambda\,a^3 b^2}{15}$$

$$\iint y\,\rho\,dx\,dy = \iint (br\sin\theta)^2\,\lambda \cdot ar\cos\theta\,ab\,r\,dr\,d\theta$$

$$= \lambda\,a^2 b^3 \int_0^{\pi/2}\sin^2\theta\cos\theta\,d\theta \int_0^1 r^4\,dr$$

$$= \frac{\lambda\,a^2 b^3}{15}$$

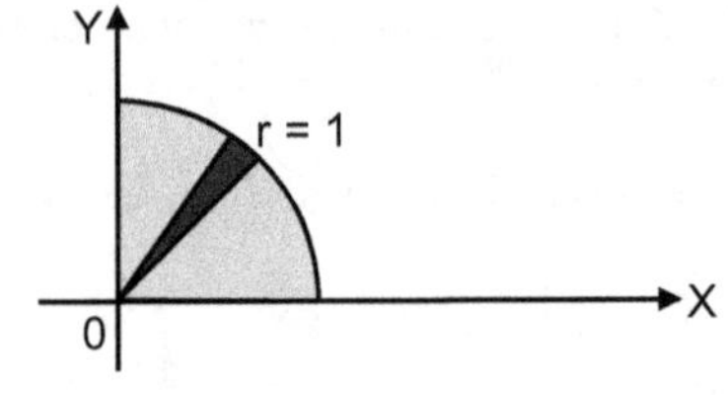
Fig. 11.18 (b)

$$\iint \rho\,dx\,dy = \iint \lambda \cdot ar\cos\theta \cdot br\sin\theta \cdot ab\,r\,dr\,d\theta$$

$$= \lambda\,a^2 b^2 \int_0^{\pi/2}\sin\theta\cos\theta\,d\theta \int_0^1 r^3\,dr = \frac{\lambda\,a^2 b^2}{8}$$

$$\bar{x} = \frac{\lambda\,a^3 b^2/15}{\lambda\,a^2 b^2/8} = \frac{8a}{15}; \quad \bar{y} = \frac{\lambda\,a^2 b^3/15}{\lambda\,a^2 b^2/8} = \frac{8b}{15}$$

> Co-ordinates of C.G. of the plate are $\left(\dfrac{8a}{15}, \dfrac{8b}{15}\right)$.

Ex. 14 : *Find the C.G. of one loop of $r = a\sin 2\theta$.* **(May 2019, May 2004, 2013, Dec. 2014, 2017)**

Sol. : $r = a\sin 2\theta$ contains 4 equal loops and lies within a circle of radius $r = a$. The first loop lies between $\theta = 0$ to $\theta = \frac{\pi}{2}$ and is symmetrical about the line $y = x \left(\text{i.e. } \theta = \frac{\pi}{4}\right)$.

$$\therefore \quad \bar{x} = \bar{y}. \quad \text{We have } \bar{x} = \frac{\iint x\,dx\,dy}{\iint dx\,dy}$$

$$\iint x\,dx\,dy = \int_0^{\pi/2}\int_0^{a\sin 2\theta} r\cos\theta\,r\,dr\,d\theta$$

$$= \int_0^{\pi/2}\cos\theta\,\frac{a^3\sin^3 2\theta}{3}\,d\theta$$

$$= \frac{a^3}{3}\int_0^{\pi/2}\cos\theta\,(2\sin\theta\cos\theta)^3\,d\theta$$

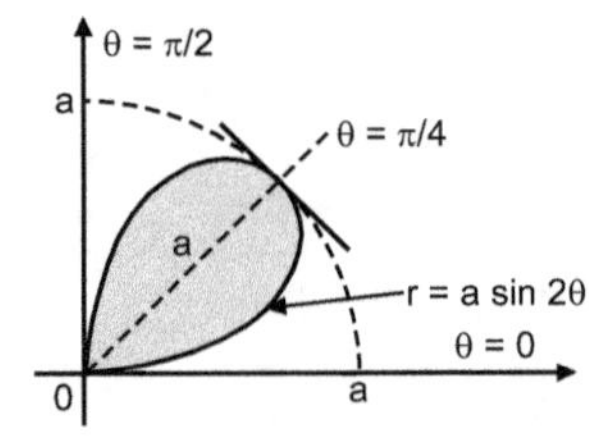
Fig. 11.19

$$= \frac{8a^3}{3}\int_0^{\pi/2}\sin^3\theta\cdot\cos^4\theta\,d\theta = \frac{8a^3}{3}\left(\frac{2\cdot 3\cdot 1}{7\cdot 5\cdot 3\cdot 1}\right) = \frac{16a^3}{105}$$

$$\iint dx\,dy = \frac{1}{2}\int_0^{\pi/2} r^2\,d\theta = \frac{1}{2}\int_0^{\pi/2} a^2\sin^2 2\theta\,d\theta = 2a^2\int_0^{\pi/2}\sin^2\theta\cos^2\theta\,d\theta = 2a^2\left(\frac{1\cdot 1}{4\cdot 2}\,\frac{\pi}{2}\right) = \frac{\pi a^2}{8}$$

$$\bar{x} = \frac{16a^3/105}{\pi a^2/8} = \frac{128\,a}{105\,\pi}. \quad \text{Also } \bar{y} = \frac{128\,a}{105\,\pi}$$

$$\therefore \quad \boxed{\text{C.G. is } \left(\frac{128\,a}{105\,\pi}, \frac{128\,a}{105\,\pi}\right)}$$

Ex. 15 : *ABCD is a square plate of side a and O is the mid point of AB. If the surface density varies as the square of distance from O, show that C.G. of the plate is at a distance $\dfrac{7a}{10}$ from AB.*

Sol. : Take O as the origin and the side OB as the x-axis, and a line through O perpendicular to OB as y-axis. The density ρ at the point P(x, y) is proportional to (PO)^2.

$$\therefore \qquad \rho \propto (x^2 + y^2) \Rightarrow \quad \rho = \lambda\,(x^2 + y^2)$$

As the distribution of the mass is symmetrical about OY, the C.G. of the plate is on the y-axis and $\bar{x} = 0$.

$$\therefore \qquad \bar{y} = \frac{\iint y\,\rho\,dx\,dy}{\iint \rho\,dx\,dy}$$

By symmetry,

$$\iint y\,\rho\,dx\,dy = 2\iint \lambda\,(x^2 y + y^3)\,dx\,dy$$

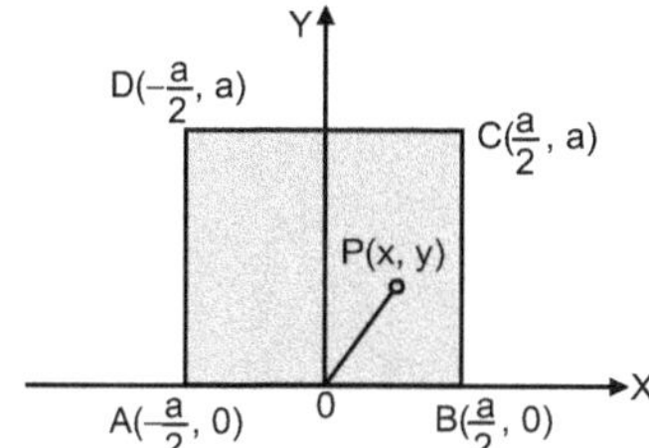

Fig. 11.20

$$= 2\lambda \int_0^{a/2} dx \int_0^a (x^2 y + y^3)\,dy$$

$$= 2\lambda \int_0^{a/2} \left(x^2\frac{a^2}{2} + \frac{a^4}{4}\right)dx = 2\lambda\left(\frac{a^5}{48} + \frac{a^5}{8}\right) = \frac{7\lambda\,a^5}{24}$$

$$\iint \rho\,dx\,dy = \iint \lambda\,(x^2 + y^2)\,dx\,dy = 2\lambda \int_0^{a/2} dx \int_0^a (x^2 + y^2)\,dy$$

$$= 2\lambda \int_0^{a/2} \left(ax^2 + \frac{a^3}{3}\right)dx = 2\lambda\left(\frac{a^4}{24} + \frac{a^4}{6}\right) = \frac{5\lambda\,a^4}{12}$$

$$\bar{y} = \frac{7\lambda\,a^5/24}{5\lambda\,a^4/12} = \frac{7a}{10}$$

Thus, the C.G. of the plate is on the y-axis, at a distance $\dfrac{7a}{10}$ from O.

Ex. 16 : *Find the C.G. of the loop of the curve $y^2\,(a + x) = x^2\,(a − x)$.* **(Nov. 2015)**

Sol. : Because of symmetry, C.G. of the loop lies on x-axis, $\bar{y} = 0$.

$$\bar{x} = \frac{\int x \cdot y \cdot dx}{\int y \cdot dx}$$

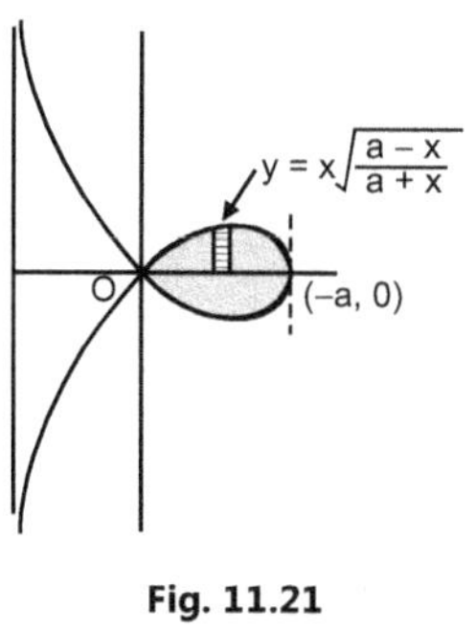

$$\int xy\,dx = \int_0^a x \cdot x\sqrt{\frac{a-x}{a+x}}\,dx = \int_0^a \frac{x^2\,(a-x)}{\sqrt{a^2 - x^2}}\,dx \quad \text{(put } x = a\sin\theta)$$

$$= \int_0^{\pi/2} \frac{a^2\sin^2\theta\,(a - a\sin\theta)\cdot a\cos\theta\,d\theta}{a\cdot\cos\theta}$$

$$= a^3 \int_0^{\pi/2} (\sin^2\theta - \sin^3\theta)\,d\theta = a^3\left(\frac{1}{2}\frac{\pi}{2} - \frac{2}{3}\right) = a^3\left(\frac{3\pi - 8}{12}\right)$$

Fig. 11.21

$$\int y \, dx = \int_0^a x \sqrt{\frac{a-x}{a+x}} \, dx = \int_0^a \frac{x(a-x)\,dx}{\sqrt{a^2-x^2}} = \int_0^{\pi/2} \frac{a\sin\theta\,(a - a\sin\theta)\,a\cos\theta\,d\theta}{a\cos\theta}$$

$$= a^2 \int_0^{\pi/2} (\sin\theta - \sin^2\theta)\,d\theta = a^2\left[1 - \frac{\pi}{4}\right] = \frac{a^2}{4}(4-\pi)$$

$$\bar{x} = \frac{a^3\,(3\pi-8)/12}{a^2\,(4-\pi)/4} = \frac{a\,(3\pi-8)}{3\,(4-\pi)} \quad \text{Required C.G. is}\left[\frac{a}{3}\left(\frac{3\pi-8}{4-\pi}\right),\, 0\right]$$

Ex. 17 : *Find the position of the centroid of the area of the curve $x^{2/3} + y^{2/3} = a^{2/3}$ lying in the positive quadrant.*

Sol. : C.G. lies on the lines $y = x$ $\therefore$ $\bar{x} = \bar{y}$. Also, $\bar{x} = \dfrac{\int x \cdot y \cdot dx}{\int y \, dx}$.

$$\int x \cdot y \, dx = \int_0^{\pi/2} a\cos^3\theta \cdot a\sin^3\theta \cdot 3a\cos^2\theta\sin\theta\,d\theta \quad \text{Putting}\begin{Bmatrix} x = a\cos^3\theta \\ y = a\sin^3\theta \end{Bmatrix}$$

$$= 3a^3 \int_0^{\pi/2} \cos^5\theta\sin^4\theta\,d\theta = 3a^3 \cdot \frac{3\cdot1\cdot4\cdot2}{9\cdot7\cdot5\cdot3\cdot1} = \frac{8a^3}{105}$$

$$\int y \, dx = \int_0^{\pi/2} a\sin^3\theta \cdot 3a\cos^2\theta\sin\theta\,d\theta$$

$$= 3a^2 \int_0^{\pi/2} \sin^4\theta\cos^2\theta\,d\theta = 3a^2\left(\frac{3\cdot1\cdot1}{6\cdot4\cdot2}\right)\frac{\pi}{2} = \frac{3\pi a^2}{32}$$

$$\bar{x} = \frac{8a^3/105}{3\pi a^2/32} = \frac{256\,a}{315\,\pi} \quad \text{and} \quad \bar{y} = \frac{256\,a}{315\,\pi}$$

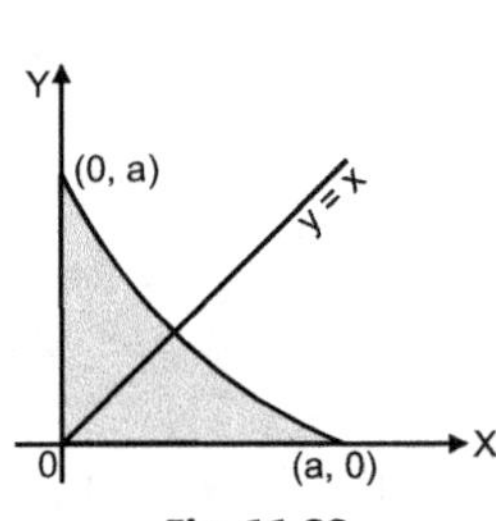

$$\boxed{\text{Required co-ordinates of C.G. are}\left(\frac{256\,a}{315\,\pi},\,\frac{256\,a}{315\,\pi}\right)} .$$

Ex. 18 : *Find the position of the centroid of the area bounded by the curve $y^2\,(2a-x) = x^3$ and its asymptote.*

Sol. : $y^2\,(2a-x) = x^3$ is known as Cissoid and is symmetrical about x-axis.   *(Dec. 2009, 2010; May 2011)*

$$\therefore \quad \bar{y} = 0 \qquad\qquad \bar{x} = \frac{\int x \cdot y \, dx}{\int y \, dx}$$

$$\int x y \, dx = \int_0^{2a} x \cdot \frac{x^{3/2}}{\sqrt{2a-x}}\,dx = \int_0^{2a} \frac{x^{5/2}}{\sqrt{2a-x}}\,dx \quad (\text{put } x = 2a\sin^2\theta)$$

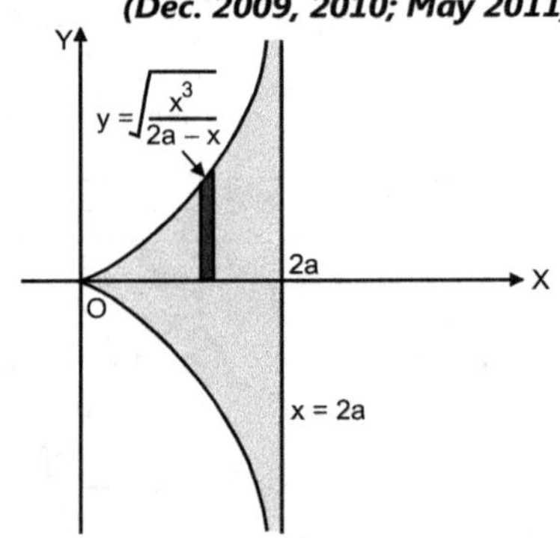

$$= \int_0^{\pi/2} \frac{(2a)^{5/2}\,\sin^5\theta \cdot 4a\sin\theta\cos\theta\,d\theta}{(2a)^{1/2}\cos\theta} = 16a^3 \int_0^{\pi/2} \sin^6\theta\,d\theta$$

$$= 16a^3\left(\frac{5}{6}\,\frac{3}{4}\,\frac{1}{2}\,\frac{\pi}{2}\right) = \frac{5\pi a^3}{2}$$

$$\int y \, dx = \int_0^{2a} \frac{x^{3/2}}{\sqrt{2a-x}}\,dx = \int_0^{\pi/2} \frac{(2a)^{3/2}\,\sin^3\theta \cdot 4a\sin\theta\cos\theta\,d\theta}{(2a)^{1/2}\cos\theta}$$

$$= 8a^2 \int_0^{\pi/2} \sin^4\theta \cdot d\theta = 8a^2\left(\frac{3}{4}\,\frac{1}{2}\,\frac{\pi}{2}\right) = \frac{3\pi a^2}{2}$$

$$\bar{x} = \frac{5\pi a^3/2}{3\pi a^2/2} = \frac{5\,a}{3}$$

$$\boxed{\text{Required co-ordinates of C.G. are}\left(\frac{5a}{3},\,0\right)} .$$

11.8 ILLUSTRATIONS ON C.G. OF SOLIDS

Ex. 19 : *Find the centroid of the region bounded by $z = 4 - x^2 - y^2$ and xy plane (assume constant density ρ).*

Sol. : As paraboloid is symmetrical about z-axis, C.G. will lie on z-axis $\bar{x} = 0$, $\bar{y} = 0$.

$$\bar{z} = \frac{\iiint\limits_V z\, dx\, dy\, dz}{\iiint\limits_V dx\, dy\, dz}$$

$$\iiint z\, dx\, dy\, dz = \iint \int_0^{4-x^2-y^2} z\, dx\, dy\, dz = \frac{1}{2} \iint (4 - x^2 - y^2)^2\, dx\, dy$$

$$= \frac{1}{2}\, 4 \cdot \int_0^{\pi/2} \int_0^2 (4 - r^2)^2\, r\, dr\, d\theta$$

$$= 2 \cdot \frac{\pi}{2} \left[\frac{(4 - r^2)^3}{(-2)\,3}\right]_0^2 = \frac{32\pi}{3}$$

$$\iiint dx\, dy\, dz = \iint \int_0^{4-x^2-y^2} dx\, dy\, dz = \iint (4 - x^2 - y^2)\, dx\, dy$$

$$= 4 \int_0^{\pi/2} \int_0^2 (4 - r^2)\, r\, dr\, d\theta = 4 \cdot \frac{\pi}{2} \cdot \left[\frac{(4 - r^2)^2}{(-2)\,2}\right]_0^2 = 8\pi$$

$$\boxed{\bar{z} = \frac{32\pi/3}{8\pi} = \frac{4}{3}}$$

C.G. is $\left(0, 0, \dfrac{4}{3}\right)$

Ex. 20 : *Find the centroid of the region in the first octant bounded by* $\dfrac{x}{a} + \dfrac{y}{b} + \dfrac{z}{c} = 1$, $(a > 0, b > 0, c > 0)$.

Sol. : We have,

$$\bar{x} = \frac{\iiint x\, dx\, dy\, dz}{\iiint dx\, dy\, dz}, \quad \bar{y} = \frac{\iiint y\, dx\, dy\, dz}{\iiint dx\, dy\, dz}, \quad \bar{z} = \frac{\iiint z\, dx\, dy\, dz}{\iiint dx\, dy\, dz}$$

Take ρ = constant

$$\iiint dx\, dy\, dz = \text{volume of tetrahedron} = \frac{abc}{6}$$

Now put $x = au$, $y = bv$, $z = cw$, $\therefore$ dx dy dz = abc du dv dw and $u + v + w = 1$

$$\iiint x\, dx\, dy\, dz = \iiint au \cdot abc \cdot du\, dv\, dw$$

$$= a^2 bc \iiint u\, du\, dv\, dw = a^2 bc \iiint u^{2-1} v^{1-1} w^{1-1}\, du\, dv\, dw$$

$$= a^2 bc\, \frac{\lceil 2 \ \lceil 1 \ \lceil 1}{\lceil 1 + 2 + 1 + 1} = \frac{a^2 bc}{4!}$$

$$= \frac{a^2 bc}{24} \quad \text{(by using Dirichlet's theorem)}$$

Similarly, $\displaystyle\iiint y\, dx\, dy\, dz = \frac{ab^2 c}{24}$ and $\displaystyle\iiint z\, dx\, dy\, dz = \frac{abc^2}{24}$

$$\bar{x} = \frac{a}{4}, \qquad \bar{y} = \frac{b}{4}, \qquad \bar{z} = \frac{c}{4}.$$

$$\boxed{\text{The required C.G. is } \left(\frac{a}{4}, \frac{b}{4}, \frac{c}{4}\right)}.$$

Ex. 21 : *If the density of a hemisphere vary as distance from the bounding plane, show that the distance from the plane of C.G. is* $\left(\dfrac{8}{15}\right)^{th}$ *of its radius.*

Sol. : Let $x^2 + y^2 + z^2 = a^2$ be the hemisphere and $\rho = \lambda z$

$$\therefore \qquad \bar{z} = \frac{\iiint z\,\rho\,dx\,dy\,dz}{\iiint \rho\,dx\,dy\,dz}$$

$$\iiint z\,\rho\,dx\,dy\,dz = \lambda \iiint z^2\,dx\,dy\,dz \qquad \text{(Transforming to spherical polar)}$$

$$= \lambda \int_0^{\pi/2} \int_0^{2\pi} \int_0^a r^2 \cos^2\theta \cdot r^2 \sin\theta \, d\theta \, d\phi \, dr$$

$$= \lambda \int_0^{\pi/2} \cos^2\theta \sin\theta \, d\theta \int_0^{2\pi} d\phi \int_0^a r^4 \, dr = \lambda \left(\frac{1}{3}\right)(2\pi)\left(\frac{a^5}{5}\right)$$

$$= \frac{2\lambda\,a^5\,\pi}{15}$$

$$\iiint \rho\,dx\,dy\,dz = \lambda \iiint z\,dx\,dy\,dz = \lambda \int_0^{\pi/2} \int_0^{2\pi} \int_0^a r\cos\theta\, r^2 \sin\theta \, d\theta \, d\phi \, dr$$

$$= \lambda \int_0^{\pi/2} \sin\theta \cos\theta \, d\theta \int_0^{2\pi} d\phi \int_0^a r^3 \, dr = \lambda\,\frac{1}{2}\cdot 2\pi \cdot \frac{a^4}{4} = \frac{\lambda\pi a^4}{4}$$

$$\bar{z} = \frac{8}{15} \ (a)$$

i.e. $\boxed{\left(\dfrac{8}{15}\right)^{th} \text{ of radius of hemisphere.}}$

Ex. 22 : *Show that C.G. of the volume common to the cylinder $x^2 + y^2 = ax$ and the sphere $x^2 + y^2 + z^2 = a^2$ (for which $z > 0$) is at a distance* $\dfrac{45\pi a}{64\,(3\pi - 4)}$ *from the plane $z = 0$.*

Sol. :

$$\bar{z} = \frac{\iiint z\,dx\,dy\,dz}{\iiint dx\,dy\,dz} \quad (\rho = \text{constant})$$

$$\iiint z\,dx\,dy\,dz = \iint \int_0^{\sqrt{a^2-x^2-y^2}} z\,dx\,dy\,dz$$

$$= \frac{1}{2} \iint (a^2 - x^2 - y^2)\,dx\,dy$$

$$= \frac{1}{2} \int_{-\pi/2}^{\pi/2} d\theta \int_0^{a\cos\theta} (a^2 - r^2)\,r\,dr$$

$$= -\frac{1}{4} \int_{-\pi/2}^{\pi/2} d\theta \left[\frac{(a^2-r^2)^2}{2}\right]_0^{a\cos\theta}$$

$$= \frac{1}{8}\,2 \int_0^{\pi/2} (a^4 - a^4 \sin^4\theta)\,d\theta = \frac{a^4}{4}\left[\frac{\pi}{2} - \frac{3}{4}\,\frac{1}{2}\,\frac{\pi}{2}\right] = \frac{5\pi a^4}{64}$$

$$\left(\because \int f\cdot f'\,dx = \frac{f^2}{2}\right)$$

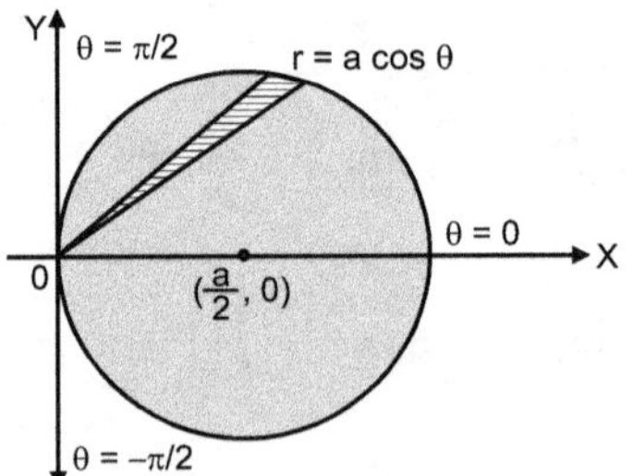

Fig. 11.24

$$\iiint dx\,dy\,dz = \iint \int_0^{\sqrt{a^2 - x^2 - y^2}} dx\,dy\,dz = \iint \sqrt{a^2 - x^2 - y^2}\ dx\,dy$$

$$= 2 \int_0^{\pi/2} d\theta \int_0^{a\cos\theta} \sqrt{a^2 - r^2}\ r\,dr = -\int_0^{\pi/2} \left[\frac{2}{3}(a^2 - r^2)^{3/2}\right]_0^{a\cos\theta} d\theta$$

$$= \frac{2}{3} \int_0^{\pi/2} (a^3 - a^3 \sin^3\theta)\,d\theta \qquad\qquad (\because \sin\theta = \pm\sqrt{1 - \cos^2\theta}\,)$$

$$= \frac{2a^3}{3}\left[\frac{\pi}{2} - \frac{2}{3}\right] = \frac{a^3(3\pi - 4)}{9}$$

$$\boxed{\overline{z} = \frac{5\pi a^4/64}{a^3(3\pi - 4)/9} = \frac{45\pi a}{64(3\pi - 4)}}$$

Ex. 23 : *A solid is cut of the cylinder $x^2 + y^2 = a^2$ by the plane $z = 0$ and that part of the plane $z = mx$ for which z is positive. The density of the solid cut off at any point varies as the height of the point above the plane $z = 0$. Find the z-co-ordinate of the centre of mass of the solid.*

Sol. : $\quad \rho \propto z \Rightarrow \rho = \lambda z$

Also, $\qquad\qquad \overline{z} = \dfrac{\iiint z\,\rho\,dx\,dy\,dz}{\iiint \rho\,dx\,dy\,dz}$

$$\iiint z\,\rho\,dx\,dy\,dz = \lambda \iint \int_0^{mx} z^2\,dx\,dy\,dz = \frac{\lambda m^3}{3} \iint_R x^3\,dx\,dy \text{ (R is circle)}$$

$$= \frac{\lambda m^3}{3}\,4 \int_0^{\pi/2} \int_0^a r^3 \cos^3\theta\ r\,dr\,d\theta$$

$$= \frac{4\lambda\,m^3}{3}\left(\frac{2}{3}\right)\left(\frac{a^5}{5}\right) = \frac{8\lambda\,m^3\,a^5}{45}$$

$$\iiint \rho\,dx\,dy\,dz = \lambda \iint \int_0^{mx} z\,dx\,dy\,dz = \frac{\lambda m^3}{2} \iint_R x^2\,dx\,dy$$

$$= \frac{\lambda m^2}{2}\,4 \int_0^{\pi/2} \int_0^a r^2 \cos^2\theta\ r\,dr\,d\theta$$

$$= 2\lambda m^2 \left(\frac{1}{2}\,\frac{\pi}{2}\right)\left(\frac{a^4}{4}\right) = \frac{\lambda\pi\,m^2\,a^4}{8}$$

$$\boxed{\overline{z} = \frac{8\lambda\,m^3\,a^5/45}{\lambda\pi\,m^2\,a^4/8} = \frac{64\,ma}{45\pi}}$$

Ex. 24 : *A solid is in the form of positive octant of the sphere $x^2 + y^2 + z^2 = a^2$. The density ρ at any point (x, y, z) is given by $\rho = \lambda\,xyz$ where λ is constant. Find the co-ordinates of C.G. of the solid.*

Sol. : $\ \overline{x} = \dfrac{\iiint x\,\rho\,dx\,dy\,dz}{\iiint \rho\,dx\,dy\,dz}\ , \quad \overline{y} = \dfrac{\iiint y\,\rho\,dx\,dy\,dz}{\iiint \rho\,dx\,dy\,dz}\ , \quad \overline{z} = \dfrac{\iiint z\,\rho\,dx\,dy\,dz}{\iiint \rho\,dx\,dy\,dz}\ .$

$$\iiint x\,\rho\,dx\,dy\,dz = \lambda \iiint x^2 y\,z\,dx\,dy\,dz$$

$$x = r\sin\theta\cos\phi$$
$$y = r\sin\theta\sin\phi$$
$$z = r\cos\theta$$

$$= \lambda \int_0^{\pi/2}\int_0^{\pi/2}\int_0^{a} r^6 \sin^4\theta\cdot\cos\theta\cdot\cos^2\phi\,\sin\phi\,d\theta\,d\phi\,dr$$

$$dx\,dy\,dz = r^2 \sin\theta\,d\theta\,d\phi\,dr$$

$$= \lambda \int_0^{\pi/2} \sin^4\theta\cos\theta\,d\theta \int_0^{\pi/2} \cos^2\phi\,\sin\phi\,d\phi \int_0^{a} r^6\,dr = \lambda\cdot\frac{1}{5}\cdot\frac{1}{3}\cdot\frac{a^7}{7} = \frac{\lambda a^7}{105}$$

Similarly, $\quad \iiint y\,\rho\,dx\,dy\,dz = \dfrac{\lambda a^7}{105}$ and $\iiint z\,\rho\,dx\,dy\,dz = \dfrac{\lambda a^7}{105}$

$$\iiint \rho\,dx\,dy\,dz = \lambda \iiint xyz\,dx\,dy\,dz$$

$$= \lambda \int_0^{\pi/2}\int_0^{\pi/2}\int_0^{a} r^5 \sin^3\theta\cdot\cos\theta\cdot\sin\phi\cos\phi\,d\theta\,d\phi\,dr$$

$$= \lambda\cdot\frac{1}{4}\cdot\frac{1}{2}\cdot\frac{a^6}{6} = \frac{\lambda a^6}{48}$$

$$\bar{x} = \frac{\lambda\,a^7/105}{\lambda\,a^6/48} = \frac{16a}{35}. \quad \text{Also } \bar{y} = \frac{16a}{35} = \bar{z}$$

Required co-ordinates of C.G. are $\left(\dfrac{16a}{35},\ \dfrac{16a}{35},\ \dfrac{16a}{35}\right)$.

EXERCISE 11.1

Problems on C.G. of an Arc of a Curve :

1.　Find the C.G. of arc of the cycloid $x = a\,(\theta + \sin\theta)$, $y = a\,(1 - \cos\theta)$ measured from cusp to cusp.　　**Ans.** $\left(0, \dfrac{2a}{3}\right)$

2.　Find the position of the centroid of the arc of the cardioide $r = a\,(1 + \cos\theta)$ lying above the initial line.

(Dec. 2006) Ans. $\left(\dfrac{4a}{5}, \dfrac{4a}{5}\right)$

3.　Find the C.G. of an arc of the catenary $y = a\cosh\dfrac{x}{a}$ from $x = -a$ to $x = a$.　　**Ans.** $\left(0, \dfrac{a}{4}\,\dfrac{2+\sinh 2}{\sinh 1}\right)$

4.　Find the C.G. of the arc of the parabola $y^2 = 4ax$ included between the vertex and the upper extremity of the latus rectum.

　　[**Hint :** C.G. of arc OA where A $(a, 2a)$, $ds = \sqrt{\dfrac{x+a}{x}}\ dx$]

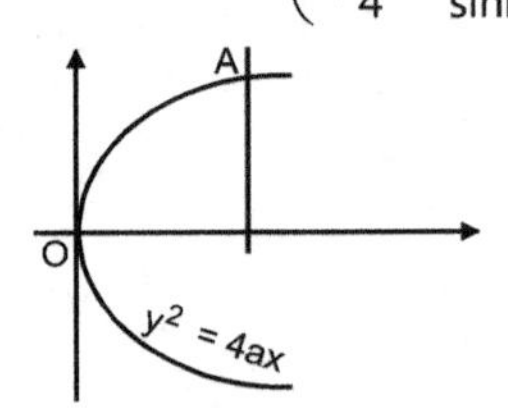

Fig. 11.25

Ans. $\left(\dfrac{a}{4}\,\dfrac{3\sqrt{2}-\log(1+\sqrt{2})}{\sqrt{2}+\log(1+\sqrt{2})},\ \dfrac{4a}{3}\,\dfrac{2\sqrt{2}-1}{\sqrt{2}+\log(1+\sqrt{2})}\right)$

Problems on C.G. of a plane area or lamina :

5.　Find the C.G. of the plate cut from the parabola $y^2 = 8x$ by its latus rectum $x = 2$, if the density is numerically equal to the distance from the latus rectum.　　**Ans.** $\left(\dfrac{6}{7}, 0\right)$

6.　Find the C.G. of a plate in the form of the upper half of the cardioide $r = 2\,(1 + \cos\theta)$ if the density is numerically equal to the distance from the pole.　　**Ans.** $\left(\dfrac{21}{10}, \dfrac{96}{25\pi}\right)$

7. Find the C.G. of mass of the first quadrant part of the disk of radius 'a' with centre at the origin, if the density function is y.

Ans. $\left(\dfrac{3a}{8}, \dfrac{3\pi a}{16}\right)$

8. Find the C.G. of the area enclosed between the curves $y^2 = ax$ and $x^2 + y^2 = 2ax$.

Ans. $\left(a\dfrac{15\pi - 44}{15\pi - 40}, 0\right)$

9. Find the C.G. of the area of the cardioide $r = a (1 + \cos \theta)$ which lies above the initial line. **(Dec. 2011)**

Ans. $\left(\dfrac{5a}{6}, \dfrac{16a}{9\pi}\right)$

10. Find the C.G. of the area of the cardioide $r = a (1 + \cos \theta)$.

(Dec. 2016, May 2017) Ans. $\dfrac{5a}{6}$

11. Find the distance of the C.G. from the initial line to that half of the area of the cardioide $r = a (1 - \cos \theta)$ which lies above the initial line.

Ans. $\bar{y} = \dfrac{16\,a}{9\pi}$

12. Find the C.G. of the area of the cardioide $r = a (1 + \cos \theta)$ if its surface density varies as its distance from the initial line.

(**Hint :** $\bar{y} = 0$, $\rho = ky = kr \sin \theta$).

Ans. $\bar{x} = \dfrac{4a}{5}$, $\bar{y} = 0$

13. Show that the C.G. of the area of the curve $r = 2a \sin^2 \dfrac{\theta}{2}$ is on the initial line at a distance $\dfrac{5a}{6}$ from the origin.

$\left(\textbf{Hint :} \ \text{Given curve is } r = a (1 - \cos \theta) \cdot \bar{y} = 0, \bar{x} = \dfrac{5a}{6}\right)$

14. Find by double integration the area which is above the initial line and bounded by the straight line $\theta = \dfrac{\pi}{2}$, the circle $r = 2a \cos \theta$ and cardioide $r = a (1 + \cos \theta)$. Determine also the position of its C.G.

Ans. Area $= (8 - \pi)\dfrac{a^2}{8}$, $\bar{y} = \left(\dfrac{14}{8 - \pi}\right)\dfrac{a}{3}$, $\bar{x} = \left(\dfrac{16 - 3\pi}{8 - \pi}\right)\dfrac{a}{2}$, $\bar{y} = \left(\dfrac{14a}{8 - \pi}\right)\dfrac{a}{3}$

15. Find the C.G. of the area outside the circle $r = a$ and inside the cardioide $r = a (1 + \cos \theta)$.

Ans. $\left(\dfrac{9\pi + 44}{\pi + 8}a, 0\right)$

16. Find the C.G. of the area bounded by $r = a \sin \theta$, $r = 2a \sin \theta$.

Ans. $\left(0, \dfrac{7a}{6\pi}\right)$

17. Find the x-co-ordinate of the C.G. of the area bounded by the parabola $y^2 = 4x$ and the line $2x - y - 4 = 0$. **(May 2014)**

Ans. $\bar{x} = \dfrac{8}{5}$

18. Find the position of the C.G. of a semi-circular lamina of radius 'a' if its surface density varies as the square of the distance from the diameter.

Hint : Take the circle above x-axis, $\rho = \lambda y^2$, $\bar{x} = 0$, $\bar{y} = \dfrac{32a}{15\pi}$.

19. Prove that the centroid of the smaller segment of the ellipse $\dfrac{x^2}{a^2} + \dfrac{y^2}{b^2}$

$= 1$ cut off by the straight line $\dfrac{x}{a} + \dfrac{y}{b} = 1$ is $\left[\dfrac{2a}{3 (\pi - 2)}, \dfrac{2b}{3 (\pi - 2)}\right]$.

[**Hint :** Limits : $x : 0$ to $a, y : b\left(1 - \dfrac{x}{a}\right)$ to $\dfrac{b}{a} \sqrt{a^2 - x^2}$]

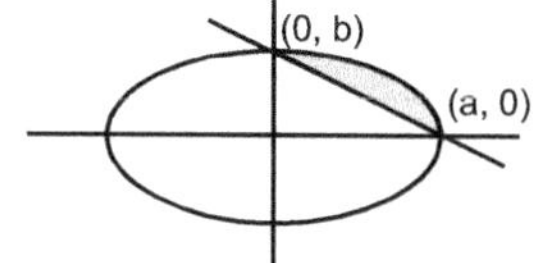

Fig. 11.26

20. Show that the C.G. of the loop of the curve $r = a \cos 3\theta$ containing the initial line lies at a distance $\dfrac{81\sqrt{3}}{80\pi} a$, from the pole.

21. Find the position of the C.G. of a loop of the curve $r = a \cos 2\theta$.

Ans. $\left(\dfrac{128 \sqrt{2}\, a}{105\pi}, 0\right)$

22. Find the C.G. of the area enclosed by the curves $y^2 = ax$, $x^2 = ay$.

Ans. $\bar{x} = \dfrac{9a}{20} = \bar{y}$

23. Find the C.G. of the area enclosed by the curves $y^2 = 4ax$, $y = 2x$. **Ans.** $\left(\dfrac{2a}{5}, a\right)$

24. Find the C.G. of the area enclosed by the curves $y = 6x - x^2$ and $y = x$. **Ans.** $\left(\dfrac{5}{2}, 5\right)$

25. Find the C.G. of the area enclosed by positive quadrant of the ellipse $\dfrac{x^2}{a^2} + \dfrac{y^2}{b^2} = 1$. **Ans.** $\left(\dfrac{4a}{3\pi}, \dfrac{4b}{3\pi}\right)$

26. Find the C.G. of the area of the cycloid $x = a\,(\theta + \sin\theta)$, $y = a\,(1 - \cos\theta)$ measured from cusp to cusp. **Ans.** $\left(0, \dfrac{7a}{6}\right)$

27. Find the C.G. of the area bounded by y-axis, the cycloid $x = a\,(\theta + \sin\theta)$, $y = a\,(1 - \cos\theta)$ and its base.

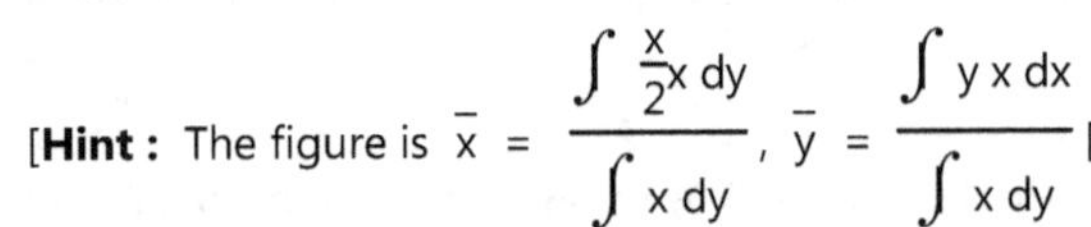

[**Hint :** The figure is $\overline{x} = \dfrac{\displaystyle\int \dfrac{x}{2}\,x\,dy}{\displaystyle\int x\,dy}$, $\overline{y} = \dfrac{\displaystyle\int y\,x\,dx}{\displaystyle\int x\,dy}$ [

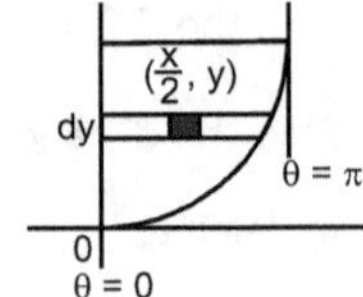

Fig. 11.27

Ans. $\left(\dfrac{a\,(9\pi^2 - 16)}{18\pi}, \dfrac{7a}{6}\right)$

28. Find the C.G. of the area of the cycloid $x = a\,(\theta - \sin\theta)$, $y = a\,(1 - \cos\theta)$ measured from cusp to cusp. **Ans.** $\left(a\pi, \dfrac{5a}{6}\right)$

29. Show that $\left(\dfrac{256a}{315\pi}, \dfrac{256b}{315\pi}\right)$ is the C.G. of the area in the first quadrant bounded by the curve $\left(\dfrac{x}{a}\right)^{2/3} + \left(\dfrac{y}{b}\right)^{2/3} = 1$.

30. Find the C.G. of the area in the first quadrant bounded by $a^4 y^2 = (a^2 - x^2)^3$. **Ans.** $\left(\dfrac{16a}{15\pi}, \dfrac{128a}{105\pi}\right)$

31. Find the distance from the origin of the C.G. of the area outside the circle $x^2 + y^2 = a^2$ and inside the circle $x^2 + y^2 = 2ax$. **Ans.** $\dfrac{a}{2} \cdot \dfrac{8\pi + 3\sqrt{3}}{2\pi + 3\sqrt{3}}$

32. Find the position of the centroid of the area under the parabola $y = 4ax^2$ from $x = 0$ to $x = c$. **(May 2006) Ans.** $\left(\dfrac{3c}{4}, \dfrac{6ac^2}{5}\right)$

33. Find the C.G. of the loop of the curve $9x^2 = (2y - 1)(y - 2)^2$. **Ans.** $\left(0, \dfrac{8}{7}\right)$

34. Find the position of the C.G. of the area of the loop of the curve $8ay^2 = x\,(2a - x)^2$. **Ans.** $\overline{x} = \dfrac{6a}{7}$

Problems on C.G. of a Solid :

35. Find the C.G. of the cube of edge 'a' with 3 faces on the co-ordinate planes, if the density is numerically equal to the sum of the distances from the co-ordinate planes. **Ans.** $\left(\dfrac{5}{9}a, \dfrac{5}{9}a, \dfrac{5}{9}a\right)$

36. Find centroid of the region bounded by $z = x^2 + y^2$, $z = 0$, $x = -a$, $x = a$, $y = -a$, $y = a$. **(Dec. 2005)** **Ans.** $\left(0, 0, \dfrac{7a^2}{15}\right)$

37. Find the position of the C.G. of the volume intercepted between the paraboloid $x^2 + y^2 = a\,(a - z)$ and the plane $z = 0$. **Ans.** $\left(0, 0, \dfrac{a}{3}\right)$

38. Find the C.G. of that half of ellipsoid $\dfrac{x^2}{a^2} + \dfrac{y^2}{b^2} + \dfrac{z^2}{c^2} = 1$, which lies above XY plane. **Ans.** $\overline{z} = \dfrac{3a}{8}$

39. Find the C.G. of a cube of side 'h' if its density at each point is proportional to the square of its distance of the point from one corner O of the base. **Ans.** $\overline{x} = \dfrac{7h}{12} = \overline{y} = \overline{z}$

40. OABC is a solid tetrahedron. The faces which meet at O are mutually at right angles and the density varies directly as the distance from the face OAB. If $OA = a$, $OB = b$, $OC = c$, find the position of its centre of gravity.

[**Hint :** Take O as origin, OA, OB, OC as x, y, z axis, $\rho = kz$. Plane ABC is $\dfrac{x}{a} + \dfrac{y}{b} + \dfrac{z}{c} = 1$.] **Ans.** $\left(\dfrac{a}{5}, \dfrac{b}{5}, \dfrac{2c}{5}\right)$

11.9 MOMENT OF INERTIA

The moment of inertia of a particle of mass m about a line (or axis) is mr^2, where r is the perpendicular distance of the particle from the line (or axis).

Next, consider a body of mass M. A body is supposed to consist of an infinite number of small particles. Let the particles composing the body be taken to be of masses m_1, m_2, m_3, ... situated at distances r_1, r_2, r_3 ...... respectively from a certain axis. Then, the moment of inertia of the body about the axis is

$$\Sigma \, mr^2 \;=\; m_1 r_1^2 \;+\; m_2 r_2^2 \;+\; m_3 r_3^2 \;+\; ......$$

In case the mass be in a continuously distributed form, consider an elementary particle of mass "dm" of the body. Let the distance of this element from the axis under consideration be p. Then the M.I. of this element about the axis = $p^2\, dm$, so that the M.I. of the whole body $= \Sigma\, p^2\, dm$. By the definition of a definite integral as the limit of a sum, we can write

$$\text{M.I.} \;=\; \lim_{dm \to 0} \Sigma\, p^2\, dm \;=\; \int p^2\, dm$$

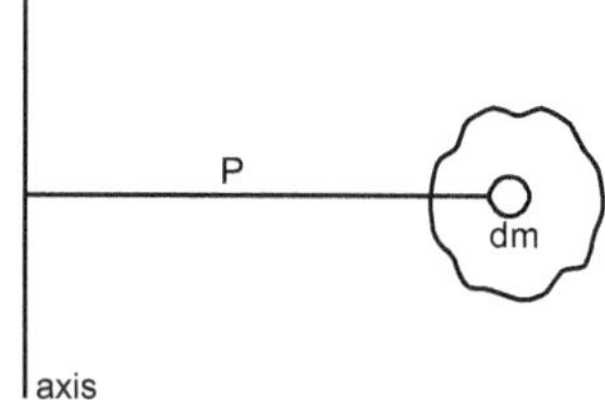

Fig. 11.28

M.I. is denoted by I :

∴   The moment of inertia I of a continuously distributed mass is $\boxed{I = \int p^2\, dm}$ where the integral is to be evaluated throughout the region of the body. (Refer Fig. 11.28).

If M is the mass of a body and the M.I. of the body about an axis is denoted by MK^2 i.e. $\boxed{I = MK^2}$ then, K is said to be the *radius of gyration* of the body about the axis.

Hence, $\boxed{\textbf{Radius of Gyration} = \textbf{K} = \sqrt{\dfrac{I}{M}}}$

Note :

1. For M.I. of an area, we take $dm = dx\, dy$

    ∴ $I = \iint p^2\, dx\, dy$ in cartesian form, $I = \iint p^2\, r\, dr\, d\theta$ in polar form

2. For M.I. of an arc, we take $dm = ds$.

    $I = \int p^2\, ds$ and convert ds into $\dfrac{ds}{dx}\, dx$ or $\dfrac{ds}{dy}\, dy$ or $\dfrac{ds}{dt}\, dt$,

    according to the nature of the curve.

3. If density is variable, say ρ at (x, y) or (r, θ) then the

    M.I. of mass $= I = \iint p^2 \rho\, dx\, dy$ or $I = \iint p^2 \rho\, r\, dr\, d\theta$ or $I = \int p^2 \rho\, ds$

4. For M.I. of a solid, $I = \iiint p^2\, \rho\, dv.$

11.10 THEOREMS ON MOMENT OF INERTIA

Two important theorems on moment of inertia are given below without proof, which will be found useful in solving problems on moment of inertia.

I.   Parallel Axes Theorem : If I_G is the M.I. of a body about an axis through the centroid G of the body and I_A, the M.I. of the body about any parallel axis through any point A, then $\boxed{I_A = I_G + M\, d^2}$, where M is mass of the body and d is the distance between the parallel axes.

II.   Perpendicular Axes Theorem : In case of a plane lamina in the XOY plane $\boxed{I_Z = I_X + I_Y}$, where I_X and I_Y are M.I. about OX and OY and I_Z, M.I. about z-axis.

Note : This theorem is applicable to plane lamina and not to three-dimensional solids.

11.11 SOME STANDARD RESULTS

1. For a Uniform Straight Rod : M.I. of a uniform straight rod of length l about an axis perpendicular to the rod through (i) the mid point, (ii) an entremity.

(i)   Let AB be rod of length l and O be the mid-point of AB, OY be axis through O perpendicular to AB. Consider a small element of length "dx" of the rod at a distance x from OY. If ρ is the mass per unit length of the rod, then mass of this element is $dm = \rho \cdot dx$ and the M.I. of this elementary mass about OY is $x^2 \cdot (\rho \cdot dx)$.

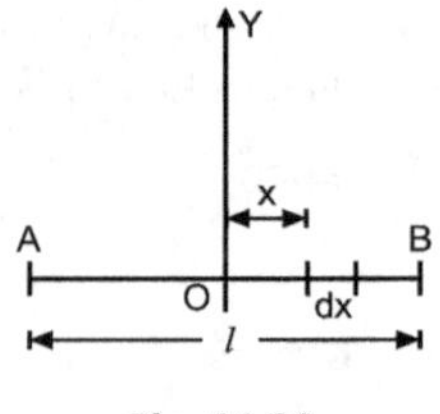

Fig. 11.29

$$\text{Hence required M.I. of the rod } = \int_{-l/2}^{l/2} x^2 \cdot (\rho\, dx) = \frac{\rho l^3}{12}$$

But $\qquad M = \text{Mass of the rod } = \rho \cdot l$

$$\text{M.I. of the rod about OY } = \frac{M \cdot l^2}{12}$$

If $l = 2a$ then, M.I. of the rod of length 2a about OY $= M \cdot \dfrac{a^2}{3}$

The radius of gyration K is given by, $K^2 = \dfrac{a^2}{3}$

(ii)   M.I. about axis through A using theorem of parallel axes,

$$I_A = I_G + M \cdot (AO)^2 = \frac{Ml^2}{12} + M \cdot \left(\frac{l}{2}\right)^2 = \frac{Ml^2}{3}$$

$$\text{If } l = 2a \text{ then, } I_A = M\left(\frac{4a^2}{3}\right)$$

2. For a Rectangle : M.I. of a rectangle of sides 'a' and 'b' about (i) a median parallel to side 'a', (ii) a median parallel to side b, (iii) an axis through centre perpendicular to the lamina.

(i)   Let EF be median parallel to side 'a'. Consider a strip of width 'dx' at a distance x from EF.

$\therefore \qquad dm = \rho \cdot a\, dx$

M.I. of this strip about EF $= x^2 \cdot (\rho\, a\, dx)$

$$\text{M.I. of a rectangle about EF} = \int_{-b/2}^{b/2} x^2 \rho\, a\, dx = \frac{\rho\, ab^3}{12}$$

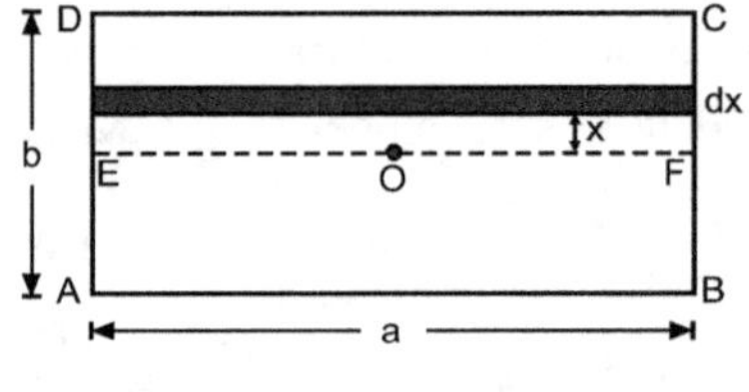

Fig. 11.30

But, $\qquad M = \text{Mass of rectangle} = \rho\,(ab)$

$$\begin{array}{l}\text{M.I. about median EF}\\ \text{parallel to side 'a'}\end{array} = \frac{Mb^2}{12}$$

(ii)   Similarly, M.I. about median parallel to side 'b' $= \dfrac{Ma^2}{12}$

(iii) M.I. about line through centre perpendicular to the plane of lamina $= \dfrac{Ma^2}{12} + \dfrac{Mb^2}{12} = \dfrac{M}{12}(a^2 + b^2)$

by using perpendicular axes theorem.

3. For a Circular Ring of Radius a : M.I. of a circular ring about (i) axes through centre perpendicular to the plane of the wire, (ii) a diameter.

(i)   Let 'a' be radius of circular ring and O be the centre. Let OZ be the axis perpendicular to the plane of the ring. Since, every element of the ring is at a distance a from the centre, the required M.I. is Ma^2, where M is the mass of the ring.

(ii)   We have :

$$I_z = I_x + I_y \quad \text{But} \quad I_x = I_y$$

$$Ma^2 = 2I_x \qquad \therefore \; I_x = \frac{Ma^2}{2}$$

M.I. of the ring about diameter $= \dfrac{Ma^2}{2}$.

4.   For a Circular Disc : M.I. of a uniform circular disc of radius 'a' and mass M about (i) the axis through its centre perpendicular to its plane, (ii) a diameter, (iii) tangent to the circular disc.

(i)   Take the centre of the disc as the pole, so that, the polar equation of the circle is $r = a$ and density $= \rho$; Mass of the disc $M = \pi a^2 \rho$. Consider element dm at a distance r from O.

$$dm = \rho \, (r \, dr \, d\theta)$$

M.I. of this element about OZ $= r^2 \, dm = r^2 \, (\rho \, r \, dr \, d\theta)$

M.I. of disc about the line through O perpendicular to plane of disc

$$= \int_0^{2\pi} \int_0^a \rho \, r^3 \, dr \, d\theta \; = \; \rho \, (2\pi) \, \frac{a^4}{4} = \frac{\rho \pi a^4}{2} = \frac{Ma^2}{2} \; (M = \pi a^2 \rho)$$

(ii)

$$I_z = I_x + I_y \quad \text{But} \quad I_x = I_y, \; \frac{Ma^2}{2} = 2I_x$$

M.I. about diameter $= \dfrac{Ma^2}{4}$

(iii)   M.I. about tangent to the circular disc $= \dfrac{Ma^2}{4} + Ma^2 = \dfrac{5Ma^2}{4}$ by using parallel axes theorem.

5.   M.I. of Uniform Solid Sphere about its Diameter : Divide the given sphere into an infinite number of circular discs, perpendicular to the given diameter and consider a typical disc at a distance 'x' from the centre and of thickness dx. The radius of this disc is $\sqrt{a^2 - x^2}$, where a is the radius of the sphere.

$$\therefore \qquad dm = \text{mass of the disc} = \rho \, \pi \, y^2 \, dx \qquad\qquad \left(\because \; y = \sqrt{a^2 - x^2} \right)$$

assuming constant density ρ.

M.I. of the disc about the line through O perpendicular to plane of the disc

$$= \frac{y^2}{2} \, dm = \frac{y^2}{2} \rho \, \pi \, y^2 \, dx$$

M.I. of the sphere about diameter $= \dfrac{\pi \rho}{2} \displaystyle\int_{-a}^{a} y^4 \, dx = \dfrac{\pi \rho}{2} \displaystyle\int_{-a}^{a} (a^2 - x^2)^2 \, dx$

$$= \frac{\pi \rho}{2} \, 2 \int_0^a (a^4 - 2a^2 x^2 + x^4) \; dx = \frac{8}{15} \pi \rho \, a^5$$

Now, mass of the sphere, $\; M = \left(\dfrac{4}{3} \pi \, a^3 \right) \cdot \rho \quad \therefore \; \rho = \dfrac{3M}{4\pi a^3}$

$\therefore$ M.I. of the sphere about a diameter $= \dfrac{8}{15} \pi \, a^5 \cdot \dfrac{3M}{4\pi a^3} = \dfrac{2}{5} Ma^2$

6.   M.I. of a Hollow Sphere about its Diameter : Here, elementary disc has a curved surface, in the form of ring of area $2\pi y \, ds$.

Let ρ be mass per unit area, then $dm = \rho \, (2\pi y \, ds)$

M.I. of the ring about line through the centre perpendicular to plane is ma^2.

Hence, M.I. of this ring about x-axis $= (\rho\, 2\pi y\, ds)\, y^2$

$$\text{M.I.} = 2\pi\rho \int_{-a}^{a} y^3\, ds$$

$$x = a\cos\theta,\ y = a\sin\theta$$
$$s = a\theta,\ ds = a\,d\theta$$

$$= 4\pi\rho \int_{0}^{\pi/2} a^3 \sin^3\theta\, a\, d\theta = 4\pi\rho\, a^4 \frac{2}{3} = \frac{8\pi\rho a^4}{3}$$

$$M = \text{Mass of hollow sphere} = 4\pi a^2 \rho$$

$$\therefore \qquad \text{M.I.} = \frac{2Ma^2}{3}$$

7. The M.I. of a uniform triangular lamina of mass M about any axis is the same as the M.I. of the particles of mass $\frac{M}{3}$ placed at the mid-points of the sides of the triangle.

11.12 ILLUSTRATIONS ON MOMENT OF INERTIA

Ex. 1 : *Find the moment of inertia of the quadrant of the ellipse $2x^2 + y^2 = 1$ in which x and y are positive about an axis through its centre perpendicular to its plane, where mass per unit area of the ellipse varies as the abscissa of the point at which it is situated.*

Sol. : Let the density be $\rho \propto x$ (abscissa) $\Rightarrow \rho = Kx$. An elementary area dx dy at the point (x, y) has then the mass $(Kx)\,(dx\,dy)$ so, $dm = Kx\,dx\,dy$. This mass is situated at a distance $r = \sqrt{x^2 + y^2}$ from the axis through the pole O at right angles to the plane XOY. So, the contribution to the total moment of inertia by it is :

$$r^2\, dm = (x^2 + y^2)\, Kx\, dx\, dy = K\, x\,(x^2 + y^2)\, dx\, dy$$

The moment of inertia of the quadrant of an ellipse is $I = K \iint x\,(x^2 + y^2)\, dx\, dy$

$$2x^2 + y^2 = 1 \ \text{ or } \ \frac{x^2}{\left(\frac{1}{\sqrt{2}}\right)^2} + \frac{y^2}{(1)^2} = 1, \quad a = \frac{1}{\sqrt{2}}, \quad b = 1$$

Put $\qquad x = ar\cos\theta = \dfrac{1}{\sqrt{2}} r\cos\theta,\ y = br\sin\theta = r\sin\theta$

and $\qquad dx\, dy = ab\, r\, dr\, d\theta = \dfrac{1}{\sqrt{2}} r\, dr\, d\theta$

$$2x^2 + y^2 = 1 \ \Rightarrow \ r = 1$$

$$I = K \int_{\theta=0}^{\pi/2} \int_{r=0}^{1} \frac{1}{\sqrt{2}} r\cos\theta \left(\frac{\cos^2\theta}{2} + \sin^2\theta\right) r^2 \cdot \frac{1}{\sqrt{2}} r\, dr\, d\theta$$

$$= \frac{K}{2} \int_{0}^{\pi/2} \left(\frac{\cos^3\theta}{2} + \sin^2\theta\cos\theta\right) d\theta \int_{0}^{1} r^4\, dr = \frac{K}{2}\left[\frac{1}{2}\cdot\frac{2}{3} + \frac{1}{3}\right]\cdot\frac{1}{5}$$

$$\boxed{I = \frac{K}{15}}$$

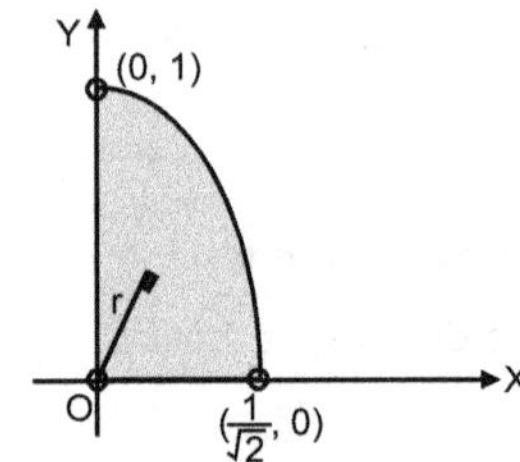

Fig. 11.31

Ex. 2 : *Find the M.I. about the line $\theta = \dfrac{\pi}{2}$ of the area enclosed by $r = a\,(1 + \cos\theta)$.*   **(May 2009, 2013; Dec. 2005, 2016, 2018)**

Sol. : Here, $p = x = r\cos\theta,\ I = \int p^2\, dA$

$$\therefore \qquad I = \iint r^2 \cos^2\theta\, r\, dr\, d\theta = 2\int_{0}^{\pi} \cos^2\theta\, d\theta \int_{0}^{a(1+\cos\theta)} r^3\, dr$$

$$= 2\int_{0}^{\pi} \cos^2\theta\, \frac{a^4(1+\cos\theta)^4}{4}\, d\theta$$

$$= \frac{a^4}{2} \int_0^\pi \cos^2\theta \,(1 + 4\cos\theta + 6\cos^2\theta + 4\cos^3\theta + \cos^4\theta)\, d\theta$$

$$= \frac{a^4}{2} \int_0^\pi (\cos^2\theta + 4\cos^3\theta + 6\cos^4\theta + 4\cos^5\theta + \cos^6\theta)\, d\theta$$

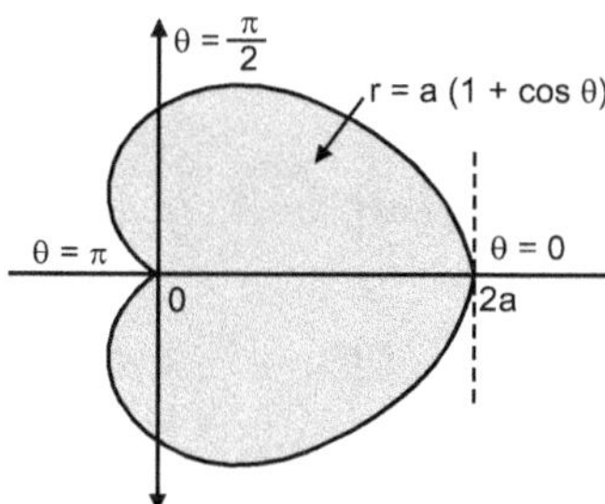

Fig. 11.32

$$= \frac{a^4}{2} \, 2 \int_0^{\pi/2} (\cos^2\theta + 6\cos^4\theta + \cos^6\theta)\, d\theta = a^4 \left[\frac{1}{2}\frac{\pi}{2} + 6 \cdot \frac{3}{4}\frac{1}{2}\frac{\pi}{2} + \frac{5}{6}\cdot\frac{3}{4}\frac{1}{2}\frac{\pi}{2} \right]$$

$$\boxed{I = \frac{49\pi a^4}{32}}$$

Ex. 3 : *Prove that the moment of inertia of the area included between the curves* $y^2 = 4ax$ *and* $x^2 = 4ay$ *about x-axis is* $\dfrac{144}{35} Ma^2$, *where, M is the mass of the area included between the curves.* **(May 2005, 2015)**

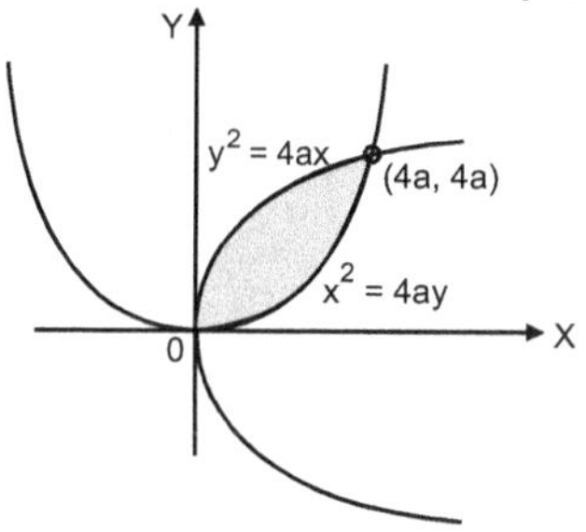

Sol. :

$$\text{M.I. about x-axis} = \int_0^{4a} \int_{\frac{x^2}{4a}}^{2\sqrt{ax}} y^2 \rho \, dx \, dy$$

$$= \frac{\rho}{3} \int_0^{4a} \left(8a^{3/2} x^{3/2} - \frac{x^6}{64a^3} \right) dx$$

$$= \frac{\rho}{3} \left[8a^{3/2}\frac{2x^{5/2}}{5} - \frac{x^7}{7.64a^3} \right]_0^{4a}$$

$$= \frac{\rho}{3} \left[\frac{2^9 a^4}{5} - \frac{2^8 a^4}{7} \right] = \frac{2^8 \, 3\rho a^4}{35}$$

Fig. 11.33

$$\text{Mass} = M = \int_0^{4a} \int_{\frac{x^2}{4a}}^{2\sqrt{ax}} \rho \, dx \, dy = \rho \int_0^{4a} \left(2\sqrt{ax} - \frac{x^2}{4a} \right) dx = \rho \left[2a^{1/2} \cdot \frac{2}{3} x^{3/2} - \frac{x^3}{12a} \right]_0^{4a} = \frac{16\rho a^2}{3}$$

$$\rho = \frac{3M}{16a^2}$$

$$\therefore \quad I = \frac{2^8 \cdot 3a^4}{35} \cdot \frac{3M}{16a^2}$$

$$\boxed{I = \frac{144\, Ma^2}{35}}$$

Ex. 4 : *Find the moment of inertia of one loop of the lemniscate* $r^2 = a^2 \cos 2\theta$ *about initial line.*

(Nov./Dec. 2019, Dec. 07, 04, Nov. 15, May 17)

Sol. : Let ρ be constant density. Consider an element of mass $dm = \rho\,(r\,dr\,d\theta)$.

Also M.I. about initial line is $\int p^2\, dm$, where $p = y = r\sin\theta$

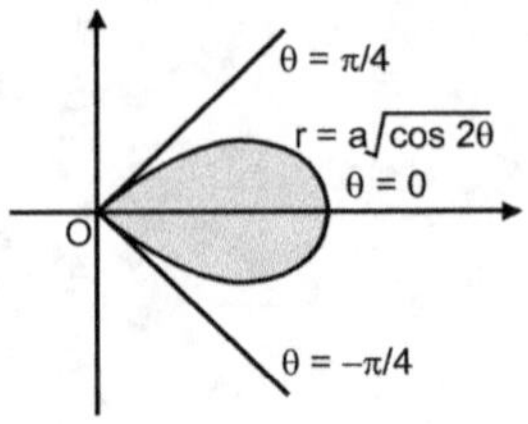

$$\text{M.I.} = \iint r^2 \sin^2 \theta \cdot \rho \, r \, dr \, d\theta$$

$$= 2\rho \int_0^{\pi/4} \sin^2 \theta \, d\theta \int_0^{a\sqrt{\cos 2\theta}} r^3 \, dr$$

$$= \frac{2\rho \, a^4}{4} \int_0^{\pi/4} \sin^2 \theta \cos^2 2\theta \, d\theta$$

$$= \frac{\rho \, a^4}{2} \int_0^{\pi/4} \left(\frac{1 - \cos 2\theta}{2}\right) \cos^2 2\theta \, d\theta$$

$$= \frac{\rho \, a^4}{4} \int_0^{\pi/2} (1 - \cos t) \, \cos^2 t \, \frac{dt}{2} = \frac{\rho \, a^4}{8} \left[\frac{1}{2}\frac{\pi}{2} - \frac{2}{3}\right] = \frac{\rho \, a^4}{96} (3\pi - 8)$$

$$\text{Mass} = \text{M} = \iint \rho \, r \, dr \, d\theta = 2\rho \int_0^{\pi/4} \int_0^{a\sqrt{\cos 2\theta}} r \, dr \, d\theta$$

$$\text{M} = \rho \, a^2 \int_0^{\pi/4} \cos 2\theta \, d\theta = \frac{\rho \, a^2}{2} \qquad \therefore \quad \rho = \frac{2M}{a^2}$$

$$\therefore \qquad \text{M.I.} = \text{I} = \frac{a^4 (3\pi - 8)}{96} \left(\frac{2M}{a^2}\right)$$

$$\boxed{\text{M.I.} = \frac{(3\pi - 8) \, Ma^2}{48}}$$

Fig. 11.34

Ex. 5 : *Find the moment of inertia of the portion of the parabola $y^2 = 4ax$, bounded by x-axis and latus rectum, about x-axis, if density at each point varies as the cube of the abscissa.* **(Dec. 2008, 2017)**

Sol. : $\rho \propto x^3 \qquad \Rightarrow \quad \rho = Kx^3$ also $p = y$

$$\text{M.I.} = \iint y^2 \, Kx^3 \, dx \, dy = K \int_0^a x^3 \, dx \int_0^{2\sqrt{ax}} y^2 \, dy = \frac{K}{3} \, 8 \cdot a^{3/2} \int_0^a x^{9/2} \, dx = \frac{16 \, K}{33} \, a^7 \qquad \ldots \text{(i)}$$

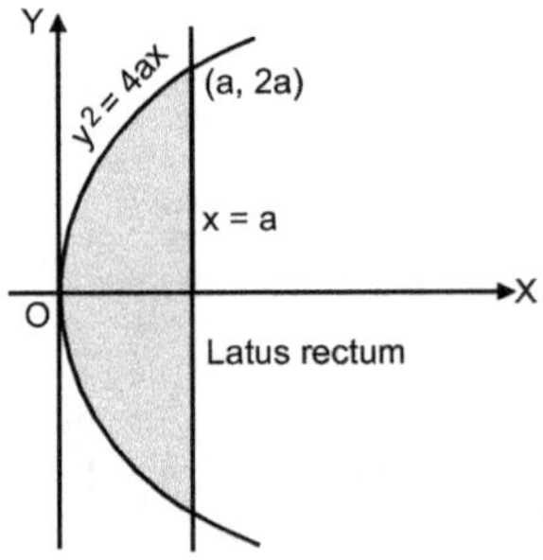

Also,
$$\text{M} = \iint \rho \, dx \, dy = \int_0^a \int_0^{2\sqrt{ax}} Kx^3 \cdot dx \, dy$$

$$= K \int_0^a x^3 \cdot 2 \sqrt{ax} \, dx = 2K \sqrt{a} \left[\frac{2}{9} x^{9/2}\right]_0^a$$

$$\text{M} = \frac{4K}{9} a^5 \quad \therefore \quad K = \frac{9M}{4a^5}$$

From (i) and (ii), $\text{I} = \frac{16a^7}{33} \cdot \frac{9m}{4a^5}$ $\boxed{\text{I} = \frac{12}{11} \, Ma^2}$

Fig. 11.35

Ex. 6 : *Find the moment of inertia about z-axis of the region bounded by $z = x^2 + y^2$, $z = 0$, $x = -a$, $x = a$, $y = -a$, $y = a$.*

Sol. :
$$p = \sqrt{x^2 + y^2}$$

$$\text{M.I.} = \iiint p^2 \rho \, dx \, dy \, dz = \rho \iiint_0^{x^2 + y^2} (x^2 + y^2) \, dx \, dy \, dz$$

$$= \rho \iint (x^2 + y^2)^2 \, dx \, dy = 4\rho \int_0^a dx \int_0^a (x^4 + 2x^2 y^2 + y^4) \, dy$$

$$= 4\rho \int_0^a \left(ax^4 + \frac{2x^2}{3} a^3 + \frac{a^5}{5}\right) dx = 4\rho \left[\frac{a^6}{5} + \frac{2}{9} a^6 + \frac{a^6}{5}\right] = \frac{112\rho \, a^6}{45}$$

$$\text{Mass} = M = \iiint \rho \, dx \, dy \, dz = \rho \iiint_0^{x^2 + y^2} dx \, dy \, dz$$

$$= 4\rho \int_0^a \int_0^a (x^2 + y^2) \, dx \, dy = 4\rho \int_0^a \left(x^2 a + \frac{a^3}{3} \right) dx$$

$$= 4\rho \left(\frac{a^4}{3} + \frac{a^4}{3} \right) = \frac{8\rho \, a^4}{3} \quad \therefore \quad \rho = \frac{3M}{8a^4}$$

$$\therefore \qquad \text{M.I.} = \frac{112a^6}{45} \left(\frac{3M}{8a^4} \right)$$

$$\boxed{\text{M.I.} = \frac{14}{15} \, Ma^2}$$

Ex. 7 : *The surface density of a circular lamina varies as the square of the distance from a point on the circumference. Find the moment of inertia of the area about an axis through O perpendicular to the plane of circle.*

Sol. : Take the fixed point O on the circumference as the pole and the diameter through O as the initial line. Then, the polar equation of the circle is $r = 2a \cos \theta$, $\rho \propto r^2$, $\rho = Kr^2$, $p = \sqrt{x^2 + y^2} = r$.

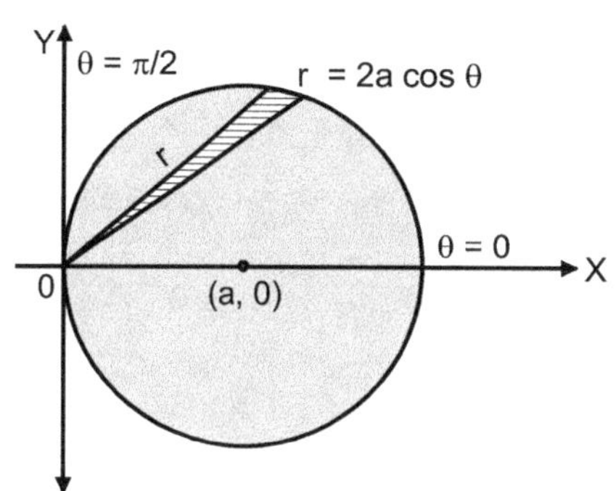

$$\text{M.I.} = \iint p^2 \rho \, dx \, dy = \iint r^2 \, Kr^2 \, r \, dr \, d\theta$$

$$= K \int_0^{\pi/2} d\theta \int_0^{2a\cos\theta} r^5 \, dr = \frac{K}{6} \int_0^{\pi/2} 64a^6 \cos^6 \theta \, d\theta$$

$$= \frac{32K}{3} a^6 \frac{5}{6} \frac{3}{4} \frac{1}{2} \frac{\pi}{2} = \frac{5\pi \, Ka^6}{3}$$

$$\text{Mass} = M = \iint \rho \, dx \, dy = K \int_0^{\pi/2} d\theta \int_0^{2a\cos\theta} r^3 \, dr$$

$$= \frac{K}{4} \int_0^{\pi/2} 16a^4 \cos^4 \theta \, d\theta = 4Ka^4 \frac{3}{4} \frac{1}{2} \frac{\pi}{2}$$

$$M = \frac{3\pi \, Ka^4}{4} \qquad \therefore \quad K = \frac{4M}{3\pi \, a^4}$$

$$\text{M.I.} = \frac{5\pi a^6}{3} \frac{4M}{3\pi a^4}$$

$$\boxed{\text{M.I.} = \frac{20}{9} \, Ma^2}$$

Ex. 8 : *Show that the M.I. of the positive octant of the ellipsoid $\dfrac{x^2}{a^2} + \dfrac{y^2}{b^2} + \dfrac{z^2}{c^2} = 1$, about the x-axis is $\dfrac{M}{5} (b^2 + c^2)$, where M is the mass of the solid.*

(May 2006)

Sol. : Here, $\quad p = \sqrt{y^2 + z^2}$ ρ be constant density

$$\text{M.I.} = \iiint p^2 \rho \, dx \, dy \, dz = \iiint (y^2 + z^2) \, \rho \, dx \, dy \, dz$$

Put $x = ar \sin\theta \cos\phi$, $y = br \sin\theta \sin\phi$, $z = cr\cos\theta$, $dx \, dy \, dz = abc \, r^2 \sin\theta \, d\theta \, d\phi \, dr$

$$\text{M.I.} = \rho \int_0^{\pi/2} \int_0^{\pi/2} \int_0^1 (b^2 \sin^2\theta \sin^2\phi + c^2 \cos^2\theta) \, r^2 \, abc \, r^2 \sin\theta \, d\theta \, d\phi \, dr$$

$$= abc \, \rho \int_0^{\pi/2} \sin\theta \, d\theta \int_0^{\pi/2} (b^2 \sin^2\theta \sin^2\phi + c^2 \cos^2\theta) \, d\phi \cdot \frac{1}{5}$$

$$= \frac{abc}{5}\,\rho \int_0^{\pi/2} \sin\theta \left[b^2 \sin^2\theta \cdot \frac{1}{2} \cdot \frac{\pi}{2} + c^2 \cos^2\theta \cdot \frac{\pi}{2} \right] d\theta$$

$$= \frac{abc\,\rho\,\pi}{10} \int_0^{\pi/2} \left(\frac{b^2}{2} \sin^3\theta + c^2 \cos^2\theta \sin\theta \right) d\theta$$

$$= \frac{abc\,\rho\,\pi}{10} \left(\frac{b^2}{2} \frac{2}{3} + \frac{c^2}{3} \right) = abc\,\rho\pi \frac{(b^2 + c^2)}{30}$$

Also, $\quad M = \iiint \rho\, dx\, dy\, dz = \rho \iiint dx\, dy\, dz = \rho\,(\text{volume of ellipsoid})$

$$= \rho \left(\frac{4}{3} \pi\, abc \right) \cdot \frac{1}{8} = \frac{\rho\,\pi\,abc}{6} \qquad \therefore \quad \rho = \frac{6M}{\pi\, abc}$$

$$\text{M.I.} = \frac{abc\,\pi\,(b^2 + c^2)}{30} \frac{6M}{\pi\,abc}$$

$$\boxed{\ \text{M.I.} = \frac{(b^2 + c^2)\,M}{5}\ }$$

Ex. 9 : *Show that the M.I. of a rectangle of sides a and b about its diagonal is $\frac{M}{6}\left(\frac{a^2 b^2}{a^2 + b^2} \right)$ where M is the mass of rectangle.*

Sol. : Consider an element of area dx dy. Let P(x, y) be any point in a rectangle. Let PM $\perp$ diagonal OB. *(May 2011)*

Equation of OB is $\frac{y}{x} = \frac{b}{a}$ i.e. $ay - bx = 0$.

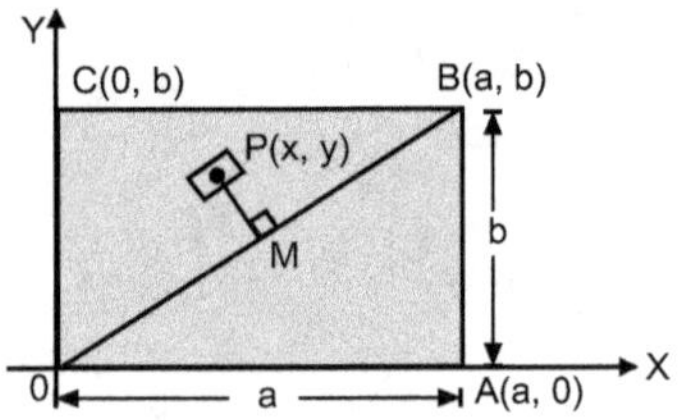

$\therefore \qquad \qquad PM = \left| \dfrac{ay - bx}{\sqrt{a^2 + b^2}} \right|$

$\therefore \qquad \qquad dm = \rho\, dx\, dy$

and $\qquad \qquad p = \dfrac{ay - bx}{\sqrt{a^2 + b^2}}$

$\qquad \qquad \text{M.I.} = \int p^2\, dm$

Fig. 11.37

$$= \int_{x=0}^{a} \int_{y=0}^{b} \frac{(ay - bx)^2}{(a^2 + b^2)}\, \rho\, dx\, dy$$

$$= \frac{\rho}{a^2 + b^2} \int_0^a dx \left[\frac{(ay - bx)^3}{a\cdot 3} \right]_0^b = \frac{\rho}{3a\,(a^2 + b^2)} \int_0^a [b^3(a - x)^3 - (- b^3 x^3)]\, dx$$

$$= \frac{\rho\, b^3}{3a\,(a^2 + b^2)} \left[\frac{(a - x)^4}{-4} + \frac{x^4}{4} \right]_0^a = \frac{\rho\, b^3}{12a\,(a^2 + b^2)}(a^4 + a^4) = \frac{\rho\,b^3 a^3}{6\,(a^2 + b^2)}$$

$$M = \text{Mass of a rectangle} = \rho\,(\text{area}) = \rho\, ab$$

$$\boxed{\ \text{M.I.} = \frac{M}{6} \frac{a^2 b^2}{(a^2 + b^2)}\ }$$

Ex. 10 : *Find the polar M.I. of the area in XY-plane bounded by $y^2 = 2x$ and $y = x$ assuming constant density ρ.* **(Dec. 2006, May 2004)**

Sol. : *"Polar M.I." means M.I. about the axis through the origin (pole) perpendicular to the plane of the region.*

$\therefore \qquad \qquad p = \sqrt{x^2 + y^2}, \quad \rho = \text{constant}$

Changing to polar,

$$\text{M.I.} = \iint p^2 \rho\, dx\, dy = \iint (x^2 + y^2)\,\rho\, dx\, dy$$

$$= \rho \iint r^2\, r\, dr\, d\theta$$

$$\left[y^2 = 2x \text{ in polar} \Rightarrow r^2 \sin^2\theta = 2r\cos\theta \Rightarrow r = \frac{2\cos\theta}{\sin^2\theta} \right]$$

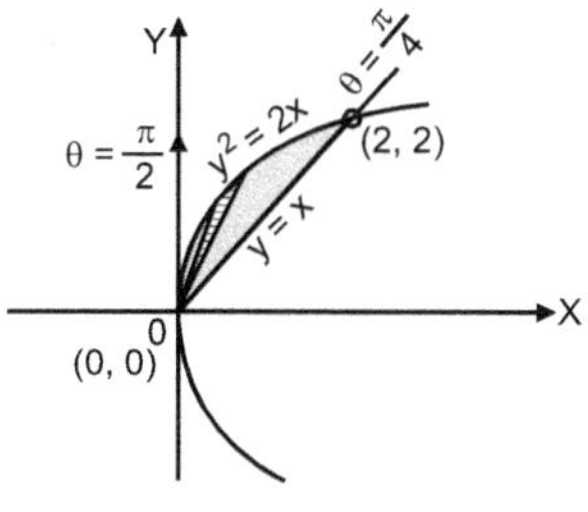

$$\text{M.I.} = \rho \int_{\pi/4}^{\pi/2} d\theta \int_{0}^{\frac{2\cos\theta}{\sin^2\theta}} r^3 \, dr$$

$$= \frac{\rho}{4} \int_{\pi/4}^{\pi/2} \frac{16\cos^4\theta}{\sin^8\theta} \, d\theta = 4\rho \int_{\pi/4}^{\pi/2} \cot^4\theta \cdot \csc^4\theta \, d\theta$$

$$= 4\rho \int_{\pi/4}^{\pi/2} \cot^4\theta \, (1 + \cot^2\theta) \csc^2\theta \, d\theta$$

$$= 4\rho \left[-\frac{\cot^5\theta}{5} - \frac{\cot^7\theta}{7} \right]_{\pi/4}^{\pi/2} = 4\rho \left[0 + \frac{1}{5} + \frac{1}{7} \right] = \frac{48\,\rho}{35}$$

$$\text{Mass M} = \iint \rho \, dx \, dy = \frac{\rho}{2} \int_{\pi/4}^{\pi/2} r^2 \, d\theta = \frac{\rho}{2} \int_{\pi/4}^{\pi/2} \frac{4\cos^2\theta}{\sin^4\theta} \, d\theta = 2\rho \int_{\pi/4}^{\pi/2} \cot^2\theta \csc^2\theta \, d\theta$$

$$\text{M} = 2\rho \left[-\frac{\cot^3\theta}{3} \right]_{\pi/4}^{\pi/2} = 2\rho \left[0 + \frac{1}{3} \right] = \frac{2\rho}{3} \quad \therefore \quad \rho = \frac{3\,M}{2}$$

$$\text{M.I.} = \frac{48}{35} \left(\frac{3M}{2} \right)$$

$$\boxed{\text{M.I.} = \frac{72}{35}\, M}$$

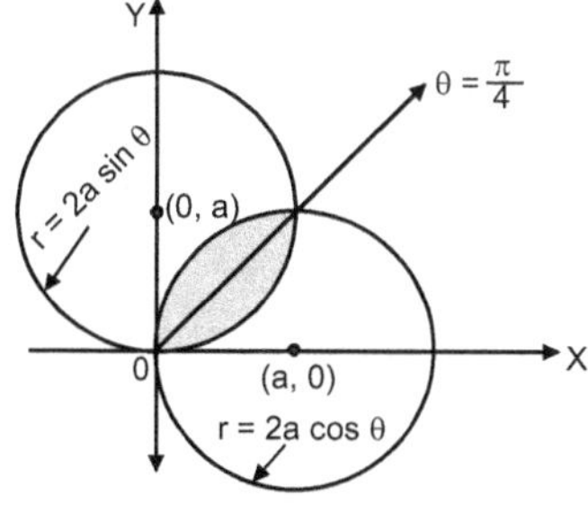

Fig. 11.38

Ex. 11 : *Prove that the moment of inertia of the area included between the smaller arcs of $r = 2a\cos\theta$ and $r = 2a\sin\theta$ about an axis through the pole perpendicular to the plane of the curve is $a^4 \left(\dfrac{3\pi}{4} - 2 \right)$.*

Sol. : $r = 2a\cos\theta$ is a circle with centre $(a, 0)$ and radius a. Similarly, $r = 2a\sin\theta$ is a circle with centre $(0, a)$ and radius a. The area included by the smaller arcs of these circles is the shaded area in the figure which is divided equally by the line $\theta = \dfrac{\pi}{4}$. An elementary area $r \, dr \, d\theta$ at (r, θ) is at a distance r from the axis through the pole O at right angles to the plane xoy; and thus the moment of inertia of the element about the axis is $r^2 (r \, dr \, d\theta) = r^3 \, dr \, d\theta$. Integrating this over the shaded area and using the fact that the line $\theta = \dfrac{\pi}{4}$ divides the region equally, we have

Fig. 11.39

$$\text{The required moment of inertia} = 2 \int_{0}^{\pi/4} \int_{0}^{2a\sin\theta} r^3 \, dr \, d\theta = 8a^4 \int_{0}^{\pi/4} \sin^4\theta \, d\theta$$

$$= 8a^4 \int_{0}^{\pi/4} \frac{(1 - \cos 2\theta)^2}{4} \, d\theta = a^4 \int_{0}^{\pi/2} (1 - \cos\phi)^2 \, d\phi, \text{ where } \phi = 2\theta$$

$$= a^4 \int_{0}^{\pi/2} [1 - 2\cos\phi + \cos^2\phi] \, d\phi = a^4 \left[\frac{\pi}{2} - 2 + \frac{1}{2}\frac{\pi}{2} \right]$$

$$\boxed{\text{M.I.} = a^4 \left[\frac{3\pi}{4} - 2 \right]}$$

Ex. 12 : *Prove that the radius of gyration of semi-circular area of radius a about the line joining one end of the bounding diameter to the mid-point of the arc is* $a\sqrt{\left(\dfrac{3}{4}-\dfrac{4}{3\pi}\right)}$.

Sol. : Take the fixed point O on the circumference as the pole, the diameter OA as the initial line and 'a' the radius of the circle. Then the polar equation of the circle is $r = 2a\cos\theta$. Let O be one end of the diameter, C be the midpoint of the arc OCA. Join OC and equation of the line OC is $x - y = 0$.

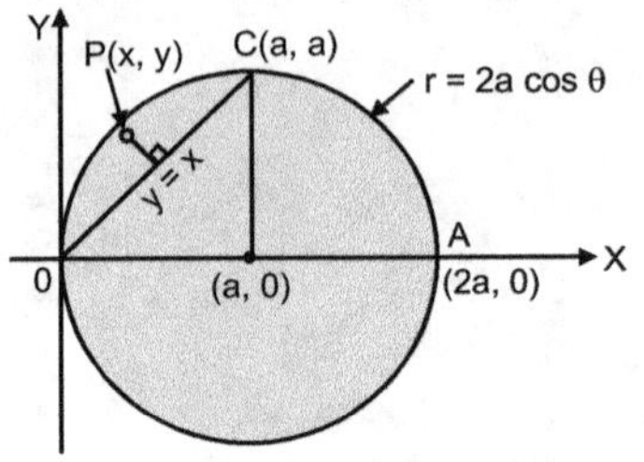

∴   The distance of any point P(x, y) of the area from OC is $p = \left|\dfrac{x-y}{\sqrt{2}}\right|$.

Fig. 11.40

$$\text{M.I.} = \iint p^2\rho\, dx\, dy = \rho\iint \frac{(x-y)^2}{2}\, dx\, dy$$

$$= \frac{\rho}{2}\iint (x^2 + y^2 - 2xy)\, dx\, dy \qquad\text{(Change to polar)}$$

$$= \frac{\rho}{2}\int_0^{\pi/2}\int_0^{2a\cos\theta} (r^2 - 2r^2\sin\theta\cos\theta)\, r\, dr\, d\theta$$

$$= \frac{\rho}{2}\int_0^{\pi/2} (1 - 2\sin\theta\cos\theta)\, d\theta \int_0^{2a\cos\theta} r^3\, dr$$

$$= \frac{\rho}{2}\int_0^{\pi/2} (1 - 2\sin\theta\cos\theta)\, \frac{16a^4\cos^4\theta}{r}\, d\theta$$

$$= 2\rho a^4\int_0^{\pi/2} (\cos^4\theta - 2\cos^5\theta\cdot\sin\theta)\, d\theta$$

$$= 2\rho a^4\left[\frac{3}{4}\,\frac{1}{2}\,\frac{\pi}{2} - 2\cdot\frac{1}{6}\right] = \rho a^4\left[\frac{3\pi}{8} - \frac{2}{3}\right]$$

$$M = \rho\,(\text{Area of the semi-circle}) = \rho\left(\frac{\pi a^2}{2}\right)$$

$$K^2 = \frac{I}{M} = \rho a^4\left(\frac{3\pi}{8} - \frac{2}{3}\right)\frac{2}{\rho\pi a^2} = a^2\left(\frac{3}{4} - \frac{4}{3\pi}\right)$$

$$\boxed{K = a\sqrt{\frac{3}{4} - \frac{4}{3\pi}}}$$

Ex. 13 : *Find the M.I. of a uniform lamina bounded by the x-axis and one arc of the cycloid* $x = a(\theta - \sin\theta)$, $y = a(1 - \cos\theta)$ *from* $\theta = 0$ *to* $\theta = 2\pi$ *about x-axis.*

Sol. : $p = y$ $\qquad$ $\text{M.I.} = \iint p^2\rho\, dx\, dy$

$$\text{M.I.} = \rho\int\int_0^y y^2\, dx\, dy = \frac{\rho}{3}\int_0^{2\pi} y^3\frac{dx}{d\theta}\cdot d\theta$$

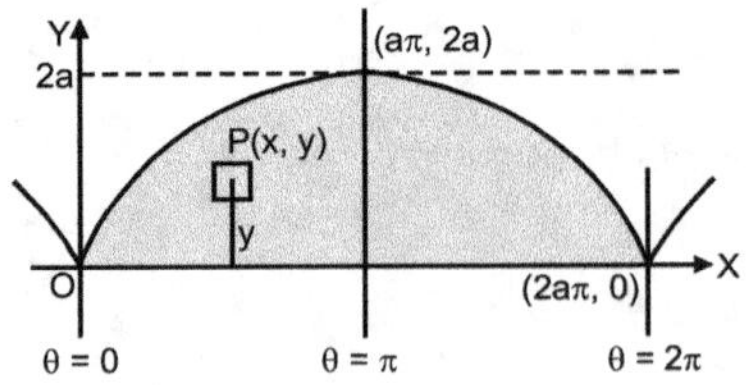

$$= \frac{\rho}{3}\int_0^{2\pi} a^3(1 - \cos\theta)^3\, a(1 - \cos\theta)\, d\theta$$

$$= \frac{\rho a^4}{3}\int_0^{2\pi} 16\sin^8\frac{\theta}{2}\, d\theta$$

$$= \frac{16\rho a^4}{3}\int_0^{\pi} \sin^8 t\, 2dt = \frac{64\rho a^4}{3}\int_0^{\pi/2} \sin^8 t\, dt = \frac{64\rho a^4}{3}\left(\frac{7}{8}\,\frac{5}{6}\,\frac{3}{4}\,\frac{1}{2}\,\frac{\pi}{2}\right)$$

$$= \frac{35\pi\rho a^4}{12}$$

Fig. 11.41

$$\text{Mass} = \iint \rho \, dx \, dy = \rho \int_0^{2\pi} y \, \frac{dx}{d\theta} \cdot d\theta$$

$$= \rho \int_0^{2\pi} a \, (1 - \cos \theta) \, a \, (1 - \cos \theta) \, d\theta = \rho \, a^2 \int_0^{2\pi} 4 \sin^4 \frac{\theta}{2} \, d\theta$$

$$= 4 \, \rho \, a^2 \int_0^{\pi} \sin^4 t \, 2 \, dt = 8 \, \rho \, a^2 \, 2 \int_0^{\pi/2} \sin^4 t \, dt$$

$$= 16 \, \rho \, a^2 \, \frac{3}{4} \frac{1}{2} \frac{\pi}{2} = 3\pi \, \rho \, a^2$$

$$\rho = \frac{M}{3\pi \, a^2}$$

$$\therefore \qquad \text{M.I.} = \frac{35\pi \, a^4}{12} \frac{M}{3\pi \, a^2}$$

$$\boxed{\text{M.I.} = \frac{35}{36} Ma^2}$$

Ex. 14 : *Find the M.I. of the area of the upper half of the circle $x^2 + y^2 = a^2$ about the line $x + y = 2a$.*

Sol. : The perpendicular distance of any point $P(x, y)$ from $x + y = 2a$ is $p = \left| \dfrac{x + y - 2a}{\sqrt{2}} \right|$.

$$\text{M.I.} = \iint p^2 \, \rho \, dx \, dy$$

$$= \rho \iint \frac{(x + y - 2a)^2}{2} \, dx \, dy$$

$$= \frac{\rho}{2} \iint (x^2 + y^2 + 2xy - 4a \, (x + y) + 4a^2) \, dx \, dy$$

$$= \frac{\rho}{2} \int_0^{\pi} \int_0^{a} [r^2 + 2r^2 \sin \theta \cos \theta - 4ar \, (\cos \theta + \sin \theta) + 4a^2] \, r \, dr \, d\theta$$

$$= \frac{\rho}{2} \int_0^{\pi} \left(\frac{a^4}{4} + \frac{a^4}{2} \sin \theta \cos \theta - \frac{4a^4}{3} (\cos \theta + \sin \theta) + 2a^4 \right) d\theta$$

$$= \frac{\rho}{2} \left[\frac{9}{4} a^4 \, (\pi) - \frac{4a^4}{3} (0 + 2) \right]$$

$$\boxed{\text{M.I.} = \rho \, a^4 \left[\frac{9\pi}{8} - \frac{4}{3} \right]}$$

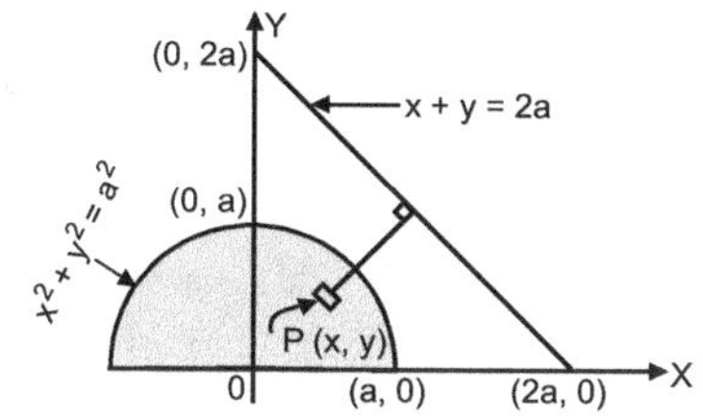

Fig. 11.42

$$\left\{ \begin{array}{l} \because \displaystyle\int_0^{\pi} \sin \theta \cos \theta \, d\theta = 0 \\[2ex] \displaystyle\int_0^{\pi} \cos \theta \, d\theta = 0 \end{array} \right.$$

Ex. 15 : *Find the M.I. and radius of gyration about z-axis of the region in the first octant bounded by* $\dfrac{x}{a} + \dfrac{y}{b} + \dfrac{z}{c} = 1$, *where a, b, c are positive.*

Sol. : Here $p = \sqrt{x^2 + y^2}$. Let ρ be density (mass per unit volume), $dm = \rho \, dx \, dy \, dz$

$$\text{M.I.} = \iiint (x^2 + y^2) \, \rho \, dx \, dy \, dz$$

Put $x = au$, $dx = a \, du$; $y = bv$, $dy = b \, dv$; $z = cw$, $dz = c \, dw$; $\dfrac{x}{a} + \dfrac{y}{b} + \dfrac{z}{c} = 1 \Rightarrow u + v + w = 1$

$$\text{M.I.} = \rho \iiint (a^2 u^2 + b^2 v^2)\, abc\, du\, dv\, dw$$

$$= \rho\, abc \iiint a^2 u^{3-1} v^{1-1} w^{1-1}\, du\, dv\, dw + \rho\, abc \iiint b^2 u^{1-1} v^{3-1} w^{1-1}\, du\, dv\, dw$$

$$= \rho\, abc \left\{ a^2 \frac{\overline{3}\ \overline{1}\ \overline{1}}{\overline{3+1+1+1}} + b^2 \frac{\overline{1}\ \overline{3}\ \overline{1}}{\overline{1+3+1+1}} \right\} \qquad \text{(by Dirichlet's theorem)}$$

$$= \rho\, abc\, \frac{2!}{5!}\, (a^2 + b^2)$$

$$= \frac{\rho\, abc}{60}\, (a^2 + b^2)$$

$$\text{Mass} = \rho\, (\text{volume of tetrahedron})$$

$$= \rho\, \frac{abc}{6}$$

$$\therefore \qquad \text{M.I.} = \frac{M}{10}\, (a^2 + b^2)\ .$$

$$\text{Also} \qquad K^2 = \frac{I}{M} = \frac{a^2 + b^2}{10}$$

$$\therefore \qquad \boxed{K = \sqrt{\frac{a^2 + b^2}{10}}}\ .$$

Ex. 16 : *Find M.I. of a sphere about a diameter.*　　　　　　　　　　*(May 2007)*

Sol. : We shall find the M.I. of the positive octant and multiply the result by 8.

Let the sphere be $x^2 + y^2 + z^2 = a^2$ and the diameter be along z-axis about which M.I. is required. If ρ is the uniform density, the required M.I. $= 8 \iiint (x^2 + y^2)\, \rho\, dx\, dy\, dz$.

$$\text{M.I.} = 8 \int_0^{\pi/2} \int_0^{\pi/2} \int_0^a r^2 \sin^2\theta \cdot \rho \cdot r^2 \sin\theta\, d\theta\, d\phi\, dr$$

$$= 8\rho \int_0^{\pi/2} \sin^3\theta\, d\theta \int_0^{\pi/2} d\phi \int_0^a r^4\, dr$$

$$= 8\rho\, \frac{2}{3}\, \frac{\pi}{2} \left(\frac{a^5}{5} \right)$$

$$= \frac{8\rho\pi a^5}{15} \qquad \text{Also mass is } M = \frac{4}{3}\pi a^3 \rho$$

$$\therefore \qquad \rho = \frac{3M}{4\pi a^3}$$

$$\text{M.I.} = \frac{8\pi a^5}{15} \cdot \frac{3M}{4\pi a^3}$$

$$\boxed{\text{M.I.} = \frac{2}{5} Ma^2}$$

Ex. 17 : *The mass of a solid right circular cylinder of radius 'a' and height 'h' is M. Find the M.I. of the cylinder about (i) its axis, (ii) a line through its C.G. perpendicular to its axis, (iii) any diameter through its base.*

Sol. : (i) Let OA be the axis of rotation. Divide the cylinder into a large number of thin circular discs parallel to the base and consider a typical disc of breadth 'dx' at distance x from the base.

The mass of disc is dm $= \pi a^2\, dx \cdot \rho$, where ρ is the density.

But M.I. of disc of mass dm about line through centre perpendicular to the plane of

disc $= dm\,\dfrac{a^2}{2}$ or $\dfrac{Ma^2}{2} = (\pi a^2 \rho\, dx)\,\dfrac{a^2}{2} = \dfrac{\pi a^4 \rho}{2}\, dx$

(Refer standard result No. 4).

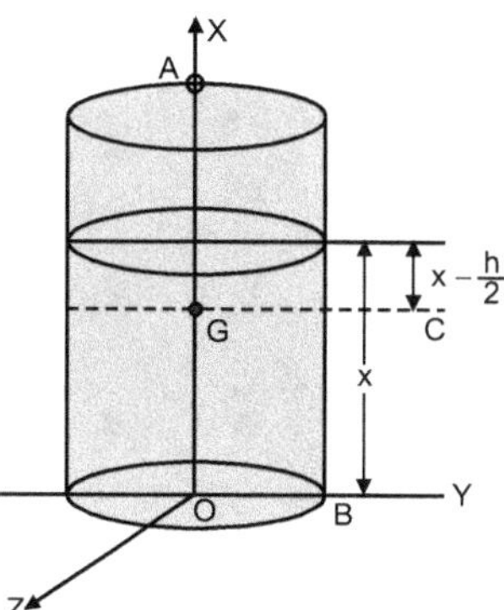

$$\text{Hence M.I. of cylinder about axis of cylinder} = \frac{\pi a^4 \rho}{2} \int_0^h dx \quad \text{(where, h is the height of cylinder)}$$

$$= \frac{\pi a^4 \rho\, h}{2}$$

Fig. 11.43

But $\qquad M = \text{mass of the cylinder} = \pi a^2 h \rho \quad \therefore\ \rho = \dfrac{M}{\pi a^2 h}$

$$\text{M.I.} = \frac{\pi a^4 h}{2}\,\frac{M}{\pi a^2 h} = \frac{Ma^2}{2}$$

(ii) Let G be the C.G. of the cylinder. Then G is at the middle point of axis OA. Divide the cylinder into circular discs as in (i). The mass dm of a typical disc at a distance x from the base is $\pi a^2 \rho\, dx$.

$\therefore$ The M.I. of the disc about a diameter is $(\pi a^2 \rho\, dx)\,\dfrac{a^2}{4}$ and so about the line GC through G parallel to base is

$$= \frac{\pi a^4 \rho}{4}\, dx + (\pi a^2 \rho\, dx)\left(x - \frac{h}{2}\right)^2 \qquad \text{(Parallel axes theorem)}$$

$$= \pi a^2 \rho\, dx \left[\frac{a^2}{4} + \left(x - \frac{h}{2}\right)^2\right]$$

Hence M.I. of the cylinder about GC

$$= \pi a^2 \rho \int_0^h \left[\frac{a^2}{4} + \left(x - \frac{h}{2}\right)^2\right] dx$$

$$= \pi a^2 \rho \left\{\frac{a^2}{4} h + \left[\frac{\left(x - \frac{h}{2}\right)^3}{3}\right]_0^h\right\} = \pi a^2 \rho\, h \left[\frac{3a^2 + h^2}{12}\right] = M\left(\frac{3a^2 + h^2}{12}\right) \qquad \left(\because\ M = \pi a^2 h \rho\right)$$

(iii) The M.I. about a diameter of the base may now be deduced from this result by the theorem of parallel axes. The distance between these axes being $\dfrac{h}{2}$,

the required $\qquad \text{M.I.} = M\left(\dfrac{3a^2 + h^2}{12}\right) + M\left(\dfrac{h}{2}\right)^2$

$$\boxed{\text{M.I.} = \frac{M}{12}\,(3a^2 + 4h^2)}$$

Ex. 18 : *The mass of a solid right circular cone is M, the radius of its base is a and its height is h. Find the moment of inertia of the cone about (i) the axis of the cone, (ii) the axis through the vertex perpendicular to the axis of the cone, (iii) a line through the centre of gravity perpendicular to the axis of the cone.*

Sol. : (i) Take a disc at a distance x from the vertex V and thickness dx. Let y be the radius of this disc.

The mass of this disc is $\rho \pi y^2\, dx$; and its moment of inertia about the line OV by the standard results is

$$(\rho \pi y^2\, dx)\,\frac{y^2}{2} = \frac{\pi \rho}{2}\, y^4\, dx \qquad\qquad \text{...(i)}$$

By similarity of the triangles,

$$\frac{y}{x} = \frac{a}{h} \quad \text{or} \quad y = \frac{a}{h}x \qquad \text{... (ii)}$$

Substituting from (ii) for y in (i), the M.I. of the elementary disc about OV is

$$\frac{\pi\rho}{2} \cdot \frac{a^4}{h^4} x^4 \, dx$$

∴ The M.I. of the cone about its axis is

$$\frac{\pi\rho}{2} \frac{a^4}{h^4} \int_0^h x^4 \, dx = \frac{\pi\rho a^4 h}{10} \qquad \text{... (iv)}$$

The mass M of the cone is

$$M = \frac{1}{3}\pi a^2 h \rho \quad \text{or} \quad \rho = \frac{3M}{\pi a^2 h}$$

Substituting this value of ρ in (iv), we have the moment of inertia I_1 of the cone about its axis as :

$$I_1 = \frac{\pi a^4 h}{10} \cdot \frac{3M}{\pi a^2 h} = \frac{3}{10} Ma^2 \qquad \text{... (vi)}$$

(ii) For this part, we use the result that the moment of inertia of a disc about its diameter is $\frac{Ma^2}{4}$ and the parallel axes theorem.

Moment of inertia of the elementary disc about its diameter

$$= (\pi y^2 \rho \, dx) \frac{y^2}{4}$$

By the parallel axes theorem, the moment of inertia of the same disc about a line through the vertex V parallel to the diameter of the disc, and hence perpendicular to the axis of the cone, which is at a distance x from the diameter is

$$(\pi y^2 \rho \, dx)\left(\frac{y^2}{4} + x^2\right) = \pi\rho \, dx \left[\frac{y^4}{4} + y^2 x^2\right] \qquad \text{... (vii)}$$

$$= \frac{\pi\rho a^2}{h^2}\left[\frac{a^2}{4h^2} + 1\right] x^4 \, dx \quad \left(\text{as } y = \frac{a}{h}x\right) \qquad \text{... (viii)}$$

The total moment of inertia I_2 of the cone about a line through the vertex perpendicular to the axis of the cone is obtained by integrating the result (viii).

Thus, $$I_2 = \frac{\pi\rho a^2}{h^2}\left[\frac{a^2}{4h^2} + 1\right] \int_0^h x^4 \, dx = \frac{\pi\rho a^2 h}{20}[a^2 + 4h^2]$$

$$\boxed{I_2 = \frac{3}{20} M (a^2 + 4h^2)} \text{ , substituting for } \rho \text{ from (v)} \qquad \text{... (ix)}$$

(iii) To find the M.I. about an axis through the centre of gravity G, perpendicular to the axis of the cone, we again use the parallel axes theorem. G divides VO in the ratio of 3 : 1.

Therefore the distance between the parallel axes through V and G is $\frac{3}{4}$ h.

By parallel axes theorem,

$$I_2 = I_G + M \cdot \frac{9}{16} h^2$$

∴ $$I_G = I_2 - M \frac{9}{16} h^2$$

$$= M \frac{3}{20} (4h^2 + a^2) - M \cdot \frac{9}{16} h^2$$

$$\boxed{I_G = \frac{3}{80} M (h^2 + 4a^2)} \qquad \text{... (x)}$$

EXERCISE 11.2

1. Find the M.I. of the area enclosed by $x^{2/3} + y^{2/3} = a^{2/3}$ about x-axis. **Ans.** $\dfrac{7}{64} Ma^2$

2. Find the M.I. of an arc in the form of cardioide $r = a (1 + \cos \theta)$ about the axis through the pole perpendicular to the plane of the arc. **(May 2014) Ans.** $\dfrac{256a^3 \rho}{15}$

3. Find the M.I. of a square lamina of side 'a' about one of its diagonals, density at any point (x, y) varies as the square of its distance from the diagonal. **(Dec. 09) Ans.** $\dfrac{Ma^2}{5}$

4. Find the M.I. of a quadrant of a circle of radius a about a line parallel to its bounding radius at a distance d (d > a) from it.

    Ans. $\rho \, a^2 \left(\dfrac{\pi d^2}{4} - \dfrac{2ad}{3} + \dfrac{a^2 \pi}{16} \right)$

5. Find the M.I. of a thin circular wire in the form of arc of a circle of a radius a which subtends an angle 2α at the centre O about :
   (i) an axis through its middle point perpendicular to its plane, (ii) the radius bisecting the angle 2α.

    Ans. (i) $2a^2 M \left(1 - \dfrac{\sin \alpha}{\alpha} \right)$, (ii) $\dfrac{Ma^2}{2} \left(1 - \dfrac{\sin 2\alpha}{2\alpha} \right)$

6. Find the M.I. about x-axis of the area under the curve $y = \sin x$ from $x = 0$ to $x = \dfrac{\pi}{2}$. **Ans.** $\dfrac{2M}{9}$

7. Find the M.I. of the area enclosed by $r = a (1 - \cos \theta)$ about the line $\theta = \dfrac{\pi}{2}$. **Ans.** $\dfrac{49\pi a^4}{32}$

8. The surface density at a point of the area of the lemniscate $r^2 = a^2 \cos 2\theta$ varies as the square of its distance from the pole. Find the moment of inertia about an axis through the pole perpendicular to its plane.

    [**Hint :** $\rho = kr^2$, $p = r$.] **Ans.** $I = \dfrac{16Ma^2}{9\pi}$

9. Find the M.I. about the polar axis of the area enclosed by $r = a (1 + \cos \theta)$ in the upper half. **Ans.** $\dfrac{35\pi\rho a^4}{32}$

10. Find the M.I. of one loop of $r^2 = a^2 \sin 2\theta$ about an axis perpendicular to its plane. **Ans.** $\dfrac{\rho \pi a^4}{16}$

11. A thin plate of uniform thickness and density is in the form of lamina bounded by the parabola $x^2 = y$ and the line $y = x + 2$. Find the M.I. of the lamina about co-ordinate axis. **(Dec. 2010) Ans.** $\dfrac{47M}{14}$, $\dfrac{7M}{10}$

12. Find the M.I. of the area bounded by $x^2 + y^2 - ax = 0$, $x^2 + y^2 - 2ax = 0$ about the axis through origin perpendicular to XOY plane if density varies as distance from the origin. **Ans.** $\dfrac{496 \lambda a^5}{75}$

13. Find the radius of gyration for the area of the cardioide $r = a (1 + \cos \theta)$ about the axis perpendicular to its plane through the pole when the density at any point varies as the distance of the point from the pole. **Ans.** $\dfrac{3\sqrt{21} \, a}{10}$

14. Prove that the M.I. of the disc of radius a about an axis through its centre and perpendicular to its plane is $\dfrac{3}{5} Ma^2$, where M is the mass of the disc, the density being proportional to the distance from the axis.

15. Prove that the M.I. of a lamina of mass M in the form of a right-angled triangle, having hypotenuse of length 'c' about the axis through the vertex containing the right angle perpendicular to the plane of the lamina is $\dfrac{M}{6} c^2$.

    Hint : $p = r$ and use $a^2 + b^2 = c^2$ Limits for $x : 0$ to a, $y : 0$ to $\dfrac{b}{a} (a - x)$

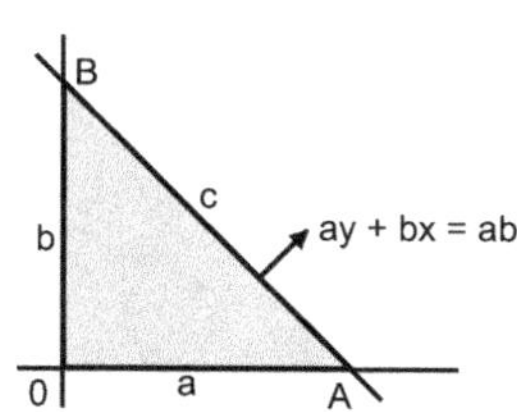

Fig. 11.45

16. Find the M.I. of a loop of the curve $r^2 = a^2 \cos 2\theta$ about a line through the pole perpendicular to its plane.   **Ans.** $\dfrac{M}{8}\pi a^2$

17. A lamina in the first quadrant is bounded by $x = 0$, $y = 0$ and $x^2 + y^2 = a^2$ and $\rho = \dfrac{\lambda x^2 y}{a^3}$ where λ is constant. Find the co-ordinates of C.G. and M.I. of the lamina about z-axis.   **(Dec. 2011) Ans.** $\left(\dfrac{5a}{8}, \dfrac{5\pi a}{32}\right)$, $\dfrac{5Ma^2}{7}$

18. Find the M.I. and radius of gyration of the area bounded by, $y = 20x\,(1-x)$ and the line $y = 0$ about (i) OX, (ii) OY.

$$\textbf{Ans. } \text{(i) } \frac{40}{7}\,M,\ \sqrt{\frac{40}{7}}\ ,\ \text{(ii) } \frac{3M}{10}\ ,\ \sqrt{\frac{3}{10}}$$

19. Find the M.I. of the area of the parabola $y^2 = 4ax$ bounded by a double ordinate at a distance h from the vertex about (i) an axis through the centroid perpendicular to the axis of the parabola, (ii) the tangent at the vertex, (iii) the axis.

    Hint :   A = vertex, BC = double ordinate, $x = h$

$$= \text{C.G. of area BAC,} \quad \bar{x} = \frac{3}{5}\,h$$

$$= \frac{8}{3}\rho\,h\sqrt{ah} \quad .I. = \rho \int_{0}^{h} \int_{-2\sqrt{ax}}^{2\sqrt{ax}} \left(x - \frac{3h}{5}\right)^2 dx\,dy$$

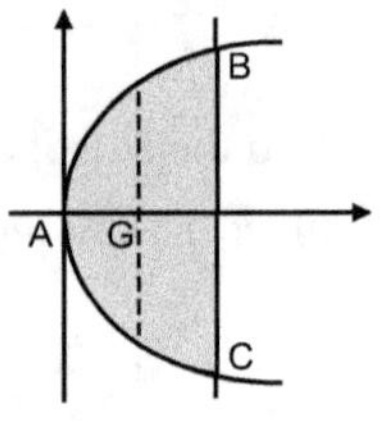

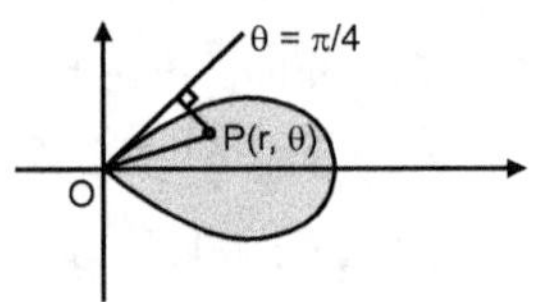

Fig. 11.46

$$\textbf{Ans. } \text{(i) } \frac{12Mh^2}{175}\ ,\ \text{(ii) tangent at vertex} = \text{y-axis}, \frac{3}{7}\,Mh^2,\ \text{(iii) } \frac{4}{5}\,Mah.$$

20. Find the M.I. of a loop of Bernoulli's Lemniscate $r^2 = a^2 \cos 2\theta$ about the tangent at the pole.

    [**Hint :** tangent at pole is $\theta = \dfrac{\pi}{4}$, $p = r\sin\left(\dfrac{\pi}{4} - \theta\right) = \dfrac{r}{\sqrt{2}}(\cos\theta - \sin\theta)$]   **Ans.** $\dfrac{\pi a^4}{32}$

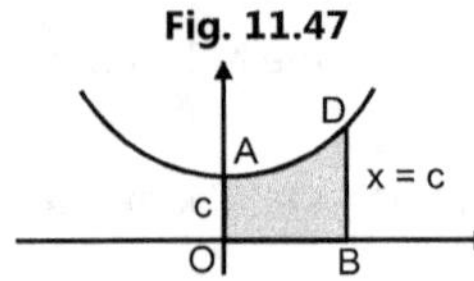

Fig. 11.47

21. An area is bounded by the curve $y = c\cosh\dfrac{x}{c}$, the axes and the ordinate $x = c$. Calculate the radius of gyration about the y-axis.

    [**Hint :** Use $\sinh 1 = \dfrac{e - e^{-1}}{2}$, $\cosh 1 = \dfrac{e + e^{-1}}{2}$]

Fig. 11.48

$$\textbf{Ans. } Mc^2\left(\frac{e^2 - 5}{e^2 - 1}\right)$$

CHAPTER 1 : FIRST ORDER ORDINARY DIFFERENTIAL EQUATIONS

Order, Degree and Formation of Differential Equation :   **Marks**

1. The differential equation $1 + \dfrac{dy}{dx} - \left(\dfrac{d^2y}{dx^2}\right)^{3/2} = 0$ is of … (1)

   (A) order 1 and degree 2   (B) order 2 and degree 3
   (C) order 3 and degree 6   (D) order 3 and degree 3

2. The differential equation $\sqrt{1 + \dfrac{dy}{dx}} = \dfrac{d^2y}{dx^2}$ is of … (1)

   (A) order 2 and degree 2   (B) order 1 and degree 2
   (C) order 2 and degree 1   (D) order 1 and degree 1

3. The differential equation $x = \dfrac{1}{\sqrt{1 + \dfrac{dy}{dx} + \dfrac{d^2y}{dx^2}}}$ is of … (1)

   (A) order 2 and degree 2   (B) order 1 and degree 2
   (C) order 2 and degree 1   (D) order 1 and degree 1

4. The differential equation $1 + \dfrac{dy}{dx} = \dfrac{y}{\dfrac{dy}{dx}}$ is of … (1)

   (A) order 2 and degree 2   (B) order 1 and degree 1
   (C) order 2 and degree 1   (D) order 1 and degree 2

5. The differential equation $(2x - y + 3)\,dx + (y - 2x - 2)\,dy = 0$ is of … (1)

   (A) order 1 and degree 1   (B) order 1 and degree 2
   (C) order 2 and degree 1   (D) order 2 and degree 2

6. The number of arbitrary constants in the general solution of ordinary differential equation is equal to … (1)

   (A) the order of differential equation
   (B) the degree of differential equation
   (C) coefficient of highest order differential coefficient
   (D) none of these

7. The general solution of nth order ordinary differential equation must involve … (1)

   (A) $(n + 1)$ arbitrary constants
   (B) $(n - 1)$ arbitrary constants
   (C) n arbitrary constants
   (D) none of these

8. The solution obtained by assigning particular values to arbitrary constants in general solution of ordinary differential equation is called … (1)

   (A) the general solution   (B) a particular solution
   (C) a singular solution   (D) none of these

9. The order of differential equation whose general solution is $y = C_1 + C_2 e^{-2x} + C_3 e^{3x} + C_4 e^{-3x}$, where C_1, C_2, C_3, C_4 are arbitrary constants, is … (1)

   (A) 1   (B) 3
   (C) 2   (D) 4

10. The order of differential equation whose general solution is $y = e^x (Ax^2 + Bx + C)$, where A, B, C are arbitrary constants, is … (1)

    (A) 2   (B) 4
    (C) 3   (D) 1

11. The differential equation whose general solution is $y = \sqrt{5x + C}$, where C is arbitrary constant, is … (1)

    (A) $2y\dfrac{dy}{dx} - 1 = 0$ (B) $2y\dfrac{dy}{dx} - 5 = 0$

    (C) $\dfrac{dy}{dx} - \dfrac{5}{2}\dfrac{1}{\sqrt{5x + C}} = 0$ (D) $y\dfrac{dy}{dx} - 5 = 0$

12. $y = Cx - C^2$, where C is arbitrary constant is the general solution of the differential equation … (1)

    (A) $\dfrac{dy}{dx} = C$ (B) $\left(\dfrac{dy}{dx}\right)^2 + xy = 0$

    (C) $\left(\dfrac{dy}{dx}\right)^2 + x\dfrac{dy}{dx} - y = 0$ (D) $\left(\dfrac{dy}{dx}\right)^2 - x\dfrac{dy}{dx} + y = 0$

13. The differential equation whose general solution is $y = C^2 + \dfrac{C}{x}$, where C is arbitrary constant is … (1)

    (A) $x^4 y_1^2 + xy_1 - y = 0$ (B) $x^4 y_1^2 - xy_1 - y = 0$

    (C) $x^2 y_1^2 - xy_1 - y = 0$ (D) $y_1 = -\dfrac{C}{x^2}$

14. By eliminating arbitrary constant A the differential equation whose general solution is $y = 4(x - A)^2$ (1)

    (A) $y_1^2 + 16y = 0$ (B) $y_1 - 2y = 0$

    (C) $y_1^2 - 16y = 0$ (D) $y_1 - 8(x - A) = 0$

15. The differential equation whose general solution is $y = A\cos(x + 3)$, where A is arbitrary constant, is … (1)

    (A) $\cot(x + 3)\,y_1 + y = 0$ (B) $\tan(x + 3)\,y_1 + y = 0$
    (C) $\cot(x + 3)\,y_1 - y = 0$ (D) $\tan(x + 3)\,y_1 - y = 0$

16. By eliminating arbitrary constant a the differential equation whose general solution is $y^2 = 4ax$ is … (1)

    (A) $xy\dfrac{dy}{dx} - y^2 = 0$ (B) $2xy\dfrac{dy}{dx} + y^2 = 0$

    (C) $2xy\dfrac{dy}{dx} - y^2 = 0$ (D) $8xy\dfrac{dy}{dx} - y^2 = 0$

17. The differential equation whose general solution is $xy = C^2$, where C is arbitrary constant, is … (1)

    (A) $xy_1 - y = 0$ (B) $xy_2 + y_1 = 0$
    (C) $xy_1 = C^2$ (D) $xy_1 + y = 0$

18. The differential equation representing the family of curves $y^2 = 2C\left(x + \sqrt{C}\right)$, where C is arbitrary constant, is … (1)

    (A) $2yy_1\left(x + \sqrt{yy_1}\right) - y^2 = 1$

    (B) $2y_1\left(x + \sqrt{yy_1}\right) - y = 0$

    (C) $y = 2y_1\left(x + \sqrt{C}\right)$ (D) $y_1\left(x + \sqrt{yy_1}\right) - y = 0$

19. By eliminating arbitrary constant A the differential equation whose general solution is $y = Ae^{-x^2}$ is ... (1)

(A) $\dfrac{dy}{dx} - 2xy = 0$

(B) $y\dfrac{dy}{dx} - 2x = 0$

(C) $\dfrac{dy}{dx} + 2xy = 0$

(D) $y\dfrac{dy}{dx} + 2x = 0$

20. $y = mx$ where m is arbitrary constant is the general solution of the differential equation is ... (1)

(A) $\dfrac{dy}{dx} = \dfrac{y}{x}$

(B) $\dfrac{dy}{dx} = \dfrac{x}{y}$

(C) $\dfrac{dy}{dx} = m$

(D) $\dfrac{dy}{dx} = -\dfrac{y}{x}$

21. The differential equation representing the family of curves $y = 3 + \sqrt{Cx}$, where C is arbitrary constant, is ... (2)

(A) $y = 3 + 2x\dfrac{dy}{dx}$

(B) $y = 3 + 2\sqrt{x}\dfrac{dy}{dx}$

(C) $y = 2x\dfrac{dy}{dx}$

(D) $\dfrac{dy}{dx} = \dfrac{\sqrt{c}}{2\sqrt{x}}$

22. The differential equation satisfied by general solution $\dfrac{x^2}{a^2} + \dfrac{y^2}{4} = 1$, where a is arbitrary constant, is ... (2)

(A) $xyy_1 - y + 4 = 0$

(B) $xyy_1 + y^2 - 4 = 0$

(C) $x^2yy_1 + y^2x - 1 = 0$

(D) $xyy_1 - y^2 + 4 = 0$

23. The differential equation representing the family of curves $x^2 + y^2 = 2Ax$, where A is arbitrary constant, is ... (2)

(A) $y_1 = \dfrac{y^2 + x^2}{2xy}$

(B) $y_1 = \dfrac{y^2 - x^2}{2xy}$

(C) $y_1 = \dfrac{y^2 - x^2}{2y}$

(D) $y_1 = \dfrac{2xy}{y^2 - x^2}$

24. $y^2 = C(4 + e^{2x})$ where C is arbitrary constant is the general solution of the differential equation ... (2)

(A) $(4 + e^{2x})\dfrac{dy}{dx} - ye^{2x} = 0$

(B) $y\dfrac{dy}{dx} - e^{2x}(4 + e^{2x}) = 0$

(C) $e^{2x}\dfrac{dy}{dx} - y^2 e^{2x} = 0$

(D) $y(4 + e^{2x})\dfrac{dy}{dx} - e^{2x} = 0$

25. $\sin(y - x) = Ce^{-\frac{x^2}{2}}$ where C is arbitrary constant is the general solution of the differential equation ... (2)

(A) $\tan(y - x)\left[\dfrac{dy}{dx} - 1\right] + x = 0$

(B) $\cot(y - x)\dfrac{dy}{dx} - 1 + x = 0$

(C) $\cot(y - x)\left[\dfrac{dy}{dx} - 1\right] + x = 0$

(D) $\cot(y - x)\left[\dfrac{dy}{dx} - 1\right] = 0$

26. By eliminating arbitrary constant A the differential equation whose general solution is $(1 + x^2) = A(1 - y^2)$ is ... (2)

(A) $\dfrac{(1 + x^2)}{(1 - y^2)}\dfrac{dy}{dx} + \dfrac{x}{y} = 0$

(B) $\dfrac{(1 - y^2)}{(1 + x^2)}\dfrac{dy}{dx} + \dfrac{x}{y} = 0$

(C) $\dfrac{(1 + x^2)}{(1 - y^2)}\dfrac{dy}{dx} + \dfrac{y}{x} = 0$

(D) $\dfrac{(1 - y^2)}{(1 + x^2)}\dfrac{dy}{dx} - \dfrac{x}{y} = 0$

27. The differential equation satisfied by general solution $x = Cy - y^2$, where C is arbitrary constants, is ... (2)

(A) $\left(\dfrac{y}{x + y^2}\right)y_1 - 2yy_1 - 1 = 0$

(B) $\left(\dfrac{x + y^2}{y}\right)y_1 - 2yy_1 - 1 = 0$

(C) $\left(\dfrac{x + y^2}{y}\right)y_1 + 2yy_1 + 1 = 0$

(D) $\left(\dfrac{x + y^2}{y}\right)y_1 + 2yy_1 = 0$

28. The differential equation satisfied by general solution $y + x^3 = Cx$, where C is arbitrary constants, is ... (2)

(A) $\dfrac{dy}{dx} + 3x^2 = C$

(B) $x\dfrac{dy}{dx} + 2x^2 - y = 0$

(C) $\dfrac{dy}{dx} + 2x^2 - y = 0$

(D) $x\dfrac{dy}{dx} + 2x^3 - y = 0$

29. $xy + y^2 - x^2 - 3y - x = C$, where C is arbitrary constant is the general solution of the differential equation ... (2)

(A) $\dfrac{dy}{dx} = \dfrac{2x - y + 1}{x + 2y - 3}$

(B) $\dfrac{dy}{dx} = \dfrac{2x + 1}{x + 2y - 3}$

(C) $\dfrac{dy}{dx} = \dfrac{y - 2x - 1}{x - 2y + 3}$

(D) $\dfrac{dy}{dx} = \dfrac{x + 2y - 3}{2x - y + 1}$

30. $y = Ce^{x/y}$, where C is arbitrary constant is the general solution of the differential equation ... (2)

(A) $yy_1 + (y - xy_1) = 0$

(B) $yy_1 - (y - xy_1) = 0$

(C) $y^2y_1 - (y - xy_1) = 0$

(D) $\dfrac{dy}{dx} - 1 = 0$

31. $\sin\left(\dfrac{y}{x}\right) = Cx$, where C is arbitrary constant is the general solution of the differential equation ... (2)

(A) $xy_1 + y = x\tan\left(\dfrac{y}{x}\right)$

(B) $xy_1 - y = x\cot\left(\dfrac{y}{x}\right)$

(C) $xy_1 - y = x\tan\left(\dfrac{x}{y}\right)$

(D) $xy_1 - y = x\tan\left(\dfrac{y}{x}\right)$

32. By eliminating arbitrary constant A the differential equation whose general solution is $y^2 = x^2 - 1 + Ax$ is ... (2)

(A) $2xy\dfrac{dy}{dx} = x^2 + y^2 + 1$

(B) $2xy\dfrac{dy}{dx} = x^2 + x + y^2 + 1$

(C) $2xy\dfrac{dy}{dx} = y^2 + 1$

(D) $2y\dfrac{dy}{dx} = 2x + A$

33. The differential equation satisfied by general solution $y = A\cos x + B\sin x$, where A and B are arbitrary constants, is ... (2)

(A) $\dfrac{d^2y}{dx^2} + \dfrac{dy}{dx} = B\sin x$

(B) $\dfrac{d^2y}{dx^2} - y = 0$

(C) $\dfrac{d^2y}{dx^2} + \dfrac{dy}{dx} = A\cos x$

(D) $\dfrac{d^2y}{dx^2} + y = 0$

34. The differential equation satisfied by general solution $y = A \cos \dfrac{2x}{3} + B \sin \dfrac{2x}{3}$, where A and B are arbitrary constants, is … (2)

(A) $\dfrac{d^2y}{dx^2} + \dfrac{9}{4}y = 0$ (B) $\dfrac{d^2y}{dx^2} - \dfrac{4}{9}y = 0$

(C) $\dfrac{d^2y}{dx^2} + \dfrac{4}{9}y = 0$ (D) $\dfrac{d^2y}{dx^2} - \dfrac{9}{4}y = 0$

35. The differential equation satisfied by general solution $y = A \cos(\log x) + B \sin(\log x)$ where A and B are arbitrary constants, is … (2)

(A) $x\dfrac{d^2y}{dx^2} + \dfrac{dy}{dx} + y = 0$ (B) $x^2\dfrac{d^2y}{dx^2} + x\dfrac{dy}{dx} + y = 0$

(C) $x^2\dfrac{d^2y}{dx^2} + x\dfrac{dy}{dx} - y = 0$ (D) $x^2\dfrac{d^2y}{dx^2} + y = 0$

36. The differential equation satisfied by general solution $y = Ae^x + Be^{-x}$, where A and B are arbitrary constants, is … (2)

(A) $y_2 - y = 0$ (B) $y_2 + y = 0$

(C) $y_2 + y = Ae^x - Be^{-x}$ (D) $y_2 - y = 2Ae^x$

37. The differential equation satisfied by general solution $xy = Ae^x + Be^{-x}$, where A and B are arbitrary constants, is … (2)

(A) $xy_2 + 2y_1 + xy = 0$ (B) $xy_2 - 2y_1 + xy = 0$

(C) $xy_2 + 2y_1 - xy = 0$ (D) $xy_2 + y_1 - xy = 0$

38. The differential equation satisfied by general solution $x = A \cos(2t + B)$, where A and B are arbitrary constants, is … (2)

(A) $\dfrac{d^2x}{dt^2} + 4x = 0$ (B) $\dfrac{d^2x}{dt^2} - 2x = 0$

(C) $\dfrac{d^2x}{dt^2} - 4x = 0$ (D) $\dfrac{d^2x}{dt^2} + x = 0$

39. By eliminating arbitrary constants A and B the differential equation whose general solution is $e^{-t} x = (A + Bt)$ is … (2)

(A) $x_2 + 2x_1 - x = 0$ (B) $x_2 - 2x_1 + x = 0$

(C) $x_2 - x_1 + x = 0$ (D) $x_2 + x = 0$

40. By eliminating arbitrary constants A and B the differential equation whose general solution is $y^2 = 4A(x - B)$ is … (2)

(A) $y_2 + y_1^2 = 0$ (B) $yy_2 + y_1 = 0$

(C) $yy_2 - y_1^2 = 0$ (D) $yy_2 + y_1^2 = 0$

41. The differential equation of family of circles having their centres at (A, 5) and radius 5, where A is arbitrary constant is … (2)

(A) $(y - 5)^2 \left\{1 + \dfrac{dy}{dx}\right\} = 5$

(B) $(y - 5)^2 \left\{1 - \left(\dfrac{dy}{dx}\right)^2\right\} = 25$

(C) $(y - 5)^2 \left\{1 + \left(\dfrac{dy}{dx}\right)^2\right\} = 25$

(D) none of these

42. The differential equation of family of circles having their centres at origin and radius a where a is arbitrary constant, is … (2)

(A) $x - y\dfrac{dy}{dx} = 0$ (B) $x + y\dfrac{dy}{dx} = 0$

(C) $x\dfrac{dy}{dx} + y = 0$ (D) $x + y\dfrac{dy}{dx} = \dfrac{a^2}{2}$

43. By eliminating arbitrary constants A and B the differential equation whose general solution is $(x - A)^2 = 4(y - B)$ is … (2)

(A) $2\dfrac{dy}{dx} - (x - A) = 0$ (B) $\dfrac{d^2y}{dx^2} + \dfrac{1}{2} = 0$

(C) $\dfrac{d^2y}{dx^2} - 2 = 0$ (D) $\dfrac{d^2y}{dx^2} - \dfrac{1}{2} = 0$

44. The differential equation satisfied by general solution $y = A \cos 4x + B \sin 4x + C$, where A, B and C are arbitrary constants, is … (2)

(A) $\dfrac{d^2y}{dx^2} - 16y = 0$ (B) $\dfrac{d^3y}{dx^3} - 16\dfrac{dy}{dx} = 0$

(C) $\dfrac{d^3y}{dx^3} + 16\dfrac{dy}{dx} = 0$ (D) $\dfrac{d^3y}{dx^3} + \dfrac{dy}{dx} = 0$

45. The differential equation satisfied by general solution $y = Ax^2 + Bx + C$, where A, B and C are arbitrary constants, is … (2)

(A) $\dfrac{d^3y}{dx^3} = 0$ (B) $\dfrac{d^2y}{dx^2} = 2A$

(C) $\dfrac{d^3y}{dx^3} = A$ (D) $\dfrac{d^4y}{dx^4} = 0$

ANSWERS

1. (B)	2. (A)	3. (C)	4. (D)	5. (A)	6. (A)	7. (C)	8. (B)
9. (D)	10. (C)	11. (B)	12. (D)	13. (B)	14. (C)	15. (A)	16. (C)
17. (D)	18. (B)	19. (C)	20. (A)	21. (A)	22. (D)	23. (B)	24. (A)
25. (C)	26. (A)	27. (B)	28. (D)	29. (A)	30. (B)	31. (D)	32. (A)
33. (D)	34. (C)	35. (B)	36. (A)	37. (C)	38. (A)	39. (B)	40. (D)
41. (C)	42. (B)	43. (D)	44. (C)	45. (A)			

•••

Differential Equations in Variable Separable Form and Reducible to Variable Separable Form :   Marks

1. The differential equation $\dfrac{dy}{dx} = e^{x-y} + 3x^2e^{-y}$ is of the form … (1)

(A) variable separable   (B) homogeneous

(C) linear   (D) exact

2. For solving the differential equation $(x + y + 1)\,dx + (2x + 2y + 4)\,dy = 0$ appropriate substitution is … (1)

(A) $x + y = 1$ (B) $x + y = u$

(C) $x - y = u$ (D) none of these

3. The differential equation $\dfrac{dy}{dx} = \dfrac{x^3 - 3xy^2}{y^3 - 3x^2y}$ is of the form ...   (1)

   (A) variable separable   (B) homogeneous

   (C) linear   (D) exact

4. The differential equation $\dfrac{dy}{dx} = \dfrac{x + 2y - 3}{3x + 6y - 1}$ is of the form ...   (1)

   (A) variable separable   (B) exact

   (C) non-homogeneous   (D) homogeneous

5. The solution of differential equation $\dfrac{dy}{dx} + y = 0$ is ...   (1)

   (A) $y = Ae^{-x}$ (B) $y = Ae^{x}$

   (C) $x = Ae^{-y}$ (D) $x = Ae^{y}$

6. The solution of differential equation $\dfrac{dy}{dx} + x = 0$ is ...   (1)

   (A) $x + y^2 = C$ (B) $x + y = C$

   (C) $x^2 + y = C$ (D) $x^2 + 2y = C$

7. The solution of differential equation $ydx + xdy = 0$ is ...   (1)

   (A) $x^2y = C$ (B) $xy = C$

   (C) $xy^2 = C$ (D) $xy + 1 = C$

8. The solution of differential equation $\dfrac{dy}{dx} + \tan x = 0$ is ...   (1)

   (A) $y + \log \sin x = C$ (B) $y + \sec^2 x = C$

   (C) $y - \log \cos x = C$ (D) $y + \log \cot x = C$

9. The solution of differential equation $\dfrac{dy}{dx} = \dfrac{1 + y}{1 + x}$ is ...   (1)

   (A) $(1 + y) = C(1 + x)$ (B) $(1 + x) = \dfrac{C}{(1 + y)}$

   (C) $xy(1 + y) = C$ (D) $(1 + y)^2 = C(1 + x)$

10. The solution of differential equation $\dfrac{dy}{dx} + \dfrac{1 + y^2}{1 + x^2} = 0$ is ...   (2)

   (A) $\tan^{-1} y - \tan^{-1} x = C$ (B) $\tan^{-1} y + \tan^{-1} x = C$

   (C) $\tan y + \tan x = C$ (D) $\cos y + \cos x = C$

11. The solution of differential equation $(4 + e^{2x}) \dfrac{dy}{dx} = ye^{2x}$ is ...   (2)

   (A) $y^2 = (4 + e^{2x}) C$ (B) $y = (4 + e^{2x}) C$

   (C) $y(4 + e^{2x}) = C$ (D) $y^2(4 + e^{2x}) = C$

12. The solution of differential equation $y - x\dfrac{dy}{dx} = 2\left(y + \dfrac{dy}{dx}\right)$ is ...   (2)

   (A) $y + (x + 2) = C$ (B) $y - (x + 2) = C$

   (C) $y = C(x + 2)$ (D) $y(x + 2) = C$

13. The solution of differential equation $xdy - ydx = 0$ is ...   (2)

   (A) $y = x + C$ (B) $x^2 - y^2 = C$

   (C) $xy = C$ (D) $y = Cx$

14. The solution of differential equation $\dfrac{dy}{dx} = e^{x-y} + 3x^2e^{-y}$ is ...   (2)

   (A) $e^y = e^x + x^3 + C$ (B) $e^y = e^x + 3x^3 + C$

   (C) $e^y = e^x + 3x + C$ (D) $e^x + e^y = 3x^3 + C$

15. The solution of differential equation is ...

   $x^3\left(x\dfrac{dy}{dx} + y\right) - \sec(xy) = 0$ by substitution $xy = u$ is ...   (2)

   (A) $\tan(xy) + \dfrac{1}{2x^2} = C$ (B) $\sin(xy) + \dfrac{1}{2x^2} = C$

   (C) $\sin(xy) - \dfrac{1}{2x^2} = C$ (D) $\sin(xy) - \dfrac{1}{4x^4} = C$

16. The solution of differential equation $\sec^2 x \tan y\,dx + \sec^2 y \tan x\,dy = 0$ is ...   (2)

   (A) $\sec^2 x \tan y = C$ (B) $\tan x \sec^2 y = C$

   (C) $\tan x \tan y = C$ (D) $\sec^2 x \sec^2 y = C$

17. The solution of differential equation $y \sec^2 x + (y + 7) \tan x \dfrac{dy}{dx} = 0$ is ...   (2)

   (A) $y + 7 \log y = -\log \tan x + C$

   (B) $y + \log(7 + y) = -\log \tan x + C$

   (C) $y - 7 \log y = \log \tan x + C$

   (D) $y + \log y = -\log \tan x + C$

18. The solution of differential equation $e^x \cos y\,dx + (1 + e^x) \sin y\,dy = 0$ is ...   (2)

   (A) $(1 + e^x) = C \sec y$ (B) $(1 + e^x) \sec y = C$

   (C) $\dfrac{\sec y}{(1 + e^x)} = C$ (D) $(1 + e^x) \cos y = C$

19. The solution of differential equation $y(1 + \log x) \dfrac{dx}{dy} - x \log x = 0$ is ...   (2)

   (A) $\log(x \log x) = yC$ (B) $\dfrac{x}{\log x} = yC$

   (C) $y(\log x) = xC$ (D) $x(\log x) = yC$

20. The solution of differential equation $3e^x \tan y\,dx + (1 + e^x) \sec^2 y\,dy = 0$ is ...   (2)

   (A) $3 \log(1 - e^x) = -\log \tan y + \log C$

   (B) $\log(1 + e^x) = \log \tan y + \log C$

   (C) $3 \log(1 + e^x) = -\log \tan y + \log C$

   (D) $\log(1 + e^x) = -\log \sin y + \log C$

21. The solution of differential equation $x(1 + y^2)\,dx + y(1 + x^2)\,dy = 0$ is ...   (2)

   (A) $(1 - x^2)(1 + y^2) = C$ (B) $\tan^{-1} x + \tan^{-1} y = C$

   (C) $(1 + x^2) = C(1 + y^2)$ (D) $(1 + x^2)(1 + y^2) = C$

22. The solution of differential equation $\dfrac{dy}{dx} = (1 + x)(1 + y^2)$ is ...   (2)

   (A) $\tan^{-1} y = x + \dfrac{x^2}{2} + C$

   (B) $\log(1 + y^2) = x + \dfrac{x^2}{2} + C$

   (C) $\tan^{-1} x = y + \dfrac{y^2}{2} + C$

   (D) $\dfrac{1}{2} \log\left(\dfrac{1 + y}{1 - y}\right) = x + \dfrac{x^2}{2} + C$

23. The solution of differential equation

 $(e^x + 1) \, y \, dy = (y + 1) \, e^x \, dx$ is ...   (2)

 (A) $y - \log (1 - y) = \log (e^x - 1) + \log C$

 (B) $y - \log (1 + y) = \log (e^x + 1) + \log C$

 (C) $y + \log (1 - y) = \log (e^x + 1) + \log C$

 (D) $y - \log (1 + y) = \log (e^x - 1) + \log C$

24. The solution of differential equation $\dfrac{dy}{dx} = e^{x+y} + e^{y-x}$ is ... (2)

 (A) $e^{-y} = e^{-x} - C$ (B) $e^y = e^x - e^{-x} + C$

 (C) $-e^{-y} = e^x - e^{-x} + C$ (D) $e^{-y} = e^x + e^{-x} + C$

25. The solution of differential equation $\dfrac{dy}{dx} + \sqrt{\dfrac{1 - y^2}{1 - x^2}} = 0$

 is ...   (2)

 (A) $\tan^{-1} x + \cot^{-1} y = C$ (B) $\sin^{-1} x + \sin^{-1} y = C$

 (C) $\sec^{-1} x + \csc^{-1} y = C$ (D) $\sin^{-1} x - \sin^{-1} y = C$

ANSWERS

1. (A)	2. (B)	3. (B)	4. (C)	5. (A)	6. (D)	7. (B)	8. (C)	9. (A)
10. (B)	11. (A)	12. (D)	13. (D)	14. (A)	15. (B)	16. (C)	17. (A)	18. (B)
19. (D)	20. (C)	21. (D)	22. (A)	23. (B)	24. (C)	25. (B)		

●●●

Exact Differential Equations and Reducible to Exact Differential Equation :   **Marks**

1. The necessary and sufficient condition that the differential equation $M(x, y) \, dx + N(x, y) \, dy = 0$ be exact is ...   (1)

 (A) $\dfrac{\partial M}{\partial y} = \dfrac{\partial N}{\partial x}$; $My + Nx \neq 0$

 (B) $\dfrac{\partial M}{\partial x} = \dfrac{\partial N}{\partial y}$; $Mx - Ny \neq 0$

 (C) $\dfrac{\partial M}{\partial y} \neq \dfrac{\partial N}{\partial x}$; $Mx + Ny \neq 0$

 (D) $\dfrac{\partial M}{\partial y} - \dfrac{\partial N}{\partial x} = 1$; $My - Nx \neq 0$

2. If homogeneous differential equation $M(x, y) \, dx + N(x, y) \, dy = 0$ is not exact then the integrating factor is ...   (1)

 (A) $\dfrac{1}{My + Nx}$; $My + Nx \neq 0$ (B) $\dfrac{1}{Mx - Ny}$; $Mx - Ny \neq 0$

 (C) $\dfrac{1}{Mx + Ny}$; $Mx + Ny \neq 0$ (D) $\dfrac{1}{My - Nx}$; $My - Nx \neq 0$

3. If the differential equation $M(x, y) \, dx + N(x, y) \, dy = 0$ is not exact and it can be written as $yf_1(xy) \, dx + xf_2(xy) \, dy = 0$ then the integrating factor is ...   (1)

 (A) $\dfrac{1}{My + Nx}$; $My + Nx \neq 0$ (B) $\dfrac{1}{Mx - Ny}$; $Mx - Ny \neq 0$

 (C) $\dfrac{1}{Mx + Ny}$; $Mx + Ny \neq 0$ (D) $\dfrac{1}{My - Nx}$; $My - Nx \neq 0$

4. If the differential equation $M(x, y) \, dx + N(x, y) \, dy = 0$ is not exact and $\dfrac{\dfrac{\partial M}{\partial y} - \dfrac{\partial N}{\partial x}}{N} = f(x)$ then the integrating factor is ... (1)

 (A) $e^{f(x)}$ (B) $e^{\int f(x) \, dy}$

 (C) $f(x)$ (D) $e^{\int f(x) \, dx}$

5. If the differential equation $M(x, y) \, dx + N(x, y) \, dy = 0$ is not exact and $\dfrac{\dfrac{\partial M}{\partial y} - \dfrac{\partial N}{\partial x}}{- M} = f(y)$ then the integrating factor is ... (1)

 (A) $e^{\int f(y) \, dy}$ (B) $e^{\int f(y) \, dx}$ (C) $f(y)$ (D) $e^{f(y)}$

6. The total derivative of $x \, dy + y \, dx$ is ...   (1)

 (A) $d\left(\dfrac{y}{x}\right)$ (B) $d\left(\dfrac{x}{y}\right)$ (C) $d(xy)$ (D) $d(x + y)$

7. The total derivative of $x \, dy - y \, dx$ with integrating factor $\dfrac{1}{x^2}$ is ...   (1)

 (A) $d\left(\dfrac{x}{y}\right)$ (B) $d\left(\dfrac{y}{x}\right)$

 (C) $d\left(\log \dfrac{x}{y}\right)$ (D) $d(x - y)$

8. The total derivative of $x \, dy + y \, dx$ with integrating factor $\dfrac{1}{xy}$ is ...   (1)

 (A) $d\left(\log \dfrac{x}{y}\right)$ (B) $d\left(\log \dfrac{y}{x}\right)$

 (C) $d[\log (x + y)]$ (D) $d(\log xy)$

9. The total derivative of $x \, dy - y \, dx$ with integrating factor $\dfrac{1}{xy}$ is ...   (1)

 (A) $d\left(\log \dfrac{x}{y}\right)$ (B) $d\left(\log \dfrac{y}{x}\right)$

 (C) $d\left[\dfrac{y}{x}\right]$ (D) $d(\log xy)$

10. The total derivative of $x \, dy - y \, dx$ with integrating factor $\dfrac{1}{x^2 + y^2}$ is ...   (1)

 (A) $d\left(\tan^{-1} \dfrac{y}{x}\right)$ (B) $d\left(\tan^{-1} \dfrac{x}{y}\right)$

 (C) $d[\log (x^2 + y^2)]$ (D) none of these

11. The total derivative of $dx + dy$ with integrating factor $\dfrac{1}{x + y}$ is ...   (1)

 (A) $d[\log (x - y)]$ (B) $d[\log x^2 - y^2)]$

 (C) $d[\log (x + y)]$ (D) none of these

12. The differential equation $(x + y - 2) \, dx + (x - y + 4) \, dy = 0$ is of the form ...   (1)

 (A) exact   (B) homogeneous

 (C) linear   (D) none of these

13. The value of λ for which the differential equation
$(xy^2 + \lambda x^2 y)\, dx + (x^3 + x^2 y)\, dy = 0$ is exact is ... (2)
(A) -3 (B) 2 (C) 3 (D) 1

14. The differential equation
$(ay^2 + x + x^8)\, dx + (y^8 - y + bxy)\, dy = 0$ is exact if ... (2)
(A) $b \neq 2a$ (B) $b = a$
(C) $a = 1, b = 3$ (D) $b = 2a$

15. The differential equation
$(3 + by \cos x)\, dx + (2 \sin x - 4y^3)\, dy = 0$ is exact if ... (2)
(A) $b = -2$ (B) $b = 3$ (C) $b = 0$ (D) $b = 2$

16. The differential equation
$(\tan y - ax^2 y - y)\, dx + (x \tan^2 y - x^3 - \sec^2 y)\, dy = 0$ is exact if ... (2)
(A) $a = 2$ (B) $a = 3$ (C) $a = -3$ (D) $a = -2$

17. The differential equation $\left(\dfrac{2x}{y^3}\right) dx + \left(\dfrac{y^2 + ax^2}{y^4}\right) dy = 0$ is exact if ... (2)
(A) $a = -3$ (B) $a = 3$ (C) $a = -2$ (D) $a = 6$

18. Integrating factor of homogeneous differential equation $(xy - 2y^2)\, dx + (3xy - x^2)\, dy = 0$ is ... (2)
(A) $\dfrac{1}{xy}$ (B) $\dfrac{1}{x^2 y^2}$ (C) $\dfrac{1}{x^2 y}$ (D) $\dfrac{1}{xy^2}$

19. Integrating factor of homogeneous differential equation $(x^2 - 3xy + 2y^2)\, dx + (3x^2 - 2xy)\, dy = 0$ is ... (2)
(A) $\dfrac{1}{xy}$ (B) $\dfrac{1}{x^3}$ (C) $\dfrac{1}{x^2 y}$ (D) $\dfrac{1}{x^2}$

20. Integrating factor of homogeneous differential equation $(y^2 - 2xy)\, dx + (2x^2 + 3xy)\, dy = 0$ is ... (2)
(A) $\dfrac{1}{x^2 y^2}$ (B) $\dfrac{1}{x^2 y}$ (C) $\dfrac{1}{4xy^2}$ (D) $\dfrac{1}{y^2}$

21. Integrating factor of homogeneous differential equation $(x^2 y - 2xy^2)\, dx - (x^3 - 3x^2 y)\, dy = 0$ is ... (2)
(A) $\dfrac{1}{x^2 y^2}$ (B) $\dfrac{1}{xy}$ (C) $\dfrac{2}{x}$ (D) $\dfrac{1}{x^2 y}$

22. Integrating factor for differential equation $(x^2 y^2 + xy + 1)\, y\, dx + (x^2 y^2 - xy + 1)\, x\, dy = 0$ is ... (2)
(A) $\dfrac{1}{2x^3 y^3}$ (B) $\dfrac{1}{xy}$ (C) $\dfrac{1}{2x^2 y^2}$ (D) $\dfrac{1}{x^2 y}$

23. Integrating factor for differential equation $(1 + xy)\, y\, dx + (1 - xy)\, x\, dy = 0$ is ... (2)
(A) $\dfrac{1}{2x^2 y^2}$ (B) $\dfrac{1}{x^2 y}$ (C) $\dfrac{1}{xy^2}$ (D) $\dfrac{1}{y}$

24. Integrating factor for differential equation $(1 + xy)\, y\, dx + (x^2 y^2 + xy + 1)\, x\, dy = 0$ is ... (2)
(A) $\dfrac{1}{x^2 y}$ (B) $-\dfrac{1}{x^3 y^3}$ (C) $\dfrac{1}{xy^2}$ (D) $\dfrac{1}{x^2 y^2}$

25. Integrating factor for differential equation $(x^2 + y^2 + x)\, dx + (xy)\, dy = 0$ is ... (2)
(A) $\dfrac{1}{x}$ (B) $\dfrac{1}{x^2}$ (C) x^2 (D) x

26. Integrating factor for differential equation $\left(y + \dfrac{y^3}{3} + \dfrac{x^2}{2}\right) dx + \left(\dfrac{x + xy^2}{4}\right) dy = 0$ is ... (2)
(A) $\dfrac{1}{x}$ (B) x^3 (C) x^2 (D) $\dfrac{1}{x^3}$

27. Integrating factor for differential equation $(2x \log x - xy)\, dy + (2y)\, dx = 0$ is ... (2)
(A) $\dfrac{1}{x}$ (B) $\dfrac{1}{x^2 y^2}$ (C) $\dfrac{1}{x^2}$ (D) $\dfrac{1}{y}$

28. Integrating factor for differential equation $(x^2 + y^2 + 1)\, dx - 2xy\, dy = 0$ is ... (2)
(A) $\dfrac{1}{x}$ (B) $\dfrac{1}{x^3}$ (C) $\dfrac{1}{x^2}$ (D) $\dfrac{1}{xy}$

29. Integrating factor for differential equation $y\,(2xy + e^x)\, dx - e^x\, dy = 0$ is ... (2)
(A) $\dfrac{1}{x}$ (B) $\dfrac{1}{y}$ (C) $\dfrac{1}{x^2}$ (D) $\dfrac{1}{y^2}$

30. Integrating factor for differential equation $y \log y\, dx + (x - \log y)\, dy = 0$ is ... (2)
(A) $\dfrac{1}{x}$ (B) $\dfrac{1}{y}$ (C) $\dfrac{1}{x^2}$ (D) $\dfrac{1}{y^2}$

31. Integrating factor for differential equation $(y^4 + 2y)\, dx + (xy^3 + 2y^4 - 4x)\, dy = 0$ is ... (2)
(A) $\dfrac{2}{x}$ (B) $\dfrac{1}{y}$ (C) $\dfrac{1}{y^3}$ (D) $\dfrac{2}{y^2}$

32. Integrating factor for differential equation $(2x + e^x \log y)\, y\, dx + (e^x)\, dy = 0$ is... (2)
(A) $\dfrac{1}{x}$ (B) $\dfrac{1}{y^2}$ (C) $\dfrac{1}{x^2}$ (D) $\dfrac{1}{y}$

33. Solution of non-exact differential equation $(x^2 - 3xy + 2y^2)\, dx + x\,(3x - 2y)\, dy = 0$ with integrating factor $\dfrac{1}{x^3}$ is ... (2)
(A) $3\dfrac{y}{x} - \dfrac{y^2}{x^2} = C$ (B) $\log x - 3\dfrac{y}{x} + \dfrac{y^2}{x^2} = C$
(C) $\log x + 3\dfrac{y}{x} - 2\dfrac{y^2}{x^2} = C$ (D) $\log x + 3\dfrac{y}{x} - \dfrac{y^2}{x^2} = C$

34. Solution of non-exact differential equation $(3xy^2 - y^3)\, dx + (xy^2 - 2x^2 y)\, dy = 0$ with integrating factor $\dfrac{1}{x^2 y^2}$ is ... (2
(A) $3 \log x - \dfrac{2y}{x^3} - 2 \log y = C$
(B) $3 \log x + \dfrac{y}{x} - 2 \log y = C$
(C) $3 \log x + \dfrac{y}{x} = C$
(D) $\log x - \dfrac{y}{x} + 2 \log y = C$

35. Solution of non-exact differential equation

$(1 + xy) \, y \, dx + (1 - xy) \, x \, dy = 0$ is integrating factor $\dfrac{1}{x^2 y^2}$ is ... (2)

(A) $\dfrac{2}{xy} - \log\left(\dfrac{x}{y}\right) = C$ (B) $-\dfrac{1}{xy} + \log\left(\dfrac{y}{x}\right) = C$

(C) $-\dfrac{1}{xy} + \log\left(\dfrac{x}{y}\right) = C$ (D) $-\dfrac{2}{x^3 y} + \log\left(\dfrac{x}{y}\right) = C$

36. Solution of non-exact differential equation

$(2 + x^2 y^2) \, y \, dx + (2 - 2x^2 y^2) \, x \, dy = 0$ with integrating factor $\dfrac{1}{x^3 y^3}$ is ... (2)

(A) $\log\left(\dfrac{x}{y^2}\right) - \dfrac{1}{x^2 y^2} = C$ (B) $\log\left(\dfrac{x}{y^2}\right) + \dfrac{1}{x^2 y^2} = C$

(C) $\log\left(\dfrac{y^2}{x}\right) - \dfrac{1}{x^2 y^2} = C$ (D) $\log x - \dfrac{1}{x^2 y^2} = C$

37. Solution of non-exact differential equation

38. Solution of non-exact differential equation

$y (2xy + e^x) \, dx - e^x \, dy = 0$ with integrating factor $\dfrac{1}{y^2}$ is ... (2)

(A) $x^2 + \dfrac{e^x}{y} - e^x \log y = C$ (B) $x^2 + \dfrac{e^x}{y} = C$

(C) $x^2 + \dfrac{2e^x}{y} = C$ (D) $x^2 - \dfrac{e^x}{y} = C$

38. Solution of non-exact differential equation

$(x^4 e^x - 2mxy^2) \, dx + (2mx^2 y) \, dy = 0$ with integrating factor $\dfrac{1}{x^4}$ is ... (2)

(A) $e^x + \dfrac{6my^2}{x^4} = C$ (B) $e^x + \dfrac{2my^2}{x^2} = C$

(C) $e^x + \dfrac{y^2}{x^2} = C$ (D) $e^x + \dfrac{my^2}{x^2} = C$

ANSWERS

1. (A)	2. (C)	3. (B)	4. (D)	5. (A)	6. (C)	7. (B)	8. (D)
9. (B)	10. (A)	11. (C)	12. (A)	13. (C)	14. (D)	15. (D)	16. (B)
17. (A)	18. (D)	19. (B)	20. (C)	21. (A)	22. (C)	23. (A)	24. (B)
25. (D)	26. (B)	27. (A)	28. (C)	29. (D)	30. (B)	31. (C)	32. (D)
33. (D)	34. (B)	35. (C)	36. (A)	37. (B)	38. (D)		

•••

Linear Differential Equations and Reducible to Linear Differential Equation :   **Marks**

1. The differential equation of the form $\dfrac{dy}{dx} + Py = Q$ where P and Q are functions of x or constants, is ... (1)
   - (A) exact differential equation
   - (B) linear differential equation in y
   - (C) linear differential equation in x
   - (D) non-homogeneous differential equation

2. The differential equation of the form $\dfrac{dx}{dy} + Px = Q$ where P and Q are functions of y or constants, is ... (1)
   - (A) exact differential equation
   - (B) linear differential equation in y
   - (C) linear differential equation in x
   - (D) non-homogeneous differential equation

3. Integrating factor of linear differential equation $\dfrac{dy}{dx} + Py = Q$ where P and Q are functions of x or constant, is ... (1)
   - (A) $e^{\int P \, dy}$ (B) $e^{\int Q \, dy}$ (C) $e^{\int Q \, dx}$ (D) $e^{\int P \, dx}$

4. Integrating factor of linear differential equation $\dfrac{dx}{dy} + Px = Q$ where P and Q are functions of y or constant, is ... (1)
   - (A) $e^{\int P \, dy}$ (B) $e^{\int P \, dx}$ (C) $e^{\int Q \, dx}$ (D) $e^{\int Q \, dy}$

5. The general solution of linear differential equation $\dfrac{dy}{dx} + Py = Q$ where P and Q are functions of x or constants, is ... (1)
   - (A) $x e^{\int P \, dy} = \displaystyle\int Q e^{\int P \, dy} \, dy + C$
   - (B) $y = \displaystyle\int Q e^{\int P \, dx} \, dx + C$
   - (C) $y e^{\int P \, dx} = \displaystyle\int Q \, dx + C$
   - (D) $y e^{\int P \, dx} = \displaystyle\int Q e^{\int P \, dx} \, dx + C$

6. The general solution of linear differential equation $\dfrac{dx}{dy} + Px = Q$ where P and Q are functions of y or constants, is ... (1)
   - (A) $x = \displaystyle\int Q e^{\int P \, dy} \, dy + C$
   - (B) $x e^{\int P \, dx} = \displaystyle\int Q e^{\int P \, dx} \, dx + C$
   - (C) $y e^{\int P \, dx} = \displaystyle\int Q e^{\int P \, dx} \, dx + C$
   - (D) $x e^{\int P \, dy} = \displaystyle\int Q e^{\int P \, dy} \, dy + C$

7. The differential equation of the form $\dfrac{dy}{dx} + Py = Qy^n$, $n \neq 1$ where P and Q are functions of x or constants, is ... (1)
   - (A) Bernoulli's differential equation
   - (B) exact differential equation
   - (C) symmetric differential equation
   - (D) linear differential equation

8. The differential equation of the form $\dfrac{dx}{dy} + Px = Qx^n,\ n \neq 1$ where P and Q are functions of y or constants, is ... (1)
   - (A) Bernoulli's differential equation
   - (B) exact differential equation
   - (C) symmetric differential equation
   - (D) linear differential equation

9. The differential equation of the form $f'(y)\dfrac{dy}{dx} + P\,f(y) = Q$ where P and Q are functions of x or constants, can be reduced to linear differential equation by the substitution ... (1)
   - (A) $f'(y) = u$ (B) $P = u$ (C) $f(y) = u$ (D) $Q = u$

10. The differential equation of the form $f'(x)\dfrac{dx}{dy} + P\,f(x) = Q$ where P and Q are functions of y or constants, can be reduced to linear differential equation by the substitution ... (1)
    - (A) $f'(x) = u$ (B) $f(x) = u$ (C) $P = u$ (D) $Q = u$

11. Integrating factor of linear differential equation $\dfrac{dy}{dx} + xy = x^3$ is ... (2)
    - (A) $e^{\log x}$ (B) e^x (C) x^2 (D) $e^{\frac{x^2}{2}}$

12. Integrating factor of linear differential equation $\dfrac{dx}{dy} + yx = y^2$ is ... (2)
    - (A) $e^{\frac{y^2}{2}}$ (B) $e^{\frac{x^2}{2}}$ (C) y^2 (D) $e^{\log y}$

13. The differential equation $\dfrac{dy}{dx} + \dfrac{y}{1+x^2} = x^2$ has integrating factor ... (2)
    - (A) $e^{\frac{1}{1+y^2}}$ (B) $e^{\tan^{-1} x}$ (C) $e^{\frac{1}{1+x^2}}$ (D) $e^{\tan^{-1} y}$

14. The differential equation $\dfrac{dx}{dy} + \dfrac{x}{1+y^2} = y^2$ has integrating factor ... (2)
    - (A) $e^{\frac{1}{1+y^2}}$ (B) $e^{\tan^{-1} x}$ (C) $e^{\frac{1}{1+x^2}}$ (D) $e^{\tan^{-1} y}$

15. The differential equation $\dfrac{dy}{dx} + \sqrt{x}\,y = x^3$ has integrating factor ... (2)
    - (A) $e^{\frac{2}{3} x\sqrt{x}}$ (B) $e^{\frac{1}{3} x\sqrt{x}}$ (C) $e^{\sqrt{x}}$ (D) e^{-x}

16. The linear differential equation
    $(1 + y^2) + (x - e^{-\tan^{-1} y})\dfrac{dy}{dx} = 0$ has integrating factor ... (2)
    - (A) $e^{\tan^{-1} x}$ (B) $e^{\frac{1}{1+y^2}}$ (C) $e^{\tan^{-1} y}$ (D) e^{2y}

17. The linear differential equation $(1 - x^2)\dfrac{dy}{dx} = 1 + xy$ has integrating factor ... (2)
    - (A) $\sqrt{1 - x^2}$ (B) $\dfrac{1}{\sqrt{1 - x^2}}$
    - (C) $e^{\tan^{-1} x}$ (D) $x\sqrt{1 - x^2}$

18. The linear differential equation $(2y + x^2)\,dx = x\,dy$ has integrating factor ... (2)
    - (A) $\dfrac{1}{x}$ (B) $\dfrac{1}{x^2}$ (C) x (D) $\dfrac{1}{y^2}$

19. The linear differential equation $y^2 + \left(x - \dfrac{1}{y}\right)\dfrac{dy}{dx} = 0$ has integrating factor ... (2)
    - (A) e^x (B) e^y (C) $\dfrac{1}{y^2}$ (D) $e^{-\frac{1}{y}}$

20. The differential equation $\dfrac{dy}{dx} + y\cot x = \sin 2x$ has integrating factor ... (2)
    - (A) $\cos x$ (B) $e^{\cot x}$ (C) $\sin x$ (D) $\sec x$

21. The differential equation $\cos x\,\dfrac{dy}{dx} + y = \sin x$ has integrating factor ... (2)
    - (A) $e^{\sec x}$ (B) $(\csc x - \cot x)$
    - (C) $(\sec x + \tan x)$ (D) $(\sec x - \tan x)$

22. The differential equation $(x^2 + 1)\dfrac{dy}{dx} + 4xy = \dfrac{1}{(x^2 + 1)^2}$ has integrating factor ... (2)
    - (A) $(x^2 + 1)^2$ (B) $(x^2 + 1)$ (C) $e^{(x^2 + 1)^{\frac{4x}{}}}$ (D) e^{4x}

23. The Bernoulli's differential equation $\dfrac{dy}{dx} - y\tan x = y^4 \sec x$ reduces to linear differential equation ... (2)
    - (A) $\dfrac{du}{dx} + (3\tan x)u = -3\sec x$ where $y^{-3} = u$
    - (B) $\dfrac{du}{dx} - (3\tan x)u = 3\sec x$ where $y^{-3} = u$
    - (C) $\dfrac{du}{dx} + (\tan x)u = -\sec x$ where $y^{-3} = u$
    - (D) none of these

24. The Bernoulli's differential equation $\dfrac{dy}{dx} - xy = -y^3 e^{-x^2}$ reduces to linear differential equation ... (2)
    - (A) $\dfrac{du}{dx} + (2x)u = 2e^{-x^2}$ where $y^{-2} = u$
    - (B) $\dfrac{du}{dx} + (x)u = e^{-x^2}$ where $y^{-2} = u$
    - (C) $\dfrac{du}{dx} - (2x)u = -2e^{-x^2}$ where $y^{-2} = u$
    - (D) none of these

25. The differential equation $\tan y \dfrac{dy}{dx} + \tan x = \cos^2 x \cos y$ reduces to linear differential equation … (2)

(A) $\dfrac{du}{dx} - \tan(x)\, u = -\cos^2 x$ where $\sec y = u$

(B) $\dfrac{du}{dx} + (\tan x)\, u = \cos^2 x$ where $\sec y = u$

(C) $\dfrac{du}{dx} + (\cot x)\, u = \cos^2 x$ where $\sec y = u$

(D) none of these

26. The differential equation

$\sin y \dfrac{dy}{dx} - 2 \cos x \cos y = -\cos x \sin^2 x$ reduces to linear differential equation … (2)

(A) $\dfrac{du}{dx} + (\cos x)\, u = \cos x \sin^2 x$ where $\cos y = u$

(B) $\dfrac{du}{dx} - (2 \cos x)\, u = -\cos x \sin^2 x$ where $\cos y = u$

(C) $\dfrac{du}{dx} + (2 \cos x)\, u = \cos x \sin^2 x$ where $\cos y = u$

(D) none of these

27. The value of α so that $e^{\alpha y^2}$ is an integrating factor of linear differential equation $\dfrac{dx}{dy} + xy = e^{-\frac{y^2}{2}}$ is … (2)

(A) -1 (B) $-\dfrac{1}{2}$ (C) 1 (D) $\dfrac{1}{2}$

28. The value of α so that $e^{\alpha x^2}$ is an integrating factor of linear differential equation $\dfrac{dy}{dx} - xy = x$ is … (2)

(A) $-\dfrac{1}{2}$ (B) $\dfrac{1}{2}$ (C) 1 (D) -2

29. If I_1, I_2 are integrating factors of the equation $x\dfrac{dy}{dx} + 2y = 1$ and $x\dfrac{dy}{dx} - 2y = 1$ then true relation is … (2)

(A) $I_1 = -I_2$ (B) $I_1 I_2 = 1$ (C) $I_1 = x^2 I_2$ (D) $I_1 I_2 = x^2$

30. The general solution of $\dfrac{dy}{dx} + \dfrac{1}{1-x} y = -x(1-x)$ with integrating factor $\dfrac{1}{1-x}$ is … (2)

(A) $y = -\dfrac{x^2}{2}\left(\dfrac{1}{1-x}\right) + C$

(B) $y\dfrac{1}{1-x} = x^2 + C$

(C) $y\dfrac{1}{1-x} = \dfrac{x^2}{2} + C$

(D) $y\dfrac{1}{1-x} = -\dfrac{x^2}{2} + C$

31. The general solution of $\dfrac{dy}{dx} + \dfrac{3}{x} y = \dfrac{e^x}{x^2}$ with integrating factor x^3 is … (2)

(A) $y\, x^3 = (x+1)\, e^x + C$

(B) $y\, x^3 = (x-1)\, e^x + C$

(C) $x\, y^3 = (x-1)\, e^x + C$

(D) none of these

32. The general solution of $\dfrac{dy}{dx} + (\cot x)\, y = \sin 2x$ with integrating factor $\sin x$ is … (2)

(A) $y \sin x = \dfrac{2}{3} \sin^2 x + C$

(B) $y \sin x = \sin^3 x + C$

(C) $y \sin x = \dfrac{2}{3} \sin^3 x + C$

(D) none of these

33. The general solution of $\dfrac{dy}{dx} + \dfrac{1}{(1-x)\sqrt{x}} y = \left(1 - \sqrt{x}\right)$ with integrating factor $\dfrac{1+\sqrt{x}}{1-\sqrt{x}}$ is … (2)

(A) $y\dfrac{1+\sqrt{x}}{1-\sqrt{x}} = -x - \dfrac{2}{3} x^{3/2} + C$

(B) $y\dfrac{1+\sqrt{x}}{1-\sqrt{x}} = x + \dfrac{2}{3} x^{3/2} + C$

(C) $y\dfrac{1+\sqrt{x}}{1-\sqrt{x}} = x + \dfrac{3}{2} x^{1/2} + C$

(D) none of these

34. The general solution of $\dfrac{dy}{dx} + \left(\tan x + \dfrac{1}{x}\right) y = \dfrac{1}{x} \sec x$ with integrating factor $x \sec x$ is … (2)

(A) $y\, (x \sec x) = \tan x + C$

(B) $y\, (x \sec x) = \dfrac{x^3}{3} + C$

(C) $x\, (y \sec y) = \tan x + C$

(D) none of these

35. The general solution of $\dfrac{dy}{dx} + \dfrac{3}{x} y = x^2$ with integrating factor x^3 is … (2)

(A) $y\, x^3 = \dfrac{x^6}{6} + C$

(B) $y\, x^3 = \dfrac{x^2}{2} + C$

(C) $y\, x^3 = \log x + C$

(D) none of these

36. The general solution of $\dfrac{dy}{dx} + \dfrac{2}{x} y = \dfrac{1}{x^3}$ with integrating factor x^2 is … (2)

(A) $y\, x^2 = \dfrac{x^2}{2} + C$

(B) $y\, x^2 = \log x + C$

(C) $y\, x^2 = \dfrac{x^6}{6} + C$

(D) none of these

37. The general solution of $\dfrac{dy}{dx} + (1 + 2x)\, y = e^{-x^2}$ with integrating factor $e^{x + x^2}$ is … (2)

(A) $y\, e^{x + x^2} = \dfrac{e^{x + x^2}}{2} + C$

(B) $y\, e^{x + x^2} = e^{x^2} + C$

(C) $y\, e^{x + x^2} = e^x + C$

(D) none of these

38. The general solution of $\dfrac{dx}{dy} + \dfrac{1}{1+y^2} x = \dfrac{e^{-\tan^{-1} y}}{1+y^2}$ with integrating factor $e^{\tan^{-1} y}$ is … (2)

(A) $x\, e^{\tan^{-1} y} = \tan^{-1} y + C$

(B) $y\, e^{\tan^{-1} y} = \tan^{-1} y + C$

(C) $e^{\tan^{-1} y} = \tan^{-1} y + C$

(D) none of these

39. The general solution of $\dfrac{dx}{dy} + (\sec y)\, x = \dfrac{2y \cos y}{1 + \sin y}$ with integrating factor $(\sec y + \tan y)$ is ... (2)

(A) $y\,(\sec y + \tan y) = y^2 + C$

(B) $x\,(\sec y + \tan y) = \dfrac{y^2}{2} + C$

(C) $x\,(\sec y + \tan y) = y^2 + C$

(D) none of these

40. The general solution of $\dfrac{dx}{dy} - \dfrac{2}{y}x = y^2\, e^{-y}$ with integrating factor $\dfrac{1}{y^2}$ is ... (2)

(A) $x\,\dfrac{1}{y^2} = -\,e^{-y} + C$

(B) $x\,\dfrac{1}{y^2} = e^{-y} + C$

(C) $y\,\dfrac{1}{x^2} = -\,e^{-y} + C$

(D) none of these

ANSWERS

1. (B)	2. (C)	3.(D)	4. (A)	5. (D)	6. (D)	7. (A)	8. (A)
9. (C)	10. (B)	11. (D)	12. (A)	13. (B)	14. (D)	15. (A)	16. (C)
17. (A)	18. (B)	19. (D)	20. (C)	21. (C)	22. (A)	23. (A)	24. (A)
25. (B)	26. (C)	27. (D)	28. (A)	29. (B)	30. (D)	31. (B)	32. (C)
33. (B)	34. (A)	35. (A)	36. (B)	37. (C)	38. (A)	39. (C)	40. (A)

•••

CHAPTER 2 : APPLICATIONS OF FIRST ORDER DIFFERENTIAL EQUATIONS

Orthogonal Trajectories :   **Marks**

1. The differential equation of orthogonal trajectories of family of straight lines $y = mx$ is ... (1)

(A) $\dfrac{dx}{dy} = -\dfrac{y}{x}$ (B) $\dfrac{dx}{dy} = -\dfrac{x}{y}$

(C) $\dfrac{dy}{dx} = \dfrac{y}{x}$ (D) $\dfrac{dy}{dx} = m$

2. If the family of curves is given by $x^2 + 2y^2 = c^2$ then the differential equation of orthogonal trajectories of family is ... (1)

(A) $x - 2y\,\dfrac{dy}{dx} = 0$ (B) $x + 2y\,\dfrac{dx}{dy} = 0$

(C) $x + 2y\,\dfrac{dy}{dx} = 0$ (D) $x - 2y\,\dfrac{dx}{dy} = 0$

3. The differential equation of orthogonal trajectories of family of curves $xy = c$ is ... (1)

(A) $x\,\dfrac{dx}{dy} + y = 0$ (B) $-x\,\dfrac{dx}{dy} + y = 0$

(C) $-x\,\dfrac{dx}{dy} - y = 0$ (D) $x\,\dfrac{dy}{dx} + y = 0$

4. If the family of curves is given by $y^2 = 4ax$ then the differential equation of orthogonal trajectories of family is ... (1)

(A) $2y\,\dfrac{dy}{dx} = 4a$ (B) $2y\,\dfrac{dy}{dx} = \dfrac{y^2}{x}$

(C) $-2y\,\dfrac{dx}{dy} = \dfrac{y^2}{x}$ (D) $2y\,\dfrac{dy}{dx} = \dfrac{x}{y^2}$

5. The differential equation of orthogonal trajectories of family of curves $2x^2 + y^2 = cx$ is ... (1)

(A) $4x + 2y\,\dfrac{dy}{dx} = \dfrac{2x^2 + y^2}{x}$ (B) $4x - 2y\,\dfrac{dx}{dy} = \dfrac{2x^2 + y^2}{x}$

(C) $4x + 2y\,\dfrac{dx}{dy} = \dfrac{2x^2 + y^2}{x}$ (D) none of these

6. The differential equation of orthogonal trajectories of family of curves $x^2 + cy^2 = 1$ is ... (1)

(A) $x - \left(\dfrac{1 - x^2}{y}\right)\dfrac{dx}{dy} = 0$ (B) $x + \left(\dfrac{1 - x^2}{y}\right)\dfrac{dy}{dx} = 0$

(C) $x + \left(\dfrac{1 - x^2}{y}\right)\dfrac{dx}{dy} = 0$ (D) none of these

7. The differential equation of orthogonal trajectories of family of curves $e^x + e^{-y} = c$ is ... (1)

(A) $e^x - e^{-y}\,\dfrac{dy}{dx} = 0$ (B) $e^x - e^{-y}\,\dfrac{dx}{dy} = 0$

(C) $e^x + e^{-y}\,\dfrac{dx}{dy} = 0$ (D) none of these

8. The differential equation of orthogonal trajectories of family of curves $r = a \cos\theta$ is ... (1)

(A) $r^2\,\dfrac{d\theta}{dr} = \tan\theta$ (B) $\dfrac{1}{r}\dfrac{d\theta}{dr} = -\tan\theta$

(C) $r\,\dfrac{d\theta}{dr} = -\tan\theta$ (D) $r\,\dfrac{d\theta}{dr} = \tan\theta$

9. The differential equation of orthogonal trajectories of family of curves $r = a \sin\theta$ is ... (1)

(A) $\dfrac{1}{r}\dfrac{d\theta}{dr} = \cot\theta$ (B) $r\,\dfrac{d\theta}{dr} = -\cot\theta$

(C) $r\,\dfrac{d\theta}{dr} = -\tan\theta$ (D) $r^2\,\dfrac{d\theta}{dr} = \tan\theta$

10. The differential equation of orthogonal trajectories of family of curves $r = a\,(1 - \cos\theta)$ is ... (1)

(A) $-r\,\dfrac{d\theta}{dr} = \dfrac{\sin\theta}{1 - \cos\theta}$ (B) $r\,\dfrac{d\theta}{dr} = \dfrac{\sin\theta}{1 - \cos\theta}$

(C) $-r^2\,\dfrac{d\theta}{dr} = \dfrac{\sin\theta}{1 - \cos\theta}$ (D) $-r\,\dfrac{d\theta}{dr} = \dfrac{1 - \cos\theta}{\sin\theta}$

11. The differential equation of orthogonal trajectories of family of curves $r^2 = a \sin 2\theta$ is … (1)

(A) $r \dfrac{d\theta}{dr} = \tan 2\theta$ (B) $r \dfrac{d\theta}{dr} = \cot 2\theta$

(C) $-r \dfrac{d\theta}{dr} = \cot 2\theta$ (D) $\dfrac{dr}{d\theta} = r \cot 2\theta$

12. The differential equation of orthogonal trajectories of family of curves $r^2 = a \cos 2\theta$ is … (1)

(A) $-r^2 \dfrac{d\theta}{dr} = \tan 2\theta$ (B) $r \dfrac{d\theta}{dr} = \cot 2\theta$

(C) $r \dfrac{d\theta}{dr} = \tan 2\theta$ (D) $\dfrac{dr}{d\theta} = -r \tan 2\theta$

13. The differential equation of orthogonal trajectories of family of curves $r = a \sec^2 \dfrac{\theta}{2}$ is … (1)

(A) $-r \dfrac{d\theta}{dr} = \tan \dfrac{\theta}{2}$ (B) $r \dfrac{d\theta}{dr} = \tan \dfrac{\theta}{2}$

(C) $-r \dfrac{d\theta}{dr} = \cot \dfrac{\theta}{2}$ (D) $\dfrac{dr}{d\theta} = 2 \tan \dfrac{\theta}{2}$

14. The differential equation of orthogonal trajectories of family of curves $r = a \cos^2 \theta$ is … (1)

(A) $\dfrac{dr}{d\theta} = -\dfrac{r \sin 2\theta}{\cos^2 \theta}$ (B) $-r \dfrac{d\theta}{dr} = \dfrac{\sin 2\theta}{\cos^2 \theta}$

(C) $r \dfrac{d\theta}{dr} = \dfrac{\cos^2 \theta}{\sin 2\theta}$ (D) $r \dfrac{d\theta}{dr} = \dfrac{\sin 2\theta}{\cos^2 \theta}$

15. The differential equation of orthogonal trajectories of family of curves $r^2 = a \sin \theta$ is … (1)

(A) $2r \dfrac{d\theta}{dr} = \cot \theta$ (B) $2r \dfrac{d\theta}{dr} = -\cot \theta$

(C) $2r \dfrac{d\theta}{dr} = -\tan \theta$ (D) $2 \dfrac{d\theta}{dr} = -r^2 \cot \theta$

16. If the differential equation of family of straight lines $y = mx$ is $\dfrac{dy}{dx} = \dfrac{y}{x}$ then its orthogonal trajectories is … (2)

(A) $xy = k$ (B) $x^2 - y^2 = k^2$

(C) $y = kx$ (D) $x^2 + y^2 = k^2$

17. If the differential equation of family of rectangular hyperbola $xy = c$ is $x \dfrac{dy}{dx} = -y$ then its orthogonal trajectories is … (2)

(A) $x^2 - y^2 = k^2$ (B) $x^2 + y^2 = k^2$

(C) $y^2 = kx$ (D) $xy = k_1$

18. Orthogonal trajectories of family of circles $x^2 + y^2 = c^2$ whose differential equation is $\dfrac{dy}{dx} = -\dfrac{x}{y}$, is equal to … (2)

(A) $x^2 - y^2 = k^2$ (B) $y = kx$

(C) $y^2 = kx$ (D) $x^2 + y^2 = k^2$

19. If the differential equation of family of rectangular hyperbola $x^2 - y^2 = c^2$ is $\dfrac{dy}{dx} = \dfrac{x}{y}$, then its orthogonal trajectories is … (2)

(A) $y^2 = kx$ (B) $x^2 + y^2 = k^2$

(C) $xy = k$ (D) $y = kx$

20. Orthogonal trajectories of family of curves $x^2 + 2y^2 = c^2$ whose differential equation is $\dfrac{dy}{dx} = -\dfrac{x}{2y}$, is … (2)

(A) $x^2 = ky$ (B) $x^2 = \dfrac{k}{y}$

(C) $x^2 + 2y^2 = k^2$ (D) none of these

21. Orthogonal trajectories of family of curves $y^2 = 4ax$, whose differential equation is $\dfrac{dy}{dx} = \dfrac{y}{2x}$, is equal to … (2)

(A) $x^2 + y^2 = k^2$ (B) $x^2 + 2y^2 = k^2$
(C) $y^2 = 4kx$ (D) $2x^2 + y^2 = k$

22. If the differential equation of family of curves $e^x + e^{-y} = c$ is $\dfrac{dy}{dx} = \dfrac{e^x}{e^{-y}}$, then its orthogonal trajectories is … (2)

(A) $e^{-x} + e^{-y} = k$ (B) $e^{-x} - e^y = k$
(C) $e^x - e^{-y} = k$ (D) $e^x + e^{-y} = k$

23. If the differential equation of family of curves $e^y - e^{-x} = c$ is $\dfrac{dy}{dx} = -\dfrac{e^{-x}}{e^y}$, then its orthogonal trajectories is … (2)

(A) $e^{-x} + e^{-y} = k$ (B) $e^x - e^{-y} = k$
(C) $e^x + e^{-y} = k$ (D) $e^y - e^{-x} = k$

24. Orthogonal trajectories of family of curves $x^2 + cy^2 = 1$ whose differential equation is $\dfrac{dy}{dx} = -\dfrac{xy}{1 - x^2}$, is … (2)

(A) $\log x + \dfrac{x^2}{2} = \dfrac{y^2}{2} + k$ (B) $\log x - \dfrac{x^2}{2} = \dfrac{y^2}{2} + k$

(C) $x^2 + y^2 = k$ (D) $x^2 + ky^2 = 1$

25. Orthogonal trajectories of family of curves $x^2 = ce^{x^2 + y^2}$ whose differential equation is $\dfrac{dy}{dx} = \dfrac{1 - x^2}{xy}$, is … (2)

(A) $\log (1 - x^2) - 2 \log y = \log k$
(B) $2 \log (1 - x^2) - \log y = \log k$
(C) $\log (1 - x^2) + 2 \log y = \log k$
(D) $\log (1 - x^2) - 2 \log y = \log k + k$

26. Orthogonal trajectories of family of curves $\dfrac{x^2}{a^2} + \dfrac{y^2}{b^2 + \lambda} = 1$, λ is an arbitrary constant, whose differential equation is $x + \left(\dfrac{a^2 - x^2}{y}\right) \dfrac{dy}{dx} = 0$, is … (2)

(A) $\dfrac{y^2}{2} = -a^2 \log x + \dfrac{x^2}{2} + k$ (B) $y^2 = b^2 \log x - x^2 +$

(C) $\dfrac{y^2}{2} = a^2 \log x - \dfrac{x^2}{2} + k$ (D) $\dfrac{y^2}{2} = -a^2 \log x + x^2 + k$

27. If the differential equation of family of curves $r = a (1 - \cos \theta)$ is $\dfrac{dr}{d\theta} = r \cot \dfrac{\theta}{2}$ then its orthogonal trajectories is given by … (2)

(A) $2 \log \sec \dfrac{\theta}{2} = \log r + \log k$

(B) $2 \log \cos \dfrac{\theta}{2} = \log r + \log k$

(C) $\dfrac{1}{2} \log \cos \dfrac{\theta}{2} = \log r + \log k$

(D) $\dfrac{1}{2} \log \sec \dfrac{\theta}{2} = \log r + \log k$

28. If the differential equation of family of curves $r = a \sec^2 \frac{\theta}{2}$ is $\frac{dr}{d\theta} = r \tan \frac{\theta}{2}$ then its orthogonal trajectories is given by ... (2)

   (A) $-2 \log \cos \frac{\theta}{2} = \log r + \log k$

   (B) $2 \log \sin \frac{\theta}{2} = \log r + \log k$

   (C) $-2 \log \sin \frac{\theta}{2} = \log r + \log k$

   (D) $2 \log \cos \frac{\theta}{2} = \log r + \log k$

29. If the differential equation of family of curves $r = a \sin \theta$ is $\frac{dr}{d\theta} = r \cot \theta$ then its orthogonal trajectories is given by ... (2)

   (A) $r = k \cos \theta$

   (B) $r = k \sec \theta$

   (C) $r = k \sin \theta$

   (D) $\log \cos \theta = rk$

30. If the differential equation of family of curves $r = a \cos \theta$ is $\frac{dr}{d\theta} = -r \tan \theta$ then its orthogonal trajectories is given by ... (2)

   (A) $\log r = -\operatorname{cosec}^2 \theta + k$ (B) $r = k \cos \theta$

   (C) $r = k \operatorname{cosec} \theta$ (D) $r = k \sin \theta$

31. If the differential equation of family of curves $r^2 = a \sin 2\theta$ is $\frac{dr}{d\theta} = r \cot 2\theta$ then its orthogonal trajectories is given by ...(2)

   (A) $r^2 = \log \sec 2\theta + k$
   (B) $r^2 = k \sin 2\theta$
   (C) $r^2 = k \cos 2\theta$
   (D) $\log r = -\frac{1}{2} \sec^2 2\theta + k$

32. If the differential equation of family of curves $r^2 = a \cos 2\theta$ is $\frac{dr}{d\theta} = -r \tan 2\theta$ then its orthogonal trajectories is given by ... (2)

   (A) $\frac{1}{2} \log \cos 2\theta = \log r + \log k$
   (B) $\frac{1}{2} \log \sin 2\theta = \log r + \log k$
   (C) $\log \sin 2\theta = -r^2 + k$
   (D) $\frac{1}{2} \log \sin 2\theta = -\log r + \log k$

33. If the differential equation of family of curves $r = a \cos^2 \theta$ is $\frac{dr}{d\theta} = -2r \tan \theta$ then its orthogonal trajectories is given by ... (2)

   (A) $\frac{1}{2} \log \sin \theta = \log r + \log k$
   (B) $\frac{1}{2} \log \sin \theta = -\log r + \log k$
   (C) $\log \sin \theta = r + k$
   (D) $\log \sec \theta = -\log r + \log k$

ANSWERS

1. (A)	2. (D)	3. (B)	4. (C)	5. (B)	6. (A)	7. (C)	8. (D)
9. (B)	10. (A)	11. (C)	12. (C)	13. (A)	14. (D)	15. (B)	16. (D)
17. (A)	18. (B)	19. (C)	20. (A)	21. (D)	22. (B)	23. (C)	24. (B)
25. (A)	26. (C)	27. (B)	28. (C)	29. (A)	30. (D)	31. (C)	32. (B)
33. (A)							

•••

Rate of Decay of Radioactive Materials :

1. Radium decomposes at the rate proportional to the amount present. The differential equation for the rate of decay of radium is ...... (1)

   (A) $\frac{du}{dt} = -\frac{k}{u}$ (B) $\frac{du}{dt} = ku$

   (C) $\frac{du}{dt} = -ku$ (D) $\frac{du}{dt} = -k(u - M)$

   Ans. (C)

•••

Newton's Law of Cooling :   **Marks**

1. Newton's law of cooling states that ... (1)

   (A) the temperature of a body changes at the rate which is proportional to the temperatures of surrounding medium.

   (B) the temperature of a body changes at the rate which is inversely proportional to the difference in temperatures between that of surrounding medium and that of body itself.

   (C) the temperature of a body changes at the rate which is proportional to the sum of temperatures of surrounding medium and that of body itself.

   (D) the temperature of a body changes at the rate which is proportional to the difference in temperatures between that of surrounding medium and that of body itself.

2. A metal ball is heated to a temperature of 100°C and at time t = 0 it is placed in water which is maintained at 40°C. By Newton's law of cooling the differential equation satisfied by temperature θ of metal ball at any time t is ...   (1)

(A) $\dfrac{d\theta}{dt} = - k\,(\theta - 100)$

(B) $\dfrac{d\theta}{dt} = - k\,(\theta - 40)$

(C) $\dfrac{d\theta}{dt} = - k\theta$

(D) $\dfrac{d\theta}{dt} = - k\theta\,(\theta - 40)$

3. According to Newton's law of cooling, the rate at which a substance cools in moving air is proportional to the difference between the temperature of the substance and that of air. A substance initially at temperature 90°C is kept in moving air at temperature 26°C, the differential equation satisfied by temperature θ of substance at any time t is ...   (1)

(A) $\dfrac{d\theta}{dt} = - k\,(\theta - 26)$

(B) $\dfrac{d\theta}{dt} = - k\,(\theta - 90)$

(C) $\dfrac{d\theta}{dt} = - k\theta$

(D) $\dfrac{d\theta}{dt} = - k\,(\theta - 64)$

4. Suppose a corpse at a temperature of 32°C arrives at mortuary where the temperature is kept at 10°C. Then by Newton's law of cooling the differential equation satisfied by temperature T of corpse t hours later is ...   (1)

(A) $\dfrac{dT}{dt} = - kT\,(T - 10)$

(B) $\dfrac{dT}{dt} = - k\,(T - 32)$

(C) $\dfrac{dT}{dt} = - k\,(T - 10)$

(D) $\dfrac{dT}{dt} = - kT\,(T - 32)$

5. A thermometer is taken outdoors where the temperature is 0°C, from a room in which the temperature is 21°C and temperature drops to 10°C in 1 minute. Then by Newton's law of cooling the differential equation satisfied by temperature T at time t is ...   (1)

(A) $\dfrac{dT}{dt} = - k\,(T - 21)$

(B) $\dfrac{dT}{dt} = - kT$

(C) $\dfrac{dT}{dt} = kT$

(D) $\dfrac{dT}{dt} = - kT\,(T - 21)$

6. If θ_0 is the temperature of the surrounding and θ is temperature of the body at any time t satisfies the differential equation $\dfrac{d\theta}{dt} = - k\,(\theta - \theta_0)$ then θ is given by ...   (2)

(A) $\theta = \theta_0 e^{-kt}$

(B) $\theta = \theta_0 + Ae^{kt}$

(C) $\theta = -k(\theta_0 + Ae^{-kt})$

(D) $\theta = \theta_0 + Ae^{-kt}$

7. Suppose a corpse at a temperature of 32°C arrives at mortuary where the temperature is kept at 10°C. If the corpse cools to 27°C in 5 minutes. If the differential equation by Newton's law of cooling is $\dfrac{dT}{dt} = - (0.05)\,(T - 10)$, then temperature T of corpse at any time t is given by ...   (2)

(A) $T = 22e^{-0.05t}$

(B) $T = 10 + 22e^{0.05t}$

(C) $T = 10 + 22e^{-0.05t}$

(D) $T = 10 - 22e^{-0.05t}$

8. A thermometer is taken outdoors where the temperature is 0°C, from a room in which the temperature is 21°C and temperature drops to 10°C in 1 minute. If the differential equation by Newton's law of cooling is $\dfrac{dT}{dt} = - (0.7419)\,T$, then temperature T of thermometer at time t is given by ...   (2)

(A) $T = 21 + 11e^{-0.7419t}$

(B) $T = 21e^{0.7419t}$

(C) $T = 10 + 21e^{-0.7419t}$

(D) $T = 21e^{-0.7419t}$

9. A body originally at 80°C cools down to 60°C in 20 minutes in a room where the temperature is 40°C. The differential equation by Newton's law of cooling is $\dfrac{d\theta}{dt} = - k\,(\theta - 40)$, then the value of k is ...   (2)

(A) $-\dfrac{1}{20}\log_e 2$

(B) $\dfrac{1}{20}\log_e 2$

(C) $20\log_e 2$

(D) $\log_e 2$

10. If the temperature of the body drops from 100°C to 60°C in 1 minute when the temperature of surrounding is 20°C satisfies the differential equation $\dfrac{d\theta}{dt} = - k\,(\theta - 20)$, then the value of k is ...   (2)

(A) $\log_e 2$ (B) $-\log_e 2$ (C) $\log_e 4$ (D) $\log_e 8$

11. The temperature of the air is 30°C and the substance cools from 100°C to 70°C in 15 minutes. If differential equation by Newton's law of cooling is $\dfrac{d\theta}{dt} = - k\,(\theta - 30)$, then the value of k is ...   (2)

(A) $\log_e \dfrac{7}{4}$

(B) $\dfrac{1}{15}\log_e \dfrac{4}{7}$

(C) $\dfrac{1}{15}\log_e \dfrac{7}{4}$

(D) $15\log_e \dfrac{7}{4}$

12. By Newton's law of cooling the differential equation of body originally at 80°C cools down to 60°C in 20 minutes in surrounding temperature of 40°C is $\dfrac{d\theta}{dt} = - (0.03465)\,(\theta - 40)$. The temperature of the body after 40 minutes is ...   (2)

(A) 60°C   (B) 50°C   (C) 35°C   (D) 85°C

13. A metal ball is heated to a temperature of 100°C and at time t = 0 it is placed in water which is maintained at 40°C. The temperature of the ball reduces to 60°C in 4 minutes. By Newton's law of cooling the differential equation is $\dfrac{d\theta}{dt} = -\left(\dfrac{1}{4}\log_e 3\right)(\theta - 40)$. Then the time required to reduce the temperature of ball to 50°C is ...   (2)

(A) 7.5 min   (B) 3.5 min   (C) 10 min   (D) 6.5 min

14. A body at temperature 100°C is placed in a room whose temperature is 20°C and cools to 60°C in 5 minutes. By Newton's law of cooling the differential equation is $\dfrac{d\theta}{dt} = -\left(\dfrac{1}{5}\log_e 2\right)(\theta - 20)$. Then the temperature after 8 minutes is ...   (2)

(A) 46.4°C   (B) 65.4°C   (C) 40.4°C   (D) 20°C

15. A copper ball is heated to a temperature of 100°C and at time t = 0 it is placed in water which is maintained at 30°C. The temperature of the ball reduces to 70°C in 3 minutes. The differential equation by Newton's law of cooling is $\dfrac{d\theta}{dt} = -\left(\dfrac{1}{3}\log_e \dfrac{7}{4}\right)(\theta - 30)$. Then the time required to reduce the temperature of ball to 31°C is ... (2)

(A) 3 min   (B) 7.78 min

(C) 22.78 min   (D) 15.78 min

ANSWERS

1. (D)	2. (B)	3. (A)	4. (C)	5. (B)	6. (D)	7. (C)	8. (D)
9. (B)	10. (A)	11. (C)	12. (B)	13. (D)	14. (A)	15. (C)	

•••

Rectilinear Motion :   Marks

1. Rectilinear motion is a motion of body along a ... (1)
(A) straight line   (B) circular path
(C) parabolic path   (D) none of these

2. According to D'Alembert's principle, algebraic sum of forces acting on a body along a given direction is equal to ... (1)
(A) velocity × acceleration   (B) mass × velocity
(C) mass × displacement   (D) mass × acceleration

3. A particle moving in a straight line with acceleration $k\left(x + \dfrac{a^4}{x^3}\right)$ directed towards the origin. The equation of motion is ... (1)
(A) $\dfrac{dv}{dx} = -k\left(x + \dfrac{a^4}{x^3}\right)$ (B) $v\dfrac{dv}{dx} = k\left(x + \dfrac{a^4}{x^3}\right)$
(C) $v\dfrac{dv}{dx} = -k\left(x + \dfrac{a^4}{x^3}\right)$ (D) $\dfrac{dv}{dx} = \left(x + \dfrac{a^4}{x^3}\right)$

4. A particle of mass m moves in a horizontal straight line OA with acceleration $\dfrac{k}{x^3}$ at a distance x and directed towards the origin O. Then the differential equation of motion is ... (1)
(A) $v\dfrac{dv}{dx} = \dfrac{k}{x^3}$ (B) $v\dfrac{dv}{dx} = -\dfrac{k}{x^3}$
(C) $\dfrac{dv}{dx} = -\dfrac{k}{x^3}$ (D) $\dfrac{dv}{dx} = \dfrac{k}{x^3}$

5. A body of mass m falling from rest is subjected to a force of gravity and air resistance proportional to square of velocity (kv^2). The equation of motion is ... (1)
(A) $m\dfrac{dv}{dx} = mg - kv^2$ (B) $mv\dfrac{dv}{dx} = mg + kv^2$
(C) $mv\dfrac{dv}{dx} = -kv^2$ (D) $mv\dfrac{dv}{dx} = mg - kv^2$

6. A particle is projected vertically upward with velocity v_1 and resistance of air produces retardation (kv^2) where v is velocity. The equation of motion is ... (1)
(A) $v\dfrac{dv}{dx} = -g - kv^2$ (B) $v\dfrac{dv}{dx} = -g + kv^2$
(C) $v\dfrac{dv}{dx} = -kv^2$ (D) $v\dfrac{dv}{dx} = g - kv^2$

7. A body starts moving from rest is opposed by a force per unit mass of value (cx) resistance per unit mass of value (bv^2) where v and x are velocity and displacement of body at that instant. The differential equation of motion is ... (1)
(A) $mv\dfrac{dv}{dx} = -cx - bv^2$ (B) $v\dfrac{dv}{dx} = cx + bv^2$
(C) $v\dfrac{dv}{dx} = -cx - bv^2$ (D) $\dfrac{dv}{dx} = -cx - bv^2$

8. A body of mass m falls from rest under gravity, in a fluid whose resistance to motion at any time t is mk times its velocity where k is constant. The differential equation of motion is ... (1)
(A) $\dfrac{dv}{dt} = -g - kv$ (B) $\dfrac{dv}{dt} = g - kv$
(C) $\dfrac{dv}{dt} = g + kv$ (D) $\dfrac{dv}{dt} = mg - mkv$

9. A particle of mass m is projected vertically upward with velocity v, assuming the air resistance k times its velocity where k is constant. The differential equation of motion is ... (1)
(A) $\dfrac{dv}{dt} = mg - kv$ (B) $m\dfrac{dv}{dt} = -mg + kv$
(C) $m\dfrac{dv}{dt} = -kv$ (D) $m\dfrac{dv}{dt} = -mg - kv$

10. Assuming that the resistance to movement of a ship through water in the form of $(a^2 + b^2v^2)$ where v is the velocity, a and b are constants. The differential equation for retardation of the ship moving with engine stopped is ... (1)
(A) $m\dfrac{dv}{dt} = -(a^2 + b^2v^2)^2$ (B) $m\dfrac{dv}{dt} = +(a^2 + b^2v^2)$
(C) $m\dfrac{dv}{dt} = -(a^2 + b^2v^2)$ (D) $m\dfrac{dv}{dx} = -(a^2 + b^2v^2)$

11. Differential equation of motion of a body of mass m falls from rest under gravity in a fluid whose resistance to motion at any time t is mk times its velocity where k is constant is $\dfrac{dv}{dt} = g - kv$ then the relation between velocity and time t is ... (2)
(A) $t = \dfrac{1}{k}\log\dfrac{g - kv}{g}$ (B) $t = \dfrac{1}{k}\log\dfrac{g}{g - kv}$
(C) $t = \dfrac{1}{k}\log\dfrac{g}{g + kv}$ (D) $t = -\dfrac{1}{k}\log\dfrac{1}{g - kv}$

12. A body of mass m falling from rest is subjected to the force of gravity and air resistance proportional to square of velocity (kv^2) satisfies the differential equation $mv\dfrac{dv}{dx} = k(a^2 - v^2)$ where $ka^2 = mg$, then the relation between velocity and displacement is ... (2)
(A) $\dfrac{2kx}{m} = \log\dfrac{a^2}{a^2 - v^2}$ (B) $\dfrac{2kx}{m} = \log\dfrac{a^2 - v^2}{a^2}$
(C) $2kx = \log\dfrac{1}{a^2 - v^2}$ (D) $\dfrac{x}{m} = \log\dfrac{a^2}{a^2 - v^2}$

13. A vehicle starts from rest and its acceleration is given by $\dfrac{dv}{dt} = k\left(1 - \dfrac{t}{T}\right)$ where k and T are constant. Then the velocity v in terms of time t is given by … (2)

(A) $v = k\left(t - \dfrac{t^2}{2}\right)$

(B) $v = k\left(t - \dfrac{t^2}{T}\right)$

(C) $v = k\left(\dfrac{t^2}{2} - \dfrac{t^3}{3T}\right)$

(D) $v = k\left(t - \dfrac{t^2}{2T}\right)$

14. A particle of unit mass moves in a horizontal straight line OA with an acceleration $\dfrac{k}{r^3}$ at a distance r and directed towards O. If initially the particle was at rest at r = a and equation of motion is $v\dfrac{dv}{dr} = -\dfrac{k}{r^3}$ then the relation between r, v is … (2)

(A) $v^2 = k\left(\dfrac{1}{r^2} + \dfrac{1}{a^2}\right)$

(B) $v^2 = k\left(\dfrac{1}{a^2} - \dfrac{1}{r^2}\right)$

(C) $v^2 = k\left(\dfrac{1}{r^2} - \dfrac{1}{a^2}\right)$

(D) $v^2 = k\left(\dfrac{1}{r^4} - \dfrac{1}{a^2}\right)$

15. A particle of mass m is projected upward with velocity V. Assuming the air resistance k times its velocity and equation of motion is $m\dfrac{dv}{dt} = -mg - kv$ then the relation between velocity v and time t is … (2)

(A) $t = \dfrac{m}{k}\log\left(\dfrac{mg + kV}{mg + kv}\right)$

(B) $t = \dfrac{m}{k}\log\left(\dfrac{mg + kv}{mg + kV}\right)$

(C) $t = m\log\left(\dfrac{mg + kV}{mg + kv}\right)$

(D) $t = \log\left(\dfrac{mg + kv}{mg + kV}\right)$

16. A body of mass m falls from rest under gravity in a fluid whose resistance to motion at any instant is mkv where k is constant. The differential equation of motion is $\dfrac{dv}{dt} = g - kv$ then the terminal velocity is … (2)

(A) $\dfrac{k}{g}$

(B) $\dfrac{g}{k}$

(C) $-\dfrac{g}{k}$

(D) none of these

17. A bullet is fired into a sand tank, its retardation is proportional to the square root of its velocity. The differential equation of motion is $\dfrac{dv}{dt} = -k\sqrt{v}$. If v_0 is initial velocity then the relation between velocity v and time t is … (2)

(A) $\sqrt{v} = -t + \sqrt{v_0}$

(B) $2\sqrt{v} = -kt$

(C) $\sqrt{v} = -kt + \sqrt{v_0}$

(D) $2\sqrt{v} = -kt + 2\sqrt{v_0}$

18. A particle moving in a straight line with acceleration $k\left(x + \dfrac{a^4}{x^3}\right)$ directed towards the origin. the equation of motion is $v\dfrac{dv}{dx} = -k\left(x + \dfrac{a^4}{x^3}\right)$. If it starts from rest at a distance x = a from the origin then the relation between velocity v and displacement x is … (2)

(A) $\dfrac{v^2}{2} = k\left(\dfrac{x^2}{2} + \dfrac{a^4}{2x^2}\right)$

(B) $\dfrac{v^2}{2} = -k\left(\dfrac{x^2}{2} - \dfrac{a^4}{2x^2}\right)$

(C) $\dfrac{v^2}{2} = -k\left(\dfrac{x^2}{2} - \dfrac{a^4}{2x^2}\right) + \dfrac{a^2}{2}$

(D) $\dfrac{v^2}{2} = -k\left(\dfrac{x^2}{2} - \dfrac{3a^4}{x^4}\right)$

ANSWERS

1. (A)	2. (D)	3. (C)	4. (B)	5. (D)	6. (A)	7. (C)	8. (B)	9. (D)
10. (C)	11. (B)	12. (A)	13. (D)	14. (C)	15. (A)	16. (B)	17. (D)	18. (B)

•••

Applications to Electrical Circuits :   **Marks**

1. A circuit containing resistance R and inductance L in series with voltage source E. By Kirchhoff's voltage law, differential equation for current i is … (1)

(A) $Li + R\dfrac{di}{dt} = E$

(B) $L\dfrac{di}{dt} + Ri = E$

(C) $L\dfrac{di}{dt} + Ri = 0$

(D) $L\dfrac{di}{dt} + \dfrac{q}{C} = E$

2. A circuit containing resistance R and capacitance C in series with voltage source E. By Kirchhoff's voltage law, differential equation for current $i = \dfrac{dq}{dt}$ is … (1)

(A) $L\dfrac{di}{dt} + \dfrac{q}{C} = E$

(B) $R\dfrac{dq}{dt} + \dfrac{q}{C} = 0$

(C) $L\dfrac{di}{dt} + Ri = 0$

(D) $R\dfrac{dq}{dt} + \dfrac{q}{C} = E$

3. A circuit containing inductance L, capacitance C in series without applied electromotive force. By Kirchhoff's voltage law, differential equation for current i is … (1)

(A) $L\dfrac{di}{dt} + \dfrac{q}{C} = 0$

(C) $L\dfrac{di}{dt} + Ri = 0$

(C) $L\dfrac{di}{dt} + Ri = E$

(D) $L\dfrac{di}{dt} + \dfrac{q}{C} = E$

4. A circuit containing inductance L, capacitance C in series with applied electromotive force E. By Kirchhoff's voltage law, differential equation for current i is … (1)

(A) $L\dfrac{di}{dt} + Ri = E$

(B) $L\dfrac{di}{dt} + Ri = 0$

(C) $L\dfrac{di}{dt} + \dfrac{q}{C} = E$

(D) $L\dfrac{di}{dt} + \dfrac{q}{C} = 0$

5. The differential equation for the current in an electric circuit containing resistance R and inductance L in series with voltage source E sin ωt is … (1)

(A) $L\dfrac{di}{dt} + \dfrac{q}{C} = E$

(B) $Li + R\dfrac{di}{dt} = E\sin \omega t$

(C) $L\dfrac{di}{dt} + Ri = 0$

(D) $L\dfrac{di}{dt} + Ri = E\sin \omega t$

6. In a circuit containing resistance R and inductance L in series with constant voltage source E, current i is given by $i = \dfrac{E}{R}\left(1 - e^{-\frac{R}{L}t}\right)$, then maximum current i_{max} is ... (1)

(A) $\dfrac{E}{R}$ (B) $\dfrac{R}{E}$ (C) ER   (D) 0

7. The differential equation for the current i in an electric circuit containing resistance 100 ohm and an inductance of 0.5 henry connected in series with battery of 20 volts is ... (1)

(A) $0.5\dfrac{di}{dt} + 100i = 0$ (B) $0.5\dfrac{di}{dt} + 100i = 20$

(C) $100\dfrac{di}{dt} + 0.5i = 20$ (D) $100\dfrac{di}{dt} + 0.5R = 0$

8. The differential equation for the current i in an electric circuit containing resistance R = 250 ohm and an inductance of L = 640 henry in series with an electromotive force E = 500 volts is ... (1)

(A) $640\dfrac{di}{dt} + 250i = 0$ (B) $250\dfrac{di}{dt} + 640i = 500$

(C) $640\dfrac{di}{dt} + 250i = 500$ (D) $250\dfrac{di}{dt} + 640i = 0$

9. A capacitor C = 0.01 farad in series with resistor R = 20 ohms is charged from battery E = 10 volts. If initially capacitor is completely discharged then differential equation for charge q(t) is given by ... (1)

(A) $20\dfrac{dq}{dt} + \dfrac{q}{0.01} = 0;\ q(0) = 0$

(B) $20\dfrac{dq}{dt} + 0.01q = 10;\ q(0) = 0$

(C) $20\dfrac{dq}{dt} + \dfrac{q}{0.01} = 10;\ q(0) = 0$

(D) $20\dfrac{dq}{dt} + 0.01q = 0;\ q(0) = 0$

10. In a circuit containing resistance R and inductance L in series with constant e.m.f. E, the current i is given by $i = \dfrac{E}{R}\left(1 - e^{-\frac{R}{L}t}\right)$, then the time required to build current half of its theoretical maximum is ... (2)

(A) $\dfrac{L}{R \log 2}$ (B) $\dfrac{L \log 2}{R}$ (C) $\dfrac{R \log 2}{L}$ (D) 0

11. In a circuit containing resistance R and inductance L in series with constant e.m.f. E, the current i is given by $i = \dfrac{E}{R}\left(1 - e^{-\frac{R}{L}t}\right)$, then the time required before current reaches its 90% of maximum value is ... (2)

(A) 0   (B) $\dfrac{L}{R \log 10}$ (C) $\dfrac{R \log 10}{L}$ (D) $\dfrac{L \log 10}{R}$

12. If the differential equation for current in an electric circuit containing resistance R and inductance L in series with constant e.m.f. E, the current i is $L\dfrac{di}{dt} + Ri = E$, then the current at any time t is given by ... (2)

(A) $i = \dfrac{E}{R} - Ae^{-\frac{R}{L}t}$; A is arbitrary constant

(B) $i = \dfrac{E}{R} + Ae^{-\frac{R}{L}t}$; A is arbitrary constant

(C) $i = \dfrac{E}{R} + Ae^{\frac{R}{L}t}$; A is arbitrary constant

(D) $i = \dfrac{E}{R} + e^{-\frac{R}{L}t}$

13. A charge q on the plate of condenser of capacity C charge through a resistance R by steady voltage V satisfies differential equation $R\dfrac{dq}{dt} + \dfrac{q}{C} = V$, then charge q at any time t is ... (2)

(A) $q = CV + Ae^{-\frac{1}{RC}t}$; A is arbitrary constant

(B) $q = CV - Ae^{\frac{1}{RC}t}$; A is arbitrary constant

(C) $q = C + Ae^{\frac{1}{RC}t}$; A is arbitrary constant

(D) $q = CV + e^{\frac{1}{RC}t}$

14. The charge q on the plate of condenser of capacity C charge through a resistance R by steady voltage V is given by $q = CV\left(1 - e^{-\frac{1}{RC}t}\right)$ then current flowing through the plate is ... (2)

(A) $i = \dfrac{V}{R}e^{-\frac{R}{L}t}$ (B) $i = \dfrac{V}{R}e^{\frac{1}{RC}t}$

(C) $i = \dfrac{V}{R}e^{-\frac{1}{RC}t}$ (D) $i = CV\left(1 - e^{-\frac{1}{RC}t}\right)$

15. A resistance R = 100 ohms, an inductance L = 0.5 henry are connected in series with a battery of 20 volts. The differential equation for the current i is $0.5\dfrac{di}{dt} + 100i = 20$, then current i at any time t is ... (2)

(A) Ae^{-200t}; A is arbitrary constant

(B) $\dfrac{1}{5} + Ae^{200t}$; A is arbitrary constant

(C) $2 + Ae^{-200t}$; A is arbitrary constant

(D) $\dfrac{1}{5} + Ae^{-200t}$; A is arbitrary constant

16. If an R-C circuit, charge q as function of time t is $q = e^{-3t} - e^{-6t}$, then time required for maximum charge on capacitor is ... (2)

(A) $3 \log 2$ (B) $-\dfrac{1}{3}\log 2$ (C) $\dfrac{1}{3}\log 2$ (D) $\dfrac{1}{2}\log 3$

17. A circuit containing resistance R and inductance L in series with voltage source E. The differential equation for current i is $L\dfrac{di}{dt} + Ri = E$. Given L = 640 H, R = 250 Ω and E = 500 volts then integrating factor of differential equation is ... (2)

(A) $e^{\frac{64}{25}t}$ (B) $e^{\frac{25}{64}t}$ (C) $e^{-\frac{25}{64}t}$ (D) e^{250t}

ANSWERS

1. (B)	2. (D)	3. (A)	4. (C)	5. (D)	6. (A)	7. (B)	8. (C)	9. (C)
10 (B)	11. (D)	12. (B)	13. (A)	14. (C)	15. (D)	16. (C)	17. (B)	

•••

Simple Harmonic Motion :
Marks

1. If a particle moves on a straight line so that the force acting on it is always directed towards a fixed point on the line and proportional to its distance from the point then the particle is said to be in … (1)
   (A) simple harmonic motion
   (B) motion under the gravity
   (C) periodic motion
   (D) circular motion

2. A particle executes simple harmonic motion then the differential equation of motion is … (1)

   (A) $\dfrac{d^2x}{dt^2} = -\omega^2 x$

   (B) $\dfrac{d^2x}{dt^2} = \omega^2 x$

   (C) $\dfrac{d^2x}{dt^2} = -\dfrac{\omega^2}{x}$

   (D) $\dfrac{dx}{dt} = -\omega^2 x$

ANSWERS

1. (A)	2. (A)

•••

Chemical Engineering Problems :
Marks

1. A tank contains 10,000 litres of brine in which 200 kg salt dissolved. Fresh water runs into the tank at the rate of 100 litres per minute and the mixture kept uniform by stirring, runs out at the same rate. If Q be the total amount of salt at any time t then the governing differential equation is … (2)

   (A) $\dfrac{dQ}{dt} = 200 - \dfrac{Q}{100}$

   (B) $\dfrac{dQ}{dt} = -\dfrac{Q}{10000}$

   (C) $\dfrac{dQ}{dt} = -\dfrac{Q}{100}$

   (D) $\dfrac{dQ}{dt} = \dfrac{Q}{100}$

2. A tank initially contains 50 litres of fresh water. Brine containing 2 grams per litre of salt, flows into the tank at the rate of 2 litres per minute and the mixture kept uniform by stirring, runs out at the same rate. If Q be the total amount of salt at any time t then the differential equation in terms of Q and t is … (2)

   (A) $\dfrac{dQ}{dt} = 4 - \dfrac{Q}{50}$

   (B) $\dfrac{dQ}{dt} = 4 - \dfrac{Q}{25}$

   (C) $\dfrac{dQ}{dt} = 2 - \dfrac{Q}{50}$

   (D) $\dfrac{dQ}{dt} = 2 - \dfrac{Q}{25}$

3. A tank initially contains 100 litres of fresh water. 2 litres of brine each containing 1 gram of dissolved salt, runs into the tank per minute and the mixture kept uniform by stirring, runs out at the rate of 1 litre per minute. Let Q be the quantity of salt present at any time t then $\dfrac{dQ}{dt}$ the rate at which salt content changing is … (2)

   (A) $\dfrac{dQ}{dt} = 1 - \dfrac{Q}{100 + t}$

   (B) $\dfrac{dQ}{dt} = 2 - \dfrac{Q}{100}$

   (C) $\dfrac{dQ}{dt} = \dfrac{Q}{100 + t}$

   (D) $\dfrac{dQ}{dt} = 2 - \dfrac{Q}{100 + t}$

4. A tank contains 1000 litres of brine in which 20 kg salt dissolved. Brine containing 0.1 kg per litre of salt, runs into the tank at the rate of 40 litres per minute and the mixture, assumed to be kept uniform by stirring, runs out at the rate of 30 litres per minute. Assuming that tank is sufficiently large to avoid overflow, the governing differential equation for rate at which the salt content changing $\dfrac{dQ}{dt}$ at any time t is … (2)

   (A) $\dfrac{dQ}{dt} = 4 - 30\,\dfrac{Q}{1000 + 10t}$

   (B) $\dfrac{dQ}{dt} = 4 - 30\,\dfrac{Q}{1000}$

   (C) $\dfrac{dQ}{dt} = 30\,\dfrac{Q}{1000 + 10t}$

   (D) $\dfrac{dQ}{dt} = 4 - \dfrac{Q}{1000 + 10t}$

5. A tank initially contains 5000 litres of fresh water. Salt water containing 100 grams per litre of salt, flows into it at the rate of 10 litres per minute and the mixture kept uniform by stirring, runs out at the same rate. If Q be the total amount of salt at any time t then the differential equation relating Q and t is … (2)

   (A) $\dfrac{dQ}{dt} = 100 - \dfrac{Q}{500}$

   (B) $\dfrac{dQ}{dt} = 1000 - \dfrac{Q}{500}$

   (C) $\dfrac{dQ}{dt} = 1000 - \dfrac{Q}{5000}$

   (D) $\dfrac{dQ}{dt} = -\dfrac{Q}{500}$

6. A tank contains 10,000 litres of brine in which 200 kg of salt are dissolved. Fresh water runs into the tank at the rate of 100 litres per minute and the mixture kept uniform by stirring, runs out at the same rate. If governing differential equation is $\dfrac{dQ}{dt} = -\dfrac{Q}{100}$, the amount of salt Q at any time t is … (2)

   (A) $\log Q = -\dfrac{\log t}{100} + \log 200$

   (B) $\log Q = -\dfrac{t}{100}$

   (C) $\log Q = -\dfrac{t}{100} + \log 200$

   (D) $\log Q = -\dfrac{t}{100} - \log 200$

7. A tank initially contains 50 litres of fresh water. Brine containing 2 grams per litre of salt, flows into the tank at the rate of 2 litres per minute and the mixture kept uniform by stirring, runs out at the same rate. The differential equation in terms of Q and t is $\dfrac{dQ}{dt} = 4 - \dfrac{Q}{25}$. The total amount of salt Q at any time t is … (2)

   (A) $t = -25 \log_e (100 - Q) + 25 \log_e 100$

   (B) $t = 25 \log_e (100 - Q) - 25 \log_e 100$

   (C) $t = -\log_e (100 - Q) + \log_e 10$

(D) none of these

8. A tank initially contains 500 litres of fresh water. Salt water containing 100 grams per litre of salt, flows into it at the rate of 10 litres per minute and the mixture kept uniform by stirring, runs out at the same rate. The differential equation in terms of Q and t is $\dfrac{dQ}{dt} = 1000 - \dfrac{Q}{500}$. The amount of salt Q at any time t is ... (2)

(A) $t = -\log_e (500000 - Q) + k$
(B) $t = 500 \log_e (500000 - Q) + k$
(C) $t = -500 \log_e (5000 - Q) + k$
(D) $t = -500 \log_e (500000 - Q) + k$

Answers

1. (C)	2. (B)	3. (D)	4. (A)	5. (B)	6. (C)	7. (A)	8. (D)

•••

One Dimensional Conduction of Heat :   **Marks**

1. Fourier's law of heat conduction states that, the quantity of heat flow across an area A cm^2 is ... (2

(A) proportional to the product of area A and temperature gradient $\dfrac{dT}{dx}$.

(B) inversely proportional to the product of area A and temperature gradient $\dfrac{dT}{dx}$.

(C) equal to sum of area A and temperature gradient $\dfrac{dT}{dx}$.

(D) equal to difference of area A and temperature gradient $\dfrac{dT}{dx}$.

2. If q be the quantity of heat that flows across an area A cm^2 and thickness δx in one second where the difference of temperature at the faces is δT, then by Fourier's law of heat conduction ... (2)

(A) $q = -k\left(A - \dfrac{dT}{dx}\right)$, where k is thermal conductivity.

(B) $q = kA\dfrac{dT}{dx}$, where k is thermal conductivity.

(C) $q = -k\left(A + \dfrac{dT}{dx}\right)$, where k is thermal conductivity.

(D) $q = -kA\dfrac{dT}{dx}$, where k is thermal conductivity.

3. The differential equation for steady state heat loss per unit time from a unit length of pipe with thermal conductivity k, radius r_0 carrying steam at temperature T_0, if the pipe is covered with insulation of thickness w, the outer surface of which remains at the constant temperature T_1, is ... (2)

(A) $Q = k(2\pi r)\dfrac{dT}{dr}$
(B) $Q = -k(2\pi r)\dfrac{dT}{dr}$

(C) $Q = -k(2\pi r^2)\dfrac{dT}{dr}$
(D) $Q = -k(\pi r^2)\dfrac{dT}{dr}$

4. The differential equation for steady state heat loss per unit time from a spherical shell with thermal conductivity k radius r_0 carrying steam at temperature T_0, if the spherical shell is covered with insulation of thickness w, the outer surface of which remains at the constant temperature T_1, is ... (2)

(A) $Q = -k(2\pi r)\dfrac{dT}{dr}$
(B) $Q = k(2\pi r)\dfrac{dT}{dr}$

(C) $Q = -k(4\pi r^2)\dfrac{dT}{dr}$
(D) $Q = -k(\pi r^2)\dfrac{dT}{dr}$

5. The differential equation for steady state heat loss Q per unit time from a unit length of pipe with thermal conductivity k, radius r_0 carrying steam at temperature T_0, if the pipe is covered with insulation of thickness W, the outer surface of which remains at the constant temperature T_1, is $Q = -k(2\pi r)\dfrac{dT}{dr}$. Then the temperature T of surface of pipe of radius r is ... (2)

(A) $T = \dfrac{Q}{2\pi k}\dfrac{1}{r} + C$
(B) $T = \dfrac{Q}{2\pi k}\log r + C$

(C) $T = -\dfrac{Q}{2\pi k}\dfrac{1}{r} + C$
(D) $T = -\dfrac{Q}{2\pi k}\log r + C$

6. The differential equation for steady state heat loss Q per unit time from a spherical shell with thermal conductivity k, radius r_0 carrying steam at temperature T_0, if the spherical shell is covered with insulation of thickness w, the outer surface of which remains at the constant temperature T_1, is $Q = -k(4\pi r^2)\dfrac{dT}{dr}$. Then the temperature T of spherical shell of radius r is ... (2)

(A) $T = -\dfrac{Q}{4\pi k}\dfrac{1}{r^2} + C$
(B) $T = \dfrac{Q}{4\pi k}\dfrac{1}{r} + C$

(C) $T = -\dfrac{Q}{4\pi k}\dfrac{1}{r} + C$
(D) $T = -\dfrac{Q}{2\pi k}\dfrac{1}{r^3} + C$

7. A pipe 20 cm in diameter contains steam at 150°C and is protected with covering 5 cm thick for which k = 0.0025. If the temperature of the outer surface of the covering is 40°C and differential equation of conduction of heat is $dT = -\dfrac{Q}{2\pi k}\dfrac{dx}{x}$. The amount of heat loss Q is ... (2)

(A) $\dfrac{110\,(2\pi k)}{\log (1.5)}$
(B) $\dfrac{\log (1.5)}{110\,(2\pi k)}$

(C) $-\dfrac{110\,(2\pi k)}{\log (1.5)}$
(D) $\dfrac{110}{\log (1.5)}$

8. A long hollow pipe has an inner diameter of 10 cm and outer diameter of 20 cm. The inner surface is kept at 200°C and outer surface at 50°C. The thermal conductivity k = 0.12. The differential equation of conduction of heat is $dT = -\dfrac{Q}{2\pi k}\dfrac{dx}{x}$. Then the amount of heat loss Q cal/sec is ... (2)

(A) $-\dfrac{150\,(2\pi k)}{\log 2}$
(B) $\dfrac{\log 2}{150\,(2\pi k)}$

(C) $\dfrac{150\,(2\pi k)}{\log 2}$
(D) $\dfrac{(2\pi k)}{\log 2}$

9. A steam pipe 20 cm in diameter is protected with covering 6 cm thick for which thermal conductivity k = 0.0003 in steady state. The inner surface of the pipe is at 200°C and outer surface of the covering is at 30°C and differential equation of conduction of heat is $dT = -\dfrac{Q}{2\pi k}\dfrac{dx}{x}$. The amount of heat loss Q is ... (2)

(A) $\dfrac{170\,(2\pi k)}{\log(1.6)}$ (B) $-\dfrac{170\,(2\pi k)}{\log(1.6)}$

(C) $\dfrac{\log(1.6)}{170\,(2\pi k)}$ (D) $\dfrac{170}{\log(1.6)}$

10. A pipe 10 cm in diameter contains steam at 100°C. It is protected with asbestos 5 cm thick for which k = 0.0006 and outer surface is at 30°C. The differential equation of conduction of heat is $dT = -\dfrac{Q}{2\pi k}\dfrac{dx}{x}$. The amount of heat loss Q is ... (2)

(A) $\dfrac{\log 2}{70\,(2\pi k)}$ (B) $\dfrac{70\,(2\pi k)}{\log 2}$

(C) $-\dfrac{70\,(2\pi k)}{\log 2}$ (D) $\dfrac{(2\pi k)}{\log 2}$

ANSWERS

1. (A)	2. (D)	3. (B)	4. (C)	5. (D)	6. (B)	7. (A)	8. (C)
9. (A)	10. (B)						

•••

Miscellaneous Examples : Differential Equations   **Marks**

1. In a certain culture of bacteria, the rate of increase is proportional to the number present. If y denote the number of bacteria at time t hours then the governing differential equation is ... (1)

(A) $\dfrac{dy}{dt} = ky$ (B) $\dfrac{dy}{dt} = -ky$

(C) $\dfrac{dy}{dt} = \dfrac{k}{y}$ (D) $\dfrac{dy}{dt} = ky^2$

2. The differential equation of the population model for natural growth of bacteria is $\dfrac{dy}{dt} = ky$. The general solution of the equation is ... (1)

(A) $y = c \log kt$ (B) $ye^{kt} = ct$

(C) $y = ce^{kt}$ (D) $y = ce^{-kt}$

3. The amount x of a substance present in certain chemical reaction at time t is given by $\dfrac{dx}{dt} + \dfrac{1}{10}x = 2 - (1.5)\,e^{-\frac{1}{10}t}$, then the amount x of substance present at time t is ... (1)

(A) $x = -\dfrac{3}{2}te^{-\frac{1}{10}t} + Ce^{-\frac{1}{10}t}$

(B) $x = 20 + \dfrac{3}{2}te^{-\frac{1}{10}t} - Ce^{-\frac{1}{10}t}$

(C) $x = 20 - \dfrac{3}{2}t + C$

(D) $x = 20 - \dfrac{3}{2}te^{-\frac{1}{10}t} + Ce^{-\frac{1}{10}t}$

4. Biotransformation of an organic compound having concentration x can be modeled using an ordinary differential equation $\dfrac{dx}{dt} + kx^2 = 0$ where k is reaction rate constant. If x = a at t = 0, the solution of equation is ... (2)

(A) $x = ae^{-kt}$ (B) $\dfrac{1}{x} = \dfrac{1}{a} + kt$

(C) $x = a\,(1 - e^{-kt})$ (D) $x = a + kt$

ANSWERS

1. (A)	2. (C)	3. (D)	4. (B)

•••

CHAPTER 3 : REDUCTION FORMULAE, BETA AND GAMMA FUNCTIONS

Reduction Formulae :   **Marks**

1. If $I_n = \displaystyle\int_0^{\pi/2} \sin^n x\, dx$ then which of the following relation is true ? (1)

(A) $I_n = \dfrac{n-1}{n} I_{n-2}$ (B) $I_n = \dfrac{n-1}{n} I_{n-1}$

(C) $I_n = \dfrac{n}{n-1} I_{n-2}$ (D) $I_n = \dfrac{n-2}{n} I_{n-1}$

2. If $I_n = \displaystyle\int_0^{\pi/2} \cos^n x\, dx$ then which of the following relation is true ? (1)

(A) $I_n = \dfrac{n-1}{n} I_{n-1}$ (B) $I_n = \dfrac{n-1}{n} I_{n-2}$

(C) $I_n = \dfrac{n}{n-1} I_{n-2}$ (D) $I_n = n(n+1) I_{n-1}$

3. If $I_n = \int_0^{\pi/2} \sin^n x \, dx$, n positive even integer then I_n is calculated from ... (1)

   (A) $\dfrac{n-1}{n} \cdot \dfrac{n-3}{n-2} \cdot \dfrac{n-5}{n-4} \cdots \dfrac{3}{4} \cdot \dfrac{1}{2}$

   (B) $\dfrac{n-1}{n} \cdot \dfrac{n-3}{n-2} \cdot \dfrac{n-5}{n-4} \cdots \dfrac{3}{4} \cdot \dfrac{1}{2} \cdot \pi$

   (C) $\left(\dfrac{n+1}{n} \cdot \dfrac{n+3}{n+2} \cdot \dfrac{n+5}{n+4} \cdots \right) \cdot \dfrac{\pi}{2}$

   (D) $\dfrac{n-1}{n} \cdot \dfrac{n-3}{n-2} \cdot \dfrac{n-5}{n-4} \cdots \dfrac{3}{4} \cdot \dfrac{1}{2} \cdot \dfrac{\pi}{2}$

4. If $I_n = \int_0^{\pi/2} \sin^n x \, dx$, n positive odd integer then I_n is calculated from ... (1)

   (A) $\dfrac{n-1}{n} \cdot \dfrac{n-3}{n-2} \cdot \dfrac{n-5}{n-4} \cdots \dfrac{4}{5} \cdot \dfrac{2}{3} \cdot \dfrac{\pi}{2}$

   (B) $\dfrac{n-1}{n} \cdot \dfrac{n-3}{n-2} \cdot \dfrac{n-5}{n-4} \cdots \dfrac{3}{4} \cdot \dfrac{1}{2} \cdot \dfrac{\pi}{2}$

   (C) $\dfrac{n-1}{n} \cdot \dfrac{n-3}{n-2} \cdot \dfrac{n-5}{n-4} \cdots \dfrac{4}{5} \cdot \dfrac{2}{3} \cdot 1$

   (D) $\dfrac{n+1}{n} \cdot \dfrac{n+3}{n+2} \cdot \dfrac{n+5}{n+4}$

5. If $I_n = \int_0^{\pi/2} \cos^n x \, dx$, n positive even integer then I_n is calculated from ... (1)

   (A) $\dfrac{n-1}{n} \cdot \dfrac{n-3}{n-2} \cdot \dfrac{n-5}{n-4} \cdots \dfrac{3}{4} \cdot \dfrac{1}{2}$

   (B) $\dfrac{n-1}{n} \cdot \dfrac{n-3}{n-2} \cdot \dfrac{n-5}{n-4} \cdots \dfrac{3}{4} \cdot \dfrac{1}{2} \cdot \dfrac{\pi}{2}$

   (C) $\left(\dfrac{n+1}{n} \cdot \dfrac{n+3}{n+2} \cdot \dfrac{n+5}{n+4} \cdots \right) \cdot \dfrac{\pi}{2}$

   (D) $\dfrac{n-1}{n} \cdot \dfrac{n-3}{n-2} \cdot \dfrac{n-5}{n-4} \cdots \dfrac{3}{4} \cdot \dfrac{1}{2} \cdot \pi$

6. If $I_n = \int_0^{\pi/2} \cos^n x \, dx$, n positive odd integer then I_n is calculated from ... (1)

   (A) $\dfrac{n-1}{n} \cdot \dfrac{n-3}{n-2} \cdot \dfrac{n-5}{n-4} \cdots \dfrac{4}{5} \cdot \dfrac{2}{3} \cdot \dfrac{\pi}{2}$

   (B) $\dfrac{n-1}{n} \cdot \dfrac{n-3}{n-2} \cdot \dfrac{n-5}{n-4} \cdots \dfrac{3}{4} \cdot \dfrac{1}{2} \cdot \dfrac{\pi}{2}$

   (C) $\dfrac{n+1}{n} \cdot \dfrac{n+3}{n+2} \cdot \dfrac{n+5}{n+4}$

   (D) $\dfrac{n-1}{n} \cdot \dfrac{n-3}{n-2} \cdot \dfrac{n-5}{n-4} \cdots \dfrac{4}{5} \cdot \dfrac{2}{3} \cdot 1$

7. The value of integral $I_{m,n} = \int_0^{\pi/2} \sin^m x \cos^n x \, dx$; m, n are positive integers ≥ 2 is ... (1)

(A) $I_{m,n} = \dfrac{[(m-1)(m-3)\ldots 2 \text{ or } 1] \cdot [(n-1)(n-3)\ldots 2 \text{ or } 1]}{[(m+n)(m+n-2)\ldots 2 \text{ or } 1]} \times P$

   where $P = \dfrac{\pi}{2}$ if m and n are both even = 1 for all other values of m and n

(B) $I_{m,n} = \dfrac{[(m-1)(m-3)\ldots 2 \text{ or } 1] \cdot (n-1)(n-3)\ldots 2 \text{ or } 1]}{[(m+n)(m+n-2)\ldots 2 \text{ or } 1]} \times P$

   where $P = 1$ if m and are both even $= \dfrac{\pi}{2}$ for all other values of m and n.

(C) $I_{m,n} = \dfrac{n-1}{n} \cdot \dfrac{n-3}{n-2} \cdot \dfrac{n-5}{n-4} \cdots \dfrac{4}{5} \cdot \dfrac{2}{3} \cdot 1$

(D) $I_{m,n} = \dfrac{n-1}{n} \cdot \dfrac{n-3}{n-2} \cdot \dfrac{n-5}{n-4} \cdots \dfrac{3}{4} \cdot \dfrac{1}{2} \cdot \dfrac{\pi}{2}$

8. $\int_0^{\pi/2} \sin^4 x \, dx$ is equal to ... (2)

   (A) $\dfrac{\pi}{2}$ (B) $\dfrac{3\pi}{8}$ (C) $\dfrac{\pi}{4}$ (D) $\dfrac{3\pi}{16}$

9. $\int_0^{\pi/2} \sin^5 x \, dx$ is equal to ... (2)

   (A) $\dfrac{8}{15} \cdot \dfrac{\pi}{2}$ (B) $\dfrac{15}{8}$ (C) $\dfrac{8}{15}$ (D) 0

10. $\int_0^{\pi/2} \cos^3 x \, dx$ is equal to ... (2)

    (A) $\dfrac{2}{3}$ (B) $\dfrac{1}{4}$ (C) $\dfrac{2}{3} \cdot \dfrac{\pi}{2}$ (D) $\dfrac{1}{3}$

11. $\int_0^{\pi/2} \cos^6 x \, dx$ is equal to ... (2)

    (A) $\dfrac{5}{16}$ (B) $\dfrac{5}{16} \cdot \dfrac{\pi}{2}$ (C) $\dfrac{16}{5} \cdot \dfrac{\pi}{2}$ (D) $\dfrac{5}{48} \cdot \dfrac{\pi}{2}$

12. $\int_0^{\pi/4} \sin^2 (2x) \, dx$ is equal to ... (2)

    (A) $\dfrac{1}{4}$ (B) $\dfrac{\pi}{2}$ (C) $\dfrac{\pi}{4}$ (D) $\dfrac{\pi}{8}$

13. $\int_0^{\pi/4} \cos^2 (2x) \, dx$ is equal to ... (2)

    (A) $\dfrac{\pi}{8}$ (B) $\dfrac{\pi}{2}$ (C) $\dfrac{\pi}{4}$ (D) $\dfrac{1}{4}$

14. $\int_0^{\pi} \sin^5 \left(\dfrac{x}{2} \right) dx$ is equal to ... (2)

    (A) $\dfrac{8\pi}{15}$ (B) $\dfrac{32}{15}$ (C) $\dfrac{16}{15}$ (D) $\dfrac{8}{15}$

15. $\int_0^{\pi} \cos^3 x \, dx$ is equal to ... (2)

    (A) 1   (B) $\dfrac{2}{3}$ (C) $\dfrac{4}{3}$ (D) 0

16. $\displaystyle\int_0^{\pi} \sin^6 t \, dt$ is equal to ... (2)

(A) $\dfrac{\pi}{8}$　　(B) $\dfrac{5\pi}{16}$　　(C) $\dfrac{5}{8}$　　(D) $\dfrac{5\pi}{8}$

17. $\displaystyle\int_0^{2\pi} \sin^6 t \, dt$ is equal to ... (2)

(A) $\dfrac{5}{4}$　　(B) $\dfrac{5\pi}{32}$　　(C) $\dfrac{5\pi}{8}$　　(D) 0

18. $\displaystyle\int_0^{2\pi} \cos^7 t \, dt$ is equal to ... (2)

(A) $\dfrac{32}{35}$　　(B) $\dfrac{32\pi}{70}$　　(C) 0　　(D) $\dfrac{16}{35}$

19. $\displaystyle\int_0^{\pi/2} \sin^6 x \cos^4 x \, dx$ is equal to ... (2)

(A) $\dfrac{3}{256}$　　(B) $\dfrac{3\pi}{512}$　　(C) $\dfrac{3}{128}$　　(D) $\dfrac{512}{3}\pi$

20. $\displaystyle\int_0^{\pi/2} \sin^4 x \cos^5 x \, dx$ is equal to ... (2)

(A) $\dfrac{4\pi}{315}$　　(B) $\dfrac{315}{8}$　　(C) $\dfrac{8\pi}{630}$　　(D) $\dfrac{8}{315}$

21. $\displaystyle\int_0^{2\pi} \sin^6 x \cos^4 x \, dx$ is equal to ... (2)

(A) $\dfrac{5\pi}{128}$　　(B) $\dfrac{3\pi}{512}$　　(C) $\dfrac{3\pi}{128}$　　(D) $\dfrac{5\pi}{64}$

22. $\displaystyle\int_0^{2\pi} \sin^5 \theta \cos^4 \theta \, d\theta$ is equal to ... (2)

(A) 0　　(B) $\dfrac{8}{315}$　　(C) $\dfrac{3\pi}{128}$　　(D) $\dfrac{\pi}{128}$

23. $\displaystyle\int_{-\pi/2}^{\pi/2} \sin^4 \theta \, d\theta$ is equal to ... (2)

(A) $\dfrac{3\pi}{16}$　　(B) $\dfrac{3\pi}{4}$　　(C) 0　　(D) $\dfrac{3\pi}{8}$

24. $\displaystyle\int_{-\pi/2}^{\pi/2} \sin^4 \theta \cos^2 \theta \, d\theta$ is equal to ... (2)

(A) $\dfrac{\pi}{32}$　　(B) $\dfrac{\pi}{16}$　　(C) $\dfrac{\pi}{8}$　　(D) 0

25. If $I_n = \displaystyle\int_0^{\pi/4} \tan^n x \, dx$ and $I_n = \dfrac{1}{n-1} - I_{n-2}$ then I_4 is equal to ... (2)

(A) $-\dfrac{2}{3}+\dfrac{\pi}{2}$　　(B) $-\dfrac{2}{3}-\dfrac{\pi}{4}$　　(C) $-\dfrac{2}{3}+\dfrac{\pi}{4}$　　(D) $-\dfrac{4}{3}+\dfrac{\pi}{4}$

26. If $I_n = \displaystyle\int_{\pi/4}^{\pi/2} \cot^n x \, dx$ and $I_n = \dfrac{1}{n-1} - I_{n-2}$ then I_4 is equal to ... (2)

(A) $-\dfrac{4}{3}+\dfrac{\pi}{4}$　　(B) $-\dfrac{2}{3}+\dfrac{\pi}{2}$　　(C) $-\dfrac{2}{3}-\dfrac{\pi}{4}$　　(D) $-\dfrac{2}{3}+\dfrac{\pi}{4}$

27. If $I_n = \displaystyle\int_0^{\pi/4} \sin^{2n} x \, dx$ and $I_n = -\dfrac{1}{n2^{n+1}} + \left(\dfrac{2n-1}{2n}\right) I_{n-1}$ then I_2 is equal to ... (2)

(A) $-\dfrac{1}{4}+\dfrac{3\pi}{32}$　　(B) $-\dfrac{3}{4}+\dfrac{3\pi}{32}$　　(C) $\dfrac{1}{4}+\dfrac{\pi}{32}$　　(D) $\dfrac{1}{4}-\dfrac{\pi}{16}$

28. If $I_n = \displaystyle\int_0^{\pi/4} \cos^{2n} x \, dx$ and $I_n = \dfrac{1}{n2^{n+1}} + \left(\dfrac{2n-1}{2n}\right) I_{n-1}$ then I_2 is equal to ... (2)

(A) $-\dfrac{1}{4}+\dfrac{3\pi}{32}$　　(B) $\dfrac{1}{4}+\dfrac{3\pi}{32}$　　(C) $\dfrac{1}{4}+\dfrac{3\pi}{16}$　　(D) $\dfrac{1}{4}+\dfrac{\pi}{16}$

29. If $I_n = \displaystyle\int_0^{\pi/3} \cos^n x \, dx$ and $I_n = \dfrac{\sqrt{3}}{n2^n} + \left(\dfrac{n-1}{n}\right) I_{n-2}$ then I_2 is equal to ... (2)

(A) $\dfrac{\sqrt{3}}{4}+\dfrac{\pi}{3}$

(B) $\left(\dfrac{\sqrt{3}}{8}+\dfrac{1}{2}\right)\dfrac{\pi}{6}$

(C) $\dfrac{\sqrt{3}}{8}+\dfrac{\pi}{6}$

(D) $\dfrac{\sqrt{3}}{16}+\dfrac{\pi}{12}$

30. If $I_n = \displaystyle\int_0^{\infty} e^{-px} \sin^n x \, dx$ and $I_n = \dfrac{n(n-1)}{n^2+p^2} I_{n-2}$ then value of $\displaystyle\int_0^{\infty} e^{-2x} \sin^2 x \, dx$ is equal to ... (2)

(A) $\dfrac{1}{4}$　　(B) $\dfrac{1}{2}$　　(C) $\dfrac{1}{8}$　　(D) 2

31. If $I_n = \displaystyle\int_0^{\pi/4} \dfrac{\sin(2n-1)x}{\sin x} \, dx$ and $I_{n+1} = \dfrac{1}{n}\sin\dfrac{n\pi}{2} + I_n$ then I_3 is equal to ... (2)

(A) $2+\dfrac{\pi}{4}$　　(B) $2-\dfrac{\pi}{2}$　　(C) $1-\dfrac{\pi}{4}$　　(D) $1+\dfrac{\pi}{4}$

32. If $I_n = \displaystyle\int (\log x)^n \, dx$ then ... (2)

(A) $I_n + nI_{n-1} = x(\log x)^n$　　(B) $I_n - nI_{n-1} = x(\log x)^n$

(C) $I_n + I_{n-1} = x(\log x)^n$　　(D) $I_n + nI_{n-1} = (\log x)$

ANSWERS

1. (A)	2. (B)	3. (D)	4. (C)	5. (B)	6. (D)	7. (A)	8. (D)
9. (C)	10. (A)	11. (B)	12. (D)	13. (A)	14. (C)	15. (D)	16. (B)
17. (A)	18. (C)	19. (B)	20. (D)	21. (C)	22.(A)	23. (D)	24. (B)
25. (C)	26. (D)	27. (A)	28. (B)	29. (C)	30. (C)	31. (D)	32.(A)

•••

Gamma Functions : **Marks**

1. Gamma function of n (n > 0), is defined as ... (1)

   (A) $\int_0^\infty e^{-x} x^{n-1} dx$ (B) $\int_0^\infty e^{x} x^{n-1} dx$

   (C) $\int_0^1 e^{-x} x^{n-1} dx$ (D) $\int_0^\infty e^{-x} x^{1-n} dx$

2. The value of equivalent form of Gamma function $\int_0^\infty e^{-kx} x^{n-1} dx$ is ... (1)

   (A) $\dfrac{\overline{\lfloor n}}{n^k}$ (B) $\dfrac{\overline{\lfloor n}}{k!}$

   (C) $\dfrac{\overline{\lfloor n}}{k^n}$ (D) $\overline{\lfloor n+k+1}$

3. Reduction formula for Gamma function is ... (1)
   (A) $\Gamma(n+1) = (n-1)\,\Gamma(n-1)$ (B) $\Gamma(n+1) = n\,\Gamma(n)$
   (C) $\Gamma(n+1) = (n-1)\,\Gamma(n)$ (D) $\Gamma(n+1) = n\,\Gamma(n-1)$

4. If n is a positive integer, then $\Gamma(n+1)$ is ... (1)
   (A) $(n+1)!$ (B) $(n-1)!$ (C) $(n+2)!$ (D) $n!$

5. $\overline{\lfloor 1}$ is equal to ... (1)

   (A) $2!$ (B) 1 (C) 0 (D) $\sqrt{\pi}$

6. $\overline{\left\lfloor \dfrac{1}{2} \right.}$ is equal to ... (1)

   (A) $\sqrt{\pi}$ (B) π (C) $\dfrac{1}{2}$ (D) 1

7. $\overline{\lfloor 7}$ is equal to ... (1)

   (A) $8!$ (B) $7!$ (C) $6!$ (D) 6

8. $\overline{\left\lfloor \dfrac{5}{2} \right.}$ is equal to ... (1)

   (A) $\dfrac{5}{2}\sqrt{\pi}$ (B) $\dfrac{15}{8}\sqrt{\pi}$ (C) $\dfrac{4}{3}\sqrt{\pi}$ (D) $\dfrac{3}{4}\sqrt{\pi}$

9. By using $\overline{\lfloor p}\ \overline{\lfloor 1-p} = \dfrac{\pi}{\sin p\pi}$, if $0 \le p \le 1$ the value of $\overline{\left\lfloor \dfrac{1}{4} \right.}\ \overline{\left\lfloor \dfrac{3}{4} \right.}$ is ... (1)

   (A) $\dfrac{\pi}{\sqrt{2}}$ (B) π (C) $\sqrt{2}\pi$ (D) 2π

10. $\int_0^\infty e^{-t} t^{3/2} dt$ is equal to ... (1)

    (A) $\dfrac{3}{4}\sqrt{\pi}$ (B) $\dfrac{15}{4}\sqrt{\pi}$ (C) $\dfrac{3}{4}\pi$ (D) $\dfrac{3}{2}\sqrt{\pi}$

11. $\int_0^\infty e^{-5x} x^4 dx$ is equal to ... (1)

    (A) $\dfrac{4!}{4^5}$ (B) $\dfrac{5!}{4^4}$ (C) $\dfrac{5!}{5^5}$ (D) $\dfrac{4!}{5^5}$

12. The appropriate substitution to reduce the given integral $\int_0^\infty \sqrt{x}\,e^{-\sqrt{x}} dx$ to Gamma function integral ... (1)

    (A) $x^3 = t$ (B) $\sqrt{x} = t$ (C) $-x^3 = t$ (D) $\log x = t$

13. The appropriate substitution to reduce the given integral $\int_0^1 (x \log x)^4 dx$ to Gamma function integral ... (1)

    (A) $\log x = -t$ (B) $x = -e^t$
    (C) $x = t^2$ (D) $\log x = t$

14. The appropriate substitution to reduce the given integral $\int_0^\infty \dfrac{x^5}{5^x} dx$ to Gamma function integral ... (1)

    (A) $\log x = -t$ (B) $5^x = e^{-t}$ (C) $5^x = e^t$ (D) $5^x = t$

15. The value of integral $\int_0^\infty \sqrt{x}\,e^{-\sqrt{x}} dx$ by using substitution $\sqrt{x} = t$ is ... (2)
    (A) 1 (B) 4 (C) 2 (D) 3

16. The value of integral $\int_0^\infty e^{-x^2} dx$ by using substitution $x^2 = t$ is ... (2)

    (A) $\sqrt{\pi}$ (B) $\dfrac{\sqrt{\pi}}{2}$ (C) $\dfrac{\sqrt{\pi}}{3}$ (D) $2\sqrt{\pi}$

17. The value of integral $\int_0^\infty e^{-x^4} dx$ by using substitution $x^4 = t$ is ... (2)

    (A) $\overline{\left\lfloor \dfrac{5}{4} \right.}$ (B) $\overline{\left\lfloor \dfrac{1}{4} \right.}$ (C) $\dfrac{1}{4}\overline{\left\lfloor \dfrac{1}{4} \right.}$ (D) $\dfrac{1}{4}\overline{\left\lfloor \dfrac{5}{4} \right.}$

18. The value of integral $\int_0^\infty \sqrt{x}\,e^{-x^3} dx$ by using substitution $x^3 = t$ is ... (2)

    (A) $\dfrac{\sqrt{\pi}}{6}$ (B) $\dfrac{\sqrt{\pi}}{2}$ (C) $3\sqrt{\pi}$ (D) $\dfrac{\sqrt{\pi}}{3}$

19. The value of integral $\int_0^\infty \sqrt[4]{x}\,e^{-\sqrt{x}} dx$ by using substitution $x^2 = t$ is ... (2)

    (A) $\dfrac{3\sqrt{\pi}}{4}$ (B) $\dfrac{3\sqrt{\pi}}{2}$ (C) $\dfrac{15\sqrt{\pi}}{4}$ (D) $\dfrac{\sqrt{\pi}}{3}$

20. The value of integral $\int_0^\infty x^9\,e^{-2x^2} dx$ by using substitution $2x^2 = t$ is ... (2)

    (A) $\dfrac{\overline{\lfloor 5}}{64}$ (B) $\dfrac{\overline{\lfloor 6}}{64}$ (C) $\dfrac{\overline{\lfloor 5}}{32}$ (D) $\dfrac{\overline{\lfloor 6}}{32}$

21. The value of integral $\int\limits_0^1 \left[\log\left(\frac{1}{y}\right)\right]^{n-1} dy$ by using substitution $\log\left(\frac{1}{y}\right) = t$ is ...   (2)

   (A) $\overline{\lfloor n}$ (B) $\overline{\lfloor n+1}$ (C) $-\overline{\lfloor n}$ (D) $\overline{\lfloor n-1}$

22. The value of integral $\int\limits_0^1 \dfrac{dx}{\sqrt{x \log\left(\frac{1}{x}\right)}}$ by using substitution $\log\left(\frac{1}{x}\right) = t$ is ...   (2)

   (A) $\dfrac{\sqrt{\pi}}{2}$ (B) $\sqrt{\pi}$ (C) $\sqrt{2\pi}$ (D) $2\sqrt{\pi}$

23. The value of integral $\int\limits_0^1 \dfrac{dx}{\sqrt{-\log x}}$ by using substitution $-\log x = t$ is ...   (2)

24. The value of integral $\int\limits_0^\infty \dfrac{x^4}{4^x} dx$ by using substitution $4^x = e^t$ is ...   (2)

   (A) $\dfrac{4}{(\log 4)^4}$ (B) $\dfrac{24}{(\log 4)^3}$ (C) $\dfrac{24}{(\log 4)^5}$ (D) $\dfrac{12}{(\log 4)^4}$

25. The value of integral $\int\limits_0^\infty \dfrac{x^2}{2^x} dx$ by using substitution $2^x = e^t$ is ...   (2)

   (A) $\dfrac{1}{(\log 2)^2}$ (B) $\dfrac{2}{(\log 2)^2}$ (C) $\dfrac{2}{(\log 2)^3}$ (D) $\dfrac{3}{(\log 2)^4}$

(A) $\dfrac{\sqrt{\pi}}{2}$ (B) $\sqrt{\pi}$ (C) $\sqrt{2\pi}$ (D) $2\sqrt{\pi}$

ANSWERS

1. (A)	2. (C)	3. (B)	4. (D)	5. (B)	6. (A)	7. (C)	8. (D)	9. (C)
10. (A)	11. (D)	12. (B)	13. (A)	14. (C)	15. (B)	16. (B)	17. (C)	18. (D)
19. (B)	20. (A)	21. (A)	22. (C)	23. (B)	24. (C)	25. (C)		

•••

Beta Functions :   **Marks**

1. B (m, n) is equal to ...   (1)

   (A) $\int\limits_0^1 x^{m-1}(1-x)^{n-1} dx$ (B) $\int\limits_0^\infty x^{m-1}(1-x)^{n-1} dx$

   (C) $\int\limits_0^1 x^m (1-x)^n dx$ (D) $\int\limits_0^\infty e^{-x} x^{n-1} dx$

2. $\int\limits_0^{\pi/2} \sin^p \theta \cos^q \theta \, d\theta$ is equal to ...   (1)

   (A) $B\left(\dfrac{p-1}{2}, \dfrac{q-1}{2}\right)$ (B) $B\left(\dfrac{p+1}{2}, \dfrac{q+1}{2}\right)$

   (C) $B(p, q)$ (D) $\dfrac{1}{2} B\left(\dfrac{p+1}{2}, \dfrac{q+1}{2}\right)$

3. $\int\limits_0^\infty \dfrac{x^{m-1}}{(1+x)^{m+n}} dx$ is equal to ...   (1)

   (A) $B\left(\dfrac{m-1}{2}, \dfrac{n-1}{2}\right)$ (B) $B(m, n)$

   (C) $B(m+1, n+1)$ (D) $B(m, m)$

4. B(m, n) is equal to ...   (1)

   (A) $\dfrac{\overline{\lfloor m}\,\overline{\lfloor n}}{\overline{\lfloor m+n}}$ (B) $\dfrac{\overline{\lfloor n}\,\overline{\lfloor m}}{\overline{\lfloor m+n}}$

   (C) $\dfrac{\overline{\lfloor m}\,\overline{\lfloor n}}{\overline{\lfloor m+n}}$ (D) $\dfrac{\overline{\lfloor m-1}\,\overline{\lfloor n-1}}{\overline{\lfloor m+n}}$

5. Duplication formula for Gamma function is ...   (1)

   (A) $\overline{\lfloor m}\,\sqrt{\overline{\left\lfloor m+\tfrac{1}{2}\right.}} = \dfrac{\sqrt{\pi}}{2^{2m-1}}\overline{\lfloor 2m-1}$

   (B) $\overline{\lfloor m}\,\overline{\left\lfloor m+\tfrac{1}{2}\right.} = \dfrac{\sqrt{\pi}}{2^2}\overline{\lfloor 2m}$

   (C) $\overline{\lfloor m}\,\overline{\left\lfloor m+\tfrac{1}{2}\right.} = \dfrac{\sqrt{\pi}}{2^{2m-1}}\overline{\lfloor 2m}$

   (D) $\overline{\lfloor m}\,\overline{\left\lfloor m+\tfrac{1}{2}\right.} = \dfrac{\pi}{2^{2m-1}}\overline{\lfloor 2m}$

6. B(4, 5) is represented by ...   (1)

   (A) $\int\limits_0^{\pi/2} \sin^3 x \cos^4 x\, dx$ (B) $\int\limits_0^1 x^3 (1-x)^4 dx$

   (C) $\int\limits_0^1 x^2 (1-x)^3 dx$ (D) $\int\limits_0^2 x^3 (1-x)^4 dx$

7. $\int\limits_0^1 x^{1/2} (1-x)^{5/2} dx$ is equal to ...   (1)

   (A) $B\left(\dfrac{3}{2}, \dfrac{7}{2}\right)$ (B) $B\left(-\dfrac{1}{2}, \dfrac{1}{2}\right)$

   (C) $\dfrac{1}{2} B\left(\dfrac{1}{2}, \dfrac{3}{2}\right)$ (D) $B\left(\dfrac{1}{2}, \dfrac{3}{2}\right)$

8. $\displaystyle\int_0^{\pi/2} \sin^{3/2}\theta\,\cos^4\theta\,d\theta$ is equal to ... (1)

    (A) $\dfrac{1}{2}B\left(\dfrac{5}{4},\dfrac{5}{2}\right)$ (B) $B\left(\dfrac{5}{2},\dfrac{3}{2}\right)$

    (C) $\dfrac{1}{2}B\left(\dfrac{3}{2},\dfrac{1}{2}\right)$ (D) $B\left(\dfrac{5}{4},\dfrac{5}{2}\right)$

9. $B(3, 5)$ is equal to ... (1)

    (A) $\dfrac{1}{105}$ (B) $\dfrac{2}{105}$ (C) $\dfrac{4}{105}$ (D) $\dfrac{1}{21}$

10. $\displaystyle\int_0^{\infty} \dfrac{x^4}{(1+x)^7}\,dx$ is equal to ... (1)

    (A) $\dfrac{1}{30}$ (B) $\dfrac{1}{20}$ (C) $\dfrac{1}{15}$ (D) $\dfrac{1}{60}$

11. $\overline{p}\;\overline{1-p} = \ldots$ (1)

    (A) $\dfrac{1}{2}\dfrac{\pi}{\sin p\pi}$ (B) $\dfrac{\pi}{\sin p\pi}$ (C) $\dfrac{1}{\sin p\pi}$ (D) $\dfrac{\dfrac{\pi}{2}}{\sin\dfrac{p\pi}{2}}$

12. The appropriate substitution to reduce the given integral $\displaystyle\int_0^2 x\sqrt[3]{8-x^3}\,dx$ to Beta function integral ... (1)

    (A) $x = t^3$ (B) $x^3 = t$ (C) $x = 8t^3$ (D) $x^3 = 8t$

13. The appropriate substitution to reduce the given integral $\displaystyle\int_0^n x^n(n-x)^m\,dx$ to Beta function integral ... (1)

    (A) $x = t^n$ (B) $x = mt$ (C) $x = t$ (D) $x = nt$

14. The appropriate substitution to reduce the given integral $\displaystyle\int_a^b (x-a)^m(b-x)^n\,dx$ to Beta function integral ... (1)

    (A) $x = (b-a)\,t$ (B) $(x-a) = (b-a)\,t$

    (C) $(x+a) = (b-a)\,t$ (D) $(x-a) = (b+a)\,t$

15. The appropriate substitution to reduce the given integral $\displaystyle\int_5^9 (x-5)^{1/4}(9-x)^{1/4}\,dx$ to Beta function integral ... (2)

    (A) $x = 4t + 5$ (B) $x = 4t$

    (C) $x = 4t - 5$ (D) $x = 14t + 5$

16. The value of integral $\displaystyle\int_0^1 x^{m-1}(1-x^2)^{n-1}\,dx$ by using substitution $x = \sqrt{t}$ is ... (2)

    (A) $\dfrac{1}{2}B\left(\dfrac{m-1}{2}, n-1\right)$ (B) $B\left(\dfrac{m}{2}, n\right)$

    (C) $\dfrac{1}{2}B\left(\dfrac{m}{2}, n\right)$ (D) $B\left(\dfrac{m-1}{2}, n-1\right)$

17. The value of integral $\displaystyle\int_0^1 x^3(1-\sqrt{x})^5\,dx$ by using substitution $\sqrt{x} = t$ is ... (2)

    (A) $2B(8, 6)$ (B) $B(8, 6)$ (C) $2B(7, 6)$ (D) $2B(6, 4)$

18. The value of integral $\displaystyle\int_0^1 (1-x^{1/n})^m\,dx$ by using substitution $x^{1/n} = t$ is ... (2)

    (A) $B(n, m+1)$ (B) $nB(n, m+1)$

    (C) $B(m, n+1)$ (D) $mB(m, n+1)$

19. The value of integral $\displaystyle\int_0^2 x(8-x^3)^{1/3}\,dx$ by using substitution $x^3 = 8t$ is ... (2)

    (A) $\dfrac{2}{3}B\left(\dfrac{2}{3},\dfrac{4}{3}\right)$ (B) $\dfrac{4}{3}B\left(-\dfrac{1}{3},\dfrac{1}{3}\right)$

    (C) $\dfrac{2}{3}B\left(-\dfrac{1}{3},\dfrac{1}{3}\right)$ (D) $\dfrac{8}{3}B\left(\dfrac{2}{3},\dfrac{4}{3}\right)$

20. The integral $\displaystyle\int_3^7 \sqrt{(x-3)(7-x)}\,dx$ by using substitution $x = 4t + 3$ transforms to ... (2)

    (A) $\displaystyle\int_0^1 t^{1/2}(1-t)^{1/2}\,dt$ (B) $4\displaystyle\int_0^1 t^{1/2}(1-t)^{1/2}\,dt$

    (C) $16\displaystyle\int_0^1 t^{1/2}(1-t)^{1/2}\,dt$ (D) $16\displaystyle\int_0^1 t^{1/4}(1-t)^{1/4}\,dt$

21. The value of $\displaystyle\int_0^{\infty}\dfrac{x^8}{(1+x)^{24}}\,dx - \int_0^{\infty}\dfrac{x^{14}}{(1+x)^{24}}\,dx$ is ... (2)

    (A) $B(8, 14) - B(9, 15)$ (B) $B(9, 15) - B(8, 14)$

    (C) 0 (D) $2B(9, 15)$

22. $\displaystyle\int_0^{\pi/2}\sqrt{\tan\theta}\,d\theta$ is equal to ... (2)

    (A) $\dfrac{1}{2}B\left(\dfrac{3}{4},\dfrac{1}{4}\right)$ (B) $B\left(\dfrac{3}{4},\dfrac{1}{4}\right)$

    (C) $B\left(\dfrac{1}{2},-\dfrac{1}{2}\right)$ (D) $\dfrac{1}{2}B\left(\dfrac{1}{2},-\dfrac{1}{2}\right)$

23. $\displaystyle\int_0^{\pi/2}\sqrt{\cot\theta}\,d\theta$ is equal to ... (2)

    (A) $\dfrac{1}{2}B\left(-\dfrac{1}{2},\dfrac{1}{2}\right)$ (B) $B\left(\dfrac{1}{4},\dfrac{3}{4}\right)$

    (C) $B\left(\dfrac{1}{2},-\dfrac{1}{2}\right)$ (D) $\dfrac{1}{2}B\left(\dfrac{1}{4},\dfrac{3}{4}\right)$

24. $\displaystyle\int_0^{\pi/2}\dfrac{1}{\sqrt{\sin\theta}}\,d\theta$ is equal to ... (2)

    (A) $B\left(\dfrac{1}{4},\dfrac{1}{2}\right)$ (B) $\dfrac{1}{2}B\left(\dfrac{1}{4},\dfrac{1}{2}\right)$

    (C) $\dfrac{1}{2}B\left(-\dfrac{1}{2},0\right)$ (D) $B\left(-\dfrac{1}{2},0\right)$

25. $\displaystyle\int_0^1 \frac{x^3 + x^2}{(1 + x)^7}\, dx$ is equal to ... (2)

    (A) $\dfrac{1}{240}$ (B) $\dfrac{1}{30}$ (C) $\dfrac{1}{60}$ (D) $\dfrac{1}{120}$

26. $B\left(\dfrac{1}{4}, \dfrac{3}{4}\right)$ is equal to ... (2)

    (A) $\dfrac{2\pi}{\sqrt{3}}$ (B) 0 (C) 2π (D) $\pi\sqrt{2}$

27. $B\left(\dfrac{5}{4}, \dfrac{5}{4}\right)$ is equal to ... (2)

    (A) $\dfrac{2}{3}\left[\overline{\left\lfloor\dfrac{1}{4}\right.}\right]^2$ (B) $\dfrac{1}{12\sqrt{\pi}}\left[\overline{\left\lfloor\dfrac{1}{4}\right.}\right]^2$

    (C) $\dfrac{2}{3\sqrt{\pi}}\left[\overline{\left\lfloor\dfrac{1}{4}\right.}\right]$ (D) $\dfrac{1}{\sqrt{\pi}}\left[\overline{\left\lfloor\dfrac{1}{4}\right.}\right]^2$

28. $B(n, n + 1)$ is equal to ... (2)

    (A) $\dfrac{1}{2}\dfrac{\left(\overline{\lfloor n}\right)^2}{\overline{\lfloor 2n}}$ (B) $\dfrac{\left(\overline{\lfloor n}\right)^2}{\overline{\lfloor 2n}}$

    (C) $\dfrac{1}{2}\dfrac{\overline{\lfloor n}\;\;\overline{\lfloor n-1}}{\overline{\lfloor 2n}}$ (D) $\dfrac{1}{2}\dfrac{\left(\overline{\lfloor n+1}\right)^2}{\overline{\lfloor 2n}}$

29. The value of $B(m, n + 1) + B(m + 1, n)$ is ... (2)

    (A) $B(m, n)$ (B) $2B(m + 1, n)$

    (C) $2B(m, n + 1)$ (D) $B(m - 1, n - 1)$

30. $B(m, n) \times B(m + n, p)$ is equal to ... (2)

    (A) $\dfrac{\overline{\lfloor m}\;\;\overline{\lfloor p}}{\overline{\lfloor m + n + p}}$ (B) $\dfrac{\overline{\lfloor m}\;\;\overline{\lfloor n}\;\;\overline{\lfloor p}}{\overline{\lfloor m + n + p}}$

    (C) $\dfrac{\overline{\lfloor m}\;\;\overline{\lfloor n}}{\overline{\lfloor m + n + p}}$ (D) $\dfrac{\overline{\lfloor m}\;\;\overline{\lfloor n}\;\;\overline{\lfloor p}}{\overline{\lfloor m - n - p}}$

ANSWERS

1. (A)	2. (D)	3. (B)	4. (C)	5. (C)	6. (B)	7. (A)	8. (A)
9. (A)	10. (A)	11. (B)	12. (D)	13. (D)	14. (B)	15. (A)	16. (C)
17. (A)	18. (B)	19. (D)	20. (C)	21. (C)	22. (A)	23. (D)	24. (B)
25. (B)	26. (D)	27. (B)	28. (A)	29. (A)	30. (B)		

•••

CHAPTER 4 : DIFFERENTIATION UNDER THE INTEGRAL SIGN AND ERROR FUNCTIONS

Differentiation under the Integral Sign (DUIS) :   **Marks**

1. If $I(\alpha) = \displaystyle\int_a^b f(x, \alpha)\, dx$, where α is parameter and a, b are constants then by DUIS rule, $\dfrac{dI(\alpha)}{d\alpha}$ is ... (1)

    (A) $\displaystyle\int_a^b \dfrac{\partial}{\partial\alpha} f(x, \alpha)\, dx$ (B) $\displaystyle\int_a^b \dfrac{\partial}{\partial x} f(x, \alpha)\, dx$

    (C) $f(b, \alpha) - f(a, \alpha)$ (D) $f(x, \alpha)$

2. If $I(\alpha) = \displaystyle\int_a^b f(x, \alpha)\, dx$ where a, b are functions of parameter α then by DUIS rule, $\dfrac{dI(\alpha)}{d\alpha}$ is ... (1)

    (A) $\displaystyle\int_a^b \dfrac{\partial}{\partial\alpha} f(x, \alpha)\, dx + f(a, \alpha)\dfrac{da}{d\alpha} - f(b, \alpha)\dfrac{db}{d\alpha}$

    (B) $\displaystyle\int_a^b \dfrac{\partial}{\partial\alpha} f(x, \alpha)\, dx + f(b, \alpha)\dfrac{db}{d\alpha} + f(a, \alpha)\dfrac{da}{d\alpha}$

    (C) $\displaystyle\int_a^b \dfrac{\partial}{\partial\alpha} f(x, \alpha)\, dx + f(b, \alpha)\dfrac{db}{d\alpha} - f(a, \alpha)\dfrac{da}{d\alpha}$

    (D) $\displaystyle\int_a^b \dfrac{\partial}{\partial\alpha} f(x, \alpha)\, dx$

3. If $\phi(a) = \displaystyle\int_0^\infty \dfrac{e^{-x}}{x}(1 - e^{-ax})\, dx$, $a > -1$ then by DUIS rule, $\dfrac{d\phi}{da}$ is ... (1)

    (A) $\displaystyle\int_0^\infty \dfrac{\partial}{\partial a} e^{-x}(1 - e^{-ax})\, dx$ (B) $\displaystyle\int_0^\infty \dfrac{\partial}{\partial a}\dfrac{e^{-x}}{x}(1 - e^{-ax})\, dx$

    (C) $\displaystyle\int_0^\infty \dfrac{\partial}{\partial x}\dfrac{e^{-x}}{x}(1 - e^{-ax})\, dx$ (D) $\dfrac{e^{-x}}{x}(1 - e^{-ax})$

4. If $\phi(a) = \displaystyle\int_0^\infty e^{-bx^2}\cos(2ax)\, dx$, $b > 0$ then by DUIS rule, $\dfrac{d\phi}{da}$ is ... (1)

    (A) $\displaystyle\int_0^\infty \dfrac{\partial}{\partial b}\left[e^{-bx^2}\cos(2ax)\right] dx$

    (B) $e^{-bx^2}\cos(2ax)$

    (C) $\displaystyle\int_0^\infty \dfrac{\partial}{\partial x}\left[e^{-bx^2}\cos(2ax)\right] dx$

    (D) $\displaystyle\int_0^\infty \dfrac{\partial}{\partial a}\left[e^{-bx^2}\cos(2ax)\right] dx$

5. If $\phi(a) = \int\limits_0^\infty \dfrac{e^{-x}}{x}\left(a - \dfrac{1}{x} + \dfrac{1}{x} e^{-ax}\right) dx$, then by DUIS rule, $\dfrac{d\phi}{da}$ is ... (1)

(A) $\dfrac{e^{-x}}{x}\left(a - \dfrac{1}{x} + \dfrac{1}{x} e^{-ax}\right)$

(B) $\int\limits_0^\infty \dfrac{\partial}{\partial a} \dfrac{e^{-x}}{x}\left(a - \dfrac{1}{x} + \dfrac{1}{x} e^{-ax}\right) dx$

(C) $\int\limits_0^\infty \dfrac{\partial}{\partial x} \dfrac{e^{-x}}{x}\left(a - \dfrac{1}{x} + \dfrac{1}{x} e^{-ax}\right) dx$

(D) $\int\limits_0^\infty \dfrac{\partial}{\partial a} e^{-x}\left(a - \dfrac{1}{x} + \dfrac{1}{x} e^{-ax}\right) dx$

6. If $\phi(a) = \int\limits_0^\infty \dfrac{e^{-x}}{x}(1 - e^{-ax}) dx$, $a > -1$ then by DUIS rule, $\dfrac{d\phi}{da}$ is ... (2)

(A) $\dfrac{e^{-x}}{x}(1 - e^{-ax})$

(B) $\int\limits_0^\infty \dfrac{a}{x} e^{-(a+1)x} dx$

(C) $\int\limits_0^\infty e^{-ax} dx$

(D) $\int\limits_0^\infty e^{-(a+1)x} dx$

7. If $\phi(a) = \int\limits_0^1 \dfrac{x^a - 1}{\log x} dx$; $a \geq 0$ then by DUIS rule, $\dfrac{d\phi}{da}$ is ... (2)

(A) $\int\limits_0^1 \dfrac{x^a \log a}{\log x} dx$

(B) $\int\limits_0^1 \dfrac{ax^{a-1}}{\log x} dx$

(C) $\int\limits_0^1 x^a dx$

(D) $\dfrac{x^a - 1}{\log x}$

8. If $\phi(\alpha) = \int\limits_0^\infty \dfrac{e^{-x} \sin \alpha x}{x} dx$ then by DUIS rule, $\dfrac{d\phi}{d\alpha}$ is ... (2)

(A) $\int\limits_0^\infty e^{-x} \sin \alpha x\, dx$

(B) $\int\limits_0^\infty e^{-x} \cos \alpha x\, dx$

(C) $\int\limits_0^\infty \dfrac{\alpha e^{-x} \sin \alpha x}{x} dx$

(D) $\dfrac{e^{-x} \sin \alpha x}{x}$

9. If $I(a) = \int\limits_0^\infty e^{-\left(x^2 + \frac{a^2}{x^2}\right)} dx$; $a > 0$ then by DUIS rule, $\dfrac{dI}{da}$ is ... (2)

(A) $\int\limits_0^\infty e^{-\left(x^2 + \frac{a^2}{x^2}\right)}\left(-\dfrac{2a}{x^2}\right) dx$

(B) $\int\limits_0^\infty e^{-\left(x^2 + \frac{a^2}{x^2}\right)}\left(\dfrac{2a}{x^2}\right) dx$

(C) $\int\limits_0^\infty e^{-\left(x^2 + \frac{a^2}{x^2}\right)}\left(-2x - \dfrac{2a^2}{x^3}\right) dx$

(D) $e^{-\left(x^2 + \frac{a^2}{x^2}\right)}$

10. If $I(a) = \int\limits_0^\pi \log (1 - a \cos x)\, dx$; $|a| < 1$ then by DUIS rule, $\dfrac{dI}{da}$ is ... (2)

(A) $\int\limits_0^\pi \dfrac{-a \sin x}{1 - a \cos x} dx$

(B) $\int\limits_0^\pi \dfrac{\cos x}{1 - a \cos x} dx$

(C) $\int\limits_0^\pi \dfrac{-\cos x}{1 - a \cos x} dx$

(D) $\int\limits_0^\pi \dfrac{1}{1 - a \cos x} dx$

11. By DUIS rule $\dfrac{d}{da}\left[\int\limits_0^\infty \dfrac{e^{-x}}{x}\left(a - \dfrac{1}{x} + \dfrac{1}{x} e^{-ax}\right) dx\right]$, a is parameter, is ... (2)

(A) $\int\limits_0^\infty \dfrac{e^{-x}}{x}(1 + e^{-ax}) dx$

(B) $\int\limits_0^\infty \dfrac{e^{-x}}{x}\left(1 - \dfrac{ae^{-ax}}{x}\right) dx$

(C) $\int\limits_0^\infty e^{-x}(1 - e^{-ax}) dx$

(D) $\int\limits_0^\infty \dfrac{e^{-x}}{x}(1 - e^{-ax}) dx$

12. If $I(x) = \int\limits_0^\infty e^{-a^2} \cos ax\, da$, x is parameter then by DUIS rule, $\dfrac{dI}{dx}$ is ... (2)

(A) $\int\limits_0^\infty x e^{-a^2} \sin (xa)\, da$

(B) $\int\limits_0^\infty a e^{-a^2} \sin (xa)\, da$

(C) $\int\limits_0^\infty -a e^{-a^2} \sin (xa)\, da$

(D) $\int\limits_0^\infty a e^{-a^2} \cos (xa)\, da$

13. If $I(a) = \int\limits_0^\infty \dfrac{e^{-x} - e^{-ax}}{x \sec x} dx$, $a > 0$ then by DUIS rule, $\dfrac{dI}{da}$ is ... (2)

(A) $\int\limits_0^\infty \dfrac{e^{-x} + e^{-ax} x}{x \sec x} dx$

(B) $\int\limits_0^\infty \dfrac{e^{-x}}{\sec x} dx$

(C) $\int\limits_0^\infty (e^{-x} - e^{-ax}) dx$

(D) $\int\limits_0^\infty \dfrac{e^{-ax}}{\sec x} dx$

14. If $\phi(a) = \int\limits_0^1 \dfrac{x^a - x^b}{\log x} dx$; $a > 0$, $b > 0$ then by DUIS rule, $\dfrac{d\phi}{da}$ is ... (2)

(A) $\int\limits_0^1 \dfrac{x^a \log a}{\log x} dx$

(B) $\int\limits_0^1 x^a dx$

(C) $\int\limits_0^1 x^b dx$

(D) $\int\limits_0^1 (x^a - x^b) dx$

15. If $\phi(b) = \int\limits_0^\infty \dfrac{e^{-ax} - e^{-bx}}{x}\,dx$; $a > 0$, $b > 0$ then by DUIS rule, $\dfrac{d\phi}{db}$ is ... (2)

  (A) $\int\limits_0^\infty e^{-bx}\,dx$

  (B) $\int\limits_0^\infty \dfrac{e^{-ax}\,(-a) - e^{-bx}\,(-b)}{x}\,dx$

  (C) $\int\limits_0^\infty e^{-ax}\,dx$

  (D) $\int\limits_0^\infty (e^{-ax} - e^{-bx})\,dx$

16. If $\phi(a) = \int\limits_0^\infty \dfrac{1}{x^2} \log(1 + ax^2)\,dx$; $a > 0$, then by DUIS rule, $\dfrac{d\phi}{da}$ is ... (2)

  (A) $\int\limits_0^\infty \dfrac{a}{x(1 + ax^2)}\,dx$ 　　(B) $\int\limits_0^\infty \dfrac{\log(1 + ax^2)}{x}\,dx$

  (C) $\int\limits_0^\infty \dfrac{2a}{x(1 + ax^2)}\,dx$ 　　(D) $\int\limits_0^\infty \dfrac{1}{1 + ax^2}\,dx$

17. If $\phi(a) = \int\limits_0^{\pi/2} \dfrac{\log(1 + a\sin^2 x)}{\sin^2 x}\,dx$; $a > 0$, then by DUIS rule, $\dfrac{d\phi}{da}$ is ... (2)

  (A) $\int\limits_0^{\pi/2} \dfrac{2\sin x \cos x}{(1 + a\sin^2 x)}\,dx$ 　　(B) $\int\limits_0^{\pi/2} \dfrac{1}{(1 + a\sin^2 x)\sin^2 x}\,dx$

  (C) $\int\limits_0^{\pi/2} \dfrac{1}{1 + a\sin^2 x}\,dx$ 　　(D) $\int\limits_0^{\pi/2} \dfrac{\sin^2 x}{(1 + a\sin^2 x)}\,dx$

18. If $\phi(a) = \int\limits_a^{a^2} \log(ax)\,dx$, then by DUIS rule II, $\dfrac{d\phi}{da}$ is ... (2)

  (A) $\int\limits_a^{a^2} \dfrac{\partial}{\partial a} \log(ax)\,dx + 2a \log a^3$

  (B) $\int\limits_a^{a^2} \dfrac{\partial}{\partial a} \log(ax)\,dx$

  (C) $\int\limits_a^{a^2} \dfrac{\partial}{\partial a} \log(ax)\,dx + 2a \log a^3 - 2\log a$

  (D) $\int\limits_a^{a^2} \dfrac{\partial}{\partial a} \log(ax)\,dx - 2a \log a^3 + 2\log a$

19. If $\phi(a) = \int\limits_0^{a^2} \tan^{-1}\left(\dfrac{x}{a}\right)\,dx$, then by DUIS rule II, $\dfrac{d\phi}{da}$ is ... (2)

  (A) $\int\limits_0^{a^2} \dfrac{\partial}{\partial a} \tan^{-1}\left(\dfrac{x}{a}\right)\,dx + 2a \tan^{-1} a$

  (B) $\int\limits_0^{a^2} \dfrac{\partial}{\partial a} \tan^{-1}\left(\dfrac{x}{a}\right)\,dx$

  (C) $\int\limits_0^{a^2} \dfrac{\partial}{\partial a} \tan^{-1}\left(\dfrac{x}{a}\right)\,dx + a^2 \tan^{-1} a$

  (D) $\int\limits_0^{a^2} \dfrac{\partial}{\partial a} \tan^{-1}\left(\dfrac{x}{a}\right)\,dx + a^2 \tan^{-1} a - \tan^{-1}\left(\dfrac{x}{a}\right)$

20. If $I(a) = \int\limits_a^{a^2} \dfrac{1}{x + a}\,dx$, then by DUIS rule II, $\dfrac{dI}{da}$ is ... (2)

  (A) $\int\limits_a^{a^2} \dfrac{\partial}{\partial a}\left(\dfrac{1}{x + a}\right)\,dx - \dfrac{1}{a^2 + a}\,(2a) + \dfrac{1}{2a}$

  (B) $\int\limits_a^{a^2} \dfrac{\partial}{\partial a}\left(\dfrac{1}{x + a}\right)\,dx + \dfrac{1}{a^2 + a}\,(2a) - \dfrac{1}{2a}$

  (C) $\int\limits_a^{a^2} \dfrac{\partial}{\partial a}\left(\dfrac{1}{x + a}\right)\,dx - \dfrac{1}{a^2 + a}$

  (D) $\int\limits_a^{a^2} \dfrac{\partial}{\partial a}\left(\dfrac{1}{x + a}\right)\,dx$

21. If $\phi(a) = \int\limits_0^a \dfrac{\log(1 + ax)}{1 + x^2}\,dx$, then by DUIS rule II, $\dfrac{d\phi}{da}$ is ... (2)

  (A) $\int\limits_0^a \dfrac{\partial}{\partial a}\left[\dfrac{\log(1 + ax)}{1 + x^2}\right]\,dx$

  (B) $\int\limits_0^a \dfrac{\partial}{\partial a}\left[\dfrac{\log(1 + ax)}{1 + x^2}\right]\,dx - \dfrac{\log(1 + a)}{1 + a^2}$

  (C) $\int\limits_0^a \dfrac{\partial}{\partial a}\left[\dfrac{\log(1 + ax)}{1 + x^2}\right]\,dx - \dfrac{\log(1 + a^2)}{1 + a^2}$

  (D) $\int\limits_0^a \dfrac{\partial}{\partial a}\left[\dfrac{\log(1 + ax)}{1 + x^2}\right]\,dx + \dfrac{\log(1 + a^2)}{1 + a^2}$

22. If $F(t) = \int\limits_{t}^{t^2} e^{tx2} \, dx$, then by DUIS rule II, $\dfrac{dF}{dt}$ is ... (2)

(A) $\int\limits_{t}^{t^2} \dfrac{\partial}{\partial t} e^{tx2} \, dx + (2t) \, e^{t4} - e^{t2}$

(B) $\int\limits_{t}^{t^2} \dfrac{\partial}{\partial t} e^{tx2} \, dx + e^{t5} - e^{t3}$

(C) $\int\limits_{t}^{t^2} \dfrac{\partial}{\partial t} e^{tx2} \, dx + (2t) \, e^{t5} - e^{t3}$

(D) $\int\limits_{t}^{t^2} \dfrac{\partial}{\partial t} e^{tx2} \, dx$

23. If $f(x) = \int\limits_{a}^{x} (x - t)^2 \, G(t) \, dt$, a is constant and x is parameter then by DUIS rule II, $\dfrac{df}{dx}$ is ... (2)

(A) $\int\limits_{a}^{x} \dfrac{\partial}{\partial x} (x - t)^2 \, G(t) \, dt + G(x)$

(B) $\int\limits_{a}^{x} \dfrac{\partial}{\partial x} (x - t)^2 \, G(t) \, dt$

(C) $\int\limits_{a}^{x} \dfrac{\partial}{\partial x} (x - t)^2 \, G(t) \, dt - (x - a)^2 \, G(a)$

(D) $(x - t)^2 \, G(t)$

24. If $y = \int\limits_{0}^{x} f(t) \sin a(x - t) \, dt$, then by DUIS rule II, $\dfrac{dy}{dx}$ is ... (2)

(A) $\int\limits_{0}^{x} af(t) \sin a \, (x - t) \, dt$

(B) $\int\limits_{0}^{x} f(t) \cos a(x - t) \, dt$

(C) $\int\limits_{0}^{x} af(t) \cos a(x - t) \, dt$

(D) $\int\limits_{0}^{x} af(t) \cos a(x - t) \, dt + f(x)$

25. Using DUIS rule the value of integral
$\phi(a) = \int\limits_{0}^{\infty} \dfrac{e^{-x}}{x} (1 - e^{-ax}) \, dx$, $a > -1$, given $\dfrac{d\phi}{da} = \dfrac{1}{a + 1}$ is ... (2)

(A) $\log (a + 1)$

(B) $-\dfrac{1}{(a + 1)^2}$

(C) $\log (a + 1) + \pi$

(D) $-\dfrac{1}{(a + 1)^2} + 1$

26. Using DUIS rule the value of integral $\phi(a) = \int\limits_{0}^{1} \dfrac{x^a - 1}{\log x} \, dx$, $a \geq 0$, given $\dfrac{d\phi}{da} = \dfrac{1}{a + 1}$ is ... (2)

(A) $\log (a + 1)$

(B) $-\dfrac{1}{(a + 1)^2}$

(C) $\log (a + 1) + \pi$

(D) $-\dfrac{1}{(a + 1)^2} + 1$

27. Using DUIS rule the value of integral $\phi(\alpha) = \int\limits_{0}^{\infty} \dfrac{e^{-2x} \sin \alpha x}{x} \, dx$, with $\dfrac{d\phi}{d\alpha} = \dfrac{2}{\alpha^2 + 4}$ is ... (2)

(a) $2 \log (\alpha^2 + 4)$

(B) $2 \tan^{-1} \left(\dfrac{\alpha}{2}\right)$

(C) $\dfrac{1}{2} \tan^{-1} \left(\dfrac{\alpha}{2}\right)$

(D) $\tan^{-1} \left(\dfrac{\alpha}{2}\right)$

28. Using DUIS rule the value of integral $\phi(\alpha) = \int\limits_{0}^{\infty} \dfrac{e^{-\alpha x} \sin x}{x} \, dx$, with $\dfrac{d\phi}{d\alpha} = -\dfrac{1}{\alpha^2 + 1}$ and assuming $\int\limits_{0}^{\infty} \dfrac{\sin x}{x} \, dx = \dfrac{\pi}{2}$ is ... (2)

(A) $\tan^{-1} \alpha + \dfrac{\pi}{2}$

(B) $-\tan^{-1} \alpha + \dfrac{\pi}{2}$

(C) $-\tan^{-1} \alpha$

(D) $\log (\alpha^2 + 1) + \dfrac{\pi}{2}$

29. Using DUIS rule the value of integral
$\phi(a) = \int\limits_{0}^{\pi/2} \dfrac{\log (1 + a \sin^2 x)}{\sin^2 x} \, dx$, with $\dfrac{d\phi}{da} = \dfrac{\pi}{2} \dfrac{1}{\sqrt{a + 1}}$ is ... (2)

(A) $\pi\sqrt{a + 1}$

(B) $\pi\sqrt{a + 1} + \pi$

(C) $\pi\sqrt{a + 1} - \pi$

(D) $3\pi \, (a + 1)^{3/2} - \pi$

30. Using DUIS rule the value of integral
$\phi(a) = \int\limits_{0}^{\infty} \dfrac{\log (1 + ax^2)}{x^2} \, dx$; $a > 0$ with $\dfrac{d\phi}{da} = \dfrac{\pi}{2} \dfrac{1}{\sqrt{a}}$ is ... (2)

(A) $\pi\sqrt{a}$

(B) $\dfrac{\pi\sqrt{a}}{4}$

(C) $-\dfrac{\pi}{4a^{3/2}}$

(D) $\dfrac{\pi}{4\sqrt{a}}$

31. Using DUIS rule the value of integral $\phi(a) = \int\limits_{0}^{\infty} \dfrac{e^{-x} - e^{-ax}}{x \sec x} \, dx$, with $\dfrac{d\phi}{da} = \dfrac{a}{a^2 + 1}$ is ... (2)

(A) $\tan^{-1} a + \dfrac{\pi}{4}$

(B) $\log \left(\dfrac{2}{a^2 + 1}\right)$

(C) $\dfrac{1}{2} \log (a^2 + 1)$

(D) $\dfrac{1}{2} \log \left(\dfrac{a^2 + 1}{2}\right)$

32. Using DUIS rule the value of integral $\phi(a) = \int_0^\infty \dfrac{1 - \cos ax}{x^2}\, dx$, with $\dfrac{d\phi}{da} = \dfrac{\pi}{2}$ is ... (2)

(A) $\dfrac{\pi}{2}$ (B) $\dfrac{\pi a}{2}$ (C) πa (D) $\dfrac{\pi a}{2} + \dfrac{\pi}{2}$

ANSWERS

1. (A)	2. (C)	3. (B)	4. (D)	5. (B)	6. (D)	7. (C)	8. (B)
9. (A)	10. (C)	11. (D)	12. (C)	13. (D)	14. (B)	15. (A)	16. (D)
17. (C)	18. (C)	19. (A)	20. (B)	21. (D)	22. (C)	23. (B)	24. (C)
25. (A)	26. (A)	27. (D)	28. (B)	29. (C)	30. (A)	31. (D)	32. (B)

•••

Error Functions :   Marks

1. Error function of x, erf(x) is defined as ... (1)

(A) $\dfrac{2}{\sqrt{\pi}} \int_0^x e^{-u^2}\, du$ (B) $\dfrac{2}{\sqrt{\pi}} \int_0^\infty e^{-u^2}\, du$

(C) $\int_0^\infty e^{-x} x^{n-1}\, dx$ (D) $\dfrac{2}{\sqrt{\pi}} \int_0^\infty e^{-u^2}\, du$

2. Complimentary error function of x, erfc (c) is defined as ... (1)

(A) $\dfrac{2}{\sqrt{\pi}} \int_0^x e^{-u^2}\, du$ (B) $\dfrac{2}{\sqrt{\pi}} \int_x^\infty e^{-u^2}\, du$

(C) $\int_0^\infty e^{-x} x^{n-1}\, dx$ (D) $\dfrac{2}{\sqrt{\pi}} \int_0^\infty e^{-u^2}\, du$

3. The value of erf (∞) is ... (1)

(A) 0   (B) ∞ (C) 1   (D) $\dfrac{2}{\sqrt{\pi}}$

4. The value of erf (0) is ... (1)

(A) -1 (B) ∞ (C) 1   (D) 0

5. The value of erfc(0) is ... (1)

(A) -1 (B) ∞ (C) 1   (D) 0

6. Which of the following is true ? (1)

(A) erf(x) – erfc(x) = 1   (B) erf(x) + erfc(x) = 1

(C) erf(x) + erfc(x) = 2   (D) erf(–x) = erf(x)

7. Error function is ... (1)

(A) a periodic function   (B) an even function

(C) a harmonic function   (D) an odd function

8. erf(–x) is equal to ... (1)

(A) –erf(x)   (B) erf(x)   (C) erfc(x)   (D) erfc(–x)

9. The proper substitution to reduce the integral $\int_0^\infty e^{-(x+a)^2}\, dx$ to complimentary error function is ... (1)

(A) $(x + a)^2 = u$ (B) $-(x + a) = u$

(C) $x + a = u$ (D) $-(x + a)^2 = u$

10. erf(x) + erf(–x) = ... (1)

(A) 2   (B) 1   (C) –1   (D) 0

11. erf(–x) + erfc(–x) = ... (1)

(A) 2   (B) 1   (C) –1   (D) 0

12. erfc(–x) – erf(x) = ... (1)

(A) 2   (B) –1   (C) 1   (D) 0

13. erfc(–x) + erfc(x) = ... (1)

(A) 2   (B) –1   (C) 1   (D) 0

14. If erf (ax) $= \dfrac{2}{\sqrt{\pi}} \int_0^{ax} e^{-u^2}\, du$ then $\dfrac{d}{dx}$ erf (ax) is ... (2)

(A) $\dfrac{2x}{\sqrt{\pi}} e^{-a^2 x^2}$ (B) $\dfrac{2a}{\sqrt{\pi}} e^{-a^2 x^2}$

(C) $ae^{-a^2 x^2}$ (D) $\dfrac{2a}{\sqrt{\pi}} e^{a^2 x^2}$

15. If erf $(\sqrt{t}) = \dfrac{2}{\sqrt{\pi}} \int_0^{\sqrt{t}} e^{-u^2}\, du$ then $\dfrac{d}{dt}$ erf $(\sqrt{t})$ is ... (2)

(A) $\dfrac{e^{-t}}{2\sqrt{t}}$ (B) $\dfrac{e^{-t^2}}{\sqrt{\pi t}}$ (C) $\dfrac{e^{-t}}{\sqrt{\pi}}$ (D) $\dfrac{e^{-t}}{\sqrt{\pi t}}$

16. If erfc (ax) $= \dfrac{2}{\sqrt{\pi}} \int_{ax}^\infty e^{-u^2}\, du$ then $\dfrac{d}{dx}$ erfc (ax) is ... (2)

(a) $-\dfrac{1}{\sqrt{\pi}} e^{-a^2 x^2}$ (B) $-\dfrac{2}{\sqrt{\pi}} e^{-a^2 x^2}$

(C) $-\dfrac{2a}{\sqrt{\pi}} e^{-a^2 x^2}$ (D) $\dfrac{2a}{\sqrt{\pi}} e^{a^2 x^2}$

17. If erfc $(\sqrt{t}) = \dfrac{2}{\sqrt{\pi}} \int_{\sqrt{t}}^\infty e^{-u^2}\, du$ then $\dfrac{d}{dt}$ erfc $(\sqrt{t})$ is ... (2)

(A) $\dfrac{e^{-t}}{2\sqrt{t}}$ (B) $-\dfrac{e^{-t}}{\sqrt{\pi t}}$ (C) $\dfrac{e^{-t}}{\sqrt{\pi t}}$ (D) $\dfrac{e^{-t^2}}{\sqrt{\pi t}}$

18. If $\dfrac{d}{dx}$ erf (ax) $= \dfrac{2a}{\sqrt{\pi}} e^{-a^2 x^2}$ then $\dfrac{d}{dx}$ erfc (ax) is ... (2)

(A) $\dfrac{2a}{\sqrt{\pi}} e^{a^2 x^2}$ (B) $1 - \dfrac{2a}{\sqrt{\pi}} e^{-a^2 x^2}$

(C) $-\dfrac{1}{\sqrt{\pi}} e^{-a^2 x^2}$ (D) $-\dfrac{2a}{\sqrt{\pi}} e^{-a^2 x^2}$

19. If $\dfrac{d}{dt}$ erf $(\sqrt{t}) = \dfrac{e^{-t}}{\sqrt{\pi t}}$ then $\dfrac{d}{dt}$ erfc $(\sqrt{t})$ is ... (2)

(A) $-\dfrac{e^{-t}}{\sqrt{\pi t}}$ (B) $1 - \dfrac{e^{-t}}{\sqrt{\pi t}}$ (C) $-\dfrac{e^{t}}{\sqrt{\pi t}}$ (D) $\dfrac{e^{-t}}{\sqrt{\pi t}}$

20. If $\dfrac{d}{dx}$ erfc (ax) $= -\dfrac{2a}{\sqrt{\pi}}\, e^{-a^2x^2}$ then $\dfrac{d}{dx}$ erf (ax) is ... (2)

 (A) $ae^{-a^2x^2}$ (B) $1 - \dfrac{2a}{\sqrt{\pi}}\, e^{-a^2x^2}$

 (C) $\dfrac{2a}{\sqrt{\pi}}\, e^{-a^2x^2}$ (D) $-\dfrac{2a}{\sqrt{\pi}}\, e^{-a^2x^2}$

21. If $\dfrac{d}{da}$ erfc (ax) $= -\dfrac{2x}{\sqrt{\pi}}\, e^{-a^2x^2}$ then $\dfrac{d}{da}$ erf (ax) is ... (2)

 (A) $xe^{-a^2x^2}$ (B) $-\dfrac{2x}{\sqrt{\pi}}\, e^{-a^2x^2}$

 (C) $\dfrac{1}{\sqrt{\pi}}\, e^{-a^2x^2}$ (D) $\dfrac{2x}{\sqrt{\pi}}\, e^{-a^2x^2}$

22. If erfc (a) $= \dfrac{2}{\sqrt{\pi}} \displaystyle\int_{a}^{\infty} e^{-u^2}\, du$ then by using substitution $x + a = u$, the integral $\displaystyle\int_{0}^{\infty} e^{-(x+a)^2}\, dx$ in terms of erfc (a) is ... (2)

 (A) $\dfrac{2}{\sqrt{\pi}}$ erfc (a)   (B) erfc (a)

 (C) $\dfrac{\sqrt{\pi}}{2}$ erfc (a)   (D) $\dfrac{\sqrt{\pi}}{2}$ erf (a)

23. $\displaystyle\int_{0}^{t}$ erf (ax) dx $+ \displaystyle\int_{0}^{t}$ erfc (ax) dx $= $ ... (2)

 (A) t (B) x (C) 0 (D) $\dfrac{t^2}{2}$

24. The integral $\displaystyle\int_{0}^{\infty} e^{-t}$ erf $(\sqrt{t})\, dt$ using $\dfrac{d}{dt}$ erf $(\sqrt{t}) = \dfrac{e^{-t}}{\sqrt{\pi t}}$ is ... (2)

 (A) $\dfrac{1}{\sqrt{\pi}} \displaystyle\int_{0}^{\infty} e^{2t}\, t^{1/2}\, dt$ (B) $\dfrac{1}{\sqrt{\pi}} \displaystyle\int_{0}^{\infty} e^{-2t}\, t^{-1/2}\, dt$

 (C) $\displaystyle\int_{0}^{\infty} e^{-2t}\, t^{-1/2}\, dt$ (D) $\dfrac{1}{\sqrt{\pi}} \displaystyle\int_{0}^{\infty} e^{-t}\, t^{1/2}\, dt$

25. Expansion of erf (x) in series is ... (2)

 (A) $\dfrac{2}{\sqrt{\pi}}\left[x - \dfrac{x^3}{3} + \dfrac{x^5}{10} - \dfrac{x^7}{42} + ... \right]$

 (B) $\dfrac{2}{\sqrt{\pi}}\left[x + \dfrac{x^3}{3} + \dfrac{x^5}{10} + \dfrac{x^7}{42} + ... \right]$

 (C) $x - \dfrac{x^3}{3!} + \dfrac{x^5}{5!} - \dfrac{x^7}{7!} + ...$

 (D) $1 + x + \dfrac{x^2}{2!} + \dfrac{x^3}{3!} + ...$

ANSWERS

1. (A)	2. (B)	3. (C)	4. (D)	5. (C)	6. (B)	7. (D)	8. (A)
9. (C)	10. (D)	11. (B)	12. (C)	13. (A)	14. (B)	15. (D)	16. (C)
17. (B)	18. (D)	19. (A)	20. (C)	21. (D)	22. (C)	23. (A)	24. (B)
25. (A)							

•••

CHAPTER 5 : CURVE TRACING AND RECTIFICATION OF CURVES

Curve Tracing :   **Marks**

1. If the powers of y in the cartesian equation are even everywhere then the curve is symmetrical about ... (1)

 (A) x-axis   (B) y-axis

 (C) both x and y axes   (D) line y = x

2. If the powers of x in the cartesian equation are even everywhere then the curve is symmetrical about ... (1)

 (A) x-axis   (B) y-axis

 (C) both x and y axes   (D) line y = x

3. If the powers of x and y both in the cartesian equation are even everywhere then the curve is symmetrical about ... (1)

 (A) x-axis only   (B) y-axis only

 (C) both x and y axes   (D) line y = x

4. On replacing x and y by $-$ x and $-$ y respectively if the cartesian equation remains unchanged then the curve is symmetrical about ... (1)

 (A) line y = x   (B) y-axis

 (C) both x and y axes   (D) opposite quadrants

5. If x and y are interchanged and cartesian equation remains unchanged then the curve is symmetrical about ... (1)

 (A) both x and y axes   (B) line y = $-$ x

 (C) line y = x   (D) opposite quadrants

6. If x is changed to $-$ y and y to $-$ x and cartesian equation remains unchanged then the curve is symmetrical about ... (1)

 (A) both x and y axes   (B) line y = $-$x

 (C) line y = x   (D) opposite quadrants

7. If the curve passes through origin then tangents at origin to the cartesian curve can be obtained by equating to zero ...(1)

 (A) lowest degree term in the equation.

 (B) highest degree term in the equation.

 (C) coefficient of lowest degree term in the equation.

 (D) coefficient of highest degree term in the equation.

8. A double point is called node if the tangents to the curve at the double point are ... (1)

 (A) real and equal   (B) imaginary

 (C) always perpendicular   (D) real and distinct

9. A double point is called cusp if the tangents to the curve at the double point are ... (1)
    (A) real and equal   (B) imaginary
    (C) always perpendicular   (D) real and distinct

10. In cartesian equation the points where $\dfrac{dy}{dx} = 0$, tangent to the curve at those points will be ... (1)
    (A) parallel to y-axis   (B) parallel to x-axis
    (C) parallel to y = x   (D) parallel to y = – x

11. In cartesian equation the points where $\dfrac{dy}{dx} = \infty$, tangent to the curve at those points will be ... (1)
    (A) parallel to y = – x   (B) parallel to x-axis
    (C) parallel to y = x   (D) parallel to y-axis

12. The asymptotes to the cartesian curve parallel to x-axis if exists is obtained by equating to zero ... (1)
    (A) coefficient of highest degree term in y.
    (B) lowest degree term in the equation.
    (C) coefficient of highest degree term in x.
    (D) highest degree term in the equation.

13. The asymptotes to the cartesian curve parallel to y-axis if exists is obtained by equating to zero ... (1)
    (A) coefficient of highest degree term in y.
    (B) lowest degree term in the equation.
    (C) coefficient of highest degree term in x.
    (D) highest degree term in the equation.

14. If the polar equation to the curve remains unchanged by changing θ to – θ then the curve is symmetrical about ... (1)
    (A) line $\theta = \dfrac{\pi}{4}$ (B) pole
    (C) line $\theta = \dfrac{\pi}{2}$ (D) initial line θ = 0

15. If the polar equation to the curve remains unchanged by changing r to – r then the curve is symmetrical about ... (1)
    (A) line $\theta = \dfrac{\pi}{4}$ (B) pole
    (C) line $\theta = \dfrac{\pi}{2}$ (D) initial line θ = 0

16. If the polar equation to the curve remains unchanged by changing θ to π – θ then the curve is symmetrical about ... (1)
    (A) initial line θ = 0
    (B) pole
    (C) line passing through pole and perpendicular to the initial line
    (D) line $\theta = \dfrac{\pi}{4}$

17. Pole will lie on the curve if for some value of θ ... (1)
    (A) r becomes zero   (B) r becomes infinite
    (C) r > 0   (D) r < 0

18. The tangents to the polar curve at pole if exist can be obtained by putting in the polar ... (1)
    (A) θ = 0   (B) θ = π
    (C) r = 0   (D) r = a, a > 0

19. For the rose curve r = a cos nθ and r = a sin nθ if n is odd then the curve consist of ... (1)
    (A) 2n equal loops   (B) (n + 1) equal loops
    (C) (n – 1) equal loops   (D) n equal loops

20. For the rose curve r = a cos nθ and r = a sin nθ if n is even then the curve consist of ... (1)
    (A) (n + 1) equal loops   (B) 2n equal loops
    (C) (n – 1) equal loops   (D) n equal loops

21. For the polar curve, angle ϕ between radius vector and tangent line is obtained by the formula ... (1)
    (A) $\cot \phi = r\dfrac{d\theta}{dr}$ (B) $\tan \phi = r\dfrac{d\theta}{dr}$
    (C) $\tan \phi = r\dfrac{dr}{d\theta}$ (D) $\sin \phi = r\dfrac{d\theta}{dr}$

22. The cartesian parametric curve x = f(t), y = g(t) is symmetrical about x-axis if ... (1)
    (A) f(t) is even and g(t) is odd.
    (B) f(t) is odd and g(t) is even.
    (C) f(t) is even and g(t) is even.
    (D) f(t) is odd and g(t) is odd.

23. The cartesian parametric curve x = f(t), y = g(t) is symmetrical about y-axis if ... (1)
    (A) f(t) is even and g(t) is odd.
    (B) f(t) is even and g(t) is even.
    (C) f(t) is odd and g(t) is even.
    (D) f(t) is odd and g(t) is odd.

24. The curve represented by the equation $x^{1/2} + y^{1/2} = a^{1/2}$ is symmetrical about ... (1)
    (A) y = – x   (B) x-axis
    (C) both x and y axes   (D) y = x

25. The curve represented by the equation $x^2 y^2 = x^2 + 1$ is symmetrical about ... (1)
    (A) y = – x   (B) x-axis only
    (C) both x and y axes   (D) y = x

26. The curve represented by the equation $r^2\theta = a^2$ is symmetrical about ... (1)
    (A) pole   (B) initial line θ = 0
    (C) line $\theta = \dfrac{\pi}{2}$ (D) line $\theta = \dfrac{\pi}{4}$

27. The curve represented by the equation r = 2a sin θ is symmetrical about ... (1)
    (A) pole   (B) initial line θ = 0
    (C) line $\theta = \dfrac{\pi}{4}$ (D) line $\theta = \dfrac{\pi}{2}$

28. The curve represented by the equation x = at^2, y = 2at is symmetrical about ... (1)
    (A) y-axis   (B) x-axis
    (C) both x and y axes   (D) opposite quadrants

29. The asymptote parallel to y-axis to the curve $xy^2 = a^2 (a - x)$ is ... (1)
    (A) y = 0   (B) x = 0   (C) x = a   (D) x = – a

30. The number of loops in the rose curve $r = a \cos 2\theta$ are ... (1)
   - (A) 4
   - (B) 2
   - (C) 3
   - (D) 8

31. The number of loops in the rose curve $r = a \sin 3\theta$ are ... (1)
   - (A) 6
   - (B) 4
   - (C) 3
   - (D) 9

32. The curve represented by the equation $y^2 (2a - x) = x^3$ is ...(2)
   - (A) symmetrical about y-axis and passing through origin.
   - (B) symmetrical about x-axis and not passing through origin.
   - (C) symmetrical about y-axis and passing through (2a, 0).
   - (D) symmetrical about x-axis and passing through origin.

33. The curve represented by the equation $x (x^2 + y^2) = a (x^2 - y^2)$ is ... (2)
   - (A) symmetrical about x-axis and passing through origin.
   - (B) symmetrical about x-axis and not passing through origin.
   - (C) symmetrical about y-axis and passing through (a, 0).
   - (D) symmetrical about y-axis and passing through origin.

34. The curve represented by the equation $a^2x^2 = y^3 (2a - y)$ is ... (2)
   - (A) symmetrical about x-axis and passing through (2a, 0).
   - (B) symmetrical about both x-axis and y-axis and passing through origin.
   - (C) symmetrical about y-axis and passing through (0, 2a).
   - (D) symmetrical about both x-axis and y-axis and passing through (2a, 0).

35. The equation of tangents to the curve at origin, if exist, represented by the equation $y^2 (2a - x) = x^3$ is ... (2)
   - (A) $y = 0, y = 0$
   - (B) $x = 0, x = 2a$
   - (C) $x = 0, x = 0$
   - (D) $y = x$

36. The equation of tangents to the curve at origin, if exist, represented by the equation $y (1 + x^2) = x$ is ... (2)
   - (A) $y = x$
   - (B) $x = 0$
   - (C) $x = 1$ and $x = -1$
   - (D) $y = 0$

37. The equation of tangents to the curve at origin, if exists, represented by the equation $3ay^2 = x (x - a)^2$ is ... (2)
   - (A) $x = a$
   - (B) $x = 0$ and $y = 0$
   - (C) $x = 0$
   - (D) $y = 0$

38. The equation of asymptotes parallel to x-axis to the curve represented by the equation $y (1 + x^2) = x$ is ... (2)
   - (A) $x = 1, x = -1$
   - (B) $x = 0$
   - (C) $y = x$
   - (D) $y = 0$

39. The equation of asymptotes parallel to y-axis to the curve represented by the equation $y^2 (4 - x) = x (x - 2)^2$ is ... (2)
   - (A) $x = 2$
   - (B) $x = 4$
   - (C) $y = 0$
   - (D) $x = 0$

40. The equation of asymptotes parallel to y-axis to the curve represented by the equation $x^2y^2 = a^2 (y^2 - x^2)$ is ... (2)
   - (A) $x = a, x = -a$
   - (B) $y = a, y = -a$
   - (C) $y = x, y = -x$
   - (D) $x = 0, y = 0$

41. The region of absence for the curve represented by the equation $x^2 = \dfrac{4a^2 (2a - y)}{y}$ is ... (2)
   - (A) $y < 0$ and $y > 2a$
   - (B) $y > 0$ and $y < 2a$
   - (C) $y > 0$ and $y > 2a$
   - (D) $y < 0$ and $y < 2a$

42. The region of absence for the curve represented by the equation $y^2 (2a - x) = x^3$ is ... (2)
   - (A) $x > 0$ and $x < 2a$
   - (B) $x < 0$ and $x > 2a$
   - (C) $x < 0$ and $x < 2a$
   - (D) $x > 0$ and $x > 2a$

43. The region of absence for the curve represented by the equation $xy^2 = a^2 (a - x)$ is ... (2)
   - (A) $x > 0$ and $x < a$
   - (B) $x < 0$ and $x < a$
   - (C) $x < 0$ and $x > a$
   - (D) $x > 0$ and $x > a$

44. The region of absence for the curve represented by the equation $y^2 = \dfrac{x^2 (a - x)}{a + x}$ is ... (2)
   - (A) $x > a$ and $x > -a$
   - (B) $x < a$ and $x < -a$
   - (C) $x < a$ and $x > -a$
   - (D) $x > a$ and $x < -a$

45. The region of absence for the curve represented by the equation $x^2 = \dfrac{a^2 y^2}{a^2 - y^2}$ is ... (2)
   - (A) $y < a$ and $y > -a$
   - (B) $y > a$ and $y < -a$
   - (C) $y > a$ and $y > -a$
   - (D) $y < a$ and $y < -a$

46. The curve represented by the equation $r = a (1 + \cos \theta)$ is ... (2)
   - (A) symmetrical about initial line and passing through pole.
   - (B) symmetrical about initial line and not passing through pole.
   - (C) symmetrical about $\theta = \dfrac{\pi}{2}$ and passing through pole.
   - (D) symmetrical about $\theta = \dfrac{\pi}{4}$ and passing through pole.

47. The curve represented by the equation $r^2 = a^2 \cos 2\theta$ is ... (2)
   - (A) symmetrical about $\theta = \dfrac{\pi}{2}$ and not passing through pole.
   - (B) symmetrical about $\theta = \dfrac{\pi}{4}$ and not passing through pole.
   - (C) symmetrical about initial line and pole.
   - (D) symmetrical about $\theta = \dfrac{\pi}{4}$ and passing through pole.

48. The curve represented by the equation $r^2 = a^2 \sin 2\theta$ is ... (2)
   - (A) symmetrical about initial line and passing through pole.
   - (B) symmetrical about initial line and not passing through pole.
   - (C) symmetrical about $\theta = \dfrac{\pi}{2}$ and passing through pole.
   - (D) symmetrical about $\theta = \dfrac{\pi}{4}$ and passing through pole.

49. The curve represented by the equation $r = \dfrac{2a}{1 + \cos \theta}$ is ... (2)
   - (A) symmetrical about initial line and passing through pole.
   - (B) symmetrical about initial line and not passing through pole.
   - (C) symmetrical about $\theta = \dfrac{\pi}{2}$ and passing through pole.
   - (D) symmetrical about $\theta = \dfrac{\pi}{4}$ and passing through pole.

50. The tangents at pole to the polar curve $r = a \sin 3\theta$ are (2)

  (A) $\theta = 0, \dfrac{\pi}{3}, \dfrac{2\pi}{3}, \pi, \dfrac{4\pi}{3}, \dfrac{5\pi}{3}, \ldots$ (B) $\theta = \dfrac{\pi}{6}, \dfrac{3\pi}{6}, \dfrac{5\pi}{6}, \dfrac{7\pi}{6}, \ldots$

  (C) $\theta = 0, \dfrac{\pi}{4}, \dfrac{\pi}{2}, \dfrac{3\pi}{4}, \pi \ldots$ (D) $\theta = 0, \dfrac{\pi}{2}, \pi, \dfrac{3\pi}{2}, \ldots$

51. The tangents at pole to the polar curve $r = a \cos 2\theta$ are ... (2)

  (A) $\theta = 0, \pi, 2\pi, 3\pi, \ldots$ (B) $\theta = \dfrac{\pi}{6}, \dfrac{3\pi}{6}, \dfrac{5\pi}{6}, \dfrac{7\pi}{6}, \ldots$

  (C) $\theta = \dfrac{\pi}{4}, \dfrac{3\pi}{4}, \dfrac{5\pi}{4}, \dfrac{7\pi}{4}, \ldots$ (D) $\theta = 0, \dfrac{\pi}{2}, \pi, \dfrac{3\pi}{2}, \ldots$

52. The curve represented by the equation $x = t^2$, $y = t - \dfrac{t^3}{3}$ is ... (2)

  (A) symmetrical about y-axis and passing through origin.

  (B) symmetrical about x-axis and not passing through origin.

  (C) symmetrical about y-axis and passing through (3, 0).

  (D) symmetrical about x-axis and passing through origin.

53. The curve represented by the equation

  $x = a (t + \sin t)$, $y = a (1 + \cos t)$ is ... (2)

  (A) symmetrical about y-axis and not passing through origin.

  (B) symmetrical about x-axis and not passing through origin.

  (C) symmetrical about y-axis and passing through origin.

  (D) symmetrical about x-axis and passing through origin.

54. The equation $r = a \cos 2\theta$ represents the curve ... (1)

  (A)

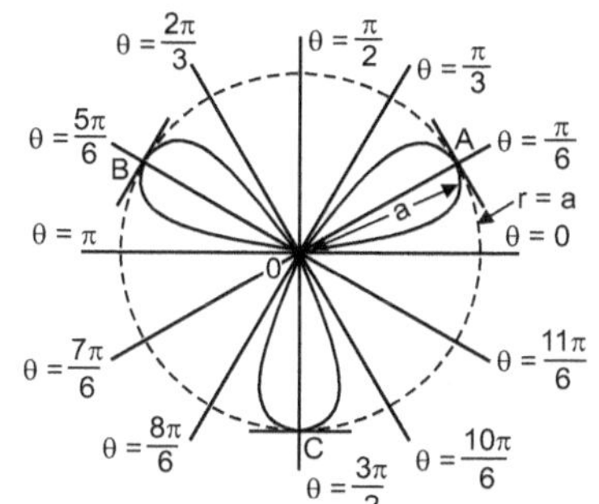

  (B)

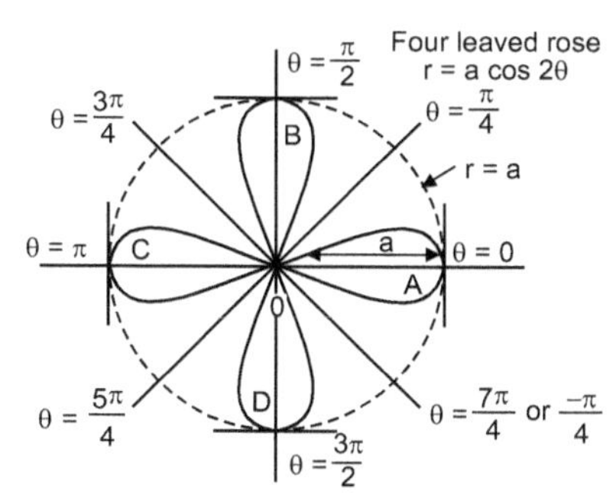

  (C)

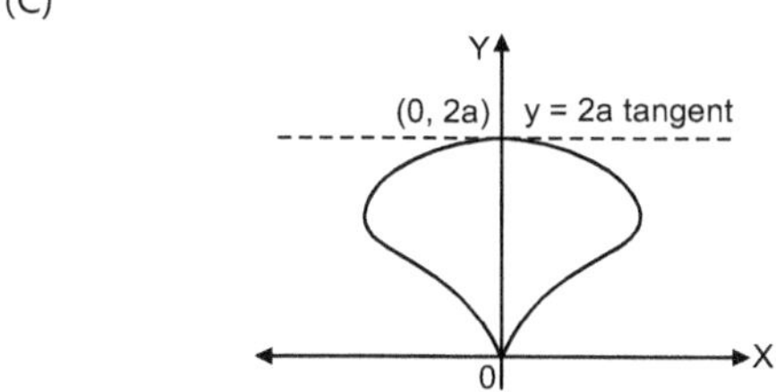

  (D)

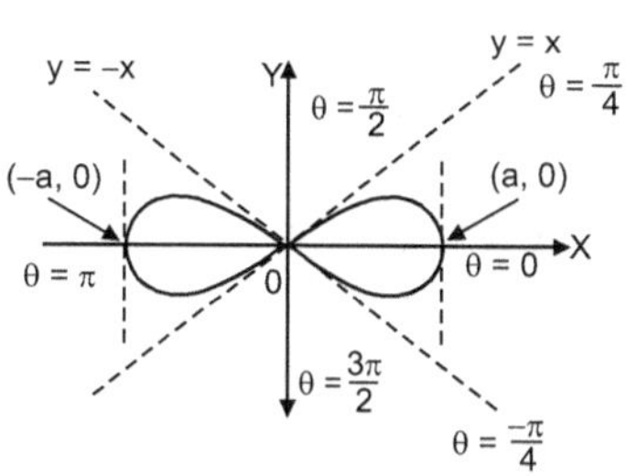

55. The equation $r = a \sin 3\theta$ represents the curve ... (1)

  (A)

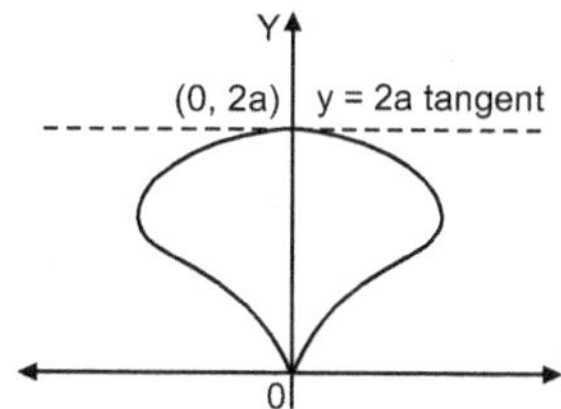

  (B)

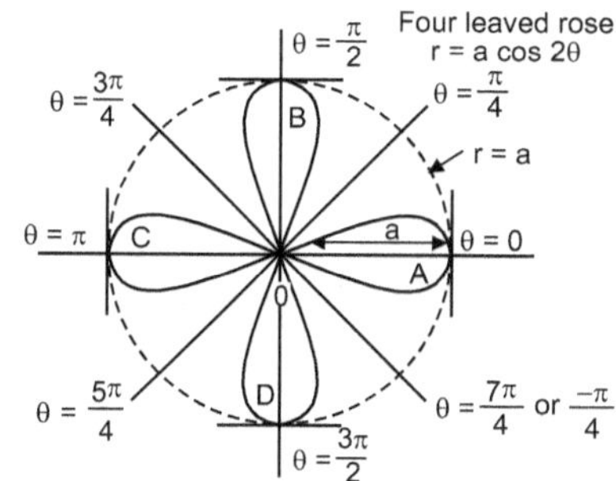

  (C)

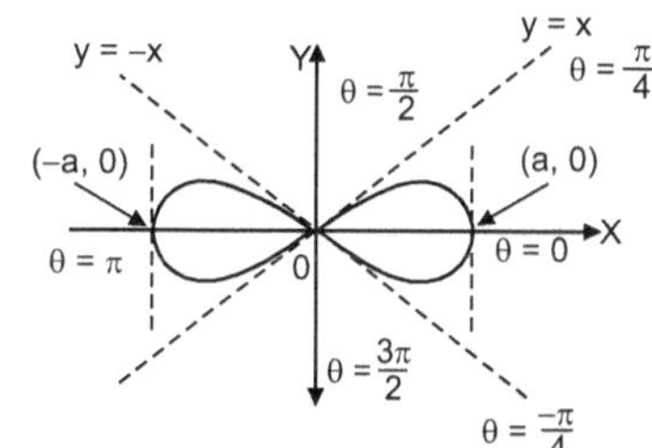

  (D)

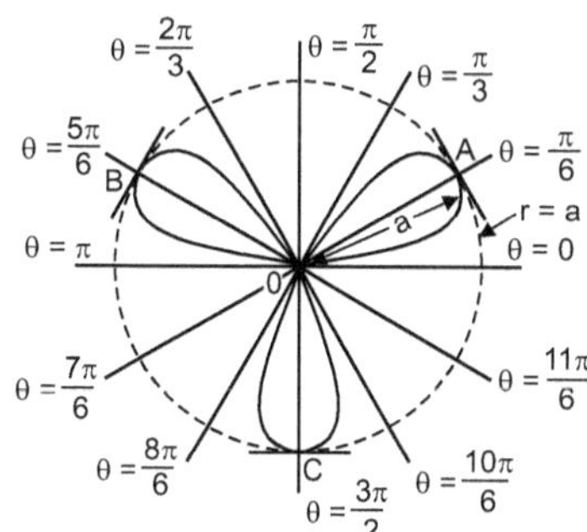

56. The equation $r^2 = a^2 \cos 2\theta$ represents the curve ... (1)

  (A)

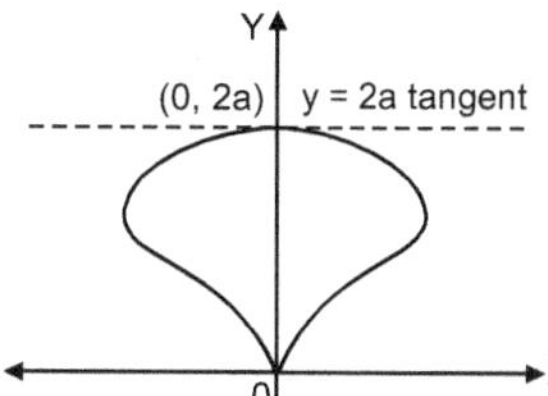

(B)

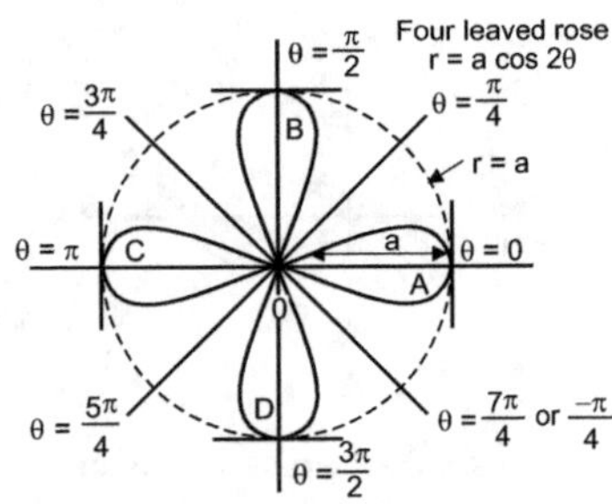

(C)

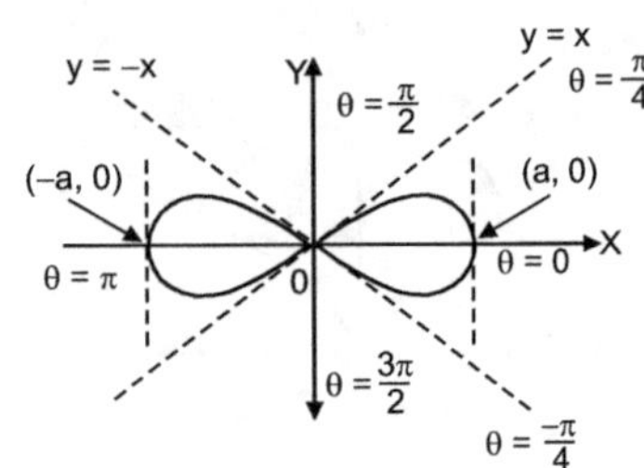

(D)

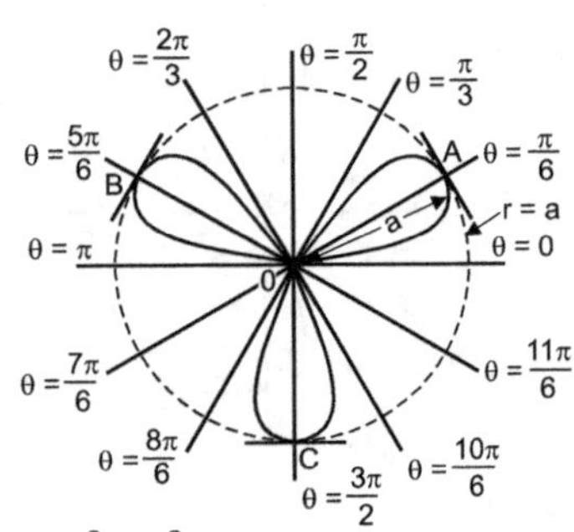

57. The equation $xy^2 = a^2 (a - x)$ represents the curve ...　(2)

(A)

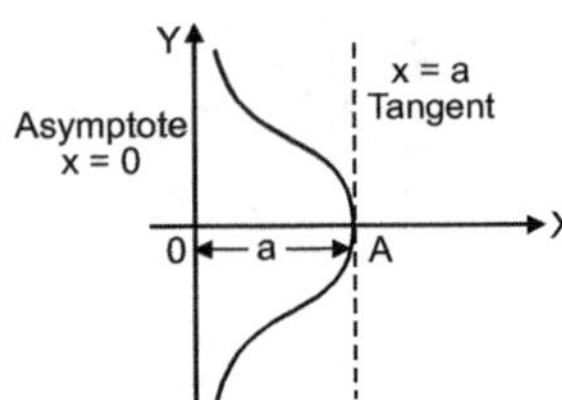

(B)

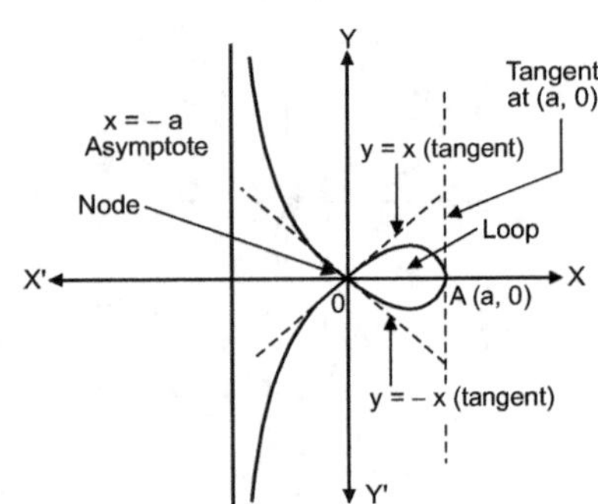

(C)

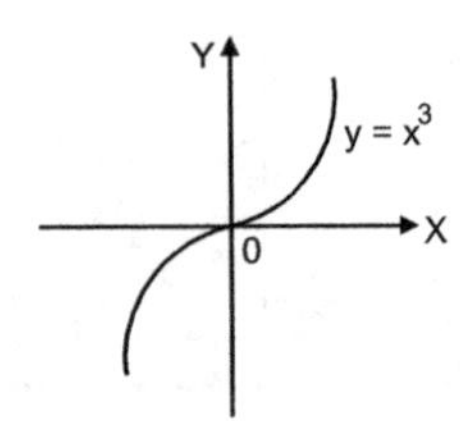

(D)

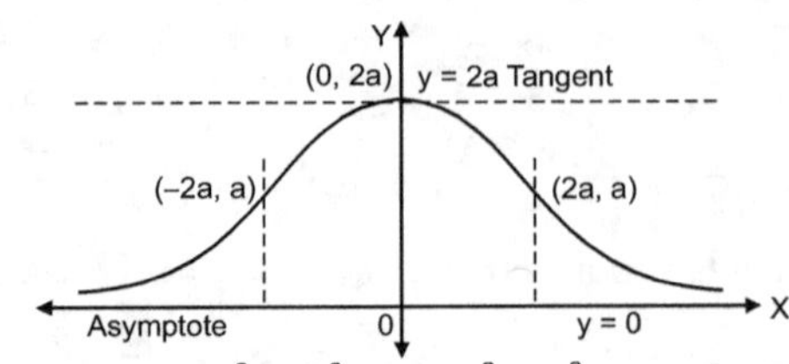

58. The equation $x (x^2 + y^2) = a (x^2 - y^2)$, $a > 0$ represents the curve ...　(2)

(A)

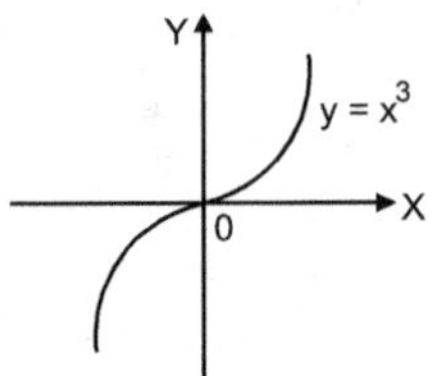

(B)

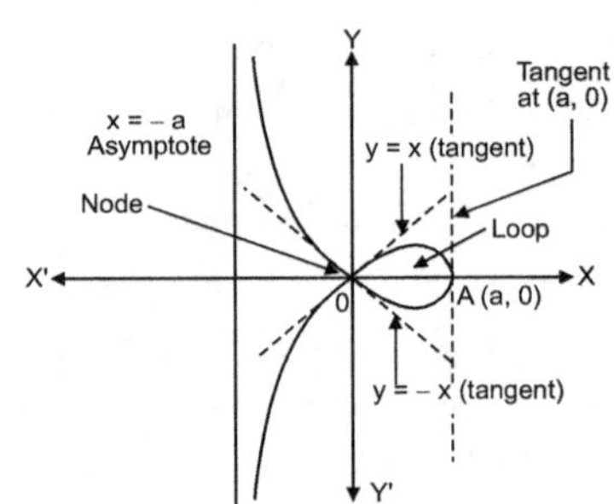

(C)

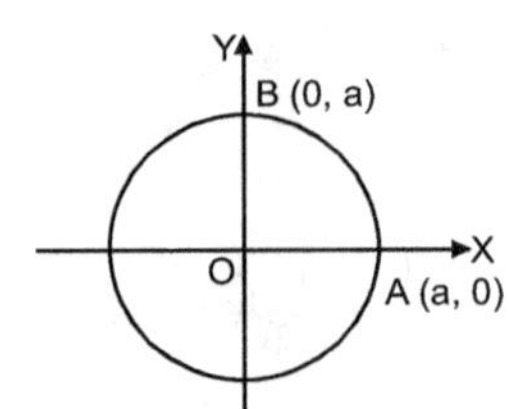

(D)

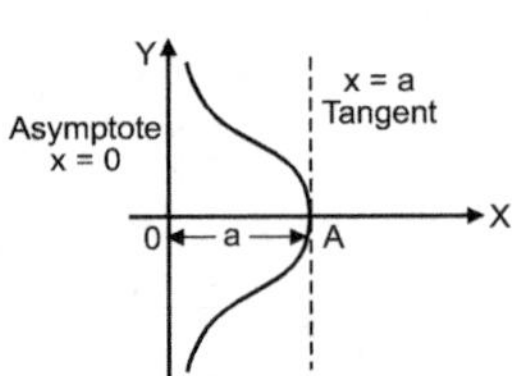

59. The equation $x^{2/3} + y^{2/3} = a^{2/3}$ represents the curve ...　(2)

(A)

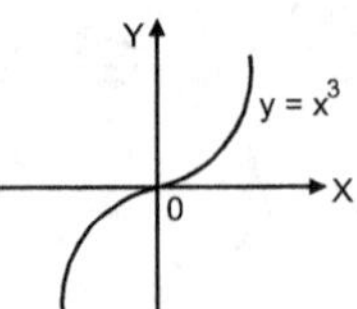

(B)

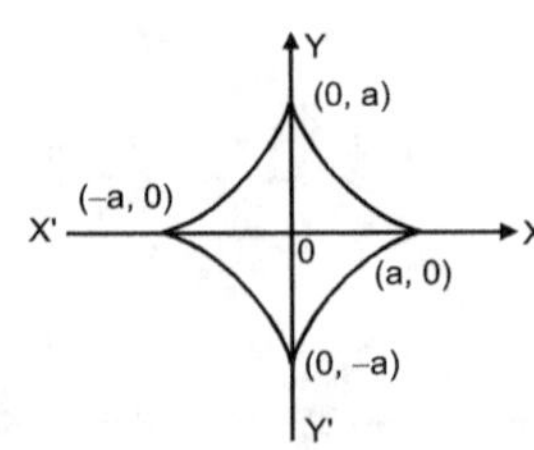

(C)

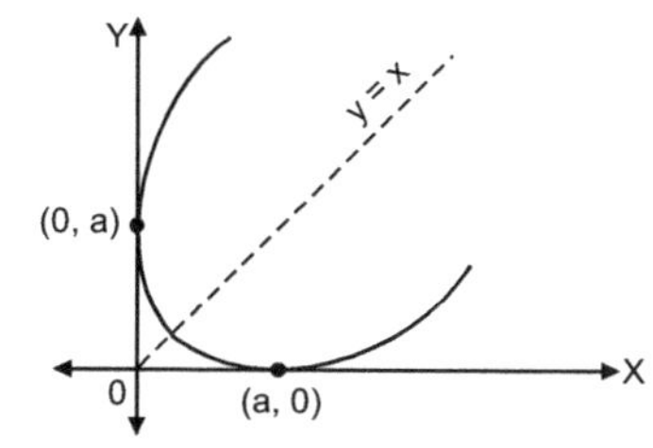

(D)

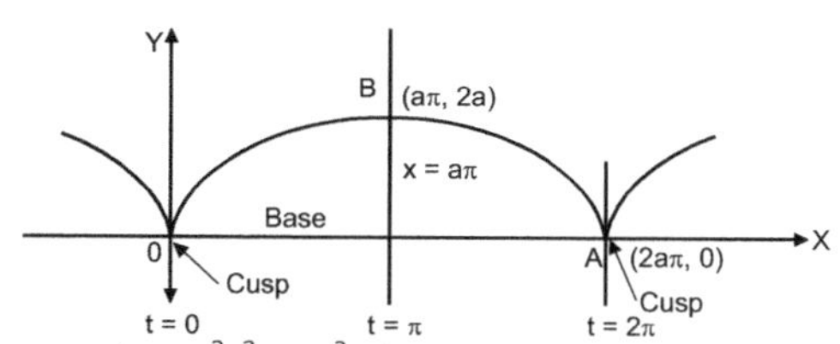

60. The equation $a^2x^2 = y^3 (2a - y)$, $a > 0$ represents the curve ... (2)

(A)

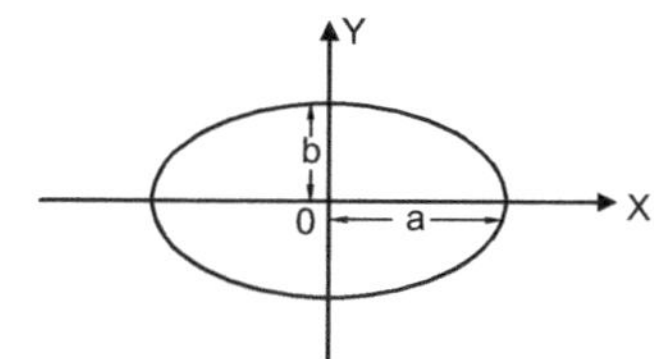

(B)

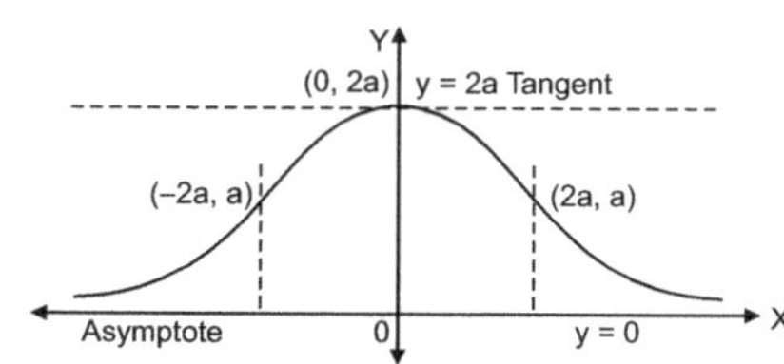

(C)

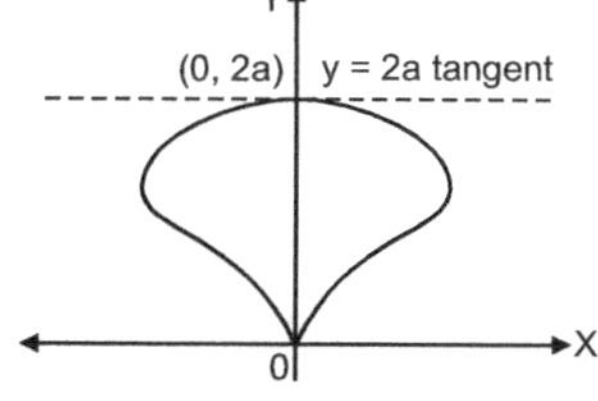

(D)

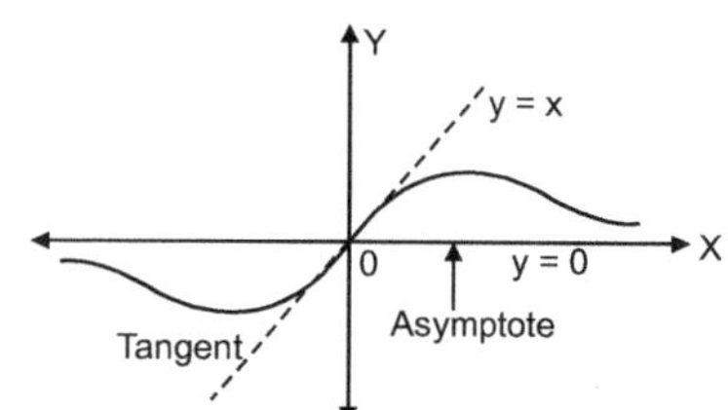

61. The following figure represents the curve whose equation is ... (2)

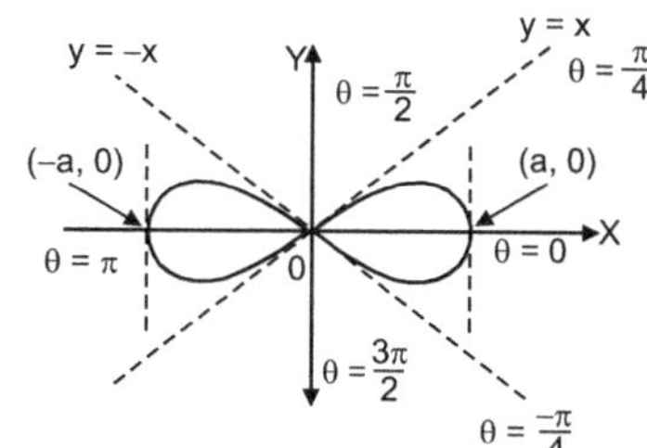

(A) $r^2 = a^2 \cos 2\theta$

(B) $r^2 = a^2 \sin 2\theta$

(C) $r = a \cos 2\theta$

(D) $r = a (1 + \cos \theta)$

62. The following figure represents the curve whose equation is ... (2)

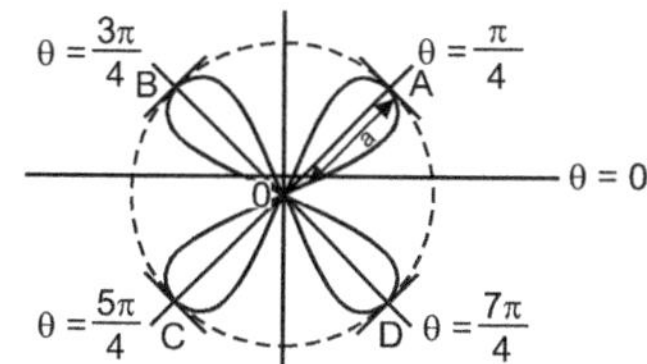

(A) $r = a \cos 3\theta$

(B) $r = a \sin 2\theta$

(C) $r = a \sin 3\theta$

(D) $r = a (1 + \cos \theta)$

ANSWERS

1. (A)	2. (B)	3. (C)	4. (D)	5. (C)	6. (B)	7. (A)	8. (D)
9. (A)	10. (B)	11. (D)	12. (C)	13. (A)	14. (D)	15. (B)	16. (C)
17. (A)	18. (C)	19. (D)	20. (B)	21. (B)	22. (A)	23. (C)	24. (D)
25. (C)	26. (A)	27. (D)	28. (B)	29. (B)	30. (A)	31. (C)	32. (D)
33. (A)	34. (C)	35. (D)	36. (A)	37. (C)	38. (D)	39. (B)	40. (A)
41. (A)	42. (B)	43. (C)	44. (D)	45. (B)	46. (A)	47. (C)	48.(D)
49. (B)	50. (A)	51. (C)	52. (D)	53. (A)	54. (B)	55. (D)	56. (C)
57. (A)	58. (B)	59. (B)	60. (C)	61. (A)	62. (B)		

•••

Rectification of Curves :   Marks

1. Formula for measuring the arc length AB where $A(x_1, y_1)$, $B(x_2, y_2)$ are any two points on the curve $y = f(x)$ is ... (1)

(A) $\int_{x_1}^{x_2} \sqrt{dx}$

(B) $\int_{x_1}^{x_2} \sqrt{1 + \left(\dfrac{dx}{dy}\right)^2} \, dx$

(C) $\int_{x_1}^{x_2} \sqrt{1 + \left(\dfrac{dy}{dx}\right)^2} \, dx$

(D) $\int_{x_1}^{x_2} \sqrt{1 - \left(\dfrac{dy}{dx}\right)^2} \, dx$

2. Formula for measuring the arc length AB where $A(x_1, y_1)$, $B(x_2, y_2)$ are any two points on the curve $x = g(y)$ is ... (1)

(A) $\displaystyle\int_{y_1}^{y_2} \sqrt{1 + \left(\frac{dy}{dx}\right)^2}\, dy$

(B) $\displaystyle\int_{y_1}^{y_2} \sqrt{1 + \left(\frac{dx}{dy}\right)^2}\, dy$

(C) $\displaystyle\int_{0}^{y_2} \sqrt{1 + \left(\frac{dx}{dy}\right)^2}\, dy$

(D) $\displaystyle\int_{y_1}^{y_2} \sqrt{1 - \left(\frac{dx}{dy}\right)^2}\, dy$

3. Formula for measuring the arc length AB where $A(r_1, \theta_1)$, $B(r_2, \theta_2)$ are any two points on the curve $r = f(\theta)$ is ... (1)

(A) $\displaystyle\int_{\theta_1}^{\theta_2} \sqrt{1 + \left(\frac{dr}{d\theta}\right)^2}\, d\theta$

(B) $\displaystyle\int_{\theta_1}^{\theta_2} \sqrt{1 + r^2\left(\frac{d\theta}{dr}\right)^2}\, dr$

(C) $\displaystyle\int_{\theta_1}^{\theta_2} \sqrt{r^2 + \left(\frac{dr}{d\theta}\right)^2}\, d\theta$

(D) $\displaystyle\int_{\theta_1}^{\theta_2} \sqrt{r^2 + r^2\left(\frac{dr}{d\theta}\right)^2}\, dr$

4. Formula for measuring the arc length AB where $A(r_1, \theta_1)$, $B(r_2, \theta_2)$ are any two points on the curve $\theta = f(r)$ is ... (1)

(A) $\displaystyle\int_{r_1}^{r_2} \sqrt{r^2 + \left(\frac{dr}{d\theta}\right)^2}\, dr$

(B) $\displaystyle\int_{r_1}^{r_2} \sqrt{1 + \left(\frac{d\theta}{dr}\right)^2}\, dr$

(C) $\displaystyle\int_{r_1}^{r_2} \sqrt{1 - r^2\left(\frac{d\theta}{dr}\right)^2}\, dr$

(D) $\displaystyle\int_{r_1}^{r_2} \sqrt{1 + r^2\left(\frac{d\theta}{dr}\right)^2}\, dr$

5. Formula for measuring the arc length AB where A, B are any two points on the parametric curve $x = f_1(t)$, $y = f_2(t)$, corresponding to parameters t_1, t_2 respectively is ... (1)

(A) $\displaystyle\int_{t_1}^{t_2} \sqrt{\left(\frac{dx}{dt}\right)^2 - \left(\frac{dy}{dt}\right)^2}\, dt$

(B) $\displaystyle\int_{t_1}^{t_2} \sqrt{\left(\frac{dx}{dt}\right)^2 + \left(\frac{dy}{dx}\right)^2}\, dt$

(C) $\displaystyle\int_{t_1}^{t_2} \left[\left(\frac{dx}{dt}\right)^2 + \left(\frac{dy}{dt}\right)^2\right]\, dt$

(D) $\displaystyle\int_{t_1}^{t_2} \sqrt{x^2(t) + y^2(t)}\, dt$

6. The arc length AB where $A(a, 0)$, $B(0, a)$ are any two points on the circle $x^2 + y^2 = a^2$, using $1 + \left(\frac{dy}{dx}\right)^2 = \frac{a^2}{a^2 - x^2}$, is ... (2)

(A) $\dfrac{\pi a}{2}$ 　　(B) $a \log a$ 　　(C) $\dfrac{\pi a}{4}$ 　　(D) a

7. The length of arc from vertex $(0, 0)$ to any point (x, y) of catenary $y = a \cos h\dfrac{x}{a}$, using $1 + \left(\frac{dy}{dx}\right)^2 = \cos h^2\dfrac{x}{a}$, is ... (2)

(A) $a \cos h\dfrac{x}{a}$ 　(B) $\sin h\dfrac{x}{a}$ 　(C) $a \sin h\dfrac{x}{a}$ 　(D) $\cos h\dfrac{x}{a}$

8. The length of arc of upper part of loop of the curve $3y^2 = x(x - 1)^2$ from $(0, 0)$ to $(1, 0)$ using $1 + \left(\frac{dy}{dx}\right)^2 = \dfrac{(3x + 1)^2}{12x}$, is ... (2)

(A) $\dfrac{4}{\sqrt{3}}$ 　(B) $\dfrac{1}{\sqrt{3}}$ 　(C) $\sqrt{3}$ 　(D) $\dfrac{2}{\sqrt{3}}$

9. Integral for calculating the length of upper arc of loop of the curve $9y^2 = (x + 7)(x + 4)^2$ is ... (2)

(A) $\displaystyle\int_{-7}^{-4} \sqrt{1 + \left(\frac{dy}{dx}\right)^2}\, dx$

(B) $\displaystyle\int_{4}^{7} \sqrt{1 + \left(\frac{dy}{dx}\right)^2}\, dx$

(C) $\displaystyle\int_{0}^{-4} \sqrt{1 + \left(\frac{dy}{dx}\right)^2}\, dx$

(D) $\displaystyle\int_{-7}^{0} \sqrt{1 + \left(\frac{dy}{dx}\right)^2}\, dx$

10. Integral for calculating the length of arc of parabola $y^2 = 4x$, cut off by the line $3y = 8x$ is ... (2)

(A) $\displaystyle\int_{0}^{16/9} \sqrt{1 + \left(\frac{dy}{dx}\right)^2}\, dx$

(B) $\displaystyle\int_{0}^{9/16} \sqrt{1 + \left(\frac{dy}{dx}\right)^2}\, dx$

(C) $\displaystyle\int_{0}^{8/3} \sqrt{1 + \left(\frac{dy}{dx}\right)^2}\, dx$

(D) $\displaystyle\int_{0}^{3/8} \sqrt{1 + \left(\frac{dy}{dx}\right)^2}\, dx$

11. The length of upper half of cardioide $r = a(1 + \cos\theta)$ where θ varies from 0 to π, using $r^2 + \left(\frac{dr}{d\theta}\right)^2 = 2a^2(1 + \cos\theta)$, is ... (2)

(A) $2a$ 　　(B) $8a$ 　　(C) $4a$ 　　(D) a

12. The length of the arc of curve $r = ae^{m\theta}$ intercepted between radius r_1 and r_2 using $1 + r^2\left(\frac{d\theta}{dr}\right)^2 = 1 + \dfrac{1}{m^2}$, is ... (2)

(A) $\dfrac{m}{\sqrt{1 + m^2}}(r_2 - r_1)$

(B) $\dfrac{\sqrt{1 + m^2}}{m}\, r_2$

(C) $\dfrac{\sqrt{1 + m^2}}{m}(r_2 + r_1)$

(D) $\dfrac{\sqrt{1 + m^2}}{m}(r_2 - r_1)$

13. Integral for calculating the length of cardioide $r = a(1 + \cos\theta)$ which lies outside the circle $r = -a \cos\theta$ is ... (2)

(A) $2\displaystyle\int_{0}^{\pi/3} \sqrt{r^2 + \left(\frac{dr}{d\theta}\right)^2}\, dr$

(B) $2\displaystyle\int_{0}^{\pi} \sqrt{r^2 + \left(\frac{dr}{d\theta}\right)^2}\, dr$

(C) $2\displaystyle\int_{0}^{2\pi/3} \sqrt{r^2 + \left(\frac{dr}{d\theta}\right)^2}\, d\theta$

(D) $2\displaystyle\int_{0}^{\pi/2} \sqrt{r^2 + \left(\frac{dr}{d\theta}\right)^2}\, dr$

14. Integral for calculating the length of upper arc of one loop of Bernoulli's lemniscate $r^2 = a^2 \cos 2\theta$ in the first quadrant is ... (2)

(A) $\displaystyle\int_{0}^{\pi/4} \sqrt{r^2 + \left(\frac{dr}{d\theta}\right)^2}\, d\theta$

(B) $\displaystyle\int_{0}^{\pi/6} \sqrt{r^2 + \left(\frac{dr}{d\theta}\right)^2}\, dr$

(C) $\displaystyle\int_{0}^{\pi/2} \sqrt{r^2 + \left(\frac{dr}{d\theta}\right)^2}\, dr$

(D) $\displaystyle\int_{0}^{\pi/3} \sqrt{r^2 + \left(\frac{dr}{d\theta}\right)^2}\, dr$

15. Integral for calculating the length of upper arc of loop of the curve $x = t^2$, $y = t\left(1 - \dfrac{t^2}{3}\right)$ is ... (2)

(A) $\displaystyle\int_0^9 \sqrt{\left(\dfrac{dx}{dt}\right)^2 + \left(\dfrac{dy}{dt}\right)^2}\, dt$ (B) $\displaystyle\int_0^3 \sqrt{\left(\dfrac{dx}{dt}\right)^2 + \left(\dfrac{dy}{dt}\right)^2}\, dt$

(C) $\displaystyle\int_0^1 \sqrt{\left(\dfrac{dx}{dt}\right)^2 + \left(\dfrac{dy}{dt}\right)^2}\, dt$ (D) $\displaystyle\int_0^{\sqrt{3}} \sqrt{\left(\dfrac{dx}{dt}\right)^2 + \left(\dfrac{dy}{dt}\right)^2}\, dt$

16. Integral for calculating the length of the arc of Astroid $x = a\cos^3\theta$, $y = a\sin^3\theta$ in the first quadrant between two consecutive cusps, is ... (2)

(A) $\displaystyle\int_0^{\pi/4} \sqrt{\left(\dfrac{dx}{d\theta}\right)^2 + \left(\dfrac{dy}{d\theta}\right)^2}\, d\theta$

(B) $\displaystyle\int_0^{\pi/2} \sqrt{\left(\dfrac{dx}{d\theta}\right)^2 + \left(\dfrac{dy}{d\theta}\right)^2}\, d\theta$

(C) $\displaystyle\int_0^{\pi/3} \sqrt{\left(\dfrac{dx}{d\theta}\right)^2 + \left(\dfrac{dy}{d\theta}\right)^2}\, d\theta$

(D) $\displaystyle\int_0^{\pi/6} \sqrt{\left(\dfrac{dx}{d\theta}\right)^2 + \left(\dfrac{dy}{d\theta}\right)^2}\, d\theta$

17. The length of arc of upper part of loop of the curve $x = t^2$, $y = t\left(1 - \dfrac{t^2}{3}\right)$ where t varies from 0 to $\sqrt{3}$, using $\left(\dfrac{dx}{dt}\right)^2 + \left(\dfrac{dy}{dt}\right)^2 = (1 + t^2)^2$, is ... (2)

(A) $2\sqrt{3}$ (B) $4\sqrt{3}$ (C) $\sqrt{3}$ (D) $\dfrac{2}{\sqrt{3}}$

18. The length of the arc of Astroid $x = a\cos^3\theta$, $y = a\sin^3\theta$ in the first quadrant between two consecutive cusps, where θ varies from 0 to $\dfrac{\pi}{2}$, using $\left(\dfrac{dx}{d\theta}\right)^2 + \left(\dfrac{dy}{d\theta}\right)^2 = 9a^2\sin^2\theta\cos^2\theta$, is ... (2)

(A) $3a$ (B) $\dfrac{3a}{2}$ (C) $\dfrac{3a}{4}$ (D) $\dfrac{3a}{8}$

19. The length of arc of the curve $x = e^\theta\cos\theta$, $y = e^\theta\sin\theta$, from $\theta = 0$ to $\theta = \dfrac{\pi}{2}$, using $\left(\dfrac{dx}{d\theta}\right)^2 + \left(\dfrac{dy}{d\theta}\right)^2 = 2e^{2\theta}$, is ... (2)

(A) $\sqrt{2}\, e^{\pi/2}$ (B) $\sqrt{2}\,(e^{\pi/2} + 1)$

(C) $\sqrt{2}\,(e^{\pi/2} - 1)$ (D) $(e^{\pi/2} + 1)$

20. The length of arc of the cycloid $x = a\,(\theta + \sin\theta)$, $y = a\,(1 - \cos\theta)$, from one cusp $\theta = -\pi$ to another cusp $\theta = \pi$, using $\left(\dfrac{dx}{d\theta}\right)^2 + \left(\dfrac{dy}{d\theta}\right)^2 = 4a^2\cos^2\dfrac{\theta}{2}$, is ... (2)

(A) $2a$ (B) a

(C) $4a$ (D) $8a$

ANSWERS

1. (C)	2. (B)	3. (C)	4. (D)	5. (B)	6. (A)	7. (C)	8. (D)
9. (A)	10. (B)	11. (C)	12. (D)	13. (C)	14. (A)	15. (D)	16. (B)
17. (A)	18. (B)	19. (C)	20. (D)				

•••

CHAPTER 6, 7 AND 8 : CO-ORDINATE SYSTEM, SPHERE, CONE AND CYLINDER

Co-ordinate System, Sphere, Cone and Cylinder :   **Marks**

1. The spherical polar co-ordinates (r, θ, ϕ) of a Carteisan point $(1, 1, 1)$ is ...

(A) $(\sqrt{3}, 54.74°, 45°)$ (B) $(\sqrt{2}, 54.74°, 45°)$

(C) $(\sqrt{2}, 45°, 54.74°)$ (D) $(\sqrt{3}, 45°, 60°)$

2. The cylindrical co-ordinates (ρ, ϕ, z) of a Cartsian point $(1, 1, 1)$ is ...

(A) $(\sqrt{3}, 45°, 2)$ (B) $(\sqrt{2}, 45°, 1)$

(C) $(\sqrt{2}, 60°, 1)$ (D) $(\sqrt{3}, 60°, 45°)$

3. The Cartesion co-ordinates (x, y, z) of a spherical polar co-ordinates $\left(2, \dfrac{\pi}{3}, \dfrac{\pi}{4}\right)$ is ...

(A) $\left(\sqrt{\dfrac{3}{2}}, \sqrt{\dfrac{3}{2}}, \dfrac{2\sqrt{3}}{1}\right)$ (B) $\left(\dfrac{\sqrt{3}}{2\sqrt{2}}, \dfrac{\sqrt{3}}{2\sqrt{2}}, 1\right)$

(C) $\left(\sqrt{\dfrac{3}{2}}, \sqrt{\dfrac{3}{2}}, 1\right)$ (D) $\left(\dfrac{9}{4}, \dfrac{2\sqrt{3}}{4}, 1\right)$

4. The Cartesion co-ordinates (x, y, z) of a cylindrical co-ordinates $(\sqrt{2}, 45°, 1)$ ...

(A) $(1, 1, -1)$ (B) $(1, -1, 1)$

(C) $(\sqrt{2}, 1, 1)$ (D) $(1, 1, 1)$

5. The d.c.'s of the lines equally inclined to the co-ordinate axes are ...

(A) $\left(\pm\dfrac{1}{\sqrt{3}}, \pm\dfrac{1}{\sqrt{3}}, \pm\dfrac{1}{\sqrt{3}}\right)$ (B) $\left(\dfrac{1}{3}, \dfrac{2}{3}, \dfrac{2}{3}\right)$

(C) $\left(\dfrac{1}{3}, \dfrac{-2}{3}, \dfrac{2}{3}\right)$ (D) $\left(\dfrac{1}{3}, \dfrac{2}{3}, -\dfrac{2}{3}\right)$

6. The centre (a, b, c) and radius 'r' form of equation of sphere is ...

(A) $(x - a)^2 + (y - b)^2 + (z - c^2) = r^2$

(B) $x^2 + y^2 + z^2 = r^2$

(C) $x + y + z = a$

(D) $(x - a)^2 + (y - b)^2 + (z - c)^2 = 1$

7. The general form of the equation of sphere with centre $(-u, -v, -w)$ and radius $r = \sqrt{u^2 + v^2 + w^2 - d}$ ...

   (A) $x^2 + y^2 + z^2 + ux + vy + wz + d = 0$

   (B) $x^2 + y^2 + z^2 + 2ux + 2vy + 2wz + d = 0$

   (C) $x^2 + y^2 + z^2 + \dfrac{u}{2}x + \dfrac{v}{2}y + \dfrac{w}{2}z + \dfrac{d}{2} = 0$

   (D) $x^2 + y^2 + z^2 + d = 0$

8. If $S = 0$ is the equation of a sphere and $U = 0$ that of a plane. The common section of sphere and the plane, is circle and it is given by ...

   (A) $S = 0, U = 0$ (B) $S + U = 0$

   (C) $S - U = 0$ (D) None of these

9. The equation of right circular cone whose vertex is (α, β, γ), semi-vertical angel A and axis $\dfrac{x - \alpha}{l} = \dfrac{y - \beta}{m} = \dfrac{z - \gamma}{n}$ is ...

   (A) $\cos A = \dfrac{l(x - \alpha) + m(y - \beta) + n(z - \gamma)}{\sqrt{l^2 + m^2 + n^2}\,\sqrt{(x - \alpha)^2 + (y - \beta)^2 + (z - \gamma)^2}}$

   (B) $\cos A = \dfrac{l(x - \alpha) + m(y - \beta) + n(z - \gamma)}{\sqrt{l^2 + m_2 + n^2}\,\sqrt{x^2 + y^2 + z^2}}$

   (C) $\cos A = \dfrac{(x - \alpha) + (y - \beta) + (z - \gamma)}{\sqrt{(x - \alpha)^2 + (y - \beta)^2 + (z - \gamma)^2}}$

   (D) $\cos A = \dfrac{l(x - \alpha) + m(y - \beta) + n(z - \gamma)}{(l^2 + m^2 + n^2)\,(x^2 + y^2 + z^2)}$

10. The equation of the right circular cylinder whose radius r and axis is the line $\dfrac{x - \alpha}{l} = \dfrac{y - \beta}{m} = \dfrac{z - \gamma}{m}$ is ...

   (A) $(x - \alpha)^2 + (y - \beta)^2 + (z - \gamma)^2 - \left\{\dfrac{l(x - \alpha) + m(y - \beta) + n(z - \gamma)}{\sqrt{l^2 + m^2 + n^2}}\right\}^2 = r^2$

   (B) $(x - \alpha)^2 + (y - \beta)^2 + (z - \gamma)^2 + \left\{\dfrac{l(x - \alpha) + m(y - \beta) + n(z - \gamma)}{\sqrt{l^2 + m^2 + n^2}}\right\}^2 = r^2$

   (C) $(x - \alpha) + (y - \beta) + (z - \gamma) - \left\{\dfrac{(x - \alpha) + (y - \beta) + (z - \gamma)}{l^2 + m^2 + n^2}\right\}^2 = r^2$

   (D) $(x - \alpha) + (y - \beta) + (z - \gamma) + \left\{\dfrac{(x + \alpha) + (y + \beta) + (z + \gamma)}{l^2 + m^2 + n^2}\right\}^2 = r^2$

ANSWERS

1. (A)	2. (B)	3. (C)	4. (D)	5. (A)	6. (A)	7. (B)	8. (A)
9. (A)	10. (A)						

•••

CHAPTER 9 AND 10 : MULTIPLE INTEGRALS AND APPLICATIONS

Multiple Integrals and Applications :   **Marks**

1. The value of $\displaystyle\int_0^2 \int_0^x e^{x+y}\, dy\, dx$ is ...

   (A) $\dfrac{1}{2}(e^2 - 1)$ (B) $\dfrac{1}{2}(e^2 - e)$

   (C) $\dfrac{1}{2}(e^2 - 1)^2$ (D) $\dfrac{1}{2}(e - e^{-1})^2$

2. The integral $\displaystyle\int_0^1 \int_0^{\sqrt{1 - x^2}} (x + y)\, dy\, dx$ after changing the order of integration is ...

   (A) $\displaystyle\int_0^2 \int_0^{\sqrt{1 - y^2}} (x + y)\, dx\, dy$ (B) $\displaystyle\int_0^1 \int_0^{\sqrt{1 - y^2}} (x + y)\, dx\, dy$

   (C) $\displaystyle\int_0^1 \int_0^{\sqrt{1 + y^2}} (x + y)\, dx\, dy$ (D) $\displaystyle\int_0^{-1} \int_0^{\sqrt{1 - y^2}} (x + y)\, dx\, dy$

3. The value of integral $\displaystyle\int_0^{\pi/2} \int_0^{\pi/2} \sin(x + y)\, dx\, dy$ is ...

   (A) 0 (B) π

   (C) $\dfrac{\pi}{2}$ (D) 2

4. Using polar transformations $x = r \cos\theta$ and $y = r \sin\theta$, the Cartesian double integral $\displaystyle\iint_R f(x, y)\, dx\, dy$ is transformed to ...

   (A) $\displaystyle\iint_{R'} f(r, \theta)\, \theta\, dr\, d\theta$ (B) $\displaystyle\iint_{R'} f(r, \theta)\, \dfrac{1}{2}\, dr\, d\theta$

   (C) $\displaystyle\iint_{R'} f(r, \theta)\, 2dr\, d\theta$ (D) $\displaystyle\iint_{R'} f(r, \theta)\, r d r\, d\theta$

5. Changing the order of the integration in the double integral $I = \displaystyle\int_0^8 \int_{x/4}^2 f(x, y)\, dy\, dx$ leads to $I = \displaystyle\int_r^s \int_p^q f(x, y)\, dx\, dy$. The value of q is ...

   (A) $4y$ (B) $16y^2$

   (C) x (D) 8

6. Considering shaded triangular region R as shown in the figure, change the order of integration in $\displaystyle\int_0^a \int_0^y f(x, y)\, dx\, dy$ gives ...

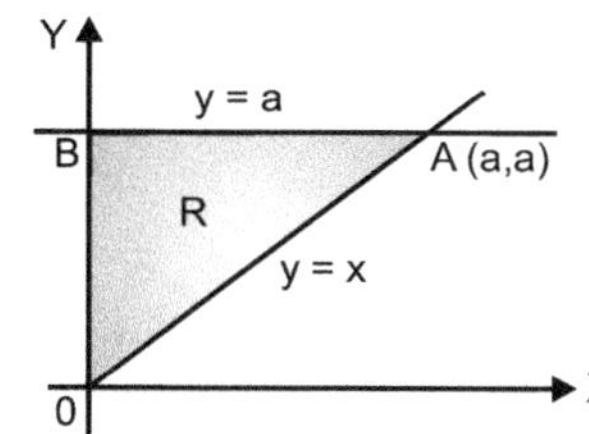

(A) $\displaystyle\int_{x=0}^{x=a}\int_{y=x}^{y=\infty} f(x,y)\,dx\,dy$ (B) $\displaystyle\int_{x=0}^{x=\infty}\int_{y=0}^{y=a} f(x,y)\,dy\,dx$

(C) $\displaystyle\int_{x=0}^{x=a}\int_{y=x}^{y=a} f(x,y)\,dy\,dx$ (D) $\displaystyle\int_{x=0}^{x=a}\int_{y=0}^{y=a} f(x,y)\,dy\,dx$

7. Considering shaded triangle region R as shown in figure, the value of $\displaystyle\iint_R x\,y\,dx\,dy$ is …

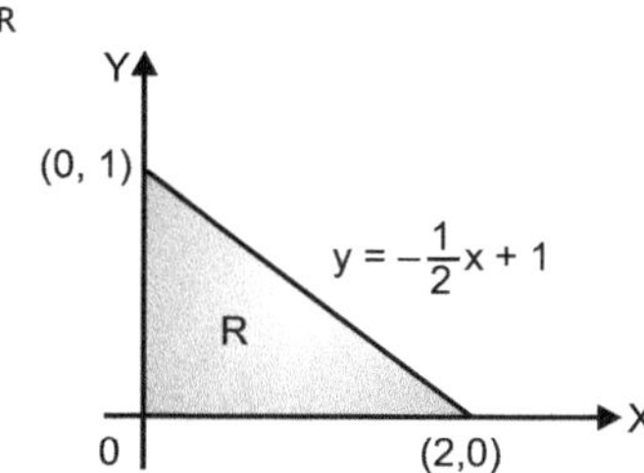

(A) $\dfrac{1}{6}$ (B) 2

(C) $\dfrac{1}{16}$ (D) 1

8. On changing the order of integration for $\displaystyle\int_0^\infty\int_x^\infty f(x,y)\,dy\,dx$, the integral becomes …

(A) $\displaystyle\int_0^\infty\int_0^\infty f(x,y)\,dy\,dx$ (B) $\displaystyle\int_0^\infty\int_y^\infty f(x,y)\,dx\,dy$

(C) $\displaystyle\int_0^\infty\int_0^y f(x,y)\,dx\,dy$ (D) $\displaystyle\int_0^\infty\int_0^x f(x,y)\,dy\,dx$

9. On changing the order of integration for $\displaystyle\int_0^a\int_{y^2/a}^y f(x,y)\,dx\,dy$, the integral becomes

(A) $\displaystyle\int_0^a\int_0^x f(x,y)\,dx\,dy$ (B) $\displaystyle\int_0^a\int_x^{\sqrt{ax}} f(x,y)\,dy\,dx$

(C) $\displaystyle\int_0^a\int_0^a f(x,y)\,dy\,dx$ (D) $\displaystyle\int_0^\infty\int_x^\infty f(x,y)\,dx\,dy$

10. Using double integration and the strip parallel to x-axis the area enclosed between the curves $y^2 = 4x$ and $x^2 = 4y$ is [Use cutting points (0, 0) and (4, 4)] …

(A) $\displaystyle\int_0^4\int_{x^2/4}^{4x} dx\,dy$ (B) $\displaystyle\int_0^4\int_0^4 dx\,dy$

(C) $\displaystyle\int_0^4\int_0^4 dy\,dx$ (D) $\displaystyle\int_{y=0}^4\int_{y^2/4}^{2\sqrt{y}} dy\,dx$

11. The area enclosed between the straight line $y = x$ and parabola $y = x^2$ in the xoy plane using double integration is calculated from …

(A) $\displaystyle\int_{x=0}^{x=1}\int_{y=x}^{y=x^2} dy\,dx$ (B) $\displaystyle\int_{x=0}^{x=\infty}\int_{y=x}^{y=\infty} dy\,dx$

(C) $\displaystyle\int_{x=0}^{x=1}\int_{y=1}^{y=1} dy\,dx$ (D) $\displaystyle\int_{x=1}^{x=\infty}\int_{y=1}^{y=\infty} dy\,dx$

12. The volume of an object expressed in spherical polar coordinate is given by $V = \displaystyle\int_0^{2\pi/3}\int_0^{\pi/3}\int_0^1 r^2 \sin\theta\,dr\,d\theta\,d\phi$. The value of integral is …

(A) $\dfrac{\pi}{3}$ (B) $\dfrac{\pi}{6}$

(C) $\dfrac{2\pi}{3}$ (D) $\dfrac{\pi}{4}$

ANSWERS

1. (C)	2. (B)	3. (D)	4. (D)	5. (A)	6. (C)	7. (A)	8. (C)
9. (B)	10. (D)	11. (A)	12. (A)				

•••

LIST OF FORMULAE

1. $\int x^n \, dx = \dfrac{x^{n+1}}{n+1} + c \ (n \neq 1)$

2. $\int \dfrac{1}{x} \, dx = \log x + c$

3. $\int e^x \, dx = e^x + c$

4. $\int a^x \, dx = \dfrac{a^x}{\log a} + c$

5. $\int \sin x \, dx = -\cos x + c$

6. $\int \cos x \, dx = \sin x + c$

7. $\int \tan x \, dx = \log \sec x + c$

8. $\int \cot x \, dx = \log \sin x + c$

9. $\int \sec^2 x \, dx = \tan x + c$

10. $\int \operatorname{cosec}^2 x \, dx = -\cot x + c$

11. $\int \sec x \tan x \, dx = \sec x + c$

12. $\int \operatorname{cosec} x \cot x \, dx = -\operatorname{cosec} x + c$

13. $\int \sec x \, dx = \log(\sec x + \tan x) + c$

14. $\int \operatorname{cosec} x \, dx = \log(\operatorname{cosec} x - \cot x) + c$

15. $\int \dfrac{1}{\sqrt{a^2 - x^2}} \, dx = \sin^{-1}\left(\dfrac{x}{a}\right) + c$

16. $\int \dfrac{1}{\sqrt{x^2 - a^2}} \, dx = \log(x + \sqrt{x^2 - a^2}) + c$

17. $\int \dfrac{dx}{\sqrt{x^2 + a^2}} = \log(x + \sqrt{x^2 + a^2}) + c \quad \text{OR} \quad \sinh^{-1}\dfrac{x}{a} + c$

18. $\int \dfrac{dx}{a^2 - x^2} = \dfrac{1}{2a} \log\left(\dfrac{a + x}{a - x}\right) + c \quad \text{OR} \quad \dfrac{1}{a} \tanh^{-1}\dfrac{x}{a} + c$

19. $\int \dfrac{1}{x^2 - a^2} \, dx = \dfrac{1}{2a} \log \dfrac{x - a}{x + a} + c$

20. $\int \dfrac{1}{a^2 + x^2} \, dx = \dfrac{1}{a} \tan^{-1}\left(\dfrac{x}{a}\right) + c$

21. $\int \sqrt{a^2 - x^2} \, dx = \dfrac{x}{a} \sqrt{a^2 - x^2} + \dfrac{a^2}{2} \sin^{-1}\left(\dfrac{x}{a}\right) + c$

22. $\int \sqrt{x^2 - a^2} \, dx = \dfrac{x}{a} \sqrt{x^2 - a^2} - \dfrac{a^2}{2} \log[x + \sqrt{x^2 - a^2}] + c$

23. $\int \sqrt{x^2 + a^2} \, dx = \dfrac{x}{2} \sqrt{x^2 + a^2} + \dfrac{a^2}{2} \log[x + \sqrt{x^2 + a^2}] + c$

24. $\int e^x [f(x) + f'(x)] \, dx = e^x f(x)$

25. $\int e^{ax} \sin bx \, dx = \dfrac{e^{ax}}{a^2 + b^2}(a \sin bx - b \cos bx)$

26. $\int e^{ax} \cos bx \, dx = \dfrac{e^{ax}}{a^2 + b^2}(a \cos bx + b \sin bx)$

Rule for Integration by Parts

27. $\int u \cdot v \, dx = u \int v \, dx - \int \left[\dfrac{du}{dx} \cdot \int v \, dx\right] dx + c$

28. $\int [f(x)]^n f'(x) \, dx = \dfrac{[f(x)]^{n+1}}{n+1}, \ n \neq -1$

29. $\int \dfrac{f'(x)}{f(x)} \, dx = \log f(x)$

30. $\int e^{f(x)} f'(x) \, dx = e^{f(x)}$

31. $\int \sin(f(x)) f'(x) \, dx = -\cos(f(x))$

32. $\int \cos(f(x)) f'(x) \, dx = \sin(f(x))$

33. $\int \dfrac{f'(x)}{\sqrt{f(x)}} \, dx = 2\sqrt{f(x)}$

34. $\int \sqrt{f(x)} \cdot f'(x) \, dx = \dfrac{2}{3}[f(x)]^{3/2}$

35. $\int_0^a f(x) \, dx = \int_0^a f(a - x) \, dx$

36. $\int_a^b f(x) \, dx = \int_a^c f(x) \, dx + \int_c^b f(x) \, dx, \ a < c < b$

37. $\int_0^{2a} f(x) \, dx = \int_0^a f(x) \, dx + \int_0^a f(2a - x) \, dx$

38. $\int_{-a}^a f(x) \, dx = 2 \int_0^a f(x) \, dx \ \text{if } f(x) \text{ is even}$

$\qquad\qquad\qquad = 0, \ \text{if } f(x) \text{ is odd}$

39. $\int_0^{\infty} e^{-ax} \cos bx \, dx = \dfrac{a}{a^2 + b^2}$

40. $\int_0^{\infty} e^{-ax} \sin bx \, dx = \dfrac{b}{a^2 + b^2}$

41. $\displaystyle\int_0^{\pi/2} \cos^n x\, dx = \int_0^{\pi/2} \sin^n x\, dx = \frac{n-1}{n} \cdot \frac{n-3}{n-2} \cdot \frac{n-5}{n-4} \cdots\cdots \frac{3}{4} \cdot \frac{1}{2} \cdot \frac{\pi}{2}$, if n is even

$$= \frac{n-1}{n} \cdot \frac{n-3}{n-2} \cdot \frac{n-5}{n-4} \cdots\cdots \frac{4}{5} \cdot \frac{2}{3} \cdot 1,\ \text{if n is odd}$$

42. $\displaystyle\int_0^{\pi} \sin^n x\, dx = 2\int_0^{\pi/2} \sin^n x\, dx \qquad$ for all n integral values of n

43. $\displaystyle\int_0^{\pi} \cos^n x\, dx \quad = \quad 2\int_0^{\pi/2} \cos^n x\, dx, \quad$ if n is even integer

$$= \quad 0, \qquad\qquad\qquad \text{if n is an odd integer}$$

44. $\displaystyle\int_0^{2\pi} \sin^n x\, dx \quad = \quad 4\int_0^{\pi/2} \sin^n x\, dx, \quad$ if n is even integer

$$= \quad 0, \qquad\qquad\qquad \text{if n is an odd integer}$$

45. $\displaystyle\int_0^{2\pi} \cos^n x\, dx \quad = \quad 4\int_0^{\pi/2} \cos^n x\, dx \quad$ if n is even integer

$$= \quad 0, \qquad\qquad\qquad \text{if n is an odd integer}$$

46. $\displaystyle\int_0^{\pi/2} \sin^m x \cos^n x\, dx = \frac{\{(m-1)\,(m-3)\,\ldots\,2\ \text{or}\ 1\}\,\{(n-1)\,(n-3)\,\ldots\,2\ \text{or}\ 1\}}{(m+n)\,(m+n-2)\,(m+n-4)\,\ldots\,2\ \text{or}\ 1} \times P$

$\qquad$ where, $\quad$ P $\quad = \quad \dfrac{\pi}{2} \qquad\qquad$ if m and n are both even

$$= \quad 1 \qquad\qquad \text{for all other values of m and n}$$

47. $\displaystyle\int_0^{\pi/2} \sin^P x \cos x\, dx = \frac{1}{P+1} = \int_0^{\pi/2} \cos^P x \cdot \sin x \cdot dx$

48. $\displaystyle\int_0^{\pi} \sin^m x \cos^n x\, dx = 2\int_0^{\pi/2} \sin^m x \cdot \cos^n x\, dx \qquad\qquad$ if n = even, m = even or odd

$$= \quad 0 \qquad\qquad\qquad\qquad\qquad \text{if n = odd, m = even or odd}$$

49. $\displaystyle\int_0^{2\pi} \sin^m x \cos^n x\, dx = 4\int_0^{\pi/2} \sin^m x \cdot \cos^n x\, dx \qquad\qquad$ if m, n = even

$$= \quad 0 \qquad\qquad\qquad\qquad\qquad \text{otherwise}$$

50. $\displaystyle\int_0^{2a} f(x)\, dx = 2\int_0^{a} f(x)\, dx \qquad$ if f(x) = f(2a − x)

$$\int_0^{2a} f(x)\, dx = 0 \qquad\qquad \text{if } f(x) = -f(2a - x)$$

•••

ADDITIONAL IMPORTANT POINTS

I. For $x^2 + y^2 = a^2$. Arc ABC represents $\quad y = \sqrt{a^2 - x^2}$ Arc ADC represents $\quad y = -\sqrt{a^2 - x^2}$	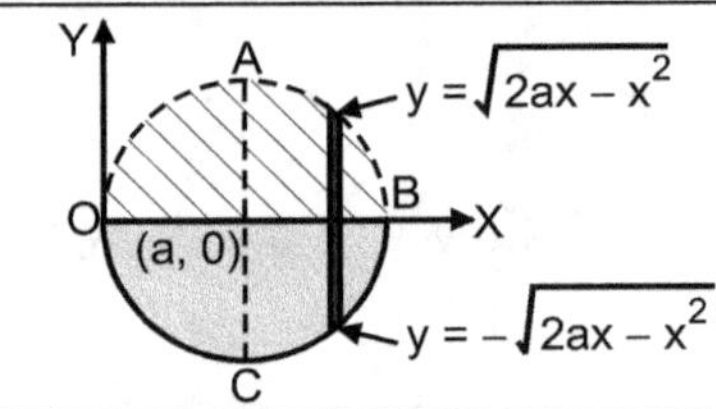
II. For $x^2 + y^2 = a^2$ Arc BAD represents $\quad x = -\sqrt{a^2 - y^2}$ Arc BCD represents $\quad x = \sqrt{a^2 - y^2}$	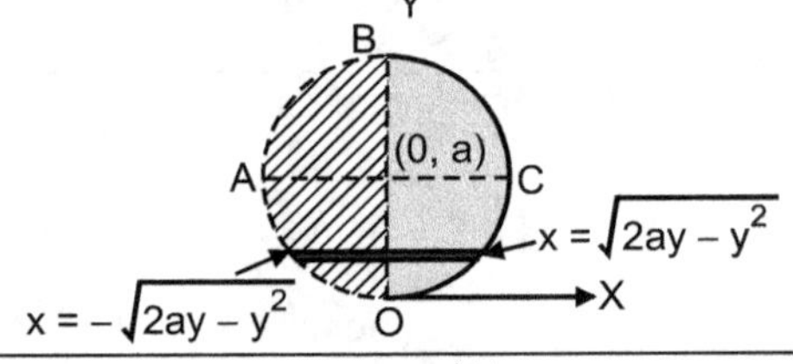
III. For $x^2 + y^2 = 2ax$ OR $(x - a)^2 + (y - 0)^2 = a^2$ Arc OAB represents $\quad y = \sqrt{2ax - x^2}$ Arc OCB represents $\quad y = -\sqrt{2ax - x^2}$	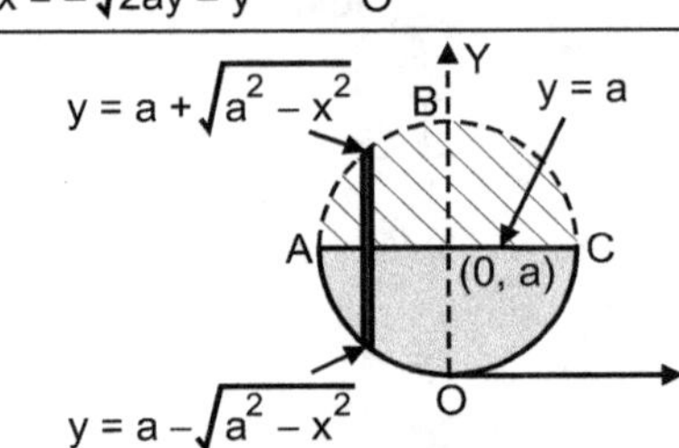
IV. For $x^2 + y^2 = 2ax$ OR $(x - a)^2 + (y - 0)^2 = a^2$ Arc AOC represents $\quad x = a - \sqrt{a^2 - y^2}$ Arc ABC represents $\quad x = a + \sqrt{a^2 - y^2}$	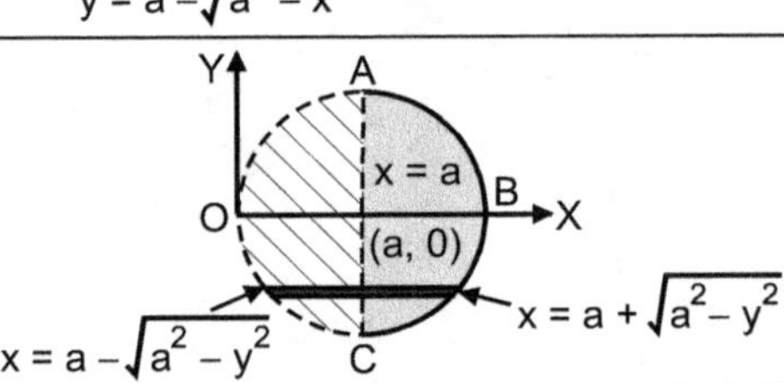
V. For $x^2 + y^2 = 2ay$ Arc OAB represents $\quad x = -\sqrt{2ay - y^2}$ Arc OCB represents $\quad x = \sqrt{2ay - y^2}$	
VI. For $x^2 + y^2 = 2ay$ OR $(x - 0)^2 + (y - a)^2 = a^2$ Arc ABC represents $\quad y = a + (a^2 - x^2)$ Arc AOC represents $\quad y = a - \sqrt{a^2 - x^2}$	

MODEL QUESTION PAPER - I

Engineering Mathematics-II (2019 Course)

Phase-I : In Semester Examination (ISE)

Time : 1 Hour   **Maxim Marks : 30**

N.B. : Attempt Q.1 or Q.2, Q. 3 or Q. 4

1. (a) $\dfrac{dy}{dx} = \dfrac{\tan y - 2xy - y}{x^2 - x\tan^2 y + \sec^2 y}$.   **(5)**

   (b) $(xy^3 + y)\,dx + 2\,(x^2 y^2 + x + y^4)\,dy = 0$.   **(5)**

   (c) $\tan y\,\dfrac{dy}{dx} + \tan x = \cos y \cos^2 x$.   **(5)**

OR

2. (a) $(x^2 + 1)\,\dfrac{dy}{dx} + 4xy = \dfrac{1}{(x^2 + 1)^2}$.   **(5)**

   (b) $\dfrac{dy}{dx} - y\tan x = y^4 \sec x$.   **(5)**

   (c) $(x^2 y^2 + xy + 1)\,y\,dx + (x^2 y^2 - xy + 1)\,x\,dy = 0$.   **(5)**

3. (a)   A body at temperature 100°C is placed in a room whose temperature is 20°C and cools to 60°C in 5 minutes. Find its temperature after a further interval of 3 minutes ?   **(5)**

   (b)   In a circuit containing inductance L, resistance R and voltage E, the current I is given by $E = Ri + L\dfrac{dI}{dt}$, where $I(0) = 0$. Given L = 640 H, R = 250 Ω and E = 500 V. Find the time.   **(5)**

   (c)   A body of mass 'm', falling from rest is subjected to the force of gravity and an air resistance proportional to the square of the velocity 'kv^2'. If it falls through a distance 'x' and possesses a velocity 'v' at that instant. prove that, $\dfrac{2kx}{m} = \log\left(\dfrac{a^2}{a^2 - v^2}\right)$ where $mg = ka^2$.   **(5)**

OR

4. (a)   Find the orthogonal trajectories of the curves given by $x^2 + 2y^2 = c^2$.   **(5)**

   (b)   A voltage Ee^{-at} is applied at $t = 0$ to a circuit containing inductance L and resistance R. Show that current at time t is $\dfrac{E}{R - aL}\left(e^{-at} - e^{-\frac{R}{L}t}\right)$.   **(5)**

   (c)   A steam pipe 20 cm in diameter is protected with a covering 6 cm thick for which the coefficient of thermal conductivity is k = 0.0003 cal./cm. deg. sec. in steady state. Find the heat loss per hour through a meter length of the pipe, if the surface of the pipe is at 200°C and the outer surface of the covering is at 30°C.   **(5)**

❏❏❏

Engineering Mathematics-II (2019 Course)
Phase-I : End Semester Examination (ESE)

Time : 3 Hour | **Maxim Marks : 70**

N.B. : Attempt Q.1 or Q. 2, Q. 3 or Q. 4, Q. 5 or Q. 6, Q.7 or Q. 8.

1. (a) If $U_n = \int_0^{\pi/2} x \cos^n x \, dx$ then show that $U_n = \frac{1}{n^2} + \frac{n-1}{n} U_{n-2}$. Hence evaluate U_3. **(6)**

   (b) Evaluate $\int_0^\infty (x-3)^{1/4} (7-x)^{1/4} \, dx$. **(5)**

   (c) Evaluate DUIS rule $\int_0^\infty \frac{e^{-x} \sin \alpha x}{x} \, dx$. **(6)**

OR

2. (a) Evaluate $\int_0^\infty \frac{x^8 - x^5}{(1 + x^3)^5} \, dx$. **(6)**

   (b) Evaluate $\int_0^1 (x \log x)^4 \, dx$. **(5)**

   (c) Show that $\int_0^\infty e^{-x^2 - 2bx} \, dx = \frac{\sqrt{\pi}}{2} e^{b^2} [1 - \operatorname{erf}(b)]$. **(6)**

3. (a) Trace the cuve : $y^2 (a^2 - x^2) = a^3 x$. **(6)**
   (b) Trace the curve : $r = a \cos 2\theta$. **(6)**
   (c) Trace the cycloid : $x = a (1 + \sin t)$, $y = a (1 - \cos t)$. **(6)**

OR

4. (a) Trace the curve : $y^2 (a^2 - x^2) = a^3 x$. **(6)**
   (b) Trace the curve : $r^2 = a^2 \cos 2\theta$. **(6)**
   (c) Find the length of cardioide $r = a (1 + \cos \theta)$ which lies outside the circle $r + a \cos \theta = 0$. **(6)**

5. (a) Find the equation of the right circular cone which passes through the point $(1, 1, 2)$ has its axis the line $6x = -3y = 4z$ and vertex at origin. **(6)**
   (b) Find the equation of right circular cylinder of radius 2 whose axis passes through $(1, 2, 3)$ and has direction cosines proportional to $2, -3, 6$. **(5)**
   (c) Find the equation of sphere passing through the circle $x^2 + y^2 = 4$, $z = 4$ and cutting the sphere $x^2 + y^2 + z^2 + 10y - 4z - 8 = 0$ orthogonally. **(6)**

OR

6. (a) Find the sphere through the circle $x^2 + y^2 + z^2 = 4$, $z = 0$ meeting the plane $x + 2y + 2z = 0$ in a circle of radius 3. **(6)**
   (b) Obtain the equation of a right circular cone which passes through the point $(2, 1, 3)$ with vertex at $(1, 1, 2)$ and axis parallel to the line $\dfrac{x-2}{2} = \dfrac{y-1}{-4} = \dfrac{z+2}{3}$. **(5)**

   (c) Find the equation of right circular cylinder of radius 2 whose axis is the line $\dfrac{x-1}{2} = \dfrac{y-2}{1} = \dfrac{z-3}{2}$. **(6)**

7. (a) Evaluate $\int_0^1 \int_0^{\sqrt{1-x^2}} \frac{dxdy}{(1 + e^y)\sqrt{1 - x^2 - y^2}}$. **(6)**

   (b) Find the area between the curve $xy^2 = a^2 (a - x)$ and its asymptote. **(6)**
   (c) Find the C.G. of one loop of the curve $r = a \sin 2\theta$. **(6)**

8. (a) Sketch the area of double integration and evaluate $\int_0^{a\sqrt{2}} \int_y^{\sqrt{a^2 - y^2}} \log_e (x^2 + y^2) \, dx \, dy$. **(6)**

   (b) Evaluate $\iiint_R \frac{z^2}{x^2 + y^2 + z^2} \, dxdydz$ where V is the volume bounded by $x^2 + y^2 + z^2 = z$. **(6)**

   (c) Find the moment of inertia of one loop of the lemniscate $r^2 = a^2 \cos 2\theta$ above initial line. **(6)**

❑❑❑